William F. Maag Library
Youngstown State University

ANNUAL REVIEW OF PHYSICAL CHEMISTRY

EDITORIAL COMMITTEE (2002)

PAUL ALIVISATOS
MARTIN GRUEBELE
FRANCES HOULE
STEPHEN R. LEONE
ANN E. MCDERMOTT
GEORGE C. SCHATZ
MOSHE SHAPIRO
SUNNEY XIE

RESPONSIBLE FOR THE ORGANIZATION OF VOLUME 53 (EDITORIAL COMMITTEE, 2000)

GERALD T. BABCOCK
LOUIS E. BRUS
JEAN-PIERRE HANSEN
STEPHEN R. LEONE
ANN E. MCDERMOTT
MALCOLM F. NICOL
MARK A. RATNER
HERBERT L. STRAUSS
GILBERT NATHANSON (GUEST)
JOHN D. WEEKS (GUEST)

Production Editor: ANNE E. SHELDON
Bibliographic Quality Control: MARY A. GLASS
Color Graphics Coordinator: EMÉ O. AKPABIO
Electronic Content Coordinator: SUZANNE K. MOSES
Subject Indexer: BRUCE TRACY

ANNUAL REVIEW OF PHYSICAL CHEMISTRY

VOLUME 53, 2002

STEPHEN R. LEONE, *Editor*
University of California, Berkeley

PAUL ALIVISATOS, *Associate Editor*
University of California, Berkeley

ANN E. McDERMOTT, *Associate Editor*
Columbia University

www.annualreviews.org   science@annualreviews.org   650-493-4400

ANNUAL REVIEWS
4139 El Camino Way • P.O. BOX 10139 • Palo Alto, California 94303-0139

ANNUAL REVIEWS
Palo Alto, California, USA

COPYRIGHT © 2002 BY ANNUAL REVIEWS, PALO ALTO, CALIFORNIA, USA. ALL RIGHTS RESERVED. The appearance of the code at the bottom of the first page of an article in this serial indicates the copyright owner's consent that copies of the article may be made for personal or internal use, or for the personal or internal use of specific clients. This consent is given on the condition that the copier pay the stated per-copy fee of $14.00 per article through the Copyright Clearance Center, Inc. (222 Rosewood Drive, Danvers, MA 01923) for copying beyond that permitted by Section 107 or 108 of the US Copyright Law. The per-copy fee of $14.00 per article also applies to the copying, under the stated conditions, of articles published in any *Annual Review* serial before January 1, 1978. Individual readers, and nonprofit libraries acting for them, are permitted to make a single copy of an article without charge for use in research or teaching. This consent does not extend to other kinds of copying, such as copying for general distribution, for advertising or promotional purposes, for creating new collective works, or for resale. For such uses, written permission is required. Write to Permissions Dept., Annual Reviews, 4139 El Camino Way, P.O. Box 10139, Palo Alto, CA 94303-0139 USA.

International Standard Serial Number: 0066-426X
International Standard Book Number: 0-8243-1053-5
Library of Congress Catalog Card Number: A-51-1658

All Annual Reviews and publication titles are registered trademarks of Annual Reviews.
⊗ The paper used in this publication meets the minimum requirements of American National Standards for Information Sciences—Permanence of Paper for Printed Library Materials, ANSI Z39.48-1992.

Annual Reviews and the Editors of its publications assume no responsibility for the statements expressed by the contributors to this *Annual Review*.

Typeset by TechBooks, Fairfax, VA
Printed and Bound in the United States of America

Preface

The field of physical chemistry today is distinguished by its depth of mathematical and quantitative rigor in a pursuit of a wide array of challenging new subjects. Physical chemistry defines a range of explorations from the most detailed small molecule spectroscopic and dynamical processes to enzyme and polymer events, with a relentless push toward more complex systems. Volume 53 of the *Annual Review of Physical Chemistry* is no exception, covering such diverse areas as exceedingly detailed gas-phase spectroscopic, scattering, and angular-distribution studies, elegant investigations of surface interfacial phenomena, liquid-phase ultrafast laser coherent dynamics, reversible polymerizations, as well as protein folding and drug design, among the many topics represented. Physical chemistry also clearly involves many technological themes, such as the development of single-molecule transistors with carbon nanotubes. Developments in sophisticated experimental tools and powerful theoretical methodologies also play key roles in defining what is approachable by the field.

This volume begins with a Prefatory chapter by Tinoco that considers the prediction of the folded structure of any RNA from its sequence using bulk- and single-molecule measurements of thermodynamics, kinetics, and circular dichroism spectroscopy. Continuing along this theme, Pratt reviews the molecular theory of hydrophobic effects relevant to biomolecular structure and assembly in aqueous solution. Oldfield reviews the use of magnetic resonance spectroscopy and ab initio methods to investigate chemical shifts in proteins with extensions of the use of NMR chemical shifts in drug design. Tse considers the applications of ab initio density functional theory to study solution reactions and biological systems, and Gao & Truhlar present a review of recent quantum mechanical theoretical methods for enzyme kinetics.

Bolhuis, Chandler, Dellago, & Geissler review the concepts of transition path sampling to study rare events, and Weeks considers a van der Waals theory to describe nonuniform liquids. Fourkas reviews experimental advances in higher-order ultrafast laser electronic and vibrational spectroscopy to study liquid-state dynamics and structure. Greer explores the reversible aggregation of monomers into polymers, comparing experiment with theory.

Surface interfacial phenomena and electron transport are also major themes of this volume. Petek & Ogawa review a detailed ultrafast laser case study of the transfer of an electron from a metal substrate to an unoccupied resonance of an atom that causes the atom to desorb. Zhu reviews the subject of electron transfer at molecule-metal interfaces by two-photon photoemission. Ouyang, Huang, & Lieber review the rapid advances in studies of the electronic properties of single-walled carbon nanotubes, as probed by sensitive scanning-tunneling microscopy. Chen,

v

Shen, & Somorjai consider the substantial body of information obtained on surface molecular chemical structures of polymer materials by sum frequency generation. Greeley, Nørskov, & Mavrikakis review the issues of the electronic structures of surfaces and their effect on the metal catalysis.

Finally, Jacobs considers the collisions of energetic hyperthermal particles with surfaces and the resulting chemical dynamics.

Detailed spectroscopic and dynamics studies of small gas phase molecules are explored in a number of reviews. Ng provides a comprehensive examination of high-resolution photoionization methods to obtain thermochemical data for many neutrals and cationic species, with unprecedented precision. Meyer studies issues related to polyatomic molecular spectroscopy by considering the comprehensive details of molecular Hamiltonians. Seideman reviews the insights that can be obtained by theoretical and experimental analyses of time-resolved photoelectron angular distributions and the use of these methods as potential probes of molecular structure. Continuing the pursuit of detailed, state-to-state phenomena, Fernández-Alonso & Zare review comprehensively the entire field of resonance phenomena in chemical dynamics.

The *Annual Review of Physical Chemistry* is an ever-changing forum, and we are pleased to welcome several new members of the Editorial Board, Martin Gruebele, George Schatz, Moshe Shapiro, and Sunney Xie. We thank them for their enthusiastic support of this venture and for providing their time for the benefit of the community.

The Editorial Committee

Annual Review of Physical Chemistry
Volume 53, 2002

Contents

Frontispiece—*Ignacio Tinoco, Jr.*	xiv
PHYSICAL CHEMISTRY OF NUCLEIC ACIDS, *Ignacio Tinoco, Jr.*	1
HIGHER-ORDER OPTICAL CORRELATION SPECTROSCOPY IN LIQUIDS, *John T. Fourkas*	17
TIME-RESOLVED PHOTOELECTRON ANGULAR DISTRIBUTIONS: CONCEPTS, APPLICATIONS, AND DIRECTIONS, *Tamar Seideman*	41
SCATTERING RESONANCES IN THE SIMPLEST CHEMICAL REACTION, *Félix Fernández-Alonso and Richard N. Zare*	67
VACUUM ULTRAVIOLET SPECTROSCOPY AND CHEMISTRY BY PHOTOIONIZATION AND PHOTOELECTRON METHODS, *Cheuk-Yiu Ng*	101
THE MOLECULAR HAMILTONIAN, *Henning Meyer*	141
REVERSIBLE POLYMERIZATIONS AND AGGREGATIONS, *Sandra C. Greer*	173
SCANNING TUNNELING MICROSCOPY STUDIES OF THE ONE-DIMENSIONAL ELECTRONIC PROPERTIES OF SINGLE-WALLED CARBON NANOTUBES, *Min Ouyang, Jin-Lin Huang, and Charles M. Lieber*	201
ELECTRON TRANSFER AT MOLECULE-METAL INTERFACES: A TWO-PHOTON PHOTOEMISSION STUDY, *X.-Y. Zhu*	221
AB INITIO MOLECULAR DYNAMICS WITH DENSITY FUNCTIONAL THEORY, *John S. Tse*	249
TRANSITION PATH SAMPLING: THROWING ROPES OVER ROUGH MOUNTAIN PASSES, IN THE DARK, *Peter G. Bolhuis, David Chandler, Christoph Dellago, and Phillip L. Geissler*	291
ELECTRONIC STRUCTURE AND CATALYSIS ON METAL SURFACES, *Jeff Greeley, Jens K. Nørskov, and Manos Mavrikakis*	319
CHEMICAL SHIFTS IN AMINO ACIDS, PEPTIDES, AND PROTEINS: FROM QUANTUM CHEMISTRY TO DRUG DESIGN, *Eric Oldfield*	349
REACTIVE COLLISIONS OF HYPERTHERMAL ENERGY MOLECULAR IONS WITH SOLID SURFACES, *Dennis C. Jacobs*	379
MOLECULAR THEORY OF HYDROPHOBIC EFFECTS: "SHE IS TOO MEAN TO HAVE HER NAME REPEATED," *Lawrence R. Pratt*	409

STUDIES OF POLYMER SURFACES BY SUM FREQUENCY GENERATION
VIBRATIONAL SPECTROSCOPY, *Zhan Chen, Y. R. Shen, and
Gabor A. Somorjai*   437

QUANTUM MECHANICAL METHODS FOR ENZYME KINETICS, *Jiali Gao
and Donald G. Truhlar*   467

SURFACE FEMTOCHEMISTRY: OBSERVATION AND QUANTUM CONTROL
OF FRUSTRATED DESORPTION OF ALKALI ATOMS FROM NOBLE
METALS, *Hrvoje Petek and Susumu Ogawa*   507

CONNECTING LOCAL STRUCTURE TO INTERFACE FORMATION: A
MOLECULAR SCALE VAN DER WAALS THEORY OF NONUNIFORM
LIQUIDS, *John D. Weeks*   533

INDEXES
   Author Index   563
   Subject Index   591
   Cumulative Index of Contributing Authors, Volumes 49–53   623
   Cumulative Index of Chapter Titles, Volumes 49–53   625

ERRATA
   An online log of corrections to *Annual Review of Physical Chemistry* chapters may be found at http://physchem.annualreviews.org/errata.shtml

Related Articles

From the *Annual Review of Astronomy and Astrophysics*, Volume 39 (2001)
  The Cosmic Infrared Background: Measurements and Implications, Michael G. Hauser and Eli Dwek
  Optical Interferometry, Andreas Quirrenbach

From the *Annual Review of Biomedical Engineering*, Volume 3 (2001)
  New DNA Sequencing Methods, Andre Marziali and Mark Akeson
  Soft Lithography in Biology and Biochemistry, George M. Whitesides, Emanuele Ostuni, Shutchi Takayama, Xingyu Jiang, and Donald E. Ingber

From the *Annual Review of Biophysics and Biomolecular Structure*, Volume 30 (2001)
  Hydrogen Bonding, Base Stacking, and Steric Effects in DNA Replication, Eric T. Kool
  Structures and Proton-Pumping Strategies of Mitochondrial Respiratory Enzymes, Brian E. Schultz and Sunney I. Chan
  Mass Spectrometry as a Tool for Protein Crystallography, Steven L. Cohen and Brian T. Chait
  A Structural View of Cre-loxP Site-Specific Recombination, Gregory D. Van Duyne
  Probing the Relation Between Force—Lifetime—And Chemistry in Single Molecular Bonds, Evan Evans
  NMR Probes of Molecular Dynamics: Overview and Comparison with Other Techniques, Arthur G. Palmer III
  Chaperonin-Mediated Protein Folding, D. Thirumalai and George H. Lorimer
  Interpreting the Effects of Small Uncharged Solutes on Protein-Folding Equilibria, Paula R. Davis-Searles, Aleister J. Saunders, Dorothy A. Erie, Donald J. Winzor, and Gary J. Pielak
  Protein Folding Theory: From Lattice to All-Atoms Models, Leonid Mirny and Eugene Shakhinovich

From the *Annual Review of Earth and Planetary Sciences*, Volume 29 (2001)
  Human Impacts on Atmospheric Chemistry, P. J. Crutzen and J. Lelieveld

Rheological Properties of Water Ice—Applications to Satellites of the Outer Planets, W. B. Durham and L. A. Stern

Hydrogen in the Deep Earth, Quentin Williams and Russell J. Hemley

The Carbon Budget in Soils, Ronald Amundson

From the ***Annual Review of Materials Research***, Volume 31 (2001)

Design and Synthesis of Energetic Materials, Laurence E. Fried, M. Riad Manaa, Philip F. Pagoria, and Randall L. Simpson

Block Copolymer Thin Films: Physics and Applications, Michael J. Fasolka and Anne M. Mayes

ANNUAL REVIEWS is a nonprofit scientific publisher established to promote the advancement of the sciences. Beginning in 1932 with the *Annual Review of Biochemistry*, the Company has pursued as its principal function the publication of high-quality, reasonably priced *Annual Review* volumes. The volumes are organized by Editors and Editorial Committees who invite qualified authors to contribute critical articles reviewing significant developments within each major discipline. The Editor-in-Chief invites those interested in serving as future Editorial Committee members to communicate directly with him. Annual Reviews is administered by a Board of Directors, whose members serve without compensation.

2002 Board of Directors, Annual Reviews

Richard N. Zare, *Chairman of Annual Reviews*
  Marguerite Blake Wilbur, Professor of Chemistry, Stanford University
John I. Brauman, *J. G. Jackson–C. J. Wood Professor of Chemistry, Stanford University*
Peter F. Carpenter, *Founder, Mission and Values Institute*
W. Maxwell Cowan, *Vice President and Chief Scientific Officer, Howard Hughes Medical Institute, Bethesda*
Sandra M. Faber, *Professor of Astronomy and Astronomer at Lick Observatory, University of California at Santa Cruz*
Eugene Garfield, *Publisher*, The Scientist
Samuel Gubins, *President and Editor-in-Chief, Annual Reviews*
Daniel E. Koshland, Jr., *Professor of Biochemistry, University of California at Berkeley*
Joshua Lederberg, *University Professor, The Rockefeller University*
Gardner Lindzey, *Director Emeritus, Center for Advanced Study in the Behavioral Sciences, Stanford University*
Sharon R. Long, *Professor of Biological Sciences, Stanford University*
J. Boyce Nute, *President and CEO, Mayfield Publishing Co.*
Michael E. Peskin, *Professor of Theoretical Physics, Stanford Linear Accelerator Ctr.*
Harriet A. Zuckerman, *Vice President, The Andrew W. Mellon Foundation*

Management of Annual Reviews

Samuel Gubins, President and Editor-in-Chief
Richard L. Burke, Director for Production
Paul J. Calvi, Jr., Director of Information Technology
Steven J. Castro, Chief Financial Officer
John W. Harpster, Director of Sales and Marketing

Annual Reviews of

Anthropology
Astronomy and Astrophysics
Biochemistry
Biomedical Engineering
Biophysics and Biomolecular Structure
Cell and Developmental Biology
Earth and Planetary Sciences
Ecology and Systematics
Energy and the Environment
Entomology
Fluid Mechanics
Genetics
Genomics and Human Genetics
Immunology
Materials Research
Medicine
Microbiology
Neuroscience
Nuclear and Particle Science
Nutrition
Pharmacology and Toxicology
Physical Chemistry
Physiology
Phytopathology
Plant Biology
Political Science
Psychology
Public Health
Sociology

SPECIAL PUBLICATIONS
Excitement and Fascination of Science, Vols. 1, 2, 3, and 4

PHYSICAL CHEMISTRY OF NUCLEIC ACIDS

Ignacio Tinoco, Jr.

Department of Chemistry, University of California, Berkeley, and Physical Biosciences Division, Lawrence Berkeley National Laboratory, Berkeley, California 94720-1460; e-mail: Intinoco@lbl.gov

Key Words DNA, RNA, circular dichroism, hypochromism, NMR

■ **Abstract** The Watson-Crick double helix of DNA was first revealed in 1953. Since then a wide range of physical chemical methods have been applied to DNA and to its more versatile relative RNA to determine their structures and functions. My major goal is to predict the folded structure of any RNA from its sequence. We have used bulk and single-molecule measurements of thermodynamics and kinetics, plus various spectroscopic methods (UV absorption, optical rotation, circular dichroism, circular intensity differential scattering, fluorescence, NMR) to approach this goal.

MADISON

In 1950, I was a 19-year-old junior at the University of New Mexico at Albuquerque when the Korean war started. The likelihood that I would soon be drafted encouraged me to get my B.S. degree a year early. Guido Daub, my organic chemistry professor, was able to get me into graduate school at the University of Wisconsin, although I had applied very late. All the teaching assistantships had already been assigned, but Professor John Ferry offered me a research assistantship, which I gladly accepted. My project was to study viscoelastic and dynamic mechanical properties of synthetic polymers. I knew nothing about viscoelasticity or polymers, and I am sure if I had been given time to pick a research director I would have ended up with someone like Paul Bender, a classical physical chemist who studied problems I understood, such as the heat of reaction of HCl with NaOH.

My first job in the lab was to clean up a water bath used for measuring viscosities. The temperature was controlled by a large bulb of mercury connected to a capillary. An adjustable wire in the capillary was the thermostat. When the mercury expanded, contact with the wire turned off the heating element in the bath. The constant making and breaking of the electrical contact at the surface of the mercury led to oxidation, which had to be removed periodically. I took all the mercury out of the bulb and washed it carefully with dilute acid, followed by water. To finish the job, I put the beaker of mercury into the drying oven for glassware to get rid of the water. I do not remember how long the mercury cooked before one of my lab partners removed it. I was told that mercury vapor was not good for you, and that it was sufficient to rinse with acetone and let the mercury dry at room temperature. At that time the proper action for a mercury spill was to sweep up the big drops and throw sulfur on the rest.

I spent the next few months measuring the viscosities of polyisobutylene solutions. But that fall one of Ferry's second-year students failed his qualifying exam, leaving a hole in the project to study the conversion of fibrinogen to fibrin. I was assigned to fill the hole. Actually, I was asked to fill in for awhile, and if I really did not like biological polymers, I could switch back.

The mechanical rigidity of a blood clot is provided by the polymerization of activated fibrinogen to form a fibrin gel. The activation is caused by the removal of a peptide from fibrinogen by the proteolytic enzyme thrombin. We wanted to characterize the mechanism of the polymerization by measuring concentrations of activated fibrinogen monomers, dimers, etc. as a function of time of polymerization. The first requirement was to slow the reaction down enough so that we could use methods like ultracentrifugation and light scattering to study it. Hexamethylene glycol in water was the preferred solvent, but I measured the effects of different chemicals on the clotting times to look for a better one. We wanted a reagent that would slow the reaction but not denature the protein. Some of Ferry's biochemical colleagues thought that studying clotting in nonbiological solvents could never yield any biologically relevant information. However, once the main intermediates in the polymerization process were identified in hexamethylene glycol, we found the same species in more physiological solvents. The nonaqueous solvent made it easier to study the reaction, but the same mechanism seemed to occur in an aqueous buffer.

Nearly 50 years later there is still discussion and disagreement about the relevance to biological function of structures of biological macromolecules measured in crystals or of mechanisms of reactions studied in nonbiological environments. We biophysical chemists are considered naive reductionists because we mainly study one or two species in vitro. The molecular biologists at least study reactions in bacteria or yeast. Of course, the real biologists insist on studying whole multicellular organisms. I am convinced that whatever we learn about biological molecules—no matter how simple the system or how unusual the environment—can be applied to understand their biological functions. Intermolecular interactions do not distinguish between biologically relevant and biologically irrelevant reactions. Whenever someone questions the biological relevance of an experiment, I respond that anything of interest to a human is clearly biologically relevant. I hasten to add that NIH requires more justification than this for funding. The correct answer for NIH is that the proposed experiment will eventually lead to a cure for cancer, AIDS, and old age.

The polymerization of fibrinogen begins after the proteolytic enzyme thrombin clips off an oligopeptide from the cigar-shaped fibrinogen. My thesis project concluded with building a transient electric birefringence apparatus to study the size and shape of fibrinogen and to learn how it changed on activation by thrombin. I applied an electric field for a few milliseconds to orient the protein molecules and measured the buildup and decay of the orientation by the appearance and disappearance of birefringence in the solution. From the slight decrease, less than 100 Debye (D), in effective dipole moment of the fibrinogen when the peptide was removed, we concluded that the peptide with its 10 negatively charged carboxylate groups must have been near the center of the molecule. Removal of 10 charges from the end of the 600-Å molecule would change the dipole moment by about 15,000 D in a vacuum. Our solutions had no added salt so we argued that the measured decrease of 100 D meant the peptide was within a few Angstroms of the center of rotation of the protein. An atomic resolution X-ray structure of fibrinogen published recently shows that the amino-terminal ends cleaved by thrombin are indeed at the center of the molecule.

YALE

After 3 years my research progress had gone well and my wife was pregnant, so I suggested to Ferry that it was time for me to move on. John Kirkwood visited about that time and Ferry persuaded him to take me on as a postdoc. Kirkwood had recently become chairman at Yale, where he had a large group of theoretical students and a small group of experimentalists, which I would join. Thus, in 1954 we drove from Madison to New Haven just in time for our daughter to be born as Hurricane Edna was hitting New England.

A postdoctoral position is the best job that I can imagine—there is little responsibility and maximum opportunity to learn new things. At Yale there was an enthusiastic group of people thinking about proteins and nucleic acids. Jon Singer (now at University of California, San Diego) was studying antigen-antibody reactions; Peter Geiduschek (also at UC San Diego) worked on DNA and was very excited about the paper by Watson and Crick that had just appeared. Kirkwood was interested in everything. I was warned to never tell him any preliminary experimental results because he could and would create a theory to explain any result—right or wrong. He had an idea about the effect of ion fluctuations on enzymatic activity that predicted maximum catalytic activity at the pK of the contributing groups. The pH dependence of catalytic activity was not consistent with this model, but it did lead to some enzyme experiments that I continued later. A Yale undergraduate who worked on this project as a "bursary boy"—a fellowship student—was Don Crothers. He is still at Yale but no longer as an undergraduate.

As an advisor in theory, Kirkwood was superb. My thesis experiments on electric orientation and birefringence of fibrinogen had suggested two questions: What are the effective electric forces on the charged protein in an external field in the presence of counterions and added ions? How does the optical activity (circular birefringence) of the protein change with orientation and thus effect the linear birefringence? Kirkwood immediately told me that the first question was very hard and I should concentrate on the second. He showed me the paper he had published in 1937 on a quantum mechanical formula for the orientation-averaged optical rotation. Bill Hammerle, a theoretical postdoc of Kirkwood's, and I derived the result that the rotational strength of an electronic transition was very sensitive to the direction of incidence of light. In fact the principal components of the optical rotation tensor generally had different signs. Kirkwood praised the work but said his name should not be on the paper as it was not his idea. His interest in optical activity had recently revived because of a paper published by Bill Moffitt from Harvard. Elkan Blout and Paul Doty at Harvard had found that the optical rotation of polypeptides showed large and characteristic changes when the polypeptide went from an α-helix to a coil. Moffitt's paper explaining the optical activity in terms of exciton interactions in the helix was incorrect; he had missed a term that did not contribute to absorption but was crucial for rotation. Kirkwood called and told Moffitt about the error and they published a correction in *PNAS* with Moffitt as first author (1). I thought it was a very considerate thing to do. Kirkwood seemed to agree with Norbert Wiener that unlike chess, in science the mistakes you make are not as important as whether you finally get it right.

After two very stimulating years at Yale, learning as much as I could from Kirkwood, Fuoss, Lyons, Onsager (his statistical mechanics course was called Norwegian I and II by the students), Singer, Sturtevant, and J. H. (Ray) Wang, I started applying for academic jobs. The University of Kansas interviewed me, but they hired Sherwood Rowland instead. As I was considering a second postdoc with John Edsall at Harvard, I received an offer of an instructorship from Berkeley. I learned that they had offered the job to Stanley Gill, who had chosen to go to the University of Colorado. Kirkwood recommended that I accept the Berkeley offer, although he assured me that I would not be given tenure. He said that they only hired their own Ph.D.s, but that after a few years at Berkeley I should be able to move on to an acceptable position. I also learned that the University of New Mexico would probably hire me after my stay at Berkeley. With no expectation of tenure, and with a backup job opportunity, the pressure to make it in my first academic job was minimal. So in 1956 we drove across the country to Berkeley in our 1949 Chevrolet; we bought a new radiator in Salt Lake City to make it across the desert and over the Sierra.

BERKELEY

Four new faculty members in physical chemistry were hired at University of California (UC), Berkeley in 1956. Frank Harris, a Berkeley Ph.D., was hired as an assistant professor. Mark (Phil) Freeman, (University of Washington), Bruce Mahan (Harvard), and I were hired as instructors. Two of us made it to tenure. Frank Harris left his pregnant wife for the wife of a graduate student; this was unacceptable in the 1950s so he was not promoted. Phil Freeman had been having a continuing argument with a senior faculty member about the interpretation of Freeman's experiments. At the end of his seminar on the subject Phil pulled a large dueling pistol out of his briefcase, set it on the table and asked "Are there any questions?" Not everyone thought this was funny.

In our first year at Berkeley the new instructors were not allowed to have graduate students. We were told that they would delay us getting started in research; this is in major contrast to the present custom of establishing a large group as quickly as possible. My Berkeley research started out as a direct continuation of my thesis and postdoctoral research. I measured the transient electric birefringence of synthetic polypeptides with the usual arrangement of light incident perpendicular to the electric field. My calculations on the optical rotation of oriented molecules had predicted large differences in sign and magnitude for the rotation of polarized light incident along a helix axis and perpendicular to the helix axis. To measure this effect I used newly available, conducting transparent windows to allow light incident parallel to the electric field. No linear birefringence occurs, so changes in circular birefringence of the polypeptide helices can be measured. The measurements were consistent with the expected results, but a direct comparison with theory was difficult. Phil Freeman and I decided that a better test was to use small copper wire helices and microwave radiation; the copper helices, inserted in styrofoam cubes, could be oriented easily by hand. We borrowed a microwave oscillator, wave guide, and detector from Bill Gwinn and measured the rotation of the oriented helices in the 2–3.5 cm wavelength region.

The paper (2) appeared in 1957 in the *Journal of Physical Chemistry* with the subheading: "I. Experimental." We planned on having the "II. Theory" paper ready shortly. It took over 20 years; the problem was that the helices were the same size as the wavelength. In 1964 Bob Woody and I derived the optical properties (dipole strengths and rotational strengths) of a free electron on a helix (3). The derivation assumes the wavelength is large compared with the helix, so no comparison with our wire helices was possible. Finally, in 1980, Dexter Moore and I made calculations to compare with the 1957 experiments, using the full transition integrals for interaction with light [exp (i k • r)] that allowed any size helix (4). The results were in reasonable agreement with the experiment, but the best part of the work was that it reminded me how much fun it had been making the measurements.

In 1958, a unique scientific conference called "Biophysical Science—A Study Program" was sponsored by the Biophysics and Biophysical Chemistry study section of NIH to educate physicists, physical chemists, biochemists, and biologists about important biological problems and potential methods to solve them. About 120 of us lived in the dorms at the University of Colorado at Boulder with our families for one month. Nearly everyone stayed for the entire conference, even the senior people. Now, in spite of e-mail and satellite phones that keep you connected continuously, if you stay for the full three to four days of a meeting you are considered either unambitious or retired. The conference covered a wide range of topics from macromolecules to the physiology of vision; the participants ranged in age from their 20s to 70s. As usual, most of the important communication occurred during meals, afternoon sports, and late-night drinking. The sports groups seemed to separate by field. Most of the afternoon tennis players were in statistical mechanics including Norman Davidson, Terrell Hill, and Walter Stockmayer. David

Davies was an exception, but his young family prevented him from going mountain hiking on weekends with the other X-ray crystallographers led by John Kendrew. David and I resumed our tennis games—on luxurious grass courts—in 1964 when we were both in Kendrew's group at the Medical Research Council Laboratory in Cambridge.

The lectures were helpful, particularly after we got instant reviews from Leo Szilard. He was a portly, very visible man who sat in the front row. After listening for a few minutes he would sometimes get up and walk slowly out to sit outside to wait for the next speaker. Eventually, when Szilard left others would follow because we knew we could learn more in the conversation outside, rather than listening inside. I felt sorry for the speaker, but it did not deter me from leaving.

A meeting like this has, I think, tremendous advantages for everyone involved. You meet your peers and future colleagues of all ages, which is of immense value, particularly for the young people. Everyone learns what the current exciting problems and possible solutions are. There is time to not only think and talk about research but to actually do some. Charles Townes in his autobiography *How the Laser Happened* (5) writes that he finished the Townes-Schawlow paper on the laser at this meeting.

One problem that attracted my attention at the meeting was the UV absorption of DNA. Concentrations of DNA were routinely measured using an extinction coefficient at 260 nm, but the extinction coefficient varied over 20% depending on preparation methods. Heating and cooling the DNA produced large changes in absorption. The puzzle was that the spectrum did not change. There were no large energy shifts, just changes in intensity. A native DNA had 40% less absorption than the sum of its constituent nucleotides with essentially no change in shape of the broad 260-nm band. The effect was called hypochromicity or hypochromism. I thought this effect could be treated by the same methods used to calculate circular dichroism. Perturbation of the wavefunctions by internucleotide interaction rather than changes in energy levels would be dominant. I derived the theory back in Berkeley that showed that the main contribution to hypochromism was interaction between the transition dipoles on stacked bases. The bands below 200 nm were borrowing intensity from the longer wavelength bands. A simple classical explanation is that the polarizability of two stacked disks is less than the sum of the individual polarizabilities. The paper was published in the *Journal of the American Chemical Society* in 1960 (6). An erratum appeared in 1961; I had included a term that should have cancelled. It did not change the conclusions, only the quantitative calculations. I was embarrassed by the mistake, but I took comfort in how the error was found. Norman Davidson at Caltech had asked Richard Feynman to explain the paper to his group at a seminar; Feynman immediately pointed out the mistake.

We were interested in experimentally testing theories for hypochromism and circular dichroism of polynucleotides and polypeptides. The simplest polymers were dinucleoside phosphates (two nucleosides, each consisting of a base and a ribose, joined by a phosphate) that could be obtained by hydrolyzing RNA. The measured hypochromism and optical rotation could be interpreted in terms of the amount of stacking of the bases; adenines stacked the most and uracils the least. Higher oligomers could be obtained by the same method, but it became much harder to separate the 64 trimers compared to the 16 dimers. This is when the Unabomber came to my aid; he suddenly resigned his position in the math department. A temporary instructorship was offered to Karen Uhlenbeck, so her husband Olke also had to obtain a job in Berkeley. He was given a Miller fellowship, which he used in my laboratory.

THERMODYNAMICS OF NUCLEIC ACIDS

Olke Uhlenbeck brought with him oligonucleotides and the enzymes to synthesize others. We started on the goal of predicting thermodynamic properties of single- and double-stranded polynucleotides from measured nearest-neighbor properties. The idea is that the free energy of a folded RNA relative to the unfolded single strand can be estimated by adding the measured free energies of its components. Base-paired helices contribute negative free energies, and single-stranded loops contribute positive free energies because of their loss of entropy. Olke's method for getting everybody in the lab involved was to take bets on the predicted properties. With real money at stake (25¢), the students made sure that the sample was pure, the experiment was done properly, and the data were analyzed rapidly. Our first paper on prediction of secondary structure—base pairing—in RNA appeared in *Nature* in 1971 (7). We underestimated the biochemists (molecular biologists had not yet been invented) who were the audience for this paper. We decided that they would not understand that negative free energies meant more stable structures, so we assigned positive stability numbers instead of free energies. Base pairs had positive stability numbers; unpaired hairpin loops, interior loops, and bulges had negative stability numbers. To predict a secondary structure we simply calculated stabilities for possible structures and looked for the maximum. In spite of the stability numbers, our method was an improvement over previous methods that only maximized numbers of base pairs.

In our next paper (8) on secondary structure (published jointly with Don Crothers' group) we switched to free energies; apparently biochemists did understand that negative values could be favorable. Prediction of secondary structure in RNA has improved considerably since the 1970s. Doug Turner and his students at the University of Rochester have measured thermodynamic contributions from the 10 possible Watson-Crick nearest neighbors, many base-base mismatches, stacking at the ends of helices, and all sorts of loops and bulges. Their parameters along with Michael Zuker's algorithm for considering all possible secondary structures provide widely used predictions of RNA secondary structure from sequence. I am particularly pleased that Doug Turner is now such an expert on thermodynamics, because when he came to Berkeley as a postdoc from Columbia he claimed that he had never even heard of Lewis and Randall.

William F. Maag Library
Youngstown State University

SPECTROSCOPY OF NUCLEIC ACIDS

We were also working on a variety of optical properties of polymers, particularly polynucleotides. The main emphasis was on circular dichroism and circular birefringence, but UV absorption—hypo- and hyper-chromism—were also considered. The effects of electric fields and magnetic fields were treated, including orientation of the molecules (electric and magnetic birefringence), as well as direct electronic effects (Faraday effect and Stark effect). Experiments, theory, and calculations were made. Doug Turner invented fluorescence-detected circular dichroism (FDCD) (9, 10), in which the chirality in the vicinity of a fluorophore can be measured from the intensity of the fluorescence emission excited by circularly polarized light. This method has developed into a very sensitive and widely used analytical method.

The usual method of measuring circular dichroism actually measures circular extinction, the sum of absorption, and scattering. As large molecules—especially aggregates of large molecules—scatter light significantly, there can be a large contribution from scattering to circular extinction. It is important to be able to separate the effects, because scattering and absorption depend on very different size scales. Absorption depends on local interactions; scattering effects are largest for distances of the order of the wavelength of the light. The most useful structural data are obtained from the angular dependence of the scattered light when circularly polarized light is incident. For the most general experimental design, the polarization of the incident light is characterized by its four-component Stokes vector, and the angular-dependent 4-by-4 Mueller matrix is used to represent the effect of the sample on the outgoing light. Marcos Maestre, Carlos Bustamante, and I derived equations for various chiral models of coupled dipoles, helices, and liquid crystals. Experimental comparison was difficult for nucleic acid double helices because the effects are small, so when we learned that sperm from a Mediterranean octopus were helical we were overjoyed. The circular differential scattering agreed with the dimensions seen in the electron microscope (11). I proudly presented this work at a Department of Energy (DOE) site visit of the Chemical Biodynamics Division of the Lawrence Berkeley Lab. David Shirley, then director of the Lawrence Berkeley Lab, could hardly keep from laughing out loud. I do not know if he thought that research on octopus sperm emphasized the multidisciplinary nature of the lab, or if he was laughing while trying to come up with a justification for DOE support of the work. In any case the site visit did not affect my DOE budget. Actually, site visits never seemed to have any effect, either positive or negative, on budgets.

The ability of circularly polarized light to distinguish chiral objects suggested to Marcos Maestre and me that it would be useful to make images of the differential absorption of the circularly polarized light. We used a digital detector and subtracted the images for right and left circularly polarized light. We applied for a patent, but the patent examiner said they would not patent a polarimeter connected to a microscope. The lawyers advised us to apply for patents on applications of the method for detecting clinically important targets in cells, but we declined.

Magnetic circular dichroism, the differential absorption of circularly polarized light, fit in well with our applications of polarized light to nucleic acids; therefore it was natural for us to apply radio frequency magnetic circular dichroism—better known as nuclear magnetic resonance. Although it is sort of a joke to lump NMR with magnetic circular dichroism, all spectroscopies are of course similar. My colleague Bob Harris and I considered the effect of circularly polarized UV light on NMR. We showed that contrary to a published report, the effect on chemical shifts of circularly polarized light was negligible (12).

As in our work with UV light, our NMR work started with RNA dinucleotides and trinucleotides. The first studies (1977) were at the Stanford Magnetic Resonance Laboratory on a 360-MHz (for protons) machine; Berkeley at that time only had a 200-MHz machine. We have most recently (2001) studied a 56-nucleotide RNA at Stanford on an 800-MHz machine; Berkeley only has 600 MHz.

RNA STRUCTURE

The number of atomic resolution structures of RNA molecules is minuscule compared with the number of protein structures. Proteins crystallize easier, and NMR determination of protein structure is easier. In proteins, scalar couplings can be used to make sequential through-bond assignments of a ^{15}N- and ^{13}C-labeled polypeptide chain. The equivalent cannot be done on nucleic acids because of the lack of a spin 1/2 oxygen nucleus to place on both sides of the phosphorus in each nucleotide. Furthermore, RNAs are easily hydrolyzed by finger nucleases, trace metal ions, and ribozymes. In spite of this, we have been able to identify and characterize several novel RNA structures by NMR.

The usual form of DNA (B-DNA) is a right-handed double helix with 10 Watson-Crick base pairs per turn. RNA is also right-handed, but with 11 base pairs per turn; this RNA structure is similar to A-DNA, the dehydrated form of B-DNA. However, alternating purine-pyrimidine sequences of DNA had been found to switch to a left-handed helix (Z-DNA) in high salt concentrations (>4 M NaCl). A *Scientific American* article (13) stated that steric hindrance from the 2′-hydroxyl group on ribose would prevent RNA from going left-handed, so we had to try it. Kathi Hall synthesized the ribo-polynucleotide with an alternating G–C sequence with the help of Mike Chamberlin, the RNA polymerase expert from the Molecular and Cell Biology Department. We needed even higher salt concentrations, but left-handed Z-RNA did form. What its function in biology is—in addition to attracting the interest of some humans—is unknown.

While on sabbatical at the University of Colorado I learned from Larry Gold about an RNA hairpin loop of four nucleotides, UUCG (a tetraloop), that blocked the enzyme reverse transcriptase that synthesizes DNA from an RNA template. Because the UUCG tetraloop's melting temperature to the single strand is higher than loops of other sequences, it became known as an extra-stable tetraloop. The obvious question was what is special about the structure? At Berkeley, Gabriele

Varani and Joon Cheong found that the guanine, G, in the loop is in a rare *syn* conformation relative to the ribose instead of the usual *anti*; it also forms an unusual type of G • U base pair. Except for the middle U, the bases make a compact structure knit together by hydrogen bonds and electrostatic attractions. The middle U is dangling outside the loop, consistent with the fact that any base can substitute for the U without decreasing stability or the reverse transcriptase stop. The structure thus actually did help in understanding the function. As a test of our NMR structure we gave some of the RNA to Steve Holbrook at the Lawrence Berkeley Lab to get a crystal structure. However, the RNA would only crystallize as a double-stranded helix with non-Watson-Crick base pairs; no hairpin loop formed in any crystal. We tried to study the double helix in solution, but it always precipitated, so we were never able to compare the NMR and X-ray structures.

Another benefit of my sabbatical at Colorado was learning about pseudoknots from Olke Uhlenbeck. Although we both heard the same lecture on pseudoknots at a meeting in the Netherlands, he paid attention. He convinced me that pseudoknots were the most important new structural motifs found in RNA. In a pseudoknot, one end of the chain in a hairpin loop (a stem and loop) forms a second loop and stem by folding back to base pair with the first loop. If each stem had 11 or more base pairs (one turn of helix), a knot could form. Jackie Wyatt and Jody Puglisi measured the thermodynamic stabilities of pseudoknots with different loop and stem sizes and determined the NMR structure of one of them. They both got theses from Olke's insight. A few years later Harold Varmus, then at University of California, San Francisco, called to ask if I was interested in studying the pseudoknot reponsible for programmed frameshifting in mouse mammary tumor virus. This is a typical retrovirus that synthesizes its essential enzymes plus a viral coat protein as one long polypeptide chain; the enzymes are reverse transcriptase to synthesize its DNA template, integrase to incorporate this DNA into its host, and protease to cut the polyprotein into its active parts. The polyprotein is made by a series of frameshifting signals in the viral RNA that synthesizes the enzymes and the coat protein as one chain. The virus needs much less of the enzymes than of the coat protein, so most of the time synthesis stops after the first protein. Ten percent of the time, synthesis continues to make the polyprotein containing the vital enzymes. In most retroviruses, but not HIV, a pseudoknot is required for this to occur.

In collaboration with the Varmus group we studied the effect of changes in sequence and structure of the pseudoknot on frameshifting efficiency (14). We learned what characteristics of the pseudoknot were required for frameshifting. I presented the work at Yale once, and afterwards Peter Moore said, "Now all you have to do to understand how this works is to determine the structure of the ribosome." He and others have since done this, so we hope that we will eventually be able to study the interaction of the pseudoknot with the ribosome and thus understand how pseudoknots cause frameshifting.

As should be clear, we and many others are determining the structures and thermodynamic stabilities of the building blocks that make up a functional RNA. I mentioned earlier the 56-nucleotide RNA molecule whose structure Ming Wu and Minxue Zheng recently determined at Berkeley. It is an independently-folding domain of the first RNA ever found to be catalytic (by Tom Cech). The complete ribozyme has over 400 nucleotides; it is presently too big for NMR. The structure of the small domain is significantly different in aqueous solution (200 mM Na^+) from its structure in a crystal as part of a larger piece of the ribozyme. As the crystalline RNA was in the presence of divalent ions, we added magnesium ions to the solution to more closely mimic the biological environment, and maybe to induce a structure similar to the crystal. Unfortunately, Mg^{2+} muddles the NMR spectrum beyond assignment; peaks split, broaden, and overlap. It is clear that multiple species are formed. This is not uncommon in RNA; it may be that RNA has evolved to be flexible and dynamic so it can change easily and respond to small changes in its environment. But it does make it difficult or impossible to determine its structure by any standard method. Multiple species produce either sums or averages of spectra depending on whether their exchange rates are slow or fast compared with the measurement time scale. In either case, or in the even worse case of intermediate exchange, determination of structure is very difficult. An obvious solution to the problem is to study a single molecule. One molecule at one time will have one conformation; a measured property will not be an average or sum of many species. It would be great to measure NMR on a single molecule, but at present we are content with using mechanical properties (force times distance) to measure thermodynamics and kinetics of single RNA molecules.

SINGLE-MOLECULE THERMODYNAMICS AND KINETICS

Carlos Bustamante, a graduate student in my group from 1976–1980, returned to Berkeley as a faculty member in 1998. He was now a leader in single-molecule studies, but he had only worked on DNA and proteins; he had ignored RNA. We agreed to rectify this. We (Jan Liphardt, Bibiana Onoa, and Steve Smith did the work) started with the 56-nucleotide domain from the first ribozyme and simpler versions of it. Micron-sized polystyrene beads are attached to the ends of the RNA by 500-base-pair DNA/RNA handles. One bead is held by a micropipet; the other is in a laser light trap. The distance (in nanometers) between the beads is measured and the force [in piconeutons (pN)] acting on the bead in the light trap is measured from its position relative to the center of the trap. Thus, the work necessary to unfold the RNA is measured; 1 pN times 1 nm (a zeptojoule) is equal to 0.6 kJ mol^{-1}. If the work is reversible at constant temperature and pressure, it is equal to the Gibbs free energy. The usual way to unfold an RNA is to heat it, add a denaturant such as 7 M urea, or remove all ions except the counterions needed to neutralize the phosphates. By using force, however, we can measure the free energy of unfolding the RNA at any temperature or solvent; we can study the effect of divalent ions, proteins, and any ligands on its unfolding and refolding (15).

The first RNA studied—a hairpin of 22 basepairs closed by a loop of 4 nucleotides (a tetraloop)—performed perfectly. It folded and unfolded (broke and reformed the base pairs) reversibly at a force of 14.5 pN with a change in length of 20 nm, thus revealing a free energy of transition of about 175 kJ mol^{-1} at room temperature. This free energy is in good agreement with that expected for unfolding the RNA without force (after correction for the loss of entropy of the stretched out single strand held by beads in our experiment). The change in length corresponds to breaking the 22 basepairs. When we hold the force constant in the folding-unfolding transition region we can see the molecule jump back and forth between folded hairpin and extended single strand. We actually watch the distance between the beads oscillate between two values. By increasing the force we shift the equilibrium toward the long single strand; decreasing the force favors the short hairpin. A plot of the logarithm of the equilibrium constant versus force gives a straight line with slope equal to the change in length of the RNA divided by kT.

$$\frac{\partial \ln K}{\partial F} = \frac{\Delta l}{kT}$$

The length from the slope agrees with the length seen in the force versus extension curves. The kinetics lessons are just as compelling. Increasing force speeds the rate of hairpin-to-single-strand reaction and slows the rate of single-strand-to-hairpin reaction. The lifetimes of the hairpin and single strand vary but have an exponential distribution of values. The logarithm of each first-order rate constant (the reciprocal of the average lifetime for each species) is linear in force. We interpret the slope (multiplied by kT) as a measure of the distance between the ends of the hairpin and the transition state, or the ends of the single strand and the transition state. The position of the transition state depends on the sequence of the hairpin, and the number of bases in the loop. We are obtaining detailed knowledge about the free energy landscapes for folding and unfolding RNA molecules with different secondary structures.

The most exciting result is that these single-molecule experiments illustrate so many concepts of freshman physical chemistry so directly:

1. Reversible mechanical work is equal to free energy.
2. At equilibrium there really are reactions going in opposite directions.
3. Force and distance are equivalent to pressure and volume; they have similar effects on equilibria as in the familiar van't Hoff equation.
4. For a first-order kinetic process the lifetimes of a species are distributed exponentially.
5. A reaction coordinate as understandable as the distance between the ends of a molecule can characterize the extent of a reaction.

Larger and more complex RNAs introduce new phenomena including irreversible unfolding, multiple unfolding paths, and so forth. However, one very pleasant property of mechanical unfolding experiments is that it does not matter how large the molecule is. Delphine Collin has unfolded a ribosomal RNA with 1500 nucleotides. It will be a while before all the transitions are assigned.

A FACULTY POSITION

I was hired at Berkeley in 1956 through the old boy network. The acting chairman, Jim Cason, telephoned his organic colleague, Jim English, at Yale, asking for a recommendation for a physical chemist. The chairman at Yale, Kirkwood, recommended me, and after a short discussion with English I was offered a job at Berkeley. As an Hispanic (both my parents were born in Mexico), I have since become a valuable addition to the diversity statistics. As far as I could tell, hiring was pretty haphazard in the 1950s and 1960s, although in biophysical chemistry, the department did very well. Dick Powell, our chairman from 1960–1966, seemed to choose people based on the cost of their interview trip. John Hearst, then a postdoc at Dartmouth, was interviewed through his research director, Jerome Vinograd, who was invited up from Caltech to serve as proxy. Ken Sauer was a postdoc in Melvin Calvin's lab at Berkeley, and Jim Wang was a postdoc at Caltech when they were hired.

In the 1970s increasing pressure was put on Berkeley to hire women and minorities. I remember one faculty meeting where one after another of my senior colleagues made impassioned pleas for obtaining a woman. Taken out of context it might have sounded like a group of sex-deprived old men begging for relief. I mention old men because it surprised me that the younger faculty were less enthusiastic; it was the older, more politically aware, men who recognized the need for women on the faculty. We hired the first faculty woman—with tenure—in 1977; the first African American—also with tenure—was hired in 1981 while I was chairman. It is very clear that the women and minorities on our faculty have been a tremendous asset to the department; it is also clear that they would not have been hired without the strong push from outside the department.

I am sometimes asked how the students and the campus have changed over the past 45 years. The answer is that the students have stayed pretty much constant, while I have changed. When I arrived in Berkeley the graduate students and I were in our twenties; they still are. We used to play tennis, rock climb on Indian Rock, run in Redwood park, backpack in the Sierra, and party—at some of the parties just breathing would get you high. Indian Rock, a 30-foot pile of rocks in Berkeley used as a practice area by beginners and expert rock climbers, provided our routine Friday afternoon outing. Curt Johnson and Arlene Blum introduced me to the sport. Curt and I moved up to bigger mountains including Mt. Hood in Oregon and the Breithorn rear Zermatt. Arlene graduated to Annapurna and Everest. We carried RNA flags—in Arlene's case a tRNA flag—up several mountains.

The students and I now, rarely, walk in Tilden park. We have both grown more conscious of funding and of the importance of publishing in high impact journals (as first or last author). Now students are more apt to choose nonacademic jobs because they fear the five-year uncertainty before a tenure decision and the recurring anxiety at each grant renewal period. Tenure and grants were not necessarily easier to get earlier, but now many students do not think the pain and effort of getting a job, getting tenure, and getting funded are worth the minor pleasures of a faculty position. Of course, I disagree.

It seems to me that by working hard the first few years you can get a grant and earn tenure while doing some good science. After that it is all fun and you cannot be fired. Students and postdocs do the research, write the papers, and most importantly of all, entertain and teach you. An ex-student, who became a patent lawyer after receiving her Ph.D., said that she liked hearing about exciting new inventions without having to spend long hours in the lab on experiments that did not work. I agree; I have not worked in a lab since my sabbatical at Oregon State in 1971. Yet I not only get to learn about exciting new results nearly every day, but I eventually get credit for them. A tenured position at a university can't be beat. A class is three 50-minute or two 80-minute lectures a week; yet my colleagues think more than one class a term is excessive. Although we all complain, correctly, of being overworked and underpaid, the fact is we can pretty much work as much or as little as we like. We get expense-paid trips all over the world, and we can significantly augment our university salary if we consult, or get patent royalties, or found a company. I love my job and plan to continue it until I get bored or die, whichever comes first.

ACKNOWLEDGMENTS

The many students, postdocs, and visitors who spent time in the lab since 1967 were welcomed, nurtured with kindness and food, and, if possible, laboratory-trained by Barbara Dengler. She preserves the group memory of procedures and protocols. David Koh has provided oligonucleotides, cold-noodle parties, and housing for visitors since 1968.

Graduate students who received a Ph.D. with me are F. Aboul-ela, A.D. Blum, P.N. Borer, C.A. Bush, C. Bustamante, Z. Cai, C.R. Cantor, D. Carroll, C.L. Cech, K.-Y. Chang, M. Chastain, C. Cheong, L. Comolli, J.D. Corbett, K.S. Dahl, P.W. Davis, R.C. Davis, S.L. Davis, A.E. Drobnies, C. Formoso, S.M. Freier, D. Glaubiger, R.L. Gonzalez, K. Hall, A.E.V. Haschemeyer, J.K. James, S.R. Jaskunas, J. Kao, D. Keller, J.S. Kieft, S.M. Landry, A.I. Levin, M.D. Levine, H. Lewis, P. Li, D.A. Lloyd, S.R. Mirmira, M. Molinaro, K.M. Morden, J.W. Nelson, J. Nowakowski, A. Pardi, J.D. Puglisi, E. Scoffone, L.X. Shen, L. Sun, B. Tomlinson, M.M. Warshaw, M.T. Watts, E. Wickstrom, B. Wimberly, S.A. Winkle, S. Wolk, R.W. Woody, J.R. Wyatt, K. Yamaoka, R. Yolles, and K. Yoon.

Postdoctoral students who spent time in the group include F.S. Allen, V.P. Antao, F.H. Arnold, R. Biltonen, K.J. Breslauer, X. Chen, L.B. Clark, D. Collin, G. Colmenarejo, P. Cruz, H. DeVoe, E. Evertsz, D.M. Gray, C.C. Hardin, J.V. Hines, D.N. Holcomb, W. Hug, M.S. Itzkowitz, J.A. Jaeger, K. Javaherian, B.B. Johnson, W.C. Johnson, H. Kang, C.-H. Kim, C.-H. Lee, J. Liphardt, K.J. Luebke, S. R. Lynch, M.F. Maestre, S. Mandeles, F.H. Martin, J. McMahon, D.W. McMullen, G. Meng, W. Mickols, D. Moore, R. Nussinov, B. Onoa, S. K. Podder, A. Pohorille, C. Reich, L.J. Rinkle, S. Ruedisser, J. SantaLucia, M. Schmitz, G. Shimer, J. Thiéry, W.N. Thurmes, D.H. Turner, O.C. Uhlenbeck, G. Varani, G.T. Walker, A.L. Williams, M. Wu, and M. Zheng.

I greatly appreciate everything I learned from everybody, and all the fun we had in and out of the lab. I trust the next few years will be as exciting as the previous ones. It was suggested a while ago in *Physics Today* that the termination of mandatory retirement might encourage the most dedicated professors to be buried with a few students and postdocs to continue their research. I have not discussed this yet with my current research group.

Visit the Annual Reviews home page at www.annualreviews.org

LITERATURE CITED

1. Moffitt W, Fitts DD, Kirkwood JG. 1957. Theory of optical activity of helical polymers. *Proc. Natl. Acad. Sci. USA* 43:723–30
2. Tinoco I Jr, Freeman MP. 1957. The optical activity of oriented copper helices. I. Experimental. *J. Phys. Chem.* 61:1196–200
3. Tinoco I Jr, Woody RW. 1964. Optical rotation of oriented helices. IV. A free electron on a helix. *J. Chem. Phys.* 40:160–65
4. Moore D, Tinoco I Jr. 1980. The circular dichroism of large helices, a free particle on a helix. *J. Chem. Phys.* 82:3396–400
5. Townes CH. 1999. *How the Laser Happened.* New York: Oxford Univ. Press
6. Tinoco I Jr. 1960. Hypochromism in polynucleotides. *J. Am. Chem. Soc.* 82:4785–90; Erratum. 1961. *J. Am. Chem. Soc.* 84:5047. The error in this article was an elementary one. The referees kept the manuscript about two months and then made only minor suggestions. I later thought that if they were going to read it so carefully, they should at least have found the mistake.
7. Tinoco I Jr, Uhlenbeck OC, Levine MD. 1971. Estimation of secondary structure in ribonucleic acids. *Nature* 230:362–67
8. Tinoco I Jr, Borer PN, Dengler B, Levine MD, Uhlenbeck OC, Crothers DM, Gralla J. 1973. Improved estimation of secondary structure in ribonucleic acids. *Nat. New Biol.* 246:40–41
9. Turner DH, Tinoco I Jr, Maestre MF. 1974. Fluorescence detected circular dichroism. *J. Am. Chem. Soc.* 96:4340–42
10. Tinoco I Jr, Turner DH. 1976. Fluorescence detected circular dichroism. Theory. *J. Am. Chem. Soc.* 98:6453–56
11. Maestre MF, Bustamante C, Hayes TL, Subirana JA, Tinoco I Jr. 1982. Differential scattering of circularly polarized light by the helical sperm head from the octopus eledone cirrhosa. *Nature* 298:773–74
12. Harris RA, Tinoco I Jr. 1994. Laser-perturbed NMR spectroscopy and the conservation of parity. *J. Chem. Phys.* 101:9289–94
13. Dickerson RE. 1983. The DNA helix and how it is read. *Sci. Am.* 249:94–98
14. Chen XY, Chamorro M, Lee SI, Shen LX, Hines JV, Tinoco I Jr, Varmus HE. 1995. Structural and functional studies of retroviral RNA pseudoknots involved in ribosomal frameshifting: Nucleotides at the junction of the two stems are important for efficient ribosomal frameshifting. *EMBO J.* 14:842–52. There was some discussion about whether the first author should be from my lab or Varmus' lab. The compromise was that the first two authors have asterisks with the statement that both contributed equally to the work. I thought that Varmus and I should also have asterisks so we could equally share the desirable last position.
15. Liphardt J, Onoa B, Smith SB, Tinoco I Jr, Bustamante C. 2001. Reversible unfolding of a single RNA molecule by force. *Science* 292:733–37

HIGHER-ORDER OPTICAL CORRELATION SPECTROSCOPY IN LIQUIDS

John T. Fourkas
Eugene F. Merkert Chemistry Center, Boston College, Chestnut Hill, Massachusetts 02467; e-mail: fourkas@bc.edu

Key Words Ultrafast spectroscopy, multidimensional spectroscopy, liquid dynamics

■ **Abstract** Linear optical spectroscopies have long been used to study the behavior of liquids. Laser technology has progressed to the point that it has become possible to perform nonlinear optical experiments that probe higher-order correlation functions in liquids, opening a new window into our understanding of the microscopic details of solution-phase processes. Here we review advances that have been made in recent years in employing higher-order electronic and vibrational spectroscopies to study liquid-state dynamics and structure.

INTRODUCTION

Liquids are inherently complex media that feature a high degree of disorder coupled with strong intermolecular interactions. As a result, dynamics in liquids span a broad range of time and distance scales. These inherent solvent dynamics couple to all chemical and physical processes that take place in solution, such as solvation, vibrational relaxation, and electron transfer, to name a few. Additionally, the local structure of a liquid is generally affected by solutes, which in turn influence local dynamics, adding yet another level of intricacy to the already rich complexity of dynamics in solution-phase processes.

Optical techniques such as electronic, Raman and infrared spectroscopies have long been powerful tools for helping to unravel the microscopic details of liquid dynamics. However, conventional (linear) spectroscopic data are often consistent with multiple interpretations. The crux of the problem is that there are a number of different sources of line broadening of absorption spectra in liquids, and it is difficult to distinguish the effects of different broadening mechanisms using linear spectroscopy. In one extreme (the homogeneous limit) the width of an absorption line may be representative of the spectrum of each absorber, while in the other extreme (the inhomogeneous limit) the width of an absorption line is far broader than the spectrum of any given single absorber due to the range of local environments possible in a liquid. Furthermore, because liquids do exhibit dynamics over such a broad range of time and distance scales, most spectra in liquids lie somewhere between these two limits. Indeed, while inhomogeneous broadening is a well-defined concept in a static medium, its definition is considerably more ambiguous in an environment that is in constant flux.

These same sorts of problems have long been attacked in nuclear magnetic resonance (NMR) spectroscopy through the use of sequences of multiple pulses to perform nonlinear spectroscopies. Feynman et al. (1) demonstrated in 1957 that there is a direct analogy between NMR and optical spectroscopy. Thus, in principle the same sorts of pulse sequences could be employed to perform nonlinear optical spectroscopy on liquids, potentially providing a great deal more information about the microscopic dynamics of liquids than can be gained directly from linear spectroscopies. However, there are also some crucial differences between NMR and optical transitions in liquids that made the implementation of multiple pulse sequences in liquids considerably more difficult. In particular, the ratio of the width of a typical solution-phase NMR line to its center frequency, $\Delta\omega/\omega_0$, is at most on the order of 10^{-7}. On the other hand, for electronic and vibrational transitions in solution typical values of $\Delta\omega/\omega_0$ lie in the range of 10^{-2} to 10^{-3}, which makes broadband, frequency-independent excitation considerably more difficult. Furthermore, optical transition frequencies are much larger than those in NMR. Optical multiple-pulse sequences must therefore employ pulses of durations in the range of 10^{-15} to 10^{-13} seconds. Until recently, the generation and phase control of such optical pulses was quite challenging.

Rapid advances in laser technology over the past two decades have now made possible the relatively routine tabletop generation of femtosecond laser pulses with center frequencies ranging from the mid-infrared to the near ultraviolet. These technologies have finally opened the door to higher-order electronic and vibrational correlation spectroscopies in liquids. Multidimensional optical spectroscopy in liquids is a field that has expanded rapidly in the last few years, and here we review recent developments in this area.

BACKGROUND

In discussing higher-order correlation spectroscopies in liquids, we will follow the spirit of Feynman et al. (1) in relying heavily on analogies with NMR via the optical Bloch equation (2). While the Bloch picture does not always provide a sufficiently detailed description of the ultrafast nonlinear spectroscopy of liquids (3, 4), it does provide an excellent, clear description of the basic physics underlying higher-order correlation spectroscopies.

Consider an ensemble of N two-level systems, each of which is comprised of a ground state $|0\rangle$ and an excited state $|1\rangle$. In general the wave function of one system j can be described by some superposition of these states, i.e., $|\psi_j\rangle = a_j|0\rangle + b_j|1\rangle$.

The density matrix for this system is then given by

$$\tilde{\rho}_j(t) = \begin{bmatrix} a_j^*(t)a_j(t) & b_j^*(t)a_j(t) \\ a_j^*(t)b_j(t) & b_j^*(t)b_j(t) \end{bmatrix}. \qquad 1.$$

The density matrix for the entire ensemble is

$$\tilde{\rho}(t) = \frac{1}{N}\sum_{j=1}^{N}\tilde{\rho}_j(t) = \begin{bmatrix} a^*(t)a(t) & b^*(t)a(t) \\ a^*(t)b(t) & b^*(t)b(t) \end{bmatrix}, \qquad 2.$$

where

$$a(t) = \frac{1}{N}\sum_{j=1}^{N} a_j^*(t)a_j(t) \qquad 3.$$

and so on. The diagonal elements of the density matrix are the relative populations of the ground and excited states and must therefore be real. The off-diagonal elements describe the coherence of the ensemble. These elements are, in general, complex numbers, so the density matrix can be characterized by three time-dependent real numbers:

$$x(t) = b^*(t)a(t) + a^*(t)b(t)$$
$$y(t) = i[b^*(t)a(t) - a^*(t)b(t)] \qquad 4.$$
$$z(t) = b^*(t)b(t) - a^*(t)a(t).$$

Taken together, x and y describe the phase and degree of the coherence between the ground and excited states, whereas z is the population difference between these states.

Note that $x^2 + y^2 + z^2 \leq 1$, so these quantities can be thought of as describing a three-dimensional vector **r** with a maximum length of unity. This is known as the "optical Bloch vector" of the ensemble, and its behavior (either under an applied field or during free evolution) is governed by the optical Bloch equations (2). The optical Bloch vector is completely analogous to the Bloch vector that describes the net magnetization in NMR (5), and the optical Bloch equations correspond directly to the magnetic resonance Bloch equations (5). By the same token, optical Bloch-vector diagrams can be used to describe coherent optical experiments. (There is a minor difference in convention that the net magnetization vector in NMR is taken to point in the positive direction along the z axis initially, whereas the optical Bloch vector is taken to point in the negative direction along this axis.)

Figure 1 (see color insert) shows examples of optical Bloch-vector diagrams after a single pulse with a flip angle of $\pi/2$. This corresponds to the simplest pulsed NMR experiment that one can perform, in which the free-induction decay (i.e., the Fourier transform of the line shape) of the in-plane magnetization after the pulse is recorded. Note that Bloch-vector diagrams are conventionally drawn in a frame of reference that rotates with a frequency corresponding to the center of the absorption line in question. Coherences of molecules with higher transition frequencies than the center frequency will rotate in a clockwise manner relative to this frame of reference and coherences of molecules with lower transition frequencies than the center frequency will rotate in a counterclockwise manner. If the absorption line width is dominated by inhomogeneous broadening, then this spreading of the different frequency components will be the major mechanism responsible for destroying the in-plane component of the net Bloch vector (see Figure 1*a*, color insert). If the line width is instead dominated by population relaxation, then there is no appreciable spreading of the in-plane component of the net Bloch vector over the course of the free-induction decay (Figure 1*b*, color insert). This is one example of a case in which the line is homogeneously broadened. Another homogeneous broadening mechanism is phase-interrupting collisions, which lead to what is known as pure dephasing (Figure 1*c*, color insert).

The above examples of line-broadening mechanisms are idealized. Because liquids are far from static, inhomogeneous broadening is not as well-defined a concept in liquid-phase spectroscopy as it is in the spectroscopy of systems that evolve on a much slower timescale (such as crystals and glasses). Thus, true inhomogeneous broadening only exists on very short (often subpicosecond) timescales in most liquid spectra, even if distinct transition frequencies can be associated with different molecules. In such a case, on longer timescales the evolution of the liquid structure causes the transition frequency of each molecule to change with time in a process known as spectral diffusion (Figure 1*d*, color insert). If spectral diffusion occurs on a fast enough timescale it can lead to what is known as motional narrowing of spectral lines (6).

In liquid-phase spectra it is generally difficult to distinguish the contributions of the different line broadening mechanisms discussed above. In the best of cases, given an isolated spectral line, it is possible to gain some information about the broadening mechanism. For instance, it is sometimes possible to estimate the amount of inhomogeneous broadening from the line shape. Even in such a favorable case, it is generally not possible to ascertain the mechanism for any homogeneous broadening. Furthermore, electronic spectra are generally congested enough that line shape analysis is not feasible.

The crux of the difficulty in making an unambiguous determination of the factors contributing to line broadening in an absorption spectrum (or, equivalently, in a free-induction decay) is that the spectrum is given by (6, 7)

$$I(\omega - \omega_0) \int_{-\infty}^{\infty} \int_{-\infty}^{\infty} e^{i(\omega - \omega_0)t} \langle \mu(t)\mu^*(0) \rangle_{\omega_0} G(\omega_0) \, dt \, d\omega_0, \qquad 5.$$

where ω_0 is the average transition frequency of a given molecule in the inhomogeneous distribution $G(\omega_0)$. This is the Fourier transform of the dipole correlation function, which is given by (6, 7)

$$\langle \mu(t)\mu^*(0) \rangle_{\omega_0} = \left\langle \exp\left(-i \int_0^t \omega(t') \, dt'\right) \right\rangle_{\omega_0}, \qquad 6.$$

where $\omega(t')$ is the time-dependent transition frequency of a molecule that started out with transition frequency ω_0 at time $t = 0$. This dipole correlation function depends on only two times (or, more to the point, on one delay time). The various line-broadening mechanisms that we have considered have time dependences that can be similar or identical, and so the dipole correlation function does not allow us to distinguish them.

The solution to this problem is to employ nonlinear spectroscopies that probe higher-order correlation functions (i.e., correlation functions that depend on three or more times). Such techniques have long been used in NMR spectroscopy (5, 8), and the first optical experiment to probe a higher-order correlation function was performed in 1964 (9). Due to the limitations of the laser technologies that were available, the early higher-order optical correlation experiments were performed on gases and solids. However, ultrafast laser technology has progressed to the point that such experiments have been performed routinely on liquids for more than a decade.

ELECTRONIC SPECTROSCOPIES

Electronic spectra in liquids differ from those in the gas phase in a number of important ways. First, both absorption and fluorescence spectra broaden considerably in solution, which is due in large part to the range of energetically different solvent environments that can exist around a guest molecule. The electronic spectra of most molecules broaden in solution to the extent that it is no longer possible to observe individual vibronic transitions. Second, the equilibrium structure of a solute may differ from that in the gas phase, which can lead to shifts in vibrational frequencies and Franck-Condon factors. Thus, electronic spectra in solution are not necessarily merely broadened versions of their gas-phase counterparts. Finally, electronic spectra exhibit solvatochromism in solution, which is to say that absorption spectra are generally shifted to the blue and fluorescence spectra to the red as compared to the gas phase, leading to a solvent-dependent equilibrium Stokes shift. This effect comes about because the solvent acts to stabilize the electronic state of a solute. Properties of the solute such as dipole moment, polarizability, shape and size change upon electronic excitation. Since electronic excitation happens on a very fast timescale, the solvent cannot adjust to the new electronic state of the solute instantaneously. A solvent configuration that stabilizes the ground state is not ideal for the excited state, and so the electronic transition occurs at a higher frequency than in the gas phase. Over time, the solvent rearranges to stabilize the new electronic state, which leads to a solvent configuration that is less than optimal when the solute returns to its ground electronic state via fluorescence. As a result, the fluorescence spectrum moves to a lower frequency than in the gas phase.

Using an ultrafast laser pulse to excite solute molecules electronically allows the time dependence of the fluorescence spectrum to be observed. Time-resolved fluorescence Stokes shift (TRFSS) experiments have been used by many groups to study the dynamics of solvation (10–15). When femtosecond laser pulses are used, such experiments generally reveal bimodal dynamics. A significant fraction of the dynamic Stokes shift occurs on a timescale of a few hundred femtoseconds. This component is generally attributed to "inertial" motions of the solvent molecules. The rest of the dynamic Stokes shift, which occurs on a timescale that can range from picoseconds to hundreds of picoseconds or longer, arises from diffusive dynamics of the solvent.

Although TRFSS experiments have taught us a considerable amount about dynamics in solution, this technique has limited time resolution, and so is not able to probe the very earliest stages of solvation. Furthermore, one cannot distinguish the various contributions to the broadening of solution-phase electronic spectra with this technique. The results of TRFSS experiments have thus provided the impetus for employing higher-order electronic correlation spectroscopies to study liquids.

Two-Pulse Photon Echoes

Perhaps the simplest higher-order optical coherence spectroscopy is the photon echo, which is directly analogous to the NMR spin echo. The Bloch-vector diagram for this technique is shown in Figure 2 (see color insert). A pulse at time $t = 0$ places the molecules in a coherence between the ground and excited states. This coherence is allowed to evolve for a time τ, after which a second pulse is used to reverse the phase of the coherence on each molecule (i.e., if a particular molecule has a phase of φ relative to the in-plane component of the net Bloch vector, this phase is converted to $-\varphi$). Any inhomogeneous broadening that remains is then rephased at time 2τ, leading to a coherent photon-echo signal.

The photon echo is most commonly implemented using two noncollinear light pulses, one with wave vector $\vec{k}_1$ and the other with wave vector $\vec{k}_2$. The signal then propagates in a unique direction in space with wave vector $\vec{k}_s = 2\vec{k}_2 - \vec{k}_1$. When set up in this noncollinear manner, the photon echo is a zero-background technique (i.e., under ideal circumstances no light should reach the detector in the absence of the signal). In the simplest implementation, the integrated signal at $\vec{k}_s$ is collected as a function of τ, although it is often worthwhile to spend the additional effort to time-resolve the echo signal for reasons that we will consider below.

The photon echo is sensitive to the correlation function (7, 16)

$$\langle \mu^*(\tau + \tau_{sig})\mu(\tau)\mu(\tau)\mu^*(0) \rangle_{\omega_0}$$

$$= \left\langle \exp\left(i \int_{\tau}^{\tau+\tau_{sig}} \omega(t')\,dt' - i \int_{0}^{\tau} \omega(t')\,dt' \right) \right\rangle_{\omega_0}, \qquad 7.$$

where $\tau + \tau_{sig}$ is the time at which the signal is collected. This technique therefore depends on a four-time correlation function, which in this case depends explicitly on two delay times. If the echo is not time-resolved, then the detected signal depends on the integral of Equation 7 from $\tau_{sig} = 0$ to ∞. In the limit in which the absorption line width is dominated by static inhomogeneous broadening (Figure 2a, color insert), $\omega(t')$ is essentially equal to ω_0, and the integral of the correlation function in Equation 7 over ω_0 will be peaked when the time intervals in the two integrals are equal, i.e., when $\tau_{sig} = \tau$. Thus, in the inhomogeneous limit the echo is generated at time 2τ, and the decay of the echo signal with τ reflects the homogeneous broadening in the system. In the homogeneous limit (Figure 2b, color insert), $\omega(t')$ displays fast fluctuations of magnitudes that can be comparable to the line width. As a result, the correlation function in Equation 7 decreases monotonically with increasing τ_{sig}, and so the "echo" signal is generated at $\tau_{sig} = 0$. As in the inhomogeneous case, the echo will decay with increasing τ if the broadening is predominantly homogeneous. In either case, for an ideal two-level system the integrated echo signal will decay at the homogeneous dephasing rate as τ is scanned.

Of course, molecules are not ideal two-level systems, and this fact can introduce complications into the echo decay. Dephasing in liquids is fast enough that to perform a photon echo experiment it is optimal to employ laser pulses that are at most a few tens of femtoseconds in duration. The bandwidth of such pulses is large enough that many vibronic coherences are excited simultaneously in the molecule being studied. As a result, phenomena such as vibrational relaxation can play an important role in the signal. Matters are complicated further by the dynamic Stokes shift, which is one reason that a Bloch-vector picture does not really provide an adequate description of photon-echo experiments in liquids (3, 4).

The two-pulse photon echo was first performed in a liquid solution in 1989 by Becker et al. (17), who used 6-fs optical pulses to study the behavior of Nile blue and malachite green in ethylene glycol. These researchers found that intramolecular vibrational coherences contribute strongly to the echo signal. In these and ensuing studies (3, 4, 18–27), it has been found in general that the integrated echo signal decays on a timescale of tens of femtoseconds and can be almost symmetric about zero delay time. Analysis of two-pulse echo data therefore requires a very accurate knowledge of the zero delay time, which can be achieved by collecting the echo signals at $2\vec{k}_2 - \vec{k}_1$ and $2\vec{k}_1 - \vec{k}_2$ simultaneously (18). As would be expected from the considerations discussed above, the time-resolved echo signal also has been found to give significantly more information about the liquid dynamics than does the integrated echo signal (28, 29). As useful as the time-resolved two-pulse echo proved to be, attention rapidly turned to the even more powerful three-pulse echo.

Three-Pulse Photon Echoes

The Bloch-vector picture for the three-pulse ("stimulated") echo in an inhomogeneously broadened ensemble is shown in Figure 3. An initial pulse of wave vector $\vec{k}_1$ prepares the ensemble in a coherent superposition of the ground and excited states at time $t = 0$. A second pulse with wave vector $\vec{k}_2$ "stores" this coherence as a population at time $t = \tau$. At time delay T after the second pulse, a third pulse with wave vector $\vec{k}_3$ is used to convert the population back into a coherence in which

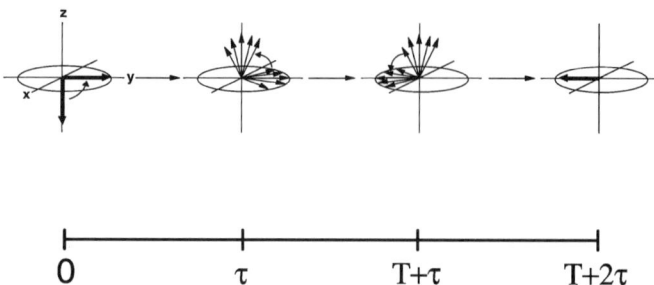

Figure 3 Optical Bloch-vector diagram for the three-pulse echo. A coherence between states $|0\rangle$ and $|1\rangle$ is created at $t = 0$. This coherence is allowed to evolve for delay time τ, after which the the net in-plane Bloch vector is stored as a population. After another waiting time T, the population is turned back into a coherence with the phases reversed from the initial coherence. If the ensemble is inhomogeneously broadened, rephasing will lead to the production of a coherent signal that will peak near time $T + 2\tau$. If the absorption line is homogeneously broadened the coherent signal will be generated at the time of the third pulse.

the phases of the inhomogeneous frequency components have been reversed from those in the original coherence. A coherent echo signal will then be generated with wave vector $\vec{k}_s = -\vec{k}_1 + \vec{k}_2 + \vec{k}_3$.

The three-pulse echo is sensitive to the correlation function (7, 16)

$$\langle \mu^*(T + \tau + \tau_{sig})\mu(T + \tau)\mu(\tau)\mu^*(0)\rangle_{\omega_0}$$

$$= \left\langle \exp\left(i \int_{T+\tau}^{T+\tau+\tau_{sig}} \omega(t')\,dt' - i \int_0^\tau \omega(t')\,dt'\right)\right\rangle_{\omega_0}. \qquad 8.$$

By using the second pulse to store the coherence as a population, frequency fluctuations can be compared over much longer timescales than in the two-pulse echo (which is the $T = 0$ limit of the three-pulse echo). In analogy to the two-pulse echo, if there exists substantial long-lived inhomogeneity at $T + \tau$, then the echo will reach its maximum at $\tau_{sig} = \tau$. If, on the other hand, the dephasing is entirely homogeneous by $T + \tau$ time, the echo signal will be at its maximum at $\tau_{sig} = \tau$. Thus, the time-resolved three-pulse echo signal provides significantly more information than the integrated three-pulse echo signal.

Three-pulse echoes were first performed in liquid solution by Ippen and coworkers (30), who found that dephasing of both Nile blue and rhodamine 640 in methanol occurred on a timescale faster than 20 fs. In 1993, Joo and Albrecht (18) studied the dynamics of cresyl violet and LD690 in ethylene glycol. These researchers found that electronic dephasing occurs in these systems on two different timescales, the faster of which is on the order of 20 fs and the slower of which is on the order of 2.5 ps in both cases. This general trend has been observed in most of the now significant number of three-pulse echo experiments that have been performed in liquids (31–43).

In 1996 it was suggested (33, 34) that if the signals at wave vectors $-\vec{k}_1 + \vec{k}_2 + \vec{k}_3$ and $\vec{k}_1 - \vec{k}_2 + \vec{k}_3$ were collected simultaneously, then the time shift between these two peaks would yield direct information about the solvation dynamics. This so-called three-pulse echo peak shift technique has since become the most common implementation of the three-pulse echo in liquids and has proven versatile enough to be employed in the study of biologically-relevant systems as well (44–49). As in the case of two-pulse photon echoes, time-resolving the three-pulse echo signal has also proven to be helpful in unraveling liquid dynamics (24, 32, 35, 38). The time-resolved technique has also proven capable of removing the contribution of individual intramolecular modes to the three-pulse echo signal (35), although in most probe molecules there are enough vibrational modes to still have a significant effect on the three-pulse echo signal (42, 43).

Cho and Fleming also proposed another three-pulse photon echo technique that relies on fifth-order (rather than the usual third-order) nonlinearity (19). This technique is capable of adding enough control to the rephasing process to distinguish between the dephasing effects of fast and slow solvent motions. The fifth-order, three-pulse echo was implemented by Joo et al. to study HITCI in ethylene glycol (31). By studying the shape of the signal along the two different delay times, they were indeed able to separate the effects of solvent motions occurring on different timescales. However, since the information content available from both methods is similar, the fifth-order echo has not been as popular as the somewhat easier three-pulse echo peak shift technique.

Transient Hole Burning

Transient hole-burning is another example of an electronic spectroscopy that depends on a higher-order time correlation. A schematic diagram of a transient hole-burning experiment on an electronic transition with some degree of inhomogeneous broadening is shown in Figure 4. At time $t = 0$ a laser pulse with a bandwidth that is less than the inhomogeneous broadening in the transition is used to excite a subset of molecules electronically, leaving a "hole" in the ground-state absorption spectrum and placing a population of molecules in the excited state within a narrow frequency range. A subsequent broadband light pulse is used to detect the absorption bleach from the ground-state hole and the stimulated emission from the excited-state population. As time progresses, the ground-state and excited-state populations both equilibrate in their respective solvation potentials, leading to shifts and broadening in the absorption bleach and stimulated-emission signals that allow the solvent dynamics to be monitored.

Transient hole-burning is sensitive to essentially the same correlation function as the three-pulse echo (16), but due to experimental constraints the systems that can be studied by these two techniques tend to be different. In order to burn a

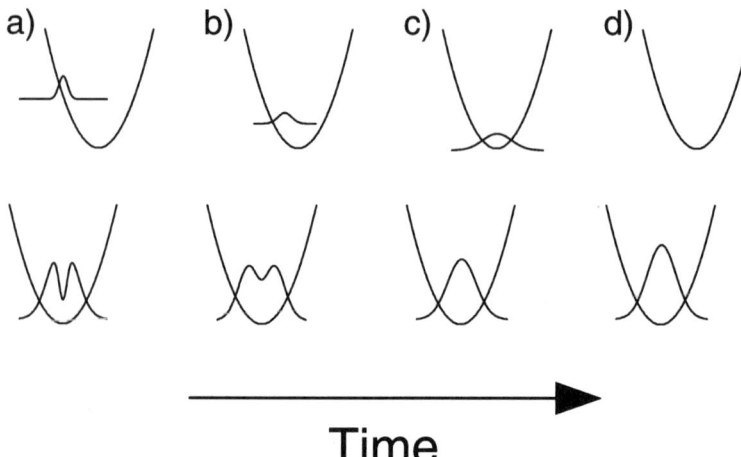

Figure 4 Schematic diagram of a transient hole-burning experiment. The lower and upper parabolas represent the ground- and excited-state solvation potentials, respectively. At time zero (*a*), a subset of the molecules in the inhomogeneous ground-state distribution is excited. As time progresses (*b*), the ground-state hole broadens and the excited-state population shifts and broadens until it reaches equilibrium (*c*). Eventually (*d*), all of the population returns to the ground state and is equlibrated there.

hole in the absorption spectrum, the bandwidth of the first pulse in the transient hole-burning experiment must not be too great, which in turn implies that the pulse cannot have an especially short duration. As a result, transient hole burning is not generally well-suited for studying subpicosecond dynamics, but is very sensitive to dynamics that occur on longer timescales.

The first liquid-phase transient hole-burning experiment was reported by Shank and coworkers in 1986 (50). These authors used 10-fs optical pulses to study Nile blue, cresyl violet and HITC in solution. The broad bandwidth of the pulses used in this experiment encompassed many vibronic levels of the absorption spectra of the dye molecules, and so only very general information about the solvent dynamics could be extracted.

Some years later, Berg and coworkers demonstrated that transient hole burning is a powerful technique for studying the solvation dynamics of solute molecules that do not undergo a change in dipole moment upon excitation to the excited state and therefore exhibit only modest solvatochromism and inhomogeneous broadening (51–59). Owing to the relatively weak interactions between the solute and solvents in these experiments, the inhomogeneous broadening was of small enough magnitude that with a picosecond burning pulse it was possible to create a spectral hole that was essentially entirely within the 0-0 transition (53). In addition, the absorption and fluorescence spectra could be modeled to good approximation by convolving the corresponding gas-phase spectra with a gaussian function (53). As a result, the dynamics of solvation could be probed in great detail without contamination from intramolecular vibrational dynamics. As in the case of polar solvation, nonpolar solvation dynamics were found to consist of a subpicosecond component due to inertial motions followed by a slower component due to diffusive solvent motion (51–56). These dynamics were followed in glass-forming solvents to temperatures below the glass transition, and notably the magnitude of the fast component extrapolated directly into the glassy phase, whereas the other component became too slow to observe (53–57, 59). This behavior was described later in terms of a viscoelastic continuum model of the system (57, 58, 60).

A number of other liquid-phase phenomena have been studied with transient hole burning as well. Yu and Berg used this technique to study the dynamics of the permanganate ion in glassy and liquid $LiCl \cdot 6H_2O$ and were able to monitor the dynamic local symmetry distortions imposed on the solute by the solvent (61). The same researchers also used this technique to study the dynamics of the breaking and formation of hydrogen bonds to resorufin in ethanol and ethylene glycol (62). In addition, Kinoshita and coworkers have used transient hole burning to study polar solvation dynamics (63–65).

Two-Dimensional Fourier-Transform Electronic Spectroscopy

One of the most powerful aspects of NMR spectroscopy is the ability to generate multidimensional spectra that, by virtue of their structure, reveal details about the system under study immediately upon visual inspection (8). Jonas and coworkers have followed this philosophy in developing techniques for obtaining three-pulse optical spectra and transforming them into the frequency domain via two-dimensional inverse Fourier transformation (66–69). In order to achieve this end they have developed a technique to measure the absolute phase of the signal generated from three noncollinear pulses, the timing of each of which can be controlled independently. This opens up the possibility of performing the inverse Fourier transforms with respect to a number of the different experimental delay times, and each of these combinations is capable of revealing unique information about the system (66, 67, 70).

Hybl et al. have used such a technique to map out the real and imaginary portions of the inverse Fourier transform of the three-pulse echo signal of the cyanine dye IR144 in methanol (68). This work is a beautiful demonstration of the potential of multidimensional Fourier-transform electronic spectroscopy in revealing the details of microscopic solvent dynamics. Such two-dimensional spectra make the connection between the excitation frequency and the emission frequency readily apparent and should prove invaluable in the study of electronic dynamics in solution.

RAMAN SPECTROSCOPIES

Vibrational spectra are not affected by solvation-induced line shape effects to the same extent as are electronic spectra. In Raman spectroscopy, line shape analysis is often useful in estimating the relative homogeneous and inhomogeneous contributions to a vibrational band. However, Raman spectra are also susceptible to the effects of Fermi resonances and other coupling to vibrations on the same or other molecules, as well as to motional narrowing and other phenomena that can distort the line shape. Furthermore, even in the absence of these effects at times it can be difficult to distinguish among the different effects that can lead to line broadening. As in the case of electronic spectroscopy, Raman spectroscopies that depend on higher-order correlation functions have the capability of resolving many of these issues.

The simplest coherent Raman technique is coherent anti-Stokes Raman spectroscopy (CARS) (71). In this technique, two time-coincident laser pulses of center frequencies $\omega_1 > \omega_2$ are used to create a coherence in a Raman-active mode with a characteristic frequency of $\omega_1 - \omega_2$. This coherence is probed time τ later by a laser pulse with center frequency ω_3, thus generating a coherent signal with a center frequency of $\omega_s = \omega_1 - \omega_2 + \omega_3$. Although this technique depends on a third-order optical nonlinearity, as do the two- and three-pulse photon echoes, it is a linear technique in the vibrational mode being probed. The CARS signal is in fact the Raman free-induction decay, which is sensitive to the two-time correlation function $\langle \tilde{\alpha}(t) \tilde{\alpha}(0) \rangle$, where $\tilde{\alpha}$ is the polarizability tensor (72, 73). Thus, to study Raman correlation functions in liquids that are higher in order in the vibrational response, it is necessary to use techniques that have fifth- or even seventh-order nonlinearity (72, 73). Despite the highly nonlinear nature of such experiments, they can be accomplished using visible light, and so for technological reasons the development of higher-order Raman techniques preceded that of the corresponding infrared techniques.

Raman-Echo Spectroscopy

The Raman echo is a direct analogue of the photon echo. As in the case of CARS spectroscopy, the pulse sequence begins with two time-coincident pulses of frequencies $\omega_1 > \omega_2$ creating a Raman coherence in the sample. Time τ later, the phases of the individual oscillators that comprise this coherence are reversed by two interactions with another time-coincident pulse pair with the same frequencies. The rephasing is then probed by a final pulse of frequency ω_1 that is incident on the sample at delay time τ_{sig} after the second pair of pulses, generating a coherent signal at frequency $\omega_{sig} = 2\omega_1 - \omega_2$. Although the Raman echo signal depends on a seventh-order optical nonlinearity, it probes the four-time correlation function $\langle \tilde{\alpha}^*(\tau + \tau_{sig}) \tilde{\alpha}(\tau) \tilde{\alpha}(\tau) \tilde{\alpha}^*(0) \rangle$ (72, 73).

The Raman echo technique was first proposed in 1968 (74), but it took more than two decades before a Raman-echo signal was observed in a liquid. This experiment is challenging owing to the high order of the optical nonlinearity as well as to the fact that the multiple colors employed make phase matching in a dispersive liquid a difficult problem. It was even suggested at one point that Raman echoes could not be observed in liquids due to complications from other nonlinear optical phenomena that occur in liquids at high laser intensities (75).

These experimental challenges were overcome by Berg and coworkers, who succeeded in observing Raman echoes from the symmetric methyl stretch of room-temperature acetonitrile in 1991 (76). The echo decays in this liquid were found to match those of the CARS signal, demonstrating that the symmetric methyl stretch is homogeneously broadened. The same conclusion was reached by the Berg group in a later Raman-echo study of the symmetric methyl stretch of ethanol-1,1-d_2 over a temperature range covering everything from the room-temperature liquid to the glassy state at 12 K (77). A similar conclusion was also reached by the Yoshihara group for the C $\equiv$ N stretch of benzonitrile (78, 79) and the S-H stretch of ethyl mercaptan (80). Berg and coworkers have suggested (81, 82) based both on these experimental results and on a theory proposed by Schweizer & Chandler (83) that Raman lines in neat liquids may generally be homogeneously broadened due to the relative stability of the local density about any given molecule in a liquid. Intramolecular vibrational energy redistribution may also be implicated, particularly in the case of the ethanol experiments (77).

Inhomogeneity was first observed in a liquid-phase Raman echo experiment in a study of the symmetric methyl stretch of methyl iodide in a 50% mixture with $CDCl_3$ (81). This system was chosen because spontaneous Raman studies had revealed concentration-dependent line widths that were highly suggestive of inhomogeneous broadening (84, 85), and the natural variations in the constitution of the first shell of molecules around any given methyl iodide represented a plausible source of long-lived inhomogeneity. At room temperature, the magnitude of this inhomogeneous broadening was found to be 5.15 cm^{-1} (81).

Fifth-Order Raman Spectroscopy

In 1993, Tanimura & Mukamel (86) proposed a new multidimensional Raman spectroscopy that depends on a fifth-order optical nonlinearity, as opposed to the seventh-order nonlinearity of the Raman echo. Fifth-order Raman spectroscopy takes advantage of two-quantum Raman transitions in order to achieve the phase reversal needed to rephase inhomogeneity (87), as is shown in Figure 5. An initial time-coincident pair of pulses creates a Raman coherence between states $|0\rangle$ and $|1\rangle$. After delay time τ_1 a second pair of pulses causes a two-quantum Raman transition, thus creating a coherence between states $|1\rangle$ and $|2\rangle$ in which the phases of any inhomogeneous components have been reversed from their previous values. As a result, rephasing occurs after a second delay of τ if the mode of interest is harmonic; if the mode is anharmonic, rephasing can occur before or after this time.

Because it explicitly involves at least three quantum states of the oscillator being studied, fifth-order Raman spectroscopy is intended to be applied to low-frequency modes (such as the intermolecular vibrations of liquids), so that the bandwidth of a single laser can span all of the necessary transitions. As a result, a number of other nonrephasing processes also contribute to the signal (87, 88); this problem can be mitigated to some extent by employing appropriate polarization schemes (90–92).

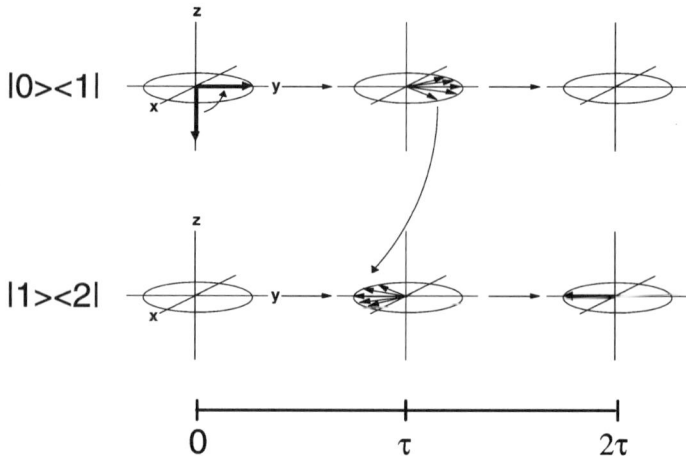

Figure 5 Optical Bloch-vector diagrams for fifth-order Raman spectroscopy. At time zero a coherence is created between states $|0\rangle$ and $|1\rangle$. At time τ this coherence is shifted to states $|1\rangle$ and $|2\rangle$ with reversed phase, leading to an echo at time 2τ.

In addition, if there is a significant distribution of anharmonicity (as may well be the case for intermolecular modes), the rephasing will not create a distinct echo. Furthermore, anharmonicity itself can serve as a source of signal, in that it breaks the selection rule against two-quantum transitions (89). All of these effects have distinct signatures, and a large body of theoretical work describes how different contributions to the signal might be distinguished (87–91, 93–119).

Tominaga & Yoshihara reported the first fifth-order Raman spectroscopic study of a liquid in 1995 (120). Based on their data they concluded that there is no appreciable inhomogeneous broadening of the intermolecular modes of CS_2 at room temperature (120). This conclusion was further supported by additional studies (121, 122), including an experiment in which CS_2 was diluted with an alkane (123). At about the same time as these later experiments were published, Steffen & Duppen reported a fifth-order Raman spectroscopic study of CS_2 that employed a different experimental geometry (124). The results of this study did not appear to be consistent with the previously reported data. This apparent discrepancy was discovered to be due to an artifact that resulted from the experimental geometry, and the results of a new study by these authors was in line with the original experiments of Tominaga & Yoshihara (125).

This was only to be the first hint of the difficulties associated with making fifth-order measurements that are not contaminated by undesired signals. Tokmakoff & Fleming used yet another experimental geometry to study the fifth-order Raman spectroscopy of liquids (126–129). These studies included experiments on CS_2 at low temperature (129) and at previously unexplored polarization conditions (126, 129), as well as experiments on other liquids (128). In all cases the data could not be fit using existing theories of the fifth-order response. Albrecht and coworkers suggested the possibility that all of the fifth-order data collected to date could be contaminated by sequential and parallel cascaded third-order signals (130). Although these contaminating signals would not be perfectly phase-matched, they also do not involve zero- or two-quantum Raman transitions and therefore could readily overwhelm the weaker but completely phase-matched fifth-order signal. This theory was subsequently confirmed by Fleming and coworkers, who found that their data could be modelled quite well by cascaded third-order signals (131).

A number of different schemes have now been devised for removing the effects of third-order cascades from the fifth-order Raman signal (132–134). The data that are being obtained now appear to be consistent with theoretical predictions of the basic shape of the signal (132–134), and more recent experiments have also largely confirmed predictions regarding the polarization dependence of the fifth-order Raman signal (135). To date, however, CS_2 is the only liquid for which data have been obtained that are clearly free of contamination from any undesired nonlinear processes, and it remains to be seen whether other liquids can be found that are amenable to study by fifth-order Raman spectroscopy.

Overtone Dephasing Spectroscopy

The transition scheme used in fifth-order Raman spectroscopy served as the inspiration for Tominaga and Yoshihara to develop a new spectroscopic technique for studying correlations among different energy levels of an intramolecular oscillator (136–138). The idea of this technique is to use two time-coincident pulses of frequencies $\omega_1 > \omega_2$ to create a Raman coherence between levels $|0\rangle$ and $|1\rangle$ of an intramolecular vibrational mode at time $t = 0$. After a delay time τ_1, a second pair of time-coincident pulses causes another one-quantum transition to create a coherence between levels $|0\rangle$ and $|2\rangle$; so long as the vibrational mode is sufficiently harmonic and the pulse bandwidth is great enough, the pulses can have the same frequencies as did the first pair. At a delay time τ_2 later, a final pulse of frequency ω_1 is used to probe this two-quantum coherence, generating a coherent signal of frequency $\omega_s = 3\omega_1 - 2\omega_2$.

The overtone dephasing signal is sensitive to the correlation function

$$\langle \tilde{\alpha}(\tau_1 + \tau_2)\tilde{\alpha}(\tau_1)\tilde{\alpha}(0) \rangle_{\omega_0}$$

$$= \left\langle \exp\left(-i \int_{\tau_1}^{\tau_1+\tau_2} \omega_{20}(t')\,dt' - i \int_0^{\tau_1} \omega_{10}(t')\,dt' \right) \right\rangle_{\omega_0}, \qquad 9.$$

where ω_{nm} is the transition frequency between levels $|n\rangle$ and $|m\rangle$. This spectroscopy thus probes the correlations between fluctuations in different quantum states in the same oscillator. In the original implementation of this technique, the C-D stretches of $CDCl_3$ and CD_3I were probed with the first delay time set to zero, such that the free-induction decay of the coherence between $|0\rangle$ and $|2\rangle$

was monitored (136). Dephasing of this overtone coherence was found to be only slightly more than twice as fast as the dephasing of the coherence between levels $|0\rangle$ and $|1\rangle$ in the same systems, although a factor of four difference would be expected in the rapid-modulation limit (136). Similar behavior was found for the decay of the coherence between states $|0\rangle$ and $|1\rangle$ in additional experiments (138). The CDCl$_3$ system was also studied further by using nonzero values of τ_1 (137). In these experiments it was found that the decay time of the overtone coherence decreased with increasing τ_1, becoming a factor of three faster than the decay time for the coherence between states $|0\rangle$ and $|1\rangle$ after 4 ps (137). This intriguing result has not yet been understood fully, but it would seem to indicate that the vibration is experiencing both rapid and slow modulations that affect the Raman line width.

INFRARED SPECTROSCOPIES

Although the optical nonlinearities upon which infrared higher-order correlation spectroscopies depend are of lower order than the nonlinearities involved in the Raman experiments discussed above, the inherent difficulty in generating short laser pulses in the mid-infrared delayed the application of higher-order infrared techniques to the study of liquids. Indeed, the earliest infrared photon-echo experiments were made possible only by access to a free-electron laser (139). Laser technology has progressed to the point that tabletop sources of pulsed mid-infrared light are now readily available, and the field of infrared vibrational higher-order coherence spectroscopy of liquids has grown rapidly as a result.

Ultrashort infrared pulses often have bandwidths that are comparable to the anharmonicity in an intramolecular vibration, which means that in higher-order experiments one needs to consider transitions between states $|1\rangle$ and $|2\rangle$ in addition to the usual transitions between states $|0\rangle$ and $|1\rangle$. This additional transition adds some interesting complications to infrared higher-order correlation spectroscopy (140, 141). In a completely harmonic oscillator, the existence of level $|2\rangle$ cancels out all nonlinear signals (73). Linear coupling of the oscillator to a bath leads to exponential energy relaxation, which is enough to restore the echo signal, albeit in a somewhat unusual form: the integrated echo signal is zero at $\tau = 0$ and then increases before decaying away (140). With the addition of anharmonicity in the vibration, beats are observed as well (141). As even broader bandwidth is employed, multiple intramolecular modes can be excited simultaneously, and the couplings between them can be investigated. It is in this capability that perhaps the greatest potential of infrared higher-order correlation spectroscopy lies.

Two- and Three-Pulse Infrared Echo Spectroscopy

The first two-pulse infrared photon echo experiments in a liquid were reported by Zimdars et al. in 1993 (139). The light source for these experiments was the Stanford free-electron laser (FEL), which produced bursts of tunable mid-infrared pulses of a few picoseconds duration (139). In these experiments, the 1960-cm^{-1}

CO stretch of W(CO)$_6$ dissolved in 2-methyltetrahydrofuran was studied from room temperature down to 16 K, well below the 88-K glass-transition temperature of the solvent (139). The homogeneous line width of this vibrational band was found to increase monotonically and continuously from the glassy state up to the room-temperature liquid (139). Following this initial report, the Fayer group used the Stanford FEL to conduct extensive studies of the temperature-dependent dynamics of metal carbonyls in solution (142–148) and of CO bound to myoglobin (151–156). In a number of these studies, the bandwidth of the laser was sufficient to create coherences between levels $|1\rangle$ and $|2\rangle$ of the vibrations under study, and the frequencies and decay times of the resultant beats allowed for the measurement of vibrational anharmonicities and level-dependent decay constants (27, 143). The behavior of these beats did not agree completely with early theoretical predictions, which led to the suggestion that fifth-order nonlinearities were playing a role in the echo decay (149). A more detailed model of the third-order response now appears to be able to fit the data quite well, however (150).

The first three-pulse infrared echo experiments were reported in 1998 by Hochstrasser and coworkers (157). In this work, the 2043-cm^{-1} band of the aqueous azide anion was studied at room temperature. Two major relaxation components were observed. The fastest relaxation took place on an 80-fs timescale and was ascribed to inertial solvation of the vibrational excitation (157). Relaxation on a 1.3-ps timescale was interpreted as long-lived inhomogeneity that decays on the timescale of the breaking of hydrogen bonds (157). These experiments were followed by three-pulse echo studies of protein fluctuations via the behavior of azide ions and carbon monoxide (158) and of the fluctuations of a small peptide via the amide I band (159).

Infrared Transient Hole Burning

Transient hole-burning experiments are best performed on absorption lines in which there is known to be long-lived inhomogeneity, a requirement that is met ideally by hydrogen-bonded systems in infrared spectroscopy. Graener et al. reported the first such experiment in 1991 (160). The dynamics of O-H bonds of HDO in D$_2$O were studied by first exciting the 3400-cm^{-1} OH stretch with a short pulse (11 ps) with a bandwidth significantly narrower than that of the vibrational transition and then probing with a second tunable pulse with a variable delay. The researchers were able to measure the anharmonic shift and the population decay time and identified three distinct spectral components that they assigned to different microscopic environments (160). This work was followed by a study of the C-H stretch vibrational dynamics of CHBr$_3$ in C$_2$D$_3$OD (161).

A number of infrared transient hole-burning investigations of hydrogen-bonded liquids have been made recently by Laenen et al. (162–168). In these newer investigations the hole-burning pulse has been reduced in duration to one or two picoseconds, which has improved the available time resolution notably. These improvements have made possible temperature- (164, 166) and concentration-dependent (163, 165, 167, 168) studies of dynamic properties of hydrogen-bonded liquids, including vibrational energy migration (163, 167), bond breaking (168), lifetime effects (165), and heating effects (168).

Two-Dimensional Infrared Echo Spectroscopy

A number of research groups are now pursuing two-dimensional infrared echo spectroscopy as a means of elucidating both the interactions between vibrational modes and molecular structure (150, 169, 170). Hochstrasser and coworkers reported a two-dimensional infrared echo experiment in which heterodyne-detected three-pulse echo signals were collected for the amide I band of small peptides as a function of the two delay times and then were inverse Fourier-transformed into the frequency domain (169). Clear coupling was seen between amide I modes in these experiments, and on this basis structural changes were observed as the solvent was changed (169). The same technique has also been used to study the behavior of the azide anion in D_2O and CO in hemoglobin (171) and of N-methylacetamide-D in D_2O (172).

A similar technique was employed by Tokmakoff and coworkers to study the anharmonic potential of the two coupled CO stretches of dicarbonylacetylacetonato rhodium (I) in hexane (170). The two-dimensional inverse Fourier transform of the two-pulse photon echo in this system clearly reveals the anharmonicity and the anharmonic coupling coefficients and shows that through-space dipole-dipole coupling is a major contributor to the carbonyl-stretching spectrum of the probe molecule (170). Fayer and coworkers used spectrally-resolved infrared photon echo spectroscopy to study the same probe molecule in poly(methyl methacrylate) and were able to use the time-dependent spectrum to elucidate the mechanism responsible for the observed anharmonic vibrational beats of the carbonyl stretching modes of this molecule (150).

The two-dimensional infrared photon echo is still in its infancy, but it is already clear that it has tremendous potential to reveal detailed microscopic structural, energetic, and dynamic information in liquids. Theoretical work has begun to reveal the wealth of information that can be extracted from such spectroscopies (173–176).

HYBRID VIBRATIONAL SPECTROSCOPIES

It is not necessary that higher-order vibrational spectroscopies involve purely Raman excitation or purely infrared excitation. Indeed, the combination of the two different types of excitation can introduce strong selectivity, as in the case of surface sum-frequency generation (177). No higher-order hybrid spectroscopy of liquids has been reported, but Cho and coworkers have investigated the potential of many such Raman/infrared and vibrational/electronic techniques theoretically (178–181). Such spectroscopies offer information that is complementary to what is available from current techniques and should not be much more difficult to implement experimentally than the spectroscopies discussed above.

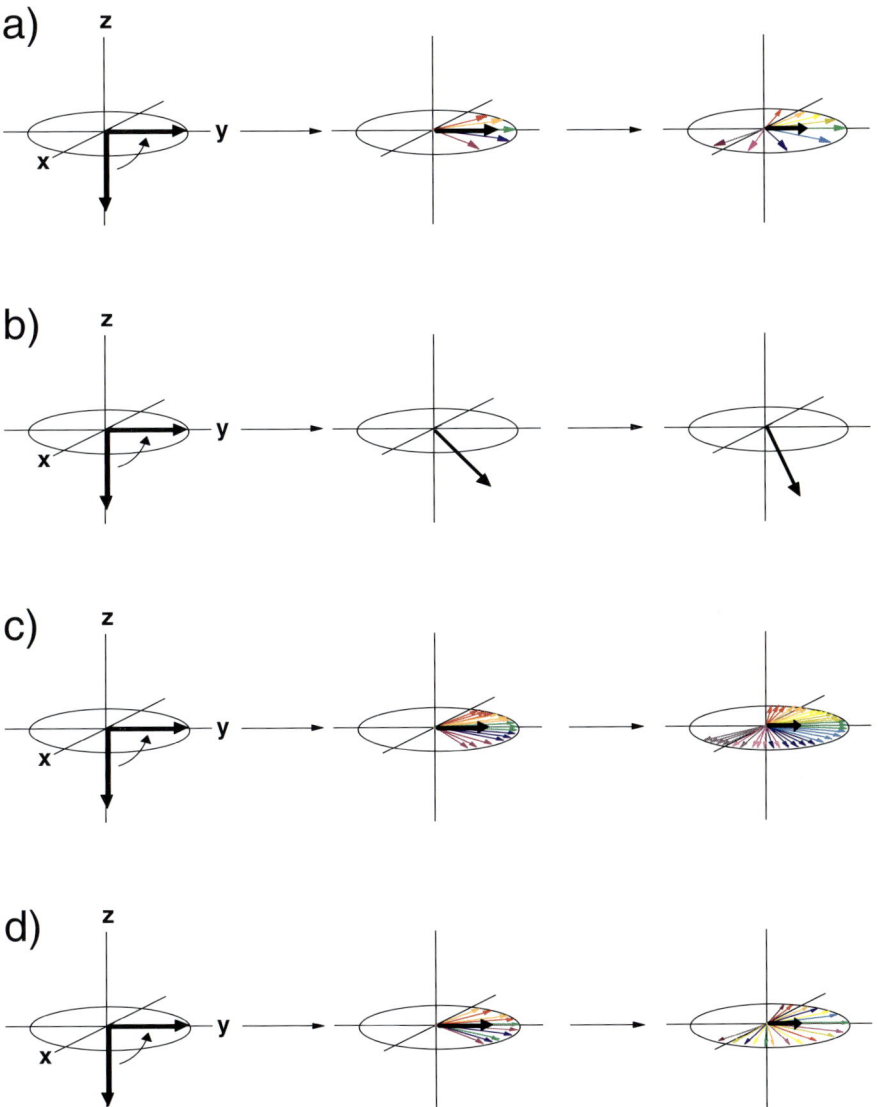

Figure 1 Optical Bloch-vector diagrams illustrating different mechanisms that affect the in-plane component of the net Bloch vector during the free-induction decay evolution after a π/2 pulse. In (*a*) the absorption line is inhomogeneously broadened, so the different spectral components spread out with time. In (*b*) the absorption line is dominated by lifetime broadening, so the in-plane component of the net Bloch vector decays by population relaxation. In (*c*) pure dephasing reduces the magnitude of the in-plane component, and in (*d*) this component is affected by spectral diffusion.

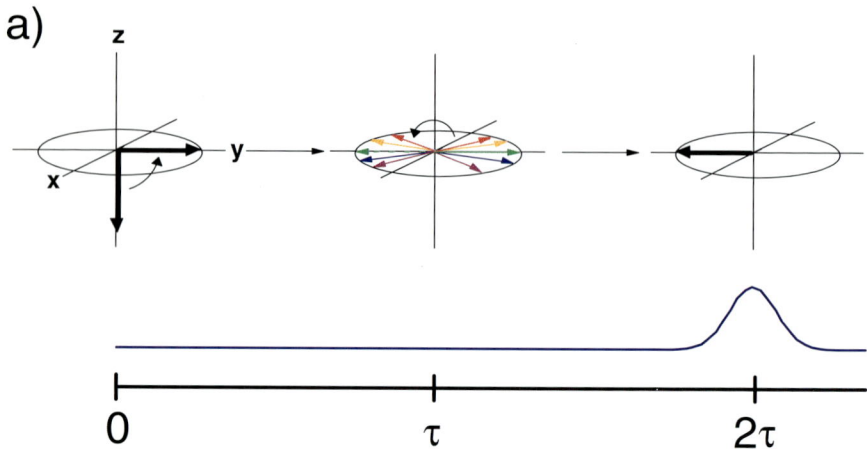
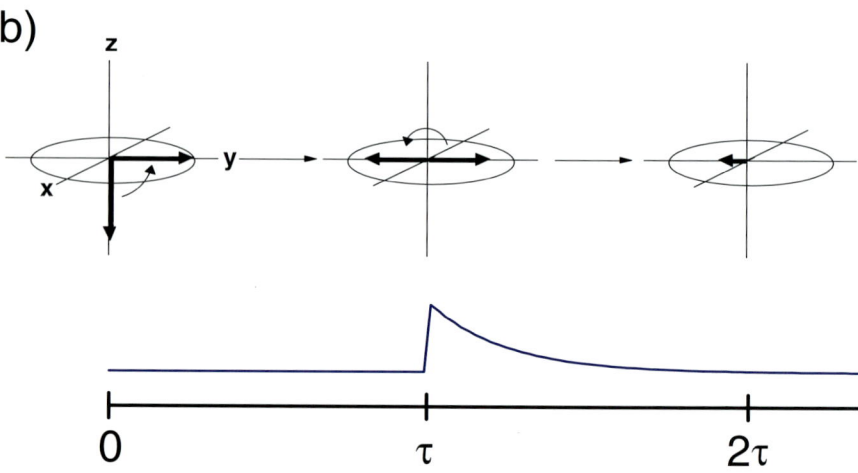

Figure 2 Optical Bloch-vector diagram for the two-pulse echo. A coherence between states $|0\rangle$ and $|1\rangle$ is created at $t=0$. This coherence is allowed to evolve for delay time τ, after which the phase of each coherent two-level system is reversed. If the ensemble is inhomogeneously broadened as in (*a*), rephasing will lead to the production of a coherent signal that will peak near time 2τ. If the absorption line is homogeneously broadened as in (*b*), the coherent signal will be generated at the time of the second pulse.

CONCLUSIONS

With advances in laser technology, optical higher-order correlation spectroscopies are becoming ever more popular tools in the optical spectroscopy of liquids. Because these techniques monitor coherent evolution over multiple time periods, they are able to provide intimate microscopic details about dynamics and structure in liquids that are not available from linear spectroscopies. Considerable experimental, instrumental, and theoretical progress remains to be made before higher-order optical spectroscopies of liquids are as readily implemented and as commonplace as multidimensional NMR methods, but applications of these optical techniques are expanding rapidly. Higher-order optical correlation spectroscopies have already affected our picture of liquid structure and dynamics significantly and promise to do so all the more as their development continues.

ACKNOWLEDGMENTS

The work described here, performed by the author's group, was supported in part by the National Science Foundation, grants CHE95-01598 and CHE00-73228. The author is a Research Corporation Cottrell Scholar, a Camille Dreyfus Teacher-Scholar, and a Sloan Research Fellow.

Visit the Annual Reviews home page at www.annualreviews.org

LITERATURE CITED

1. Feynman RP, Vernon FL Jr, Hellwarth RW. 1957. *J. Appl. Phys.* 28:49–52
2. Allen L, Eberly JH. 1975. *Optical Resonance and Two-Level Atoms*. New York: Dover. 233 pp.
3. Nibbering ETJ, Wiersma DA, Duppen K. 1991. *Phys. Rev. Lett.* 66:2464–67
4. de Boeij WP, Pshenichnikov MS, Duppen K, Wiersma DA. 1994. *Chem. Phys. Lett.* 224:243–52
5. Becker ED. 1980. *High Resolution NMR*. Orlando: Academic. 354 pp.
6. Kubo R. 1961. In *Fluctutation, Relaxation and Resonance in Magnetic Systems*, ed. D Ter Haar, pp. 23–68. Edinburgh: Oliver & Boyd
7. Bai YS, Fayer MD. 1989. *Comments Cond. Matter Phys.* 14:343–64
8. Benn R, Gunther H. 1983. *Agnew. Chem. Int. Ed. Eng.* 22:350–80
9. Kurnit NA, Abella ID, Hartmann SR. 1964. *Phys. Rev. Lett.* 13:567–68
10. Castner EW Jr, Fleming GR, Bagchi B, Maroncelli M. 1988. *J. Chem. Phys.* 89:3519–34
11. Castner EW Jr, Bagchi B, Maroncelli M, Webb SP, Ruggiero AJ, Fleming GR. 1988. *Ber. Bunsenges. Phys. Chem.* 92:363–72
12. Maroncelli M, Fleming GR. 1988. *J. Chem. Phys.* 89:875–81
13. Kahlow MA, Kang TJ, Barbara PF. 1988. *J. Chem. Phys.* 88:2372–87
14. Barbara PF, Kang TJ, Jarzeba W, Fonseca T. 1990. In *Perspectives in Photosynthesis*, ed. J Jortner, B Pullman, pp. 273–92. Deventer: Kluwer. 444 pp.
15. Horng ML, Gardecki JA, Papazyan A, Maroncelli M. 1995. *J. Phys. Chem.* 99:17311–37

16. Berg M, Walsh CA, Narasimhan LR, Littau KA, Fayer MD. 1988. *J. Chem. Phys.* 88:1564–87
17. Becker PC, Fragnito HL, Bigot JY, Cruz CHB, Fork RL, Shank CV. 1989. *Phys. Rev. Lett.* 63:505–7
18. Joo T, Albrecht AC. 1993. *Chem. Phys.* 176:233–47
19. Cho M, Fleming GR. 1993. *J. Chem. Phys.* 98:2848–59
20. Nibbering ETJ, Wiersma DA, Duppen K. 1994. *Chem. Phys.* 183:167–85
21. Vohringer P, Arnett DC, Westervelt RA, Feldstein MJ, Scherer NF. 1995. *J. Chem. Phys.* 102:4027–36
22. Yang T-S, Vohringer P, Arnett DC, Scherer NF. 1995. *J. Chem. Phys.* 103:8346–59
23. Cho M, Yu J-Y, Joo T, Nagasawa Y, Passino SA, Fleming GR. 1996. *J. Phys. Chem.* 100:11944–53
24. de Boeij WP, Pshenichnikov MS, Wiersma DA. 1996. *J. Phys. Chem.* 100:11806–23
25. Fleming GR, Cho M. 1996. *Annu. Rev. Phys. Chem.* 47:109–34
26. Fleming GR, Joo T, Cho M. 1997. *Adv. Chem. Phys.* 101:141–83
27. Zimdars D, Francis RS, Ferrante C, Fayer MD. 1997. *J. Chem. Phys.* 106:7498–511
28. Pshenichnikov MS, Duppen K, Wiersma DA. 1995. *Phys. Rev. Lett.* 74:674–77
29. Vohringer P, Arnett DC, Yang T-S, Scherer NF. 1995. *Chem. Phys. Lett.* 237:387–98
30. Weiner AM, De Silvestri S, Ippen EP. 1985. *J. Opt. Soc. Am. B* 2:654–62
31. Joo T, Jia Y, Fleming GR. 1995. *J. Chem. Phys.* 102:4063–68
32. de Boeij WP, Pshenichnikov MS, Wiersma DA. 1995. *Chem. Phys. Lett.* 238:1–8
33. Joo T, Jia Y, Yu J-Y, Lang MJ, Fleming GR. 1996. *J. Chem. Phys.* 104:6089–108
34. de Boeij WP, Pshenichnikov MS, Wiersma DA. 1996. *Chem. Phys. Lett.* 253:53–60
35. de Boeij WP, Pshenichnikov MS, Wiersma DA. 1996. *J. Chem. Phys.* 105:2953–60
36. Passino SA, Nagasawa Y, Joo T, Fleming GR. 1997. *J. Phys. Chem. A* 101:725–31
37. Passino SA, Nagasawa Y, Fleming GR. 1997. *J. Chem. Phys.* 107:6094–108
38. de Boeij WP, Pshenichnikov MS, Wiersma DA. 1998. *Chem. Phys.* 233:287–309
39. de Boeij WP, Pshenichnikov MS, Wiersma DA. 1998. *Annu. Rev. Phys. Chem.* 49:99–123
40. Lang MJ, Jordanides XJ, Song X, Fleming GR. 1999. *J. Chem. Phys.* 110:5884–92
41. Larsen DS, Ohta K, Fleming GR. 1999. *J. Chem. Phys.* 111:8970–79
42. Larsen DS, Ohta K, Xu Q-H, Cyrier M, Fleming GR. 2001. *J. Chem. Phys.* 114:8008–19
43. Ohta K, Larsen DS, Yang M, Fleming GR. 2001. *J. Chem. Phys.* 114:8020–39
44. Joo T, Jia Y, Yu J-Y, Jonas DM, Fleming GR. 1996. *J. Phys. Chem.* 100:2399–409
45. Yu J-Y, Nagasawa Y, van Grondelle R, Fleming GR. 1997. *Chem. Phys. Lett.* 280:404–10
46. Groot M-L, Yu J-Y, Agarwal R, Norris JR, Fleming GR. 1998. *J. Phys. Chem. B* 102:5923–31
47. Homoelle BJ, Edington MD, Diffey WM, Beck WF. 1998. *J. Phys. Chem. B* 102:3044–52
48. Salverda JM, van Mourik F, van der Zwan G, van Grondelle R. 2000. *J. Phys. Chem. B* 104:11395–408
49. Bursing H, Ouw D, Kundu S, Vohringer P. 2001. *Phys. Chem. Chem. Phys.* 3:2378–87
50. Cruz CHB, Fork RL, Knox WH, Shank CV. 1986. *Chem. Phys. Lett.* 132:341–44
51. Kang TJ, Yu J, Berg M. 1990. *Chem. Phys. Lett.* 174:476–80
52. Kang TJ, Yu J, Berg M. 1991. *J. Chem. Phys.* 94:2413–24
53. Yu J, Kang TJ, Berg M. 1991. *J. Chem. Phys.* 94:5787–95

54. Yu J, Berg M. 1992. *J. Chem. Phys.* 96: 8741–49
55. Fourkas JT, Berg M. 1993. *J. Chem. Phys.* 98:7773–85
56. Fourkas JT, Benigno A, Berg M. 1993. *J. Chem. Phys.* 99:8552–58
57. Fourkas JT, Benigno A, Berg M. 1994. *J. Non-Cryst. Solids* 172:234–40
58. Berg M. 1994. *Chem. Phys. Lett.* 228: 317–22
59. Ma J, Vanden Bout D, Berg M. 1995. *J. Chem. Phys.* 103:9146–57
60. Berg M. 1998. *J. Phys. Chem. A* 102:17–30
61. Yu J, Berg M. 1993. *J. Phys. Chem.* 97:1758–64
62. Yu J, Berg M. 1993. *Chem. Phys. Lett.* 208:315–20
63. Kinoshita S. 1989. *J. Chem. Phys.* 91: 5175–84
64. Kinoshita S, Itoh H, Murakami H, Miyasaka H, Okada T, Mataga N. 1990. *Chem. Phys. Lett.* 166:123–27
65. Murakami H, Kinoshita S, Hirata Y, Okada T, Mataga N. 1992. *J. Chem. Phys.* 97:7881–88
66. Hybl JD, Albrecht AW, Faeder SMG, Jonas DM. 1998. *Chem. Phys. Lett.* 297: 307–13
67. Faeder SMG, Jonas DM. 1999. *J. Phys. Chem. A* 103:10489–505
68. Hybl JD, Faeder SMG, Albrecht AW, Tolbert CA, Green DC, Jonas DM. 2000. *J. Lumin.* 87–89:126–29
69. Hybl JD, Christophe Y, Jonas DM. 2001. *Chem. Phys.* 266:295–309
70. Okumura K, Tokmakoff A, Tanimura Y. 1999. *Chem. Phys. Lett.* 314:488–95
71. Zheltikov AM. 2000. *J. Raman Spectrosc.* 31:653–67
72. Loring RF, Mukamel S. 1985. *J. Chem. Phys.* 83:2116–28
73. Mukamel S. 1995. *Principles of Nonlinear Optical Spectroscopy*. New York: Oxford Univ. Press. 543 pp.
74. Hartmann SR. 1968. *IEEE J. Quantum Electron.* QE-4:802–7
75. Muller M, Wynne K, Van Voorst JDW. 1988. *Chem. Phys.* 128:549–53
76. Vanden Bout D, Muller LJ, Berg M. 1991. *Phys. Rev. Lett.* 67:3700–3
77. Vanden Bout D, Freitas JE, Berg M. 1994. *Chem. Phys. Lett.* 229:97–102
78. Inaba R, Tominaga K, Tasumi M, Nelson KA, Yoshihara K. 1993. *Chem. Phys. Lett.* 211:183–818
79. Yoshihara K, Inaba R, Okamoto H, Tasumi M, Tominaga K, Nelson KA. 1994. In *Femtosecond Reaction Dynamics*, ed. D Wiersma, pp. 299–310. Amsterdam: North-Holland. 320 pp.
80. Tominaga K, Inaba R, Kang TJ, Naitoh Y, Nelson KA, et al. 1994. In *Proceedings of the XIV International Conference on Raman Spectroscopy*, ed. N-T Yu, X-Y Li, 1:444–45. New York: Wiley. 1123 pp.
81. Muller LJ, Vanden Bout D, Berg M. 1993. *J. Chem. Phys.* 99:810–19
82. Berg M, Vanden Bout D. 1997. *Acc. Chem. Res.* 30:65–71
83. Schweizer KS, Chandler D. 1982. *J. Chem. Phys.* 76:2296–314
84. Doege G, Arndt R, Buhl H, Bettermann G. 1980. *Z. Naturforsch. Teil A* 35:468–70
85. Knapp EW, Fischer SF. 1982. *J. Chem. Phys.* 76:4730–35
86. Tanimura Y, Mukamel S. 1993. *J. Chem. Phys.* 99:9496–511
87. Khidekel V, Mukamel S. 1995. *Chem. Phys. Lett.* 240:304–14
88. Steffen T, Fourkas JT, Duppen K. 1996. *J. Chem. Phys.* 105:7364–82
89. Okumura K, Tanimura Y. 1997. *J. Chem. Phys.* 107:2267–83
90. Murry RL, Fourkas JT. 1997. *J. Chem. Phys.* 107:9726–40
91. Murry RL, Fourkas JT. 1998. *J. Chem. Phys.* 109:7913–22
92. Kaufman LJ, Blank DA, Fleming GR. 2001. *J. Chem. Phys.* 114:2312–31
93. Palese S, Buontempo JT, Schilling L, Lotshaw WT, Tanimura Y, et al. 1994. *J. Phys. Chem.* 98:12466–70

94. Okumura K, Tanimura Y. 1997. *J. Chem. Phys.* 106:1687–98
95. Okumura K, Tanimura Y. 1997. *Chem. Phys. Lett.* 278:175–83
96. Okumura K, Tanimura Y. 1997. *Chem. Phys. Lett.* 277:159–66
97. Tanimura Y, Okumura K. 1997. *J. Chem. Phys.* 106:2078–95
98. Steffen T, Duppen K. 1997. *Chem. Phys. Lett.* 273:47–54
99. Chernyak V, Mukamel S. 1998. *J. Chem. Phys.* 108:5812–25
100. Cho M, Okumura K, Tanimura Y. 1998. *J. Chem. Phys.* 108:1326–34
101. Tanimura Y. 1998. *Chem. Phys.* 233:217–29
102. Steffen T, Duppen K. 1998. *Chem. Phys. Lett.* 290:229–36
103. Saito S, Ohmine I. 1998. *J. Chem. Phys.* 108:240–51
104. Cho M. 1998. *J. Chem. Phys.* 109:5327–37
105. Cho M. 1998. *J. Chem. Phys.* 109:6227–36
106. Hahn S, Park K, Cho M. 1999. *J. Chem. Phys.* 111:4121–30
107. Park K, Cho M, Hahn S, Kim D. 1999. *J. Chem. Phys.* 111:4131–39
108. Okumura K, Tokmakoff A, Tanimura Y. 1999. *J. Chem. Phys.* 111:492–503
109. Jansen TIC, Snijders JG, Duppen K. 2000. *J. Chem. Phys.* 113:307–11
110. Keyes T, Fourkas JT. 2000. *J. Chem. Phys.* 112:287–93
111. Steffen T, Tanimura Y. 2000. *J. Phys. Soc. Jpn.* 69:3115–32
112. Tanimura Y, Steffen T. 2000. *J. Phys. Soc. Jpn.* 69:4095–106
113. Ma A, Stratt RM. 2000. *Phys. Rev. Lett.* 85:1004–7
114. Hahn S, Kwak K, Cho M. 2000. *J. Chem. Phys.* 112:4553–56
115. Sung J, Cho M. 2000. *J. Chem. Phys.* 113:7072–83
116. Denny RA, Reichman DR. 2001. *Phys. Rev. E* 57:5101–4
117. Jansen TIC, Snijders JG, Duppen K. 2001. *J. Chem. Phys.* 114:10910–21
118. Okumura K, Jonas DM, Tanimura Y. 2001. *Chem. Phys.* 266:237–50
119. Piryatinski A, Chernyak V, Mukamel S. 2001. *Chem. Phys.* 266:311–22
120. Tominaga K, Yoshihara K. 1995. *Phys. Rev. Lett.* 74:3061–64
121. Tominaga K, Keogh GP, Naitoh Y, Yoshihara K. 1995. *J. Raman Spectrosc.* 26:495–501
122. Tominaga K, Yoshihara K. 1996. *J. Chem. Phys.* 104:4419–26
123. Tominaga K, Yoshihara K. 1996. *J. Chem. Phys.* 104:1159–62
124. Steffen T, Duppen K. 1996. *Phys. Rev. Lett.* 76:1224–27
125. Steffen T, Duppen K. 1997. *J. Chem. Phys.* 106:3854–64
126. Tokmakoff A, Fleming GR. 1997. *J. Chem. Phys.* 106:2569–82
127. Tokmakoff A, Lang MJ, Larsen DS, Fleming GR. 1997. *Chem. Phys. Lett.* 272:48–54
128. Tokmakoff A, Lang MJ, Larsen DS, Fleming GR, Chernyak V, Mukamel S. 1997. *Phys. Rev. Lett.* 79:2702–5
129. Tokmakoff A, Lang MJ, Jordanides XJ, Fleming GR. 1998. *Chem. Phys.* 233:231–42
130. Ulness DJ, Kirkwood JC, Albrecht AC. 1998. *J. Chem. Phys.* 108:3897–902
131. Blank DA, Kaufman LJ, Fleming GR. 1999. *J. Chem. Phys.* 111:3105–14
132. Blank DA, Kaufman LJ, Fleming GR. 2000. *J. Chem. Phys.* 113:771–78
133. Golonzka O, Demirdoven N, Khalil M, Tokmakoff A. 2000. *J. Chem. Phys.* 113:9893–96
134. Astinov V, Kubarych KJ, Milne CJ, Miller RJD. 2000. *Chem. Phys. Lett.* 327:334–42
135. Kaufman LJ, Heo J, Fleming GR, Sung J, Cho M. 2001. *Chem. Phys.* 266:251–71
136. Tominaga K, Yoshihara K. 1996. *Phys. Rev. Lett.* 76:987–90
137. Tominaga K, Yoshihara K. 1997. *Phys. Rev. A* 55:831–34

138. Tominaga K, Yoshihara K. 1998. *J. Phys. Chem. A* 102:4222–28
139. Zimdars D, Tokmakoff A, Chen S, Greenfield SR, Fayer MD, et al. 1993. *Phys. Rev. Lett.* 70:2718–21
140. Fourkas JT, Kawashima H, Nelson KA. 1995. *J. Chem. Phys.* 103:4393–407
141. Fourkas JT. 1995. *Laser Phys.* 5:661–66
142. Tokmakoff A, Zimdars D, Sauter B, Francis RS, Kwok AS, Fayer MD. 1994. *J. Chem. Phys.* 101:1741–44
143. Tokmakoff A, Kwok AS, Urdahl RS, Francis RS, Fayer MD. 1995. *Chem. Phys. Lett.* 234:289–95
144. Tokmakoff A, Fayer MD. 1995. *J. Chem. Phys.* 103:2810–26
145. Tokmakoff A, Zimdars D, Urdahl RS, Francis RS, Kwok AS, Fayer MD. 1995. *J. Phys. Chem.* 99:13310–20
146. Rector KD, Kwok AS, Ferrante C, Tokmakoff A, Rella CW, Fayer MD. 1997. *J. Chem. Phys.* 106:10027–36
147. Rector KD, Kwok AS, Ferrante C, Francis RS, Fayer MD. 1997. *Chem. Phys. Lett.* 276:217–23
148. Rector KD, Fayer MD. 1998. *J. Chem. Phys.* 108:1794–803
149. Hamm P, Lim M, Asplund M, Hochstrasser RM. 1999. *Chem. Phys. Lett.* 301:167–74
150. Merchant KA, Thompson DE, Fayer MD. 2001. *Phys. Rev. Lett.* 86:3899–902
151. Rella CW, Kwok A, Rector K, Hill JR, Schwettman HA, et al. 1996. *Phys. Rev. Lett.* 77:1648–51
152. Hill JR, Dlott DD, Rella CW, Peterson KA, Decatur S, et al. 1996. *J. Phys. Chem.* 100:12100–7
153. Hill JR, Ziegler CJ, Suslick KS, Dlott DD, Rella CW, Fayer MD. 1996. *J. Phys. Chem.* 100:18023–32
154. Rella CW, Rector KD, Kwok A, Hill JR, Schwettman HA, et al. 1996. *J. Phys. Chem.* 100:15620–29
155. Rector KD, Rella CW, Hill JR, Kwok AS, Sligar SG, et al. 1997. *J. Phys. Chem. B* 101:1468–75
156. Rector KD, Engholm JR, Hill JR, Myers DJ, Hu R, et al. 1998. *J. Phys. Chem. B* 102:331–33
157. Hamm P, Lim M, Hochstrasser RM. 1998. *Phys. Rev. Lett.* 81:5326–29
158. Lim M, Hamm P, Hochstrasser RM. 1998. *Proc. Natl. Acad. Sci. USA* 95:15315–20
159. Hamm P, Lim M, DeGrado WF, Hochstrasser RM. 1999. *J. Phys. Chem. A* 103:10049–53
160. Graener H, Seifert G, Laubereau A. 1991. *Phys. Rev. Lett.* 66:2092–95
161. Graener H, Seifert G. 1991. *Chem. Phys. Lett.* 185:68–74
162. Laenen R, Rauscher C. 1997. *J. Chem. Phys.* 106:8974–80
163. Laenen R, Rauscher C, Laubereau A. 1997. *J. Phys. Chem. A* 101:3201–6
164. Laenen R, Rauscher C, Laubereau A. 1998. *J. Phys. Chem. B* 102:9304–11
165. Laenen R, Simeonidis K. 1998. *J. Phys. Chem. A* 102:7207–10
166. Laenen R, Rauscher C, Laubereau A. 1998. *Phys. Rev. Lett.* 80:2622–25
167. Laenen R, Rauscher C. 1999. *Laser Chem.* 19:389–92
168. Laenen R, Rauscher C, Simeonidis K. 1999. *J. Chem. Phys.* 110:5814–20
169. Asplund MC, Zanni MT, Hochstrasser RM. 2000. *Proc. Natl. Acad. Sci. USA* 97:8219–24
170. Golonzka O, Khalil M, Demirdoven N, Tokmakoff A. 2001. *Phys. Rev. Lett.* 86:2154–57
171. Asplund MC, Lim M, Hochstrasser RM. 2000. *Chem. Phys. Lett.* 323:269–77
172. Zanni MT, Asplund MC, Hochstrasser RM. 2001. *J. Chem. Phys.* 114:4579–90
173. Tominaga K, Maekawa H. 2001. *Bull. Chem. Soc. Jpn.* 74:279–86
174. Hochstrasser RM. 2001. *Chem. Phys.* 266:273–84
175. Khalil M, Tokmakoff A. 2001. *Chem. Phys.* 266:213–30
176. Scheurer C, Piryatinski A, Mukamel

S. 2001. *J. Am. Chem. Soc.* 123:3114–24
177. Eisenthal KB. 1996. *Chem. Rev.* 96:1343–60
178. Cho M. 1999. *J. Chem. Phys.* 111:4140–47
179. Park K, Cho M. 2000. *J. Chem. Phys.* 112:5021–36
180. Cho M. 2000. *J. Chem. Phys.* 112:9978–85
181. Cho M. 2001. *J. Chem. Phys.* 114:8040–47

TIME-RESOLVED PHOTOELECTRON ANGULAR DISTRIBUTIONS: Concepts, Applications, and Directions

Tamar Seideman
Steacie Institute for Molecular Sciences, National Research Council of Canada, Ottawa, Ontario K1A 0R6, Canada; e-mail: tamar.seideman@nrc.ca

Key Words femtochemistry, angle-resolved photoionization, photoelectron imaging, radiationless transitions, time-dependent alignment

■ **Abstract** The use of photoelectron angular distributions (PADs) as a probe in short-pulse, pump-probe scenarios is reviewed. We focus on concepts, on the insight that can be gained through theoretical analysis, on applications, and on future opportunities. Time-resolved PADs are sensitive to both the time-evolving rotational composition of wavepackets and their time-evolving electronic symmetry. The former feature renders this observable a potential probe of molecular structure, intensity effects, and rotational perturbations. The latter feature renders the PAD a potential probe of radiationless transitions.

1. PREFACE

The information content of the photoelectron angular distribution, as measured in short-pulse excitation-ionization experiments, is a young subject of research but one which is expected to develop rapidly within the next few years. The present review therefore differs in nature from articles that survey mature areas of research; it focuses less on review of published works and more on concepts and future opportunities.

In introducing photoelectron angular distributions (PADs) as a potential probe in femtosecond-resolved pump-probe experiments (1–3) in Section 2, we first briefly review two well-established fields of research; time-resolved photoelectron spectroscopy (PES) (4–9) and frequency-domain, angle-resolved ionization (10–12). The concluding paragraphs of Section 2 outline recent research on the problem of time-resolved PADs, most of which is discussed in more detail in the subsequent sections. In Section 3 we outline the theory of time-resolved PADs. We do not reproduce the complete formalism (13, 14), but rather present a much simplified view that contains only (but hopefully all) the elements required for reference in Section 4, where we focus on potential applications of time-resolved

PADs. The last section contains an inevitably biased view of opportunities for future theoretical research. Although the sections of this review are related, each can be read independently of the others.

Before concluding this preface we refer the reader to several review articles on closely related topics that are not addressed in the following sections and which we found particularly helpful. The previous volume of the *Annual Review of Physical Chemistry* includes a comprehensive review of time-resolved photoelectron spectroscopies by Neumark (4), who discusses the application of these techniques to study bound state and dissociative dynamics in neutral and in negatively-charged species. The advantages of time-resolved PES of negative ions as a probe of ground state dynamics as well as solvation and relaxation dynamics in clusters are highlighted (4). In earlier review articles (5), Hayden & Stolow and Stolow et al. (6) discuss the technique of femtosecond time-resolved PES and its application to study nonadiabatic dynamics in polyatomic molecules. Gerber and coworkers (7, 8) review experimental work on time-resolved PES, focusing primarily on the application of this technique as a tool in cluster research (7) and in coherent control (8). Domcke & Stock (9) present a comprehensive review of the theory of vibronic dynamics in polyatomic excited states, including a discussion and illustrations of the use of time-resolved photoelectron energy distributions to unravel the dynamics of internal conversions.

Coincidence spectroscopy was reviewed earlier this year by Continetti (15), who discusses in detail both energy-domain and time-domain techniques and points out the new opportunities offered by the use of angular correlation between photoelectrons and photofragments. In a recent review of the technology and application of photoelectron and photoion imaging, Suzuki & Whitaker (16) include a comprehensive discussion of the technique of femtosecond-resolved photoelectron imaging. A forthcoming article by Hayden reviews recent advances in the field of time-resolved photoelectron-photoion coincidence, focusing predominantly on bound-free dynamics. The application of time- and angle-resolved PES to the study of electron dynamics at interfaces was reviewed in several recent articles, of which we note the works of Harris et al. (17) and Petek et al. (18). Finally, the information content of frequency-resolved PADs has been surveyed in a large number of review articles during the past three decades, several of which are collected in (10–12).

The reader is referred also to a series of articles by Fielding and coworkers (19) on the related topic of time- and angle-resolved autoionization spectroscopy of atomic Rydberg states, and to Reference (20), where we present a briefer and much more qualitative review of time-resolved PADs than what the present article aims to provide.

2. BACKGROUND

The time-domain pump-probe approach to molecular dynamics (1–3) continues to expand in scope and sophistication after more than two decades of intensive research. The basic principle of the pump-probe methodology is simple and extremely general in application (1–3, 21). A first (pump) pulse prepares a nonstationary superposition of excited eigenstates. A second (probe) pulse interrogates the wavepacket at a series of instances during its evolution. This concept has been applied, in the last three decades, to the study of a vast variety of problems in physics, chemistry, biology and material science (1–3, 21).

Whereas the pump stage is straightforward and essentially universal, the design of a successful probe presents an interesting challenge and is often system- and application-dependent. Ideally one desires that the probe scheme be specifically sensitive to those properties of the wavepacket that carry the information sought. In general this requires resolving the detected particle (or particles) with respect to one or more attributes. Often one would like a probe that would filter out part of the frequency content of the wavepacket and focus on a selected subset of frequencies—that subset that carries the physics probed. This notion will be clarified in the following sections.

Photoelectron spectroscopies are becoming increasingly popular as the probe stage in pump-probe experiments, owing to both technical and conceptual advantages (4–8, 16). [See also the discussions of (22, 23), where zero electron kinetic energy (ZEKE) detection is used in femtosecond-resolved experiments]. By contrast to other popular probes, such as laser-induced fluorescence and resonance multiphoton ionization, PES does not rely on a resonance condition. Consequently the probe pulse need not be tunable and, more interestingly, the excited state evolution can be followed to regions far from the equilibrium configuration. Other advantages are the sensitivity of charged particle detection and the fact that optically dark states are as readily observed as bright ones. Similar to its energy-domain counterpart, time-domain PES is sensitive to the rotational and electronic dynamics as well as to the vibrational motion. Finally, of relevance to the present article, either or both the photoelectron and the photoion can be resolved with respect to energy, angle, or spin.

The information content of angle-integrated, energy-resolved pump-probe ionization signals has been extensively discussed in the recent theoretical and experimental literature. Applications range from bound-state problems (5, 7, 24–38)[1] and wavepacket control (8, 39) through dissociation (40–42) and reaction (4, 43, 44) dynamics to solvation and relaxation in clusters (4, 45–47), collective excitation in nanoparticles (48) and electron dynamics in metals (17, 18, 49). We refer the reader to References (4–9) for reviews and for a complete bibliographical list.

Angular resolution in time-domain PES offers new opportunities while introducing new challenges to both theoretical and experimental research. Several of the new opportunities are discussed in Section 4. The PAD is extremely sensitive to strong field effects—a feature common to femtosecond experiments whose range of applications is expected to grow (13, 26, 50). It provides potentially a quantitative

---

[1]Wavepacket vibration is studied in (24–26). Intramolecular vibrational energy redistribution is studied in (27) and (28), in the second experiment using ZEKE detection. Internal conversions are studied in (29–36). An intersystem crossing is studied in (37). A rearrangement process is observed through negative ion–neutral–positive ion spectroscopy in (38).

measure of coupling mechanisms that perturb the rotational spectrum (51–53) as well as of the unperturbed evolution of rotational wavepackets (54, 55); the latter contains structural information. Furthermore, recent studies (14, 56) suggest that femtosecond-resolved PADs can serve to unravel the dynamics of nonradiative transitions.[2] Several of the challenges to theoretical research will become apparent in Section 3. Exact treatment of angular momentum algebra is essential, because the observable is specifically sensitive to the rotational content of the wavepacket. Nonperturbative treatment of the pump field is likewise important in the femtosecond domain. In those respects angle-integrated, energy-resolved observables are more forgiving.

In the continuous wave (CW) domain, the ability of PADs to convey information that is not available from angle-integrated observables has long been recognized (10–12, 57). Rotationally-selected and -resolved PAD measurements have reached the stage where a fit of the data to an appropriately formulated theory could generate a complete description of the ionization dynamics, consisting of parameters that specify the bound-free transition as well as parameters pertaining to the final core + electron state (57). Preceding this level of detail are over three decades of theoretical and experimental research on angle-resolved ionization, starting with the 1968 work of Tully et al. (58) and including a number of historical landmarks in the development of quantum defect theory (59), several formally gratifying applications of group theory (60), and substantial effort to gain new insights by comparing theory with experiments. Several reviews on molecular ionization that include work on photoelectron angular distributions are collected in References (10–12).

To the best of our knowledge, the first measurement of time-resolved PADs was reported by Berry's group as early as 1978 (61). Here, a laser pulse of 4 ns duration was used to excite a superposition of hyperfine components of the $3p^2 P_{3/2}$ level of atomic sodium. A time-delayed nanosecond pulse served to ionize the excited superposition. Photoelectrons emitted in the plane perpendicular to the propagation axis of the light beams were recorded as a function of the angle between the scattering vector and the probe polarization vector for different delay times and different angles between the excitation and the ionization polarization vectors. Analysis of the data yielded microscopic information about the ionization continuum and pointed out the sensitivity of the PAD to the hyperfine coupling in the excited state. The information about the ionization continuum results from the (relatively) high frequency resolution of the nanosecond probe pulse and is in common with time-independent PADs. The information about the excited state results from the fact that the nanosecond pump is sufficiently broad to excite coherently two levels and is in common with short-pulse pump-probe PAD measurements.

Figure 1 reproduces the results of a similar experiment, reported in the following year by Leuchs and coworkers (62). Here the time delay between the

---

[2]For an early review of the problem of nonradiative transitions see, Jortner J, Rice SA, Hochstrasser RM. 1969. *Adv. Photochem.* 7:149–309. Angle-integrated, energy-resolved studies of radiationless transitions are reported in References (29–37).

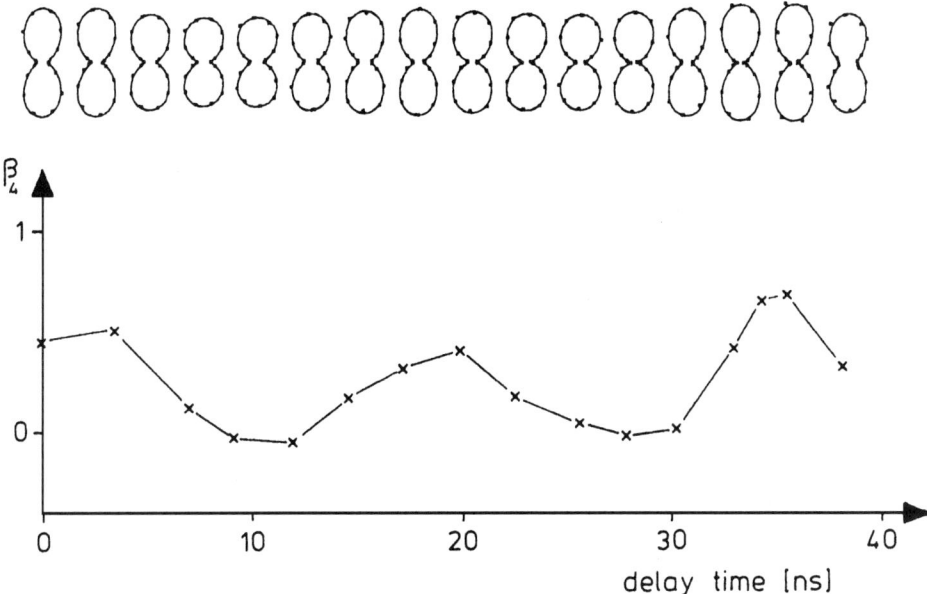

Figure 1 Experimental PADs from nanosecond–resolved ionization of the $3^2 P_{3/2}$ state of atomic sodium. The upper part shows polar diagrams of the photoelectron distribution and the lower part gives the ionization asymmetry parameter (Equation 37) as a function of the delay between the excitation and the ionization pulses. Reprinted from Reference (62) with permission from the authors and from Elsevier Science.

excitation and ionization pulses was varied continuously at parallel orientation of the polarization vectors, yielding quantitative information about the hyperfine coupling. The upper panel of Figure 1 shows polar diagrams of the PAD and the lower panel gives the ionization asymmetry parameter as a function of time (see Section 3.2.2). Fourier analysis of the signal yielded a global maximum at 59.8 MHz, in reasonable agreement with the level spacing of the $F' = 2$ and $F' = 3$ hyperfine components of the $3^2 P_{3/2}$ state of Na. Contemporaneously with the experiments of (61, 62), quantum beats in two-pulse atomic ionization were observed in the numerical studies of Zygan-Maus & Wolter (63) and Georges & Lambropoulos (64).

In the molecular domain, measurement of a nanosecond-resolved PAD was reported in 1994 by Reid et al. (65), who used a pair of time-delayed 5 ns pulses to excite and subsequently ionize NO molecules, recording the time-evolution of the photoelectron flux at fixed angle of emission [see (66) for analysis of the experiment]. As in the studies of (61, 62), the signal reported in (65) is sensitive to coherent interference between different hyperfine components of the single rotational level accessed by the pump field. It is thus modulated on a 10^{-8} sec time scale. A picosecond-resolved experiment was reported by the same group in 1999 (52). Here the time-evolution of PADs following excitation of a vibrational band of

$S_1 p$-difluorobenzene was monitored at 20 ps intervals. Polar plots of the electron distribution revealed marked variation that could not be explained in terms of pure rotational motion and was attributed to Coriolis coupling (52). The physical origin of the sensitivity of time-resolved PADs to coupling mechanisms that perturb the rotational spectrum is discussed in Section 4.1 (20, 53).

The recent work of Suzuki et al. (16, 55, 67, 68) and Hayden and colleagues (69, 70) demonstrates the ability of current technology to record PADs with femtosecond time-resolution. Invaluable in achieving this exciting goal has been the introduction and development of time-resolved photoelectron imaging technology as a subfield of charged-particle imaging. We refer the reader to the review articles of Heck & Chandler (71) for the concept, technical details and application of charged particle imaging. Comprehensive discussions of time-resolved photoelectron imaging are included in the recent review of Suzuki & Whitaker (16) and in a forthcoming review article by Hayden. A beautiful illustration of the capability of state-of-the-art time-resolved photoelectron imaging technology (C. C. Hayden, unpublished results) is reproduced in Figure 3 (see color insert) and discussed in Section 4.2.

On the theoretical side, the framework for calculation of time-resolved PADs was developed by Seideman, earlier on for linear systems (13) and more recently for nonlinear molecules (14). A first, fully nonperturbative application of the formalism explored the effects of the pump and the probe intensity on the angle-resolved signal (50). The reason for the significant pump intensity effect but minor probe intensity effect found in (50) will become apparent in Section 3. A potential application of the sensitivity of angle-resolved signals to intensity effects in femtochemistry is proposed in Section 5.

More recently, several numerical studies explored the utility of PADs as a probe in time-domain pump-probe experiments. Arasaki et al. (72) used the split-operator method within a rotationless ($J = 0$) approximation to study the time-resolved PADs of Na_2 and NaI. Althorpe & Seideman (53) used the formalism of (13) to illustrate the ability of time-resolved PADs to measure the magnitude of rotation-vibration coupling. The theory of (14) is used in (56) to illustrate the possibility of applying PADs to probe the dynamics of radiationless transitions in polyatomic excited states. Further support of the sensitivity of time-resolved PADs to nonadiabatic coupling (56) is provided by the recent work of Underwood & Reid (73), who presented model calculations of time-independent PADs for different electronic orbital symmetries of the C_{3v} point group.

We return to these applications in Section 4 after presenting a qualitative, but hopefully useful, review of the theory in Section 3. Several of the challenges to future theoretical and numerical work are discussed in Section 5.

3. THEORY

In order to keep this section as brief and transparent as possible, we do not reproduce the general formalism of time-resolved PADs (13, 14) but rather provide a simplified exposition that contains only (but hopefully all) the essential ingredients.

In Section 3.1 we derive general, formally exact expressions for the observables in a pump-probe ionization experiment. In Section 3.2 we focus on the physical interpretation of the results. To that end we first introduce two approximations of a different nature and use each to derive a closed-form expression for the observable with specific physical content. Section 3.2.1 investigates the limit of a perturbative field, leading to an analytical form for the temporal structure of the PAD of an arbitrary system, Equation 24. In Section 3.2.2 we return to the nonperturbative formalism of Section 3.1, limit attention to the simplest system that can be prepared experimentally, and derive a closed-form expression for the angular structure of the time-resolved PAD, Equation 37. Although derived here under simplifying approximations, both Equation 24 and Equation 37 apply quite generally.

3.1. Time-Evolution and Observables

We consider a molecular eigenstate subject to two sequential light pulses,

$$\vec{\varepsilon}_j(t) = \hat{\varepsilon}_j f_j(t) e^{-i\omega_j t} + \text{c.c.}, \quad j = \text{pu, pr}, \qquad 1.$$

where $\hat{\varepsilon}_j$ is a unit vector along the polarization direction, ω_j is the central frequency, and $f_j(t)$ is a smooth envelope; in the case of a Gaussian pulse,

$$f_j(t) = \frac{1}{2}\varepsilon_j \exp\left[-(t-t_j)^2/\tau_j^2\right]. \qquad 2.$$

The first (pump) pulse, $\vec{\varepsilon}_{\text{pu}}$, projects an initial eigenstate onto a superposition of rovibrational eigenstates of an electronically excited state. The second (probe) pulse, $\vec{\varepsilon}_{\text{pr}}$, couples the electronically excited state with the ionization continuum. The complete Hamiltonian is thus

$$H = H_M - \vec{\mu} \cdot \vec{\varepsilon}(t), \quad \vec{\varepsilon}(t) = \vec{\varepsilon}_{\text{pu}}(t) + \vec{\varepsilon}_{\text{pr}}(t), \qquad 3.$$

where H_M is the field-free Hamiltonian and

$$\vec{\mu} = e \sum_\nu \vec{q}_\nu, \qquad 4.$$

e being the electronic charge and $\vec{q}_\nu$ the position vectors of the electrons.

The time-dependent wavepacket is expanded in the stationary eigenstates of the field-free Hamiltonian as

$$|\Psi(t)\rangle = \sum_{\xi=0,1} \sum_{n_\xi M_\xi} C_\xi^{n_\xi M_\xi}(t) |\xi \, n_\xi \, M_\xi\rangle \exp\left(-i E_\xi^{n_\xi} t\right)$$

$$+ \sum_{n_c M_c} \int d\epsilon \int d\hat{k} \, C^{n_c M_c}(\epsilon \, \hat{k} \, t) |\epsilon \, \hat{k} \, n_c \, M_c\rangle \exp[-i(E^{n_c} + \epsilon)t], \qquad 5.$$

where $\xi = 0, 1$ is an electronic index; $\xi = 0$ is the initial electronic state, and $\xi = 1$ is the excited electronic state whose properties are the subject of the pump-probe experiment. The subscript c denotes the core indices; we reserve the label $\xi = 2$ to attributes of the ion + electron system. The collective index n_ξ denotes the energy quantum numbers, including the rovibrational indices and the electron spin, M_ξ is the magnetic quantum number, the projection of the matter angular momentum J_ξ onto the space-fixed z-axis, and $E_\xi^{n_\xi}$ is the eigenenergy. We denote by ϵ the photoelectron energy and by $\hat{k} = (\theta_k, \phi_k)$ the photoelectron ejection direction in the space-fixed frame. Atomic units are used in Equation 5 and throughout this review.

Substituting Equation 5 in the time-dependent Schrödinger equation and using the orthogonality of the field-free eigenstates, one obtains a set of coupled differential equations for the expansion coefficients,

$$i\dot{C}_0^{n_0 M_0}(t) = \sum_{n_1 M_1} C_1^{n_1 M_1}(t) \langle 0\, n_0\, M_0 | \vec{\mu} \cdot \vec{\varepsilon}_{\text{pu}}(t) | 1\, n_1\, M_1 \rangle \exp\left[i\left(E_0^{n_0} - E_1^{n_1}\right)t\right], \quad 6.$$

$$i\dot{C}_1^{n_1 M_1}(t) = \sum_{n_0 M_0} C_0^{n_0 M_0}(t) \langle 1\, n_1\, M_1 | \vec{\mu} \cdot \vec{\varepsilon}_{\text{pu}}(t) | 0\, n_0\, M_0 \rangle \exp\left[i\left(E_1^{n_1} - E_0^{n_0}\right)t\right]$$

$$+ \sum_{n_c M_c} \int d\epsilon \int d\hat{k}\, C^{n_c M_c}(\epsilon\, \hat{k}\, t) \langle 1\, n_1\, M_1 | \vec{\mu} \cdot \vec{\varepsilon}_{\text{pr}}(t) | \epsilon\, \hat{k}\, n_c\, M_c \rangle$$

$$\times \exp\left[i\left(E_1^{n_1} - E^{n_c} - \epsilon\right)t\right] \quad 7.$$

and

$$i\dot{C}^{n_c M_c}(\epsilon\, \hat{k}\, t) = \sum_{n_1 M_1} C_1^{n_1 M_1}(t) \langle \epsilon\, \hat{k}\, n_c\, M_c | \vec{\mu} \cdot \vec{\varepsilon}_{\text{pr}}(t) | 1\, n_1\, M_1 \rangle$$

$$\times \exp\left[i\left(E^{n_c} + \epsilon - E_1^{n_1}\right)t\right], \quad 8.$$

supplemented by a set of initial conditions, $C_\xi^{n_\xi M_\xi}(t=0) = \delta_{\xi,0}\delta_{n_\xi,n_i}\delta_{M_\xi,M_i}$. (We distinguish between the subscript 0, which labels the energy levels and magnetic states of the ground electronic state, and the subscript i, which labels the energy level and magnetic state of the initial rovibrational level, because a nonperturbative field excites in general a wavepacket of n_0 and M_0 levels.) Typically the two pulses do not overlap in time and hence Equations 6–8 separate into two sets of equations, one describing the dynamics during the pump stage ($\vec{\varepsilon}_{\text{pr}}(t) = 0$) and the second corresponding to the dynamics during the probe stage ($\vec{\varepsilon}_{\text{pu}}(t) = 0$).

As they stand Equations 6–8 are notoriously difficult to solve numerically because of the double continuum integration in Equation 7. A numerically practical but yet accurate method of circumventing this difficulty is derived in Reference (13) within a variant of the slowly varying continuum approximation. We do not discuss the technical details here but note only that the integrals on the right hand side of Equation 7 are carried out analytically, leading to a more complicated but temporally-smooth summand.

The state-, time-, and angle-resolved cross section is given as the squared modulus of the corresponding continuum amplitude,

$$\sigma(\bar{\epsilon}\, \hat{k}\, n_c\, M_c | \Delta t | n_i\, M_i) = \lim_{t \to \infty} \int d\epsilon |C^{n_c M_c}(\epsilon\, \hat{k}\, t)|^2, \quad 9.$$

where Δt is the delay time, the difference between the centers of the pump and probe pulses, and integration is over the range of photoelectron energies defined by the probe band-width. We denote by $\bar{\epsilon}$ an averaged photoelectron energy, defined to within the energy resolution, $\sim 2/\tau_{\text{pr}}$, and upper-limited by energy conservation, $\bar{\epsilon} \leq E_0^{n_i} + \omega_{\text{pu}} + \omega_{\text{pr}} - E^{n_c}$.

Formally the solution of Equation 8 is given as

$$C^{n_c M_c}(\epsilon \, \hat{k} \, t) = \frac{\sqrt{2\pi}}{i} \sum_{n_1 M_1} \langle \epsilon \, \hat{k} \, n_c \, M_c | \vec{\mu} \cdot \hat{\varepsilon}_{\text{pr}} | 1 \, n_1 \, M_1 \rangle \tilde{C}_1^{n_1 M_1}(t; E^{n_c} + \epsilon - E_1^{n_1}), \qquad 10.$$

where

$$\tilde{C}_1^{n_1 M_1}(t; \omega) \equiv \frac{1}{\sqrt{2\pi}} \int^t dt' \, \varepsilon_{\text{pr}}(t') C_1^{n_1 M_1}(t') e^{i\omega t'} \qquad 11.$$

is a finite-time Fourier transform of the product $\varepsilon_{\text{pr}} C_1^{n_1 M_1}$ and $\vec{\varepsilon}_{\text{pr}}(t) = \hat{\varepsilon}_{\text{pr}} \varepsilon_{\text{pr}}(t)$ (see Equation 1). With Equation 10, the cross section of Equation 9 takes the form

$$\sigma(\bar{\epsilon} \, \hat{k} \, n_c \, M_c | \Delta t | n_i \, M_i) = 2\pi \int d\epsilon \left| \sum_{n_1 M_1} \langle \epsilon \, \hat{k} \, n_c \, M_c | \vec{\mu} \cdot \hat{\varepsilon}_{\text{pr}} | 1 \, n_1 \, M_1 \rangle \right.$$

$$\left. \times \tilde{C}_1^{n_1 M_1}(E^{n_c} + \epsilon - E_1^{n_1}) \right|^2, \qquad 12.$$

where

$$\tilde{C}_1^{n_1 M_1}(\omega) = \lim_{t \to \infty} \tilde{C}_1^{n_1 M_1}(t | \omega). \qquad 13.$$

Equation 12 is the basic ingredient in the calculation of all observables of a pump-probe ionization experiment. Observables of practical interest are summed over the dipole allowed magnetic (M_c) and rotational (J_c) states of the core and averaged over the initial magnetic (M_i) levels. In general [although not always (74)] the initial rovibronic state is not selected and the final vibrational state of the core is not resolved. Hence, the cross section needs to be further averaged over a Boltzmann distribution of initial rovibrational states and summed over the vibrational core levels spanned by the probe band-width. Because the field-free Hamiltonian does not couple magnetic sublevels, summation over the magnetic indices can be generally carried out analytically (see Section 3.2.2).

Integration over the Euler angles yields the photoelectron energy spectrum as,

$$\sigma(\bar{\epsilon} \, n_c \, M_c | \Delta t | n_i \, M_i) = \int d\hat{k} \, \sigma(\bar{\epsilon} \, \hat{k} \, n_c \, M_c | \Delta t | n_i \, M_i)$$

$$= 2\pi \int d\epsilon \sum_{n_1, M_1, n_1', M_1'} \tilde{C}_1^{n_1 M_1}(E^{n_c} + \epsilon - E_1^{n_1})$$

$$\times \tilde{C}_1^{n_1' M_1' *}(E^{n_c} + \epsilon - E_1^{n_1'}) F(n_1' \, M_1' | \epsilon \, n_c \, M_c | n_1 \, M_1), \qquad 14.$$

where the dependence of the right hand side on the initial condition $\{n_i, M_i\}$ is implicit in the $\tilde{C}_1^{n_1 M_1}$ and

$$F(n_1' M_1' | \epsilon n_c M_c | n_1 M_1) \equiv \int d\hat{k} \langle 1 n_1' M_1' | \vec{\mu} \cdot \hat{\varepsilon}_{\text{pr}} | \epsilon \hat{k} n_c M_c \rangle$$
$$\times \langle \epsilon \hat{k} n_c M_c | \vec{\mu} \cdot \hat{\varepsilon}_{\text{pr}} | 1 n_1 M_1 \rangle. \qquad 15.$$

The integral over the scattering angles in Equation 15 is readily evaluated analytically once an explicit expression for the $\hat{k}$-dependence of the scattering wavefunction is derived. Analytical integration over energy requires the introduction of an approximation, but a reliable one (vide infra). Finally, the total ion signal is obtained by numerically integrating Equation 14 over $\bar{\epsilon}$ from zero up to $E_0^{n_i} + \omega_{\text{pu}} + \omega_{\text{pr}} - E^{n_c}$.

3.2. Physical Interpretation

In this subsection we focus on the physical interpretation of the expressions of Section 3.1. To that end we introduce several assumptions that simplify the derivation of closed-form expressions for the temporal and angular dependencies of the signal. Our approximations are often not valid but are useful in providing insight into the content of the observables and can serve in the interpretation of exact calculations and future experiments. In Section 3.2.1 we derive a closed-form expression for the time-dependence of the PAD. In Section 3.2.2 we express explicitly its angular dependence.

3.2.1. TEMPORAL STRUCTURE   In the limit of a weak probe field Equation 10 reduces to

$$C^{n_c M_c}(\epsilon \hat{k}, t) = \frac{\sqrt{2\pi}}{i} \sum_{n_1 M_1} \langle \epsilon \hat{k} n_c M_c | \vec{\mu} \cdot \hat{\varepsilon}_{\text{pr}} | 1 n_1 M_1 \rangle$$
$$\times C_1^{n_1 M_1}(T) \tilde{\varepsilon}_{\text{pr}}(t; E^{n_c} + \epsilon - E_1^{n_1}), \qquad 16.$$

where $T \approx t_{\text{pu}} + 3\tau_{\text{pu}}$ is time subsequent to which the pump field essentially vanishes and $\tilde{\varepsilon}_{\text{pr}}(t; \omega)$ is a finite-time Fourier transform of the probe pulse,

$$\tilde{\varepsilon}_{\text{pr}}(t; \omega) = \frac{1}{\sqrt{2\pi}} \int^t dt' \, \varepsilon_{\text{pr}}(t') e^{i\omega t'}. \qquad 17.$$

The corresponding differential cross section is

$$\sigma^0(\bar{\epsilon}\hat{k} n_c M_c | \Delta t | n_i M_i) = 2\pi \int d\epsilon \left| \sum_{n_1 M_1} \langle \epsilon \hat{k} n_c M_c | \vec{\mu} \cdot \hat{\varepsilon}_{\text{pr}} | 1 n_1 M_1 \rangle \right.$$
$$\left. \times C_1^{n_1 M_1}(T) \tilde{\varepsilon}_{\text{pr}}(E^{n_c} + \epsilon - E_1^{n_1}) \right|^2, \qquad 18.$$

where

$$\tilde{\varepsilon}_{\text{pr}}(\omega) = \lim_{t \to \infty} \tilde{\varepsilon}_{\text{pr}}(t|\omega). \qquad 19.$$

In the limit of a weak pump pulse the excited wavepacket rotational composition is restricted to $J_1 = J_i, J_i \pm 1, M_1 = M_i, M_i \pm 1, C_0^{n_0 M_0} \approx \delta_{n_0,n_i}\delta_{M_0,M_i}$, and $C_1^{n_1 M_1}(t)$ is given explicitly as

$$C_1^{n_1 M_1}(t) \approx \frac{\sqrt{2\pi}}{i} \langle 1\, n_1\, M_1 | \vec{\mu} \cdot \hat{\varepsilon}_{\text{pu}} | 0\, n_i\, M_i \rangle \, \tilde{\varepsilon}_{\text{pu}}\left(t; E_1^{n_1} - E_0^{n_i}\right). \qquad 20.$$

Assuming both pump and probe fields to be perturbative, we find, by substituting the long time ($t \gtrsim T$) limit of Equation 20 in Equation 18,

$$\sigma^{00}(\vec{\varepsilon}\,\hat{k}\,n_c\,M_c|\Delta t|n_i\,M_i) = 4\pi^2 \int d\epsilon \left| \sum_{n_1 M_1} \langle \epsilon\, \hat{k}\, n_c\, M_c | \vec{\mu} \cdot \hat{\varepsilon}_{\text{pr}} | 1\, n_1\, M_1 \rangle \right.$$

$$\times \langle 1\, n_1\, M_1 | \vec{\mu} \cdot \hat{\varepsilon}_{\text{pu}} | 0\, n_i\, M_i \rangle$$

$$\left. \times \tilde{\varepsilon}_{\text{pr}}\left(E^{n_c} + \epsilon - E_1^{n_1}\right) \tilde{\varepsilon}_{\text{pu}}\left(E_1^{n_1} - E_0^{n_i}\right) \right|^2. \qquad 21.$$

In order to express explicitly the temporal dependence of the PAD, we separate out the dependence of the $\tilde{\varepsilon}_j(\omega)$ in Equation 21 on t_j, the pulse center in Equation 2, by defining real-arithmetic amplitudes $f_j(\omega)$ through

$$f_j(\omega) e^{i(\omega - \omega_j)t_j} = \frac{1}{\sqrt{2\pi}} \int_{-\infty}^{\infty} dt\, f_j(t) e^{i(\omega - \omega_j)t} \qquad j = \text{pu, pr.} \qquad 22.$$

Substituting Equation 22 in Equation 21 and Fourier transforming the probe envelope to the frequency domain we have

$$\sigma^{00}(\vec{\varepsilon}\,\hat{k}\,n_c\,M_c|\Delta t|n_i\,M_i) = \sqrt{\frac{\pi^5}{2}}\,\varepsilon_{\text{pr}}^2 \tau_{\text{pr}} \sum_{n_1 n_1'} \sum_{M_1 M_1'} f_{\text{pu}}\left(E_1^{n_1} - E_0^{n_i}\right) f_{\text{pu}}\left(E_1^{n_1'} - E_0^{n_i}\right)$$

$$\times \langle \vec{\varepsilon}\,\hat{k}\,n_c\,M_c | \vec{\mu} \cdot \hat{\varepsilon}_{\text{pr}} | 1 n_1 M_1 \rangle \langle 1 n_1 M_1 | \vec{\mu} \cdot \hat{\varepsilon}_{\text{pu}} | 0 n_i M_i \rangle$$

$$\times \langle 0 n_i M_i | \vec{\mu} \cdot \hat{\varepsilon}_{\text{pu}} | 1 n_1' M_1' \rangle \langle 1 n_1' M_1' | \vec{\mu} \cdot \hat{\varepsilon}_{\text{pr}} | \vec{\varepsilon}\,\hat{k}\,n_c\,M_c \rangle$$

$$\times \exp\left[-\frac{\tau_{\text{pr}}^2}{8}\left(E_1^{n_1} - E_1^{n_1'}\right)^2\right] \exp\left[-i\left(E_1^{n_1} - E_1^{n_1'}\right)\Delta t\right],$$

$$\qquad 23.$$

where, consistent with our perturbative treatment of the fields, we introduced a rotating wave approximation. The energy integration over the pulse band-width has been performed analytically, using the method of Reference (13).

Equation 23 takes the generic sinusoidal Δt-dependence, common to all pump-probe signals,

$$\sigma \propto \sum_{\nu,\nu'\leq\nu} |A_{\nu\nu'}| \cos\left[\left(E_1^{n_1} - E_1^{n_1'}\right)\Delta t + \Phi_{\nu\nu'}\right], \qquad 24.$$

where ν denotes all excited state indices, and $\Phi_{\nu\nu'}$ is the relative phase of the two complex transition dipole matrix elements connecting the excited state levels ν and ν' with the ionization continuum [see Reference (75)]. This phase has often been overlooked in theoretical and experimental studies of pump-probe ionization spectroscopies. It vanishes if the ionization continuum satisfies a set of well-defined conditions (75) but is important to account for in general. The temporal dependence of Equation 24 applies (with different definitions of the $A_{\nu\nu'}$ and $\Phi_{\nu\nu'}$) also to the angle-, and the angle- and energy-integrated pump-probe signals.

3.2.2. ANGULAR STRUCTURE   In this section we return to a nonperturbative treatment of the radiation field (13, 14), as outlined in Section 3.1, but restrict attention to the simplest molecular system that can be experimentally prepared. We consider a $^1\Sigma \rightarrow {}^1\Sigma \rightarrow {}^2\Sigma$ transition of a diatomic molecule, such as the Li$_2$ $A(^1\Sigma_u^+) \rightarrow E(^1\Sigma_g^+) \rightarrow X(^2\Sigma_g^+)$ transition, studied in the pump-probe experiment of Leone and coworkers (74) or the Na$_2$ $X(^1\Sigma_g^+) \rightarrow A(^1\Sigma_u^+) \rightarrow X(^2\Sigma_g^+)$ of Reference (53), taking the pump and probe fields to be linearly polarized with common polarization direction. To further simplify the notation we assume that the system has been prepared in the ground rotational state, $J_i = 0$; hence the space-fixed projection of the total angular momentum is zero, $M_i = 0$. For explicit solution of the general polarization case, we refer the reader to Reference (13). The extension to systems where the spin plays a role is discussed in Reference (50). The case of nonlinear molecules is formulated in Reference (14).

In general, the pump and probe fields do not overlap in time and hence Equations 6–8 break into two sets of coupled differential equations separated by field-free evolution of duration determined by the pump-probe delay time. The pump stage is described by

$$i\dot{C}_0^{n_0}(t) = \sum_{n_1} C_1^{n_1}(t) \langle 0\,n_0 |\vec{\mu}\cdot\vec{\varepsilon}_{\text{pu}}(t)| 1\,n_1 \rangle \exp\left[i\left(E_0^{n_0} - E_1^{n_1}\right)t\right],$$

$$i\dot{C}_1^{n_1}(t) = \sum_{n_0} C_0^{n_0}(t) \langle 1\,n_1 |\vec{\mu}\cdot\vec{\varepsilon}_{\text{pu}}(t)| 0\,n_0 \rangle \exp\left[i\left(E_1^{n_1} - E_0^{n_0}\right)t\right], \qquad 25.$$

where we omit the magnetic indices, because $M_i = 0$ is conserved in a linearly polarized field, and the collective energy level index reduces to $n_\xi = \{\nu_\xi, J_\xi\}$. Equation 25 describes the dynamics of sequential rotational excitation accompanied by creation of a ground state wavepacket. Under nonperturbative conditions, the system cycles several times between the two electronic states during the pump stage, exchanging another unit of angular momentum with the field on each transition, and producing ro-vibrational wavepackets in both electronic states (76). The degree of rotational excitation is determined in the short pulse limit by the product of the pulse-duration and field-amplitude of the pump. A rough estimate of the number of rotations significantly populated is the number of Rabi cycles per pulse, $\tau_{\text{pu}}/\Omega_R^{-1}$, where $\Omega_R \propto \varepsilon_{\text{pu}}$ is the Rabi coupling and Ω_R^{-1} is the corresponding period. A complete discussion is given in (76). Rotational excitation and ground state wavepacket creation are the two strong field phenomena most relevant to femtochemistry experiments because they appear at lower intensities than other strong field effects.[3]

Once the molecule is set free ($t \gtrsim T$), the ro-vibrational composition of the wavepacket is fixed and only the phases evolve:

$$|\Psi(t > T)\rangle = \sum_{\xi=0,1} \sum_{n_\xi} C_\xi^{n_\xi}(T) |\xi n_\xi\rangle \exp\left(-i E_\xi^{n_\xi} t\right). \qquad 26.$$

The information about the field-free Hamiltonian is imprinted onto the time evolution through the eigenenergies, $E_\xi^{n_\xi}$ in Equation 26.

The dynamics during the probe stage are described by

$$i\dot{C}_1^{n_1}(t) = \sum_{n_c M_c} \int d\epsilon \int d\hat{k}\, C^{n_c M_c}(\epsilon\,\hat{k}\,t) \langle 1\, n_1 | \vec{\mu} \cdot \vec{\varepsilon}_{\text{pr}}(t) | \epsilon\,\hat{k}\, n_c\, M_c \rangle$$
$$\times \exp\left[i(E_1^{n_1} - E^{n_c} - \epsilon)t\right],$$
$$i\dot{C}^{n_c M_c}(\epsilon\,\hat{k}\,t) = \sum_{n_1} C_1^{n_1}(t) \langle \epsilon\,\hat{k}\, n_c\, M_c | \vec{\mu} \cdot \vec{\varepsilon}_{\text{pr}}(t) | 1\, n_1 \rangle \exp\left[i(E^{n_c} + \epsilon - E_1^{n_1})t\right].$$
$$27.$$

In principle one might envision rotational excitation induced by the moderately intense probe field further enriching the excited state ro-vibrational wavepacket and building up a rotationally broad wavepacket also in the ionic state. In practice we find that Rabi-type cycling between the excited and continuum states takes place only at intensities well above the range relevant to femtosecond experiments. The reason is that direct ionization takes place on a short timescale with respect to the Rabi period under relevant intensities. Thus, the set (27) is well approximated by its Golden Rule limit.

We proceed by expressing explicitly the abstract state vectors in Equations 25–27. For the simple system under consideration the stationary H_M eigenstates are given as products of rotational, vibrational, and electronic eigenstates,

$$\langle \hat{R}\, R\, \mathbf{Q} | \xi\, n_\xi \rangle = \langle \hat{R} | J_\xi\, 0 \rangle \langle R | n_\xi \rangle \langle \mathbf{Q}; R | \xi \rangle \qquad 28.$$

$$\langle \hat{R}\, R\, \mathbf{Q} | \epsilon\, \hat{k}\, n_c\, M_c \rangle = \langle \hat{R} | J_c\, M_c \rangle \langle R | n_c \rangle \langle \mathbf{Q}; R | \epsilon\, \hat{k} \rangle, \qquad 29.$$

where $\mathbf{Q}$ denotes collectively the electronic coordinates, defined with respect to the body-fixed frame, R is the internuclear separation, and $\hat{R} = (\phi, \theta, \nu)$ are the Euler angles of rotation of the body fixed with respect to the space-fixed frame.

---

[3]Effects such as ac-Stark shifting of excited state levels, above threshold ionization and dissociation, and nonresonant (field) ionization obtain at higher intensities than typical of femtochemistry experiments and are not discussed here.

In Equations 28 and 29 $\langle \hat{R} | J M \rangle$ are spherical harmonics, $\langle R | n_\xi \rangle$ are vibrational functions, and $\langle \mathbf{Q}; R | \xi \rangle$ and $\langle \mathbf{Q}; R | \epsilon \hat{k} \rangle$ are parametrically R-dependent bound and continuum electronic functions, respectively.

Partial-wave expansion of the continuum state in Equation 29 gives

$$\langle \mathbf{Q}; R | \epsilon \hat{k} \rangle = \sqrt{\frac{2}{\pi}} \sum_{l k_l} i^l Y^*_{l k_l}(\hat{K}) \phi_{l k_l}(\epsilon \mathbf{Q}|R), \qquad 30.$$

where $\hat{K}$ denotes the direction of ejection of the electron in the body-fixed frame and $\phi_{l k_l}(\epsilon \mathbf{Q}|R)$ includes the wavefunction describing the ionized electron and the electronic wavefunction of the core. In terms of the space-fixed coordinates,

$$Y_{l k_l}(\hat{K}) = \sum_{m_l} D^l_{m_l k_l}(\hat{R}) Y_{l m_l}(\hat{k}), \qquad 31.$$

where $D^l_{mm'}$ are rotation matrices and we use the notation of Zare (77).

Expanding the dipole vector and the field polarization vector in Equations 25 and 27 in spherical unit vectors, we find

$$\vec{\mu} \cdot \hat{\varepsilon} = \sum_s \mu_s D^{1*}_{0s}(\hat{R}). \qquad 32.$$

With Equations 28–32, the bound-bound matrix elements in Equation 25 are given as (13)

$$\langle \xi n_\xi | \vec{\mu} \cdot \vec{\varepsilon}_{\text{pu}}(t) | \xi' n'_{\xi'} \rangle = \varepsilon_{\text{pu}}(t) W(J_\xi | J'_{\xi'}) T(\xi n_\xi | \xi' n'_{\xi'}), \quad (s=0), \qquad 33.$$

and the bound-free matrix elements in Equation 27 are given as

$$\langle \epsilon \hat{k} n_c M_c | \vec{\mu} \cdot \vec{\varepsilon}_{\text{pr}}(t) | 1 n_1 \rangle = \varepsilon_{\text{pr}}(t) \sum_{l m_l k_l} Y_{l m_l}(\hat{k}) W(J_c M_c | l m_l k_l | J_1)$$
$$\times T(n_c | \epsilon l k_l | n_1), \quad (s = -k_l), \qquad 34.$$

where $\vec{\varepsilon}_i(t) = \hat{\varepsilon}_i \varepsilon_i(t)$. Equations 33 and 34 separate the transition dipole matrix elements into products of geometric (W) and dyamical (T) functions. The former are analytically solvable integrals over products of functions of the Euler angles of rotation. These integrals couple the field and matter angular momenta and give rise to the various excitation and ionization selection rules. Their explicit form is given in Reference (13) for the general-polarization–linear-molecule case and in Reference (14) for nonlinear systems. The reduction to the case considered here is straightforward and omitted. The T are matrix elements of the spherical components of the transition dipole operator (μ_s in Equation 32) in the vibronic basis. These functions contain the details of the bound-bound and bound-free dynamics and need be computed numerically for a specific system in mind.

Substituting Equation 34 in Equation 12 we find

$$\sigma(\bar{\epsilon}\hat{k}n_c M_c|\Delta t|n_i) = \sqrt{\pi}\int d\epsilon \sum_{n_1 n_1'} \tilde{C}_1^{n_1}(E^{n_c}+\epsilon - E_1^{n_1})\tilde{C}_1^{n_1'*}(E^{n_c}+\epsilon - E_1^{n_1'})$$
$$\times \sum_{lm_l k_l}\sum_{l'k_l'} W(J_c M_c|lm_l k_l|J_1)W(J_c M_c|l'm_l k_l'|J_1')$$
$$\times T(n_c|\epsilon lk_l|n_1)T^*(n_c|\epsilon l'k_l'|n_1')$$
$$\times (-1)^{m_l} \sum_L \sqrt{(2l+1)(2l'+1)(2L+1)}$$
$$\times \begin{pmatrix} l & l' & L \\ m_l & -m_l & 0 \end{pmatrix}\begin{pmatrix} l & l' & L \\ 0 & 0 & 0 \end{pmatrix} Y_{L0}^*(\hat{k}), \qquad 35.$$

where the properties of the W [Reference (13), Equation 12] and the series expansion of a pair of spherical harmonics [Reference (17) Equation 3.116] have been used, and we recall that energy integration is over the narrow range of photoelectron energies defined by the probe band-width. The time-resolved PAD is thus of the form,

$$\sigma(\bar{\epsilon}\,\hat{k}\,n_c\,M_c|\Delta t|n_i) = \sigma(\bar{\epsilon}\,\theta_k\,n_c\,M_c|\Delta t|n_i)$$
$$= \sum_L A_L P_L(\cos\theta_k), \quad L=0,1,2,\ldots, \qquad 36.$$

where P_L are Legendre polynomials, and the cylindrical symmetry results from our choice of linearly polarized fields with common polarization direction.

Equation 35 is not quite the observable of interest as it is resolved with respect to the magnetic states of the core. Because only the (analytically known) geometric functions in Equation 35 depend on the magnetic indices, the sum over the core (M_c) and electronic (m_l) magnetic indices (and in the general case also the sum over the initial and excited magnetic indices) can be carried out analytically. For the simple system considered here this procedure is straightforward. Using Equation 12 of Reference (13) and Equation 4.15 of Reference (77) one finds,

$$\sigma(\bar{\epsilon}\,\theta_k\,n_c\,|\Delta t|n_i) = \frac{\sigma_{\text{tot}}(\Delta t)}{4\pi}[1+\beta_2(\Delta t)P_2(\cos\theta_k)+\beta_4(\Delta t)P_4(\cos\theta_k)$$
$$+\beta_6(\Delta t)P_6(\cos\theta_k)+\cdots+\beta_{L_m}(\Delta t)P_{L_m}(\cos\theta_k)] \qquad 37.$$

where $L_m = 2(J_m+1)$, J_m being the largest significantly populated angular momentum state in the excited electronic state. Equation 37 is derived here for a particularly simple transition and initial condition but holds quite generally. In the limit of perturbative pump and probe fields $J_m = 1$, $L_m = 4$, and Equation 37 reduces to the form of a two photon ionization cross section. The same result is obtained by summing Equation 23 over all magnetic indices. This result may have been anticipated because, in the weak field limit, a pump-probe experiment is equivalent to a two-photon experiment in as far as angular momentum algebra is concerned. The exact pump-probe signal of Equation 37 may also seem familiar; it has the same functional form as the time-independent PAD resulting from a j-photon ionization experiment with $j = L_m/2$. The physical origin of the high ($L > 4$) moments of the PAD is nevertheless different; in the present case it arises from sequential excitation of rotational levels that accompanies Rabi cycling between the initial and excited states during the nonperturbative pump pulse. We note that Equation 37 treats the probe field nonperturbatively and hence accounts, in principle, for the possibility of further enrichment of the rotational content of the wavepacket through cycling between the excited and continuum states. In practice, however, we find that such cycling does not take place in direct ionization at relevant intensities; it requires the Rabi period $\Omega_R^{-1} \propto \varepsilon_{pr}^{-1}$ to be short with respect to the ionization time scale (50).

4. APPLICATIONS

Currently we are aware of two different types of phenomena that are mirrored in time-resolved PADs; phenomena that change the rotational composition of the wavepacket and phenomena that change its electronic symmetry. The sensitivity of the time-resolved PAD to the time-evolving rotational composition of the wavepacket renders this observable a useful probe of coupling mechanisms that perturb the rotational spectrum as well as of the unperturbed evolution of rotational wavepackets. The physical origin of this sensitivity and its application are discussed in Section 4.1. The sensitivity of the same observable to the electronic symmetry of the wavepacket makes it a potentially powerful probe of electronically nonadiabatic and spin-orbit induced processes (internal conversions and intersystem crossings, referred to collectively as nonradiative transitions). The origin and applications of this sensitivity are discussed in Section 4.2.

4.1. Time-Evolving Rotational Composition

The possibility of using time-resolved PADs to probe Coriolis coupling was proposed and illustrated theoretically by Fujimura and coworkers (51). An experimental illustration of this possibility is given in the work of Reid et al. (52), discussed in Section 2. Here strong variation of the PAD on a 10^{-11} sec time-scale following excitation of a vibrational band of p-difluorobenzene is attributed to Coriolis coupling. A mapping of the wavepacket property that carries the information about rotational perturbations onto the ionization asymmetry parameters is identified in the work of Althorpe & Seideman (53) and used to propose a quantitative measure of rotation-vibration coupling. The sensitivity of the PAD to the unperturbed evolution of rotational wavepackets, which manifests both strong field effects and the molecular structure, is illustrated numerically in (50). An experimental demonstration of this sensitivity is given in the recent study of Tsubouchi et al. (55), who used time-resolved PADs to probe the revival structure of excited state pyrazine.

To understand the origin of the responsiveness of the PAD to internal (rotation-vibration) and external (field-induced) interactions we return to Equation 26 and consider the structure of the energy phases, which contain the information about the field-free Hamiltonian. For the diatomic system considered in Equation 26, the early stages of the evolution are simple and nicely conveyed by any integrated signal; on a time scale of order $2\pi/\omega_{vib}$ the wavepacket oscillates on the vibrational time-scale, approaching the motion of a spatially well-defined classical particle as the wavepacket broadens in the conjugate quantum number space. Vibrational dephasing and revivals do not spoil this simple picture in the diatomic case as they appear on a long time-scale as compared to vibrational periods, imposing a slow envelope upon the rapid oscillations. On the time-scale of rotational periods and energy exchange between rotations and vibrations, the wavepacket of Equation 26 exhibits a broad range of energy-level spacings and the integral cross section no longer tracks the dynamics (see Figure 2a).

In order to filter out the dominating high frequency modes and focus on the frequencies that carry the information sought, we inspect the time-evolution of the wavepacket alignment.[4] The alignment, a measure of the degree of localization of the wavepacket with respect to the Euler angles, contains only the spectral information pertaining to the quantum numbers conjugate to these angles. In the case of a linearly polarized pump field and a linear molecule, one centers attention on the polar Euler angle θ and the conjugate quantum number J.

A convenient measure of the alignment is its first moment or (simply related and more commonly used) the expectation value of $\cos^2 \theta$ in the wavepacket (see Figure 2b). In the nearly classical limit—the limit of a broad rotational wavepacket, attained at nonperturbative pump intensities—$\langle \cos^2 \theta \rangle$ describes the rotation of the molecular axis in the space-fixed frame. In the fully quantum mechanical limit of a two-level system, corresponding to a perturbative pump field, $\langle \cos^2 \theta \rangle$ beats at the inverse of the rotational level spacing (54). Whereas in the absence of rotational perturbations, the alignment moments are determined solely by the rotational level spacing content of the wavepacket, coupling between rotations and vibrations introduces an additional time-scale into the time evolution of $\langle \cos^2 \theta \rangle$, $\tau_{int} = 2\pi/E_{int}^{v_1,J_1}$, $E_{int}^{v_1 J_1}$ being the coupling energy. Typically the rotation-vibration interaction is small compared with the rotational spacings. Consequently the alignment moment oscillates on the rotational time-scale with a slow envelope of time-scale τ_{int}, see Figure 2b.

The time evolution of the moments of alignment is nicely mapped onto the time evolution of the moments of the PAD, namely the time-dependent ionization asymmetry parameters in Equation 37. The origin of this correspondence can be appreciated by inspection of Equation 23. In the absence of coupling, the energy phase in that equation factorizes into a product of a vibrational and a rotational

---

[4]The notion of a time-resolved observable that serves as a "spectral filter" by exhibiting only a specific subset of the frequencies contained in the wavepacket is further explored in: Resch K, Blanchet V, Stolow A, Seideman T. 2001. *J. Phys. Chem. A* 105:2756–63.

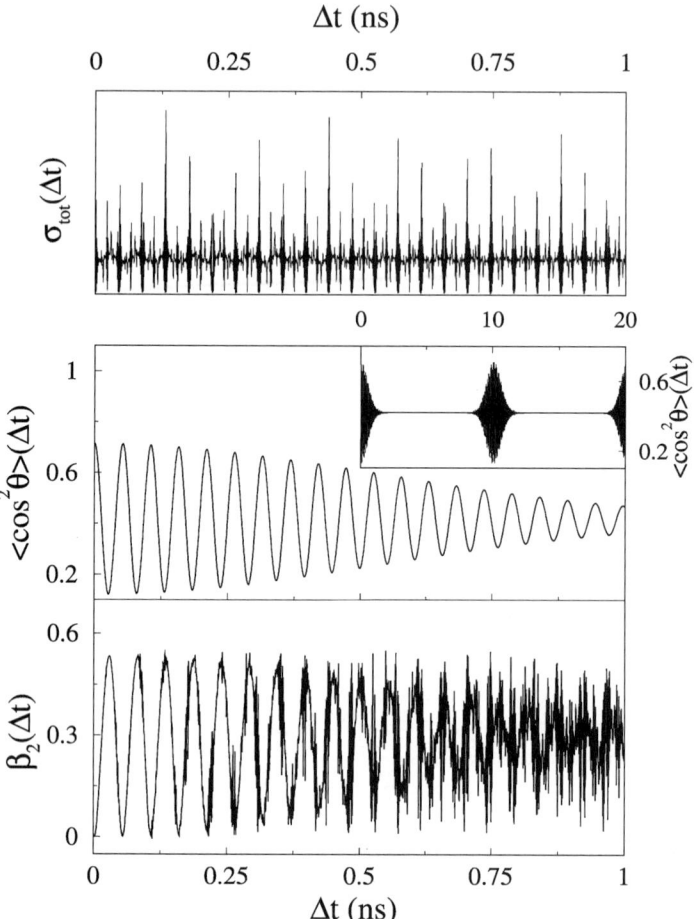

Figure 2 The effect of rotation-vibration coupling on the wavepacket alignment and the ionization signals. The wavepacket is produced by excitation from the ground vibronic state into the A state of Na_2 with a 20 fs pulse and, after a time Δt, ionized into the ground state of the ion with a 20 fs probe pulse. (*a*) The total ionization cross section [$\sigma_{tot}(\Delta t)$ in Equation 37] as a function of the pump-probe time delay. The integrated signal contains all time scales in the problem and is dominated by the high frequency modes, conveying no information about the rotational motion. (*b*) The expectation value of $\cos^2\theta$ in the wavepacket. $\langle\cos^2\theta\rangle$ provides an average measure of the alignment and hence of the time-evolving rotational composition of the wavepacket. (*c*) The lowest order asymmetry parameter [$\beta_2(\Delta t)$ in Equation 37]. β_2 maps the alignment moment of Figure 2*b*, hence providing a probe of rotational motions and rotational perturbations. Reproduced from Reference (53) with permission from the American Institute of Physics.

SEIDEMAN C-1

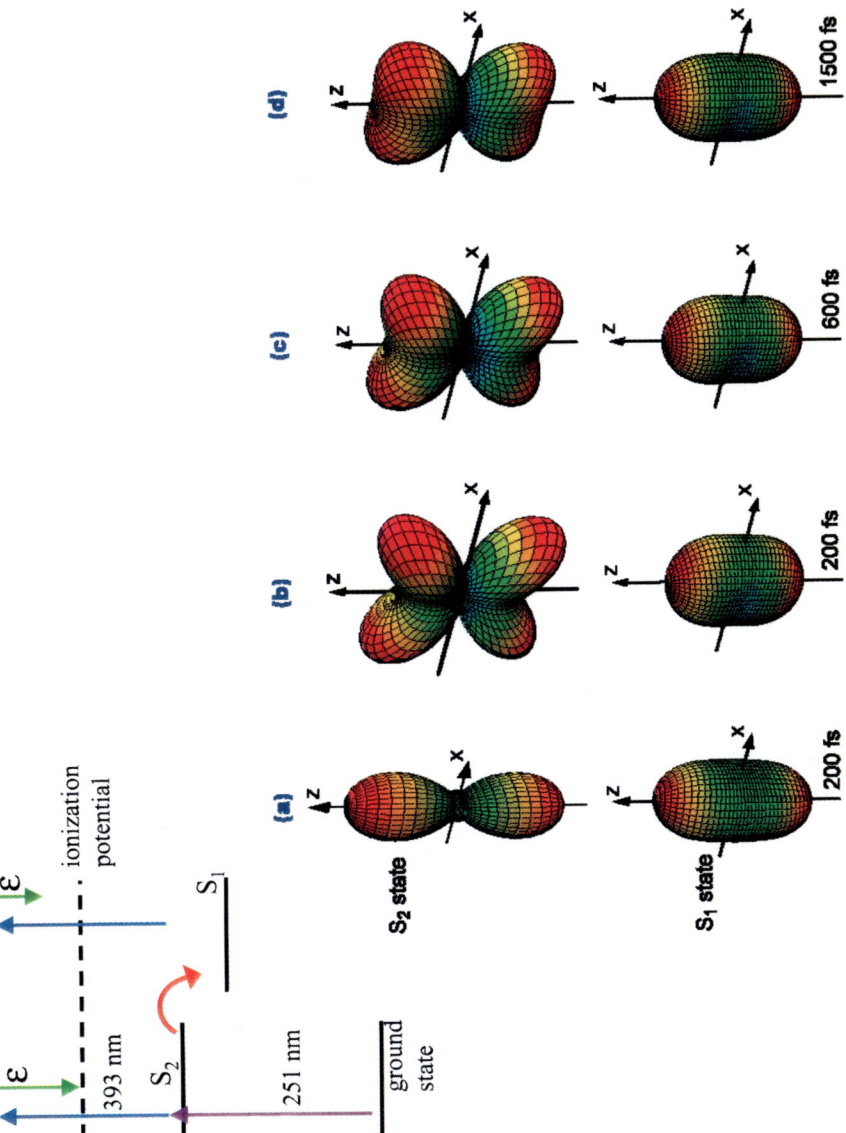

See text page C-2

Figure 3 (page C-1) Experimental PADs from femtosecond ionization of excites triethylenediamine (DABCO). The *inset* shows the energy levels involved in the experiment. The molecule is excited to the optically bright $^1E'(S_2)$ state that internally converts to a dark $^1A_1'(S_1)$ state. After a variable delay the vibronic superposition is ionized into the ground state of the ion. Panel (*a*) shows the resultant PADs at short (200 fs) delay for parallel polarization vectors of the excitation and ionization fields. The difference in electronic symmetry between the optically bright and dark states is seen to translate into a qualitative change of the PAD upon conversion. Panel (*b*), corresponding to perpendicular polarization of the excitation and ionization fields, illustrates the role of molecular alignment. Panels (*c*) and (*d*), taken at increasing time delays, show the evolution of the molecular alignment with time, as reflected in the time-resolved PAD (C. C. Hayden, Sandia National Laboratories, unpublished results).

functions, $\exp[-i(E_1^{n_1}-E_1^{n'_1})\Delta t] = \exp[-i(E^{v_1}-E^{v'_1})\Delta t]\exp[-i(E^{J_1}-E^{J'_1})\Delta t]$. Consequently the rapid vibrations are contained in the total cross section, σ_{tot} in Equation 37, whereas the β are slowly varying functions that respond only to the rotations. In the presence of rotation-vibration interactions, an energy phase, $\exp[-iE_{int}^{v_1 J_1}\Delta t]$ in Equation 23, gives rise to a modulated fine structure in the asymmetry parameter (see Figure 2c). The structure is periodic in Δt with period $t_{vib} = 2\pi/\omega_{vib}$ because the PAD probes the wavepacket within a restricted Franck-Condon window; the signal peaks each time the wavepacket traverses the Franck-Condon region. The vibrational structure sets in when Δt becomes non-negligible compared to $1/E_{int}^{v_1 J_1}$, disappears as the rotational components rephase, at $\Delta t \sim 2\pi/E_{int}^{v_1 J_1}$, and reappears periodically. Whereas the rapid oscillations exhibit a trivial time scale (that of the fundamental vibration), the beat envelope of $\beta(\Delta t)$ follows the beat pattern of $\langle\cos^2\theta\rangle$ and measures directly the strength of the rotation-vibration coupling. The generalization to polyatomic molecules, where both centrifugal and Coriolis interactions couple the vibrations with the rotations, is discussed in (53).

The conclusions of this subsection can be summarized as follows. The pump field produces a rotationally coherent wavepacket that is narrow (a 2-level system) at perturbative, and broader at nonperturbative intensities. The time-resolved PAD responds to changes in the relative phases of the rotational components of the wavepacket. This is the origin of its ability to probe coupling mechanisms that perturb the rotational spectra as well as the unperturbed evolution of rotational wavepackets. The latter evolution contains information about the molecular structure and the pump intensity.

4.2. Time-Evolving Electronic Symmetry

The possibility of using time-resolved PADs to probe nonradiative transitions in polyatomic excited states was proposed and illustrated theoretically by Seideman (56). A first experimental demonstration of this possibility is given in the recent work of Hayden and coworkers on the $S_2 \to S_1$ internal conversion of triethylenediamine (DABCO), already mentioned in Section 2 (C. C. Hayden, unpublished results). Before proceeding to discuss the theory and the experiment it is appropriate to recall the intensive research carried out on nonradiative transitions by use of time-resolved photoelectron energy spectra (see Section 2).

Work initiated by Domcke and coworkers (9) and followed by several experimental and theoretical studies (5, 7, 29–37) demonstrated the ability of time-resolved photoelectron energy distributions to probe not only the rates of nonradiative transitions but also the coherent vibronic dynamics induced (35). The origin of this capability is readily understood. Given an energy gap between the optically bright and dark states, electronic energy is converted into vibrational excitation upon transition. If the excited and ionic states have similar equilibrium configurations, vibrational energy is conserved in the ionization, and the energy gap translates into a change of the photoelectron energy spectrum in the course of the transition. In case the coupled electronic states correlate with different states of the ion, the condition of similar equilibrium configurations is relaxed (35). Recent experiments (32–37) illustrate the beauty of the energy-resolved probe while noting its lack of generality—often in molecular systems neither condition is met.

Time-resolved PADs offer a complementary approach to unraveling the dynamics of nonradiative transitions (14, 56). Provided that the coupled electronic states differ in electronic symmetry, the free electronic waves corresponding to their ionization differ in symmetry (as the product of irreducible representations of the neutral electronic state, the dipole operator, the free electronic state, and the ion-core electronic state must contain the totally symmetric representation). In the case of internal conversions, this holds true unless the mode promoting conversion is totally symmetric. The distribution of the departing electron in the body-fixed frame is thus different in the two electronic states, more so the higher the symmetry of the molecule. It is due to the alignment induced by the pump field (see Section 4.1), that the change of symmetry with respect to the molecular axes translates into a laboratory frame observable.

This concept is illustrated numerically in (14) for the case of a trans-linear polyene converting between the bright $S_2(1B_u)$ state and a dark $S_1(2A_g)$ state. Here the pump field induces a plane-polarized transition and thus mildly aligns the molecular C_2 axis perpendicular to the polarization vector. Because the internal conversion is fast as compared to the rotational periods, the pump-induced alignment is effectively constant on relevant time-scales, and the information about the internal conversion is carried through to the laboratory frame.

An impressive experimental realization of the theory (14), from the work of Hayden et al., is reproduced in Figure 3 (see color insert). The inset shows the energy levels involved in the pump-probe experiments. Ground state DABCO molecules were excited to the optically bright $S_2(^1E)$ state and subsequently ionized at variable time delays. Internal conversion to the dark $S_1(^1A_1)$ state takes place with a time constant of about 1 ps. Fully three-dimensional images of the ejected photoelectrons were collected as a function of the pump-probe delay time and analyzed with respect to energy. The molecule under consideration benefits from a large energy gap and essentially equal equilibrium displacements of the excited and ionic states. Consequently, the photoelectrons could be assigned to a parent electronic state based on their energy. The main frame of Figure 3 illustrates several general features of time-resolved PADs. Panel (a) shows the predicted qualitative difference between PADs following ionization of the bright and dark states of DABCO. Panel (b) shows that the pump-induced alignment is a key element in transforming the change of symmetry into a laboratory frame observable (56). When the probe polarization is rotated with respect to that of the pump, the PAD records the wavepacket alignment with respect to a different space-fixed coordinate system. Panels (c) and (d) show the change of the PAD as Δt becomes nonnegligible with respect to the rotational time-scale (see Figure 2 in the previous subsection).

It is important to note that, like the energy-resolved signal, the PAD is not a universal probe of radiationless transitions. It requires that the coupled states would differ in electronic symmetry and, in general, that the ionization would be a one-photon process. As discussed below, care may be needed to ensure that the body-fixed change of electron angular distribution would survive the transformation to the laboratory frame.

Research on the application of time-resolved PADs to probe nonradiative dynamics is, as yet, in a preliminary stage and much remains to be accomplished. We return to this problem and propose future directions in Section 5.

5. OUTLOOK

In the previous sections we discussed the current status of time- and angle-resolved photoelectron spectroscopy. Our goal in the concluding section is to note several of the possible avenues for future research.

1. Section 4.2 illustrated the crucial role played by the pump-induced alignment in the application of time-resolved PADs to probe radiationless transitions. In the limit of a perturbative field the alignment is mild, as it arises from the interference of just two or three levels. The possibility of using a moderately intense pump field to enhance the alignment and thereby also the change of the PAD upon conversion (56) is inviting; in short-pulse experiments high intensity is much easier to produce than to avoid. We remark that only a moderate increase of the intensity is necessary (below thresholds of competing processes), as enrichment of the rotational content of the wavepacket by one level significantly sharpens the alignment.

    It is important to point out that the (linearly polarized) pump field induces one-dimensional alignment (mild in the weak field limit and sharper at nonperturbative intensities); the molecule remains free to rotate about both space-fixed and body-fixed z-axes. Depending on the type of the pump-induced transition, this rotation might severely degrade the sensitivity of the space-fixed PAD to the change of the electronic symmetry that accompanies transition. An exciting possibility is that of using an external, moderately intense far-off-resonance pulse of appropriate polarization to align the molecules prior to the pump-probe experiment. An elliptically polarized pre-aligning field is then of advantage because the utility of a linearly or a circularly polarized field depends on the molecular symmetry and its polarizability tensor. The ability of elliptically polarized, far-off-resonance fields to hinder the rotation in all three Euler angles of molecules was recently illustrated (78).

    We note that only symmetry changing transitions that take place on a timescale short with respect to rotational periods can benefit from the pump-induced alignment. Means of inducing long-term post-pulse alignment, using moderately intense laser fields remain to be investigated (54).

2. Time-resolved techniques of measuring angular correlations between photoelectrons and photofragments (as well as between multiple fragments) in

photodissociation and three-body break-up are rapidly advancing (15). An interesting problem for theoretical research is thus the investigation of the type and quality of new (not available from more conventional observables) information that could be gained from angle-resolved photoelectron-photoion coincidence measurements and from vector correlation measurements in the time domain. This may be expected to go beyond the determination of a molecule-fixed PAD (70).

3. From a numerical view-point, the most expensive ingredient in the time-resolved PAD are the matrix elements of the transition dipole operator between the bound and free electronic functions. In case the PAD serves to probe the time-evolving rotational composition of the wavepacket (as, e.g., in studies of rotation-vibration coupling, field effects or molecular structure) these elements are not necessary because the information sought is contained in the nuclear dynamics (see Section 4.1). If, however, the PAD serves to probe the time-evolving electronic symmetry of wavepackets, the electronic dipole elements are essential, as the information sought is contained in the electronic states. Experimentally one is primarily interested in large polyatomic molecules. Such systems offer interesting [perhaps potentially useful (14, 35)] vibronic dynamics (5, 7, 32–37) as well as large absorption and ionization cross sections in experimentally convenient regimes. In that respect time-domain ionization experiments present a challenge to numerical research. On the other hand, time-resolved observables are forgiving as compared to their energy-resolved counterparts because they are significantly more averaged in the frequency sense. Current experiments therefore call for the development of methods for evaluating bound-free electronic dipole matrix elements that are approximate and apply to large polyatomic molecules.

4. Energy-resolved detachment of negative ions has yielded a wealth of information about problems that are not readily accessed by PES of neutral molecules (4, 38). We expect that PADs from detachment of negative ions will nicely complement the angle-integrated detachment data (4, 38), by analogy to PADs from ionization of neutral species.

Future experimental and theoretical work will likely point to new directions, beyond what can be envisioned at present.

ACKNOWLEDGMENTS

The author is indebted to many postdoctoral fellows and colleagues who contributed directly or indirectly to the time-resolved photoelectron spectroscopy work of her group; Stuart Althorpe, Saman Alavi, Yoshi-ichi Suzuki, Albert Stolow, Marek Zgierski and Valérie Blanchet. Special thanks are owed to Carl Hayden for authorizing the author to include in this review images from his work prior to publication, and to Albert Stolow, Phil Bunker, and Yoshi-ichi Suzuki for their careful reading of the manuscript. The Natural Science and Engineering Research Council of Canada is acknowledged for partial support.

Visit the Annual Reviews home page at www.annualreviews.org

LITERATURE CITED

1. Zewail AH. 1994. *Femtochemistry: Ultrafast Dynamics of the Chemical Bond*. Singapore: World Sci.
2. Polanyi JC, Zewail AH. 1995. *Acc. Chem. Res.* 28:119–32
3. Zhong D, Zewail AH. 1998. *J. Phys. Chem. A* 102:4031–58
4. Neumark DM. 2001. *Annu. Rev. Phys. Chem.* 52:255–77
5. Hayden CC, Stolow A, Ng CY, eds. 1999. *Adv. Phys. Chem. Photoionization and Photodetachment*, 10:91–126. Singapore: World Sci.
6. Lochbrunner S, Larsen JJ, Shaffer JP, Schmitt M, Schultz T, et al. 2000. *J. Elect. Spect. Relat. Phenom.* 112:183–98
7. Baumert T, Thalweiser R, Weiss V, Gerber G. See Ref. 21, Chapter 12
8. Baumert T, Helbing J, Gerber G. 1997. *Adv. Chem. Phys.* 1001:47–77
9. Domcke W, Stock G. 1997. *Adv. Chem. Phys.* 100:1–169
10. Warren TD. 1970. *Molecular Photoelectron Spectroscopy: A Handbook of He 584 Spectra*. New York: Wiley
11. Carlson TA. 1975. *Annu. Rev. Phys. Chem.* 26:211–33
12. Dill D, Wuilleumier FJ, eds. 1976. *Photoionization and Other Probes of Many-Electron Interactions*, NATO-ASI Ser. B 18:387–94. New York: Plenum
13. Seideman T. 1997. *J. Chem. Phys.* 107:7859–68
14. Seideman T. 2001. *Phys. Rev. A.* 64:042504-1–18
15. Continetti RE. 2001. *Annu. Rev. Phys. Chem.* 52:165–92
16. Suzuki T, Whitaker BJ. 2001. *Int. Rev. Phys. Chem.* 20:313–56
17. Harris CB, Ge N-H, Lingle RL Jr, McNeill JD, Wong CM. 1997. *Annu. Rev. Phys. Chem.* 48:711–44
18. Petek H, Ogawa S. 1997. *Prog. Surf. Sci.* 56:239–310
19. Ramswell JA, Fielding HH. 1998. *J. Chem. Phys.* 108:7653–61
20. Seideman T, Althorpe SC. 2000. *J. Elect. Spect. Relat. Phenomena* 108:99–108
21. Manz J, Wöste L, eds. 1995. *Femtosecond Chemistry*. Basel: Weinheim
22. Baumert T, Thalweiser R, Gerber G. 1993. *Chem. Phys. Lett.* 209:29–34
23. Fischer I, Villeneuve DM, Vrakking MJJ, Stolow A. 1995. *J. Chem. Phys.* 102:5566–69
24. Meier C, Engel V. 1993. *Chem. Phys. Lett.* 212:691–96
25. Assion A, Geisler M, Helbing J, Seyfried V, Baumert T. 1996. *Phys. Rev. A* 54: R4605–7
26. Frohnmeyer T, Hofmann H, Strehel M, Baumert T. 1999. *Chem. Phys. Lett.* 312: 447–54
27. Smith JM, Lakshminarayan C, Knee JL. 1990. *J. Chem. Phys.* 93:4475–76
28. Song XB, Wilkerson W, Lucia J, Pauls S, Reilly JP. 1990. *Chem. Phys. Lett.* 174: 377–83
29. Seel M, Domcke W. 1991. *J. Chem. Phys.* 95:7806–22
30. Seel M, Domcke W. 1991. *Chem. Phys.* 151:59–72
31. Mahapatra S, Köppel H, Cederbaum LS, Stampfuss P, Wenzel W. 2000. *Chem. Phys.* 259:211–26
32. Thantu N, Weber PM. 1993. *Chem. Phys. Lett.* 214:276–80
33. Cyr DR, Hayden CC. 1996. *J. Chem. Phys.* 104:771–74
34. Radloff W, Stert V, Freudenberg T, Hertel

IV, Jouvet C, et al. 1997. *Chem. Phys. Lett.* 281:20–26
35. Blanchet V, Zgierski M, Seideman T, Stolow A. 1999. *Nature* 401:52–54
36. Blanchet V, Lochbrunner S, Schmitt M, Shaffer JP, Larsen JJ, et al. 2000. *Faraday Discuss.* 115:33–48
37. Kim B, Schick CP, Weber PM. 1995. *J. Chem. Phys.* 103:6903–12
38. Boo DW, Ozaki Y, Andersen LH, Lineberger WC. 1997. *J. Phys. Chem. A* 101:6688–96
39. Frohnmeyer T, Baumert T. 2000. *Appl. Phys. B* 71:259–66
40. Dobber MR, Buma WJ, de Lange CA. 1993. *J. Chem. Phys.* 99:836–53
41. Ludowise P, Blackwell M, Chen Y. 1997. *Chem. Phys. Lett.* 273:211–18
42. Farmanara P, Stert V, Ritze HH, Radloff W. 2000. *J. Chem. Phys.* 113:1705–13
43. Lopez-Martens R, Long P, Solgadi D, Soep B, Syage J, Millie Ph. 1997. *Chem. Phys. Lett.* 273:219–26
44. Farmanara P, Radloff W, Stert V, Ritze HH, Hertel IV. 1999. *J. Chem. Phys.* 111:633–42
45. Davis AV, Zanni MT, Frischkorn C, Neumark DM. 2000. *J. Elect. Spect. Relat. Phenom.* 108:203–11
46. Greenblatt BJ, Zanni MT, Neumark DM. 1997. *Science* 276:1675–78
47. Lehr L, Zanni MT, Frischkorn C, Weinkauf R, Neumark DM. 1999. *Science* 284:635–38
48. Lehmann J, Meerschdorf M, Pfeiffer W, Thon A, Voll S, Gerber G. 2000. *J. Chem. Phys.* 112:5428–34
49. Hofer U, Shumay IL, Reuss C, Thomann U, Wallauer W, Fauster T. 1997. *Science* 277:1480–82
50. Althorpe SC, Seideman T. 1999. *J. Chem. Phys.* 110:147–55
51. Noguchi T, Sato S, Fujimura Y. 1989. *Chem. Phys. Lett.* 155:177–82
52. Reid KL, Field TA, Towrie M, Matousek P. 1999. *J. Chem. Phys.* 111:1438–45
53. Althorpe SC, Seideman T. 2000. *J. Chem. Phys.* 113:7901–10
54. Seideman T. 1999. *Phys. Rev. Lett.* 83:4971–74
55. Tsubouchi M, Whitaker BJ, Wang L, Kohguchi H, Suzuki T. 2001. *Phys. Rev. Lett.* 86:4500–3
56. Seideman T. 2000. *J. Chem. Phys.* 113:1677–80
57. Park H, Zare RN. 1996. *J. Chem. Phys.* 104:4554–80
58. Tully JC, Berry RS, Dalton BJ. 1968. *Phys. Rev.* 176:95–105
59. Jungen Ch, ed. 1996. *Molecular Applications of Quantum Defect Theory*. Bristol: Inst. Phys.
60. Chandra N. 1987. *J. Phys. B* 20:3405–14
61. Strand MP, Hansen J, Chien RL, Berry RS. 1978. *Chem. Phys. Lett.* 59:205–9
62. Leuchs G, Smith SJ, Khawaja E, Walther H. 1979. *Opt. Commun.* 31:313–16
63. Zygan-Maus R, Wolter HH. 1978. *Phys. Lett. A* 64:351–53
64. Georges AT, Lambropoulos P. 1978. *Phys. Rev. A* 18:1072–78
65. Reid KL, Duxon SP, Towrie M. 1994. *Chem. Phys. Lett.* 228:351–56
66. Reid KL. 1993. *Chem. Phys. Lett.* 215:25–30
67. Suzuki T, Wang L, Kohguchi H. 1999. *J. Chem. Phys.* 111:4859–61
68. Wang L, Kohguchi H, Suzuki T. 1999. *Faraday Discuss.* 113:37–46
69. Davies JA, LeClaire JE, Continetti RE, Hayden CC. 1999. *J. Chem. Phys.* 111:1–4
70. Davies JA, Continetti RE, Chandler DW, Hayden CC. 2000. *Phys. Rev. Lett.* 84:5983–86
71. Heck AJR, Chandler DW. 1995. *Annu. Rev. Phys. Chem.* 46:335–72
72. Arasaki Y, Takatsuka K, Wang K, McKoy V. 2000. *J. Chem. Phys.* 112:8871–84
73. Underwood JG, Reid KL. 2000. *J. Chem. Phys.* 113:1067–74
74. Williams RM, Papanikolas JM, Rathje J, Leone SR. 1997. *J. Chem. Phys.* 106:8310–23

75. Seideman T. 1999. *Faraday Discuss.* 113: 465–66
76. Seideman T. 1995. *J. Chem. Phys.* 103: 7887–96
77. Zare RN. 1988. *Angular Momentum. Understanding Spatial Aspects in Chemistry and Physics.* New York: Wiley
78. Larsen JJ, Hald K, Bjerre N, Stapelfeldt H, Seideman T. 2000. *Phys. Rev. Lett.* 85: 2470–73

SCATTERING RESONANCES IN THE SIMPLEST CHEMICAL REACTION

Félix Fernández-Alonso[1] and Richard N. Zare[2]

[1]*Istituto di Struttura della Materia—Consiglio Nazionale delle Ricerche, Area della Ricerca di Roma—Tor Vergata, 00133 Rome, Italy; e-mail: felix@ism.rm.cnr.it; and* [2]*Department of Chemistry, Stanford University, Stanford, California 94305-5080; e-mail: zare@stanford.edu*

Key Words transition state, quasibound states, scattering resonance, $H + H_2$ reaction family, reactive scattering, reaction dynamics

■ **Abstract** Recent studies of state-resolved angular distributions show the participation of reactive scattering resonances in the simplest chemical reaction. This review is intended for those who wish to learn about the state-of-the-art in the study of the $H + H_2$ reaction family that has made this breakthrough possible. This review is also intended for those who wish to gain insight into the nature of reactive scattering resonances. Following a tour across several fields of physics and chemistry where the concept of resonance has been crucial for the understanding of new phenomena, we offer an operational definition and taxonomy of reactive scattering resonances. We introduce simple intuitive models to illustrate each resonance type. We focus next on the last decade of $H + H_2$ reaction dynamics. Emphasis is placed on the various experimental approaches that have been applied to the search for resonance behavior in the $H + H_2$ reaction family. We conclude by sketching the road ahead in the study of $H + H_2$ reactive scattering resonances.

1. INTRODUCTION: FROM NUCLEAR PHYSICS TO CHEMICAL REACTIONS

Resonance behavior is ubiquitous in a wide range of physical phenomena. One of the most familiar examples comes from spectroscopy, where the energy of a photon $\hbar\omega$ is tuned to match the energy difference ΔE_{12} between two levels of an atom or molecule. The resulting energy spectrum is characterized by the appearance of sharp peaks (resonances) that obey the Bohr energy condition $\Delta E_{12} = \hbar\omega$. Likewise, the energy widths of these resonances may be associated with the lifetime of the excited state. These two quantities, line position and line shape, represent the most fundamental observables in atomic (1) and molecular (2) spectroscopy.

The behavior described above is not restricted to the use of photons as projectiles. In the 1930s, Fermi and collaborators noticed that slow neutron scattering had anomalously high cross sections at certain energies (3–6). These large capture cross sections were interpreted by Bethe (7) and Bohr (8) in terms of the existence of quasistable states of the compound neutron-nucleus system at the resonance energies with lifetimes τ directly related to the peak energy widths Γ, i.e., $\tau = \hbar/\Gamma$. The energy dependence of the cross section in the neighborhood of an isolated narrow resonance followed the Breit-Wigner (9) formula

$$\sigma(E) \propto |A_{nr} + A_r|^2 = \left| A_{nr} + \frac{\Gamma}{(E - E_o) + i\frac{\Gamma}{2}} \right|^2. \qquad 1.$$

Here E_o and Γ are the energy position and width of the resonance, and A_{nr} and A_r represent the nonresonant (direct) and resonant (indirect) contributions to the cross section. If direct scattering is negligible, the cross section has a Lorentzian energy profile with a width that is directly related to the lifetime of the compound state. This behavior, however, is not general, and the actual profile may show a variety of shapes, so-called Fano line shapes (10), depending on the destructive or constructive interference between the direct and indirect terms in Equation 1.

In the years following these initial findings, elementary particle resonances have been crucial for the understanding of the fundamental forces of nature. For example, pion-nucleon elastic scattering at low energies $\sim$100–300 MeV is dominated by the so-called P_{33} resonance, a touchstone for the understanding of the strong nuclear force. This resonance has been entirely attributed to the formation of an excited nucleon state (11–13).

Some years had to pass between the firm establishment of resonance behavior in nuclear reactions and the experimental and theoretical verification of similar phenomena in atomic and molecular physics. In electron-atom/molecule scattering, resonances were discovered almost simultaneously by experiment and theory at the end of the 1950s, even though older measurements had already given some evidence of unexpected structure in scattering cross sections. In this sense, some authors (14, 15) consider the discovery of the Auger effect (16) as the first observation of a resonance in electron/heavy-particle scattering. Haas (17) and Schulz (18) measured significant structure in e^-–N_2 collision cross sections associated with vibrational excitation of the molecular target. These results were in good agreement with the theoretical expectations of Herzenberg & Mandl (19) using resonance theories. Similarly, experiments of Schulz & Fox (20) on e^-–He collisions were successfully explained using a one-level Breit-Wigner formula by Baranger & Gerjuoy (21). Following these findings, a multitude of atoms and small molecules were investigated yielding numerous examples of the interaction of an incident electron with the target, resulting in temporary capture and subsequent decay (14, 15, 22–27).

It is worth emphasizing that such resonances are not simply of academic interest alone. The powerful infrared carbon dioxide laser, for example, is powered by efficient vibrational energy transfer from resonantly excited $N_2(v = 1)$ to CO_2 (28, 29). Recently, experimental evidence has been presented that double-strand breaks in DNA, caused by ionizing radiation incident on cells, arises from transient molecular resonances induced by the secondary electrons at energies lower than the ionization potential (30). Thus, resonant scattering processes not only affect how some stars twinkle, but they also affect life processes.

In the late 1960s, several groups began the study of the role of scattering resonances and quasibound states in elastic and inelastic heavy-particle collisions using realistic models and specific molecular systems amenable to experimental investigation (30a–k). Using a statistical model for molecular collisions, Miller (30l) estimated the direct and resonant contributions to the cross section for the Cs + RbCl reaction. This chemical reaction was known to proceed via a long-lived intermediate from the scattering experiments of Herschbach and coworkers (30m). A more clear indication that scattering resonances would be important in chemical reactions came in the early 1970s when Truhlar & Kuppermann (31, 32) performed the first exact quantum-mechanical H + H_2 collinear computations. They observed that the total reactive and nonreactive cross sections displayed pronounced oscillations as a function of collision energy. The authors ascribed these features to the effect of interfering amplitudes for different semiclassical paths between reagents and products (32). Based on the results of close-coupling collinear calculations for the same reaction system, Levine & Wu (33) explicitly demonstrated the presence of resonances in reactive collisions. The surprising feature was the absence of a local minimum in the H + H_2 potential energy surface (PES) that could readily explain trapping during the course of reaction. The early H + H_2 classical trajectory studies of Hirshfelder, Eyring, & Topley (34) using the crude London-Eyring-Polanyi (35, 36) H_3 potential energy surface (PES) had noted the existence of long-lived trajectories owing to temporary trapping in a local minimum at the saddle point region (which has been called "Lake Eyring"). Such a minimum was later shown to be an artifact in the earlier H + H_2 PES arising from a sign error (37–39). It was not present in the one used for these calculations. Levine & Wu (33) explained these resonances in terms of the strong coupling between the relative motion along the reaction coordinate and the internal degrees of freedom of the compound triatomic molecule. Owing to such couplings, the structure in the reaction cross section could be accounted for in terms of quasibound states that arise from internal excitation with lifetimes on the order of 30 fs. Subsequent work by Wu et al. (40–42) on collinear H + H_2 and model A + BC reactions provided strong evidence for the existence of the dynamical Lake Eyring under a wide range of circumstances. In addition, resonances were clearly observed whenever the effect of changing the vibrational frequency along the reaction coordinate gave rise to adiabatic potential wells. This oversimplified picture of reactive resonances has been extremely useful in further investigations and has served as a basis for many approximate treatments that followed. Concurrent with these efforts, Schatz & Kuppermann (43) analyzed in detail the oscillations in their calculated reactive cross sections for collinear H + H_2 at total energies of 0.90 eV and 1.276 eV. A calculation of scattering phase shifts, delay times, and Argand diagrams (a plot of the imaginary and real parts of the reactive amplitudes as a function of collision energy) enabled an interpretation of these features in terms of compound-state

(Feshbach) resonances. This work appears to be the first time that the term Feshbach resonance, originally coined for nuclear processes (44, 45, 45a), was utilized in the context of chemical reactions to denote quasibound states associated with a PES that possesses no local minima and thus is not capable of supporting bound states.

The world of collinear collisions of course does not describe accurately the real world. Would resonance scattering behavior still be prominent for three-dimensional scattering? Schatz & Kuppermann (46) answered this question in the affirmative. They showed that resonance peaks were reduced in magnitude but persisted in two- and three-dimensional (total angular momentum $J = 0, 1$) calculations. The energy shifts observed in going from lower to higher dimensionality were consistent with the addition of the zero-point-energies of the bending degrees of freedom of the H_3 complex. Further details of the developments on the dynamics for the $H + H_2$ reaction in this early period may be found in the review by Truhlar & Wyatt (47). These early studies showed that scattering resonances were not as uncommon in chemical reactions as initially thought, even in the three-dimensional world. Unfortunately, carrying out fully converged, three-dimensional calculations in a reliable manner had to wait more than a decade owing to the need to develop efficient algorithms for their calculation and the need for more powerful computers. The early work had established, however, that a study of the resonance spectrum of a chemical reaction could provide a means of investigating the structure of the transition state region of elementary chemical reactions [see the review by Kuppermann (48)].

During this initial period, a number of approximate methods for describing reactive resonance scattering were developed and tested on the two benchmark reactions, $H + H_2$ and $F + H_2$. The goal of this work was to develop simplified yet realistic models that allowed for the solution of a lower-dimensionality problem via quantal and semiclassical adiabatic treatments in natural (49, 50), hyperspherical (51–53), and Jacobi collision coordinates (54–56) [see also the reviews by Bowman (57, 58)]. By adiabatic we mean that the motions of the slow modes are frozen, and the resulting equations of motion solved with this constraint. This work met with some success. It was able to reproduce and provide a physical explanation for previous (and often obscure) computational results. The classical approach to reaction dynamics of Child & Pollak (59–63) stimulated the study of the correspondence between quantum mechanical resonances and classical mechanics in terms of resonant periodic orbits (RPOs). An RPO corresponds to an unstable periodic path along the PES. Trajectories starting close to an RPO have a tendency to be guided along it, thereby causing temporary trapping of the system and enhanced energy transfer between the reactant and product valleys. The location of these RPOs for collinear $H + H_2$ collisions matched with quantitative accuracy the quantal resonance energies (50, 64). Pollak and coworkers (53, 55, 65) tested the approach further for isotopic variants of the same reaction and showed the reliability of the method for the prediction of resonance energies. Resonances appeared near the energetic thresholds of new vibrational channels. The RPO approach was successfully extended to three-dimensional calculations (66, 67). The method, however, has been the subject of some criticism; see, for example, Hipes & Kuppermann (68) as well as Pollak (67).

A new impetus for theoretical calculations arose at the end of the 1980s with the experimental report by Valentini and coworkers of Feshbach resonances in the state-resolved integral cross sections (ICS) for the $H + H_2$ (69, 70) and $D + H_2$ (71) reactions. These experiments produced fast H or D atoms by photolysis and detected the resulting molecular reaction product by coherent anti-Stokes Raman scattering (CARS). [See also Reference (72) for a review of $H + H_2$ experiments carried out prior to the 1990s]. The resonances reported in these experiments appeared to be in agreement with previous quantum mechanical calculations for low values of the total angular momentum J (46, 56, 68, 73–75). The first fully converged (J up to 31) three-dimensional calculations on the $H + H_2$ reaction by Zhang & Miller (76, 77) showed, however, a dramatic blurring of the resonance peaks once all partial waves were computed. The same effect was observed for the isotopic variant of this reaction $D + H_2$ (78). Calculations by a number of other groups (79, 80) corroborated the theoretical results of Zhang & Miller. Moreover, photoinitated experiments using resonance enhanced multiphoton ionization (REMPI) for molecular product detection by Zare and coworkers (81), showed no structure in the state-resolved integral cross sections. Miller (82) has presented an excellent review of this brief yet intense period in $H + H_2$ reaction dynamics. In subsequent work, Miller & Zhang (83) made detailed calculations of state-resolved integral and differential cross sections (DCS) for the $H + H_2$ and $D + H_2$ reactions for total energies up to 1.4 eV. Although resonances were not discernable in the integral cross sections, they distinctly appeared in the energy dependence of state-resolved differential cross sections in the form of a ridge in the energy-angle (E–θ) plane. Such a ridge was related to the dependence on J of the resonance energy and was used to provide the rotational constants and lifetimes of the compound triatomic complex. The resonances seen in these calculations were broad in energy, thereby providing an explanation for their absence in state-resolved integral cross sections. It is interesting to note that quasiclassical trajectory (QCT) calculations for the $D + H_2$ reaction by Aoiz and coworkers (84–86) also showed resonance ridges similar to the ones previously ascribed to a purely quantum mechanical effect. These findings suggest that these features may have an important semiclassical interpretation, one that needs further explication.

By the early 1990s, two important conclusions were reached for further studies on the $H + H_2$ reaction: (*a*) It is not advisable to draw strong conclusions from a comparison between scattering experiments and the results of partial calculations, that is, the need exists to perform an extensive database of fully converged quantum mechanical calculations on this reaction system. (*b*) The signatures of scattering resonances are likely to be most easily found in experiments that measure the energy dependence of fully state-resolved differential cross sections.

The $H + H_2$ reaction family invites detailed study because it is for this reaction system that the PES can be most accurately calculated. Beginning in the 1960s with the appearance of the Porter-Karplus PES (37), outstanding progress has been made. To date, a number of accurate PESs have appeared: the Liu-Siegbahn-Truhlar-Horowitz PES (LSTH) (87–89), the double-many-body-expansion PES (DMBE) of Varandas and coworkers (90), and more recently the Boothroyd-Keogh-Martin-Peterson PESs (BKMP and BKMP2) (91, 92) as well the Exact-Quantum-Monte-Carlo PES (EQMC) of Wu et al. (93), of which the latter is estimated to have an accuracy of better than 4 cm^{-1}. At the same time, time-dependent and time-independent scattering calculations have advanced where it is possible to obtain all scattering attributes in the context of motion on a single potential energy surface.

It is important to stress that from an experimental point of view the operational definition of a resonance is a scattering feature that changes sharply as a function of the total energy of the reaction system and which may be associated with metastability of the compound system. This is the sense in which experimentalists search for resonances, as has been seen in past work on nuclear and electron scattering. On the other hand, theoretical calculations are needed to understand what causes the resonance in the scattering process. In this sense, theory and experiment are complementary. To illustrate this definition we refer to the recent crossed-beam experiments and QM calculations of Liu and coworkers on the F + HD → HF + D reaction (94–97) that have firmly established the existence of resonance features in the ICS of a chemical reaction. F + HD ICSs have also been complemented with energy- and angle-resolved measurements (95) that display sharp backward-forward variations with collision energy. In addition, similar studies by Dong et al. (98), Chao & Skodje (99), and Sokolovski & Castillo (100) have once more revived the long-standing issue regarding resonance signatures in the angular distributions of the F + H$_2$ reaction (101), whose origin goes back to the ground-breaking experiments of Lee and coworkers in the mid-1980s (102, 103). At present, the F + H$_2$ reaction is far less well understood than the H + H$_2$ reaction and a fully quantitative comparison between experiment and theory has been hampered by deficiencies in our present knowledge of the underlying PES. The resonances reported by Liu and coworkers provide a benchmark measurement to tune the electronic structure calculations for the FH$_2$ reaction system. For a review, see Liu (97).

This realization of how to search for resonances has inspired three independent efforts in the H + H$_2$ reaction family in the past decade:

1. Wrede & Schnieder (104) used photoinitiated crossed-beam techniques to search for predicted resonance structure over a narrow energy range. They were able to obtain state-resolved integral and differential cross sections for the H + D$_2$(v = 0, j = 0) → HD(v' = 0, 1; j') + D between 1.27 eV and 1.30 eV. They achieved outstanding resolution but they found no resonances.
2. Kendrick et al. (105, 106) also examined the H + D$_2$ reaction in which HD(v' = 0, j' = 7) was detected in a velocity-sensitive manner, from which the integral cross section was estimated between a collision energy of 0.73 eV

and 1.02 eV. They observed a bump (∼50% variation) in the integral cross section that agreed well with their calculations (partial waves up to J = 6). Subsequently, the calculations have been extended to higher partial waves, and the resonance feature is no longer present (107–109).

3. Zare and coworkers (110, 111; B. D. Bean, F. Fernández-Alonso, J. D. Ayers, A. E. Pomerantz, R. N. Zare, unpublished results) have measured state-resolved differential cross sections for the reaction of H + D$_2$ → HD(v' = 3, j') + D between 1.39 eV and 1.85 eV using photoinitiation in which the angular distribution is derived from the law of cosines (113, 114). They observed strong forward scattering for the HD(v' = 3, j' = 0) product but not for HD(v' = 3, j' > 2). Moreover, the amount of forward scattering varied sharply with collision energy.

Following a discussion of the nature of reactive scattering resonances, we shall return to possible interpretations of these experiments.

2. CLASSIFICATION OF REACTIVE SCATTERING RESONANCES

This section provides a taxonomy of reactive resonances. We make a special effort to connect formal definitions of scattering resonances with approximate treatments of reactive scattering.

2.1. Shape Resonances

Our starting point for a discussion of reactive resonances is the reaction coordinate s, by which we mean the motion along the minimum energy path (MEP) in a given PES. Motion along a direction ρ transverse to s can be assigned to the internal coordinate of our reactive system. These concepts can be put on a firmer basis for a collinear collision between three atoms using natural collision coordinates as first introduced by Marcus (115, 116) and extensively discussed in References (117–120). In natural collision coordinates, s asymptotically approaches the asymmetric stretch motion of the three-atom complex. It ultimately leads to the breakage of chemical bonds. Similarly, motion along ρ corresponds to the symmetric-stretch mode. We assume that reactions can be thought to proceed primarily along the MEP.

Let us forget momentarily about the existence of ρ motion and concentrate on the one-dimensional MEP potential energy profile. Such a profile may look like what is shown in Figure 1a (see color insert) with a well around $s = 0$ indicating the existence of a quasibound state at a particular energy below the barrier top. The situation depicted in Figure 1a is similar to what the first PESs predicted for the H + H$_2$ reaction (Lake Eyring). It is known to be present in other reactive systems as, for example, the ion-molecule reaction He + H$_2^+$, isoelectronic with H + H$_2$ (121–123). The one-dimensional Schrödinger equation describing the motion along the energy profile $V_{MEP}(s)$ in Figure 1a (see color insert) is given by

$$\frac{\partial^2 \psi}{\partial s^2} + \frac{2\mu}{\hbar^2}[E - V_{MEP}(s)]\psi = 0, \qquad 2.$$

where is μ the reduced mass and E the total energy of the system. A plane wave with positive energy incident from the left gives rise to a reflected wave (back to reagents) and a transmitted wave (forward to reaction products). The corresponding boundary conditions for the solution of Equation 2 may then be written as

$$\psi(s = -\infty) = e^{iks} + S_R e^{-iks} \qquad 3a.$$

and

$$\psi(s = \infty) = S_T e^{iks}, \qquad 3b.$$

where $k = \sqrt{2\mu E/\hbar}$ is the wavevector, and S_R and S_T are the (complex) energy-dependent reflection and transmission amplitudes. The transmission coefficients (reaction probabilities) are obtained by computing $|S_T|^2$. Figure 1b (see color insert) shows the resulting transmission coefficient as a function of incident energy. Owing to the presence of the well, we observe a sharp peak below the maximum of the potential caused by a close match in energy between the incident energy and a quasibound (quantized) level. In the context of reactive scattering we attribute this peak to what is called a shape resonance, i.e., the shape of the interaction potential is responsible for the presence of structure in the reaction cross section. In formal scattering theory, S_R and S_T are elements of the scattering S matrix. All observables related to the scattering process may be computed from knowledge of this quantity (124). In particular, resonances are associated with poles in the S matrix in the fourth quadrant of the complex energy plane. Referring back to Equation 1, we can associate a pole in such an expression with the complex energy $E = E_0 - i\frac{\Gamma}{2}$ that causes the denominator of the resonant contribution to the cross section to vanish. This feature is perhaps the most discriminating definition of a scattering resonance. From a theoretical viewpoint, a study of resonance scattering reduces to the search for such poles in the complex plane. In the language of complex analysis, the S matrix is an analytical function. Consequently, knowledge of the location of the poles of the S matrix completely defines the S matrix and hence, completely characterizes the scattering process (124, 125). In this light, all scattering may be regarded as arising from scattering resonances. But this way of regarding the scattering process may not be useful or insightful if the resonance widths overlap appreciably.

Other ways exist for ascertaining the resonance character of the sharp peak shown in Figure 1a. In Figure 1b we also show the time delay associated with the passage along the MEP. The desired time delay may be obtained from the S matrix by use of the expression (126)

$$\tau = \mathrm{Im}\left(\hbar \frac{1}{S_T} \frac{dS_T}{dE}\right), \qquad 4.$$

where Im denotes the imaginary part. The time delay τ as a function of energy shows a pronounced delay at the resonance energy indicative of a long residence time inside the potential well. Depending on the depth and width of the well shown in Figure 1a, the reaction cross section may show a number of peaks, indicative of the presence of several quasibound (quantized) levels. The depth of the well could also be below the asymptotic energy of products and reagents, indicative of the existence of a stable molecule (bound states). Such is the case of the $O + H_2$ reaction that has the water molecule as its intermediate (stable) state (127). It is known through unimolecular decomposition experiments that in this case structure persists even above the well, at energies associated with virtual levels of the intermediate (128, 129).

2.2. Barrier Resonances

In a recent study, Friedman & Truhlar (130) have posed a very intriguing question: Are chemical reaction barriers resonances? Using a methodology very similar to the one we have used for our description of shape resonances, they were able to show that symmetric energy barriers are also associated with poles in the S matrix. To illustrate this fact, they demonstrated the possibility of effecting a continuous change from a barrier resonance to a shape resonance (see Figures 1a and 1b, color insert). A hint for this behavior can already be observed in Figure 1b. It shows a small bump in the time delay near the top of the barrier but smaller and broader than the one we ascribed to a shape resonance. In Figures 1c and 1d, (see color insert) we show the solution to the one-channel scattering problem for an Eckart barrier (131) that closely resembles the $H + H_2$ MEP, as calculated by Friedman & Truhlar (130). The transmission probability rises smoothly, but we can also observe a broad peak in the time delay associated with energies close to the potential barrier. Classically, it corresponds to metastability associated with passage through a potential maximum, but more interestingly, it is associated quantum mechanically with an S-matrix pole far removed from the real energy axis. Such a pole possesses a large imaginary energy component, which implies that it corresponds to a short-lived metastable state. Subsequent work by Truhlar and coworkers (132) on asymmetric one-dimensional potentials corroborated the above conclusions for a number of asymmetric potential functions. A similar pole structure of the S matrix for parabolic and Eckart potential barriers has also been discussed by Seideman & Miller (133), and Ryabov & Moiseyev (134) in the context of transition state theory and the calculation of reaction probabilities. In their approach, reaction rate expressions are given in terms of Siegert eigenvalues, that is, the complex eigenvalues of the Schrödinger equation with outgoing boundary conditions at energies where the poles occur (135). These Siegert eigenstates have been traditionally associated with scattering resonances (15, 136, 137). Seideman & Miller (133) have contended, however, that the progressions of poles associated with barriers do not bear a direct relationship with conventional (isolated) scattering resonances as they are used to describe direct dynamics. We note, however, that a distinction between isolated and overlapped resonances is not a sharp one, and as will be further explained below, barrier resonances do enjoy many of the characteristics of conventional resonances.

Let us consider the properties of a harmonic barrier (a completely soluble problem by analytical methods) (133, 134, 138). Similar conclusions apply to other analytically solvable problems such as Eckart barriers (133, 134, 139). The poles associated with the barrier are infinite in number and accumulate at the energy of the barrier top. Their precise values are given by

$$E_r = E_o - i\hbar\omega(2n + 1), \quad n = 0, 1, 2 \ldots, \qquad 5.$$

where E_o is the position of the barrier and ω its frequency. As shown by Equation 5, the leading pole occurs at $n = 0$, and it is the longest-lived (smallest imaginary part). Atabek et al. (138) have shown that these simple harmonic barriers can lead to localization effects similar to those found in the usual resonance phenomenon.

Skodje and coworkers (140–144) have presented extensive evidence for the existence of such barrier resonances in collinear and three-dimensional calculations for the H + H$_2$ reaction family. In their computational method, a wave packet is launched in the transition state region. The Fourier transform of the wave packet temporal decay shows structure directly attributable to the participation of metastable states. Collinear studies on D + H$_2$ displayed a complicated resonance spectrum (143). Three different types of peaks were found and identified according to the following categories: conventional reactive resonances, barrier resonances, and threshold anomalies (Wigner cusps). The last of these were easily identified in their method because they became arbitrarily small as the resolution of the calculation was increased. The other two cases, however, corresponded to the formation of true metastable states. A total of ten barrier states dominated the lower energy part of the spectrum, between 0.525 eV and 2.268 eV, and formed two progressions along the reactant and product channels. Similar to conventional resonances in this reaction system (Feschbach resonances; see Section 2.3), their stability also increased with energy. Semiclassically, they were associated with maxima in the vibrationally adiabatic potential curves, that is, repulsive periodic orbit dividing surfaces (PODS), in contrast with the RPOs related to conventional resonances.

Motivated by the need to interpret these barrier resonances, Sadeghi & Skodje (143, 144) have derived an analytic expression for parabolic barrier resonances using a time-dependent formalism. For the lowest even wavepacket, it takes the form

$$S(\Omega) \propto \prod_{k=0}^{\infty} \left(\frac{(\frac{1}{2} + 2k)\omega}{\Omega + i(\frac{1}{2} + 2k)\omega} \right), \qquad 6.$$

where $\Omega = E/\hbar$, and ω is the barrier frequency. The denominator of this expression is reminiscent of the pole positions for the harmonic barrier shown in Equation 5: There is a contribution of an infinite number of poles located at $\Omega = -i(\frac{1}{2} + 2k)\omega$. These line shapes have been successfully applied to the barrier resonances appearing in the collinear D + H$_2$ reaction (143, 144). Varandas & Yu have also observed these barrier resonances in time-dependent wavepacket calculations for the H + H$_2$ (145), Mu + H$_2$/D$_2$ (146, 147), and H + DH/D + HD (148) reactions.

2.3. Feshbach Resonances

Our previous discussion on shape and barrier resonances was strictly one-dimensional in nature, although even the simplest atom-diatom reaction involves more than one degree of freedom. It is precisely the energy exchange between various collective modes of the compound molecule that are responsible for what is commonly called Feshbach resonances. We have already mentioned their importance in chemical reactions and, in particular, for the explanation of resonances in the H + H$_2$ reaction. In this section, we provide a simple model that serves as an illustration of the origin of this effect.

Let us consider a one-dimensional potential like the one used in our discussion of shape resonances (Section 2.1). In the present example, this interaction potential still represents the passage between reagents and products, e.g., the minimum energy path with a potential energy profile $V_{MEP}(s)$. If we add a second degree of freedom ρ perpendicular to the reaction path whose interaction potential depends parametrically on the position along s, the Hamiltonian becomes

$$H = -\frac{\hbar^2}{2\mu_s}\left(\frac{\partial^2}{\partial s^2}\right) - \frac{\hbar^2}{2\mu_\rho}\left(\frac{\partial^2}{\partial \rho^2}\right) + V_{MEP}(s) + V(\rho;s), \qquad 7.$$

where the μ_s and μ_ρ represent the reduced masses for s and ρ motions, respectively. For simplicity, we assume a harmonic potential $V(\rho;s) = \frac{1}{2}\mu_\rho\omega(s)\rho^2$. The harmonic frequency $\omega(s)$ is taken to be smallest at $s = 0$ as an indication of bond weakening at the point of switching between reagents ($s < 0$) and products ($s > 0$). We proceed to solve this very crude model for a chemical reaction taking the motion along s to be slow compared to the one along ρ (in the spirit of the Born-Oppenheimer separation of electronic and nuclear motions). Such an approximation amounts to neglecting the first kinetic energy term in Equation 7. The resulting Schrödinger equation is

$$\left(-\frac{\hbar^2}{2\mu_\rho}\left(\frac{\partial^2}{\partial \rho^2}\right) + V(\rho;s)\right)\varphi(\rho,s) = (\varepsilon_n(s) - V_{MEP}(s))\varphi(\rho,s), \qquad 8.$$

where $\varepsilon_n(s) = \hbar\omega(s)(n + \frac{1}{2}) + V_{MEP}(s)$ are the eigenenergies associated with the harmonic motion perpendicular to the reaction path. The total wave function can be calculated from a series expansion in terms of the solutions of Equation 8,

$$\Psi(\rho,s) = \sum_m \Phi_m(s)\varphi_m(\rho,s). \qquad 9.$$

We seek to solve the Schrödinger equation

$$\left(-\frac{\hbar^2}{2\mu_s}\left(\frac{\partial^2}{\partial s^2}\right) - \frac{\hbar^2}{2\mu_\rho}\left(\frac{\partial^2}{\partial \rho^2}\right) + V_{MEP}(s) + V(\rho;s)\right)\Psi(\rho,s) = E\Psi(\rho,s). \qquad 10.$$

After substitution of Equations 8 and 9 into 10, we find

$$\left(-\frac{\hbar^2}{2\mu_s}\left(\frac{\partial^2}{\partial s^2}\right) + \varepsilon_n(s) + \sum_m A_{nm}\right)\Phi_n(s) = E\Phi_n(s), \qquad 11.$$

where the matrix elements of A_{nm} are given by

$$A_{nm} = -\frac{\hbar^2}{2\mu_\rho}\left(2\left\langle\varphi_n\left|\frac{\partial}{\partial s}\right|\varphi_m\right\rangle\frac{\partial}{\partial s} + \left\langle\varphi_n\left|\frac{\partial^2}{\partial s^2}\right|\varphi_m\right\rangle\right). \qquad 12.$$

For a harmonic potential describing motion transverse to the reaction coordinate, the matrix elements in Equation 12 are given by (40)

$$\left\langle\varphi_n\left|\frac{\partial}{\partial s}\right|\varphi_m\right\rangle = \left(\frac{1}{2}\frac{d\ln\omega(s)}{ds}\right) \cdot \left\{\left[\frac{m(m-1)}{4}\right]^{1/2}\delta_{nm-2}\right.$$

$$\left. -\left[\frac{(m+1)(m+2)}{4}\right]^{1/2}\delta_{nm+2}\right\}, \qquad 13.$$

and

$$\left\langle\varphi_n\left|\frac{\partial^2}{\partial s^2}\right|\varphi_m\right\rangle = \left(\frac{1}{2}\frac{d\ln\omega(s)}{ds}\right)^2 \cdot \left\{\left[\frac{m(m-1)(m-2)(m-3)}{16}\right]^{1/2}\delta_{nm-4}\right.$$

$$\left. -\frac{(m^2+m+1)}{2}\delta_{nm} + \left[\frac{(m+1)(m+2)(m+3)(m+4)}{16}\right]^{1/2}\delta_{nm+4}\right\}$$

$$+ \left(\frac{1}{2}\frac{d^2\ln\omega(s)}{ds^2}\right) \cdot \left\{\left[\frac{m(m-1)}{4}\right]^{1/2}\delta_{nm-2} - \left[\frac{(m+1)(m+2)}{4}\right]^{1/2}\delta_{nm+2}\right\}, \qquad 14.$$

Equations 13 and 14 show that, in the spirit of this simple model, the A_{nm} matrix can effect transitions between vibrational adiabatic states differing by two and four quanta.

By considering only the diagonal elements of the A_{nm} matrix, we arrive at the following first-order approximation

$$\left(-\frac{\hbar^2}{2\mu_s}\frac{\partial^2}{\partial s^2} + \varepsilon_n(s) + \frac{\hbar^2}{16\mu_\rho}(n^2+n+1)\left(\frac{d\ln\omega(s)}{ds}\right)^2\right)\Phi_n(s) = E\Phi_n(s). \qquad 15.$$

Equation 15 has a very simple interpretation: motion along s takes place with an effective potential

$$V_{eff}(s) = V_{MEP}(s) + \hbar\omega(s)\left(n + \frac{1}{2}\right) + \frac{\hbar^2}{16\mu_\rho}(n^2 + n + 1)\left(\frac{d\ln\omega(s)}{ds}\right)^2. \quad 16.$$

The first term in Equation 16 is the original MEP along the reaction coordinate. The presence of a second (coupled) degree of freedom has the effect of adding a vibrational energy (vibrationally adiabatic correction) plus a diagonal correction to the energy. The interplay between these two terms can be strong enough to modify greatly the shape of the potential. Also, the energy corrections to the original MEP increase with vibrational quantum number. In this manner, new Lake Eyrings may appear once the interaction of the various degrees of freedom of the system are taken into account. Such is the case of the $H + H_2$ reaction. We note that the interaction between internal degrees of freedom can be sufficiently large to yield deep wells lying below the ground state adiabatic potential curve (that is, they are true bound states of the system). In this situation we speak of vibrational bonding. This issue has been extensively discussed in the literature in the context of heavy-light-heavy triatomics where the rapid motion of the light atom acts as the binding force (149–156).

At this point, we can begin to use the concepts discussed in the context of shape and barrier resonances. Feshbach resonances associated with adiabatic potentials may have shape and/or barrier character. This behavior depends on which features of the effective interaction potential are responsible for the temporary trapping of the system. As an example of a realistic situation, we show in Figure 2 the first three vibrationally adiabatic potentials for the $H + H_2$, $D + H_2$, and $H + D_2$ reactions using the computer program of Truhlar and coworkers, ABCRATE (157) on the DMBE PES (90). These curves represent the $J = 0$ adiabatic potential curves, taking into account the energies of all the internal modes of the triatomic species in three dimensions (one symmetric stretch v_{str} and two degenerate bends v_{bend}). In all three cases, we can clearly discern the appearance of potential wells in the vibrationally adiabatic potentials near the transition state region ($s = 0$). Also, the potentials for the isotopic variants $D + H_2$ and $H + D_2$ are not symmetric owing to the different reagent and product mass combinations. This asymmetry will likely manifest itself as a preference for metastable states to be localized in the reagent or product side, depending on the nature of the resonance. The $H + D_2$ reaction does not show potential wells up through $v_{str} = 2$, which suggests that the observation of metastable states will occur at higher collision energies. Indeed, Garrett et al. (50) have predicted the absence of low-energy reactive resonances for this system owing to the lack of resonance energy levels for $v_{str} = 0, 1, 2$. The adiabatic potential curves shown in Figure 2 correspond to $J = 0$. It is possible to compute similar adiabatic potential curves for an arbitrary total angular momentum J by addition of the rotational energy of

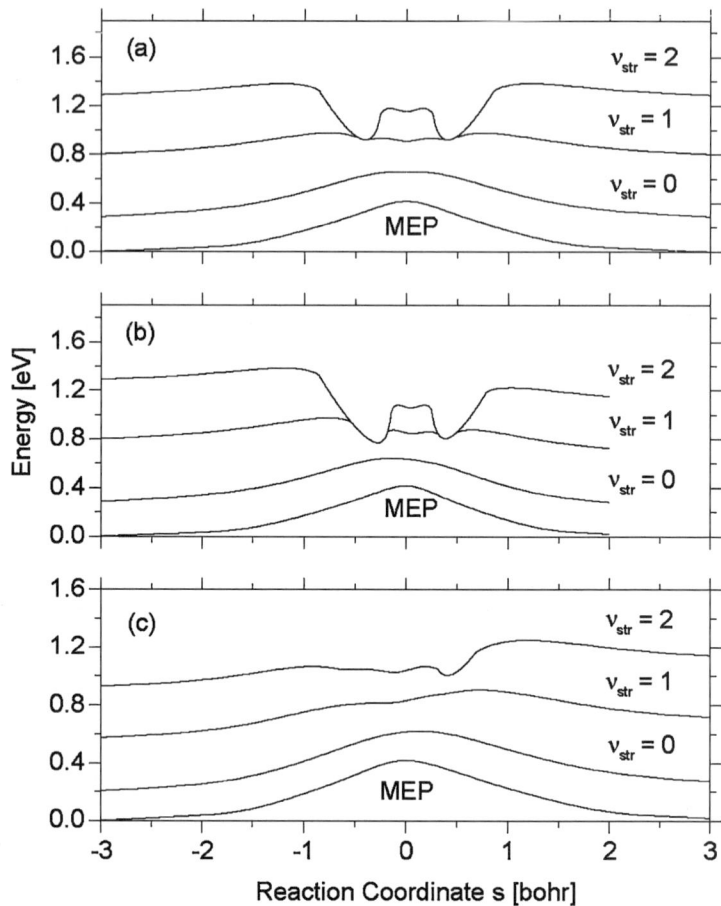

Figure 2 Three-dimensional (J = 0) vibrationally adiabatic potentials for (a) H + H_2; (b) D + H_2; and (c) H + D_2.

the triatomic complex. Under the rigid rotor approximation it takes the form

$$E_{rot} = B_{rot}(s)J(J+1) = \frac{\hbar^2}{2I(s)}J(J+1),\qquad 17.$$

where $I(s)$ is the moment of inertia of the triatomic complex.

We show in Figure 3 the effect of adding the energy given by Equation 17 to the $v_{str} = 3$ adiabatic potential for the H + D_2 reaction for values of the total angular momentum that are known to participate in reactive scattering below 2.0 eV (J < 35). Aside from the more pronounced appearance of metastable potential wells, we observe the presence of potential barriers both on the reagent and product sides of the reaction. Both features may be responsible for the presence of Feshbach resonances associated with adiabatic wells and barriers at energies

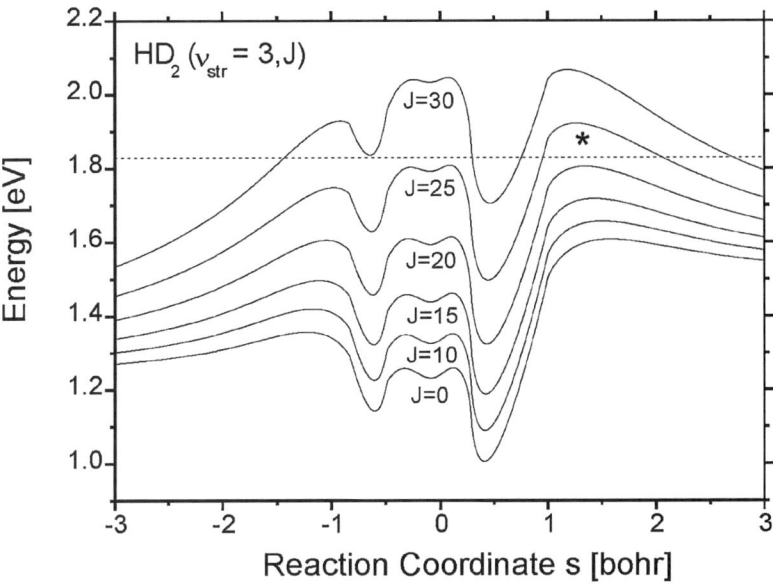

Figure 3 Adiabatic potential for H + D$_2$ (v_{str} = 3) as a function of the total angular momentum J = 0, 10, 15, 20, 25, and 30 using Equation 17. The horizontal dashed line marks the total energy of the experiments E$_{tot}$ = 1.83 eV, and the asterisk the location of the energy barrier thought to be responsible for trapping (see Section 3.3).

between ~1.4 eV and ~2.0 eV. This example is particularly relevant to developing a simple picture for resonances in the H + D$_2$ → HD(v' = 3, j') + D reaction (see next section).

To illustrate the importance of Feshbach resonances associated with barriers, we comment on their relation to new formulations of transition state theory. Recent theoretical work strongly suggests that the overall reactivity of a chemical reaction (the reaction rate) is directly related to the presence of quantized bottlenecks that control the passage of reactive flux to products (158–160). This approach to rate theories has been successfully tested using accurate three-dimensional quantal calculations for an extensive number of atom-diatom systems including H + H$_2$ (161–165), D + H$_2$ (166), F + H$_2$ (167, 168), O + H$_2$ (164, 165, 169–171), H + O$_2$ (172–174), Cl + HCl, I + HI, I + DI (175), Cl + H$_2$ (176), He + H$_2^+$ (177, 178), and Ne + H$_2^+$ (179, 180). In this context, reaction thresholds associated with maxima of vibrationally adiabatic curves have been related to barrier resonances. Although isolated narrow resonances arising from conventional trapped states (wells) are assigned using a full set of triatomic quantum numbers (v_1, v_2^K, v_3) corresponding to the symmetric v_1, bend v_2, vibrational angular momentum K, and asymmetric stretch motion v_3 (56, 68, 73, 181, 182), threshold or barrier transition states have been consistently labeled by (v_1, v_2^k), that is, they are missing the asymmetric stretch quantum number v_3 associated with the reaction coordinate. Truhlar and coworkers (159) have suggested that such a missing degree of freedom may be recovered if it is identified with the dominant barrier pole as, for example, by setting n = 0 in Equation 5. More rigorously, Zhao & Rice (183) have identified transition states with scattering resonances using complex scaling techniques.

3. RESONANCES IN THE H + D_2 REACTION: RECENT PROGRESS, CURRENT STATUS, AND FUTURE PROSPECTS

Only in the last decade have experiments been capable of measuring state-resolved differential cross sections as a function of collision energy. This task has been made possible by the use of lasers. H or D atoms are generated by laser photolysis of suitably chosen precursors. The reaction products are detected again with lasers, using either atomic (184) or molecular (185) Rydberg-tagging or, alternatively, by REMPI of the HD product (186–188).

For the H + H_2 family, special attention has been given to the H + D_2 reaction. Theoretical efforts to carry out extensive and accurate calculations on this particular reaction system have been more or less synchronous with experiment. Prior to the experiments to be described in this section, quantal calculations for the H + D_2 reaction were scarce. D'Mello et al. (189) were the first to calculate well-converged integral cross sections at collision energies of 0.55 and 1.3 eV. Their results compared favorably with the photoinitiated and state-resolved experiments of Zare and coworkers (190). Prior to this work, calculations for the H + D_2 system had been restricted to low values of the total angular momentum (75, 191, 192).

Moreover, theoretical calculations had neglected the incorporation of geometric phase (GP) effects [for a review of GP effects in molecular systems, see Yarkony (193) and Mead (194)]. In the H + H_2 reaction system, GP effects arise from the existence of a conical intersection between its two lowest adiabatic PESs (195–200). As the nuclei perform a closed loop around this intersection, the electronic wave function is forced to change sign if it is to remain real-valued. Such a purely quantum mechanical effect, which may be called a dynamical Jahn-Teller effect, is expected to influence the reaction dynamics in a fundamental way even at energies that are well below the energy at which the two PESs intersect ($\sim$2.7 eV). Whenever reaction products can be scattered into the same solid angle element by direct scattering and by a scattering mechanism that involves a pseudorotation about the conical intersection, it is necessary to sum the two scattering amplitudes and square them to obtain the reaction probability. In such circumstances, constructive and destructive interference occurs in the DCS. Moreover, such GP effects can persist in the integral cross section even after a partial-wave summation. Calculations by Kuppermann and coworkers showed this behavior to hold for the H + H_2 (195–197), D + H_2 (197, 198), and H + D_2 (199, 200) reactions. As Baer (201) and others have stressed, GP effects are part of the nature of the scattering process. Their inclusion is not a matter of choice. GP effects are expected to play an important

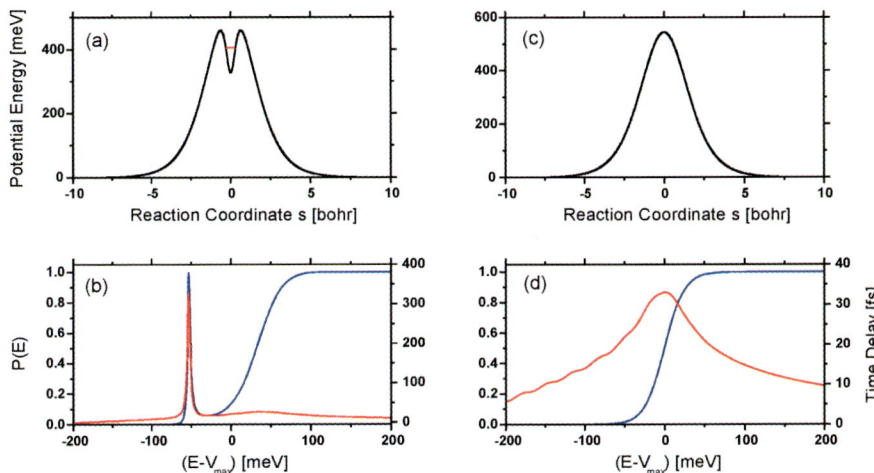

Figure 1 (*a*) Potential energy profile along the reaction coordinate displaying a well at the center. (*b*) Transmission probability P(E) (*blue curve*, left axis) and time delay (*red curve*, right axis) as a function of (E–V_{max}), where E is the total energy and V_{max} is the maximum value of the potential energy. The resonance peak below threshold at approximately –50 meV corresponds to the energy level of the well shown by the horizontal red line in (*a*). Panels (*c*) and (*d*) correspond to an Eckart barrier displaying no well along the potential energy profile. The potential parameters have been chosen so as to mimic as closely as possible the MEP for the H + H_2 reaction.

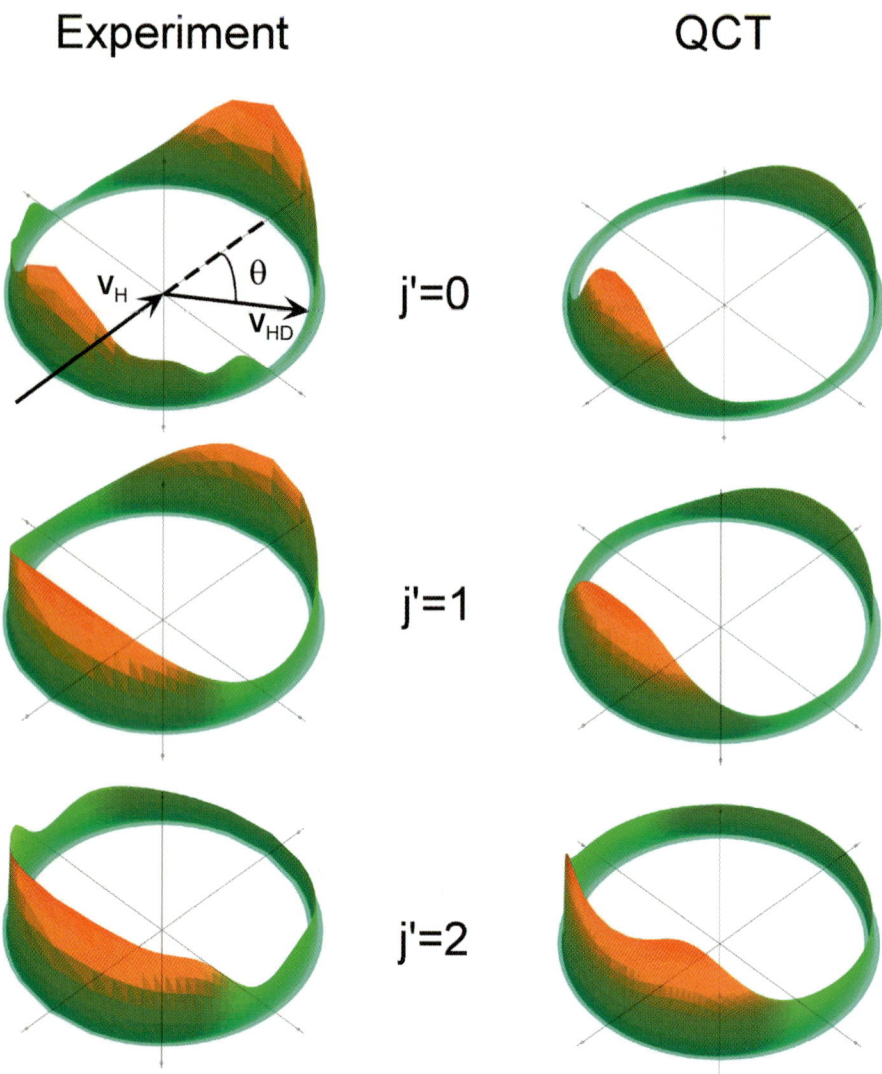

Figure 4 Polar plots of the HD(v' = 3, j' = 0–2) center-of-mass differential cross sections from experiment (*left*) and QCT calculations (*right*) at E_{col} = 1.64 eV. The top polar plot on the left indicates with arrows the directions of reagents and products.

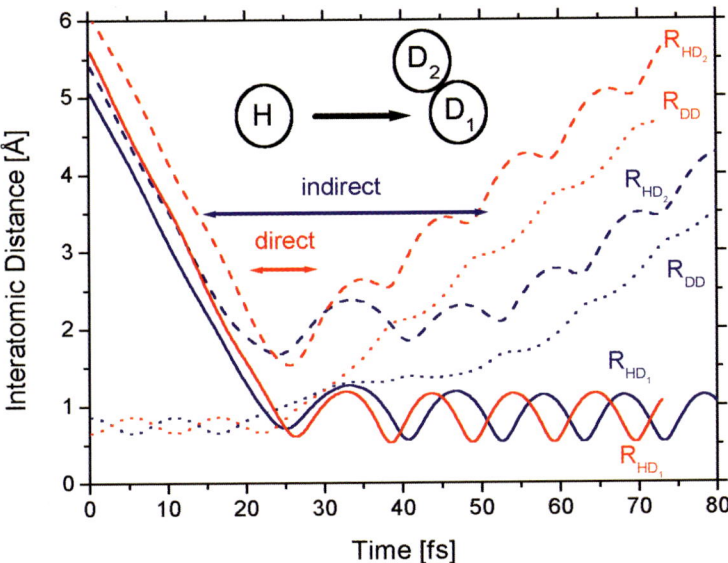

Figure 5 Temporal evolution of interatomic distances for H + D_2(v = 0, j = 0) → HD(v′ = 3; j′ = 0) + D at 1.64 eV obtained from QCT calculations. The *blue curves* represent an indirect trajectory that is forward scattered; the *red curves* represent a direct trajectory that is back scattered. The slope of the curves is the distance covered per unit time, that is, their relative velocity. Bars mark the approximate interaction time during which the complex hangs together.

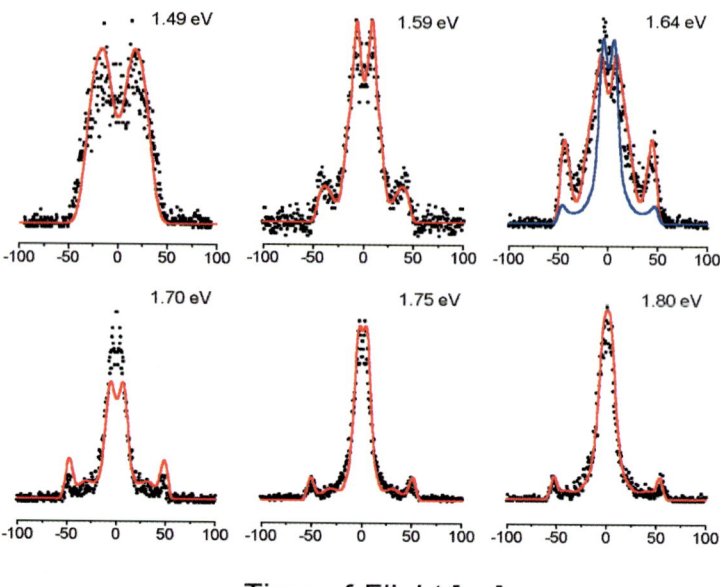

Figure 6 HD(v′ = 3, j′ = 0) experimental time-of-flight profiles (*black circles*) as a function of collision energy. The *solid red lines* are the result of a forward convolution of the QM DCS calculations. At E_{col} = 1.64 eV, the *blue line* corresponds to the forward-convoluted time-of-flight profile for the QCT DCS.

role in the dynamics of many bimolecular reactions as well as unimolecular decompositions (202–206).

3.1. Crossed-Beam Experiments in Search of Resonances; The Role of Geometric Phase

In the past few years, Schnieder and collaborators have carried out a series of very elegant high-resolution crossed-beam experiments for the H + D_2 reaction. Their experimental setup consists of two parallel, spatially separated molecular beams. One beam contains HI as the photolytic precursor for fast H atoms. The other beam contains the diatomic reagent D_2. Laser photolysis of HI generates H atoms with a well-defined laboratory spatial distribution. The laser polarization is chosen so as to direct this H-atom beam toward the D_2 beam. D-product detection is achieved with meV resolution by use of Rydberg atom tagging (184). An analysis of the kinetic energy spectrum of the D product as a function of scattering angle provides the desired HD(v′, j′) product state resolution. This experimental arrangement trades off signal production for energy resolution. More details about the experimental technique can be found in recent reviews by Liu (97) and Casavecchia and coworkers (207, 208).

Kuppermann & Wu (199) predicted for the LSTH PES a pronounced reactive scattering resonance at a total energy of 1.481 eV (1.29 eV collision energy) for the state-specific exchange H + D_2(v = 0, j = 0) → HD(v′ = 0, j′ = 4, 5) + D only if GP was included in the calculation. Guided by these predictions, Wrede & Schnieder (104) measured D-atom angular distributions between 1.27 and 1.30 eV, but they found no evidence for the predicted resonance structure. Instead, their experimental results were in good agreement with QM calculations on the LSTH PES at 1.30 eV by D'Mello et al. (189) that did not include GP effects. It was argued (104) that such a lack of agreement could be caused by inaccuracies of the LSTH PES used in the calculations of Wu & Kuppermann, which could shift the resonance energy by as much as 13 meV, that is, outside of the collision energy range investigated in the experiments.

Based on the results of the experiments of Wrede & Schnieder (104) and the calculations of Kuppermann (199), it appears that the influence of geometric phase on H + H_2 scattering dynamics is a very sensitive function of collision energy as well as the form of the PES. It seems true, however, that at some collision energies GP effects essentially make no contribution to the scattering dynamics. This absence of GP effects might result from the lack of collision trajectories that make a closed loop around the conical intersection. At present, the experiments of Schnieder and coworkers for the H + D_2 reaction have been compared with converged NGP (NGP = not including GP effects) QM calculations on the LSTH and on the newer BKMP2 surfaces at collision energies of 0.5 eV (209, 210), 1.28 eV (209, 211), 2.2 eV (212, 213), and 2.67 eV (214). The agreement between the two is very good, including the last collision energy of 2.67 eV, which lies slightly above the conical intersection for the H + H_2 system.

Recent QM calculations by Kendrick (215, 216) have shown that for the H + D_2 reaction, GP effects beautifully cancel out for the first six partial waves and for total energies between 0.4 and 2.4 eV. These conclusions are valid for both the BKMP2 and LSTH PESs. It is unlikely that higher partial waves will behave in a different way, but to date no fully converged GP calculations have been completed. The cancellation of GP effects calculated by Kendrick is so astonishing that we wonder whether a simple explanation exists to account for it. To date, no answer to this important question has been put forward. Xu & Varandas (217) and Mahapatra et al. (218) reported a similar absence of GP effects in their calculations.

In addition, Kendrick's calculations showed broad transition-state resonances in the rotationally resolved integral and differential cross sections obtained from a consideration of these first few partial waves. It is still possible, but we believe not likely, that the true PES will show more pronounced GP effects. Wu et al. (93) have developed a new Exact Quantum Monte Carlo (EQMC) PES that the authors claim to be one order of magnitude more accurate than previous ones. A comparison of converged QM mechanical calculations on the EQMC PES with experiment has not yet been performed. It would be very desirable, however, to have a calculation of cross sections and resonance positions including GP effects on this new PES and to compare them with previous work. At this point, we conclude that no general consensus has been reached on the effects of GP and their impact on the reaction dynamics and resonance spectrum for this chemical reaction. It seems clear, however, that GP effects are minor for the H + D_2 reaction over the collision energies so far investigated.

We expect GP to be important under some reaction conditions, and past experiments on the integral cross section for D + $H_2(v' = 1, j' = 1)$ (219) suggest this may be the case (198), but more work, experimental and theoretical, is needed before this effect can be claimed to be quantitatively understood. We note that the computational effort required to incorporate GP effects in scattering calculations is considerable. This difficulty may be overcome once recent and efficient timedependent methods for the solution of the reactive scattering problem (220) are utilized.

3.2. An Old Problem Revisited: Does Resonance Structure in the Integral Cross Section Survive Partial-Wave Summation?

Recent photoinitiated experiments by Shafer-Ray and coworkers (105) have indicated the existence of resonance structure in the integral cross section for the state-specific reaction H + $D_2 \rightarrow$ HD($v' = 0, j' = 7$) + D at a collision energy of 0.94 eV. The GP QM calculations on which the comparison with theory had been made included only the first seven partial waves (J < 7). Conventional wisdom based on a large body of calculations for this reaction system, as well as other isotopic variants, has brought these results into question. It now seems clear that after summation of partial waves up to J = 30–35, no structure remains in the ICS energy dependence for the HD($v' = 0, j' = 7$) product state (106). This conclusion has been independently obtained from the NGP QM calculations of Chao et al. (107), Aoiz et al. (108), and Kendrick (109).

Kendrick (109) has analyzed in detail how partial wave summation affects state-resolved ICSs and DCSs for the H + D$_2$ reaction. Owing to its heavier mass, the H + D$_2$ reaction system is a more favorable case for the survival of resonance features in the ICS than its isotopic cousins H + H$_2$ and D + H$_2$, which Miller & Zhang (83) showed a decade ago to be devoid of them. The most distinct peaks in H + D$_2$ state-resolved ICSs occur for the vibrationless and rotationless product state. Oscillations in the ICS become very faint and difficult to detect experimentally for $v' > 1$ and/or $j' > 3$. Much more interesting are the ICSs for the vibrationally excited reaction H + D$_2$(v = 1, j = 0) → HD(v', j') + D that display very clear resonance bumps. Experimental work along these lines would be very timely in order to corroborate these predictions. Vibrationally excited D$_2$ may be obtained by stimulated-Raman-pumping (SRP) techniques (219, 221–226). Moreover, ICSs are far less difficult to measure than DCSs, and the wealth of information to gain from them appears to make efforts in this direction worthwhile.

It is tempting to offer some reassessment of the ICS measurements by Valentini and coworkers on H + p-H$_2$ (69, 70) and D + H$_2$ (71). Our confidence in fully converged theoretical calculations has advanced to the point that the features they observed cannot be attributed to structure in the ICS as a function of collision energy. We do know, however, that resonances occur for low partial waves in the energy range they investigated. Their observations might be explained as arising from a partial selection of scattering angles. It is very easy to imagine that the experiments discriminated in favor of slow-moving, backward-scattered molecular products. This discrimination could be caused by the time delay between photo-initiation and CARS detection as well as by the spatial overlap of photolysis and probe laser beams. Theory teaches us that these backward-scattered products correlate to a large extent with low-J partial waves (small impact parameters). Thus, the observation of such products defeats the blurring of resonance structure arising from partial-wave summation. Possibly, the first observations of scattering resonances were made by Valentini and coworkers, but unfortunately, it is not possible to quantify with ease what exactly was observed.

3.3. Forward Scattering in the H + D$_2$ → HD(v' = 3, j') + D Reaction

The photoloc approach, as recently developed by Zare and coworkers (113, 114, 227, 228), represents an alternative and promising approach to crossed-beam experiments for the measurement of state-resolved integral and differential cross sections for the hydrogen exchange reaction. Its name stems from the two major elements of the technique: first, laser photolysis initiates the chemical reaction in a free-jet expansion of a photolytic precursor (such as HBr or HI) and reagent (D$_2$); second, the law of cosines is used to relate the product laboratory velocity distribution to the center-of-mass differential cross section. Using REMPI detection of the HD(v', j') product and the core-extraction technique (229), Fernández-Alonso et al. (228, 230, 231) have measured product-state-resolved DCSs for the HD(v' = 1) and HD(v' = 2) vibrational manifolds at collision energies of 1.70 eV and 1.55 eV, respectively. Comparison with converged QM calculations on the BKMP2 PES shows good agreement with experiment (232). The shift from backward scattering toward sideways scattering observed with increasing product rotational angular momentum was interpreted as a tendency for the reagent orbital angular momentum to be channeled into product rotation. Such a trend is consistent with a direct reaction mechanism and is also very clear in the crossed-molecular-beam data of Schnieder and coworkers (104, 209–214).

Subsequent measurements of $H + D_2 \rightarrow HD(v' = 3, j') + D$ DCSs at a collision energy of 1.64 eV (110) have shown for the first time clear deviations from the direct behavior previously observed for other $HD(v', j')$ product states. A large forward-scattering peak was observed for low-j' product states as shown in Figure 4 (see color insert). Quasiclassical trajectory (QCT) calculations at this collision energy (110) indicated some forward scattering but failed to reproduce its magnitude. An analysis of the classical trajectories associated with forward scattering features revealed a very different underlying reaction mechanism involving large values of the total angular momentum of approximately $J = 20$ (impact parameters of 0.70–0.80 Å) and time delays on the order of 26 fs (see Figure 5, color insert). These time delays are still well below the rotational period of the HD_2 complex, thereby explaining why the angular distribution lacks forward-backward symmetry. Analysis of single forward-scattered trajectories demonstrated a preference for an early elongation of the D_2 chemical bond causing the appearance of potential wells about the HDD collinear configuration that leads to temporary trapping of the complex. These observations were in agreement with a classical model for $H + H_2$ resonances previously proposed by Muga & Levine (233).

These experimental findings and first attempts to explain forward scattering in this reaction have stimulated further theoretical studies. Truhlar and coworkers (234) have used a simplified vibrationally adiabatic model in the same spirit as that discussed in Section 2.3, to provide a quantum mechanical interpretation of the resonance signatures found in the $HD(v' = 3, j' = 0)$ angular distributions. The resonance was associated with an $HD_2(v_1 = 3, v_2 = 0, J = 20)$ complex localized on a vibrationally adiabatic barrier on the product side (see Figure 3). The predictions of this reduced-dimensionality quantum mechanical study were in very good agreement with the previous conclusions reached from the analysis of quasiclassical trajectories (110). Based on previous theoretical predictions for the $H + H_2$ reaction (162, 165), the assignment $v_2 = 0$ was consistent with the observation of the largest amount of forward scattering for the rotationless state $HD(v' = 3, j' = 0)$.

As a continuation of this line of work, Zare and coworkers (111; B.D. Bean, F. Fernández-Alonso, J.D. Ayers, A.E. Pomerantz, R.N. Zare, unpublished results) have measured $HD(v' = 3, j')$ DCS's between 1.39 eV and 1.85 eV in order to characterize in more detail forward-scattering features. Figure 6 (see color insert) presents the observed $HD(v' = 3, j' = 0)$ time-of-flight profiles as a function of collision energy along with a comparison with NGP QM predictions (F.J. Aoiz, L. Bañares, J.F. Castillo, B.D. Bean, F. Fernández-Alonso, et al., unpublished results). The agreement between the two is very satisfactory. In particular, at $E_{col} =$

1.64 eV the NGP QM results seem to reproduce the experiments well, whereas the QCT method underestimates the amount of forward scattering by at least a factor of three. The comparison between experiment and theory has been performed by forward convolution of the theoretical results. Such a forward convolution involves a simulation of the time-of-flight profile that would be expected for a given theoretical center-of-mass differential cross section. The validity of this procedure for comparing the results of photoloc experiments with theory has already been assessed for HD($v' = 1, 2$) product-state-resolved DCSs (232). Because photoloc experiments measure directly a velocity distribution, the transformation between product laboratory speed and center-of-mass scattering angle implies a lower angular resolution to forward scattering and an increasing difficulty in separating the effects of height and width for narrow forward scattering peaks. Despite this limitation, both the experimental data and NGP QM calculations agree on the fact that the largest (total) amount of forward scattering relative to backward scattering occurs around 1.64 eV. We furthermore note that the crossed-beam experiments of Schnieder and coworkers have not been able to access the forward scattering region below 50° owing to geometrical constraints in the apparatus (backward-scattered D product is obscured by the presence of the reagent D_2 beam). From this quantitative comparison of experimental data and NGP QM calculations, it is possible to ascribe HD($v' = 3$, low-j') forward scattering (see Figure 4, color insert) to a primarily quantum-mechanical phenomenon also present in lower vibrational levels at lower collision energies. A partial-wave analysis of the QM calculations showed that a narrow range of high-J partial waves mainly caused the forward scattering. Based on the J-shifting approximation (58), it was possible to assign the strongest forward feature at 1.64 eV to the lowest bending level of the complex, a finding consistent with preferential decay into the rotationless product state HD($v' = 3$, $j' = 0$). By invoking the concept of vibrational adiabaticity, we can use the vibrational quantum label of the HD product to assign the symmetric stretch motion of the HD_2 complex to $v_1 = 3$. This full assignment of the quantum numbers of the HD_2 complex agrees with what Truhlar and coworkers (234) found.

Kendrick (109) has carried out further theoretical analyses of forward scattering in the H + D_2 reaction using converged NGP QM calculations. The appearance of forward-scattering features as a function of J led Kendrick to the general conclusion that it is unlikely that this feature can be interpreted as arising from resonances owing to the presence of significant classical and quantum mechanical nonresonant contributions at these scattering angles. These conclusions do not agree with NGP QM (F.J. Aoiz, L. Bañares, J.F. Castillo, B.D. Bean, F. Fernández-Alonso, et al., unpublished results) and QCT (110) results that indicate a relatively narrow contribution of partial waves to forward scattering as well as a marked difference between QCT and QM results. Furthermore, we note that the experimental data meet the two requirements needed to fulfill the operational definition of scattering resonances (see Introduction), namely: (*a*) the amount of HD($v' = 3$, j') forward scattering varies rapidly with collision energy and product rotational excitation; and (*b*) reactive flux leading to forward-scattering features has been associated with metastability of the triatomic complex. It is still unclear from a theoretical viewpoint whether the trapping of the system in the transition-state region is caused by one or more conventional resonances or by the presence of a threshold effect (see Sections 2.2 and 2.3). An explanation along the latter lines would necessarily need to account for the preferential decay into HD($v' = 3$, $j' = 0$) and, to a lesser extent, into HD($v' = 3$, $j' = 1$) observed in the experiments. We once more note that whether threshold effects may be considered scattering resonances in their own right is still a matter of discussion in the chemical physics community. It is beyond the scope of this review to provide an answer to this pending yet important question. Of course, identifying resonance features either in the laboratory or by accurate quantum mechanical calculations is very different from understanding what causes them.

Recently, Althorpe (220) has made a breakthrough in being able to perform full scattering calculations using time-dependent wave packet propagation methods for the H + H$_2$ reaction family. This approach is considerably faster than time-independent methods and will probably find its best use in more complex reaction systems. Figure 7 shows a series of time-elapsed snapshots of the H + D$_2 \rightarrow$ HD($v' = 3$, $j' = 0$) + D reaction over an energy range of 0.9–2.4 eV (S.C. Althorpe, unpublished results). This "movie" shows that the amount of forward-scattered HD($v' = 3$, $j' = 0$) product constitutes a considerable fraction of the overall flux. Moreover, the forward-scattered component lags the backward-scattered component by about 25 fs owing to temporary trapping in the transition-state region. The major features shown in Figure 7 are in qualitative agreement with previous QCT results and indicate a very different and more indirect underlying mechanism leading to forward scattering. A detailed analysis of these very interesting theoretical results, however, needs to be completed before we can fully understand what causes the observed resonance features.

The product of HD($v' = 3$, $j' = 0$) is truly a small fraction of the product yield. Specifically, it is estimated to have a total cross section of $\sim 10^{-3}$ Å^2 whereas the total reaction cross section is on the order of ~ 1 Å^2. Most scattering is direct and can be simply described using billiard-ball models. Nevertheless, it is remarkable how much detailed information can be gained from a study of the tiny fraction of the overall cross section showing signatures ascribable to reactive scattering resonances.

4. BEYOND STATE-RESOLVED DIFFERENTIAL CROSS SECTIONS

4.1. Vector Correlations in Scattering Experiments

What is to be measured beyond state-, energy-, and angle-resolved cross sections? If we think in terms of the vector properties or stereodynamics of a chemical reaction, a DCS constitutes what is called a two-vector (k-k') correlation: it relates the relative direction of reagents (k) to that of products (k'). Other vector correlations are possible, as for example, k-j' (where j' is the rotational angular momentum vector of the product), k-k'-j', etc. For more details about vector properties in photodissociation and chemical reactions, the reader is referred to the original

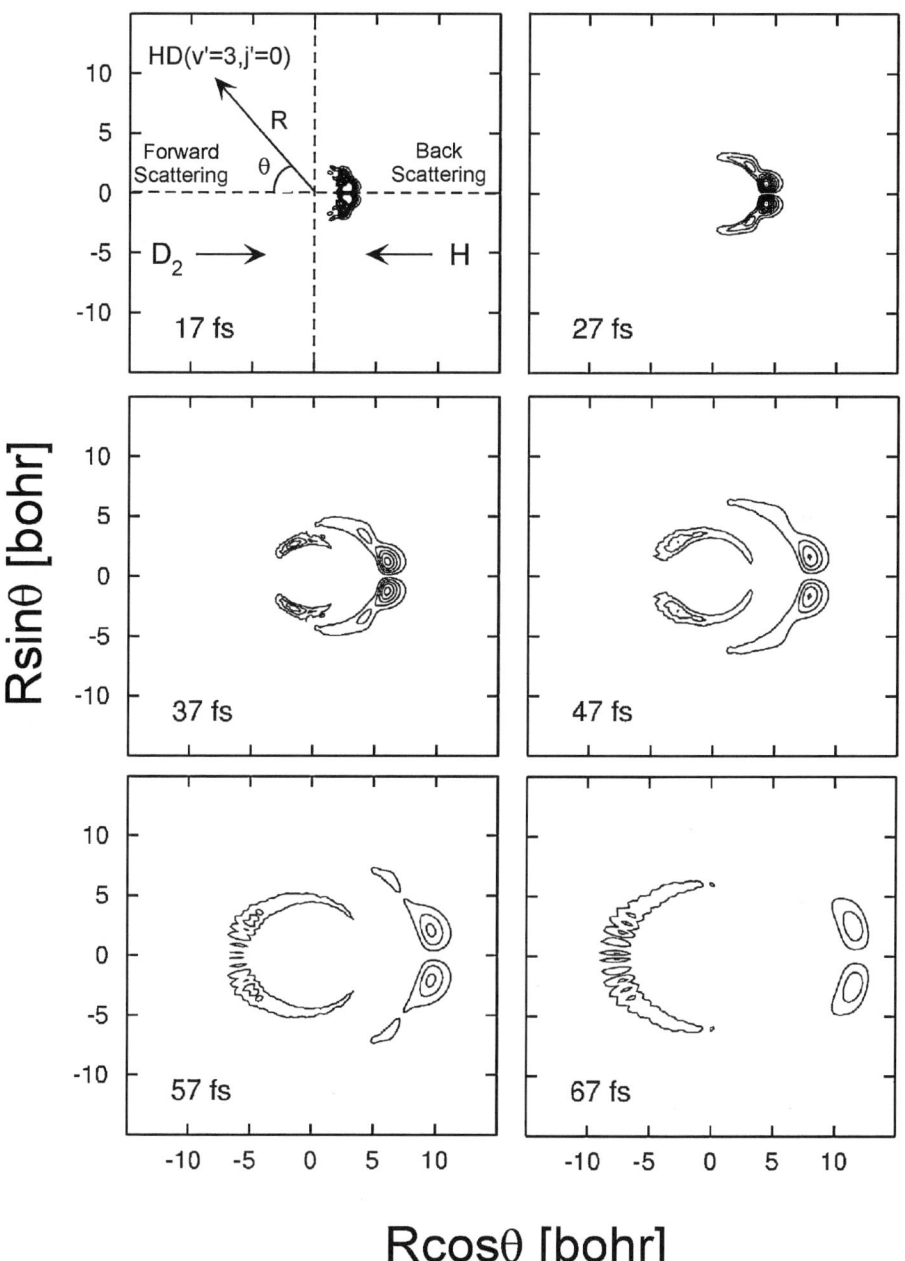

Figure 7 Wave packet snapshots of the reaction H + D$_2$(v = 0, j = 0) → HD(v' = 3; j' = 0) + D over a total energy range of 0.9–2.4 eV as calculated by a time-dependent method (see text for details) and starting at an H-D$_2$ separation of 6 bohr. The first snapshot, at a time of 17 fs, shows the directions of the incoming reagents and outgoing HD(v' = 3, j' = 0) products.

papers (237), several reviews (238–242), devoted journal volumes (243, 244), and more recent work not yet reviewed (114, 245–248). The stereodynamics of a few chemical reactions, including Cl + HD (249), Cl + CH_4 (250, 251), and Cl + C_2H_6 (252–255), have been studied using the photoloc technique by Zare and coworkers. In these experiments, state-selective detection of HCl, CH_3, and C_2H_5 molecules allows for a simultaneous measurement of the angle-resolved spatial distribution of product angular momenta. In addition, SRP techniques have also permitted the study of k-j-k' correlations for vibrationally excited reagents, which can be readily associated with the steric requirements of a chemical reaction. Kandel et al. (249) have used the SRP technique to study the stereochemistry of the Cl + HD(v = 1; j = 1, 2) reaction.

As for the H + H_2 reaction, work along similar lines has been mainly hampered by the lack of suitable multiphoton detection schemes sensitive to the angular momentum anisotropy. Recently, Zare and coworkers (256) have developed a novel multiphoton ionization detection scheme for the H_2 molecule that may prove useful in future investigations of product polarization in the hydrogen exchange reaction. Such progress on the experimental side has been also complemented by recent classical and quantal descriptions of the stererodynamics of elementary chemical reactions by de Miranda et al. (257). The benchmark system used in this work has been the prototypical H + D_2 reaction at a collision energy of 1.29 eV. It was shown that a study of the vector correlations would deepen significantly our knowledge about the PES and reaction mechanisms. The search for resonance features using these novel tools is not an exception to these conclusions as evidenced, for example, by the magnetic-sublevel-resolved QM DCS calculated by Miller & Zhang for the H + H_2 and D + H_2 reactions (83). Moreover, the study of stereodynamical effects offers the experimentalist additional means of investigating dynamical features that other and more highly averaged observables may tend to obscure.

As for the use of experimental techniques to prepare reagents with an anisotropic spatial distribution, SRP techniques are mandatory for homonuclear diatomics like H_2 and isotopologs. There has been some work in this direction for ICS measurements in D + H_2 (81, 222–224) and H + D_2 (258) but not for state-resolved angular distributions. Given the propensity for the H + H_2 reaction to be collinear, we would expect significant changes in the angular distributions as the reagent relative velocity vector is preferentially aligned parallel or perpendicular to the diatom bond axis. If our current picture of quasibound states in the H + D_2 reaction leading to forward scattering features is to stand the test of time, these experiments will undoubtedly be extremely insightful.

4.2. Direct Observation of the H_3 Transition-State Region: Electron Attachment/Detachment Experiments

An ever-present concern in this review was related to the effect of partial-wave summation over the outcome of asymptotic reactive scattering experiments. This necessary evil in collision experiments has hampered the unambiguous detection of quasibound states in reactive scattering. Experimentally, several ways exist to avoid this averaging process. The central idea is based on the preparation of a stable precursor, for example, the positive or negative ion of the compound species of interest. Either light to photodetach a negative ion or charge neutralization in the case of the positive ion may be used to launch almost instantaneously the stable precursor onto the transition-state region of a chemical reaction. Examples of this approach are the famous photodetachment experiments of Neumark and coworkers on the transition state spectroscopy of various triatoms including FH_2 and IHI (259). This unconventional way of probing the transition state region has given strong evidence for the existence of transition-state resonances in the IHI system.

At the moment, all experiments on the dissociation of H_3 states use the H_3^+ ion as precursor. Charge neutralization in alkali-metal vapors leads to two- and three-body breakup of the triatomic complex [see, for example, Krause et al. (260) for an early quantum mechanical treatment and simulation of experiments]. The UV spectra of H_3, D_2H, and H_2D have been studied using both experimental and theoretical methods (261–265). The bimodal structure observed in these spectra in the range of 200–400 nm has been interpreted as arising from radiative decay into the two Jahn-Teller sheets of the ground-state PES (also responsible for GP effects). Mahapatra & Köppel (265) have found in their time-dependent calculations distinct differences between resonance states associated with each adiabatic sheet, the upper one leading to a much more pronounced and more easily assigned structure.

Cold ion storage rings have also been used in the last decade to achieve relative collision energies for recombination as low as 0.001 eV (266–268). These experiments have measured recoil energy as well as angular distributions for two- and three-body breakup, but they still lack product-state selectivity. Very recently, Kokoouline et al. (269) have made significant progress in the theoretical understanding of the recombination of H_3^+ by electron impact at low energies (<1 eV). In their theoretical treatment, it is mandatory to incorporate the Jahn-Teller-symmetry-distortion effect in order to explain the large dissociation recombination rates, the (H + H + H)/(H_2 + H) branching fraction, and the vibrational distribution of H_2 products already measured using cold ion storage rings. In addition, H_3 excited states can be sufficiently long-lived to allow initial-state selection by laser pumping. Helm, Müller, and coworkers have embarked on a detailed characterization of the fragmentation dynamics of state-selected Rydberg H_3 states into H + H + H (270–272) and H + H_2(v', j') (273, 274), and they are at a point to achieve product-state resolution in very impressive and technically involved two- and three-particle coincidence experiments.

Photodetachment experiments akin to the ones pioneered by Neumark and coworkers are also possible for the H_3 triatomic system. As early as 1975, Aberth, Schnitzer, & Anbar (275) performed the mass-spectrometric detection of H_3^-, H_2D^-, HD_2^-, and D_3^- anions using a hollow-cathode duoplasmatron negative-ion source. The half-lives for these triatomic species were estimated to be greater than 10 μs. Sadeghi & Skodje (140) have noted that Franck-Condon overlaps between the negative and neutral species will be significant for vibrationally excited H_3^- and will make possible experiments that can be compared against their time-dependent wave packet calculations (140–143).

CONCLUDING REMARKS

Resonances are exquisitely sensitive probes of the nature of the scattering process. It is for the simplest of chemical reactions, the $H + H_2$ reaction family, that we expect theory to be most advanced. These reactions are not simple from an experimental viewpoint. Nevertheless, experiments have reached a level of sophistication that makes possible the observation of resonance features in spite of the overwhelming dominance of direct dynamics. Both theory and experiment are needed. It is not enough to measure only scattering features. It is also not enough to calculate them. What is needed is to develop simple pictures and accurate approximations that can be transferred to other reactions that display a richer chemistry.

ACKNOWLEDGMENTS

This work was supported at Stanford University by the U.S. National Science Foundation under grant number CHE-99-00305. F.F.A. acknowledges financial support in the form of a Marie Curie Fellowship of the European Program "Improving Human Research Potential and the Socioeconomic Knowledge Base" under contract number HPMFCT-2000-00683. We thank our coworkers B.D. Bean, J.D. Ayers, and A.E. Pomerantz for their contribution to the experimental work carried out at Stanford University as well as for useful discussions and suggestions. We also thank F.J. Aoiz, L. Bañares, J.F. Castillo, and S.C. Althorpe for sharing the results of their calculations prior to publication.

Visit the Annual Reviews home page at www.annualreviews.org

LITERATURE CITED

1. Condon EU, Shortley GH. 1977. *The Theory of Atomic Spectra*. New York: Cambridge Univ. Press
2. Graybeal JD. 1988. *Molecular Spectroscopy*. New York: McGraw-Hill
3. Fermi E. 1934. *Nature* 133:757
4. Fermi E. 1934. *Nature* 133:898–99
5. Fermi E, Amaldi E, D'Agostino O, Rasetti F, Segré E. 1934. *Proc. R. Soc. London Ser. A* 133:483–500
6. Amaldi E, D'Agostino O, Fermi E, Pontecorvo B, Rasetti F, Segré E. 1935. *Proc. R. Soc. London Ser. A* 149:522–58
7. Bethe HA. 1935. *Phys. Rev.* 47:747–59
8. Bohr N. 1936. *Nature* 137:344–48
9. Breit G, Wigner EP. 1936. *Phys. Rev.* 49:519–31
10. Fano U. 1961. *Phys. Rev.* 124:1866–78
11. Frauenfelder H, Henley EM. 1974. *Subatomic Physics*. Englewood Cliffs, NJ: Prentice-Hall
12. Blatt JM, Weisskopf VF. 1979. *Theoretical Nuclear Physics*. New York: Springer
13. Feshbach H. 1992. *Theoretical Nuclear Physics: Nuclear Reactions*. New York: Wiley
14. Burke PG. 1968. *Adv. At. Mol. Phys.* 4:173–19

15. Golden DE. 1978. *Adv. At. Mol. Phys.* 14:1–85
16. Auger P. 1925. *J. Phys. Radium* 6:205–8
17. Haas R. 1957. *Z. Phys.* 148:177–91
18. Schulz GJ. 1959. *Phys. Rev.* 116:1141–47
19. Herzenberg A, Mandl F. 1962. *Proc. R. Soc. London Ser. A* 270:48–71
20. Schulz GJ, Fox RE. 1957. *Phys. Rev.* 106:1179–81
21. Baranger E, Gerjouy E. 1957. *Phys. Rev.* 106:1182–85
22. Phelps AV. 1968. *Rev. Mod. Phys.* 40:399–410
23. Schulz GJ. 1973. *Rev. Mod. Phys.* 45:378–422
24. Schulz GJ. 1973. *Rev. Mod. Phys.* 45:423–86
25. Biondi MA, Herzenberg A, Kuyatt CE. 1979. *Phys. Today* 32:44–49
26. Morrison MA. 1987. *Adv. At. Mol. Phys.* 24:51–56
27. Buckman SJ, Clark CW. 1994. *Rev. Mod. Phys.* 66:539–655
28. Patel CKN. 1964. *Phys. Rev. Lett.* 13:617–19
29. Cheo PK. 1971. In *Lasers, A Series of Advances*, ed. AK Levine, AJ DeMaria, pp. 111–267. New York: Marcel Dekker
30. Boudaïffa B, Cloutier P, Hunting D, Huels MA, Sanche L. 2000. *Science* 287:1658–60
30a. Eu BC, Ross J. 1966. *J. Chem. Phys.* 44:2467–75
30b. Levine RD. 1967. *J. Chem. Phys.* 46:331–45
30c. Micha DA. 1967. *Chem. Phys. Lett.* 1:139–42
30d. Micha DA. 1967. *Phys. Rev.* 162:88–97
30e. O'Malley TF. 1967. *Phys. Rev.* 162:98–104
30f. Levine RD. 1968. *J. Chem. Phys.* 49:51–55
30g. Levine RD, Johnson BR, Muckerman JT, Bernstein RB. 1968. *Chem. Phys. Lett.* 1:517–20
30h. Levine RD, Johnson BR, Muckerman JT, Bernstein RB. 1968. *J. Chem. Phys.* 49:56–64
30i. Johnson BR, Shapiro M, Levine RD. 1969. *Chem. Phys. Lett.* 3:131–33
30j. Levine RD. 1970. *Acc. Chem. Res.* 3:273–80
30k. Secrest D. 1973. *Annu. Rev. Phys. Chem.* 24:379–406
30l. Miller WH. 1970. *J. Chem. Phys.* 52:543–51
30m. Miller WB, Safron SA, Herschbach DR. 1967. *Faraday Discuss. Chem. Soc.* 44:108–22
31. Truhlar DG, Kuppermann A. 1970. *J. Chem. Phys.* 52:3841–43
32. Truhlar DG, Kuppermann A. 1972. *J. Chem. Phys.* 56:2232–52
33. Levine RD, Wu S-F. 1971. *Chem. Phys. Lett.* 11:557–61
34. Hirschfelder JO, Eyring H, Topley B. 1936. *J. Chem. Phys.* 4:170–77
35. London F. 1929. *Z. Elektrochem.* 35:552–55
36. Ering H, Polanyi M. 1931. *Z. Phys. Chem. Abt. B* 12:279–311
37. Porter RN, Karplus M. 1964. *J. Chem. Phys.* 40:1105–15
38. Shavitt I, Stevens RM, Minn FL, Karplus M. 1968. *J. Chem. Phys.* 48:2700–13
39. Truhlar DG, Wyatt RE. 1977. *Adv. Chem. Phys.* 36:141–204
40. Wu S-F, Levine RD. 1971. *Mol. Phys.* 22:881–97
41. Wu S-F, Johnson BR, Levine RD. 1973. *Mol. Phys.* 25:609–30
42. Wu S-F, Johnson BR, Levine RD. 1973. *Mol. Phys.* 25:839–56
43. Schatz GC, Kuppermann A. 1973. *J. Chem. Phys.* 59:964–65
44. Feshbach H. 1958. *Ann. Phys.* 5:357–90
45. Feshbach H. 1962. *Ann. Phys.* 19:287–313
45a. Feshbach H. 1967. *Ann. Phys.* 43:410–20
46. Schatz GC, Kuppermann A. 1975. *Phys. Rev. Lett.* 35:1266–69
47. Truhlar DG, Wyatt RE. 1976. *Annu. Rev. Phys. Chem.* 47:1–43
48. Kuppermann A. 1981. In *Potential Energy Surfaces and Dynamics Calculations*, ed.

DG Truhlar, pp. 375–420. New York: Plenum
49. Garrett BC, Truhlar DG. 1982. *J. Phys. Chem.* 86:1136–41
50. Garrett BC, Schwenke DW, Skodje RT, Thirumalai D, Thompson TC, Truhlar DG. 1984. In *Resonances in Electron-Molecule Scattering, Van der Waals Complexes, and Reactive Chemical Dynamics. ACS Symp. Ser.*, ed. DG Truhlar, pp. 375–400. Washington, DC: ACS
51. Kuppermann A, Kaye JA, Dwyer JP. 1980. *Chem. Phys. Lett.* 74:257–62
52. Launay JM, LeDourneuf M. 1982. *J. Phys. B: At. Mol. Phys.* 15:L455–61
53. Römelt J. 1983. *Chem. Phys.* 79:197–209
54. Marston CC, Wyatt RE. 1984. *Chem. Phys.* 81:1819–24
55. Pollak E, Wyatt RE. 1984. *J. Chem. Phys.* 81:1801–12
56. Bowman JM. 1986. *Chem. Phys. Lett.* 124:260–63
57. Bowman JM. 1985. *Adv. Chem. Phys.* 61:115–63
58. Bowman JM. 1991. *J. Phys. Chem.* 95:4960–68
59. Child MS, Pollak E. 1980. *J. Chem. Phys.* 73:4365–72
60. Pollak E, Child MS. 1980. *J. Chem. Phys.* 73:4373–80
61. Pollak E, Child MS, Pechukas P. 1980. *J. Chem. Phys.* 72:1669–78
62. Pollak E. 1981. *J. Chem. Phys.* 74:5586–94
63. Pollak E, Wyatt RE. 1983. *J. Chem. Phys.* 78:4464–76
64. Pollak E, Child MS. 1981. *Chem. Phys.* 60:23–32
65. Manz J, Pollak E, Römelt J. 1982. *Chem. Phys. Lett.* 86:26–32
66. Pollak E, Wyatt RE. 1982. *J. Chem. Phys.* 77:2689–91
67. Pollak E. 1987. *Chem. Phys. Lett.* 137:171–74
68. Hipes PG, Kuppermann A. 1987. *Chem. Phys. Lett.* 133:1–7
69. Nieh J-C, Valentini JJ. 1988. *Phys. Rev. Lett.* 60:519–22
70. Nieh J-C, Valentini JJ. 1990. *J. Chem. Phys.* 92:1083–97
71. Phillips DL, Levene HB, Valentini JJ. 1989. *J. Chem. Phys.* 90:1600–9
72. Buchenau H, Toennies JP, Arnold J, Wolfrum J. 1990. *Ber. Bunsenges. Phys. Chem.* 94:1231–48
73. Colton MC, Schatz GC. 1986. *Chem. Phys. Lett.* 124:256–59
74. Schatz GC. 1988. *Annu. Rev. Phys. Chem.* 39:317–40
75. Webster F, Light JC. 1986. *J. Chem. Phys.* 85:4744–55
76. Zhang JZH, Miller WH. 1988. *Chem. Phys. Lett.* 153:465–70
77. Zhang JZH, Miller WH. 1989. *Chem. Phys. Lett.* 159:130–33
78. Zhang JZH, Miller WH. 1989. *J. Chem. Phys.* 91:1528–47
79. Manolopoulos DE, Wyatt RE. 1989. *Chem. Phys. Lett.* 159:123–29
80. Launay JM, LeDourneuf M. 1989. *Chem. Phys. Lett.* 163:178–88
81. Kliner DAV, Adelman DE, Zare RN. 1991. *J. Chem. Phys.* 94:1069–80
82. Miller WH. 1990. *Annu. Rev. Phys. Chem.* 41:245–81
83. Miller WH, Zhang JZH. 1991. *J. Phys. Chem.* 95:12–19
84. Aoiz FJ, Herrero VJ, Rábanos VS. 1991. *J. Chem. Phys.* 95:7767–68
85. Aoiz FJ, Herrero VJ, Rábanos VS. 1994. *J. Chem. Phys.* 97:7423–36
86. Aoiz FJ, Bañares L, Herrero VJ. 1998. *Adv. Class. Trajectory Methods* 3:121–82
87. Siegbahn P, Liu B. 1978. *J. Chem. Phys.* 68:2457–65
88. Truhlar DG, Horowitz CJ. 1978. *J. Chem. Phys.* 68:2466–76
89. Truhlar DG, Horowitz CJ. 1979. *J. Chem. Phys.* 71:1514
90. Varandas AJC, Brown FB, Mead CA, Truhlar DG, Blais NC. 1987. *J. Chem. Phys.* 86:6258–69
91. Boothroyd AI, Keogh WJ, Martin PG, Peterson MR. 1991. *J. Chem. Phys.* 95:4343–59

92. Boothroyd AI, Keogh WJ, Martin PG, Peterson MR. 1996. *J. Chem. Phys.* 104: 7139–52
93. Wu Y-SM, Kuppermann A, Anderson JB. 1999. *Phys. Chem. Chem. Phys.* 1:929–37
94. Skodje RT, Skouteris D, Manolopoulos DE, Lee S-H, Dong F, Liu K. 2000. *J. Chem. Phys.* 112:4536–52
95. Skodje RT, Skouteris D, Manolopoulos DE, Lee S-H, Dong F, Liu K. 2000. *Phys. Rev. Lett.* 85:1206–9
96. Schatz GC. 2000. *Science* 288:1599–600
97. Liu K. 2001. *Annu. Rev. Phys. Chem.* 52:139–64
98. Dong F, Lee S-H, Liu K. 2000. *J. Chem. Phys.* 113:3633–40
99. Chao SD, Skodje RT. 2000. *J. Chem. Phys.* 113:3487–91
100. Sokolovski D, Castillo JF. 2000. *Phys. Chem. Chem. Phys.* 2:507–12
101. Castillo JF, Manolopoulos DE, Stark K, Werner H-J. 1996. *J. Chem. Phys.* 104:6531–46
102. Neumark DM, Wodtke AM, Robinson GN, Hayden CC, Lee YT. 1985. *J. Chem. Phys.* 82:3045–66
103. Neumark DM, Wodtke AM, Robinson GN, Hayden CC, Shobatake K, et al. 1985. *J. Chem. Phys.* 82:3045–66
104. Wrede E, Schnieder L. 1997. *J. Chem. Phys.* 107:786–90
105. Kendrick BK, Jayasinghe L, Moser S, Auzinsh M, Shafer-Ray N. 2000. *Phys. Rev. Lett.* 84:432–58
106. Kendrick BK, Jayasinghe L, Moser S, Auzinsh M, Shafer-Ray N. 2001. *Phys. Rev. Lett.* 86:2482
107. Chao SD, Skodje RT. 2001. *Chem. Phys. Lett.* 336:364–70
108. Aoiz FJ, Bañares L, Castillo JF. 2001. *J. Chem. Phys.* 114:823–79
109. Kendrick BK. 2001. *J. Chem. Phys.* 114: 8796–819
110. Fernández-Alonso F, Bean BD, Ayers JD, Pomerantz AE, Zare RN, et al. 2000. *Angew. Chem. Int. Ed.* 39:2748–52
111. Bean BD. 2000. *The hydrogen atom, hydrogen molecule exchange reaction. Experimental evidence for dynamical resonances in chemical reactions.* PhD thesis. Stanford Univ. 106 pp.
112. Deleted in proof
113. Shafer NE, Orr-Ewing AJ, Simpson WR, Xu H, Zare RN. 1993. *Chem. Phys. Lett.* 212:155–62
114. Shafer-Ray NE, Orr-Ewing AJ, Zare RN. 1995. *J. Phys. Chem.* 99:7591–603
115. Marcus RA. 1966. *J. Chem. Phys.* 45: 4493–99
116. Marcus RA. 1966. *J. Chem. Phys.* 45: 4500–4
117. Levine RD. 1969. *Quantum Mechanics of Molecular Rate Processes.* Oxford: Oxford Univ. Press
118. Light JC. 1970. *Adv. Chem. Phys.* 19:1–31
119. Child MS. 1974. *Molecular Collision Theory.* London/New York: Academic
120. Levine RD, Bernstein RB. 1987. *Molecular Reaction Dynamics and Chemical Reactivity.* Oxford: Oxford Univ. Press
121. Chapman FM, Hayes EF. 1975. *J. Chem. Phys.* 62:4400–3
122. Chapman FM, Hayes EF. 1976. *J. Chem. Phys.* 65:1032–33
123. Aquilanti V, Capecchi G, Cavalli S, Fazio DD, Palmieri P, et al. 2000. *Chem. Phys. Lett.* 318:619–28
124. Newton RG. 1982. *Scattering Theory of Waves and Particles.* Heidelberg: Springer-Verlag
125. Taylor JR. 1972. *Scattering Theory.* New York: Wiley
126. Smith FT. 1960. *Phys. Rev.* 118:349–56
127. Simons JP. 1997. *J. Chem. Soc. Faraday Trans.* 93:4095–105
128. Green WH, Moore CB, Polik WF. 1992. *Annu. Rev. Phys. Chem.* 43:591–626
129. Reid SA, Reisler H. 1996. *Annu. Rev. Phys. Chem.* 47:495–525
130. Friedman RS, Truhlar DG. 1991. *Chem. Phys. Lett.* 183:539–46
131. Eckart C. 1930. *Phys. Rev.* 35:1303–9
132. Friedman RS, Hullinger VD, Truhlar DG. 1995. *J. Phys. Chem.* 99:3184–94
133. Seideman T, Miller WH. 1991. *J. Chem. Phys.* 95:1768–80

134. Ryabov V, Moiseyev N. 1993. *J. Chem. Phys.* 98:9618–23
135. Siegert AJF. 1939. *Phys. Rev.* 56:750–52
136. Bohm A. 1986. *Quantum Mechanics: Foundations and Applications.* New York: Springer-Verlag
137. Junker BR. 1982. *Adv. At. Mol. Phys.* 18:207–63
138. Atabek O, Lefebvre R, Sucre MG, Gomez-Llorente J, Taylor H. 1991. *Int. J. Quant. Chem.* 40:211–24
139. Landau LD, Lifshitz EM. 1977. *Quantum Mechanics (Non-relativistic Theory).* New York: Pergamon
140. Sadeghi R, Skodje RT. 1993. *J. Chem. Phys.* 98:9208–10
141. Sadeghi R, Skodje RT. 1993. *J. Chem. Phys.* 99:5126–40
142. Skodje RT, Sadeghi R, Köppel H, Krause JL. 1994. *J. Chem. Phys.* 101:1725–29
143. Sadeghi R, Skodje RT. 1995. *J. Chem. Phys.* 102:193–213
144. Sadeghi R, Skodje RT. 1995. *Phys. Rev. A* 52:1996–2010
145. Varandas AJC, Yu HG. 1996. *Chem. Phys. Lett.* 259:336–41
146. Varandas AJC, Yu HG. 1996. *Chem. Phys.* 209:31–40
147. Yu HG, Varandas AJC. 1996. *J. Phys. Chem.* 100:14598–601
148. Varandas AJC, Yu HG. 1999. *J. Mol. Struct. Theochem.* 493:81–88
149. Manz J, Meyer R, Pollak E, Römelt J. 1982. *Chem. Phys. Lett.* 93:184–87
150. Manz J, Meyer R, Römelt J. 1983. *Chem. Phys. Lett.* 96:607–12
151. Clary DC, Connor JNL. 1983. *Chem. Phys. Lett.* 94:81–84
152. Pollak E. 1983. *Chem. Phys. Lett.* 94:85–89
153. Pollak E. 1983. *J. Chem. Phys.* 78:1228–36
154. Atabek O, Lefebvre R. 1983. *Chem. Phys. Lett.* 98:559–62
155. Clary DC, Connor JNL. 1984. *J. Phys. Chem.* 88:2758–64
156. Manz J, Meyer R, Pollak E, Römelt J, Schor HHR. 1984. *Chem. Phys.* 83:333–43
157. Garrett BC, Lynch GC, Allison TC, Truhlar DG. 1998. *Comp. Phys. Commun.* 109:47–54
158. Truhlar DG, Garrett BC. 1992. *J. Phys. Chem.* 96:6515–18
159. Chatfield DC, Friedman RS, Mielke SL, Lynch GC, Allison TC, et al. 1996. In *Dynamics of Molecules and Chemical Reactions*, ed. RE Wyatt, JZH Zhang, pp. 323–86. New York: Marcel Dekker
160. Truhlar DG, Garrett BC, Klippenstein SJ. 1996. *J. Phys. Chem.* 100:12771–800
161. Chatfield DC, Friedman RS, Truhlar DG, Garrett BC, Schwenke DW. 1991. *J. Am. Chem. Soc.* 113:486–94
162. Chatfield DC, Friedman RS, Truhlar DG. 1991. *Faraday Discuss. Chem. Soc.* 91:289–304
163. Truhlar DG. 1991. *Faraday Discuss. Chem. Soc.* 91:395–98
164. Chatfield DC, Truhlar DG, Schwenke DW. 1992. *J. Chem. Phys.* 96:4313–23
165. Chatfield DC, Friedman RS, Schwenke DW, Truhlar DG. 1992. *J. Phys. Chem.* 96:2414–21
166. Chatfield DC, Mielke SL, Allison TC, Truhlar DG. 2000. *J. Chem. Phys.* 112:8387–408
167. Lynch GC, Halvick P, Zhao M, Truhlar DG, Yu C-h, et al. 1991. *J. Chem. Phys.* 94:7150–58
168. Kress JD, Hayes EF. 1992. *J. Chem. Phys.* 97:4881–89
169. Bowman JM. 1987. *Chem. Phys. Lett.* 141:545–47
170. Haug K, Schwenke DW, Truhlar DG, Zhang Y, Zhang JZH, Kouri DJ. 1987. *J. Chem. Phys.* 87:1892–94
171. Chatfield DC, Friedman RS, Lynch GC, Truhlar DG. 1993. *J. Chem. Phys.* 98:342–62
172. Pack RT, Butcher EA, Parker GT. 1993. *J. Chem. Phys.* 99:9310–13
173. Leforestier C, Miller WH. 1994. *J. Chem. Phys.* 100:733–35

174. Zhang DH, Zhang JZH. 1994. *J. Chem. Phys.* 101:3671–78
175. Chatfield DC, Friedman RS, Lynch GC, Truhlar DG. 1992. *J. Phys. Chem.* 96:57–63
176. Srinivasan J, Allison TC, Schwenke DW, Truhlar DG. 1999. *J. Phys. Chem. A* 103:1487–503
177. Darakjian Z, Hayes EF, Parker GA, Butcher EA, Kress JD. 1991. *J. Chem. Phys.* 95:2516–22
178. Klippenstein SJ, Kress JD. 1992. *J. Chem. Phys.* 96:8164–70
179. Kress JD. 1991. *J. Chem. Phys.* 95:8673–74
180. Kress JD, Walker RB, Hayes EF, Pendergast P. 1994. *J. Chem. Phys.* 100:2728–42
181. Cuccaro SA, Hipes PG, Kuppermann A. 1989. *Chem. Phys. Lett.* 157:440–46
182. Zhao M, Mladenovic M, Truhlar DG, Schwenke DW, Sharafeddin O, et al. 1989. *J. Chem. Phys.* 91:5302–9
183. Zhao M, Rice SA. 1994. *J. Phys. Chem.* 98:3444–49
184. Schnieder L, Meier W, Welge KH, Ashfold MNR, Western CM. 1990. *J. Chem. Phys.* 92:7027–37
185. Merkt F, Xu H, Zare RN. 1996. *J. Chem. Phys.* 104:950–61
186. Rinnen K-D, Kliner DAV, Zare RN, Huo WM. 1989. *Isr. J. Chem.* 29:369–82
187. Huo WM, Rinnen K-D, Zare RN. 1991. *J. Chem. Phys.* 95:205–13
188. Rinnen K-D, Buntine MA, Kliner DAV, Zare RN, Huo WM. 1991. *J. Chem. Phys.* 95:214–25
189. D'Mello M, Manolopoulos DE, Wyatt RE. 1991. *J. Chem. Phys.* 94:5985–93
190. Rinnen K, Kliner D, Zare RN. 1989. *J. Chem. Phys.* 91:7514–29
191. Zhao M, Truhlar DG, Blais NC, Schwenke DW, Kouri DJ. 1990. *J. Chem. Phys.* 94:6696–706
192. Webster F, Light JC. 1989. *J. Chem. Phys.* 90:300–21
193. Yarkony DR. 1996. *Rev. Mod. Phys.* 68:985–1013
194. Mead CA. 1992. *Rev. Mod. Phys.* 64:51–85
195. Lepetit B, Kuppermann A. 1990. *Chem. Phys. Lett.* 166:581–88
196. Wu Y-SM, Kuppermann A, Lepetit B. 1991. *Chem. Phys. Lett.* 186:319–28
197. Wu Y-SM, Kuppermann A. 1993. *Chem. Phys. Lett.* 201:178–86
198. Kuppermann A, Wu Y-SM. 1993. *Chem. Phys. Lett.* 205:577–36
199. Kuppermann A, Wu Y-SM. 1995. *Chem. Phys. Lett.* 241:229–40
200. Wu Y-SM, Kuppermann A. 1995. *Chem. Phys. Lett.* 235:105–10
201. Baer M. 2000. *Chem. Phys.* 259:123–47
202. Yarkony DR. 2001. *J. Phys. Chem. A* 105:6277–93
203. Mordaunt DH, Ashfold MN, Dixon RN. 1996. *J. Chem. Phys.* 104:6460–71
204. Mordaunt DH, Dixon RN, Ashfold MN. 1996. *J. Chem. Phys.* 104:6472–81
205. Dixon RN. 1996. *Mol. Phys.* 88:949–77
206. Mordaunt DH, Ashfold MN, Dixon RN. 1998. *J. Chem. Phys.* 109:7659–62
207. Casavecchia P, Balucani N, Volpi GG. 1999. *Annu. Rev. Phys. Chem.* 50:347–76
208. Casavecchia P. 2000. *Rep. Prog. Phys.* 63:355–414
209. Schnieder L, Seekamp-Rahn K, Wrede E, Welge KH. 1997. *J. Chem. Phys.* 107:6175–95
210. Bañares L, Aoiz FJ, Herrero VJ, D'Mello MJ, Niederjohann B, et al. 1998. *J. Chem. Phys.* 108:6160–69
211. Schnieder L, Seekamp-Rahn K, Borkowski J, Wrede E, Welge KH, et al. 1995. *Science* 269:207–10
212. Wrede E, Schnieder L, Welge KH, Aoiz FJ, Bañares L, Herrero VJ. 1997. *Chem. Phys. Lett.* 265:129–36
213. Wrede E, Schnieder L, Welge KH, Aoiz FJ, Bañares L, et al. 1999. *J. Chem. Phys.* 110:9971–81
214. Wrede E, Schnieder L, Welge KH, Aoiz FJ, Bañares L, et al. 1997. *J. Chem. Phys.* 106:7862–64
215. Kendrick BK. 2000. *J. Chem. Phys.* 112:5679–704

216. Kendrick BK. 2001. *J. Chem. Phys.* 114:4335–42
217. Xu ZR, Varandas AJC. 2001. *J. Phys. Chem. A* 105:2246–50
218. Mahapatra S, Köppel H, Cederbaum LS. 2001. *J. Phys. Chem. A* 105:2321–29
219. Kliner DAV, Adelman DE, Zare RN. 1991. *J. Chem. Phys.* 95:1648–62
220. Althorpe SC. 2001. *J. Chem. Phys.* 114:1601–16
221. Farrow RL, Chandler DW. 1988. *J. Chem. Phys.* 89:1994–98
222. Adelman DE, Shafer NE, Kliner DAV, Zare RN. 1992. *J. Chem. Phys.* 97:7323–41
223. Kliner DAV, Zare RN. 1990. *J. Chem. Phys.* 92:2107–9
224. Neuhauser D, Judson RS, Kouri DJ, Adelman DE, Shafer NE, et al. 1992. *Science* 257:519–22
225. Gostein M, Parhikhteh H, Sitz GO. 1995. *Phys. Rev. Lett.* 75:342–45
226. Gostein M, Watts E, Sitz GO. 1997. *Phys. Rev. Lett.* 79:2891–94
227. Shafer NE, Xu H, Tuckett RP, Springer M, Zare RN. 1993. *J. Phys. Chem.* 98:3369–78
228. Fernández-Alonso F, Bean BD, Zare RN. 1999. *J. Chem. Phys.* 111:1022–34
229. Simpson WR, Orr-Ewing AJ, Rakitzis TP, Kandel SA, Zare RN. 1995. *J. Chem. Phys.* 103:7299–312
230. Fernández-Alonso F, Bean BD, Zare RN. 1999. *J. Chem. Phys.* 111:2490–98
231. Fernández-Alonso F, Bean BD, Zare RN. 1999. *J. Chem. Phys.* 111:1035–42
232. Fernández-Alonso F, Bean BD, Zare RN, Aoiz FJ, Bañares L, Castillo JF. 2001. *J. Chem. Phys.* 114:4534–45
233. Muga JG, Levine RD. 1989. *Chem. Phys. Lett.* 162:7–13
234. Allison TC, Friedman RS, Kaufman DJ, Truhlar DG. 2000. *Chem. Phys. Lett.* 327:439–45
235. Deleted in proof
236. Deleted in proof
237. Case DA, McClelland GM, Herschbach DR. 1978. *Mol. Phys.* 35:541–73
238. Hall GE, Houston PL. 1989. *Annu. Rev. Phys. Chem.* 40:375–405
239. Orr-Ewing AJ, Zare RN. 1994. *Annu. Rev. Phys. Chem.* 45:315–66
240. Orr-Ewing AJ, Zare RN. 1995. In *Advanced Series in Physical Chemistry*, ed. K Liu, A Wagner, pp. 936–1063. Singapore: World Sci.
241. Brouard M, Simons JP. 1995. In *Advanced Series in Physical Chemistry*, ed. K Liu, A Wagner, pp. 795–841. Singapore: World Sci.
242. Loesch HJ. 1995. *Annu. Rev. Phys. Chem.* 46:555–94
243. 1991. *J. Phys. Chem.* 95:7961–8422
244. 1997. *J. Phys. Chem. A* 101:7461–90
245. Aoiz FJ, Brouard M, Enriquez PA. 1996. *J. Chem. Phys.* 105:4964–82
246. Rakitzis TP, Kandel SA, Zare RN. 1997. *J. Chem. Phys.* 107:9382–91
247. de Miranda MP, Clary DC. 1997. *J. Chem. Phys.* 106:4509–21
248. de Miranda MP, Clary DC, Castillo JF, Manolopoulos DE. 1998. *J. Chem. Phys.* 108:3142–53
249. Kandel SA, Alexander AJ, Kim ZH, Zare RN, Aoiz FJ, et al. 2000. *J. Chem. Phys.* 112:670–85
250. Simpson WR, Rakitzis TP, Kandel SA, Lev-On T, Zare RN. 1996. *J. Phys. Chem.* 100:7938–47
251. Orr-Ewing AJ, Simpson WR, Rakitzis TP, Kandel SA, Zare RN. 1997. *J. Chem. Phys.* 106:5961–71
252. Kandel SA, Rakitzis TP, Lev-On T, Zare RN. 1996. *J. Chem. Phys.* 105:7550–59
253. Kandel SA, Rakitzis TP, Lev-On T, Zare RN. 1996. *Chem. Phys. Lett.* 265:121–28
254. Kandel SA, Rakitzis TP, Lev-On T, Zare RN. 1998. *J. Phys. Chem. A* 102:2270–73
255. Rakitzis TP, Kandel SA, Lev-On T, Zare RN. 1997. *J. Chem. Phys.* 107:9392–405
256. Fernández-Alonso F, Bean BD, Ayers JD, Pomerantz AE, Zare RN. 2000. *Z. Phys. Chem.* 214:1167–86
257. de Miranda MP, Aoiz FJ, Bañares L, Sáez Rábanos V. 1999. *J. Chem. Phys.* 111:5368–83

258. Lanziserra DV, Valentini JJ. 1995. *J. Chem. Phys.* 103:607–17
259. Metz RB, Bradforth SE, Neumark DM. 1992. *Adv. Chem. Phys.* 81:1–61
260. Krause JL, Kulander KC, Light JC, Orel AE. 1992. *J. Chem. Phys.* 96:4283–92
261. Bruckmeier R, Wunderlich C, Figger H. 1994. *Phys. Rev. Lett.* 72:2550–53
262. Azinovic D, Figger H. 1997. *Z. Phys. D* 42:105–12
263. Azinovic D, Bruckmeier R, Wunderlich C, Figger H, Theodorakopoulos G, Petsalakis ID. 1998. *Phys. Rev. A* 58:1115–28
264. Mahapatra S, Köppel H. 1998. *Phys. Rev. Lett.* 81:3116–19
265. Mahapatra S, Köppel H. 1998. *J. Chem. Phys.* 109:1721–33
266. Datz S. 2001. *J. Phys. Chem. A* 105:2369–73
267. Jensen MJ, Pedersen HB, Safvan CP, Seiersen K, Urbain X, Andersen LH. 2001. *Phys. Rev. A* 63:(5)2701–5
268. Strasser D, Lammich L, Krohn S, Lange M, Kreckel H, et al. 2001. *Phys. Rev. Lett.* 86:779–82
269. Kokoouline V, Greene CH, Esry BD. 2001. *Nature* 412:891–94
270. Müller U, Cosby PC. 1999. *Phys. Rev. A* 59:3632–42
271. Müller U, Eckert T, Braun M, Helm H. 1999. *Phys. Rev. Lett.* 83:2718–21
272. Mistrík I, Reichle R, Helm H, Müller U. 2000. *Phys. Rev. A* 63:042711–10
273. Cosby PC, Helm H. 1988. *Phys. Rev. Lett.* 61:298–301
274. Müller U, Cosby PC. 1996. *J. Chem. Phys.* 105:3532–50
275. Aberth W, Schnitzer R, Anbar M. 1975. *Phys. Rev. Lett.* 34:1600–3

VACUUM ULTRAVIOLET SPECTROSCOPY AND CHEMISTRY BY PHOTOIONIZATION AND PHOTOELECTRON METHODS

Cheuk-Yiu Ng

Department of Chemistry, University of California, Davis, California 95616; e-mail: cyng@chem.ucdavis.edu

Key Words vacuum ultraviolet laser, synchrotron radiation, pulsed field ionization, photoelectron-photoion coincidence

■ **Abstract** The recent developments of vacuum ultraviolet (VUV) laser and third generation synchrotron radiation sources, together with the introduction of pulsed field ionization (PFI) schemes for photoion-photoelectron detection, have had a profound impact on the field of VUV spectroscopy and chemistry. Owing to the mediation of near-resonant autoionizing states, rovibronic states of ions with negligible Franck-Condon factors for direct photoionization can be examined by VUV-PFI measurements with rotational resolutions. The VUV-PFI spectra thus obtained have provided definitive ionization energies (IEs) for many small molecules. The recent synchrotron-based PFI-photoelectron-photoion coincidence experiments have demonstrated that dissociative photoionization thresholds for a range of molecules can be determined to the same precision as in PFI-photoelectron measurements. Combining appropriate dissociation thresholds and IEs measured in PFI studies, thermochemical data for many neutrals and cations can be determined with unprecedented precision. The further development of two-color excitation-ionization schemes promises to expand the scope of spectroscopic and chemical applications using the photoionization-photoelectron method.

INTRODUCTION

Photoionization and photoelectron spectroscopy is a major technique for research in the physical sciences (1, 2). Single-photon ionization usually occurs in the vacuum ultraviolet (VUV) region and is the most general and cleanest photoionization method (3, 4). The photoionization efficiency (PIE) measurement, which involves the detection of photoions as a function of VUV photon energy, provides a direct measure of the cross sections for the parent and fragment ions formed in photoionization. When a Rydberg series with sufficiently high-n levels is resolved in a PIE study, the Rydberg series analysis can yield the most accurate ionization energy (IE) for an ionic state. The line shape of an autoionizing Rydberg state observed in a PIE measurement contains pertinent information about the coupling between the

Rydberg state and ionization continua (3). However, owing to strong competition of neutral predissociation processes, autoionizing Rydberg features are seldom observable in PIE measurements of polyatomic molecules. Thus, photoelectron measurements for polyatomic species are more useful in providing spectroscopic information for the parent ion.

Limited by the space charge effect, ions cannot be prepared with high concentrations. Consequently, spectroscopic studies of ions using traditional optical excitation and fluorescence techniques have been difficult. Much information available in the literature on vibrational frequencies of molecular ions has been obtained by VUV photoelectron measurements (1, 2). Because photoelectrons ejected in a photoionization process carry finite angular momenta with specific symmetry representations, metastable states for ions can be produced by VUV photoionization (3). Furthermore, owing to the mediation of nearby resonance autoionizing states, the detection of threshold (or zero kinetic energy) photoelectron (TPE) bands for vibronic states using a tunable VUV source is not totally subject to the Franck-Condon factor (FCF) restriction for direct photoionization (5, 6). It has been shown in previous experiments that TPE bands for vibronic states with near zero FCFs can be observed, making VUV threshold photoionization a unique method for spectroscopic studies of ions.

The photoelectron-photoion coincidence (PEPICO) scheme involves the measurement of correlated photoelectron-photoion pairs (7, 8). Knowing the photon energy and the photoelectron kinetic energy in a PEPICO measurement, the state or energy of photoions can be selected for dissociation and reactivity studies. By virtue of the high collection efficiency for TPEs, the threshold-PEPICO (TPEPICO) method has been widely adopted for studies of ion dissociation dynamics. The use of this technique for the study of state-selected ion-molecule reactions dynamics, which is referred to as the TPE-secondary ion coincidence (TPESICO) technique (9), has also been demonstrated. Despite many advantages offered by these methods, their applications have been limited by the relatively low-achievable TPE resolution.

In the past decade the most significant technical development in the field of photoionization and photoelectron studies has been the introduction of pulsed field ionization (PFI) schemes, including PFI-zero kinetic energy photoelectron spectroscopy (6, 10, 11), mass analyzed threshold ion spectroscopy (12, 13), and threshold ion pair production spectroscopy (14–17). To unify these acronyms, these respective techniques are referred to here as PFI-photoelectron (PFI-PE), PFI-photoion (PFI-PI), and PFI-ion pair (PFI-IP) spectroscopy. These techniques were originally developed using pulsed lasers with repetition rates in the range of 10–100 Hz. Laser PFI experiments have demonstrated energy resolutions close to the laser optical bandwidth.

Recently the PFI techniques have been successfully implemented at the Chemical Dynamics Beamline of the Advanced Light Source (ALS) (18–25). The ALS is a third generation synchrotron source. Equipped with a 6.65-m scanning monochromator, the high-resolution VUV synchrotron source of the Chemical

Dynamics Beamline achieves a resolution of 1 cm^{-1} [full width at half maximum (FWHM)] at 12.12 eV (22), which is comparable to that achieved by common VUV lasers. The ease of tunability (20) over a wide VUV range (8–30 eV) of this high-resolution source has considerably expanded the usefulness of PFI-PE measurements as a general, effective high-resolution method for spectroscopic studies of molecular ions. Because the ALS is a pseudo-continuum VUV source with repetition rates of 3–488 MHz, it is most suitable for PEPICO studies. The recent success in developing the PFI-PE–photoion coincidence (PFI-PEPICO) method (23–34) is an important step, promising to bring about PFI techniques that are more useful for chemical dynamics studies. The PFI-PEPICO method is a variant of the PFI-PI scheme (12), in which ions of known internal states or energies formed in PFI are detected.

Photoionization mass spectrometry and photoelectron spectroscopy have traditionally played an essential role in providing energetic data, such as 0°K heats of formation (ΔH_{f0}°s) and 0°K bond dissociation energies (D_0s) for neutrals and ions (35, 36). Recent PFI-PEPICO studies (26–34) have shown that 0°K dissociative photoionization thresholds or appearance energies (AEs) for a range of molecules can be determined with precision limited only by PFI-PE measurements. These AE measurements, along with IE measurements using the PFI-PE method, promise to provide highly accurate energetic data for a range of molecules and their ions with well-founded error limits, which are 10–100-fold smaller than those for existing literature values (36). These 0°K energetic data, which measure differences between well-defined molecular energy levels, are most appropriate for direct comparison with theoretical calculations. The availability of these new energetic data is expected to have a significant impact on development of the next generation of computational procedures (37, 38).

This review is intended to survey recent developments in the field VUV spectroscopy and chemistry using the photoionization and photoelectron methods. After an overview on the development of VUV light sources, selected results on recent PFI-PE, PFI-IP, and PFI-PEPICO measurements using VUV lasers and the high-resolution VUV synchrotron source at the ALS are discussed. Armed with many new experimental techniques, the field of VUV photoionization mass spectrometry and photoelectron spectroscopy is ready for the next level of developments and applications. A few possible new directions for VUV photoionization-photoelectron studies are speculated upon here.

VACUUM ULTRAVIOLET LIGHT SOURCES

Laboratory Discharge Lamps

The developments in the field of VUV spectroscopy and chemistry have been limited by the available VUV light sources. Up to the mid-1980s, VUV photoionization and photoelectron studies relied mostly on laboratory discharge lamps and second generation synchrotron sources (4). Using the He-Hopfield continuum and Ar continuum formed in condensed discharges and the H_2 many-lined pseudo-continuum produced in a dc discharge coupled with a windowless VUV monochromator, tunable VUV light in the range of ≈580–1650 Å (7.5–21.4 eV) can be produced (3, 4, 39). This energy range covers the first IEs of all atoms and molecules except that of He. The H_2 many-lined pseudo-continuum is not suitable for high-resolution studies. At a resolution of 1.4 Å (FWHM), VUV intensities in the range of 10^9–10^{11} photons · s^{-1} can be obtained using these discharge sources (41). The classic photoionization study of H_2 by Dehmer & Chupka (40) achieved an optical resolution of 1.5 cm^{-1} (FWHM) using a monochromatized He-Hopfield continuum as the VUV source. Owing to the limited VUV ranges of individual discharge sources, the problem of photoionization background caused by second- and higher-order VUV lights diffracted from the grating does not exist.

Vacuum Ultraviolet Lasers

The use of coherent VUV lasers for photoionization and photoelectron studies was firmly established in the late 1980s (42–47). The principles and experimental arrangements for coherent VUV generation have been discussed in several reviews (47–51). Kung and coworkers (42–44) pioneered the application of a rare-gas pulsed jet as a nonlinear medium for the generation of coherent VUV laser radiation. When a more extensive mixing volume is needed, Kung and coworkers also suggested the attachment of a T-shaped adapter onto the nozzle for directing the gas into a 0.3-mm diameter, 30-mm long tube (43). For a negatively dispersive medium such as Ar in the region of 969–1047 Å, this has produced a significant increase in conversion efficiency and substantially relaxes alignment constraints. For VUV mass spectrometric measurements, it is necessary to separate the unwanted fundamental, harmonic, and nonlinearly mixed frequencies from the VUV frequency of interest. This is usually achieved by using a VUV monochromator (48, 49). If the unwanted laser frequencies do not present a problem, it is useful to use a quartz capillary for guiding the VUV light from the nonlinear mixing chamber to the photoionization region (46, 51). The small conductance of the capillary also serves to isolate the high-vacuum photoionization chamber from the low-vacuum jet-mixing chamber.

The most common VUV laser sources are generated by two-photon resonance-enhanced sum- and difference-frequency mixings of dye lasers in rare gases and metal vapors as the nonlinear medium (45, 48, 49). Using commercial UV and visible lasers with optical bandwidths of 0.04–0.2 cm^{-1} (FWHM), coherent VUV radiation up to 19 eV (52) with bandwidths of ≈0.12–0.8 cm^{-1} (FWHM) and intensities of 10^{-7}–10^{12} photons/pulse can be generated. Kung et al. (47) and Hollenstein et al. (53) have shown that VUV laser radiation with diffraction-limited resolutions of 210–250 MHz can be obtained by mixing two single-mode laser beams produced by an Ar^+ ion or diode laser (frequency-doubled Nd:YVO$_4$ laser at 532 nm) pumped ring dye lasers. To double and triple the ring dye laser outputs, the amplification by Nd:YAG pumped dye stages is necessary. The cost for this high-resolution VUV laser system is about twice that for the lower-resolution version constructed using common pulsed Nd:YAG pumped dye lasers.

Although the two-photon resonance-enhanced sum- and difference-frequency mixing schemes are more efficient than frequency tripling, Hennen and coworkers (54) have shown that usable VUV laser intensities in the range of 905–1093 Å can be obtained by nonresonance frequency tripling in Ar, Kr, Xe, H_2, CO, N_2, O_2, and C_2H_2. This tripling scheme takes advantage of the fact that the corresponding UV outputs (2700–3300 Å) with high pulse energies up to 20 mJ can be obtained by commercial dye lasers. Tunable femtosecond VUV radiation in the range of 1240–1020 Å has also been generated by two-photon-resonant and near-resonant four-wave difference-frequency mixings in Kr and Ar, respectively (55, 56).

In principle, VUV laser radiation based on four-wave mixing schemes in rare gases and metal vapors, such as Hg and Mg, can cover the full range of 6–19 eV (48, 49). However, small gaps of low intensities exist in the tuning curve for VUV lasers. The use of a tunable VUV laser source to cover a wide energy range as required in many experiments remains difficult and very time consuming.

Vacuum Ultraviolet Synchrotron Radiation

Photoionization experiments performed using laboratory VUV discharge lamps were competitive with synchrotron radiation in both sensitivities and optical resolutions until the early 1990s, when the high-resolution monochromatized third-generation VUV synchrotron source at the Chemical Dynamics Beamline of ALS was established. The greatest advantage of using a synchrotron radiation source is its ease of tunability. The tunable range of a synchrotron source depends on the monochromator used. The higher brightness of a third-generation synchrotron source, which is achieved by a tighter focusing of the orbiting electron bunches, allows a more effective coupling with an insertion device, such as an undulator, for further amplification of the VUV intensities. An undulator consists of a periodic array of permanent magnets, which forces the electrons to oscillate with a period of a few centimeters over a length of several meters. Each wiggle of the electron beam emits synchrotron radiation in the same forward direction. Owing to interference effects, an amplification factor up to the square of the number of periods is observed in an undulator. Using a 10-cm period undulator, the first order of the undulator harmonics can be tuned in the photon energy range of 8–30 eV by varying the undulator gap (20). A novel harmonic gas filter was designed to suppress the unwanted higher-order undulator harmonics (18, 20). When He, Ne, or Ar is used as the filter gas, VUV undulator harmonics at energies higher than the IE(He) (24.59 eV), IE(Ne) (21.56 eV), or IE(Ar) (15.76 eV) are eliminated. The first-order undulator harmonic radiation (bandwidth $\approx$2.5%, photon intensity $\approx 10^{16}$ photons $\cdot$ s^{-1}) emerging from the gas filter is focused on the entrance slit of a 6.65-m monochromator. Equipped with a 4800–lines mm^{-1} grating, the monochromator has demonstrated an optical resolution of 1 cm^{-1} (FWHM) at 12.12 eV (22). The dispersed VUV beam from the monochromator is further focused into the photoionization/photoexcitation (PI/PEX) center of the photoion-photoelectron apparatus, attaining a spot size of 0.24 mm in height and 0.36 mm in width. At a resolving power of 25,000, the measured VUV intensities at the PI/PEX center are in the range of 10^9–10^{11} photons $\cdot$ s^{-1} (20). We note that the ALS radiation is 99% polarized in the horizontal plane perpendicular to the direction of the VUV beam.

The ALS storage ring is capable of filling 328 electron buckets in a period of 656 ns. Each electron bucket emits a light pulse of 50 ps with a time separation of 2 ns between successive bunches. In each storage period, a dark gap of 16–144 ns consisting of 8–72 consecutive unfilled buckets exists for the ejection of ions from the ring orbit. Thus, in the multi-bunch mode, the synchrotron ring has 256–320 electron bunches in its orbit, corresponding to a repetition rate of 390–488 MHz. The ALS storage ring also operates in the two-bunch mode in which the temporal separation between successive bunches is 328 ns, corresponding to the repetition rate of 3.04 MHz.

For absorption measurements, the photon energy calibration can be made by known absorption resonances of He, Ne, and Ar recorded using the gas filter as an absorption cell (20). For PFI-PE measurements, the photon energy calibration was achieved using the known IEs of the rare gases (He, Ne, Ar, Kr, and Xe) and diatomic molecules (O_2, N_2, NO, and H_2) measured under the same experimental conditions (25). This calibration scheme assumes that the Stark shifts for the molecule of interest and the calibration gases are identical. On the basis of previous experiments, we estimate that the energy calibration has an uncertainty of ± 0.5 meV (57–59).

PHOTOION-PHOTOELECTRON STUDIES

Photoionization Mass Spectrometric Measurements

Owing to experimental simplicity, the majority of IE and AE values available in the literature were obtained by PIE measurements (35, 36). However, AE values determined based on PIE measurements can suffer large uncertainties because of the hot band and kinetic shift effects, particularly for ion dissociation processes involving polyatomic ions (35). The resolution of a PIE measurement is limited not only by the VUV optical bandwidth, but also by the electric field applied at the PI/PEX region. The electric field used for the ion extraction can induce Stark mixings of Rydberg levels associated with different orbital angular momenta, resulting in broadening of autoionizing Rydberg peaks. In order to obtain the highest possible resolution in PIE measurements, it is necessary to minimize the residual electric field at the PI/PEX region.

PHOTOIONIZATION EFFICIENCY SPECTRUM FOR CS_2 Figure 1 shows the PIE spectra for CS_2 (60) in the region of 81,100–81,800 cm^{-1} measured using VUV laser

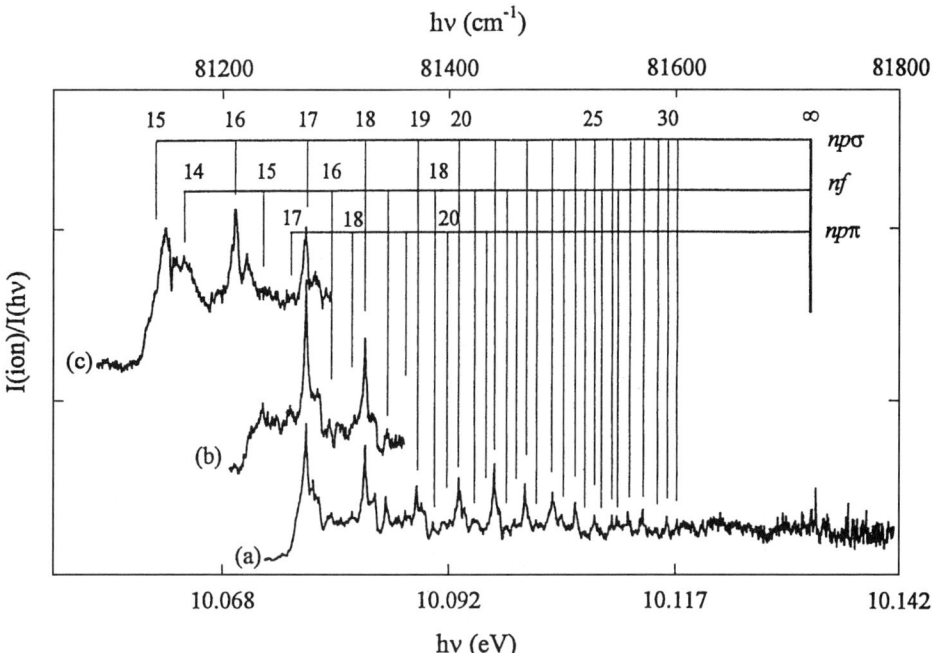

Figure 1 Photoionization efficiency spectra for CS_2 in the range of 81,200–81,800 cm^{-1} (60) measured using vacuum ultraviolet (VUV) laser and dc Stark fields (a) $F = 23$ V · cm^{-1}, (b) $F = 158$ V · cm^{-1}, and (c) $F = 787$ V · cm^{-1}. VUV optical bandwidth ≈ 1 cm^{-1} (FWHM).

radiation [optical bandwidth $= 1$ cm^{-1} (FWHM)] and the dc Stark fields (a) $F = 23$ V · cm^{-1}, (b) $F = 158$ V · cm^{-1}, and (c) $F = 787$ V · cm^{-1}. Three Rydberg series, $np\sigma$, $np\pi$, and nf (n = 14–30), are resolved in the PIE spectra. Autoionizing peaks for higher-n (n > 30) Rydberg levels may be partly broadened owing to Stark mixings resulting from the relatively high F-field used for ion extracting. The analysis of these Rydberg series yields the IE[$CS_2^+(^2\Pi_{1/2})$] $= 81,727.1 \pm 0.5$ cm^{-1} (60). It can be seen from these spectra that lower-n Rydberg levels for the $np\sigma$, $np\pi$, and nf series, which lie below the IE[$CS_2^+(^2\Pi_{3/2})$] $= 81,285.7 \pm 2.8$ cm^{-1} (60), become observable in the PIE spectra as F is increased. The accurate IE determination using PIE measurements requires the extrapolation of the ionization onset to $F = 0$.

PHOTOIONIZATION EFFICIENCY SAMPLING OF PHOTOFRAGMENTS   As a mass selective technique, an important application of photoionization mass spectrometry has been in the study of transient radicals (61). Owing to the reactive nature of radicals, it is difficult to prepare a pure sample of radicals in abundance. The impure nature of a radical source does not present any problems for mass spectrometric samplings. By preparing radicals in pyrolysis, electrical discharge, chemical reactions, and photodissociations, previous PIE studies have provided valuable energetic and spectroscopic data for many radicals (41, 61–67), of relevance to plasma, atmospheric, and combustion chemistry.

Because polyatomic radicals can exist in many isomeric forms, one important issue is how to control the isomeric structure of radicals under photoionization studies. It has been demonstrated that radicals of a specific structure can be prepared in abundance by excimer laser photodissociation (66, 67). The IEs for several radicals in selected isomeric structures thus prepared have been determined in PIE measurements (68–78). These studies suggest that the isomeric structure of a neutral polyatomic photofragment can be identified by IE measurements (70, 73, 75, 76). Although the IEs for most polyatomic radicals are unknown, they can be predicted with an accuracy of <0.15 eV using standard computation packages, such as Gaussian-2/Gaussian-3 procedures (37, 38). By comparing theoretical IE predictions with IEs determined for photofragments in PIE measurements (see Table 1), we have identified the isomeric structures of photofragments formed in several photodissociation reactions (69–76, 79). Selective photodissociation to produce CH_3S are found from the photodissociation of CH_3SH (74, 79), CH_3SCH_3 (69, 74), and CH_3SSCH_3 (73, 74). The structures of major photoproducts formed in the 193-nm photodissociation of $CH_3CH_2SCH_2CH_3$ (70, 76), C_4H_4S (thiophene) (75), and CH_2Br_2 (71) have also been identified using this

TABLE 1 Direct identification of nascent product structures formed in photodissociation or fast atom reactions

Neutral precursor	Nascent photoproduct structure	G2 IE(eV)[a]	Experimental IE(eV)[b]
CH_3SH	CH_3S	9.24	9.2649[c] (74)
CH_3SCH_3	CH_3S	9.24	9.23 (69)
	CH_3SCH_2[d]	6.85	6.85[c,d] (77)
$CH_3CH_2SCH_2CH_3$	CH_3CH_2S	9.10	8.97 (70)
			9.077[c] (76)
CH_3SSCH_3	CH_3S	9.24	9.2649[c] (74)
	CH_3SS	8.55	8.67 (73)
	CH_3SSCH_2[d]	6.90	7.00[c,d] (78)
C_4H_4S (thiophene)	$CH_2=CH-C\equiv CH$	(9.58)[e] (36)	9.57 (75)
	$CH_2=C=S$	8.77	8.91 (75)
	$HC\equiv CH$	(11.40)[e] (36)	11.39 (75)
CH_2Br_2	CH_2Br	8.47	8.61 (71)

[a]Gaussian-2 predictions unless specified.
[b]Ionization energies of photoproducts by photoionization efficiency measurements unless specified.
[c]Determined by photoelectron spectroscopy.
[d]Products formed in fast atom reactions.
[e]Experimental ionization energies.

photodissociation/photoionization scheme. These experiments were conducted using a laboratory H_2 discharge source. The sensitivity of this scheme should be significantly improved by using a third generation synchrotron source for photoionization sampling.

PHOTOFRAGMENT ION IMAGING It is known that prompt dissociation to form $SF_5^+ + F$ occurs upon the photoionization of SF_6 and that the parent SF_6^+ ion is unstable (80). Recently we employed the photofragment ion-imaging technique at the Chemical Dynamics Beamline to probe the angular distributions of SF_5^+ and SF_4^+ fragment ions formed in the dissociative photoionization of SF_6 (81). This experiment represents the first application of the fragment ion-imaging technique in conjunction with VUV synchrotron radiation. The analysis of the fragment ion images yields the anisotropy β parameters, which characterize the angular distributions of the fragment ions. Although this is not a state- or energy-selected experiment, it provides additional information that cannot be easily obtained in a traditional PEPICO study. The dynamics of many dissociative photoionization processes involving small molecules can be examined using this method.

Threshold Photoelectron Spectroscopy

Photoelectron measurements were first introduced using a fixed photon energy source, such as the HeI (584 Å) resonance line (1). The vibrational bands for a molecule observed in a HeI photoelectron spectrum usually reflect the Franck-Condon pattern for direct photoionization transitions. Vibrational levels with negligible FCFs are usually not observable in a HeI photoelectron study. When TPEs are detected as a function of VUV photon energy, the FCF restriction may not apply because of the perturbation by nearby resonance intermediate autoionizing states (5). This, together with the high collection efficiency of TPEs, allows photoelectron bands of highly vibrationally excited levels with near zero FCFs to be observed in TPE measurements. Assuming that energetic electrons are ejected into a large solid angle, TPEs can be detected with only a minor background of energetic electrons by using a tube-like steradiancy analyzer, which defines a small acceptance solid angle for the electron detector (82, 83). Energetic background photoelectrons ejected into the solid angle sustained by the steradiancy analyzer are detected and manifested as a hot tail in the TPE transmission function, which is the most undesirable feature and has limited the achievable resolution in TPE measurements. The resolution of TPE measurements also depends on the repeller field for TPE extraction and the VUV optical bandwidth. The highest resolution (0.8 meV, FWHM) is demonstrated in the TPE spectrum for $Ar^+(^2P_{3/2, 1/2})$ (Figure 2a) obtained using a VUV bandwidth of ≈ 2 cm^{-1} (FWHM) (19). The hot-electron tails associated with the $Ar^+(^2P_{3/2,1/2})$ TPE bands are evident in Figure 2a, resulting in false structures from autoionizing Rydberg states converging to the $Ar^+(^2P_{1/2})$ limit. It has been demonstrated that when a pulsed VUV source has a pulse interval >100 ns, the hot-tail problem can be alleviated

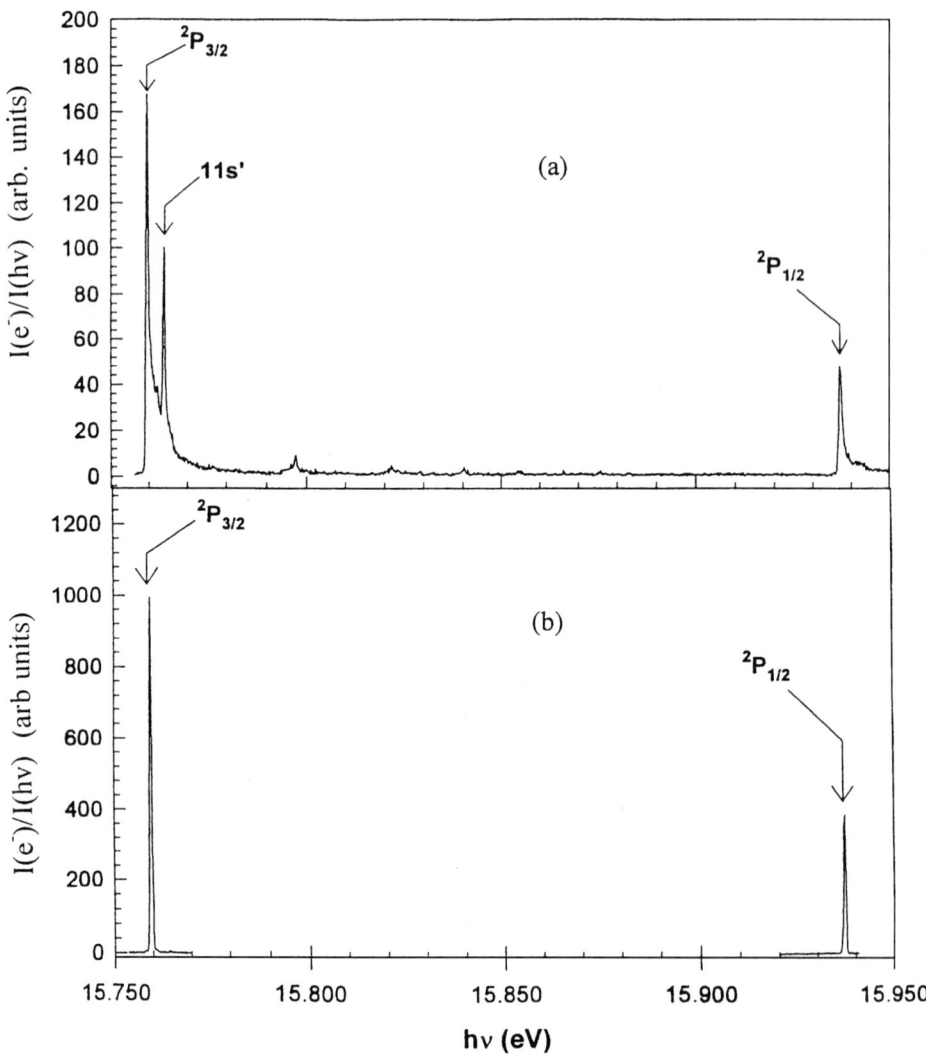

Figure 2 (*a*) Threshold photoelectron spectrum for $Ar^+(^2P_{3/2})$ and $Ar^+(^2P_{1/2})$ observed using the multibunch synchrotron radiation (19). The false structure owing to to the Ar(11s') autoionizing resonance is marked. (*b*) Pulsed field ionization–photoelectron spectrum for $Ar^+(^2P_{3/2})$ and $Ar^+(^2P_{1/2})$ using the multibunch synchrotron radiation with a dark gap of 40 ns.

by employing time-of-flight (TOF) selection along with steradiancy detection (84, 85).

A variation of the TPE method is the penetration field technique (86–89), which replaces the steradiancy analyzer with a specially designed electron analyzer. Because the collection for TPEs formed at the PI/PEX region is optimized by a penetrating electrostatic field, a higher TPE resolution can be attained in penetration field measurements. The fact that background hot electrons that travel toward the electron detector are filtered by both the chromatic aberration of the electrostatic lens system and the differential energy analyzer makes possible a significant reduction of the hot-tail problem. The highest resolution demonstrated (88) using this method is ≈1.0 meV (FWHM), which was mainly limited by the bandwidth of the VUV photoionization source used. Thus, a higher TPE resolution for the penetration technique is to be expected by using a higher resolution VUV source in the future. By employing the penetration technique, together with the free jet method to cool the gas samples, Morioka and coworkers have resolved vibrational TPE bands for H_2^+, N_2^+, and O_2^+ with vibrational v^+ levels close to the dissociation limits of these ions (90).

Laser-Based Pulsed Field Ionization–Photoelectron Studies

Although Müller-Dethlefs et al. discovered the basic procedures for laser-based PFI-PE measurements in 1984 (10), the detailed mechanism involved was only suggested in 1988 (11, 91). This technique involves the detection of PFI-PEs formed in the delayed PFI of long-lived high-n ($n \geq 100$) Rydberg species initially formed by pulsed laser excitation. Owing to the existence of small stray electric fields at the PI/PEX region, background electrons produced by direct photoionization and/or prompt autoionization can be effectively dispersed in a few hundred nanoseconds. Thus, the PFI-PE detection can be made free from background prompt electrons by delaying the application of the pulsed electric field for PFI by ≈1–5 μs with respect to the excitation laser pulse. We note that PFI-PEs are near zero kinetic energy photoelectrons formed slightly (a few cm^{-1}) below the true IE. The observed PFI-PE peak is lower than the true IE by the Stark shift (Δ) as predicted by the equation $\Delta = A\sqrt{F}$ cm^{-1}, where F is the magnitude of the pulsed field in V/cm. Experiments have shown that the value for A is 4–6. The true IE of a gas sample can thus be obtained by extrapolating the ionization onset to $F = 0$. Based on the continuity of oscillator strength density, relative intensities for photoelectron bands observed in a PFI-PE spectrum are expected to be identical to those resolved in a TPE spectrum, provided that perturbations by nearby autoionizing states are ignored.

The key requirement for the success of delayed PFI-PE measurements is the long lifetimes for high-n Rydberg species prepared in optical excitation. Radiative decay is not an important loss mechanism for high-n Rydberg states. However, autoionization of a neutral Rydberg species lying above the lowest IE can always occur by transferring sufficient energy from the ion core to the Rydberg electron. Based on the type of core energies transferred to the departing electron in molecular autoionization, the mechanisms involved can be classified as rotational, vibrational, and electronic autoionization. Simple consideration based on the atomic model predicts that the lifetime of a Rydberg state is proportional to n^3 (92, 93). This prediction is found to be much too low compared with lifetimes observed in PFI experiments. Owing to dipole selection rules, the high-n Rydberg molecules initially formed by optical excitation should have low l quantum numbers. It is generally accepted that stray electric fields and/or electric fields arising from nearby prompt ions at the PI/PEX region induce l- and m_l-mixings, where l and m_l are the orbital angular momentum quantum number and the magnetic quantum number, respectively (94–97). Because high-l states have nearly circular orbits and thus the Rydberg electrons have little interaction with the ion core, the lifetimes of high-l states are longer because they are more stable against autoionization (and predissociation, in the case of molecules). The lifetime lengthening effect by m_l-mixings is statistical in nature. The average lifetime for a statistical mixture of (l, m_l) states of the same n is predicted to scale as $n^{4.5}$, a prediction in accord with experimental observations.

The PFI-PE resolutions achieved using VUV lasers are in the range of 0.2–4 cm^{-1} (FWHM), which are determined by both the Stark field F and the bandwidth of the excitation VUV laser (51–53). These resolutions are sufficient for resolving rotational structures of many diatomic and triatomic species and simple polyatomic hydrides. Using an ingenious pulsing scheme for selective field ionization of high-n Rydberg states, together with the transform-limited VUV laser, Merkt and coworkers recently reported a PFI-PE resolution of 0.05 cm^{-1} (FWHM) (98). For a more in-depth discussion of VUV laser PFI-PE studies, readers are referred to reviews by Hepburn (48, 49), Wiedmann & White (51), Lipson et al. (99), and Ng (25).

ROTATIONALLY RESOLVED PULSED FIELD IONIZATION–PHOTOELECTRON STUDY OF CH$_4$ AND ITS ISOTOPOMERS   As a highlight of the current state-of-the-art experimental studies using the VUV laser PFI-PE method, we shown in Figures 3a–d the first rotationally resolved PFI-PE spectra (upper traces) for CH$_4$, CD$_4$, CDH$_3$, and CD$_2$H$_2$ near their ionization onsets obtained with a PFI-PE resolution of 0.7 cm^{-1} (FWHM) (100). The CH$_4^+$ ion originally formed by photoionization is subject to a Jahn-Teller distortion to structures of lower symmetry. This results in small FCFs for the formation of CH$_4^+$ at the photoionization onset of CH$_4$. Partly for this reason, the IE determinations for CH$_4$ in previous PIE and photoelectron studies were inconsistent. In addition to providing insight into the dynamics of interconversion between various structures of CH$_4^+$, the analysis of the PFI-PE spectra has yielded accurate IEs for CH$_4$ and its isotopomers.

Synchrotron-Based Pulsed Field Ionization–Photoelectron Studies

The (pulsed) VUV laser PFI-PE technique traditionally requires a delay of $\approx$1–3 μs for the dispersion of prompt background electrons and thus is not directly applicable to PFI-PE measurements using synchrotron radiation, which is essentially a continuous light source and has insufficient dead time for the delay. We have overcome this difficulty by making use of the dark gap in a storage ring period for PFI-PE measurements (19–25). Two schemes have been developed and successfully

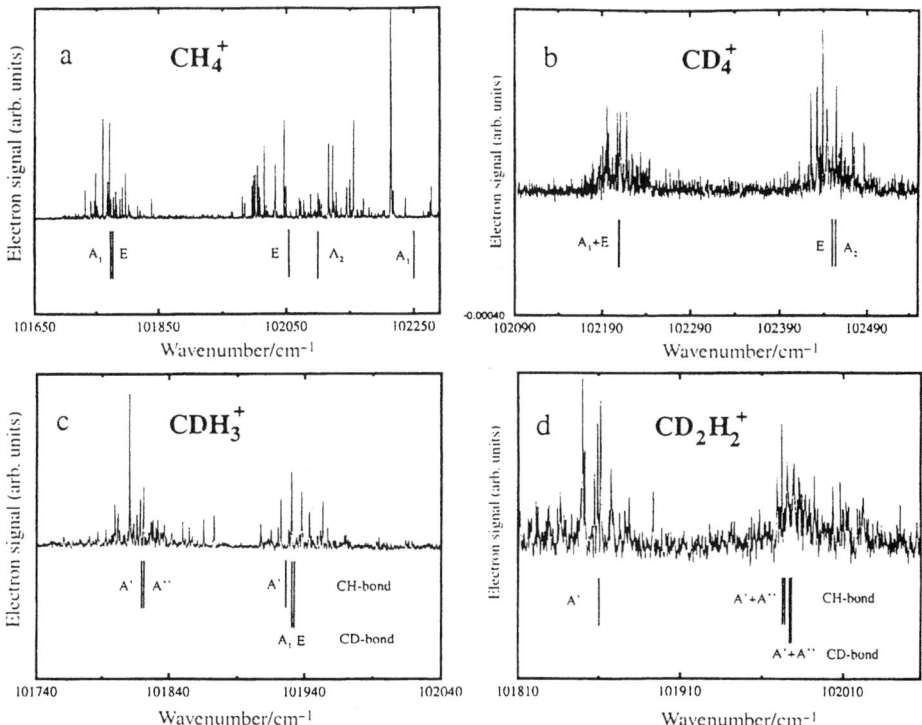

Figure 3 The pulsed field ionization–photoelectron (PFI-PE) spectra of CH_4, CD_4, CDH_3, and CD_2H_2 near their ionization onsets are depicted in the upper traces (*a* to *d*), respectively (100). The lower sticks spectra are based on ab initio predictions of the vibrational transitions. PFI-PE resolution = 0.7 cm^{-1} (full width at half maximum).

applied to PFI-PE studies of many atomic and molecular systems. The earlier method involves the use of an electron spectrometer (19), which is equipped with a steradiancy analyzer and a hemispherical energy analyzer arranged in tandem. Using such an electron spectrometer, we have shown that PFI-PEs can be detected with little background from prompt electrons after only an 8-ns delay with respect to the beginning of the dark gap. Figure 2*b* shows the PFI-PE bands for $Ar^+(^2P_{3/2,1/2})$ obtained using this method with a dark gap of 40 ns. These bands achieve a PFI-PE resolution of 4 cm^{-1} (FWHM) and are completely free from hot tails manifested in the TPE bands depicted in Figure 2*a*, which are measured using the same VUV optical bandwidth. The highest PFI-PE resolution achieved using this method with an electron spectrometer is 2 cm^{-1} (FWHM) (21).

The most recent PFI-PE method employs the TOF selection scheme (22) and is superior to that using the electron spectrometer in all aspects, including sensitivity, achievable resolution, and background prompt electron rejection. However, the TOF scheme requires a larger synchrotron dark gap (>80 ns). By varying the height, width, and delay of the electric field pulse with respect to the beginning of the dark gap, PFI-PEs can be adjusted to arrive within the dark gap, allowing them to be detected free from background prompt electrons. The success of this TOF scheme can be attributed to the small dispersion of electron TOFs resulting from the small VUV spot size at the PI/PEX region. Figures 4a, b show the

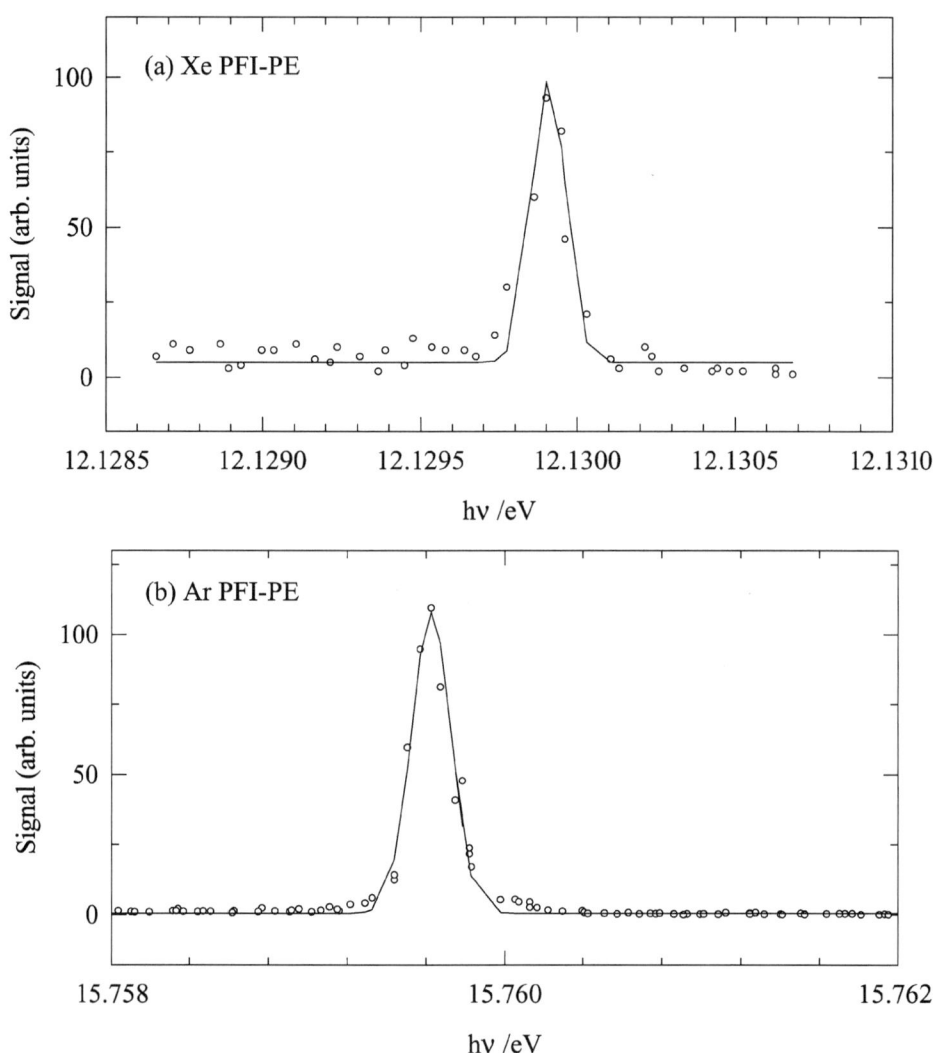

Figure 4 Pulsed field ionization–photoelectron (PFI-PE) spectra for (a) $Xe^+(^2P_{3/2})$ [PFI-PE resolution = 1.0 cm^{-1} (FWHM)] and (b) $Ar^+(^2P_{3/2})$ [PFI-PE resolution = 1.9 cm^{-1} (FWHM)] obtained using the 4800-lines/mm grating and monochromator entrance/exit slits of 10/10 μm (22).

PFI-PE bands for $Xe^+(^2P_{3/2})$ and $Ar^+(^2P_{3/2})$ obtained using the TOF selection scheme, attaining resolutions of 1.0 cm^{-1} (FWHM) and 1.9 cm^{-1} (FWHM), respectively (22). These resolutions are essentially limited by the achievable VUV optical bandwidth.

The ease of tunability of the ALS synchrotron source has made rotationally resolved PFI-PE measurements for many molecules a routine operation. By employing these synchrotron-based PFI-PE schemes, in the past few years we have obtained rotationally resolved or partially rotationally resolved PFI-PE spectra for $H_2^+(v^+ = 0–18)$ (57), $HD^+(v^+ = 0–21)$, $D_2^+(v^+ = 0–26)$, $O_2^+(X^2\Pi_g, v^+ = 0–38;$ $a^4\Pi_u, v^+ = 0–18; A^2\Pi_u, v^+ = 0–12; b^4\Sigma_g^-; v^+ = 0–9; 2^2\Pi_u; B^2\Sigma_g^-, v^+ = 0–7;$ $^2\Sigma_u^-;$ and $c^4\Sigma_u^-, v^+ = 0–1)$ (21, 58, 101–106), $NO^+(X^1\Sigma^+, v^+ = 0–32; a^3\Sigma^+,$ $v^+ = 0–16;$ and $A'^1\Sigma^-; v^+ = 0–17)$ (59, 107, 108), $CO^+(X^2\Sigma^+, v^+ = 0–42; A^2\Pi,$ $v^+ = 0–41; B^2\Sigma^+, v^+ = 0–15; D^2\Pi, v^+ = 0–8;$ and $3^2\Sigma^+, v^+ = 0–4)$ (109–112), HCl^+, and HF^+ (113). High-resolution PFI-PE spectra for CS_2^+ (60, 114), OCS^+ (115), CO_2^+ (116), N_2O^+, NO_2^+ (117), and H_2S^+, NH_3^+, ND_3^+, and $C_2H_2^+$ in the VUV photon energy range of 9–25 eV have also been recorded. The analyses of the PFI-PE spectra have not only allowed more accurate spectroscopic measurements for these ions, but also yielded useful information on photoionization dynamics of the corresponding neutral molecules by determining the photoelectron angular momentum states (25). Accurate spectroscopic data thus obtained have provided rigorous tests for state-of-the art ab initio calculations (60, 111, 112, 114–120). For several diatomic ions, the PFI-PE measurements cover vibrational levels close to the dissociation limits of the corresponding ionic states, such as $H_2^+(v^+ = 0–18)$ (57), $HD^+(v^+ = 0–21)$, $D_2^+(v^+ = 0–26)$, $O_2^+(X^2\Pi_{3/2,1/2g}; v^+ = 0–38)$ (58), $NO^+(X^1\Sigma^+; v^+ = 32)$ (59), $CO^+(X^2\Sigma^+; v^+ = 0–42)$ (109), and $CO^+(A^2\Pi_{3/2,1/2};$ $v^+ = 0–41)$ (112).

PULSED FIELD IONIZATION–PHOTOELECTRON SPECTRUM FOR $O_2^+(X^2\Pi_{3/2,1/2g}, v^+ = 0–38)$ To illustrate the detailed information obtained in these spectroscopic studies, we show in Figure 5 the rotationally resolved PFI-PE bands for $O_2^+(X^2\Pi_{3/2,1/2g},$ $v^+ = 0–38)$ in the energy range of 11.8–18.2 eV (58). These bands exhibit doublet structures, corresponding to the spin-orbit states $O_2^+(X^2\Pi_{3/2g})$ and $O_2^+(^2\Pi_{1/2g})$. The rotationally resolved PFI-PE bands for $O_2^+(X^2\Pi_{3/2,1/2g}, v^+ \leq 24)$ have been obtained using tunable VUV lasers (121). The fact that PFI-PE bands for $O_2^+(X^2\Pi_{3/2,1/2g}, v^+ \geq 21)$ are in serious overlaps with PFI-PE bands of the $O_2^+(a^4\Pi_u, v^+ = 0–18)$ and $O_2^+(A^4\Pi_u, v^+ = 0–12)$ states (see Figures 6a,b) makes the analysis of these PFI-PE bands difficult (58). However, owing to the high resolution achieved in this experiment, we have been able to deconvolute these overlapping bands by simulation of the PFI-PE spectra obtained using O_2 samples at 20°K and 220°K (105, 106).

The spin-orbit splitting constants A_{v+} and rotational constants B_{v+} in cm^{-1} for $O_2^+(X^2\Pi_{3/2,1/2g}, v^+ = 0–38)$ obtained by spectral simulation are plotted in Figure 7 (58). The $v^+ = 0–38$ levels cover up to $\approx 90\%$ of the well depth. The availability of these B_{v+} values should be useful for the construction of an accurate potential for $O_2^+(X^2\Pi_{3/2,1/2g})$. The A_{v+} values have been used to test a newly developed

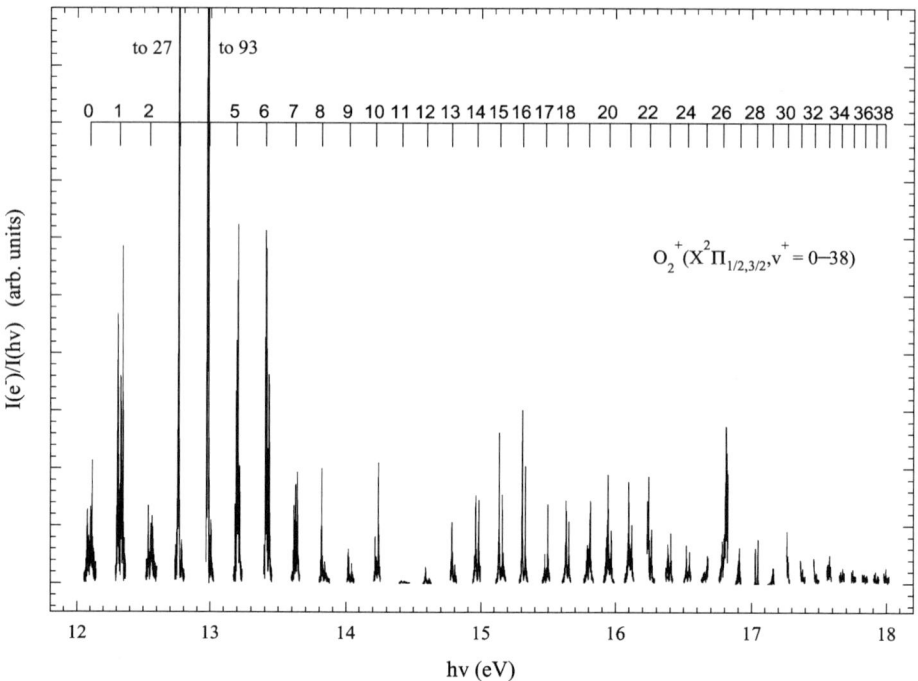

Figure 5 The pulsed field ionization–photoelectron bands for $O_2^+(X^2\Pi_g, v^+ = 0-38)$ in the energy range of 11.8–18.2 eV (58). The positions for these vibrational bands are marked. The maximum intensities of the $v^+ = 3$ and 4 bands are 27 and 93, respectively. The band intensities for $v^+ = 27-29$ are based on simulations.

ab initio computation code (112, 118) for the calculation of spin-orbit coupling constants.

LIFETIME MEASUREMENTS OF HIGH-N RYDBERG STATES A major difference between PFI-PE measurements using a pulsed VUV laser and a pseudo-continuum synchrotron source is in the density of ions produced at the PI/PEX region. Because the number of ions produced per micro-light pulse using synchrotron radiation is much less than one, the l- and m_l-mixings induced by prompt ions are unimportant.

Figure 6 Pulsed field ionization–photoelectron spectra for $O_2^+(X^2\Pi_{1/2,3/2g}, a^4\Pi_u,$ and $A^2\Pi_u)$ in the regions of (a) 15.90–16.96 eV and (b) 16.96–18.16 eV (58). The positions for the $O_2^+(X^2\Pi_{1/2,3/2g}, v^+ = 20-38)$, $O_2^+(a^4\Pi_u v^+ = 0-18)$, and $O_2^+(A^2\Pi_u v^+ = 0-12)$ are marked. The spectra were obtained using an O_2 molecular beam with an estimated rotational temperature of 220° K.

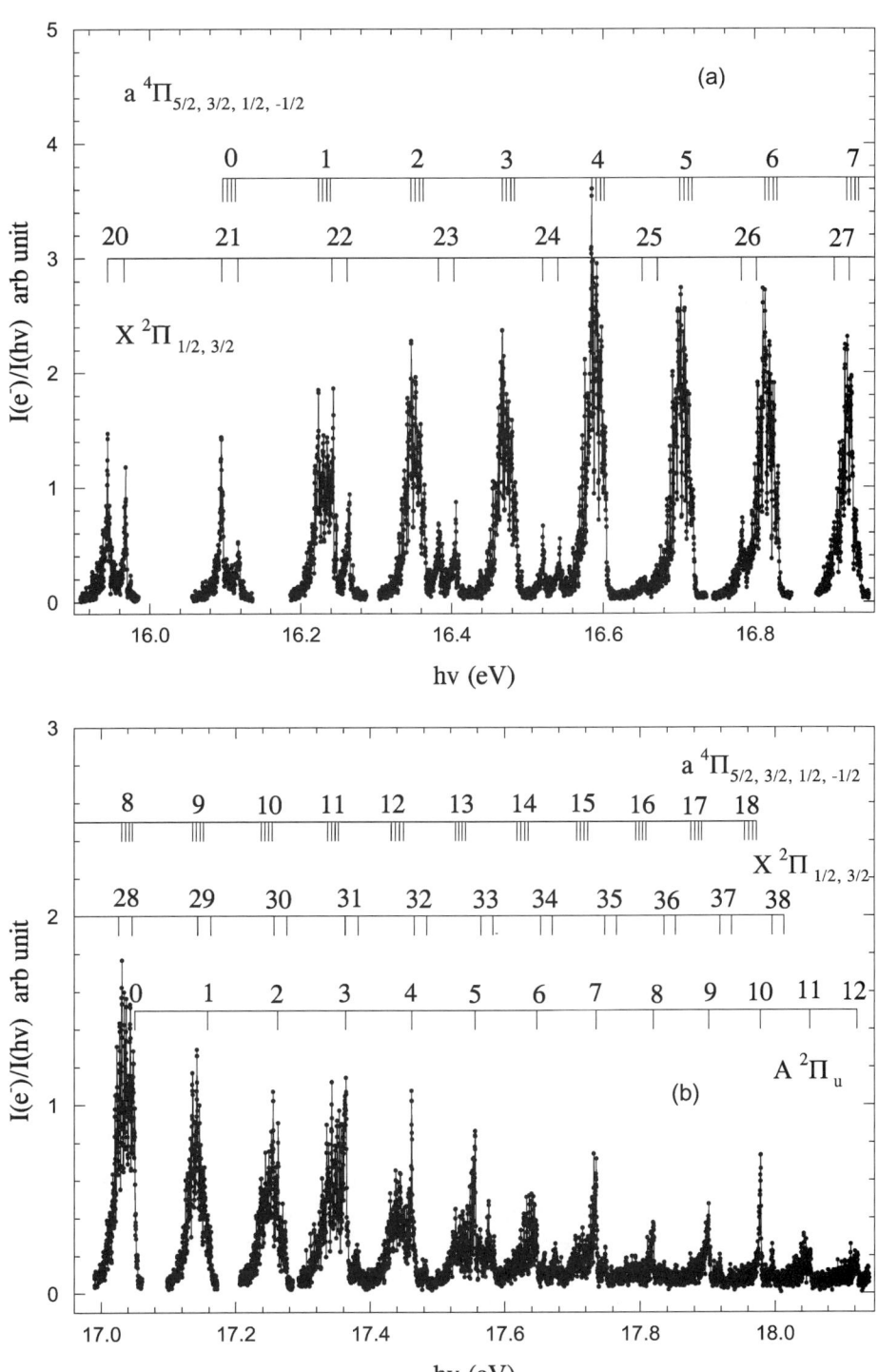

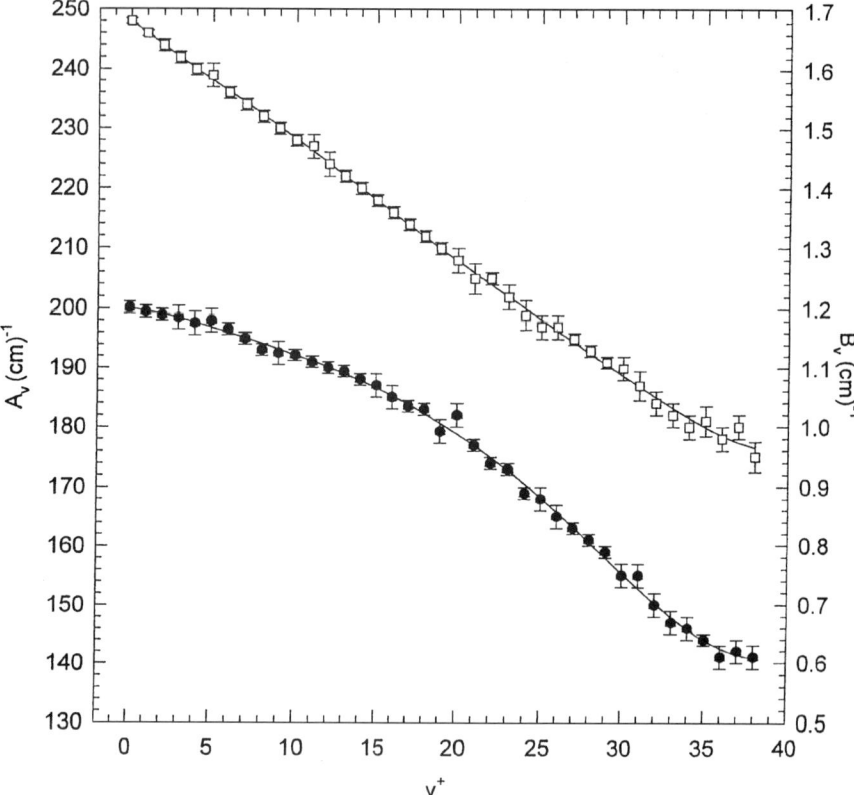

Figure 7 Plots of the spin-orbit splitting constants A_{v+} (*filled circles*) and rotational constants B_v^+ (*open squares*) versus v^+ for $O_2^+(X^2\Pi_g, v^+ = 0$–$38)$ (58).

Thus, the lifetimes for high-n Rydberg species prepared by synchrotron excitation are expected to be less than those prepared by pulsed VUV lasers.

The effective lifetime (τ) of high-n ($n \approx 100$) Rydberg states can be examined by measuring the PFI-PE intensity, i.e., applying the pulsed electric field for PFI, as a function of the number of synchrotron ring periods (104). Shiell et al. have measured the τ values for high-n Rydberg states converging to the ... $5s^25p^4(^3P)6s$ and ... $5s^25p^4(^3P)5d$ satellite states (121, 122). It is interesting that the τ values observed for the ... $5s^25p^5(^2S_{1/2})$ nl and ... $5s^25p^46s$ $(^{2S+1}L_J)$ nl states are generally greater than 5 μs. In contrast all the Rydberg states that converge onto the ... $5s^25p^46d(^{2S+1}L_J)$ core except the $^2D_{5/2}$ state show shorter τ values, falling in the range of 0.8–4 μs. The effective radius of the 5d-orbital is expected to be larger than that of the 6s- and 6p-orbitals. Thus, to the first approximation the interaction between the high-n Rydberg electron with electrons in the ... $5s^25p^45d(^{2S+1}L_J)$ ion core is greater than that between the Rydberg electron and the ... $5s5p^5(^{2S+1}L_J)$

[or ... $5s^25p^46s(^{2S+1}L_J)$] ion core. This greater electron-electron interaction may lead to faster decay by autoionization and thus shorter τ values for high-n Rydberg states associated with the ... $5s^25p^45d(^{2S+1}L_J)$ nl configuration.

DISSOCIATION MECHANISM FOR O_2 IN HIGH-N RYDBERG STATES   When the dissociation lifetime (τ_d) of a molecular ion core is $\leq 10^{-12}$ s, it can be derived by the rotational linewidth resolved in the PFI-PE measurement. We have determined the τ_ds for $O_2^+(B, v^+ = 0-6)$ (104), and $O_2^+(c, v^+ = 0, 1)$ (103) to be <0.3 ps in recent synchrotron-based PFI-PE experiments. The τ_ds for $O_2^+(b, v^+ = 4, 5)$ determined in previous laser dissociation studies are <4 ns (123). By comparing the τ_d and nominal τ values for high-n Rydberg states converging to these dissociative ionic states, we have concluded that the decay of dissociative high-n (n > 100) Rydberg states of O_2 [$O_2^*(n)$] initially prepared by VUV excitation (Equation 1a) follows the stepwise mechanism

$$O_2 + h\nu \xrightarrow{(a)} O_2^*(n) \xrightarrow{(b)} O^*(n') + O \xrightarrow{(c)} O^+ + e^- + O. \qquad 1.$$

Equation 1b represents the prompt dissociation to form $O^*(n') + O$, where $O^*(n')$ is an excited oxygen atom in a high-n' Rydberg state. The PFI-PEs observed at energies above the ion dissociation thresholds result from PFI of $O^*(n')$ (Equation 1c). A previous fluorescence study concerning excited O_2 in low-n Rydberg states has provided evidence that the n' for O^* formed in Equation 1b is identical to the n for O_2^*, i.e., the principal quantum number is conserved in the dissociation process. This dissociative-PFI mechanism is expected to be valid for high-n Rydberg species converging to a dissociative ion core with a dissociative lifetime $\leq 10^{-7}$ s.

The lowest dissociation threshold for O_2^+ is known to be just below $O_2^+(b^4\Sigma_g^-, v^+ = 4, N^+ = 9)$. At this threshold $O^*(n') + O(^3P)$ are produced in Equation 1b, where $O^*(n')$ converges to $O^+(^4S_{3/2})$. At photon energies below this dissociation threshold, the PFI-PE signal originates from the PFI of $O_2^*(n)$, whereas the PFI-PE signal at photon energies above the dissociation threshold results from the PFI of $O^*(n')$. Hence, the ratio of the PFI-PE intensity at an energy below the dissociation threshold to that above the dissociation threshold should be proportional to the ratio of the intensity for $O_2^*(n)$ to that for $O^*(n')$. Because $O^*(n')$ formed at the dissociation threshold cannot autoionize, whereas $O_2^*(n)$ is well above the IE of O_2 and can decay by autoionization as well as neutral predissociation, the lifetime of $O^*(n')$ is likely longer than that for $O_2^*(n)$. Therefore, we expect enhancements of PFI-PE intensities for rotational levels $O_2^+(b^4\Sigma_g^-, v^+ = 4, N^+ \geq 9)$. The PFI-PE bands for $O_2^+(b^4\Sigma_g^-, v^+ = 0, 3)$ are compared to those for $O_2^+(b^4\Sigma_g^-, v^+ = 4)$ in Figure 8a–c (21). The O, Q, and S branches labeled in these figures correspond to $\Delta N = N^+ - N'' = -2, 0$, and $+2$, respectively, where N^+ and N'' are the respective rotational quantum numbers for O_2^+ and O_2. The N'' values are marked in Figures 8a–c. Without any perturbations, we expect that the rotational intensity distributions of the PFI-PE vibrational bands for $O_2^+(b^4\Sigma_g^-, v^+ = 0-3)$ are

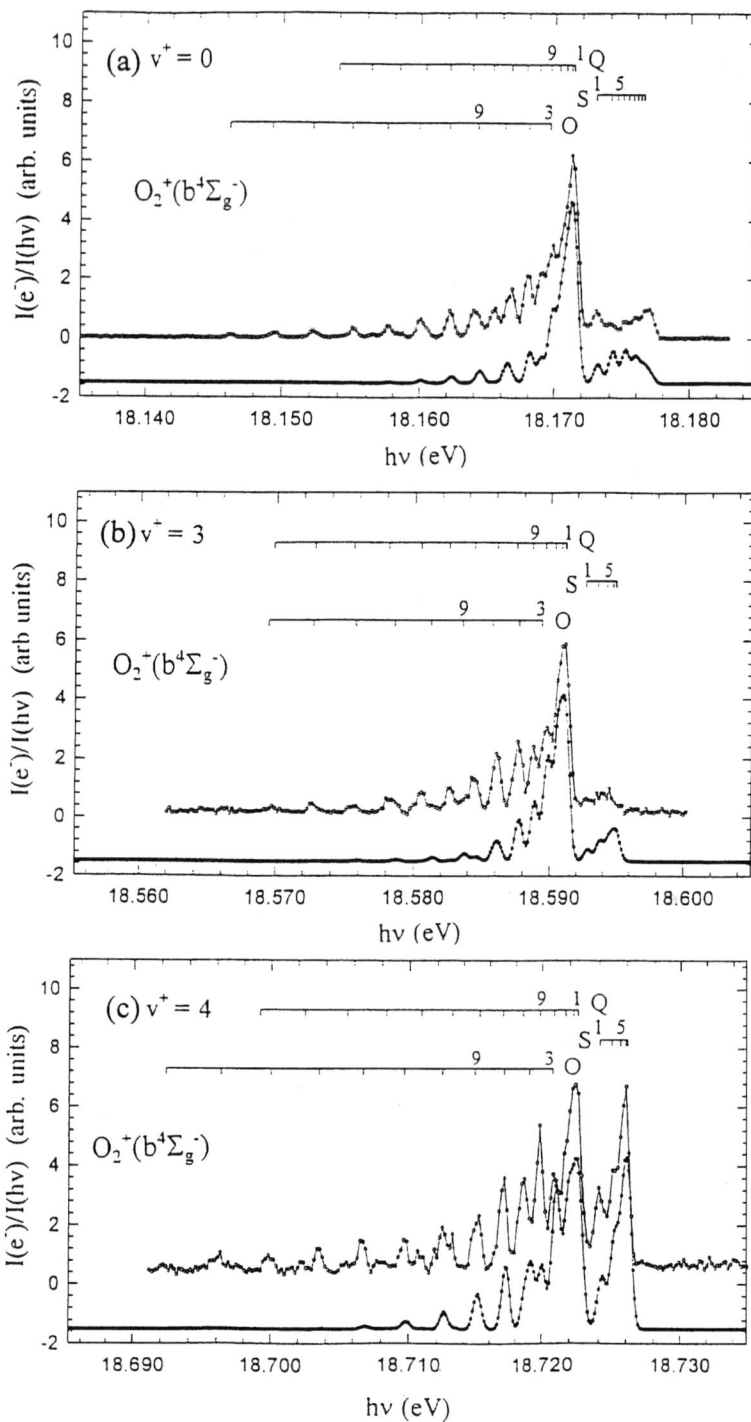

similar, as confirmed by the experimental PFI-PE measurement (21). The PFI-PE bands for $O_2^+(b^4\Sigma_g^-, v^+ = 1$ and 2) are shown here. Interestingly, the rotational intensities for ionization transitions to form $O_2^+(b^4\Sigma_g^-, v^+ = 4, N^+ \geq 9)$ are significantly enhanced (see Figure 8c) compared with those of the $v^+ = 0$–3 states. This observation is consistent with the expectation that the lifetime for $O^*(n')$ is longer than that for $O_2^*(n)$.

Pulsed Field Ionization–Ion Pair Spectroscopy

Hepburn and coworkers (14–17) demonstrated the formation of ion pairs in PFI experiments. Pratt and coworkers first reported the ion pair formation of H_2 induced by a dc Stark field (124). The discrimination of background prompt ions is achieved by a separation dc electric field maintained at the PI/PEX region. The electric field pulse for ion pair formation is applied after an appropriate delay with respect to the excitation laser pulse.

The interaction potential for an ion pair state is analogous to that for a high-n Rydberg state and is Coulombic in nature. Thus, the ion pair potential is expected to sustain an infinite number of bound vibrational levels. In addition to spectroscopy information for the ions produced, the measured ion pair threshold can be used for determining the D_0 value for the neutral parent molecule provided that the electron affinity and IE values associated with the ion pair produced are known. The PFI-IP technique has been successfully applied to the study of H_2, D_2, O_2, HCl, HF, and H_2S (14–17). Because the long-range Coulomb potential for the ion pair is stronger than the effective centrifugal contribution, no centrifugal barrier to dissociation is expected. Thus, the ion pair threshold determined in PFI-IP experiments should provide highly accurate D_0 values for the molecules involved.

PULSED FIELD IONIZATION–ION PAIR SPECTRUM FOR O_2 Figure 9 shows the first PFI-IP spectrum for O_2 obtained near its ion pair threshold (14). Each of the peaks in the energy range of 139,160–139,320 cm^{-1} corresponds to a range of unresolved ion pair vibrational levels converging on the first ion pair dissociation asymptote, and the individual lines arises from the different initial thermally populated rotational levels of $O_2(X^2\Sigma_g^-, v'' = 0, N = 1, 3, 5$, etc.). Based on energetic consideration, O^+ ions are produced in the ground $O^+(^4S_{3/2})$ state. However, O^- ions have two spin-orbit states, $^2P_{3/2}$ (ground) and $^2P_{1/2}$ (excited), separated by 177 cm^{-1}. The opening of the second ion pair dissociation asymptote $O^+(^4S_{3/2}) + O^-(^2P_{1/2})$

Figure 8   Comparison of simulated (●) experimental (○) PFI-PE bands for O_2^+ $(b^4\Sigma_g^-)$ (a) $v^+ = 0$, (b) $v^+ = 3$, and (c) $v^+ = 4$ (21). The markings of the $\Delta N = N^+ - N'' = -2, 0$, and $+2$ (or O, Q, and S, respectively) rotational branches are shown, where N^+ and N'' are rotational quantum numbers for O_2^+ and O_2, respectively. The numbers given in these figures are N'' values. Note the enhancement of rotational transitions to form $O_2^+(b^4\Sigma_g^-, v^+ = 4, N^+ \geq 9)$.

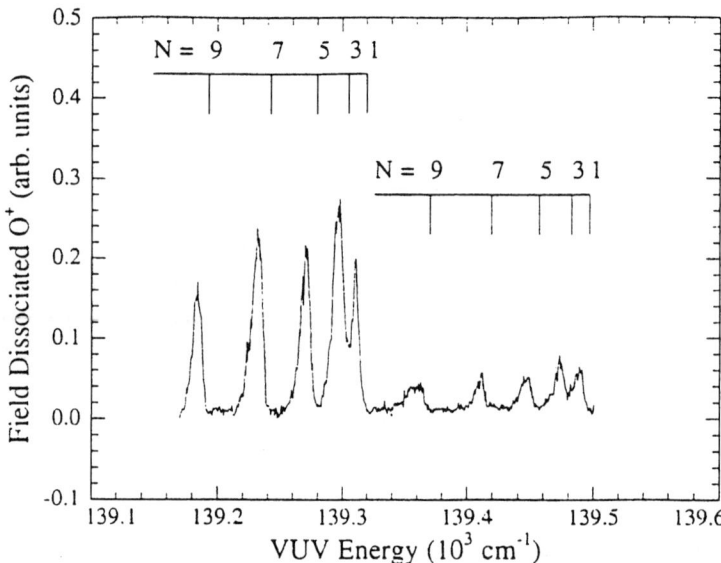

Figure 9 Pulsed field ionization–ion pair spectrum for $O^+(^4S_{3/2}) + O^-(^2P_{3/2,1/2})$ from $O_2(X^3\Sigma_g^-, v'' = 0, N)$ (14). The field-free ion pair formation thresholds corresponding to different initial rotational N states and the two final $O^-(^2P_{3/2,1/2})$ states are marked.

produces the set of peaks at the energies from 139,320 to 139,500 cm^{-1}. By extrapolating the ion pair onset to F = 0, the field-free ion pair dissociation threshold for the N = 1, J = 2 level of O_2 is determined to be 139,319.1 ± 0.7 cm^{-1}, compared with the value of 139,318.9 ± 1.1 cm^{-1} calculated using known spectroscopic constants (14).

PULSED FIELD IONIZATION–ION PAIR SPECTRUM FOR H$_2$S Shiell et al. have recorded the PFI-IP spectrum (lower spectrum of Figure 10) for the formation of $H^+ + SH^-(X^1\Sigma^+, v' = 0, J')$ from $H_2S(\tilde{X}^1A_1, J''_{Ka,Kc})$ (16). Because the rotational constants of both $SH^-(X^1\Sigma^+)$ and $H_2S(\tilde{X}^1A_1)$ are known, it is possible to simulate the PFI-IP spectrum for H$_2$S. The simulated spectrum (upper spectrum of Figure 9b) reproduces the structure of the PFI-IP spectrum well, yielding a value of 122,458 ± 3 cm^{-1} for the field-free thermodynamics threshold for the formation of $H^+ + SH^-(X^1\Sigma^+, v' = 0, J' = 0)$ from $H_2S(\tilde{X}\ ^1A_1, J''_{Ka,Kc} = 0_{00})$. After taking into account the known IE(H) and EA(SH), a highly precise value for $D_0(H-SH) = 31,451 \pm 4$ cm^{-1} is deduced.

Generally, the FCFs for the formation of an ion pair state from a neutral ground state are poor. This may limit the number of ion pair systems that can be examined by this method. When the ion pair threshold is accurately known, the PFI-IP spectrum should yield spectroscopic information about the cation and/or anion produced. However, if neither the ion pair threshold nor the spectroscopic constants

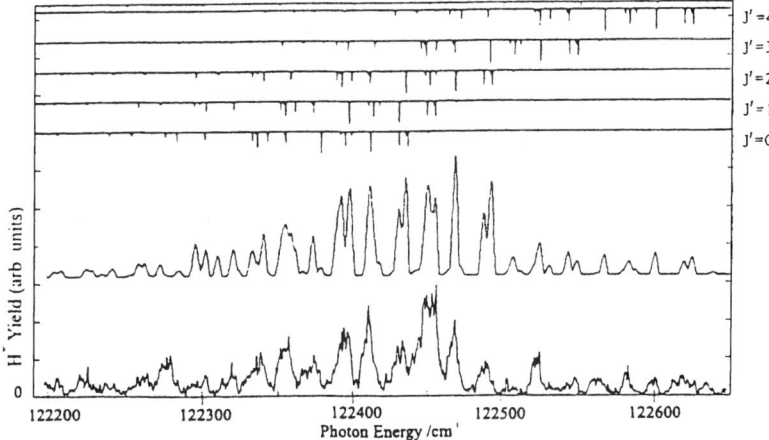

Figure 10 Experimental (*lower trace*) and simulated (*upper trace*) pulsed field ionization–ion pair spectrum for $H^+(^2S_{1/2}) + HS^-(X^1\Sigma^+, v' = 0, J')$ from $H_2S(\tilde{X}\ ^1A_1, J''_{Ka,Kc})$ in the region of 122,200–122,650 cm^{-1} (16).

for the fragment ions are known, it may be difficult to analyze the PFI-IP spectrum for a polyatomic system. In principle, a PFI-IP spectrum can be measured in a coincidence study, in which correlated anion-cation fragments induced by PFI are measured.

STATE- OR ENERGY-SELECTED STUDIES OF ION DYNAMICS

The internal state or energy of a photoion can be prepared by the PEPICO method. The resolution for state- or energy-selection of photoions depends critically on the accompanying photoelectron method. If the ion and electron are formed from the same precursor, the difference between the electron and ion TOFs to their respective detectors should be fixed. Because the TOF for an electron is significantly shorter than that for an ion, the electron signal is usually used to trigger a multichannel scaler for photoion detection. In such an arrangement, correlated ions can be found to arrive at the ion detector at a fixed time, whereas uncorrelated ions would arrive randomly in time, giving rise to a constant background.

One of the most valuable thermochemical data that can be obtained in VUV photoionization-photoelectron measurements is the 0°K AE of a dissociative photoionization process (35, 36). In principle the most reliable method for AE determinations is based on the analysis of the breakdown curves measured in PEPICO experiments (35). For a 0°K molecular sample, the fractional abundance of the parent (daughter) ion determined in a PEPICO experiment with infinitely narrow energy resolution should exhibit a step function behavior, i.e., switch from unity (zero) to zero (unity), at the $0°K$ AE. A major difficulty in AE determinations is due to the hot band effect, which causes the parent ion at a finite temperature to dissociate well below the $0°K$ AE. Thus, the fractional abundance of the parent (daughter) ion decreases (increases) gradually. However, regardless of the ion temperature, the true $0°K$ AE is always marked by the disappearance energy of the parent ion found in a PEPICO study. The disappearance energy of the parent ion is the energy at which even the coldest part of the energy distribution of the parent neutral exceeds the dissociation threshold.

Threshold Photoelectron-Photoion Coincidence Studies

The TPEPICO scheme is the combination of the PIE and TPE spectroscopic methods for the detection of correlated mass-selected ions and TPEs. The TPEPICO scheme requires the continuous production of electron-ion pairs and is traditionally the preferred coincidence method when a tunable, continuous, or pseudo-continuous ionizing VUV light source is used (7, 8, 41). Because TPEs are selected in a TPEPICO study, the internal energy for the molecular ion produced is thus equal to the difference between the photon energy and the IE for the molecule of interest. The TPEPICO technique has been successfully employed for studies of state- or energy-selected unimolecular dissociation dynamics and for measuring the TPE spectrum for a radical or a size-selected cluster produced in an impure radical or cluster source (41, 67). However, due to the hot-tail problem associated with TPE detection, most AE values determined in previous TPEPICO studies have uncertainties comparable to those of PIE measurements.

The successful application of the penetration field method for TPEPICO measurements has made possible the measurement of high-resolution TPE spectra for heterogeneous rare gas dimers, achieving resolutions of 2–3 meV (90). The penetration field method does not allow the use of an ion extraction field. In order to attain high ion collection efficiency, the ion flight tube used in the penetrating field TPEPICO experiment is very short, which in effect sacrifices the TOF mass resolution. By nature of the design of the penetration field technique, its use in TPEPICO measurements is not expected to yield accurate kinetic energy release information for fragment ions.

Synchrotron-Based Pulsed Field Ionization–Photoelectron-Photoion Coincidence Measurements

Javis et al. have recently developed a synchrotron-based PFI-PEPICO method (23), which involves the coincidence detection of PFI-PE and PFI-PI. The highest PFI-PE resolution achieved so far involved the use a shaped pulse for PFI and ion extraction. The shaped pulse consists of a low field for PFI immediately followed by a higher pulse for ion extraction. By collecting only PFI-PEs from the low field pulse, the resolution is expected to be high. An important experimental consideration is that the PFI-PEs formed by PFI owing to the low field pulse must exit the PI/PEX region prior to the employment of the high field pulse. As the PFI-PEs enter the electron TOF spectrometer, they are shielded from the high field pulse for ion extraction and thus the TOFs of PFI-PEs produced by the low field pulse are not disturbed. Figure 11 compares the PFI-PE and PFI-PEPICO bands for $Ar^+(^2P_{3/2,1/2})$ that were obtained using the shaped pulse scheme and achieved a resolution of 0.5–0.6 meV (FWHM) (23). The shaped pulse used is shown in the inset of Figure 11. The PFI-PEPICO intensity is ~25% of the PFI-PE intensity, indicating that good collection efficiencies were achieved for both PFI-PEs and PFI-PIs. The shaped pulse basically solves the dilemma of achieving a high photoelectron resolution (which requires a low electric field pulse) and a high ion transmission (which requires a high electric field pulse).

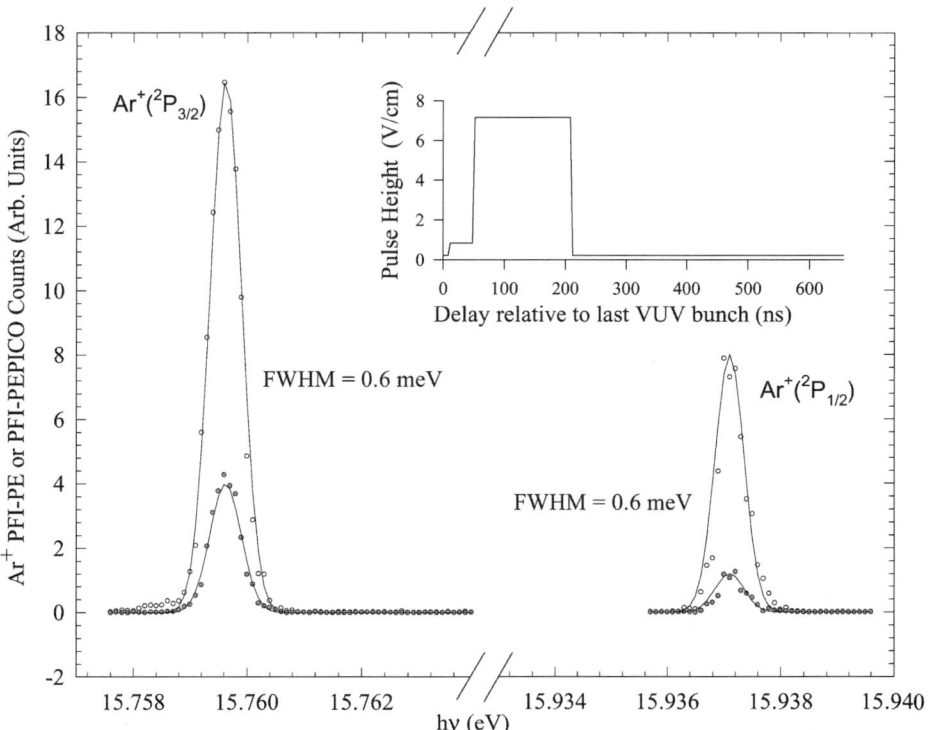

Figure 11 Comparison of the pulsed field ionization–photoelectron (○) and pulsed field ionization–photoelectron-photoion coincidence (PFI-PEPICO) (·) bands for $Ar^+(^2P_{3/2,1/2})$ obtained using the shaped pulse coincidence scheme (23). The shaped pulse is shown in the inset. It consists of a $0.5\ V \cdot cm^{-1}$ low field pulse (duration = 40 ns) followed by a $7\ V \cdot cm^{-1}$ high field pulse (duration = 150 ns). PFI-PEPICO resolution = 0.6 meV (FWHM).

Most PFI-PEPICO measurements (26–34) reported recently used a simpler scheme, in which a dc field of ≤ 0.20 V·cm^{-1} was applied across the PI/PEX region as in the PFI-PE detection. The application of a PFI electric field pulse (height = 7.0 V·cm^{-1}, width = 200 ns) was delayed by $\approx$10 ns with respect to the beginning of a 112-ns dark gap. This same electric field pulse also serves to extract the PFI-PIs toward the ion detector. The use of a relatively high pulse electric field makes this PFI-PEPICO scheme more sensitive. However, it also limits the achievable PFI-PEPICO resolution to $\approx$1.0 meV (FWHM). Nevertheless, this resolution is sufficient for the preparation of diatomic ions in specific rotational states. The unimolecular dissociation of rotationally selected $O_2^+(b^4\Sigma_g^-$, $v^+ = 4$, N^+) has been examined using this PFI-PEPICO scheme (23).

To illustrate the merits of this newly developed PFI-PEPICO technique for state-selected ion dissociation studies and accurate energetic measurements of simple polyatomic species, we summarize below the PFI-PEPICO studies on CH_4 (26), CD_4 (33), and CH_3X (X = Br and I) (32).

ENERGY-SELECTED DISSOCIATION STUDY OF CH_4^+ For dissociation studies of polyatomic ions, the resolution of the PFI-PEPICO method is insufficient for the selection of specific rotational levels. In the PFI-PEPICO study of CH_4 (26) we have established that highly accurate 0°K AE values for some dissociative photoionization reactions can be determined by the disappearance energy of the parent ion. Figure 12 depicts selected PFI-PEPICO TOF spectra measured near the 0°K AE(CH_3^+) from CH_4 (26). At 13.9225 eV only parent CH_4^+ ions are observed, whereas only daughter CH_3^+ ions are found at 14.3240 eV. At 13.9225 eV the CH_4^+ TOF peak is composed of a narrow and a broad component owing to photoionization of cold ($\approx$30°K) CH_4 in the supersonic beam and thermal (298°K) background CH_4 in the photoionization chamber, respectively. As the photon energy increased to 14.3044 eV, a broad TOF peak for CH_3^+ was observed, concomitant with the disappearance of the broad thermal component for the CH_4^+ TOF peak. The TOF peak structures resolved in Figure 12 unambiguously show that CH_3^+ ions formed at 14.3044 eV are mostly produced by dissociation of rotationally excited CH_4^+ formed in the photoionization of thermal CH_4. The narrow components of the TOF peak for CH_3^+ resulting from the dissociation of cold CH_4^+ are observed with increasing intensity as the photon energy is increased from 14.3162 to 14.3240 eV.

Because the PFI-PEPICO TOF spectra resolve the dissociation owing to cold CH_4 from that of thermal CH_4, we have constructed in Figure 13b the breakdown data for CH_3^+ and CH_4^+ based only on the cold ion signals (26). As expected, these data form steep breakdown curves, which represent the dissociation of cold CH_4^+ formed by photoionization of supersonically cooled CH_4 (33°K).

Figure 12 Pulsed field ionization–photoelectron-photoion coincidence TOF spectra of CH_4 at hν = 13.9225, 14.3044, 14.3162, 14.3201, and 14.3240 eV (26). The TOF peaks centered at 14.10 and 14.35 μs are due to CH_3^+ and CH_4^+, respectively.

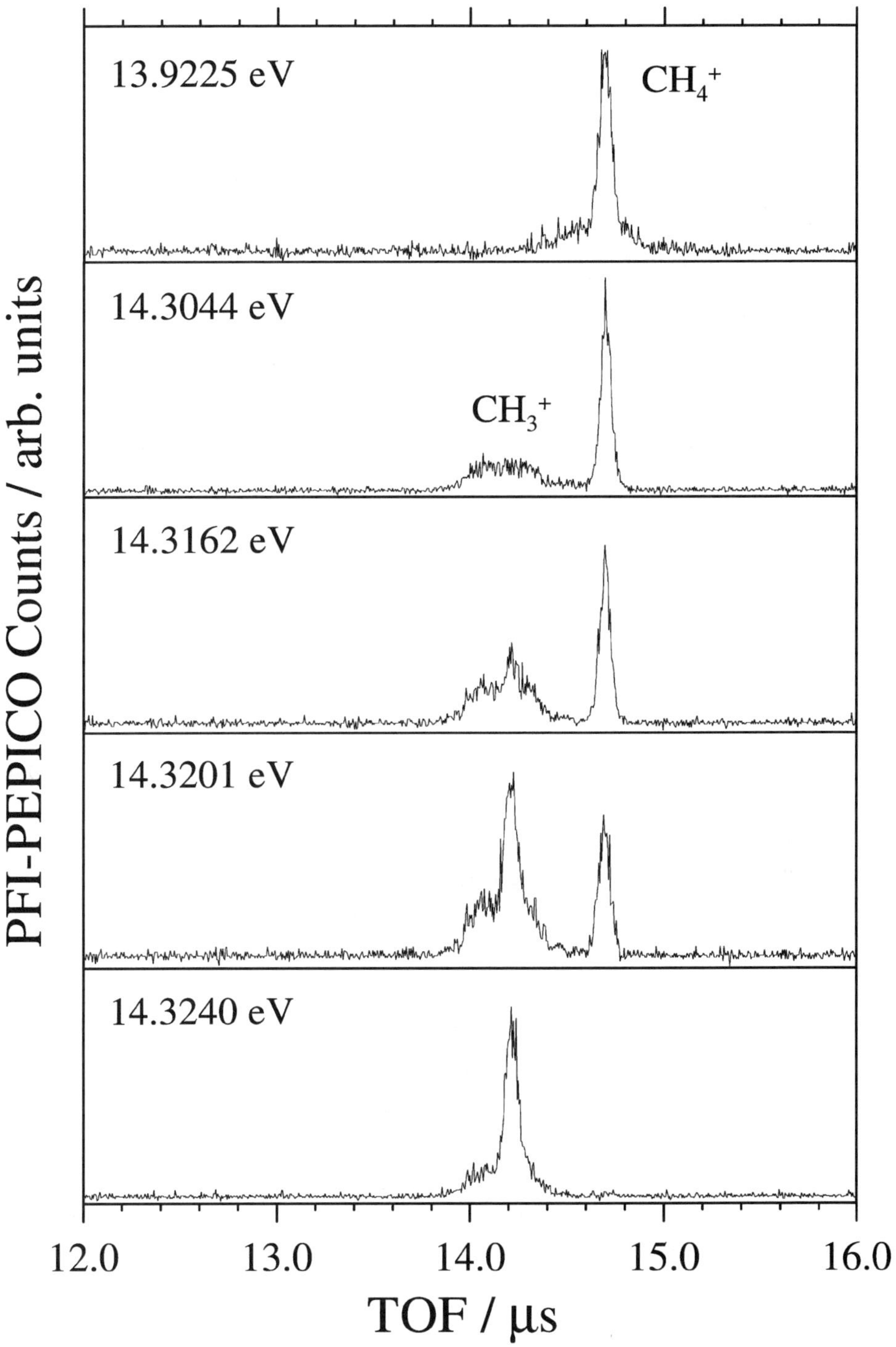
13.9225 eV
CH_4^+
14.3044 eV
CH_3^+
14.3162 eV
14.3201 eV
14.3240 eV
PFI-PEPICO Counts / arb. units
12.0
13.0
14.0
15.0
16.0
TOF / μs

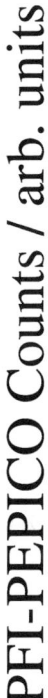

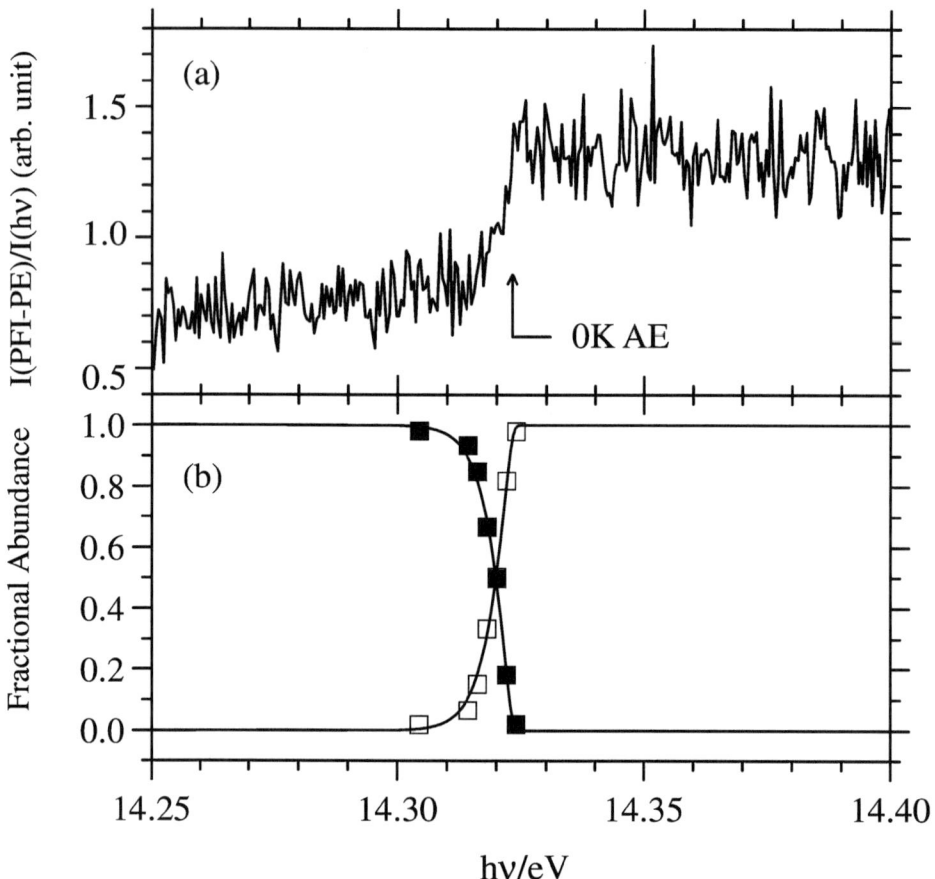

Figure 13 Comparison of the pulsed field ionization–photoelectron (PFI-PE) spectrum and breakdown diagram for CH_4 in the energy range of 14.25–14.44 eV (127). (*a*) PFI-PE spectrum for CH_4. The arrow marks the 0°K AE for CH_3^+. (*b*) The breakdown curves for CH_3^+ (*open squares*) and CH_4^+ (*filled squares*).

The 0°K AE(CH_3^+) determined by the disappearance energy of the parent CH_4^+ ion is 14.323 ± 0.001 eV. Using this value, along with the IE(CH_3) = 9.8357 ± 0.00037 eV (125) and IE(CH_4) = 12.618 ± 0.004 eV (100) determined from rotationally resolved PFI-PE measurements, we have obtained $D_0(H-CH_3) = 4.487 \pm 0.001$ eV and $D_0(H-CH_3^+) = 1.705 \pm 0.004$ eV (26). We note that in TPEPICO studies, where the TPE measurements were affected by the hot-tail problem, the fractional abundance for the parent ion is not zero at the 0°K AE (126).

DISSOCIATION MECHANISM FOR CH_4 IN HIGH-N RYDBERG STATES   The PFI-PE spectrum for CH_4 (127) measured in the region of 14.25–14.40 eV is depicted in Figure 13a, revealing a distinct step-like feature at 14.323 eV, which is identical to the 0°K AE(CH_3^+) from CH4 determined in the PFI-PEPICO study. The excellent correlation observed between the step and the 0°K AE(CH_3^+) indicates that a dissociative-PFI mechanism (Equation 2) similar to Equation 1 is operative (103, 104) for the formation of CH_3^+ from excited CH_4 in high-n (n > 100) Rydberg states (CH_4^*). The dissociative-PFI mechanism for CH_4^* is also consistent with the PFI-PEPICO data. Because the density of states for CH_4^+ is much higher than that for a diatomic ion (21), a step is observed at the 0°K AE(CH_3^+), instead of the enhancement of discrete transitions as observed in the dissociative-PFI of O_2^*(n) (21).

$$CH_4 + h\nu \xrightarrow{(a)} CH_4^* \xrightarrow{(b)} CH_3^* + H$$

$$\downarrow (c) \qquad \downarrow (d)$$

$$CH_4^{+*} + e^- \qquad CH_3^+ + e^- \qquad\qquad 2.$$

Here, CH_3^* represents excited CH_3 in long-lived high-n Rydberg states and CH_4^{+*} stands for excited CH_4^+. Equations 2c and 2d are PFI processes. At energies below the AE(CH_3^+), the PFI-PE signal originates from Equation 2c and is proportional to the concentration of CH_4^* species that have survived the decay for a time longer than the delay (Δt) of the PFI pulse relative to the excitation VUV pulse. The CH_3^* species formed by Equation 2b at the AE converge to the ground state of CH_3^+ and have energies below the IE(CH_3). Consequently autoionization is not accessible to these CH_3^* fragments. Thus, the PFI-PE signal derived from Equation 2d at the AE is expected to be higher than that obtained from Equation 2c below the AE. The step marking the 0°K AE in the PFI-PE spectrum can be attributed to the "lifetime switching" effect at the AE, where CH_4^* species with shorter lifetimes are converted into CH_3^* fragments with longer lifetimes. The observation of the sharp step-like feature in the PFI-PE spectrum is consistent with the conclusion that the conversion from CH_4^* to CH_3^* is complete prior to Equation 4d and that the dissociation process has a rate constant $\gg 1/\Delta t$ ($\approx 10^7$ s^{-1}).

Similar step-like features have been observed in the PFI-PE spectra of CD_4 (33), C_2H_2 (127), NH_3 (31), H_2O, and CH_3Br (32) at the respective 0°K AEs for the formation of CD_3^+, C_2H^+, NH_2^+, OH^+, and CH_3^+. The observation of the step in PFI-PE measurements, together with the breakdown curves obtained in PFI-PEPICO studies, can provide unambiguous 0°K AEs for the dissociation reactions involved, which in turn can yield highly accurate energetic information for simple neutrals and ions. The step marking the 0°K AE observed in synchrotron-based PFI-PE measurements might not be observable in VUV laser PFI-PE studies, in which excited CH_4^* species formed by pulsed VUV laser excitation are fully stabilized by l- and m_l-mixings and thus have significantly longer lifetimes.

For a slower dissociation reaction in which the dissociation lifetimes for excited parents in high-n Rydberg states are longer than Δt, the PFI-PE signal should originate entirely from PFI of excited parent species at photon energies both below and above the AE. Hence, the PFI-PE spectrum should be smooth across the AE. However, if excited parent ions thus formed completely dissociate within the time scale of the PFI-PEPICO experiment ($\approx 10^{-5}$ s), we still expect to observe complete dissociation at the 0°K AE, i.e., the disappearance energy of the parent ion can still be used to identify the 0°K AE. This can be considered as an intermediate case, in which the 0°K AE can be determined in a PFI-PEPICO experiment but not in a PFI-PE study.

The reaction $C_2H^4 + h\nu \rightarrow C_2H_2^+ + H_2 + e^-$ is known to have dissociation rates of 10^3–10^5 s^{-1} near its 0°K AE. As expected, we found that the PFI-PE spectrum for C_2H_4 is smooth across the 0°K AE for $C_2H_2^+$ (127). This observation can be taken as strong support for the dissociative-PFI mechanism described above.

ENERGY-SELECTED DISSOCIATION STUDY OF CD_4^+ We have also conducted a PFI-PE and PFI-PEPICO study of CD_4. Figure 14 shows a magnified view of the simulated and experimental breakdown data for CD_4^+ (33), together with their error bars, in the region of 14.410–14.447 eV. This figure consists of two breakdown curves for CD_4^+, one based on the total (thermal and cold beam; open circles in Figure 13) ion signals and the other (solid dots in Figure 13) constructed using only the cold beam ion signal. Owing to a finite coincidence background resulting from prompt electrons dispersed into the dark gap, the disappearance energy becomes a sharp break (i.e., the lowest energy at which the breakdown curve for the parent ion reaches its lowest value) of the breakdown curve for the parent ion as marked in Figure 15. It is clear that the break marking the 0°K AE(CD_3^+) at 14.4184 eV can be determined unambiguously to within ±0.0010 eV.

The PFI-PE spectrum for CD_4 obtained using the cold beam sample also reveals a sharp step at the 0°K AE, indicating the dissociation of CD_4^* occurs in a time scale $\leq 10^{-7}$ s. Furthermore, this step resolved in the PFI-PE spectrum for CD_4 renders a confirmation for the 0°K AE(CD_3^+) = 14.4184 ± 0.0010 eV determined in the PFI-PEPICO study (33). This value, along with the known IE(CD_4) = 12.6708 ± 0.0002 eV (100) and IE(CD_3) = 9.8303 ± 0.0006 eV (128) determined in recent PFI studies, has allowed us to calculate the 0°K D_0(D–CD_3^+) = 1.748 ± 0.001 eV and D_0(D–CD_3) = 4.5881 ± 0.0012 eV (33). These experimental IE, D_0, and AE values for CD_3/CD_3^+ and CD_4/CD_4^+ are compared with those for CH_3/CH_3^+ and CH_4/CH_4^+ (26) in Table 2.

The energetic data for CH_3/CH_3^+ and CH_4/CH_4^+ can be converted into those for CD_3/CD_3^+ and CD_4/CD_4^+ by using the zero point vibrational energies (ZPVEs) of these species. Because not all the vibrational frequencies for CH_4^+, CD_4^+, CH_3^+, and CD_3^+ are known, accurate experimental ZPVEs for these cations cannot be obtained. We have calculated the ZPVEs at the MP2(Full)/6-311++G(3d2f,2pd) level of theory (33). Comparing the experimental ZPVEs and theoretical ZPVEs for the neutral species, we obtain an average scaling factor of 0.957 for the theoretical ZPVEs. Using the scaled theoretical ZPVEs (given in Table 2), we have

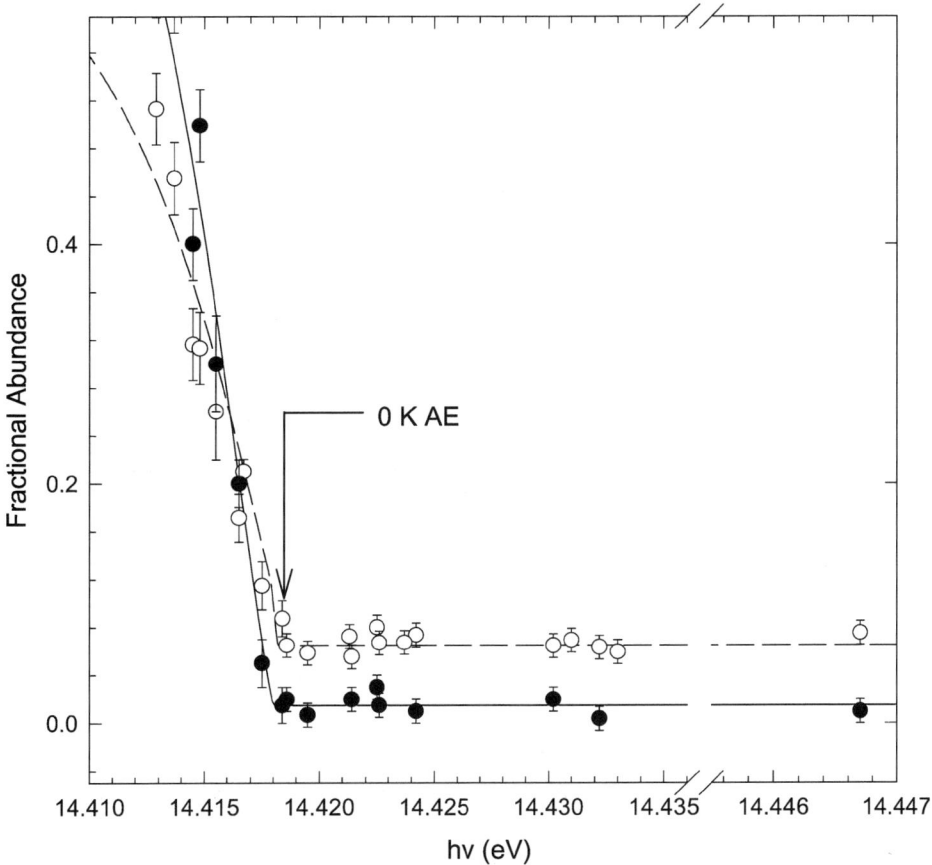

Figure 14 A magnified view of the breakdown curve for CD_4^+ based on the entire ion signals (○) and the cold ion signals (●) in the $h\nu$ range of 14.410–14.447 eV (33). Both curves show a sharp break at 14.4184 ± 0.0010 eV, which is taken to be the 0°K $AE(CD_3^+)$.

calculated values (given in parentheses in Table 2) for the $IE(CD_3)$, $IE(CD_4)$, $D_0(D-CD_3)$, $D_0(D-CD_3^+)$, and $AE(CD_3^+)$ based on the corresponding experimental $IE(CH_3)$, $IE(CH_4)$, $D_0(H-CH_3)$, $D_0(H-CH_3^+)$, and $AE(CH_3^+)$. Because the error bars assigned to these calculated values have ignored the uncertainties associated with the theoretical ZPVEs, they represent lower limits. As shown in Table 2, the calculated (values in parentheses) and experimental values are in good agreement, indicating that the available experimental IE, D_0, and AE values for CH_4/CH_4^+, CH_3/CH_3^+, CD_4/CD_4^+, and CD_3/CD_3^+ are highly reliable and have well-founded error limits.

In similar PFI-PE and PFI-PEPICO studies, we have compared the energetic data between the OH/OH^+ and H_2O/H_2O^+ and OD/OD^+ and D_2O/D_2O^+ systems

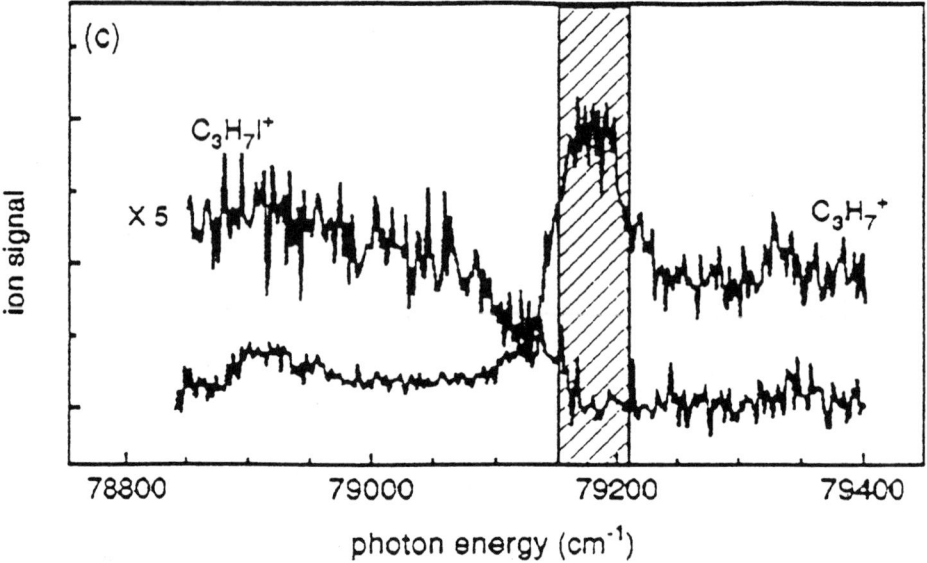

Figure 15 Single-photon pulsed field ionization–photoion spectra for $C_3H_7^+$ and $C_3H_7I^+$ from 2-iodopropane (130). The 0°K AE($C_3H_7^+$) is marked by the disappearance energy of the parent $C_3H_7I^+$ ion. The shaded region indicates the error limit for the 0°K AE($C_3H_7^+$).

(34) and those between the NH_2/NH_2^+ and NH_3/NH_3^+ and ND_2/ND_2^+ and ND_3/ND_3^+ systems. On the basis of these comparisons, we conclude that the energetic data for OX/OX^+ and X_2O/X_2O^+, NX_2/NX_2^+ and NX_3/NX_3^+, and CX_3/CX_3^+ and CX_4/CX_4^+ systems are reliable with error limits within ±0.004 eV.

ENERGY-SELECTED DISSOCIATION STUDY OF CH_3X^+ (X = Br AND I) The uncertainties of ΔH_{f0}° values for some radicals and ion fragments derived from recent PFI-PEPICO studies, such as those for CH_3 and CH_3^+ (26), are now limited by the error limit for the $\Delta H_{f0}^\circ(CH_4)$ (36), which is recognized to be among the most precisely measured values. Using these accurate ΔH_{f0}° values for small radicals and ions, together with appropriate AE determinations using the PFI-PEPICO method, it is possible to determine accurate ΔH_{f0}° values for larger neutral molecules and their ions. As an example (32), we have obtained the 0°K AEs for the formation of CH_3^+ from CH_3Br (12.834 ± 0.002 eV) and CH_3I (12.269 ± 0.003 eV) using the PFI-PE and PFI-PEPICO methods. Combining these AE(CH_3^+) values and the known $\Delta H_{f0}^\circ(Br)$ (28.186 ± 0.031 kcal/mol), $\Delta H_{f0}^\circ(I)$ (25.612 ± 0.010 kcal/mol), and $\Delta H_{f0}^\circ(CH_3^+)$ (262.79 ± 0.08 kcal/mol), we have obtained more precise values for $\Delta H_{f0}^\circ(CH_3Br) = -5.09 \pm 0.10$ kcal/mol and $\Delta H_{f0}^\circ(CH_3I) = 5.36 \pm 0.12$ kcal/mol (32). The error limits of these values are significantly smaller than the maximum discrepancies (0.4–0.8 kcal/mol) among previously reported ΔH_{f298}° values for CH_3X (X = Br and I).

TABLE 2 Zero point vibrational energies (ZPVEs), IE, D_0, and AE values for the CH_4/CH_4^+, CH_3/CH_3^+, CD_4/CD_4^+, and CD_3/CD_3^+ systems (33)[a]

	ZPVE (eV)[b] (33)			IE (eV)		D_0 (eV)		AE (eV)
X	CX_4	CX_3	CX_3^+	CX_4	CX_3	CX_3^+-X	CX_3-X	CX_3^+
H	1.183	0.790	0.832	12.618 ± 0.004 (100)	9.8380 ± 0.0004 (125)	1.705 ± 0.004 (26)	4.485 ± 0.001 (26)	14.323 ± 0.001 (26)
D	0.868	0.580	0.615	12.6708 ± 0.0002 (100) (12.664 ± 0.004)	9.8303 ± 0.0006 (128) (9.8291 ± 0.0004)	1.748 ± 0.001 (33) (1.754 ± 0.004)	4.5881 ± 0.0012 (33) (4.591 ± 0.001)	14.4184 ± 0.0010 (33) (14.421 ± 0.001)

[a] The values in parentheses are converted from corresponding IE or D_0 or AE values of CH_n or CH_n^+ (n = 3, 4) using the scaled theoretical ZPVEs.
[b] Scaled (scaling factor = 0.957) ZPVEs calculated at the MP2(Full)/6-311++G(3d2f,2pd) level of theory.

Pulsed Field Ionization–Photoion Measurements

Instead of detecting the photoelectrons formed in the PFI of high-n Rydberg species, the same PFI process can be probed by detecting the PFI-PIs. Because both the internal energy and mass of a PFI-PI are known, the PFI-PI detection scheme is equivalent to PFI-PEPICO measurements, which employ a continuous or pseudo-continuous photoionization source. Zhu and Johnson reported the first PFI-PI or mass analyzed threshold ion measurement in 1991 (12, 13). For a detailed account, readers are referred to the recent review by Johnson (13).

The experimental arrangements for PFI-PI and the PFI-IP detections are similar and require the separation of background prompt ions from PFI ions by the TOF separation scheme. Because the dispersion of background prompt ions is more difficult than the dispersion of background prompt electrons in PFI-PE detection, PFI-PI and PFI-IP measurements are more challenging experiments than PFI-PE studies. Potentially, the PFI-PI method is useful for the measurements of photoelectron spectra of size-selected clusters and radicals, where the preparation of pure cluster and radical samples cannot be made (41).

PULSED FIELD IONIZATION–PHOTOION STUDIES OF IODOPROPANES AND IODOBUTANES Because the PFI-PI technique is equivalent to the PFI-PEPICO method, it can also be used for ion dissociation threshold measurements. Previous measurements have been concerned with the measurements of D_0s for van der Waals complexes and their ions (129). Recently Kim and coworkers have successfully applied the VUV laser PFI-PI method for the determination of $0°K$ $AE(C_3H_7^+)$ from 1-iodopropane (9.8332 ± 0.0017 eV) and 2-iodopropane (9.8180 ± 0.0017 0.0037 eV) (130). The $0°K$ $AE(C_4H_9^+)$ values from isomeric iodobutanes have also been reported with error limits of ± 4 meV (131). Figure 15 depicts the PFI-PI spectra for $C_3H_7^+$ and $C_3H_7I^+$ from 2-iodopropane near its $0°K$ AE. As shown in this figure, the $0°K$ $AE(C_3H_7^+)$ from 2-iodopropane can be determined by the disappearance energy of the parent $C_3H_7I^+$ ion. The shaded region indicates the uncertainty range for the $0°K$ $AE(C_3H_7^+)$.

SYNCHROTRON-BASED PULSED FIELD IONIZATION–PHOTOION MEASUREMENTS The difficulty in PFI-PI detection using synchrotron radiation is due to the small dark gap available in a synchrotron ring period. We have demonstrated that the synchrotron-based TOF selection method of PFI-PE detection (22) can also be applied to PFI-PI detection of H_2^+ ($v^+ = 0$, $N^+ = 0$–5) by using the supersonic beam method and the two-bunch synchrotron operation at the ALS, which has a dark gap of 328 ns (24). The separation of PFI-PIs from background prompt ions of higher masses using the TOF selection scheme would require a significantly larger dark gap than 328 ns. For a heavier ion such as Ar^+ (mass 40), a dark gap of ≈ 1 μs is needed (24). This can be accomplished by using a fast chopper wheel with an appropriate slot pattern to chop the VUV beam.

TWO-COLOR PHOTOINDUCED RYDBERG IONIZATION MEASUREMENTS

Long-lived molecules in high-n (n ≥ 100) Rydberg states lying a few cm^{-1} below the first IE can easily be prepared by single-photon VUV excitation (132). The lifetimes of ≈1–10 μs for these high-n Rydberg species are sufficiently long to allow their further excitation by a second laser. Upon absorption of the second laser photon by the ion core, an excited high-n Rydberg level with an excited ion core is formed. The autoionization lifetimes of such Rydberg states with excited ion cores are likely shorter than that of the high-n Rydberg level with an ion core in its ground state. By detecting the ions or electrons produced from autoionization of the excited-ion-core high-n Rydberg states as a function of the second excitation laser frequency, we essentially produce an absorption spectrum of the ion core. This two-color ionization scheme is referred to as photo-induced Rydberg ionization spectroscopy (PIRI) (13, 132–136). As long as the quantum defects for the lower and upper Rydberg states are similar, we expect that the PIRI spectrum is similar to that of the ion. In PFI experiments the energy resolution is limited by the F value of the Stark field. However, the PIRI method does not require a PFI field. The energy resolution for PIRI measurements should be governed by the optical resolution of the second laser, the quantum defect spread, and the natural absorption line-width. For this reason, the PIRI method is expected to provide better resolution than those of PFI-PE and PFI-PI measurements. The resolution capability for PIRI measurements has not been tested because the molecules investigated to date have natural absorption line-widths greater than the rotational spacings.

In principle there is no limit on the frequency of the second excitation laser so long as autoionization is induced by such an excitation. Johnson and coworkers have performed PIRI measurements by excitation of the ion core to rovibrational levels of an excited electronic state using a dye laser in the visible frequency range (13, 132–136). Recent PIRI measurements using infrared (IR) lasers have also been made, resulting in an IR spectrum for the ion core (137, 138).

CONCLUSIONS AND OUTLOOK

The establishment of an array of new experimental techniques, including PFI-PE, PFI-PI, PFI-PEPICO, PFI-IP, and PIRI schemes, and the availability of VUV lasers and high-resolution third-generation VUV synchrotron sources, has laid a sound foundation for the next level of developments and applications. A profitable direction for photoionization mass spectrometry is to extend its application to the study of biomolecules. Several studies (139) have demonstrated the advantages of using VUV lasers for the photoionization sampling of biomolecules. However, the full potential of VUV laser mass spectrometry for the study of biomolecules has not yet been realized. Learning about the chemical structures of polyatomic photofragments is most fundamental to photochemical research. By virtue of its simplicity, we expect that the photodissociation/photoionization scheme (75) would become more popular for probing the chemical structures of nascent photoproducts. More high-resolution PFI studies of radicals are needed (63, 64, 125). Without doubt, the PFI-PEPICO and PFI-PI schemes can be further perfected by improving the sensitivity and by relaxing the restriction on the dissociation lifetimes. For results to be useful for the development of the new generation of computational codes, it is necessary to perform accurate thermochemical measurements for a sufficiently large number of chemical systems using the PFI-PE, PFI-PEPICO, and PFI-PI methods. The further development of the PFI-PESICO (9) and PFI-PI (140) methods for the preparation of state-selected reactant ions for ion-molecule reaction dynamics studies are to be anticipated. Owing to the high resolution achieved in these PFI schemes, it should be possible to prepare reactant ions in specific rotational states using these methods. The PIRI technique that combines the use of VUV synchrotron or VUV laser radiation to prepare intermediate long-lived high-n Rydberg species, along with tunable (IR, visible, UV, or VUV) laser radiation to induce ionization, holds promise for further improvement in the energy resolution and the energy range in PFI studies. By using the synchrotron-based VUV excitation method to prepare intermediate high-n Rydberg species, we may use a very high-resolution continuum wave–tunable dye laser as the photoionization source. In addition to the higher resolution offered by a continuum-wave laser, its use would also avoid the fragmentation of parent ions resulting from multiphoton processes (13) in pulsed laser PIRI studies. It is natural to combine the high-resolution VUV PFI-PEPICO method with the fragment ion-imaging scheme (81, 141). We may look forward to high-resolution ion-imaging PEPICO studies in the near future with the selection of parent ion internal states to the rotational level.

The development of the two-color scheme involving IR excitation and VUV ionization is highly profitable. With current IR laser technologies, the resolution and pulse energy obtainable from a broadly tunable IR optical parametric oscillator can saturate the IR transition for exciting a range of molecular vibrational modes. Because the lifetime of an IR excited species is long (usually in the millisecond range), the IR-excitation and VUV-photoionization schemes should allow photoionization-photoelectron studies of molecular samples in well-defined rovibrational states.

ACKNOWLEDGMENTS

This work was supported by the Director, Office of Energy Research, Office of Basic Energy Sciences, Chemical Science Division of the U.S. Department of Energy under Contract No. W-7405-Eng-82 for the Ames Laboratory and Contract No. DE-AC03-76SF00098 for the Lawrence Berkeley National Laboratory. Support from the AFOSR Grant No. F49620-99-1-0234 and NSF ATM 001644 is also acknowledged.

Visit the Annual Reviews home page at www.annualreviews.org

LITERATURE CITED

1. Turner DW, Baker C, Baker AD, Brundle CR. 1970. *Molecular Photoelectron Spectroscopy*. London: Wiley
2. Kimura K, Katsumata S, Achibi Y, Yamazaki T, Iwata S. 1981. *Handbook of HeI Photoelectron Spectra of Fundamental Organic Molecules*. Tokyo: Halsted
3. Berkowitz J. 1979. *Photoabsorption, Photoionization, and Photoelectron Spectroscopy*. New York: Academic
4. Ng CY, ed. 1991. *Vacuum Ultraviolet Photoionization and Photodissociation of Molecules and Clusters*. Singapore: World Sci.
5. Baer T, Guyon PM. 1995. See Ref. 142, pp. 1–17
6. Schlag EW. 1996. *ZEKE Spectroscopy*. Cambridge: Cambridge Univ. Press
7. Baer T. 1979. In *Gas Phase Ion Chemistry*, ed. MT Bowers, 1:153–96. New York: Academic
8. Baer T, Booze J, Weitzel KM. 1991. In *Vacuum Ultraviolet Photoionization and Photodissociation of Molecules and Clusters*, ed. CY Ng, pp. 259–96. Singapore: World Sci.
9. Koyano I, Tanaka K. 1992. *Adv. Chem. Phys.* 82:263–307
10. Müller-Dethlefs K, Sander M, Schlag EW. 1984. *Chem. Phys. Lett.* 112:291–94
11. Reiser G, Habenicht W, Müller-Dethlefs K, Schlag EW. 1988. *Chem. Phys. Lett.* 152:119–23
12. Zhu L, Johnson P. 1991. *J. Chem. Phys.* 94:5769–71
13. Johnson P. 2000. See Ref. 143, pp. 296–346
14. Martin JDD, Hepburn JW. 1997. *Phys. Rev. Lett.* 79:3154–57
15. Martin JDD, Hepburn JW. 1998. *J. Chem. Phys.* 109:8139–42
16. Shiell RC, Hu XK, Hu QJ, Hepburn JW. 2000. *J. Phys. Chem.* 104:4339–42
17. Shiell RC, Hu XK, Hu QJ, Hepburn JW. 2000. *Faraday Discuss. Chem. Soc.* 115:331–44
18. Suits AG, Heimann P, Yang XM, Evans M, Hsu CW, et al. 1995. *Rev. Sci. Instrum.* 66:4841–44
19. Hsu CW, Heimann P, Evans M, Ng CY. 1997. *Rev. Sci. Instrum.* 68:1694–1702
20. Heimann P, Kioke M, Hsu CW, Evans M, Lu KT, et al. 1997. *Rev. Sci. Instrum.* 68:1945–51
21. Hsu CW, Heimann P, Evans M, Stimson S, Ng CY. 1998. *Chem. Phys.* 231:121–43
22. Jarvis GK, Song Y, Ng CY. 1999. *Rev. Sci. Instrum.* 70:2615–21
23. Jarvis GK, Weitzel KM, Malow M, Baer T, Song Y, Ng CY. 1999. *Rev. Sci. Instrum.* 70:3892–906
24. Jarvis GK, Shiell RC, Hepburn JW, Song Y, Ng CY. 2000. *Rev. Sci. Instrum.* 71:1325–31
25. Ng CY. 2000. See Ref. 143, pp. 394–538
26. Weitzel KM, Malow M, Jarvis GK, Baer T, Song Y, Ng CY. 1999. *J. Chem. Phys.* 111:8267–70
27. Jarvis GK, Weitzel KM, Malow M, Baer T, Song Y, Ng CY. 1999. *Phys. Chem. Chem. Phys.* 1:5259–62
28. Baer T, Song Y, Liu JB, Chen WW, Ng CY. 2000. *Faraday Discuss. Chem. Soc.* 115:137–46
29. Baer T, Song Y, Ng CY, Liu JB, Chen WW. 2000. *J. Phys. Chem.* 104:1959–64
30. Ng CY. 2000. *J. Electron Spectrosc. Relat. Phenom.* 112:31–46
31. Song Y, Qian XM, Lau KC, Ng CY, Liu JB, Chen WW. 2001. *J. Chem. Phys.* 115:2582–89
32. Song Y, Qian XM, Lau KC, Ng CY, Liu JB, Chen WW. 2001. *J. Chem. Phys.* 115:4095–104
33. Song Y, Qian XM, Lau KC, Ng CY. 2001. *Chem. Phys. Lett.* 347:51–58
34. Qian XM, Song Y, Lau KC, Ng CY, Liu J, et al. 2002. *Chem. Phys. Lett.* In press

35. Rosenstock HM, Draxl MK, Steiner BW, Herron JT. 1977. *J. Phys. Ref. Data* 6(Suppl. 1)
36. *The NIST Chemistry WebBook.* http://webbook.nist.gov/chemistry/
37. Curtiss LA, Raghavachari K, Truck GW, Pople JA. 1991. *J. Chem. Phys.* 94:7221–30
38. Curtiss LA, Raghavachari K, Redfern PC, Rassolov V, Pople JA. 1998. *J. Chem. Phys.* 109:7764–76
39. Samson JAR. 1967. *Techniques in Vacuum Ultraviolet Spectroscopy.* New York: Wiley
40. Dehmer PM, Chupka WA. 1976. *J. Chem. Phys.* 65:2243–73
41. Ng CY. 1991. See Ref. 4, pp. 169–257
42. Kung AH. 1983. *Opt. Lett.* 8:24
43. Marinero EE, Rettner CT, Zare RN, Kung AH. 1983. *Chem. Phys. Lett.* 95:486–91
44. Rettner CT, Marinero EE, Zare RN, Kung AH. 1984. *J. Phys. Chem.* 88:4459–65
45. Hilber R, Largo A, Wolff B, Wallenstein R. 1986. *Comments At. Mol. Phys.* 18:157
46. Tonkyn RG, White MG. 1989. *Rev. Sci. Instrum.* 60:1245–51
47. Kung AH, Lee YT. 1991. See Ref. 4, pp. 487–502
48. Hepburn JW. 1991. See Ref. 4, pp. 435–85
49. Hepburn JW. 1994. In *Laser Techniques in Chemistry,* ed. A Meyers, TR Rizzo, pp. 149–83. New York: Wiley
50. Vidal CR. 1987. In *Tunable Lasers,* ed. LF Mollenauer, JC White, p. 57. Berlin: Springer-Verlag
51. Wiedmann RT, White MG. 1995. See Ref. 142, pp. 79–112
52. Palm H, Merkt F. 1998. *Phys. Rev. Lett.* 81:1385–88
53. Hollenstein U, Palm H, Merkt F. 2000. *Rev. Sci. Instrum.* 71:4023–28
54. Hennen PC. 1997. *XUV-laser spectroscopy of H_2 and the mystery of the diffuse intersteller bands,* PhD thesis. Vrije Univ. The Netherlands. 142 pp.
55. Kittlemann O, Ringling J, Korn G, Nazarkin A, Hertel IV. 1996. *Opt. Lett.* 21:1159–61
56. Nazarkin A, Korn G, Kittlemann O, Ringling J, Hertel IV. 1997. *Phys. Rev. A* 56:671–84
57. Stimson S, Chen YJ, Evans M, Liao CL, Ng CY, et al. 1998. *Chem. Phys. Lett.* 289:507–15
58. Song Y, Evans M, Ng CY, Hsu CW, Jarvis GK. 1999. *J. Chem. Phys.* 111:1905–16
59. Jarvis GK, Evans M, Ng CY, Mitsuke K. 1999 *J. Chem. Phys.* 111:3058–64
60. Huang JC, Cheung YS, Evans M, Liao CX, Ng CY, et al. 1997. *J. Chem. Phys.* 106:864–77
61. Berkowitz J, Ruscic B. 1991. See Ref. 4, pp. 1–42
62. Cockett MCR, Dyke JM, Zamampour HP. 1991. See Ref. 4, pp. 43–99
63. Gilbert T, Pfab R, Fischer I, Chen P. 2000. *J. Chem. Phys.* 112:1306–15
64. Gilbert T, Fischer I, Chen P. 2000. *J. Chem. Phys.* 113:561–66
65. Schultz T, Clarke JS, Gilbert T, Deyerl HJ, Fischer I. 2000. *Faraday Discuss. Chem. Soc.* 115:17–32
66. Ng CY. 1997. *Adv. Photochem.* 22:1–116
67. Ng CY. 1996. In *The Structure, Energetics and Dynamics of Organic Ions,* ed. T Baer, CY Ng, I Powis, pp. 35–124. Chichester, UK: Wiley
68. Norwood K, Nourbakhsh S, He GZ, Ng CY. 1991. *Chem. Phys. Lett.* 184:147–51
69. Nourbakhsh S, Norwood K, He GZ, Ng CY. 1991. *J. Am. Chem. Soc.* 113:6311–12
70. Ma ZX, Liao CL, Yin HM, Ng CY, Chiu SW, et al. 1993. *Chem. Phys. Lett.* 213:250–56
71. Ma ZX, Liao CL, Ng CY, Ma NL, Li WK. 1993. *J. Chem. Phys.* 99:6470–73
72. Hsu CW, Baldwin DP, Liao CL, Ng CY. 1994. *J. Chem. Phys.* 100:8047–54
73. Ma ZX, Liao CL, Ng CY, Cheung YS, Li WK, Baer T. 1994. *J. Chem. Phys.* 100:4870–75

74. Hsu CW, Ng CY. 1994. *J. Chem. Phys.* 101:5596–603
75. Hsu CW, Liao CL, Ma ZX, Ng CY. 1995. *J. Phys. Chem.* 99:1760–67
76. Cheung YS, Hsu CW, Ng CY. 1998. *J. Electron Spectrosc. Relat. Phenom.* 97:115–20
77. Baker J, Dyke JM. 1993. *Chem. Phys. Lett.* 213:257–61
78. Baker J, Dyke JM. 1994. *J. Phys. Chem.* 98:757–64
79. Jensen E, Keller JS, Waschewsky GCG, Stevens JE, Graham RL, et al. 1993. *J. Chem. Phys.* 98:2882–90
80. Evans M, Ng CY, Hsu CW, Heimann P. 1997. *J. Chem. Phys.* 106:978–81
81. Peterka DS, Ahmed M, Ng CY, Suits AG. 1999. *Chem. Phys. Lett.* 312:108–14
82. Peatman WB, Borne TB, Schlag EW. 1969. *Chem. Phys. Lett.* 3:492–97
83. Baer T, Peatman WB, Schlag EW. 1969. *Chem. Phys. Lett.* 4:243–47
84. Weitzel KM, Mahnert J, Penno M. 1994. *Chem. Phys. Lett.* 224:371–80
85. Güthe F, Weitzel KM. 1997. *Ber. Bunsenges. Phys. Chem.* 104:484
86. Hall RI, McConkey A, Ellis K, Dawber G, Avaldi L, et al. 1992. *Meas. Sci. Technol.* 3:316–24
87. Morioka Y, Tanaka T, Yoshii H, Hayaishi T. 1998. *J. Chem. Phys.* 109:1324–28
88. Morioka Y, Lu Y, Matsui T, Tanaka T, Yoshii H, et al. 1996. *J. Chem. Phys.* 104:9357–61
89. Lu Y, Morioka Y, Matsui T, Tanaka T, Yoshii H, et al. 1995. *J. Chem. Phys.* 102:1553–60
90. Morioka Y. 2000. See Ref. 143, pp. 347–93
91. Chupka WA. 1993. *J. Chem. Phys.* 98:4520–30
92. Stebbings R, Dunning F. 1983. *Rydberg States of Atoms and Molecules*. Cambridge: Cambridge Univ. Press
93. Gallagher T. 1994. *Rydberg Atoms*. Cambridge: Cambridge Univ. Press
94. Pratt ST. 1993. *J. Chem. Phys.* 98:9241–50
95. Muhlpfordt A, Even U. 1995. *J. Chem. Phys.* 103:4427–30
96. Merkt F. 1994. *J. Chem. Phys.* 100:2623–28
97. Vrakking MJJ, Lee YT. 1995. *J. Chem. Phys.* 102:8818–32
98. Hollenstein U, Seiler R, Schmutz H, Andrist M, Merkt F. 2001. *J. Chem. Phys.* 115:5461–69
99. Lipson RH, Dimov SS, Wang P, Shi YJ, Mao DM, et al. 2000. *Instrum. Sci. Technol.* 28:85–118
100. Signorell R, Merkt F. 2000. *Faraday Discuss. Chem. Soc.* 115:205–28
101. Hsu CW, Heimann P, Evans M, Stimson S, Fenn PT, Ng CY. 1997. *J. Chem. Phys.* 106:8931–34
102. Hsu CW, Evans M, Stimson S, Ng CY. 1998. *J. Chem. Phys.* 108:4701–4
103. Evans M, Stimson S, Ng CY, Hsu CW. 1998. *J. Chem. Phys.* 109:1285–92
104. Evans M, Stimson S, Ng CY, Hsu CW, Jarvis GK. 1999. *J. Chem. Phys.* 110:315–27
105. Song Y, Evans M, Ng CY, Hsu CW, Jarvis GK. 2000. *J. Chem. Phys.* 112:1271–78
106. Song Y, Evans M, Ng CY, Hsu CW, Jarvis GK. 2000. *J. Chem. Phys.* 112:1306–15
107. Jarvis GK, Song Y, Ng CY. 1999. *J. Chem. Phys.* 111:1937–46
108. Song Y, Ng CY, Jarvis GK, Dressler R. 2001. *J. Chem. Phys.* 115:2101–8
109. Evans M, Ng CY. 1999. *J. Chem. Phys.* 111:8879–92
110. Shiell RC, Evans M, Ng CY, Hepburn JW. 1999. *Chem. Phys. Lett.* 315:390–96
111. Lefebvre-Brion H, Ng CY. 2000. *Chem. Phys. Lett.* 327:404–8
112. Fedorov D, Evans M, Song Y, Gordon M, Ng CY. 1999. *J. Chem. Phys.* 111:6413–21
113. Yencha A, Lopes MCA, King GC, Hochlaf M, Song Y, Ng CY. 2000. *Faraday Discuss. Chem. Soc.* 115:355–62
114. Liu JB, Hochlaf M, Chambaud G,

Rosmus P, Ng CY. 2001. *J. Phys. Chem.* 105:2183–91

115. Stimson S, Evans M, Ng CY, Hsu CW, Heimann P, et al. 1998. *J. Chem. Phys.* 108:6205–14
116. Liu JB, Chen WW, Hsu CW, Hochlaf M, Evans M, et al. 2000. *J. Chem. Phys.* 112:10767–77
117. Jarvis K, Song Y, Ng CY, Grant ER. 1999. *J. Chem. Phys.* 111:9568–73
118. Fedorov DG, Gordon MS, Song Y, Ng CY. 2001. *J. Chem. Phys.* 115:7393–400
119. Okada K, Iwata S. 2000. *J. Chem. Phys.* 112:1804–8
120. Okada K, Iwata S. 2000. *J. Electr. Spectrosc. Relat. Phenom.* 108:225–34
121. Shiell RC, Evans M, Stimson S, Hsu CW, Ng CY, Hepburn JW. 1998. *Phys. Rev. Lett.* 80:472–75
122. Shiell RC, Evans M, Stimson S, Hsu CW, Ng CY, Hepburn JW. 1999. *Phys. Rev. A* 59:2903–9
123. Moseley JT, Cosby PC, Ozenne JB, Durup J. 1979. *J. Chem. Phys.* 70:1474–81
124. Pratt ST, McCormack EF, Dehmer JL, Dehmer PM. 1992. *Phys. Rev. Lett.* 68:584–87
125. Blush JA, Chen P, Wiedmann RT, White MG. 1993. *J. Chem. Phys.* 98:3557–59
126. Weitzel KM, Mähnert J, Baumgärtel H. 1993. *Ber. Bunsenges. Phys. Chem.* 97:134
127. Weitzel KM, Jarvis G, Malow M, Baer T, Song Y, Ng CY. 2001. *Phys. Rev. Lett.* 86:3526–29
128. Dickenson H, Chelmick T, Softley TP. 2001. *Chem. Phys. Lett.* 338:37–45
129. Krause H, Neusser H. 1993. *J. Chem. Phys.* 99:6278–86
130. Park ST, Kim SY, Kim MS. 2001. *J. Chem. Phys.* 114:5568–76
131. Park ST, Kim SY, Kim MS. 2001. *J. Chem. Phys.* 115:2492–98
132. Hofstein JD, Johnson PM. 2000. *Chem. Phys. Lett.* 316:229–37
133. Taylor DP, Goode JG, LeClaire JE, Johnson PM. 1995. *J. Chem. Phys.* 103:6293–95
134. Goode JG, LeClaire JE, Johnson PM. 1996. *Int. J. Mass Spectrom. Ion Proc.* 159:49–64
135. Goode JG, Hofstein JD, Johnson PM. 1997. *J. Chem. Phys.* 107:1703–16
136. LeClaire JE, Anand R, Johnson PM. 1997. *J. Chem. Phys.* 106:6785–94
137. Fujii A, Iwasaki A, Ebata T, Mikami N. 1997. *J. Phys. Chem.* 101:5963–65
138. Gerhards M, Schiwek M, Unterberg C, Kleinermanns K. 1998. *Chem. Phys. Lett.* 297:515–22
139. Nir E, Hunziker HE, de Vries MS. 1999. *Anal. Chem.* 71:1674–78
140. Softley T, Mackenzie S, Merkt F, Rolland D. 1997. *Adv. Chem. Phys.* 101:667–99
141. Parker DH. 2000. See Ref. 143, pp. 3–46
142. Powis I, Baer T, Ng CY, eds. 1995. *High Resolution Laser Photoionization and Photoelectron Studies*. Chichester, UK: Wiley
143. Ng CY, ed. 2000. *Photoionization and Photodetachment*. Singapore: World Sci.

THE MOLECULAR HAMILTONIAN

Henning Meyer
Department of Physics and Astronomy and Department of Chemistry, The University of Georgia, Athens, Georgia 30602-2451; e-mail: hmeyer@hal.physast.uga.edu

Key Words spectroscopy, reaction dynamics, quantization, normal modes, internal coordinates

■ **Abstract** The molecular Hamiltonian represents one of the most basic concepts in spectroscopy and molecular reaction dynamics. Its derivation is notoriously difficult owing to the use of a rotating reference frame which, in turn, is necessary to define the concept of vibration and rotation. In this article, we review the construction of the molecular Hamiltonian in normal mode and in internal coordinates. For normal mode coordinates, the Watson Hamiltonian including its modification for linear molecules is derived using an approach based on classical mechanics and the Podolsky transformation. The method is subsequently used to derive the molecular Hamiltonian in terms of Jacobi and valence coordinates. Results are presented for the triatomic system and for the extension toward N-atom systems with $N \geq 3$.

1. INTRODUCTION

The theoretical study of many topics in the area of physical chemistry or chemical physics is based on the existence of a complete molecular Hamiltonian (1). This is especially true for the areas of spectroscopy (2–5) and molecular reaction dynamics (6, 7). While the first deals exclusively with the bound states, the latter typically involves bound and continuum states of the molecular system under consideration. In any case, the attempt to determine the eigenvalues and eigenfunctions must begin with the construction of an appropriate representation of the Hamiltonian. It is thus not surprising that the derivation of an exact molecular Hamiltonian has a long history dating back more than 70 years (8–16). Even more surprising considerable efforts are still aimed at improving or simplifying our methods for the construction of exact molecular Hamiltonians (17–24).

Within the framework of nonrelativistic quantum mechanics the Hamiltonian for a system of N_n nuclei and N_e electrons is easily written down in cartesian coordinates. The resulting Schroedinger Equation is usually far too complicated to allow even approximate solutions. In practice, the problem of finding the energy eigenvalues and eigenfunctions of the molecular system is broken down into an electronic and a nuclear part. Due to the disparity in mass between the electrons and the nuclei, it is possible to separate the motion of electrons and nuclei assuming the validity of the Born-Oppenheimer approximation (25). Within this approximation, the nuclear motion is controlled by a multidimensional potential energy surface representing the electronic energy. Unless otherwise stated, the discussion in this article is restricted to the dynamics of N_n nuclei moving under the influence of a Born-Oppenheimer potential. The associated Hamiltonian for the N_n-body system is again written out easily in terms of cartesian nuclear coordinates referring to a laboratory (lab) reference frame. Exploiting the translational invariance of the Hamiltonian, it is straightforward to separate off the motion of the center of mass (cm) of the complete system. Therefore the task reduces to the description of the nuclear motion with respect to a translation-free space-fixed (sf) frame. By definition, the origin of this frame is located in the cm of the molecule while its axes are parallel to the axes of the lab frame.

The major difficulty in deriving an exact, but also tractable coordinate representation of the complete nuclear Hamiltonian results from the introduction of a rotating reference frame referred to as the body-fixed (bf) frame. In most cases, it is defined through the introduction of three Euler angles (three-angle embedding). For some special cases, e.g., linear molecules or van der Waals complexes undergoing large amplitude motion, a bf frame based on two-angle embedding has been used (26, 27). Ideally, the orientation of the bf frame for a given sf configuration of the nuclei should be specified in such a way that the angular momentum of the nuclei as viewed from the bf frame is minimized (28); a condition met for example by the Eckart condition (29). The introduction of a rotating frame results inevitably in a nonlinear relation between sf and bf coordinates. Although this does not constitute a serious problem for the potential energy function, the transformation of the kinetic energy operator to the set of new coordinates can be a formidable problem. Essentially two alternative strategies have been developed to accomplish this task: In the chain rule approach, the kinetic energy operator is set up in cartesian sf coordinates and the Laplacian is transformed directly using differential calculus. In the Podolsky approach, an expression for the bf representation of the classical kinetic energy is derived first. In a second step, the appropriate quantum mechanical kinetic energy operator is derived by applying the Podolsky transformation (30). It is important to emphasize that both methods are equivalent and it is a matter of taste or convenience which method is applied. Both methods have been applied to the construction of Hamiltonians for different sets of coordinates yielding identical results!

In this article, we follow the Podolsky approach to derive different coordinate representations of the molecular Hamiltonian: Normal mode coordinates and internal coordinates. We discuss the use of Jacobi and valence type coordinates for triatomic systems and their extension towards N-atom systems parameterized through $(N-1)$ internal vectors. The approach chosen here makes extensive use of two very basic properties known from vector algebra. First, the invariance of the dot product between vectors $\vec{A}$ and $\vec{B}$ under an orthogonal transformation $\mathbf{C}$: $(\vec{A} \cdot \vec{B}) = (\mathbf{C}\vec{A}) \cdot (\mathbf{C}\vec{B})$. This property allows us to calculate the kinetic energy defined in the sf frame using the bf representation of the sf velocity vectors. Second, the transformation behavior of the vector product under an orthogonal transformation $\mathbf{C}$: $\mathbf{C}(\vec{A} \times \vec{B}) = (\mathbf{C}\vec{A}) \times (\mathbf{C}\vec{B})$. Because of this property, it is possible to calculate directly the total angular momentum vector defined in the sf frame but projected onto bf axes in terms of the transformed sf velocities. In this way, it is possible not only to obtain compact expressions for physically relevant classical observables and their quantum mechanical counterparts, but also to keep track of the different coordinate representations facilitating the identification of symmetry properties. We hope that by applying consistently a single method the interested reader will gain a better understanding of the Hamiltonians and their properties presently used in different areas of spectroscopy as well as molecular reaction dynamics. Therefore the emphasis is on the practical derivation of these Hamiltonians rather than on providing a complete review of the historical developments in this area.

Because of its fundamental importance in spectroscopy, we derive in the next section the Watson Hamiltonian. The third section is devoted to the construction of various Hamiltonians in internal coordinates that are employed presently in most theoretical studies. The last section describes and applies in some detail the new powerful method of rovibrational S-vectors proposed by Lukka (19).

2. THE HAMILTONIAN IN NORMAL MODE COORDINATES: THE WATSON HAMILTONIAN

2.1. Nonlinear Molecules

The normal mode representation of the Hamiltonian is now usually referred to as the Watson Hamiltonian in honor of Watson's contributions in deriving its compact form (14, 15). It is particularly useful at low vibrational energies for semi-rigid molecules that are characterized by a well-defined equilibrium structure. Consequently, it is this Hamiltonian that serves as the basis for most spectroscopic applications. In this section, we discuss the Hamiltonian for nonlinear molecules and the extension toward linear molecules is presented in Section 2.3.

We start the derivation by introducing a laboratory fixed frame that defines the position $\vec{R}_i^{(lab)}$ of the nuclei of the system under consideration. Because of the translational invariance of the Hamiltonian, it is convenient to reference the nuclear positions to a space-fixed (sf) frame with origin in the cm $\vec{R}_{cm}$ as shown in Figure 1. With respect to this frame, the instantaneous nuclear positions are determined by the set of vectors $\vec{R}_i = \vec{R}_i^{(lab)} - \vec{R}_{cm}$. In a similar way, we can exploit the rotational invariance of the Hamiltonian to define nuclear position vectors with respect to a rotating coordinate frame. Formally, the transformation to this body fixed (bf) frame is defined by the direction cosine matrix $\mathbf{C}$ (31). Its elements are given in terms of the three Euler angles α, β and γ that represent rotations around the z-axis of the sf frame, the y-axis of an intermediate frame and the z-axis of the bf frame,

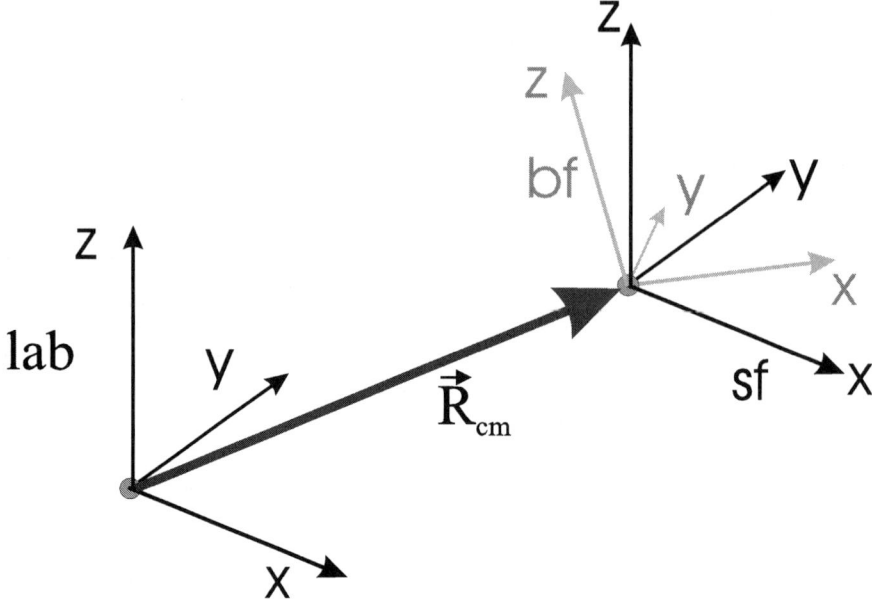

Figure 1 Definition of the lab, sf and bf frames used for the specification of the nuclear positions. The lab and sf frames are connected by the vector $\vec{R}_{cm}$.

respectively:

$$\mathbf{C}(\alpha, \beta, \gamma) = \mathbf{F}(\gamma)\mathbf{B}(\beta)\mathbf{A}(\alpha) \quad \text{with} \quad \mathbf{A}(\alpha) = \begin{pmatrix} \cos\alpha & \sin\alpha & 0 \\ -\sin\alpha & \cos\alpha & 0 \\ 0 & 0 & 1 \end{pmatrix},$$

$$\mathbf{B}(\beta) = \begin{pmatrix} \cos\beta & 0 & -\sin\beta \\ 0 & 1 & 0 \\ \sin\beta & 0 & \cos\beta \end{pmatrix}, \quad \text{and} \quad \mathbf{F}(\gamma) = \begin{pmatrix} \cos\gamma & \sin\gamma & 0 \\ -\sin\gamma & \cos\gamma & 0 \\ 0 & 0 & 1 \end{pmatrix} \quad 1.$$

The nuclear positions in the laboratory frame are thus expressible in terms of the bf vectors $\vec{r}_i$ and the cm position:

$$\vec{R}_i^{(lab)} = \vec{R}_{cm} + \vec{R}_i = \vec{R}_{cm} + \mathbf{C}^{-1}(\alpha, \beta, \gamma)\vec{r}_i. \quad 2.$$

In order to derive the kinetic energy in terms of these new coordinates, we exploit the invariance of the dot product that allows us to work with velocity vectors projected onto the axes of the rotating frame, i.e., with $\mathbf{C}\dot{\vec{R}}_i$ rather than the velocities $\dot{\vec{R}}_i$ derived from Equation 2. In this way, we avoid the necessity of determining explicitly the Euler angles in terms of a given sf nuclear configuration.

$$\mathbf{C}\dot{\vec{R}}_i^{(lab)} = \mathbf{C}\dot{\vec{R}}_{cm} + \dot{\vec{r}}_i + \mathbf{C}\dot{\mathbf{C}}^{-1}\vec{r}_i = \mathbf{C}\dot{\vec{R}}_{cm} + \dot{\vec{r}}_i + \vec{\omega} \times \vec{r}_i. \quad 3.$$

At this point, it is worthwhile to mention that we also make use of this property in the calculation of the total angular momentum defined with respect to the sf frame.

For the actual calculation, it is advantageous to introduce its bf representation: $\vec{J} = \mathbf{C}\vec{J}^{(sf)}$. Using the definition of the angular momentum and the transformation properties of the vector product, we find the following prescription for the calculation of $\vec{J}$:

$$\vec{J} = \mathbf{C}\vec{J}^{(sf)} = \sum_i m_i \mathbf{C}\vec{R}_i \times \mathbf{C}\dot{\vec{R}}_i. \qquad 4.$$

The components of the angular velocity vector $\vec{\omega}$ are related to the time derivatives of the Euler angles by the transpose of a matrix $\mathbf{W}$:

$$\vec{d} = \mathbf{W}^t \vec{\omega} \quad \text{with} \quad \vec{d} = \begin{pmatrix} \dot{\alpha} \\ \dot{\beta} \\ \dot{\gamma} \end{pmatrix}$$

and

$$\mathbf{W}^t = \begin{pmatrix} -\csc\beta\cos\gamma & \csc\beta\sin\gamma & 0 \\ \sin\gamma & \cos\gamma & 0 \\ \cot\beta\cos\gamma & -\cot\beta\sin\gamma & 1 \end{pmatrix} \qquad 5.$$

with the inverse relation:

$$\vec{\omega} = \mathbf{W}^{-1^t}\vec{d} \quad \text{with} \quad \mathbf{W}^{-1^t} = \begin{pmatrix} -\cos\gamma\sin\beta & \sin\gamma & 0 \\ \sin\gamma\sin\beta & \cos\gamma & 0 \\ \cos\beta & 0 & 1 \end{pmatrix}. \qquad 6.$$

Note that the matrix $\mathbf{W}$ is not orthogonal because its determinant is equal to $-\csc\beta$.

Next we express the kinetic energy in terms of the set of sf coordinates $\{\vec{R}_{cm}, \vec{R}_i\}$ or the set of bf coordinates $\{\vec{R}_{cm}, \alpha, \beta, \gamma, \vec{r}_i\}$:

$$T = \frac{1}{2}\sum_i m_i \dot{\vec{R}}_i^{(lab)2} = T_{cm} + \frac{1}{2}\sum_i m_i \dot{\vec{R}}_i^2 = T_{cm} + T_{vib} + T_{rot} + T_{cor}$$

with

$$T_{cm} = \frac{1}{2}M\dot{\vec{R}}_{cm}^2, \; T_{vib} = \frac{1}{2}\sum_i m_i \dot{\vec{r}}_i^2, \; T_{rot} = \frac{1}{2}\sum_i m_i \left(\vec{\omega} \times \vec{r}_i\right)^2,$$

and

$$T_{cor} = \vec{\omega} \cdot \sum_i m_i (\vec{r}_i \times \dot{\vec{r}}_i). \qquad 7.$$

In Equation 7, the different parts of the equation refer to the laboratory, the space fixed and the bf representation of the kinetic energy, respectively. The bf representation is made up of the cm contribution and three contributions due to the motion of the nuclei with respect to the sf frame. The latter describe the nuclear vibrations, the overall rotation of the system and their coupling, respectively. Obviously, the new sets of coordinates are not linearly independent since the position vectors must fulfill the cm condition $\sum m_i \vec{R}_i = \sum m_i \vec{r}_i = 0$. For the bf representation, three additional constraints are imposed by the definition of the rotating frame. As a convenient choice, Eckart defined the bf frame in such a way that the nuclei do not generate angular momentum at a particular reference configuration $\{\vec{a}_i\}$: $\sum m_i \vec{a}_i \times \vec{r}_i = 0$ (29). For a nonlinear molecule, the bf coordinates $r_{i\kappa}$ are thus related to $(3N_n - 6)$ normal mode coordinates Q_k by an orthogonal transformation defined by the matrix **L**:

$$r_{i\kappa} = a_{i\kappa} + m_i^{-\frac{1}{2}} \sum_k L_{(i\kappa)k} Q_k \quad \text{and} \quad \dot{r}_{i\kappa} = m_i^{-\frac{1}{2}} \sum_k L_{(i\kappa)k} \dot{Q}_k. \qquad 8.$$

Note that the cartesian vector components are indexed with the atom number i and the cartesian component $\kappa = x, y, z$ while the index k ranges from 1 to $(3N - 6)$. In addition to the coordinates $\{Q_k\}$, the two Eckart conditions can be used to define six external coordinates associated with the overall translation and rotation of the system. In this case, **L** becomes a square matrix of dimension $3N_n$, but the external coordinates make no contribution to Equation 8 because of the Eckart conditions. The orthogonality of the matrix **L** gives rise to several sum rules which have been discussed in detail by Watson (14). In combination with the Eckart conditions, the orthogonality ensures the simple form of the vibrational contribution in Equation 7: $T_{vib} = \frac{1}{2} \sum_k \dot{Q}_k^2$. The other two contributions involve the vector product, which is conveniently evaluated with the help of the anti-symmetric tensor $e_{\kappa\delta\nu}$. The resulting expression for the Coriolis coupling energy is then cast into a compact form by introducing the antisymmetric ζ-matrix:

$$T_{cor} = \sum_{i\kappa\delta\nu} e_{\kappa\delta\nu} m_i r_{i\delta} \dot{r}_{i\nu} \omega_\kappa = \sum_{\kappa kl} \zeta_{lk}^\kappa Q_l \dot{Q}_k \omega_\kappa$$

with

$$\zeta_{lk}^\kappa = \sum_{i\delta\nu} e_{\kappa\delta\nu} L_{(i\delta)l} L_{(i\nu)k}. \qquad 9.$$

The rotational contribution is usually written in a form that involves the instantaneous moment of inertia tensor **I**:

$$T_{rot} = \frac{1}{2} \sum_{\kappa\delta} I_{\kappa\delta} \omega_\delta \omega_\kappa \quad \text{with} \quad I_{\kappa\delta} = \sum_{i\delta'\nu\nu'} m_i e_{\delta'\kappa\nu} e_{\delta'\delta\nu'} r_{i\nu} r_{i\nu'}. \qquad 10.$$

Employing the properties of the antisymmetric tensor, the sum rules for the L-matrix and the ζ-matrix, Watson could show that the moment of inertia tensor can be partitioned into two terms one of which defines the modified moment of inertia tensor **I**':

$$I_{\kappa\delta} = I'_{\kappa\delta} + \sum_{kll'} \zeta_{lk}^\delta \zeta_{l'k}^\kappa Q_l Q_{l'}$$

with

$$I'_{\kappa\delta} = I_{\kappa\delta}^0 + \sum_l a_l^{\kappa\delta} Q_l + \frac{1}{4} \sum_{\nu\nu'll'} a_l^{\kappa\nu} I_{\nu\nu'}^{0-1} a_{l'}^{\chi\delta} Q_l Q_{l'}$$

and

$$a_l^{\kappa\delta} = 2 \sum_{i\delta'\nu\nu'} e_{\delta'\kappa\nu} e_{\delta'\delta\nu'} (m_i)^{+\frac{1}{2}} a_{i\nu} L_{(i\nu')l}. \qquad 11.$$

The tensor $\mathbf{I}'$ involves the moment of inertia tensor $\mathbf{I}_0$ associated with the reference configuration. Although it is quadratic in the normal mode coordinates, it can be factored into a product of three tensors that are at most linear in the normal coordinates:

$$I'_{\kappa\delta} = \sum_{\nu\nu'} I''_{\kappa\nu} I''^{0^{-1}}_{\nu\nu'} I''_{\nu'\delta} \quad \text{and} \quad I''_{\kappa\nu} = I^0_{\kappa\nu} + \frac{1}{2}\sum_l a_l^{\kappa\nu} Q_l. \qquad 12.$$

Combining the different contributions, we find the kinetic energy as a function of the generalized coordinates $\{\vec{R}_{cm}, \alpha, \beta, \gamma, Q_1, \ldots, Q_{3N-6}\}$:

$$T = \frac{1}{2} M \dot{\vec{R}}^2_{cm} + \frac{1}{2}\sum_k \left(\dot{Q}_k + \sum_{\delta l} \zeta^\delta_{lk} Q_l \omega_\delta\right)^2 + \frac{1}{2}\sum_{\kappa\delta} I'_{\kappa\delta} \omega_\kappa \omega_\delta \qquad 13.$$

Because of the translational invariance of the interaction potential, the cm motion is identical to the motion of a free particle, and we disregard its contribution in the following. From Equation 13, the momenta conjugate to Q_k and ω_κ are determined in the usual way:

$$P_k = \frac{\partial T}{\partial \dot{Q}_k} = \dot{Q}_k + \sum_{\delta l} \zeta^\delta_{lk} Q_l \omega_\delta \quad \text{and} \quad J_\kappa = \frac{\partial T}{\partial \omega_\kappa} = \sum_\delta I'_{\kappa\delta} \omega_\delta + j_\kappa. \qquad 14.$$

Here we have introduced the components of the vibrational angular momentum $j_\kappa = \sum_{kl} \zeta^\kappa_{lk} Q_l P_k$. Using the definition of the total angular momentum (see Equation 4, it is straightforward to show that J_κ represent the components of the total angular momentum vector $\vec{J}$ projected onto bf axes. Obviously the Hamiltonian form of the kinetic energy requires us to express the momenta J_κ in terms of the momenta conjugate to the Euler angles. The latter are easily calculated by applying the chain rule and using the definition of $\vec{\omega}$. In this way, the following relation between the bf total angular momentum vector and the true conjugate momenta J_α, J_β, and J_γ is found:

$$\vec{J} = \mathbf{W}\vec{f}, \quad \text{and} \quad \vec{f} = \mathbf{W}^{-1}\vec{J} \quad \text{with} \quad \vec{f} = \begin{pmatrix} J_\alpha \\ J_\beta \\ J_\gamma \end{pmatrix}. \qquad 15.$$

Expressing the velocities in Equation 13 in terms of the conjugate momenta given in Equation 14, we arrive at the classical Hamilton function first derived by Howard & Wilson (9).

$$H = \frac{1}{2}(\vec{J}-\vec{j})^t \mathbf{I}'^{-1}(\vec{J}-\vec{j}) + \frac{1}{2}\sum_k P_k^2 + \frac{1}{2}\sum_k \Lambda_k Q_k^2 + \sum_{rst} \Phi_{rst} Q_r Q_s Q_t + \cdots.$$

$$16.$$

Here, the Born-Oppenheimer potential $V(Q_k)$ is expanded around the reference configuration of the molecule. By definition, the set of normal mode coordinates diagonalizes its quadratic contribution. Higher order contributions, e.g., cubic and quartic terms, are responsible for anharmonic couplings between different modes.

QUANTIZATION Before proceeding we mention that the quantum mechanical operators for the bf components of $\vec{J}$ are also given by Equation 15 if the order is preserved and if we apply the quantization rule, $J_\alpha = \frac{\hbar}{i}\frac{\partial}{\partial \alpha}$, $J_\beta = \frac{\hbar}{i}\frac{\partial}{\partial \beta}$, and $J_\gamma = \frac{\hbar}{i}\frac{\partial}{\partial \gamma}$ (31).

$$J_x = -\csc\beta \cos\gamma\, J_\alpha + \sin\gamma\, J_\beta + \cot\beta \cos\gamma\, J_\gamma$$
$$J_y = \csc\beta \sin\gamma\, J_\alpha + \cos\gamma\, J_\beta - \cot\beta \sin\gamma\, J_\gamma$$
$$J_z = J_\gamma \qquad 17.$$

In contrast to the sf components of $\vec{J}$, the bf components fulfill the anomalous commutation rules, i.e., $[J_\kappa, J_\delta] = \frac{\hbar}{i}\sum_\epsilon e_{\kappa\delta\epsilon} J_\epsilon$ (32). For the volume element $s = \sin\beta$, the operators J_α and J_γ are hermitian although clearly J_β is not. Using the hermitian conjugate operator $J_\beta^\dagger = \frac{1}{\sin\beta}\frac{\hbar}{i}\frac{\partial}{\partial \beta}\sin\beta$, it is straightforward to show that the operators in Equation 17 are indeed hermitian (15).

According to Podolsky, the quantum mechanical operator for the kinetic energy can be derived from the corresponding classical expression given as a function of the generalized coordinates q_ϵ and their conjugate momenta p_ϵ by applying the following transformation (30).

$$T_{qm} = \frac{1}{2}s^{-\frac{1}{2}}g^{\frac{1}{4}}\sum_{\epsilon\delta} p_\epsilon g^{-\frac{1}{2}} G_{\epsilon\delta} p_\delta g^{\frac{1}{4}}s^{\frac{1}{2}} \quad \text{and} \quad p_\epsilon = \frac{\hbar}{i}\frac{\partial}{\partial q_\epsilon}. \qquad 18.$$

Here s represents the volume element used in the normalization of the complete wavefunction while g denotes the determinant of the matrix $\mathbf{G}$. The following discussion is simplified if we regard the conjugate momenta $J_\alpha, J_\beta, J_\gamma$, and P_k as the components of the generalized momentum vector $\vec{p}$. An inspection of Equation 16 reveals that the matrix $\mathbf{G}$ can be factored into a product of three matrices: $\mathbf{G} = \widetilde{\mathbf{W}}^t \tilde{\mu} \widetilde{\mathbf{W}}$. The matrix $\tilde{\mu}$ represents the generating matrix for the quadratic form in the momenta P_k and the pseudo-conjugate momenta $J_\kappa - j_\kappa$:

$$\tilde{\mu} = \begin{pmatrix} \mathbf{I}'^{-1} & 0 \\ 0 & 1 \end{pmatrix}, \qquad 19.$$

where 1 denotes a 3N − 6 dimensional unit matrix. The matrix $\widetilde{\mathbf{W}}$ is used to express the pseudo conjugate momenta in terms of the true conjugate momenta. Since the P_k are already true conjugate momenta, we can regard the matrix $\widetilde{\mathbf{W}}$ as being made up of four sub-matrices.

$$\widetilde{\mathbf{W}} = \begin{pmatrix} \mathbf{W} & \mathbf{K} \\ 0 & 1 \end{pmatrix}. \qquad 20.$$

As diagonal blocks, we find the 3 × 3 matrix $\mathbf{W}$ defined in Equation 6 and a $(3N - 6)$ dimensional unit matrix. The block in the lower left corner is made up of zeros since there are no contributions from J_α, J_β and J_γ to the momenta P_k. On the other hand, the upper off-diagonal block, denoted $\mathbf{K}$, gives rise to the contribution from the vibrational angular momentum $\vec{j}$ (see Equation 14). $\mathbf{K}$ is a

$3 \times (3N-6)$ dimensional matrix whose elements are given by $K_{\kappa k} = -\sum_l \zeta_{lk}^{\kappa} Q_l$. For the following, we use the volume element $s = \sin \beta$ and the determinant

$$g = (det(\widetilde{W}))^2 \, det(\tilde{\mu}) = \mu \csc^2 \beta \quad \text{with} \quad \mu = det(\mathbf{I}'^{-1}). \qquad 21.$$

Note that g depends only on the Euler angle β while its dependence on the normal mode coordinates is given by the determinant μ of the inverse modified moment of inertia tensor.

In order to recover the different rotational and vibrational contributions to the kinetic energy, it is convenient to investigate directly the individual matrix elements of **G**. Since $\tilde{\mu}_{\chi\kappa}$ is a block diagonal matrix, we distinguish two sets of indices: $\chi \in \{x, y, z\}$ and $\chi = k \in \{1, \ldots, 3N-6\}$. If the indices are from different sets, the corresponding matrix element $\tilde{\mu}_{\chi\kappa}$ vanishes.

$$G_{\epsilon\delta} = \sum_{\chi\kappa} \widetilde{W}_{\epsilon\chi}^t \tilde{\mu}_{\chi\kappa} \widetilde{W}_{\kappa\delta} = \sum_{\chi,\kappa=x,y,z} \widetilde{W}_{\epsilon\chi}^t I'^{-1}_{\chi\kappa} \widetilde{W}_{\kappa\delta} + \sum_{k=1,\ldots,3N-6} \widetilde{W}_{\epsilon k}^t \widetilde{W}_{k\delta}. \qquad 22.$$

Therefore the matrix elements of **G** give rise to two types of contributions which are associated with the rotational and the vibrational part of the kinetic energy operator:

$$T_{qm} = T_{rot} + T_{vib}$$

where

$$T_{rot} = \frac{1}{2} s^{-\frac{1}{2}} g^{\frac{1}{4}} \sum_{\epsilon\delta} p_\epsilon g^{-\frac{1}{2}} \sum_{\chi,\kappa=x,y,z} \widetilde{W}_{\epsilon\chi}^t I'^{-1}_{\chi\kappa} \widetilde{W}_{\kappa\delta} p_\delta g^{\frac{1}{4}} s^{\frac{1}{2}}$$

and

$$T_{vib} = \frac{1}{2} s^{-\frac{1}{2}} g^{\frac{1}{4}} \sum_{\epsilon\delta} p_\epsilon g^{-\frac{1}{2}} \sum_{k=1,\ldots,3N-6} \widetilde{W}_{\epsilon k}^t \widetilde{W}_{k\delta} p_\delta g^{\frac{1}{4}} s^{\frac{1}{2}}. \qquad 23.$$

Since there are no contributions of the angular momenta to the momenta conjugate to the normal coordinates, the corresponding block of the matrix $\widetilde{W}_{\kappa\delta}$ vanishes and contributions arise only for $\kappa, \delta = 1, \ldots, (3N-6)$. Furthermore, the block associated with these indices is a unit matrix, and we immediately recover the following vibrational operator:

$$T_{vib} = \frac{1}{2} \mu^{\frac{1}{4}} \sum_k P_k \mu^{-\frac{1}{2}} P_k \mu^{\frac{1}{4}} \quad \text{with} \quad P_k = \frac{\hbar}{i} \frac{\partial}{\partial Q_k}. \qquad 24.$$

The rotational contribution is rearranged as follows:

$$T_{rot} = \frac{1}{2} s^{-\frac{1}{2}} g^{\frac{1}{4}} \sum_{\chi,\kappa=x,y,z} \sum_\epsilon p_\epsilon g^{-\frac{1}{2}} \widetilde{W}_{\epsilon\kappa}^t I'^{-1}_{\kappa\chi} \left(\sum_\delta \widetilde{W}_{\chi\delta} p_\delta \right) g^{\frac{1}{4}} s^{\frac{1}{2}}$$

$$= \frac{1}{2} \frac{\mu^{\frac{1}{4}}}{\sin \beta} \sum_{\chi,\kappa=x,y,z} \sum_\epsilon p_\epsilon \mu^{-\frac{1}{2}} \sin \beta \widetilde{W}_{\epsilon\kappa}^t I'^{-1}_{\kappa\chi} (J_\chi - j_\chi) \mu^{\frac{1}{4}}. \qquad 25.$$

In the last line, we have applied the definition of the elements of $\widetilde{W}$ for the upper block and its action on the vector $\vec{p}$. In principle, we would like to consider the action of $\widetilde{W}^t$ on $\vec{p}$ in an analogous manner. This requires us to calculate the commutators of p_ϵ with $(\mu^{-1/2}\sin\beta)$ and $\widetilde{W}^t_{\epsilon\gamma}$. In order to calculate these commutators, we distinguish operators p_ϵ with $\epsilon \in \{\alpha,\beta,\gamma\}$ and $\epsilon = k \in \{1,\ldots,(3N-6)\}$. Remembering that κ is restricted to the set $\{x,y,z\}$, we find:

$$\frac{1}{\sin\beta}\sum_\epsilon p_\epsilon \sin\beta\,\mu^{-\frac{1}{2}}\widetilde{W}_{\kappa\epsilon} = \sum_{\epsilon=\alpha,\beta,\gamma} W_{\kappa\epsilon}J_\epsilon\mu^{-\frac{1}{2}} + \sum_k K_{\kappa k}P_k\mu^{-\frac{1}{2}} + \sum_k(P_k K_{\kappa k})\mu^{-\frac{1}{2}}$$

$$+\mu^{-\frac{1}{2}}\left(-\delta_{\kappa x}\frac{\hbar}{i}\sin\gamma\cot\beta - \delta_{\kappa y}\frac{\hbar}{i}\cos\gamma\cot\beta + \frac{\hbar}{i}\cot\beta\, W_{\kappa\beta}\right) \qquad 26.$$

The last three terms on the right-hand-side of Equation 26 cancel each other and we can effectively commute J_ϵ with the different matrix elements of **W**. Since ζ-matrix elements with identical lower indices vanish, also the third term vanishes and P_k commutes with the elements $K_{\kappa k}$. Using the identity

$$\sum_{\epsilon=\alpha,\beta,\gamma} W_{\kappa\epsilon}J_\epsilon\mu^{-\frac{1}{2}} + \sum_k K_{\kappa k}P_k\mu^{-\frac{1}{2}} = \sum_\epsilon \widetilde{W}_{\kappa\epsilon}p_\epsilon\mu^{-\frac{1}{2}} = (J_\kappa - j_\kappa)\mu^{-\frac{1}{2}}, \qquad 27.$$

we find the rotational part of the kinetic energy operator:

$$T_{rot} = \frac{1}{2}\mu^{\frac{1}{4}}\sum_{\chi,\kappa=x,y,z}(J_\kappa - j_\kappa)\mu^{-\frac{1}{2}}I'^{-1}_{\kappa\chi}(J_\chi - j_\chi)\mu^{\frac{1}{4}}. \qquad 28.$$

The resulting molecular Hamiltonian was first reported by Howard & Wilson (9) and, in modified form, by Darling & Dennison (10):

$$H = \frac{1}{2}\mu^{\frac{1}{4}}\sum_{\delta,\kappa=x,y,z}(J_\kappa - j_\kappa)\mu^{-\frac{1}{2}}I'^{-1}_{\kappa\delta}(J_\delta - j_\delta)\mu^{\frac{1}{4}}$$

$$+\frac{1}{2}\mu^{\frac{1}{4}}\sum_k P_k\mu^{-\frac{1}{2}}P_k\mu^{\frac{1}{4}} + V(Q_k). \qquad 29.$$

Up to this point, only the commutators of J_κ with those parts of the G-matrix and its determinant g that depend on the Euler angles have been taken into account explicitly. A similar treatment regarding the normal coordinates results in a further simplification of the Hamiltonian as pointed out first by Watson (14). Using the commutator $[P_k,\mu^a] = a\mu^{a-1}(P_k\mu)$, the vibrational part of the Hamiltonian is easily simplified:

$$H_{vib} = -\frac{\hbar^2}{2}\sum_k \frac{\partial^2}{\partial Q_k^2} + V(Q_k) + U_3 + U_4$$

with

$$U_3 = \sum_k \frac{\hbar^2}{8}I''^{-2}\left(\frac{\partial I''}{\partial Q_k}\right)^2 \quad \text{and} \quad U_4 = \sum_k \frac{\hbar^2}{4}I''^{-1}\left(\frac{\partial^2 I''}{\partial Q_k^2}\right). \qquad 30.$$

Since the terms U_3 and U_4 depend only on the normal coordinates, they represent additional mass-dependent contributions to the potential energy.

The rotational part of the Hamiltonian (see Equation 29) involves the different angular momentum operators. Because the operators for the total angular momentum only act on the Euler angles, they commute with the inverse moment of inertia tensor and with the components of the vibrational angular momentum. Although the latter involve the momenta P_k, it can be shown that the sum of the commutators of j_κ with the elements of the inverse moment of inertia tensor $I'^{-1}_{\kappa\delta}$ vanishes, i.e., $\sum_\kappa [j_\kappa, I'^{-1}_{\kappa\delta}] = 0$ (14). In combination with the commutators $[j_\kappa, \mu^a] = -2a\mu^a I''^{-1}[j_\kappa, I'']$, we can remove the dependence on the Podolsky factors μ^a from the rotational Hamiltonian:

$$H_{rot} = \frac{1}{2}\sum_{\kappa\delta}(J_\kappa - j_\kappa)I'^{-1}_{\kappa\delta}(J_\delta - j_\delta) + U_1 + U_2$$

with

$$U_1 = \frac{1}{4}\sum_{\kappa\delta} I'^{-1}_{\kappa\delta} I''^{-1}[[j_\kappa, I''], j_\delta] \quad \text{and} \quad U_2 = \frac{1}{8}\sum_{\kappa\delta} I'^{-1}_{\kappa\delta} I''^{-2}[j_\kappa, I''][j_\delta, I''].$$

31.

The mass-dependent contributions U_1 and U_2 involve commutators between the vibrational angular momentum components and the determinant of the matrix $\mathbf{I}''$. Applying the rules for the differentiation of determinants, Watson could show that the sum of all four terms U_i is proportional to the trace of the inverse moment of inertia tensor:

$$U(Q_k) = U_1 + U_2 + U_3 + U_4 = -\frac{\hbar^2}{8}\sum_\kappa I'^{-1}_{\kappa\kappa}.$$

32.

In the case of semi-rigid molecules, the mass dependent extra potential term U results in an almost constant shift of the energy levels, which can only be detected spectroscopically in ultra high resolution experiments. On the other hand, these terms can become important for molecules with large amplitude motion (33).

It is noteworthy that the Watson Hamiltonian, $H = H_{vib} + H_{rot}$ with H_{vib} and H_{rot} defined in Equations 30 and 31, has been derived also using the chain rule approach (34).

2.2. Effective Hamiltonian

Traditionally, the molecular eigenvalues and eigenfunctions are determined via perturbation theory. A zeroth order Hamiltonian is derived from the Watson Hamiltonian by expanding the inverse moment of inertia tensor around the reference configuration. Factoring out the matrix $\mathbf{I}_0^{\frac{1}{2}}$ in Equation 12, $\mathbf{I}'$ is cast into a form suitable to a Taylor expansion around $Q_k = 0$:

$$\mathbf{I}' = \mathbf{I}_0^{+1/2}\left(1 + \frac{1}{2}\tilde{\mathbf{b}}\right)^2 \mathbf{I}_0^{+1/2} \quad \text{with} \quad \tilde{\mathbf{b}} = \sum_k \mathbf{I}_0^{-1/2}\mathbf{a}_k \mathbf{I}_0^{-1/2} Q_k,$$

33.

where the elements of the 3×3 matrices $\mathbf{a}_k$ have been defined in Equation 11. Applying the Taylor series expansion of the function $(1 + x)^{-2}$, the following expansion of $\mathbf{I}'$ is found:

$$\mathbf{I}'^{-1} = \mathbf{I}_0^{-1} - \sum_k \mathbf{I}_0^{-1} \mathbf{a}_k \mathbf{I}_0^{-1} Q_k + \frac{3}{4} \sum_{kl} \mathbf{I}_0^{-1} \mathbf{a}_k \mathbf{I}_0^{-1} \mathbf{a}_l \mathbf{I}_0^{-1} Q_k Q_l + \cdots. \qquad 34.$$

Adapting the notation $\mu_{\kappa\delta} = I'_{\kappa\delta}$, the individual matrix elements are of the form:

$$\mu_{\kappa\delta} = \mu^0_{\kappa\kappa}\delta_{\kappa\delta} - \sum_k \sum_{\nu\epsilon} \mu^0_{\kappa\nu} a^k_{\nu\epsilon} \mu^0_{\epsilon\delta} Q_k + \frac{3}{4} \sum_{kl} \sum_{\nu\epsilon\rho\gamma} \mu^0_{\kappa\nu} a^k_{\nu\epsilon} \mu^0_{\epsilon\rho} a^l_{\rho\gamma} \mu^0_{\gamma\delta} Q_k Q_l + \cdots.$$

$$35.$$

As a result, the Hamiltonian is partitioned into a zeroth order Hamiltonian H_0 and different correction terms: $H = H_0 + H_{cent} + H_{cor} + H_{anh}$. The corrections represent centrifugal distortion H_{cent}, Coriolis coupling H_{cor} and the anharmonicity of the potential surface H_{anh}. The centrifugal distortion terms are quadratic in the total angular momentum components and of nth power in the normal mode coordinates. Coriolis coupling terms involve the normal coordinates, the vibrational angular momentum and the total angular momentum components. Anharmonic perturbations depend solely on the normal coordinates and their conjugate momenta. Note that this also includes terms quadratic in the vibrational angular momentum.

$$H_0 = \frac{1}{2}\sum_\kappa \mu^0_{\kappa\kappa} J_\kappa^2 + \frac{1}{2}\sum_k \{P_k^2 + \Lambda_k Q_k^2\} = H_{02} + H_{20}$$

$$H_{cent} = -\frac{1}{2}\sum_k \sum_{\kappa\chi\delta\epsilon} \mu^0_{\kappa\delta} a^k_{\delta\epsilon} \mu^0_{\epsilon\chi} Q_k J_\kappa J_\chi$$

$$+ \frac{3}{4}\sum_{kl}\sum_{\kappa\chi\delta\epsilon\rho\gamma} \mu^0_{\kappa\delta} a^k_{\delta\epsilon} \mu^0_{\epsilon\rho} a^l_{\rho\gamma} \mu^0_{\gamma\chi} Q_k Q_l J_\kappa J_\chi + \cdots$$

$$= H_{12} + H_{22} + \cdots$$

$$H_{cor} = -\frac{1}{2}\sum_\kappa \mu^0_{\kappa\kappa}[j_\kappa J_\kappa + J_\kappa j_\kappa] + \sum_k\sum_{\kappa\chi\delta\epsilon} \mu^0_{\kappa\delta} a^k_{\delta\epsilon} \mu^0_{\epsilon\chi}[j_\kappa Q_k J_\kappa + J_\kappa Q_k j_\kappa] + \cdots$$

$$= H_{21} + H_{31} + \cdots$$

$$H_{anh} = \frac{1}{6}\sum_{klm} \Phi_{klm} Q_k Q_l Q_m + \frac{1}{2}\sum_\kappa \mu^0_{\kappa\kappa} j_\kappa^2 + \frac{1}{24}\sum_{klmn} \Phi_{klmn} Q_k Q_l Q_m Q_n + \cdots$$

$$= H_{20} + H_{40} + \cdots. \qquad 36.$$

As indicated in Equation 36, it is customary to label a correction term by the different powers of the vibrational and angular momentum operators contributing to it. Clearly different vibrational zeroth order states can be coupled in different ways by the higher order corrections. As a result, the true molecular eigenstates are linear combinations of different basis states. Especially, at higher vibrational energies, the mixing of zeroth order states can manifest itself in the phenomenon of intramolecular vibrational redistribution (35–37). According to the expansion of the Hamiltonian in Equation 36, we can distinguish anharmonic, centrifugal, and Coriolis coupling mechanisms, which differ by their rotational dependence. While the anharmonic coupling is independent of the rotational state, we expect a linear and a quadratic J-dependence in case of Coriolis and centrifugal couplings, respectively (38). As a consequence, the importance of centrifugal coupling will be enhanced at higher temperatures and for larger molecules.

THE WATSONIAN   The expansion of the Hamiltonian in Equation 36 provides also the basis for various perturbation theory schemes. Of the different schemes employed, we mention here the method of contact or Van Vleck transformations (4, 39, 40). The application of a contact transformation involving a hermitian operator S generates a transformed Hamiltonian $\tilde{H} = e^{iS} H e^{-iS}$ with the same eigenvalue spectrum as the original Hamiltonian. The operator S is chosen in such a way as to remove certain correction terms. As a result, the diagonal elements of the transformed Hamiltonian in the basis of zeroth order states, now contain corrections to the zeroth order energies. Applying subsequent contact transformations, a diagonal representation within the zeroth order basis set is achieved. The first contact transformation is usually defined as a sum of operators, e.g., $S_1 = S_{30} + S_{12} + S_{21}$, which results in the removal of those terms from the original Hamiltonian responsible for the direct coupling of different zeroth order vibrational states. One such term is the cubic potential contribution. For a harmonic oscillator basis set, it does not have diagonal matrix elements and, consequently, it is removed from the Hamiltonian by applying the contact transformation S_{30}. Low-order couplings result further from the centrifugal distortion term H_{12} that defines the operator S_{12}. Similarly, the lowest order Coriolis term H_{21}, which is linear in the vibrational operators P_k and Q_l with $k \neq l$, can be removed by another transformation defined by the operator S_{21}. Note that the diagonal matrix elements of H_{21} do not necessarily vanish in the presence of degenerate vibrations. The transformed Hamiltonian $\tilde{H}^{(1)}$ has been discussed in great detail in Reference (4). It contains three correction terms in addition to the zeroth order terms H_{20} and H_{02}. The purely vibrational term $\tilde{H}_{40}$ describes the major anharmonic corrections. Since the term is of fourth order in the vibrational operators, the corrections to the energy are quadratic in the vibrational quantum numbers, i.e., proportional to $(v_k + \frac{1}{2})(v_l + \frac{1}{2})$. Similarly the purely rotational contribution $\tilde{H}_{04}$ describes the leading centrifugal corrections in terms of the quartic centrifugal distortion constants. Finally the diagonal elements of the interaction term $\tilde{H}_{22}$ give rise to the vibrational state dependence of the rotational constants. In conclusion, the effective Hamiltonian when averaged over the vibrational zeroth order basis states, can serve as an effective rotational Hamiltonian that incorporates the major distortion effects.

The application of additional contact transformations generates higher order rotational terms. Therefore, the effective Hamiltonian can be regarded as a power series in the different angular momentum components J_κ. Successive application of the commutators for the angular momentum operators reduces it to the standard form: $\tilde{H}_{rot} = \sum_{pqr} \tau_{pqr} J_x^p J_y^q J_z^r$. A careful investigation of the symmetry properties of this Hamiltonian (41, 42) reveals that for nonplanar asymmetric top molecules only even powers of p, q and r contribute defining a total of six quartic and ten quintic centrifugal distortion constants (43). Watson could also show that the different distortion constants are not completely independent requiring the further reduction of the Hamiltonian. Since a contact transformation does not change the eigenvalue spectrum of the Hamiltonian, it is possible to use such a transformation to eliminate certain parameters from the Hamiltonian. This procedure leads to the so-called reduced Hamiltonian or Watsonian, which is characterized by five quartic and seven sextic distortion constants (40, 44).

2.3. Modifications for Linear Molecules

For molecules with a linear reference configuration the bf frame is already completely determined by specifying a single direction; for example through its polar angles α and β with respect to the sf frame. Choosing $\gamma = 0$, the angular velocity vector in Equation 6 reduces to:

$$\vec{\omega} = \begin{pmatrix} -\sin\beta\dot{\alpha} \\ \dot{\beta} \\ \cos\beta\dot{\alpha} \end{pmatrix}. \qquad 37.$$

Since in the reference configuration the nuclei lie along the bf z-axis, we have $a_{i\kappa} = a_i \delta_{z\kappa}$. Therefore the second Eckart condition reduces to $\sum_i m_i e_{\chi zx} a_i r_{ix} = 0$ and $\sum_i m_i e_{\chi zy} a_i r_{iy} = 0$ from which α and β must be determined. Since the third equation is trivially fulfilled, there are only two rotational and thus $3N - 5$ vibrational degrees of freedom for a linear molecule. The normal coordinates are related to the bf positions through the matrix $\mathbf{L}$:

$$r_{i\kappa} = a_i \delta_{z\kappa} + (m_i)^{-\frac{1}{2}} \sum_k L_{(i\kappa)k} Q_k. \qquad 38.$$

Substituting this expression into the definition of the moment of inertia tensor (see Equation 10), it can be shown that the related modified moment of inertia tensor $\mathbf{I}'$ can again be factored (45).

$$I_{\kappa\kappa'} = I'_{\kappa\kappa'} + \sum_{kll'} \zeta_{lk}^{\kappa'} \zeta_{l'k}^{\kappa} Q_{l'} Q_l \quad \text{with} \quad I'_{\kappa\kappa'} = \epsilon_{\kappa\kappa'} I',$$

$$I' = I_0 + \sum_l b_l Q_l + \frac{1}{4} \sum_{ll'} b_l b_{l'} I_0^{-1} Q_{l'} Q_l = I''^2 I_0^{-1}$$

and

$$I'' = I_0 + \frac{1}{2} \sum_l b_l Q_l. \qquad 39.$$

Here we have defined also the quantities $\epsilon_{\kappa\kappa'} = \sum_\alpha e_{\kappa z\alpha} e_{\kappa' z\alpha} = \delta_{\kappa\kappa'} - \delta_{z\kappa'}\delta_{z\kappa}$ and $b_l = 2\sum_i (m_i)^{\frac{1}{2}} a_i L_{izl}$. $\zeta_{l'k}^{\kappa}$ and $I_0 = \sum_i m_i a_i^2$ represent ζ-matrix elements and the moment of inertia for the reference configuration, respectively. The modified moment of inertia tensor is thus diagonal even for deviations from the reference configuration. Since $I'_{zz} = 0$, its determinant vanishes and its inverse does not exist.

Because the classical Coriolis energy for the linear case is identical to the expression given in Equation 9, the analogue of Equation 13 becomes:

$$T = \frac{1}{2}M\dot{R}_{cm}^2 + \frac{1}{2}\sum_k \left(\dot{Q}_k + \sum_{\alpha l}\zeta_{lk}^\alpha Q_l \omega_\alpha\right)^2 + \frac{1}{2}I'(\omega_x^2 + \omega_y^2). \qquad 40.$$

The translation-free part of the kinetic energy is next brought into the canonical form

$$T = \frac{1}{2}\sum_k P_k^2 + \frac{1}{2}I'^{-1}\left((J_x - j_x)^2 + (J_y - j_y)^2\right), \qquad 41.$$

where j_κ denotes a component of the vibrational angular momentum. In order to derive the quantum mechanical operator for the kinetic energy, it is convenient to express Equation 41 directly in terms of the conjugate momenta J_α and J_β:

$$T = \frac{1}{2}\sum_k P_k^2 + \frac{1}{2}I'^{-1}\left((-\csc\beta J_\alpha - j_x + \cot\beta j_z)^2 + (J_\beta - j_y)^2\right). \qquad 42.$$

In order to apply the Podolsky transformation, we identify the generating matrix $\mathbf{G} = \widetilde{\mathbf{W}}^t \tilde{\mu} \widetilde{\mathbf{W}}$ with matrices

$$\widetilde{\mathbf{W}} = \begin{pmatrix} -\csc\beta & 0 & K_{x1} & \cdots & K_{x3N-5} \\ 0 & 1 & K_{y1} & \cdots & K_{y3N-5} \\ 0 & 0 & 1 & \cdots & 0 \\ \cdot & \cdot & \cdot & \cdots & 0 \\ 0 & 0 & 0 & \cdots & 1 \end{pmatrix}$$

and

$$\tilde{\mu} = \begin{pmatrix} I'^{-1} & 0 & 0 & \cdots & 0 \\ 0 & I'^{-1} & 0 & \cdots & 0 \\ 0 & 0 & 1 & \cdots & 0 \\ \cdot & \cdot & \cdot & \cdots & 0 \\ 0 & 0 & 0 & \cdots & 1 \end{pmatrix}. \qquad 43.$$

The upper right $2 \times (3N - 5)$ dimensional block of the matrix $\widetilde{\mathbf{W}}$ is responsible for the vibrational angular momentum contributions. Its elements are defined as follows:

$$K_{xk} = \sum_l (\cot\beta \zeta_{lk}^z - \zeta_{lk}^x) Q_l \quad \text{and} \quad K_{yk} = -\sum_l \zeta_{lk}^y Q_l. \qquad 44.$$

Using the determinant $g = det(\mathbf{G}) = I'^{-2}\csc^2\beta$ and the volume element $s = \sin\beta$, the quantum mechanical version of Equation 42 is derived:

$$T = \frac{1}{2}I'^{-\frac{1}{2}}(-\csc\beta J_\alpha + \cot\beta j_z - j_x)(-\csc\beta J_\alpha + \cot\beta j_z - j_x)I'^{-\frac{1}{2}}$$

$$+ \frac{1}{\sin\beta}I'^{-\frac{1}{2}}(J_\beta - j_y)\sin\beta(J_\beta - j_y)I'^{-\frac{1}{2}} + \frac{1}{2}\sum_k I'^{-\frac{1}{2}} P_k I' P_k I'^{-\frac{1}{2}}. \quad 45.$$

The Podolsky transformation affects only those terms that depend either on the normal coordinates or on the angle β. Therefore, the last term is only modified by factors involving I'. This also holds true for the energy contribution due to the x-component of the rotational angular momentum $J_x - j_x$, while the contribution due to the y-component is affected due to the factors $\sin\beta$ and I'. Taking into account the dependence of I' on the normal mode coordinates, Equation 45 is easily simplified to yield:

$$T = \frac{1}{2}\sum_k P_k^2 + \frac{1}{2}I'^{-1}\left((J_x - j_x)^2 + (J_\beta - j_y)^2 + \frac{\hbar}{i}\cot\beta(J_\beta - j_y)\right), \quad 46.$$

where we have introduced the hermitian operator $J_x = -\csc\beta J_\alpha + \cot\beta j_z$. In this case no extra potential terms result. Obviously the operator $J_y = J_\beta$ is not hermitian since the volume element contains the factor $s = \sin\beta$. The corresponding hermitian operator $\tilde{J}_y$ is constructed according to $\tilde{J}_y = \frac{1}{2}(J_\beta + J_\beta^\dagger) = J_\beta + \frac{\hbar}{2i}\cot\beta$. It is important to realize that the components of $\vec{J}$ obey normal commutation rules while the components of $\vec{J}$ do not follow either normal or anomalous commutation rules. This unsatisfactory situation is remedied by introducing an isomorphic Hamiltonian that is derived from the original Hamiltonian by applying the unitary transformation $U = e^{\frac{i}{\hbar}j_z\chi}$ (15, 26, 51). The transformation introduces the additional angle χ causing the isomorphic Hamiltonian to have a larger number of eigenfunctions than the original. For example, among the possible eigenfunctions of the isomorphic Hamiltonian, only those represent transformed eigenfunctions of the true Hamiltonian for which the actions of the operators j_z and J_χ are identical (46). Finally, the isomorphic Hamiltonian can be cast into a form that involves a total angular momentum vector $\vec{J}'$ whose components obey the usual anomalous commutation rules (15, 51).

For completeness, we mention the situation for which the non-Euclidian volume element $s = 1$ is used. Now, the operator J_β is hermitian while the Podolsky factor $g^{\frac{1}{4}}s^{\frac{1}{2}} = \mu^{\frac{1}{2}}(\csc\beta)^{\frac{1}{2}}$ is responsible for a mass dependent extra potential term (15, 47).

$$T = \frac{1}{2}\sum_k P_k^2 + \frac{1}{2}\mu(J_x - j_x)^2 + \frac{1}{2}\mu\left(J_y - j_y + \frac{\hbar}{2i}\cot\beta\right)\left(J_y - j_y - \frac{\hbar}{2i}\cot\beta\right).$$

$$47.$$

2.4. Modifications Due to Electronic Structure

As a consequence of the breakdown of the Born-Oppenheimer approximation, nonadiabatic effects can occur in the dynamics of electronically excited states including nonradiative relaxation, reactive scattering and photodissociation (48, 49). But there are also spectroscopic problems such as the Jahn-Teller or the Renner effects that cannot be treated properly within the Born-Oppenheimer approximation. In order to provide an adequate description, it is necessary to include the electronic degrees of freedom. Such a Hamiltonian was derived by Howard & Moss who also treated the interaction with external electric and magnetic fields (50, 51).

In treating the electronic degrees of freedom explicitly, we must distinguish the cm of all particles, $\vec{R}_{cm}$, and the cm of the nuclei $\vec{X}_{cm}$. If we denote the position of the nuclei in the sf frame associated with $\vec{R}_{cm}$ by $\vec{\xi}_i$ and the position of the electrons with $\vec{\xi}_j$, the kinetic energy in the sf representation becomes:

$$T = \frac{1}{2}\sum_i m_i \dot{\vec{R}}_i^{(lab)2} + \frac{1}{2}m_e \sum_j \dot{\vec{R}}_j^{(lab)2} = T_{cm} + T_n + T_{el}$$

with

$$T_{cm} = \frac{1}{2}M\dot{\vec{R}}_{cm}^2, \quad T_n = \frac{1}{2}\sum_i m_i \dot{\vec{\xi}}_i^2, \quad \text{and} \quad T_{el} = \frac{1}{2}m_e \sum_j \dot{\vec{\xi}}_j^2. \qquad 48.$$

The nuclear cm is introduced by defining the new position vectors $\vec{R} = \vec{\xi} - \vec{X}_{cm}$, which are indexed with i and j for nuclei and electrons, respectively. In combination with the cm condition $\sum_i m_i \vec{\xi}_i + m_e \sum_j \vec{\xi}_j = 0$, it is straightforward to express $\vec{X}_{cm}$ in terms of the new coordinates: $\vec{X}_{cm} = -\frac{m_e}{M}\sum_j \vec{R}_j$. Substituting these expressions into Equation 48 yields the kinetic energy in a representation for which the individual positions are referenced to a space fixed frame with origin in the nuclear cm.

$$T(sf) = T_{cm} + T_n + T_{el} \quad \text{with} \quad T_{cm} = \frac{1}{2}M\dot{\vec{R}}_{cm}^2, \quad T_n = \frac{1}{2}\sum_i m_i \dot{\vec{R}}_i^2,$$

and

$$T_{el} = \frac{1}{2}m_e \sum_j \dot{\vec{R}}_j^2 - \frac{1}{2}\frac{m_e^2}{M}\left(\sum_j \dot{\vec{R}}_j\right)^2. \qquad 49.$$

In comparison with the first equation in Equation 7, the sf representation contains now the additional contribution T_{el} made up of the kinetic energy of the electrons and a contribution due to the mass polarization. Because of the difference in mass between the electrons and the nuclei, the latter term is very small and usually it is neglected. Since the bf frame is defined exclusively through the nuclear configuration, the form of the electronic contribution is not affected by the transformation. In the bf frame, the electron positions are denoted by vectors $\vec{\eta}_j = \mathbf{C}^{-1}\vec{R}_j$ while the associated conjugate momenta are given by $\vec{p}_j$. The electronic contribution to Equation 49 becomes (52):

$$T_{el} = \frac{1}{2} m_e \sum_j (\dot{\vec{\eta}}_j + \vec{\omega} \times \vec{\eta}_j)^2 - \frac{1}{2} \frac{m_e^2}{M} \left(\sum_j (\dot{\vec{\eta}}_j + \vec{\omega} \times \vec{\eta}_j) \right)^2. \qquad 50.$$

Since no other terms in T depend on the velocities of the electrons, the conjugate momenta are found to be: $\vec{p}_j = \nabla_{\dot{\vec{\eta}}_j} T_{el} = m_e(\dot{\vec{\eta}}_j + \vec{\omega} \times \vec{\eta}_j) - \frac{m_e^2}{M} \sum_{j'} (\dot{\vec{\eta}}_{j'} + \vec{\omega} \times \vec{\eta}_{j'})$. Therefore the bf representation of the operator T_{el} has the same form as its sf counterpart. Also the relation between the conjugate momenta J_α, J_β, and J_γ and the components of the bf representation of the total angular momentum vector is still given by Equation 17. On the other hand, the individual components of the latter now contain a contribution due to the electrons, $\vec{l} = \nabla_{\vec{\omega}} T_{el} = \sum_{j_e} \vec{\eta}_{j_e} \times \vec{p}_{j_e}$. As a consequence, we must replace in Equation 14 the components of the internal angular momentum j_κ with $j_\kappa + l_\kappa$. This yields the classical kinetic energy of the system in the canonical form:

$$T = \frac{1}{2}(\vec{J} - \vec{j} - \vec{l})^t \mathbf{I}'^{-1}(\vec{J} - \vec{j} - \vec{l}) + \frac{1}{2} \sum_k P_k^2 + \frac{1}{2m_e} \sum_j \vec{p}_j^2 + \frac{1}{2M_n} \left(\sum_j \vec{p}_j \right)^2.$$

$$51.$$

Since the other conjugate momenta do not contribute to the electronic momenta, T_{el} represents also the quantum mechanical operator with $\vec{p}_j = \frac{\hbar}{i} \nabla_{\vec{\eta}_j}$. On the other hand, the K-block in the matrix $\widetilde{\mathbf{W}}$ of Equation 20 is modified due to the electronic orbital angular momentum. Nevertheless, the general structure of $\widetilde{\mathbf{W}}$ is preserved and thus the determinant g is not altered. Since the components of $\vec{l}$ commute with the components of $\vec{J}$ and $\vec{j}$, the simplifications resulting from the evaluation of the various commutators are not affected by the presence of the electronic terms. As a result the rotational kinetic energy operator has exactly the form derived by Watson if $\vec{j}$ is replaced by $\vec{j} + \vec{l}$.

3. THE HAMILTONIAN IN INTERNAL COORDINATES

While the Watson Hamiltonian provides an adequate description for semi-rigid molecules, perturbation theory approaches based on this Hamiltonian must fail at higher vibrational energies or for systems exhibiting large amplitude vibrations. In these cases, the bound state energies and eigenfunctions have to be determined by directly solving the Schroedinger equation. This approach relies on the use of curvilinear coordinates as opposed to normal mode coordinates since the potential energy function is nearly separable in internal coordinates. One of the major difficulties has been the derivation of the kinetic energy operator in these coordinates. In the past, several approaches have been advocated to derive the kinetic energy operator including the overall rotation in internal coordinates. A first step in this direction was undertaken by Meyer & Gunthardt who generalized the G-matrix technique for a rotating frame defined as the instantaneous principal axis system (53). The influence of the choice of different rotating frames was investigated by Pickett who, starting from the kinetic energy expression given in Equation 7, derived the Hamiltonian without applying the Eckart condition (54). His Hamiltonian in combination with Van Vleck perturbation theory was used recently by Burleigh et al. in their investigation of the rotation-vibration mixing in the dynamics of polyatomic molecules at higher energies and J-values (55).

More recently, the emphasis has shifted towards the use of simple Hamiltonians in curvilinear coordinates whose eigenvalue spectrum is then determined directly using numerical methods. Two classes of Hamiltonians can be distinguished depending on the use of a single vector or a two-vector embedded bf frame. The single vector embedded frame uses two rotations to define the z-direction of the bf frame. It is used to describe molecules with a linear reference configuration (26), but it is also applied extensively to molecular systems, which are best described in terms of two fragments. Examples include scattering and photodissociation problems (56–59) or van der Waals complexes (27, 60).

The other class is based on a bf frame defined through two internal vectors. Typical applications include semi-rigid molecules, but also floppy molecules at energies close to the dissociation limit. Early work concentrated on the treatment of triatomic systems using different types of coordinates: Three bond distances (61), two bond lengths and one bond angle (62–66) or Jacobi coordinates (67). While in these contributions, the rotating frame is defined purely geometrically, Natanson employed directly the Eckart condition to the triatomic Hamiltonian in valence coordinates (68). Explicit expressions for the complete kinetic energy operator in terms of bond, Jacobi and Radau coordinates were reported later by Carrington and Wei (69–71).

The extension towards larger systems was pioneered by Handy and coworkers who used computer algebra programs to derive the Hamiltonian via the chain rule method (17, 72). The method was successfully applied to construct the Hamiltonian for triatomic (17), tetra-atomic (17, 72), penta-atomic systems (73) and rotationfree hexatomic molecules (74). The resulting expressions are fairly complicated and, consequently, the identification of different angular momentum operators in the final result is not straightforward. Furthermore, the procedure must be adapted to each individual molecule.

A more general approach was recently advocated by Chapuisat, Iung, Gatti and coworkers (18, 75, 76). These authors propose a vector parametrization scheme in combination with polyspherical coordinates to describe the N-atom molecular system. This set of internal coordinates consists of $(N-1)$ distances, $(N-2)$ polar angles, and $(N-3)$ azimuthal angles. The method is based on the sf representation of the kinetic energy. Expressing the conjugate momentum vectors in terms of associated orbital angular momentum vectors and transforming the result to bf coordinates yield the desired canonical form of the kinetic energy. The appropriate quantum mechanical operator is finally found by applying the Podolsky transformation. Alternatively, the same Hamiltonian has been derived directly through application of the chain rule (22).

In the approach of Chapuisat et al., the internal configuration of the molecule is defined by $N - 1$ internal vectors $\vec{q}_i$ (18). Incorporating these vectors into a set of generalized coordinates results in an expression for the kinetic energy, which in general contains also cross terms in the velocity vectors $\dot{\vec{q}}_i$. These cross terms are absent for an orthogonal set of vectors resulting in a greatly simplified kinetic energy operator. The implications resulting from the use of a nonorthogonal set of internal vectors, e.g., bond vectors, have been discussed as well (77, 78). The great advantage of this method is the parametrization of the internal vectors in terms of polar coordinates so that the commutation properties and matrix elements of the standard orbital angular momentum operator can be applied directly to these Hamiltonian. The matrix representation of the complete Hamiltonian thus requires only numerical integration over the stretch coordinates.

A set of polyspherical coordinates was also used to parametrize the purely vibrational Hamiltonians of a polyatomic molecule (20, 21). Frederick & Woywood gave explicit results for the G-matrix elements associated with the polar angles between adjacent Jacobi vectors and the dihedral angles between sequentially or centrally connected Jacobi vetors. These authors also present expressions for the extra potential terms resulting from the use of a non-Euclidian volume element for the normalization of the wavefunction (21).

3.1. Jacobi Coordinates

To illustrate the use of an orthogonal set of internal vectors, we discuss first the three-atom system described by Jacobi vectors. It turns out that the results obtained for this system can be applied directly towards the extension to molecules with N atoms. The idea of building up the kinetic energy operator of a large system from the results obtained for a smaller system has also been explored in a similar context (23, 79). For the three-atom system, we use the Jacobi vectors $\vec{q}_1$ and $\vec{q}_2$ as defined in Figure 2:

$$\vec{q}_1(sf) = \vec{R}_1 - \vec{R}_2 \quad \text{and} \quad \vec{q}_2(sf) = \vec{R}_3 - \vec{X}_2 = \vec{R}_3 - \frac{1}{M_2}(m_1\vec{R}_1 + m_2\vec{R}_2), \quad 52.$$

where the total mass of the diatomic fragment (atoms 1 and 2) is denoted M_2 and its cm vector is $\vec{X}_2$. In the following, we adopt the convention $M_k = \sum_{i=1}^{k} m_i$ and $\vec{X}_k = \frac{1}{M_k}\sum_{i=1}^{k} m_i \vec{R}_i$. In combination with the cm position $\vec{X}_3$ of the complete three atom system, $\vec{q}_1(sf)$ and $\vec{q}_2(sf)$ define a new set of generalized coordinates. As expected for orthogonal internal vectors, the resulting sf representation of the kinetic energy does not contain any cross terms:

$$2T = M_3\dot{\vec{X}}_3^2 + \mu_1\dot{\vec{q}}_1^2(sf) + \mu_2\dot{\vec{q}}_2^2(sf) \quad \text{with} \quad \mu_1 = \frac{m_1 m_2}{M_2}, \mu_2 = \frac{M_2 m_3}{M_3} \quad 53.$$

In a next step, we introduce a bf system whose z-axis points along the direction of $\vec{q}_2(sf)$. Therefore, the first Euler rotations are defined by the polar angles of

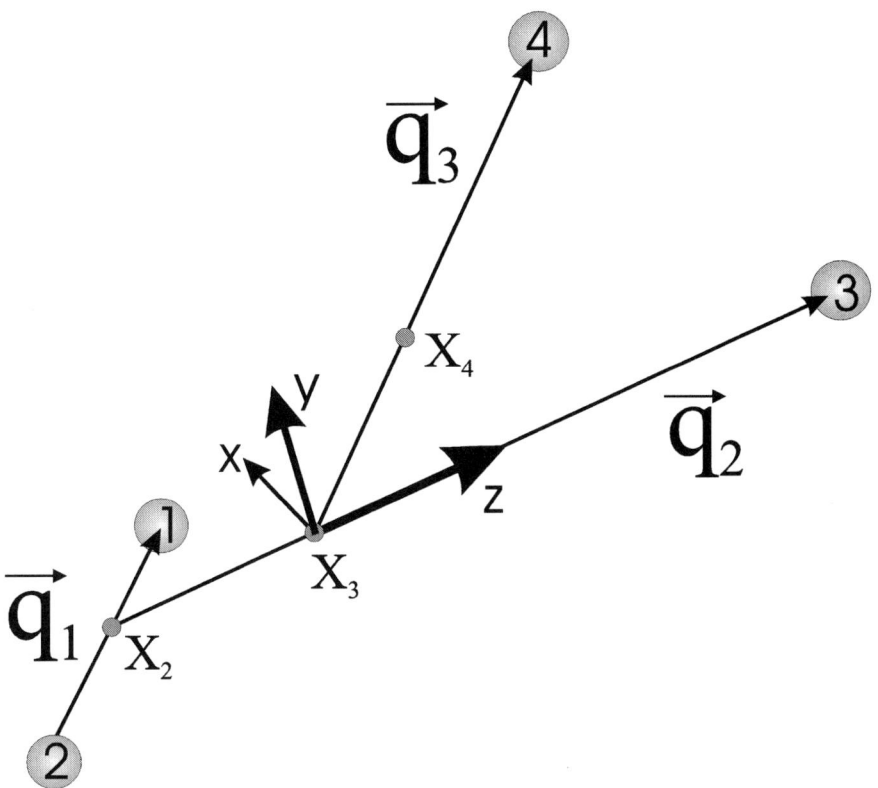

Figure 2 Definition of Jacobi vectors for a four-atom system, The orientation of the bf frame for the three-atom system is defined through the vectors $\vec{q}_1$ and $\vec{q}_2$. Its origin is located in the cm $\vec{X}_3$ of this system.

$\vec{q}_2(sf)$ with respect to the sf frame, i.e., $\alpha = \Phi_2$ and $\beta = \Theta_2$. The third Euler rotation is defined by requiring that the three atoms lie in the x,z-plane of the bf frame. This implies that the y-component of the bf representation of $\vec{q}_1$ vanishes. Thus, the Euler angle γ can be expressed in terms of the components of $\vec{q}_1(sf)$:

$$\tan \gamma = \frac{-\sin \alpha\, q_{1x}(sf) + \cos \alpha\, q_{1y}(sf)}{\cos \alpha \cos \beta\, q_{1x}(sf) + \sin \alpha \cos \beta\, q_{1y}(sf) - \sin \beta\, q_{1z}(sf)}. \qquad 54.$$

As before, we calculate $(C\dot{\vec{q}}_i(sf))^2$ rather than $(\dot{\vec{q}}_i(sf))^2$, which yields the translation-free contribution to the kinetic energy in the bf representation:

$$2T = \mu_1(\dot{\vec{q}}_1 + \vec{\omega} \times \vec{q}_1)^2 + \mu_2(\dot{\vec{q}}_2 + \vec{\omega} \times \vec{q}_2)^2. \qquad 55.$$

Because of the definition of the rotating frame, the bf representations of $\vec{q}_1$ and $\vec{q}_2$ are very restricted. Only the length of $\vec{q}_2$ can vary while for $\vec{q}_1$ the length and its angle with the direction of $\vec{q}_2$ are free to change.

$$\vec{q}_2 = \mathbf{C}(\alpha, \beta, \gamma)\vec{q}_2(sf) = \begin{pmatrix} 0 \\ 0 \\ q_2 \end{pmatrix} \quad \text{and} \quad \vec{q}_1 = \mathbf{C}(\alpha, \beta, \gamma)\vec{q}_1(sf) = \begin{pmatrix} q_1 \sin\theta_1 \\ 0 \\ q_1 \cos\theta_1 \end{pmatrix}.$$

56.

Substituting these definitions into Equation 55, we find the kinetic energy in terms of the generalized coordinates and their time derivatives:

$$2T = \mu_2 \dot{q}_2^2 + \mu_2 q_2^2 (\omega_x^2 + \omega_y^2) + \mu_1 \dot{q}_1^2$$
$$+ \mu_1 q_1^2 \big((\omega_y + \dot{\theta}_1)^2 + (\omega_z \sin\theta_1 - \omega_x \cos\theta_1)^2\big).$$

57.

This form is suitable to determine the conjugate momenta $P_i = \frac{\partial T}{\partial \dot{q}_i}$, $j_{\theta_1} = \frac{\partial T}{\partial \dot{\theta}_1}$, and the total angular momentum in its bf representation: $\vec{J} = \nabla_{\vec{\omega}} T$. Eliminating the velocities from Equation 57 finally yields the canonical form of the kinetic energy:

$$T = \frac{1}{2\mu_1} P_1^2 + \frac{1}{2\mu_2} P_2^2 + \frac{1}{2}\left(\frac{1}{\mu_1 q_1^2} + \frac{1}{\mu_2 q_2^2}\right)\left(j_{\theta_1}^2 + \csc^2\theta_1 J_z^2\right)$$
$$+ \frac{1}{2\mu_2 q_2^2}\left(J^2 - 2J_z^2 + \cot\theta_1(J_x J_z + J_z J_x) - 2J_y j_{\theta_1}\right).$$

58.

Using the determinant $g = \mu_1^{-3} \mu_2^{-3} q_1^{-4} q_2^{-4} \csc^2\theta_1 \csc^2\beta$ and the volume element $s = q_1^2 q_2^2 \sin\theta_1 \sin\beta$ in combination with the Podolsky transformation, we finally derive the quantum mechanical operator for the kinetic energy:

$$T = -\frac{\hbar^2}{2\mu_1} \frac{1}{q_1^2} \frac{\partial}{\partial q_1} q_1^2 \frac{\partial}{\partial q_1} - \frac{\hbar^2}{2\mu_2} \frac{1}{q_2^2} \frac{\partial}{\partial q_2} q_2^2 \frac{\partial}{\partial q_2}$$
$$- \frac{1}{2}\left(\frac{1}{\mu_1 q_1^2} + \frac{1}{\mu_2 q_2^2}\right)\left(\frac{\hbar^2}{\sin\theta_1} \frac{\partial}{\partial \theta_1} \sin\theta_1 \frac{\partial}{\partial \theta_1} - \csc^2\theta_1 J_z^2\right)$$
$$- \frac{1}{2\mu_2 q_2^2} \frac{\hbar}{i}\left(2\frac{\partial}{\partial \theta_1} + \cot\theta_1\right) J_y + \frac{1}{2\mu_2 q_2^2}\left(J^2 - 2J_z^2 + \cot\theta_1(J_x J_z + J_z J_x)\right),$$

59.

where the operators J_κ ($\kappa = x, y, z$) are defined in Equation 17. This Hamiltonian was first reported by Sutcliff & Tennyson using the chain rule approach (67). The presence of the anti-commutator in J_x and J_z results from the use of an axis system in which the moment of inertia tensor is not diagonal. In terms of a symmetric top basis set, this term is responsible for off-diagonal matrix elements with $\Delta K = \pm 1$. On the other hand, the Coriolis coupling term involves the bending motion and the y-axis rotation. Through this term, different rovibrational basis states are coupled. As pointed out by Carrington, choosing the z-axis of the bf frame perpendicular to the molecular plane results in a Coriolis coupling term that is diagonal in the quantum number K (71). In a crude adiabatic approximation, the pure bending problem is solved first for fixed values of K. In a second step, the overall rotation problem is treated including the K-dependent bending energy eigenvalues. While the decoupling of the bending motion from the overall rotation represents an important advantage, the rotational problem is slightly more complicated due to the presence of complex valued off-diagonal matrix elements. A similar approach, called the ρ-axis method, was developed by Herbst et al. for the treatment of the coupling between overall rotation and internal rotation of symmetric top internal rotors (80–82).

EXTENSION TOWARDS N ATOMS At this point, the extension towards systems with N atoms ($N \geq 3$) becomes straightforward. The system will be described by (N − 1) Jacobi vectors $\vec{q}_i$. The sf representation of the kinetic energy is of the same form given in Equation (53):

$$2T = M_N \dot{\vec{X}}_N^2 + \mu_2 \dot{\vec{q}}_2^2(sf) + \mu_1 \dot{\vec{q}}_1^2(sf) + \sum_{i=3}^{N-1} \mu_i \dot{\vec{q}}_i^2(sf) \quad \text{with} \quad \mu_i = \frac{M_i m_{i+1}}{M_{i+1}}.$$
60.

Here we have separately listed the contributions due to the first two Jacobi vectors that define the rotating frame in exactly the same way as discussed above for the three-atom system. Therefore the bf representation of the translation-free kinetic energy can be derived simply in extension of Equation 55.

$$T = T_1 + T_2 + T_3 \quad \text{with} \quad T_1 = \frac{1}{2}\mu_1(\dot{\vec{q}}_1 + \vec{\omega} \times \vec{q}_1)^2,$$

$$T_2 = \frac{1}{2}\mu_2(\dot{\vec{q}}_2 + \vec{\omega} \times \vec{q}_2)^2, \quad \text{and} \quad T_3 = \frac{1}{2}\sum_{i=3}^{N-1}\mu_i(\dot{\vec{q}}_i + \vec{\omega} \times \vec{q}_i)^2. \quad 61.$$

While the motion of the frame defining vectors $\vec{q}_1$ and $\vec{q}_2$ is restricted (see Equation 56), the vectors $\vec{q}_i$ for $i \geq 3$ are completely free. Therefore, their cartesian components can serve as generalized coordinates exactly in the same manner as the electronic coordinates discussed in Section 2.4. Expressed in terms of the appropriate conjugate momenta $\vec{P}_i = \nabla_{\dot{\vec{q}}_i} T_3$, their contribution to the kinetic energy is of the simple form $T_3 = \sum_{i=3}^{N-1} \frac{\vec{P}_i^2}{2\mu_i}$, while the contribution due to the first two terms is similar to the expression given in Equation 59. Since these terms include the rotational contribution, we must investigate the modification of the total angular momentum due to the extension towards more atoms. Similar to the discussion of Section 2.4, the rotation of the additional Jacobi vectors results in the contribution $(\nabla_{\vec{\omega}} T_3) = \sum_{i=3}^{N-1} \vec{q}_i \times \vec{P}_i = \sum_{i=3}^{N-1} \vec{l}_i = \vec{L}$ to $\vec{J}$. Therefore the total angular momentum for the N-atom system can be expressed in terms of the orbital angular momentum $\vec{L}$ and the total angular momentum for the three atom system $\vec{J}(3)$: $\vec{J} = \vec{J}(3) + \vec{L}$. This allows us to find the kinetic energy contribution due to $T_1 + T_2$ by replacing $\vec{J}$ with $(\vec{J} - \vec{L})$ in Equation 59:

$$T = -\frac{\hbar^2}{2\mu_1 q_1^2}\frac{1}{\partial q_1}q_1^2\frac{\partial}{\partial q_1} - \frac{\hbar^2}{2\mu_2 q_2^2}\frac{1}{\partial q_2}q_2^2\frac{\partial}{\partial q_2} - \frac{1}{2\mu_2 q_2^2}\frac{\hbar}{i}$$
$$\times \left(2\frac{\partial}{\partial\theta_1} + \cot\theta_1\right)(J_y - L_y) - \frac{1}{2}\left(\frac{1}{\mu_1 q_1^2} + \frac{1}{\mu_2 q_2^2}\right)$$
$$\times \left(\frac{\hbar^2}{\sin\theta_1}\frac{\partial}{\partial\theta_1}\sin\theta_1\frac{\partial}{\partial\theta_1} - \csc^2\theta_1(J_z - L_z)^2\right) + \sum_{i=3}^{N-1}\frac{\vec{P}_i^2}{2\mu_i}$$
$$+ \frac{1}{2\mu_2 q_2^2}((\vec{J} - \vec{L})^2 - 2(J_z - L_z)^2 + \cot\theta_1((J_x - L_x)(J_z - L_z)$$
$$+ (J_z - L_z)(J_x - L_x))). \qquad 62.$$

The Hamiltonian in Equation 62, is essentially the one obtained by Mladenovic who used the chain rule approach (22) and Gatti et al. whose method is also based on classical mechanics (76). Since the orbital angular momenta $\vec{l}_i$ are associated with the unrestricted motion of the vectors $\vec{q}_i$ (with $i \geq 3$) in the bf frame, they fulfill normal commutation rules in contrast to the bf components of $\vec{J}$: $[l_\kappa, l_\delta] = -\frac{\hbar}{i}\sum_\epsilon e_{\kappa\delta\epsilon}l_\epsilon$. If the vectors $\vec{q}_i$ are described in terms of spherical polar coordinates, the well known properties of the orbital angular momentum operator and its matrix elements can be employed to treat the internal rotation angles analytically greatly simplifying the problem of finding the eigenvalue spectrum and its eigenfunctions.

3.2. Valence Coordinates

Besides Jacobi or Radau coordinates, valence type coordinates like bond lengths and bond angles are frequently used because of the efficient decoupling of vibrational degrees of freedom. For a triatomic system, the introduction of valence coordinates is facilitated by defining the internal vectors $\vec{q}_1(sf) = \vec{R}_1 - \vec{R}_3$ and $\vec{q}_2(sf) = \vec{R}_2 - \vec{R}_3$, which yields the following expression for the kinetic energy:

$$T = \frac{1}{2}M\vec{R}_{cm}^2 + \frac{1}{2}\mu_{11}\dot{\vec{q}}_1^2(sf) + \frac{1}{2}\mu_{22}\dot{\vec{q}}_2^2(sf) + \mu_{12}\dot{\vec{q}}_1(sf)\dot{\vec{q}}_2(sf),$$

with

$$\mu_{11} = \frac{m_1(m_2 + m_3)}{M}, \quad \mu_{22} = \frac{m_2(m_1 + m_3)}{M} \quad \text{and} \quad \mu_{12} = -\frac{m_1 m_2}{M}. \qquad 63.$$

It is common to define the bf frame either with the z-axis along the bisector (65) of $\vec{q}_1$ and $\vec{q}_2$ or along $\vec{q}_2$ (17). For the z-bond system, we use the coordinate representation

$$\vec{q}_2 = \begin{pmatrix} 0 \\ 0 \\ q_2 \end{pmatrix} \quad \text{and} \quad \vec{q}_1 = q_1 \begin{pmatrix} \sin\theta_1 \\ 0 \\ \cos\theta_1 \end{pmatrix}. \qquad 64.$$

Substituting the analogue of Equation 3 into Equation 63, the bf representation of the translation-free kinetic energy becomes:

$$\begin{aligned}
T =\ & \tfrac{1}{2}\mu_{11}\dot{q}_1^2 + \tfrac{1}{2}\mu_{22}\dot{q}_2^2 + \mu_{12}\cos\theta_1 \dot{q}_1\dot{q}_2 + \tfrac{1}{2}\mu_{11}q_1^2\dot{\theta}_1^2 - \mu_{12}\sin\theta_1 q_1\dot{q}_2\dot{\theta}_1 \\
& + \left(\mu_{11}q_1^2 + \mu_{12}q_1q_2\cos\theta_1\right)\omega_y\dot{\theta}_1 - \mu_{12}\sin\theta_1(q_1\dot{q}_2 - q_2\dot{q}_1)\omega_y \\
& + \tfrac{1}{2}\left(\mu_{11}q_1^2 + \mu_{22}q_2^2 + 2\mu_{12}q_1q_2\cos\theta_1\right)\omega_y^2 \\
& + \tfrac{1}{2}\left(\mu_{22}q_2^2 + \mu_{11}q_1^2\cos^2\theta_1 + 2\mu_{12}q_1q_2\cos\theta_1\right)\omega_x^2 + \tfrac{1}{2}\mu_{11}q_1^2\sin^2\theta_1\,\omega_z^2 \\
& - \left(\mu_{12}q_1q_2\sin\theta_1 + \mu_{11}q_1^2\sin\theta_1\cos\theta_1\right)\omega_x\omega_z.
\end{aligned} \qquad 65.$$

In comparison with the kinetic energy in Jacobi coordinates, the use of valence coordinates results in a more complex expression demonstrating very clearly the advantage of using an orthogonal set of internal vectors. Nevertheless, the above expression can be used to calculate the conjugate momenta P_1, P_2, j_{θ_1}, J_y, J_x, J_z. Introducing these momenta into Equation 65, yields the canonical form of the kinetic energy.

$$\begin{aligned}
T =\ & \frac{1}{2\mu_2}P_2^2 + \frac{1}{2\mu_1}P_1^2 + \frac{\cos\theta_1}{m_3}P_1 P_2 - \frac{\sin\theta_1}{m_3}\left(\frac{P_2}{q_1} + \frac{P_1}{q_2}\right)j_{\theta_1} + \frac{\sin\theta_1}{m_3 q_2}J_y P_1 \\
& + \frac{1}{2}\left(\frac{1}{\mu_1 q_1^2} + \frac{1}{\mu_2 q_2^2} - \frac{2\cos\theta_1}{m_3 q_1 q_2}\right)j_{\theta_1}^2 + \left(\frac{\cos\theta_1}{m_3 q_1 q_2} - \frac{1}{\mu_2 q_2^2}\right)j_{\theta_1} J_y \\
& + \frac{1}{2\mu_2 q_2^2}(J_x^2 + J_y^2) + \frac{1}{2}\left(\frac{\csc^2\theta_1}{\mu_1 q_1^2} + \frac{\cot^2\theta_1}{\mu_2 q_2^2} - \frac{2\csc\theta_1\cot\theta_1}{m_3 q_1 q_2}\right)J_z^2 \\
& + \left(\frac{\cot\theta_1}{\mu_2 q_2^2} - \frac{\csc\theta_1}{m_3 q_1 q_2}\right)J_x J_z \quad \text{with} \quad \mu_1 = \mu_{11} - \frac{\mu_{12}^2}{\mu_{22}},
\end{aligned}$$

$$\mu_2 = \mu_{22} - \frac{\mu_{12}^2}{\mu_{11}}, \quad \text{and} \quad m_3 = \frac{\mu_{11}\mu_{22}}{\mu_{12}} - \mu_{12}. \qquad 66.$$

Here, the canonical form is given in terms of the original coefficients μ_{ij}. Thus the result can be formally reduced to the Jacobi case in considering the limit $\mu_{12} \to 0$, i.e., $m_3 \to \infty$. The corresponding quantum mechanical operator is readily derived using the determinant $g = (\frac{M}{m_1 m_2 m_3})^3 q_1^{-4} q_2^{-4} \csc^2\theta_1 \csc^2\beta$ and the volume elements $s = q_1^2 q_2^2 \sin\theta_1 \sin\beta$ or $s = \sin\theta_1 \sin\beta$ as reported by Sutcliffe and coworkers (65, 71).

EXTENSION TOWARD N ATOMS   It is also straightforward to extend the results of this section toward molecules with N atoms although the results are not quite as compact as the ones for the Jacobi vectors. If the molecule is described by a nonorthogonal set of (N − 1) vectors $\vec{q}_i$, the translation-free kinetic energy in the sf representation will be generated by a symmetric and constant matrix **M**:

$$T = \frac{1}{2} \sum_{i,j} M_{ij} \vec{\dot{q}}_i(sf) \vec{\dot{q}}_j(sf). \qquad 67.$$

Since the rotating frame is assumed to be defined through the vectors $\vec{q}_1$ and $\vec{q}_2$, we can reduce the N-atom (N ≥ 3) problem to the triatomic form if we separate out the contributions due to the motion of $\vec{q}_1$ and $\vec{q}_2$. For this purpose, it is convenient to introduce the inverse to the sub-matrix M_{kl} with $k, l \geq 3$. Using modified matrix elements $\tilde{M}_{ij}$ for $i, j = 1, 2$, it is possible to express the canonical form of the kinetic energy in terms of the corresponding triatomic energy $T(3, \tilde{M}_{ij})$ as given in Equation 66:

$$T(N) = T(3, \tilde{M}_{ij}) + \frac{1}{2} \sum_{k,l \geq 3} M_{kl}^{-1} \vec{P}_k \vec{P}_l$$

and

$$\tilde{M}_{ij} = M_{ij} - \sum_{kl \geq 3} M_{ik} M_{kl}^{-1} M_{lj} \quad \text{for} \quad i, j = 1, 2. \qquad 68.$$

Furthermore, the conjugate momenta P_1, P_2, j_{θ_1} and $\vec{J}$ in $T(3, \tilde{M}_{ij})$ are modified by contributions due to the momenta $\vec{P}_k$ resulting in the following substitutions:

$$P_2 \to P_2 - \sum_{kl \geq 3} M_{2k} M_{kl}^{-1} P_{lz}$$

$$P_1 \to P_1 - \sin\theta_1 \sum_{kl \geq 3} M_{1k} M_{kl}^{-1} P_{lx} - \cos\theta_1 \sum_{kl \geq 3} M_{1k} M_{kl}^{-1} P_{lz}$$

$$j_{\theta_1} \to j_{\theta_1} - q_1 \cos\theta_1 \sum_{kl \geq 3} M_{1k} M_{kl}^{-1} P_{lx} + q_1 \sin\theta_1 \sum_{kl \geq 3} M_{1k} M_{kl}^{-1} P_{lz}$$

$$\vec{J} \to \vec{J} - \vec{L} - \sum_{i=1,2} \sum_{kl \geq 3} M_{ik} M_{kl}^{-1} \vec{q}_i \times \vec{P}_l \quad \text{and} \quad \vec{L} = \sum_{k \geq 3} \vec{q}_k \times \vec{P}_k. \qquad 69.$$

Because of the nonorthogonality of the internal vectors, additional couplings between the stretch and bending coordinates R_i, θ_1 and the free vectors $\vec{q}_k$ exist. This coupling manifests itself also in a contribution to the rotational angular momentum that takes the role of the total angular momentum in Equation 66.

3.3. Rovibrational S-Vectors

In the discussion presented so far, it was not necessary to specify explicitly how the nuclear sf configuration defines the three Euler angles. On the other hand, the use of a two-vector embedded frame as described in the previous sections results in simple expressions for the Euler angles. This fact was exploited recently by Lukka in Reference (19) who developed an extension of the vibrational S-vector method (83). This new approach towards constructing molecular Hamiltonians has been applied already towards molecules containing four and five atoms (19, 84, 85).

In order to treat the overall rotation of the bf frame in a manner similar to the motion along internal coordinates, it is necessary to start with the sf representation of the kinetic energy. We seek a relation between the sf positions $\{R_{i\alpha}\}$ and the set of coordinates $\{t_x, t_y, t_z, \epsilon_x, \epsilon_y, \epsilon_z, q_1, \ldots, q_{3N-6}\}$. Here $\{t_x, t_y, t_z\}$ represent three translational coordinates that must be set equal to zero because we consider only the motion of the nuclei with respect to the sf frame. The coordinates $\{\epsilon_x, \epsilon_y, \epsilon_z\}$ represent three rotational coordinates that specify the location of the bf frame relative to the sf frame. The q_i represent the 3N − 6 internal coordinates. In the following, we regard ϵ_x, ϵ_y, and ϵ_z as defined through the Euler angles α, β, and γ, respectively. Considering infinitesimal coordinate changes, we determine a linear relationship between the cartesian sf displacements $R_{i\alpha}$ and the resulting variation of the internal coordinates:

$$dq_j = \sum_{i=1}^{N} \sum_{\alpha=x,y,z} \frac{\partial f_j(R_{i\alpha})}{\partial R_{i\alpha}} dR_{i\alpha} = \sum_{i=1}^{N} \sum_{\alpha=x,y,z} S_{i\alpha}^{q_j} dR_{i\alpha} = \sum_i \vec{S}_i^{q_j} \cdot d\vec{R}_i$$

$$d\epsilon_\kappa = \sum_{i=1}^{N} \sum_{\alpha=x,y,z} \frac{\partial f_\kappa(R_{i\alpha})}{\partial R_{i\alpha}} dR_{i\alpha} = \sum_{i=1}^{N} \sum_{\alpha=x,y,z} S_{i\alpha}^{\epsilon_\kappa} dR_{i\alpha} = \sum_i \vec{S}_i^{\epsilon_\kappa} \cdot d\vec{R}_i. \quad 70.$$

Note that the variation of the translational coordinates vanishes for the sf frame. Substituting Equation 70 into the sf representation of the kinetic energy, we find:

$$T = \frac{1}{2}\dot{\vec{R}}^\dagger \mathbf{M} \dot{\vec{R}} = \frac{1}{2}\dot{\vec{q}}^\dagger \mathbf{S}^{-1\dagger} \mathbf{M} \mathbf{S}^{-1} \dot{\vec{q}}. \quad 71.$$

Here we have introduced the 3N-dimensional vectors $\vec{R}$ and $\vec{q}$. From this expression, the momenta conjugate to $\vec{q}$ are calculated in the usual way. This leads to the canonical form of the kinetic energy:

$$T = \frac{1}{2}\vec{P}^\dagger \mathbf{G} \vec{P} = \frac{1}{2}\sum_{ij} p_i G_{ij} p_j \quad \text{with} \quad \mathbf{G} = \mathbf{S}\mathbf{M}^{-1}\mathbf{S}^\dagger, \quad 72.$$

where the indices i, j refer to the rotational and the internal coordinates. Remembering that the matrix $\mathbf{M}$ is diagonal, the elements of the G-matrix are easily expressed in terms of S-vectors:

$$G_{ij} = (\mathbf{S}\mathbf{M}^{-1}\mathbf{S}^\dagger)_{ij} = \sum_{(l\alpha)} \frac{1}{m_l} S_{i(l\alpha)} S_{j(l\alpha)} = \sum_l \frac{1}{m_l} \vec{S}_l^i \cdot \vec{S}_l^j. \quad 73.$$

In this way, the G-matrix gives rise to the vibrational as well as the rotational contributions to the kinetic energy and their interactions. The important contribution of Lukka consists in the realization that, because of the rotational invariance of the molecular Hamiltonian, we can determine the rotational S-vectors for a fixed orientation of the bf frame. Most conveniently, this is chosen to coincide with the particular Euler angles $\alpha = \gamma = 0$ and $\beta = \frac{\pi}{2}$. From Equation 17 it follows that under these conditions the momenta conjugate to the Euler angles are identical to the components of the total angular momentum projected onto bf axes: $J_x = -J_\alpha$, $J_y = J_\beta$, and $J_z = J_\gamma$. Similarly, the components of $\vec{\omega}$ are identical to the angular velocities $-\dot{\alpha}$, $\dot{\beta}$, and $\dot{\gamma}$. Introducing the associated S-vectors, the kinetic energy is expressed directly in terms of the angular momentum components J_x, J_y, and J_z.

Let us assume that the bf frame is defined through the vectors $\vec{v}$ and $\vec{u}$. The direction of $\vec{v}$ defines the z-axis of the bf frame while its y-axis is parallel to $\vec{v} \times \vec{u}$. Under these conditions, the polar angles of the sf representation of $\vec{v}$ define the Euler angles α and β:

$$\cos\beta = \frac{v_z^{(sf)}}{v} \quad \text{and} \quad \tan\alpha = \frac{v_y^{(sf)}}{v_x^{(sf)}}. \qquad 74.$$

The third Euler angle γ is determined from Equation 54 with $\vec{u} = \vec{q}_1$. It ensures the condition that $\vec{u}$ lies in the x, z-plane of the bf frame. The rotational S-vectors are determined from the total differential of the Euler angles as a function of the sf components of $\vec{v}$ and $\vec{u}$ for the special orientation $\alpha = \gamma = 0$ and $\beta = \frac{\pi}{2}$:

$$d\alpha = \frac{dv_y^{(sf)}}{v}, \quad d\beta = -\frac{dv_z^{(sf)}}{v}, \quad \text{and} \quad d\gamma = \frac{u_x^{(sf)}}{u_z^{(sf)}} d\alpha - \frac{du_y^{(sf)}}{u_z^{(sf)}}. \qquad 75.$$

To illustrate the method, we apply it towards the triatomic system in Jacobi coordinates, i.e., we choose $\vec{v} = \vec{q}_2$ and $\vec{u} = \vec{q}_1$ where $\vec{q}_1$ and $\vec{q}_2$ are defined in Equation 56. It is important to realize that Equation 75 involves sf components, e.g., $u_x^{(sf)} = q_1 \cos\theta_1$ and $u_z^{(sf)} = -q_1 \sin\theta_1$:

$$d\alpha = \frac{1}{q_2} dy_3 - \frac{m_1}{M_2 q_2} dy_1 - \frac{m_2}{M_2 q_2} dy_2$$

$$d\beta = -\frac{1}{q_2} dz_3 + \frac{m_1}{M_2 q_2} dz_1 + \frac{m_2}{M_2 q_2} dz_2$$

$$d\gamma = -\frac{\cot\theta_1}{q_2} dy_3 + \left(\frac{m_1 \cot\theta_1}{M_2 q_2} + \frac{1}{q_1 \sin\theta_1}\right) dy_1 + \left(\frac{m_2 \cot\theta_1}{M_2 q_2} - \frac{1}{q_1 \sin\theta_1}\right) dy_2.$$

$$76.$$

The resulting rotational and vibrational S-vectors are listed in Table 1; the G-matrix elements calculated according to Equation 73 are consistent with the Hamilton function derived in Section 3.1.

TABLE 1 S-vectors for the triatomic system in Jacobi coordinates. Note that $\hat{e}_\kappa$ with $\kappa = x, y, z$ represents unit vectors in the sf frame for the special orientation $\alpha = \gamma = 0$ and $\beta = \frac{\pi}{2}$

	1	2	3
ϵ_x	$\dfrac{m_1}{M_2 q_2}\hat{e}_y$	$\dfrac{m_2}{M_2 q_2}\hat{e}_y$	$-\dfrac{1}{q_2}\hat{e}_y$
ϵ_y	$\dfrac{m_1}{M_2 q_2}\hat{e}_z$	$\dfrac{m_2}{M_2 q_2}\hat{e}_z$	$-\dfrac{1}{q_2}\hat{e}_z$
ϵ_z	$\left(\dfrac{m_1 \cot\theta_1}{M_2 q_2} + \dfrac{1}{\sin\theta_1}\right)\hat{e}_y$	$\left(\dfrac{m_2 \cot\theta_1}{M_2 q_2} - \dfrac{1}{\sin\theta_1}\right)\hat{e}_y$	$-\dfrac{\cot\theta_1}{q_2}\hat{e}_y$
θ_1	$-\dfrac{\sin\theta_1}{q_1}\hat{e}_x - \left(\dfrac{m_1}{M_2 q_2} + \dfrac{\cos\theta_1}{q_1}\right)\hat{e}_z$	$\dfrac{\sin\theta_1}{q_1}\hat{e}_x - \left(\dfrac{m_2}{M_2 q_2} - \dfrac{\cos\theta_1}{q_1}\right)\hat{e}_z$	$\dfrac{1}{q_2}\hat{e}_z$
q_1	$\cos\theta_1 \hat{e}_x - \sin\theta_1 \hat{e}_z$	$-\cos\theta_1 \hat{e}_x + \sin\theta_1 \hat{e}_z$	0
q_2	$-\dfrac{m_1}{M_2}\hat{e}_x$	$-\dfrac{m_2}{M_2}\hat{e}_x$	$\hat{e}_x$

4. CONCLUSIONS

The construction of a molecular Hamiltonian is the prerequisite for any theoretical study in spectroscopy as well as molecular reaction dynamics. It is hoped that the derivation of several of the most commonly used Hamiltonians with a single approach makes this extremely basic but also difficult topic more transparent and thus accessible to the nonspecialist, but interested researcher.

Depending on the type of system or the conditions of interest, different types of coordinates are selected. One of the most important aims in choosing a set of coordinates is the efficient decoupling of zeroth order vibrations and of vibration and overall rotation. Several alternative methods for the construction of exact molecular Hamiltonians including rotation have been developed recently. The availability of these methods can shift the emphasis from avoiding difficulties and mistakes in the construction of the Hamiltonian towards deriving a Hamiltonian in a representation that results in minimal coupling between different degrees of freedom. These advances should stimulate the theoretical study of larger systems and of systems at higher energies.

ACKNOWLEDGMENTS

I would like to thank Professor A. van der Avoird and Professor P. R. Bunker for several fruitful and stimulating discussions on the subject of this article. Professor T. Heil deserves special thanks for stimulating the author's interest in the application of concepts of classical mechanics towards the derivation of quantum mechanical Hamiltonians. Financial support from the National Science Foundation (grant CHE-0097189) and the donors of The Petroleum Research Fund, administered by the ACS is gratefully acknowledged.

Visit the Annual Reviews home page at www.annualreviews.org

LITERATURE CITED

1. Sutcliffe BT. 1982. In *Current Aspects of Quantum Chemistry*, ed. R Carbo, pp. 99–125. New York: Elsevier
2. Herzberg G. 1966. *Molecular Spectra and Molecular Structure*, Vols. 1–3. New York: Van Nostrand
3. Kroto HW. 1975. *Molecular Rotation Spectra*. New York: Wiley
4. Papousek D, Aliev MR. 1982. *Molecular Vibrational-Rotational Spectra*. New York: Elsevier
5. Bunker PR, Jensen P. 1998. *Molecular Symmetry and Spectroscopy*. Ottawa: NRC
6. Child MS. 1974. *Molecular Collision Theory*. New York: Academic
7. Zhang JZH. 1999. *Theory and Application of Quantum Molecular Dynamics*. Singapore: World Sci.
8. Kronig RdeL. 1930. *Band Spectra and Molecular Structure*. New York: Macmillan
9. Wilson EB Jr, Howard JB. 1936. *J. Chem. Phys.* 4:260–68
10. Darling BT, Dennison DM. 1940. *Phys. Rev.* 87:128–39
11. Nielsen HH. 1951. *Rev. Mod. Phys.* 23:90–136
12. Nielsen HH. 1959. *Handb. Phys.* 37:173–313
13. Lin CC, Swalen JD. 1959. *Rev. Mod. Phys.* 31:841–92
14. Watson JKG. 1968. *Mol. Phys.* 15:479–90
15. Watson JKG. 1970. *Mol. Phys.* 19:465–87
16. Hougen JT, Bunker PR, Johns JWC. 1970. *J. Mol. Spectrosc.* 34:136–72
17. Handy NC. 1987. *Mol. Phys.* 61:207–23
18. Chapuisat X, Iung C. 1992. *Phys. Rev. A* 45:6217–35
19. Lukka TJ. 1995. *J. Chem. Phys* 102:3945–55
20. Makarewicz J, Skalozub A. 1999. *J. Chem. Phys.* 306:352–56
21. Frederick JH, Woywood C. 1999. *J. Chem. Phys.* 111:7255–71
22. Mladenovic M. 2000. *J. Chem. Phys.* 112:1070–81
23. Wang XG, Carrington T Jr. 2000. *J. Chem. Phys.* 113:7097–101
24. Mitchell KA, Littlejohn RG. 1999. *Mol. Phys.* 96:1305–15
25. Born M, Oppenheimer R. 1928. *Ann. Phys.* 84:457–62
26. Hougen JT. 1962. *J. Chem. Phy.* 36:519–34
27. Brocks G, van der Avoird A, Sutcliffe BT, Tennyson J. 1983. *Mol. Phys.* 50:1025–43
28. Casimir HBG. 1931. *The Rotation of a Rigid Body in Quantum Mechanics*. Den Haag: JB Walters
29. Eckart C. 1935. *Phys. Rev.* 47:552–58
30. Podolsky B. 1928. *Phys. Rev.* 32:812–16
31. Zare RN. 1988. *Angular Momentum*. New York: Wiley
32. Van Vleck JH. 1951. *Rev. Mod. Phys.* 23:213–27
33. Epa VC, Bunker PR. 1991. *J. Mol. Spectrosc.* 150:511–20
34. Louck JD. 1976. *J. Mol. Spectrosc.* 61:107–37
35. Lehmann KK, Scoles G, Pate BH. 1994. *Annu. Rev. Phys. Chem.* 45:241–74
36. Nesbitt DJ, Field RW. 1996. *J. Phys. Chem.* 100:12735–56

37. Perry DS. 1997. In *Highly Excited Molecules, ACS Symp. Ser.*, ed. AS Mullin, GC Schatz, 678:70–80. Washington, DC: ACS
38. Lawrence WD, Knight AEW. 1988. *J. Chem. Phys.* 92:5900–8
39. Kemble EC. 1937. *The Fundamental Principles of Quantum Mechanics.* New York: McGraw-Hill
40. Aliev MR, Watson JKG. 1985. In *Molecular Spectroscopy: Modern Research*, ed. KN Rao, III:1–67. New York: AP
41. Hougen JT. 1962. *J. Chem. Phys.* 37:1433–41
42. Hougen JT. 1963. *J. Chem. Phys.* 39:358–65
43. Watson JKG. 1967. *J. Chem. Phys.* 46:1935–49
44. Watson JKG. 1977. In *Vibrational Spectroscopy and Structure*, ed. JR Durig, 6:1–89. New York: Dekker
45. Amat G, Henry L. 1958. *Cah. Phys.* 12:273
46. Sayvetz A. 1939. *J. Chem. Phys.* 7:383–89
47. Chapuisat X, Belafhal A, Nauts A. 1991. *J. Mol. Spectrosc.* 149:274–304
48. Tully JC. 1976. In *Dynamics of Molecular Collisions*, Part B, ed. WH Miller, pp. 217–67. New York: Plenum
49. Butler LJ. 1998. *Annu. Rev. Phys. Chem.* 49:125–71
50. Howard BJ, Moss RE. 1970. *Mol. Phys.* 19:433–50
51. Howard BJ, Moss RE. 1971. *Mol. Phys.* 20:147–59
52. Kim Y, Meyer H. 2001. *Int. Rev. Phys. Chem.* 20:219–82
53. Meyer R, Gunthard HsH. 1968. *J. Chem. Phys.* 49:1510–20
54. Pickett HM. 1971. *J. Chem. Phys.* 56:1715–23
55. Burleigh DC, McCoy AB, Sibert EL III. 1995. In *Molecular Dynamics and Spectroscopy by Stimulated Emission Pumping*, ed. HL Dai, RW Field, pp. 999–1036. Singapore: World Sci.
56. Pack RT. 1974. *J. Chem. Phys.* 60:633–39
57. Alexander MH. 1982. *J. Chem. Phys.* 76:5974–88
58. Alexander MH. 1985. *Chem. Phys.* 92:337–44
59. Balint-Kurti GG, Shapiro M. 1981. *Chem. Phys.* 61:137–55
60. Hutson JM. 1991. In *Advances in Molecular Vibration and Collision Dynamics*, ed. JM Bowman, MA Ratner, IA:1–45. Greenwich, CT: JAI
61. Diehl H, Flügge S, Schroder U, Volkel A, Weiguny A. 1961. *Z. Phys.* 162:1–14
62. Freed KF, Lombardi JR. 1966. *J. Chem. Phys.* 45:591–98
63. Bardo RD, Wolfsberg M. 1977. *J. Chem. Phys.* 67:593–603
64. Carney GD, Sprandel LL, Kern CW. 1978. *Adv. Chem. Phys.* 37:305–79
65. Sutcliffe BT. 1983. *Mol. Phys.* 48:561–66
66. Carter S, Handy NC, Sutcliffe BT. 1984. *Mol. Phys.* 48:745–48
67. Tennyson J, Sutcliffe BT. 1982. *J. Chem. Phys.* 77:4061–72
68. Natanson GA. 1989. *Mol. Phys.* 66:129–41
69. Wei H, Carrington T Jr. 1997. *J. Phys. Chem.* 107:2813–18
70. Wei H, Carrington T Jr. 1997. *J. Phys. Chem.* 107:9493–501
71. Wei H, Carrington T Jr. 1998. *Chem. Phys. Lett.* 287:289–300
72. Bramley MJ, Green WH Jr, Handy NC. 1991. *Mol. Phys.* 73:1183–1208
73. Csaszar AG, Handy NC. 1995. *Mol. Phys.* 86:959–79
74. Rempe SB, Watts RO. 1998. *J. Chem. Phys.* 108:10084–95
75. Iung C, Gatti F, Viel A, Chapuisat X. 1999. *Phys. Chem. Chem. Phys.* 1:3377–85
76. Gatti F, Iung C, Menou M, Justum Y, Nauts A, Chapuisat X. 1998. *J. Chem. Phys.* 108:8804–20
77. Gatti F, Munoz C, Iung C. 2001. *J. Chem. Phys.* 114:8275–81
78. Mladenovic M. 2000. *J. Chem. Phys.* 112:1082–95
79. Xantheas SS, Sutcliffe BT. 1995. *J. Chem. Phys.* 103:8022–30

80. Herbst E, Messer JK, DeLucia FC. 1984. *J. Mol. Spectrosc.* 108:42–57
81. Hougen JT, Kleiner I, Godefroid M. 1994. *J. Mol. Spectrosc.* 163:559–86
82. Kim Y, Fleniken J, Meyer H. 1998. *J. Chem. Phys.* 109:3401–8
83. Wilson EB, Decius JC, Cross PC. 1955. *Molecular Vibrations.* New York: McGraw–Hill
84. Colwell SM, Handy NC. 1997. *Mol. Phys.* 92:317–30
85. Wang XGN, Sibert EL III, Child MS. 2000. *Mol. Phys.* 98:317–26

REVERSIBLE POLYMERIZATIONS AND AGGREGATIONS

Sandra C. Greer
Department of Chemical Engineering and Department of Chemistry and Biochemistry, The University of Maryland, College Park, College Park, Maryland 20742-2111; e-mail: sg28@umail.umd.edu

Key Words self-assembly, living polymer, supramolecular

■ **Abstract** The aggregation of monomers into polymers, whether by covalent or noncovalent interactions, is often reversible and frequently occurs with the entropy and enthalpy of the aggregation sharing the same sign. In such a case, the aggregation goes forward or reverses, depending on such variables as temperature and composition, rather like a phase transition. We explore the physical chemistry of three such systems: an organic monomer (α-methylstyrene), an inorganic monomer (sulfur), and a biopolymer (actin). We compare the available theories and experiments and list issues still open.

INTRODUCTION AND TERMINOLOGY

The thrill of science comes when we can find common features among disparate systems and are able to construct a common model to describe those various systems. Between 1970 and 1990, the theory of phase transitions and critical phenomena matured into a powerful model to describe changes of physical phase in magnets and in fluids. In 1980 the theory of phase transitions was suggested as a model for a particular kind of chemical reaction: reversible polymerization (1). Many polymerization reactions have an initiation step that starts monomers coming together to make polymers, followed by a reversible propagation step that adds monomers to the growing polymer, but lack a termination step to fix the size of the polymers. If, in addition, the enthalpy and entropy of the propagation step have the same sign, then the propagation will be thermodynamically allowed only in certain temperature ranges. The propagation will commence at a polymerization temperature, T_p, the polymers will grow in average degree of polymerization as the temperature is taken further into the polymerizing region, and the polymers will revert to monomer if the polymerization temperature is retraversed and the system is taken back into the nonpolymerizing region. Thus the onset of the propagation with the change of temperature is like the onset of a phase transition.

Earlier work on these systems focused on cases for which covalent bonds connect the monomers to form polymers, hence the term polymerizations. More recently, it has become clear that the formation of polymeric entities by noncovalent attachments can also be considered in the same theoretical framework, hence the term aggregations. For simplicity, we refer to all cases as polymerizations, whether covalent or noncovalent. The polymerizations of interest are all fundamentally reversible, even though some issues about the conditions for that reversibility are still unresolved. The term living polymer was used first (2, 3) and is still used (4) to refer to the polymerization of reversible organic systems but usually under nonequilibrium conditions. "Living polymer" has also been used to refer to micellar systems under equilibrium conditions (5). Here we use living polymer to mean any polymer molecule that is not terminated, but is free to change its degree of polymerization (DP) as conditions vary, whether at equilibrium or not at equilibrium. A "dead polymer" is a polymer that is terminated and fixed in DP.

Previous reviews of this subject may be helpful to the reader (6–8).

THEORETICAL FRAMEWORK

Thermodynamics

We assume that the reaction mechanisms have at least two steps:

$$\text{Initiation} \quad M \rightleftharpoons M^*, \qquad 1.$$

$$\text{Propagation} \quad M^* + M \rightleftharpoons M_2^*,$$

$$M_i^* + M \rightleftharpoons M_{i+1}^*, \qquad 2.$$

where M is an uninitiated monomer, M^* is an initiated monomer, and M_i^* is a polymer with a DP of i. The initiation can involve an initiator species. Each step will have associated rate constants (forward, k_{init} and k_{prop}, and reverse, k'_{init} and k'_{prop}) and an equilibrium constant (K_{init} and K_{prop}). In some cases, the initiation step may go irreversibly to completion, fixing the concentration of initiated species; in other cases, the concentration of initiated species will be determined by K_{init}. At full thermodynamic equilibrium, there will be a distribution of the DP. However, the polydisperse equilibrium polymer in solution is still just one component in terms of proper thermodynamic components: There are many polymer species, but there is only one polymer component because the species are all related by the chemical equilibrium constants, K_{prop}. In principle, these rate and equilibrium constants can be functions of DP.

In order for the propagation at constant temperature and pressure to proceed, $\Delta G_{prop} = \Delta H_{prop} - T\Delta S_{prop}$ must be negative. If ΔH_{prop} and ΔS_{prop} are both negative, then ΔG_{prop} is negative only at temperatures low enough for the enthalpy term to be larger than the entropy term. Such enthalpically driven reactions occur only below "ceiling temperatures" that are functions (polymerization lines) of pressure, concentration of solvent, and nature of solvent. This is the case for the organic system we discuss below. If ΔH_{prop} and ΔS_{prop} are both positive, then ΔG_{prop} is negative only at temperatures high enough for the entropy term to be larger than the enthalpy term. Such entropically driven systems react only above "floor temperatures" that are functions of pressure, concentration of solvent, and nature of solvent. As we discuss below, the propagation step for sulfur polymerization and the initiation and propagation steps for actin aggregation show floor temperatures.

Let us take as the standard state the pure liquid monomer at one atmosphere converting to liquid polymer (9). In the case of the pure monomer,

$$T_p^0 = \Delta H_{prop}^0 / \Delta S_{prop}^0. \qquad 3.$$

The quantities are given per mole of monomer converted to polymer. When a solvent is added to the monomer, T_p decreases. T_p as a function of the mole fraction of initial monomer (x_m^0) is the polymerization line. The equilibrium constant for propagation, in which a polymer molecule of i monomers combines with another monomer molecule to make a polymer molecule of (i + 1) monomers, is $K_{prop} = a_{i+1}/a_i a_m$, where a is the activity of each species. If $a_i \approx a_{i+1}$, then $K_p \approx 1/a_m$. If the solution is ideal and the standard state is the pure monomer, then $K_{prop} \approx x_m^{-1} \approx \exp[(\Delta S_{prop}^0/R) - (\Delta H_{prop}^0/RT)]$, where x_m is the mole fraction of monomer at equilibrium. Then $T = \Delta H_{prop}^0 / (\Delta S_{prop}^0 + R \ln x_m)$, for any T. At T_p, $x = x_m^0$, so the polymerization line is:

$$T_p = \Delta H_{prop}^0 / (\Delta S_{prop}^0 + R \ln x_m^0). \qquad 4.$$

In fact, if the solution is ideal, x_m depends only on T, so a measure of x_m at some T is a measure of x_m^0 in the case $T_p = T$. Note that Equation 4 assumes that solution is ideal and that $a_i \approx a_{i+1}$. Equation 4 is the Dainton & Ivin Equation (10) equivalent to the Van't Hoff Equation (11).

Statistical Mechanics

Researchers in the 1960s achieved considerable success in predicting the thermodynamics of the reversible polymerizations by setting up the equilibrium constants for the coupled initiation and propagation reactions and then solving for residual unreacted monomer as a function of temperature. Tobolsky and coworkers considered organic monomers and the inorganic monomer, elemental sulfur (12–17). Oosawa and coworkers considered biological monomers (18–20).

The connection of polymer physics to phase transition physics (21) began in 1972 with the realization by de Gennes (22, 23) that the partition function for a very long lone polymer chain in a good solvent is equivalent to the partition function for a magnet that has a magnetic moment with n dimensions, if n is formally allowed to go to zero (n → 0). In 1975, des Cloizeaux (24) showed that the partition function for a solution of polymer chains is equivalent to the partition function of that same n → 0 magnet, but in an external magnetic field. In the 1980s, Wheeler and collaborators treated "equilibrium polymerization as a critical phenomenon" (1) by combining the n → 0 magnet model with a parametric equation of state taken from fluid-critical phenomena (25). They applied their theory first to sulfur (1, 26) and sulfur solutions (27–30) and later to organic living polymers and their solutions (30–32).

The "chemical equilibrium" approach of the 1960s is mathematically equivalent to the "mean field" limit of the n → 0 magnet models (31). Mean field models treat the microscopic system as having an average intermolecular interaction, devoid of correlated interactions between or among molecules. Flory-Huggins lattice models (33, 34) of polymerization are also mean field models, and interest in their application to equilibrium polymerization has been renewed by recent work of Dudowicz et al. (35–38).

Why bother to treat reversible polymerizations as phase transitions if simpler chemical equilibrium models work fairly well? There are two reasons. First, the chemical equilibrium models fail when fluctuations and correlations in the polymer-excluded volume become important. Second, the use of the phase transition approach is a far more general framework, within which not only mean field and nonmean field calculations may be made, but such issues as polymer ring formation may be explicitly addressed. The phase transition approach allows the application of a formalism developed for magnets, simple fluids, and terminated polymers to be applied to polymerizing systems ranging from sulfur to biopolymers.

We focus here on the nonmean field n → 0 magnet model of Wheeler and coworkers and on the recent mean field lattice model of Dudowicz et al. The basic features of the n → 0 model are (1, 26–32): (a) It is a nonmean field in that it allows for correlations; (b) cases of infinitely small or finite initiator concentrations may be treated; (c) the physical parameters required from experiments or from fitting the model to data are the enthalpies and entropies of each step of the reaction mechanism; (d) the physical parameters are assumed to be independent of chain length; (e) the proximity to a critical point is such that asymptotic critical behavior pertains and parametric equations of state may be used; (f) the polymers are chains, not rings. The formation of rings alters the universality class to n = 1 (39–42) and leads to bicritical phenomena (43, 44): the confluence of a (second-order) polymerization to form chains, a (second-order) polymerization to form rings, and a (first-order) conversion between rings and chains. (g) If the polymers are in a solvent, then flexible living linear polymers will be in the dilute n = 0 universality class (31, 32), and living ring polymers will fall in the dilute n = 1 universality class (30). The theory can predict the behavior of various thermodynamic properties near the polymerization transition: the heat capacity, the fraction of monomer converted to polymer, etc.

The basic features of the Dudowicz-Freed-Douglas (DFD) lattice model (35–38) are (a) It is a Flory-Huggins incompressible lattice model, inherently mean field in nature; (b) It emphasizes the case of finite concentration of initiator species. In the language of magnetic phase transitions, the finite initiator concentration is equivalent to the presence of an external magnetic field, which moves the system away from the true second order phase transition. The critical fluctuations are thus suppressed and the mean field approximation should become more valid as the initiator concentration increases. (c) The physical parameters required from experiments or from fitting the model to data are the enthalpies and entropies of each step of the reaction mechanism. (d) A polymer flexibility parameter allows for inclusion of chain stiffness, but this feature turns out to be irrelevant for the thermodynamic predictions, affecting the magnitudes of the entropy and enthalpy of propagation, but not affecting the equilibrium constant for propagation (37). (e) A monomer-solvent interaction parameter allows prediction of phase equilibria.

The DFD theory can predict the same thermodynamic properties as the nonmean field model. Additional results include: (a) Three kinds of polymerization lines are predicted. Initially, there is a "crossover line," corresponding to the temperature, T_p^x, where the polymerization first becomes detectable. Second, there is the true polymerization temperature, T_p, corresponding to the points at which the specific heat shows a maximum and the extent of polymerization, Φ, shows an inflection point. Third, there is the saturation temperature, T_s, which is the point at which the DP reaches a plateau for cases for which the number of initiator species is fixed. There is not necessarily such plateau for cases where the number of initiated species is determined by K_{init}. (b) The dependence of $\hat{L}$ (the average DP, excluding the free monomer from the collection of polymeric species) or of Φ on the initial monomer concentration, x_m^0, will be linear ($\hat{L} \sim \Phi \sim x_m^0$) and not of the form ($\hat{L} \sim \Phi \sim (x_m^0)^\alpha$), where $\alpha = 0.5$ for a mean field case, $\alpha = 0.46$ in a dilute polymer solution, and $\alpha = 0.6$ in a semidilute polymer solution. The latter form is expected from scaling arguments for systems for which the number of polymer chains is determined by an equilibrium constant (45–48), but fixing the number of chains changes the dependence to linear.

If the living polymer is in a poor solvent, then there can be phase separation into coexisting liquid phases, with such features as upper or lower critical solution points, closed miscibility gaps, and higher-order critical points, depending on the polymer/solvent system. The published theoretical treatments of these phase equilibria have all been mean field treatments and thus do not account for either excluded volume correlations or liquid-liquid critical fluctuations. The first such mean field theory was that of Scott (34), who developed a Flory-Huggins lattice model for sulfur polymerizing in solution.

Later Wheeler and colleagues made extensive calculations of phase equilibria in sulfur solutions (27, 28), predicting symmetric tricritical points analogous to that in ^{3}He + ^{4}He (49) at low values of K_{init}. They predict the appearance of nonsymmetric tricritical points when K_{init} is larger (29, 30). In the presence of both chain and ring polymers, they predict a tetracritical point (50). Wheeler and collaborators also presented studies aimed at organic monomers polymerizing in solution at ceiling temperatures, for very small concentrations of initiated monomer, [M*] (31), and also for fixed finite [M*] (51). From these models, organic systems are predicted to have a number of interesting features, including higher-order critical points (52). These predictions have been summarized by Greer (7).

DFD (36) have used their lattice model to calculate the effects of changing ΔH_{prop} on liquid-liquid phase equilibrium for a system with a fixed number of initiated species and a ceiling temperature. Their most interesting result is a prediction of a critical value of ΔH_{prop}, below which the miscibility gap is insensitive to ΔH_{prop}.

Chemical Kinetics and Molecular Weight Distribution

When a system of initiated monomer in solvent is quenched into the polymerizing region of the phase diagram, all initiated monomers are equally likely sites for polymerization. The immediate but nonequilibrium result is the formation of a narrow Poisson distribution of polymers (stage 1). If the living polymer is allowed to continue toward equilibrium, the next stage is the attainment of the equilibrium concentration of residual monomer and the completion of the enthalpic relaxation of the system (stage 2). Full thermodynamic equilibrium is not attained until the final equilibrium MWD is developed, when the entropic relaxation is completed (stage 3). These stages were pointed out in 1958 by Brown & Szwarc (53), and there have been many other theoretical papers in the ensuing years [see recent review (54)]. The detailed time development has been addressed by Miyake & Stockmayer (55), Taganov (56), and Milchev (47). Computer simulations for the equilibrium formation of cyclic polymers are also of interest (57). The nature of the final equilibrium MWD has been considered by both mean field and nonmean field theories. The mean field prediction is that the equilibrium MWD will be of the Flory-Schulz form (12, 37, 58):

$$n_x^{FS} = p^{x-1}(1-p), \qquad 5.$$

where $p = 1 - (1/\hat{L})$. When $p \to 1$, Equation 5 becomes (59)

$$n_x^E = (1/\hat{L}) \exp(-x/\hat{L}), \qquad 6.$$

and thus the Flory-Schulz distribution becomes the "exponential distribution." For the nonmean field case, scaling arguments predict the exponential distribution in the semidilute regime, and a power law times an exponential in the dilute regime (48, 60, 61):

$$n_x^{SCALING} = [\gamma^\gamma / L\Gamma(\gamma)](x/L)^{(\gamma-1)} \exp(-\gamma x/L), \qquad 7.$$

where $\gamma (= 1.16)$ (62) is the susceptibility exponent for the $n \to 0$ model. Bouchaud et al. (63) and Rouault (64) suggest that the polydispersity of living polymer systems allows the smaller polymers to swell the larger ones, resulting in yet another form for a living polydisperse system:

$$n_x^{POLY} = cx^{-2\sigma} \exp(-x/L), \qquad 8.$$

where c is a normalization constant and the exponent σ is found to be 0.25 by an indirect experiment (63) and by a computer simulation (64). Nonmean field effects were also studied by Schäfer (65) by renormalization group methods to obtain a complicated function which, in the semidilute limit, "reduces to an exponential distribution, modified by power-law prefactors in the extreme wings." Schäfer's theory is valid only up to 10 weight percent polymer and for i ≥ 500.

ORGANIC POLYMERIZATIONS

Many organic monomers have ceiling temperatures (9), but those temperatures are often well above room temperature. For example, the ceiling temperature of styrene is somewhere (not yet measured) above 508 K. The polymerization involves initiation by an initiator species; the initiation step (Equation 1) generally goes to completion and fixes the number of propagating species. Depending on the choice of initiator, the mechanism for the polymerization can produce a propagating species with one, two, or even more active sites (4); the equations for the thermodynamics change if the functionality changes (66).

Model: Poly(α-methylstyrene)

One organic monomer with a convenient polymerization line has been the subject of much study: α-methylstyrene. Poly(α-methylstyrene) is a fully flexible linear polymer chain joined by covalent bonds. One mechanism for the polymerization of α-methylstyrene is anionic polymerization with sodium naphthalide initiator (67), in which the sodium naphthalide transfers an electron to the α-methylstyrene to form a radical ion, which immediately dimerizes to form the propagating species. This mechanism results in a living polymer that propagates from both ends; the ends are anionic and the sodium counterions are nearby. Other mechanisms (other initiators for anionic polymerization, cationic polymerization, radical polymerization) are possible (3, 68, 69), but they would have different experimental considerations in terms of undesirable side reactions, solubility in the solvent, etc., and may form species with a different number of propagating sites.

Experimental Considerations

Ionic polymerizations are extremely sensitive to air and water, which terminate the living polymers. Samples of living polymers in solution must be prepared in vacuum lines and high-purity dry boxes, with careful attention to technique (70, 71). When the solution can be prepared above the ceiling temperature, initiation takes place but not polymerization; initiator molecules can react with any residual impurities and thus "clean" the sample still further before it is quenched into the polymerizing temperature regime. A key sample property is the concentration of initiator, which we denote as $r = [I]/[M_0]$, where [I] is the initiator concentration and $[M_0]$ is the initial monomer concentration. Organic monomers must be studied in solution because the "neat" monomers form glasses when they polymerize.

Thermodynamics

If the reversible polymerization is to be seen as a phase transition, then we should have a phase diagram. Figure 1a shows the phase diagram for poly(α-methylstyrene). If we have a sample at a fixed initial monomer concentration (x_m^0) and initially at high temperature, then as we lower the temperature, the monomer will begin to polymerize at T_p. The polymerization line in Figure 1a, which is T_p as a function of x_m^0, was constructed from determinations of T_p from a variety of experiments, as indicated in the legend; we have not tried to distinguish T_p from T_p^x. The solvent for the experiments in Figure 1a is tetrahydrofuran (THF), except

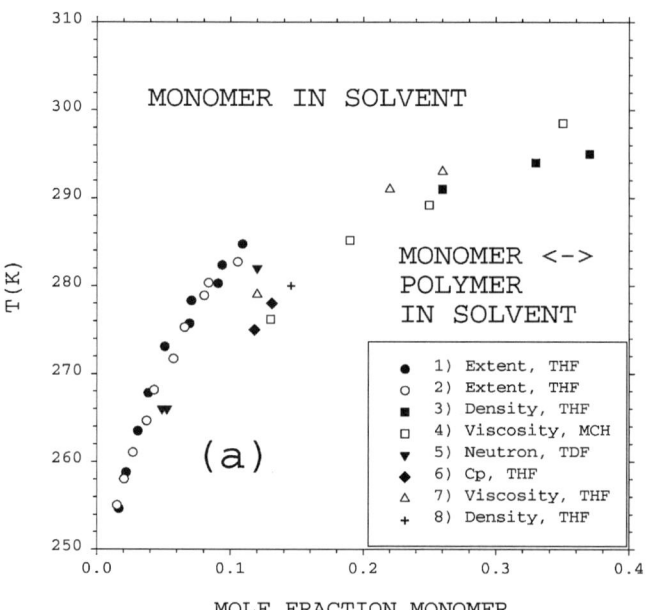

Figure 1 The equilibrium polymerization of α-methylstyrene: (a) the phase diagram, (b) the extent of polymerization (66), (c) a Dainton-Ivin/Van't Hoff plot of the logarithm of the mole fraction of residual monomer as a function of (1/T), (d) the heat capacity (78), and (e) the correlation length, ξ (81). In (a), the legend indicates various determinations of $T_p(x_m^0)$, where the References are: 1 = (66), 2 = (66), 3 = (82), 4 = (72), 5 = (81), 6 = (78), 7 = (85), and 8 = (74). For (b) through (e), the symbols represent the data (66), the dotted lines represent the mean field model, and the solid lines represent the nonmean field (dilute n → 0) model. In (e), the extent of polymerization (Φ) and the degree of polymerization (DP) were calculated from the mean field model; the experimental data are a first cooling run (91-1-A) and a second cooling run (91-1-B) on the same sample. Parts (b) and (d) are from Reference (6). Part (e) is from Reference (81).

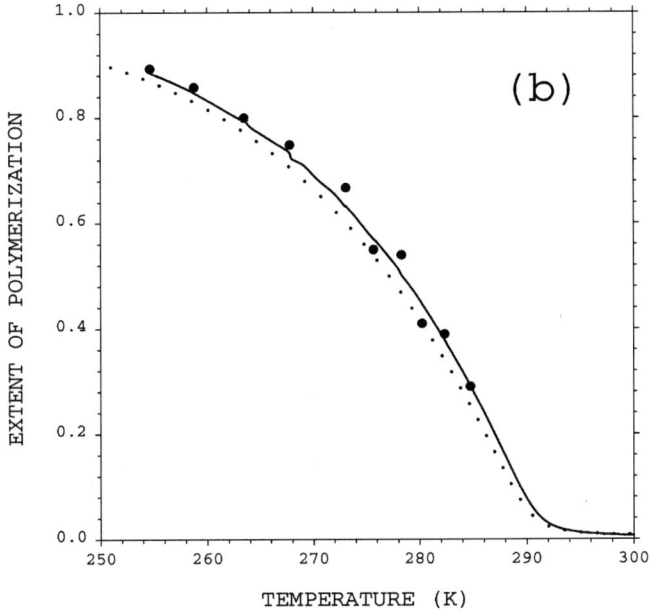

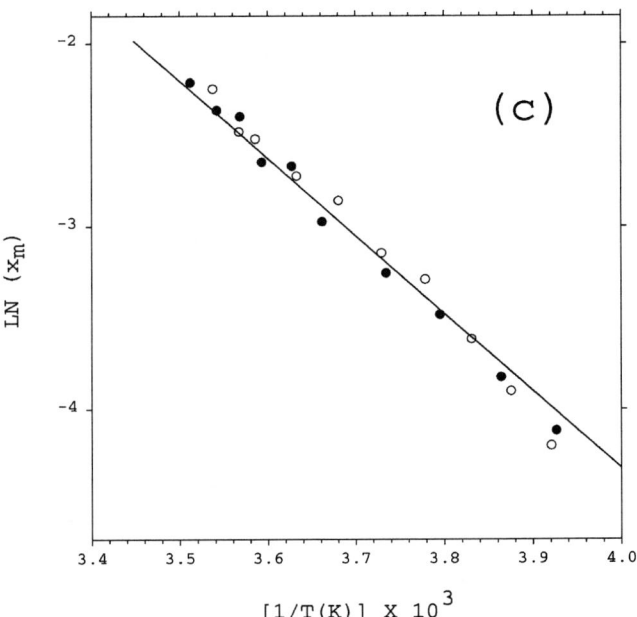

Figure 1 (*Continued*)

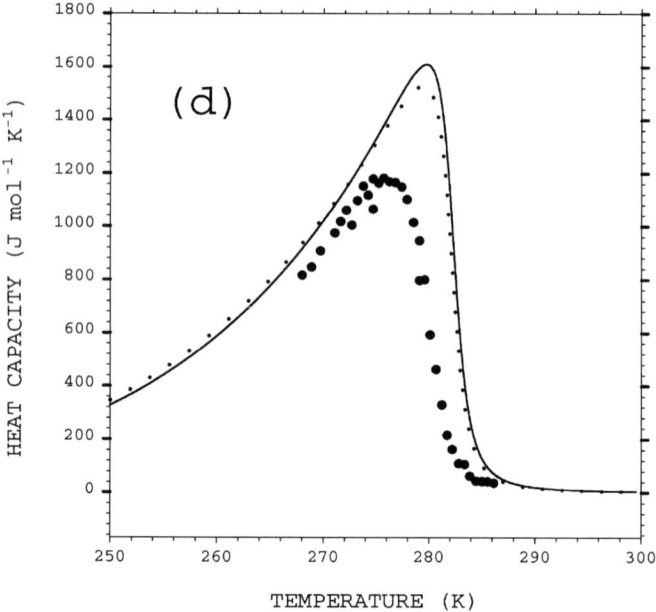

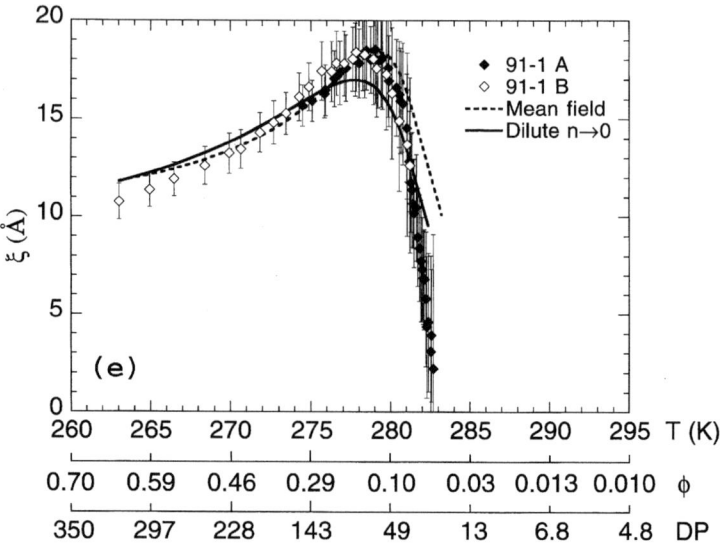

Figure 1 (*Continued*)

for one set of points in deuterated THF and one set of points in methylcyclohexane (MCH) (see legend). THF is a good solvent for poly(α-methylstyrene), and MCH is a poor solvent (72). Figure 1a indicates that the polymerization line for α-methylstyrene may not be very sensitive to the nature of the solvent, even though the MCH points seem slightly different from the THF points. If we have a sample at fixed temperature and then add monomer until polymerization commences, we can determine the critical polymerization concentration (37), which is just another way of considering the same polymerization line [i.e., $T_p(x_m^0)$ versus $x_m^0(T_p)$].

Most of the T_p determinations in Figure 1a were based on the observed initial deviations of some experimental variable (e.g., the viscosity) from its background behavior in the unpolymerized region, so they are actually T_p^x determinations. The exceptions are the data from experiments on the extent of polymerization (see discussion below) and on C_p, which should yield T_p. DFD (37) predict that T_p should decrease slightly as r increases and that $T_p^x = 1.21\ T_p$. The value of r varies from 0.0018 to 0.04 among the data in Figure 1a, but there is no dependence of T_p on r outside the considerable scatter. There is also no indication that T_p^x is 21% higher than T_p.

The second property of interest is the extent of polymerization, or conversion fraction of monomer to polymer, Φ, as a function of T. $\Phi(T)$ is proportional to $\hat{L}:\hat{L}(T) = 2\Phi(T)/r$ (13, 37). Figure 1b shows $\Phi(T)$ for poly(α-methylstyrene)/THF/sodium naphthalide system, as determined from gas chromatographic analysis for residual monomer at each temperature (66). These same data allow us to make a Dainton-Ivin/Van't Hoff plot for this system, as shown in Figure 1c. Equation 4 predicts a straight line for this plot, and it is nearly so. If we fit these data by Equation 4, we obtain $\Delta H_{prop} = (-39 \pm 2)$ kJ mol^{-1} and $\Delta S_{prop} = (-119 \pm 8)$ J mol^{-1} K^{-1}, where the uncertainty is given at the 99% confidence level. ΔH_{prop} has been directly determined by combustion calorimetry for poly(α-methylstyrene) (73): $\Delta H_{prop} = (-35.2 \pm 1.0$ kJ mol^{-1}) for DP above about 50, but becomes larger in magnitude at lower DP (7). If we fix $\Delta H_{prop} = -35.2$ kJ mol^{-1}, then we obtain $\Delta S_{prop} = (104.8 \pm 0.2$ J mol^{-1} K^{-1}), and these values lead to the line shown in Figure 1c. The line deviates from the data at lower T. Best fits to other data on this system have found [-110 J mol^{-1} K$^{-1} < \Delta S_{prop} < -105$ J mol^{-1} K^{-1}] (74). Equation 4 works fairly well for this system, which suggests that it is not extremely nonideal.

The data on $\Phi(T)$ can be inverted to give the mole fraction of residual monomer as a function of temperature, $x_m(T)$. If Equation 4 is valid, then a plot of $T(x_m)$ is the polymerization line. The extent data are plotted this way in Figure 1a, and agree reasonably well with other measurements of the polymerization line, which indicates again that the solution is surprisingly near to ideal. DFD (37) note that the inflection point in $\Phi(T)$ marks T_p, but there are insufficient experimental data in Figure 1b to determine the inflection point.

The calculation of $\Phi(T)$ by the mean field theory requires the simple solution of a quadratic equation (37, 66), whereas the n $\rightarrow$ 0 model requires the numerical solution of parametric equations. Figure 1b shows the comparison of the two theories (66): The n → 0 model is slightly better, but the mean field theory is very good. Both theories capture the "rounding" of the transition: the gradual onset of the polymerization with temperature. The only other published $\Phi(T)$ data for an organic polymerization are for THF as the monomer (75–77). Those data have been analyzed twice (31, 66) and twice were found to be slightly better described by the n → 0 model than by the mean field model.

In the study of phase transitions, the heat capacity at constant pressure, C_p, is a telling thermodynamic parameter. The only organic equilibrium polymerization for which C_p has been measured is poly(α-methylstyrene)/THF/sodium naphthalide (78). The data and theoretical predictions are shown in Figure 1d. The mean field and n → 0 models both show the same qualitative behavior as the data, but are both too rounded. Zhuang et al. (78) argue that the rounding could be caused by the dependence of ΔH_{prop} on DP at low DP (73) because of steric effects and Coulombic interactions.

Phase equilibrium data for an organic equilibrium polymerization in a poor solvent are very rare. For poly(α-methylstyrene) initiated by n-butyllithium in methylcyclohexane, the liquid-liquid coexistence curve with an upper critical solution point has been determined and is consistent with mean field expectations (72). The coexistence curve has also been roughly determined for poly(styrene) initiated by n-butyllithium in cyclohexane (7, 79, 80).

The osmotic compressibility (or concentration susceptibility), κ, near T_p for poly(α-methylstyrene) in deuterated THF, initiated by sodium naphthalide, has been measured by small-angle neutron scattering (81). The measured $\kappa(T)$ is zero above T_p, rises as the temperature is decreased below T_p, then reaches a maximum at about 10 K below T_p, and declines. The mean field model does not show such a maximum (32, 35). A dilute n → 0 model that assumes no interaction between monomer and solvent does show the maximum in $\kappa(T)$ (32), but a full nonmean field calculation has not been made. $\kappa(T)$ is one property for which the dilute n → 0 theory seems to model an experimental feature that is not modeled by the mean field theory and that can be attributed to fluctuations. It has been suggested from scaling arguments (32a) that the maximum in $\kappa(T)$ is due to the crossover from dilute to semi-dilute regime.

Other thermodynamic parameters that have been measured are the mass density and the surface tension (74, 82).

Chemical Kinetics and Molecular Weight Distribution

To understand fully the approach to equilibrium in these systems, including the development of the molecular weight distribution (MWD), we need information on the rate constants for propagation and depropagation. The functional groups of ionic polymers can exist in at least two forms, as "tight" contact ion pairs and as "loose" solvent-separated or free ion pairs (3). Then the effective rate constant for propagation, k_{prop}, has at least two terms:

$$k_{prop} = (1-f)k_\pm + fk_- \approx k_\pm + fk_-, \qquad 9.$$

where f is the fraction of ion pairs dissociated into free ions and depends on the nature of the solvent, the initial mole fraction of monomer, and the temperature, and where $k_\pm$ is the rate constant for propagation of the ion pair and k_- is the rate constant for the solvent-separated ion. The terms $k_\pm$ and k_- can be determined by studying samples in which ion pair dissociation is prevented by the presence of a salt with a common ion (83). The effective rate constant for the system poly(α-methylstyrene)/tetrahydrofuran/sodium naphthalide has been recently measured (84) and collected with earlier measurements to show k_{prop} approaching zero as T approaches T_p.

The only experimental studies of the time and temperature development of the MWD of a living organic polymer system are on poly(α-methylstyrene)/tetrahydrofuran/sodium naphthalide (54, 84). For this system, stage 1 is complete in less than 30 min, stage 2 requires about 30 min, and stage 3 requires more than 20 h (depending on T). There were some complicating features in the data, for which we refer the reader to the original paper, but the conclusion was that the final MWD is exponential (Equations 5 or 6, indistinguishable from Equation 7) at $\hat{L} > 66$ (further from T_p), but shows nonexponential behavior (Equation 8) at lower $\hat{L}$, nearer T_p. In principle, an equilibrium polymerization should be fully reversible. In practice, once formed, polymeric species are slow to depolymerize when taken back into the region of the phase diagram where the monomer is the stable state. This effect has been reported in several kinds of experiments on poly(α-methylstyrene) (78, 81, 85–87). This same metastability has been seen in actin (see below). The cause of the metastability is not understood.

Transport Properties

We can expect the changing structure of a polymerizing system to be reflected in its transport properties. The shear viscosity of poly(α-methylstyrene)/tetrahydrofuran/sodium naphthalide has been measured as a function of x_m^0, T, and r (74, 85). The qualitative behavior of the viscosity is as expected, increasing as DP increases, but the data show no signal at the overlap concentration (see next section) and do not extend to high enough DP to reach the expected behavior at entanglement. More data at higher DP (i.e., lower r and lower T) would be of interest.

The diffusion constant for poly(α-methylstyrene)/tetrahydrofuran/sodium naphthalide solutions has been studied as a function of T by means of dynamic light scattering (86). This experiment shows a diffusion constant that corresponds to the growing living polymer. However, this and similar experiments on solutions of living poly(styrene) in hydrocarbons by light and small-angle neutron scattering at a fixed T below T_p (88–91) indicate that living organic polymers can assemble into various micelle-like forms owing to the Coulombic interactions of the ionic end groups and their associated counter-ions. In a sense, these aggregations are "meta-living polymers": reversible aggregations of reversible aggregations. This complexity in the structure is interesting in itself and has implications for the development of the MWD (54) and for other properties.

Structure

The polymerization of the organic polymers can be followed by small-angle neutron scattering, as has been done for poly(α-methylstyrene) initiated by sodium naphthalide in deuterated tetrahydrofuran (81). The correlation length, ξ, is the characteristic size measured by the neutron scattering: For dilute polymer chains in a solvent, it corresponds to the diameter of the polymer coil; for semidilute solution of polymer chains, it corresponds to the mesh size of the overlapping polymer chains. Figure 1e shows ξ(T) for living poly(α-methylstyrene). Above T_p, the neutron beam detects nothing. As the temperature is decreased and reaches T_p, ξ begins to increase, owing to the scattering from the growing dilute polymer coils. At about 10 K below T_p and at DP = 100, ξ reaches a maximum and starts to decrease. The maximum is the overlap concentration, and the values of ξ at temperatures below that maximum represent the size of the semidilute polymer mesh. The data can be described well (see Figure 1e) by either a mean field (81) or a nonmean field theory (32).

INORGANIC POLYMERIZATIONS

The inorganic molecules, sulfur and selenium, show equilibrium polymerizations (92). The polymerization transition in pure sulfur occurs well above the melting point, but the transition in selenium occurs below its melting point. Although some attention has been given to selenium (93–96), we focus here on sulfur.

Model: Sulfur

The eight-atom rings of elemental sulfur can open to form diradical chains, which rapidly react with one another to form very long, flexible, covalently bonded chains (97, 98). The ring-opening is a thermally activated initiation step, so the number of polymer species is determined by the equilibrium constant for this initiation. K_{init} for sulfur is extremely small ($\sim 10^{-12}$) (12, 15), so the sulfur polymerization would be expected to be more like a second-order phase transition than is either poly(α-methylstyrene) or actin. The enthalpy and entropy of propagation are both positive ($\Delta H_{prop} = +13.3$ kJ mol^{-1}, $\Delta S_{prop} = +31$ J mol^{-1} K^{-1}) (12, 15, 26, 99, 100), and pure sulfur has a floor temperature at 1 atm and 432 K.

In addition to linear chains, polymeric sulfur can form rings, "tadpoles," and other nonlinear species (101–103). The formation of rings is a nontrivial difference from poly(α-methylstyrene) or actin. The n $\rightarrow$ 0 model specifically excludes the formation of rings, and the universality class becomes n = 1 when rings are present. Thus, although sulfur was the first system to which the n → 0 model was applied (1, 26), it probably does not fit in this universality class, and a nonmean field model for equilibrium polymerization in the n = 1 class is not available. The mean field theory is still applicable.

Experimental Considerations

Sulfur is difficult to study because the diradicals react easily with many vessels and solvents and thereby contaminate the sample (104, 105). Very pure sulfur can be prepared by sublimation (100) and remains yellow even in the polymerized region, whereas impure sulfur turns red in the polymerized region. "Neat" sulfur remains nonglassy upon polymerization.

Thermodynamics

The thermodynamics of the polymerization of pure sulfur was developed in the mean field model by Powell & Eyring (106), Tobolsky & Eisenberg (15), and Anisimov et al. (107), and in the nonmean field model by Wheeler and collaborators, as discussed above.

The extent of polymerization as a function of temperature for pure sulfur, $\Phi(T)$, has been determined from an analysis of quenched sulfur samples (108) and is plotted in Figure 2a, along with theoretical predictions (6). Neither theory gives a good description of the data. The n → 0 model works better near T_p, and the mean field model works better overall. Values for $\Phi(T)$ recently obtained from the excess sound velocity (109) reach the same conclusion.

Because the number of initiated species for sulfur is set by K_{init}, mean field theories (15) predict a maximum in average DP(T) at about 440 K. DP(T) has not been measured for sulfur, but the viscosity of sulfur does show a maximum at about 453 K (110), and the width of the central Brillouin peak shows a maximum at 423 K (111). As seen in Figure 2a, $\Phi(T)$ does not show such a maximum in either theory or experiment, nor does the optical absorption (112), which should be proportional to $\Phi(T)$.

The heat capacity at constant pressure, C_p, for pure sulfur has been measured by two different groups (113, 114). Figure 2b compares the experimental data [minus the background contribution (6)] to the theoretical prediction for the contribution to C_p from the polymerization process. Again, the mean field theory works better than the n → 0 model.

Sulfur in solvents shows interesting phase diagrams, including miscibility gaps with upper and lower critical solution points (115). Larkin et al. (116) explored experimentally the phase diagrams of sulfur in 10 organic solvents and found good qualitative agreement with Scott's mean field theory for sulfur solutions (34). Anderson & Greer (105) studied the sulfur + biphenyl system (shown in Figure 2c) in more detail and concluded that the data support the nonsymmetric tricritical point predicted by Wheeler (29, 30). Many other predictions of Wheeler and colleagues have not been tested experimentally.

Other thermodynamic parameters of interest for sulfur are the mass density (100), the dielectric constant (117), and the speed of sound (109).

Chemical Kinetics and Molecular Weight Distribution

The rate constant for ring-chain equilibrium in sulfur has been studied (118), but there are no published experimental measurements of the development of the MWD in sulfur polymerization.

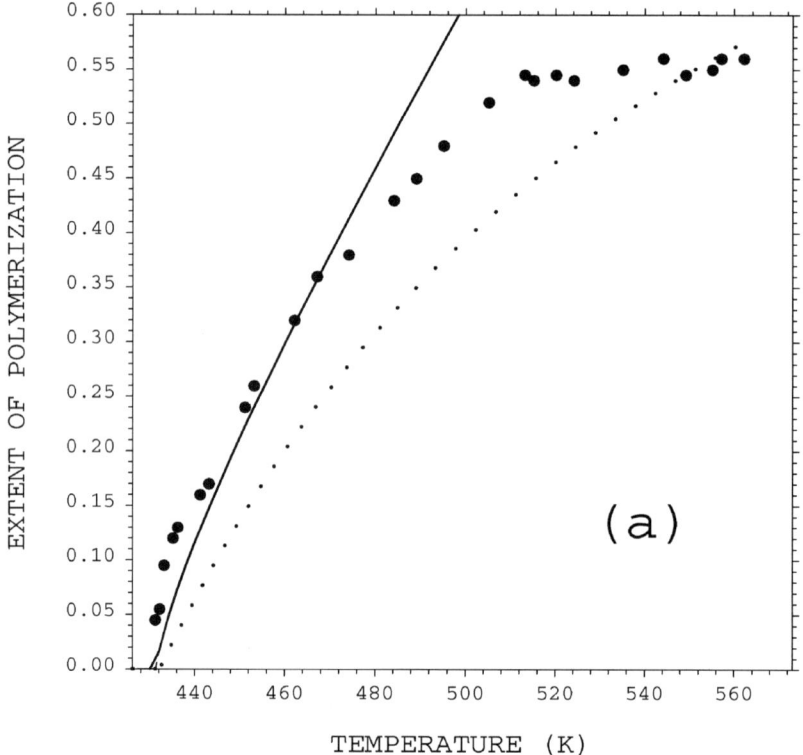

Figure 2 The equilibrium polymerization of sulfur: (*a*) the extent of polymerization (108), (*b*) the heat capacity, (*c*) the phase diagram in a solvent (biphenyl) (105, 116). For (*b*), the solid circles (●) are the data of West (113) and the open squares (□) are the data of Feher et al. (114). For (*c*), the solid triangles (▲) are points on the polymerization line; the open squares (□) are points on the coexistence curve; and the solid square (■) is the experimentally determined liquid-liquid critical point. In (*a*) and (*b*), the symbols represent the data, the dotted lines represent the mean field model, and the solid lines represent the nonmean field (n → 0) model. Parts (*a*) and (*b*) are from reference (6).

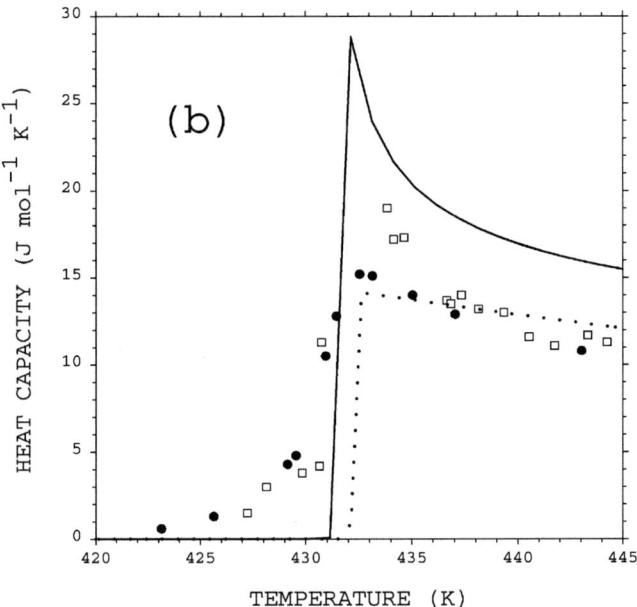

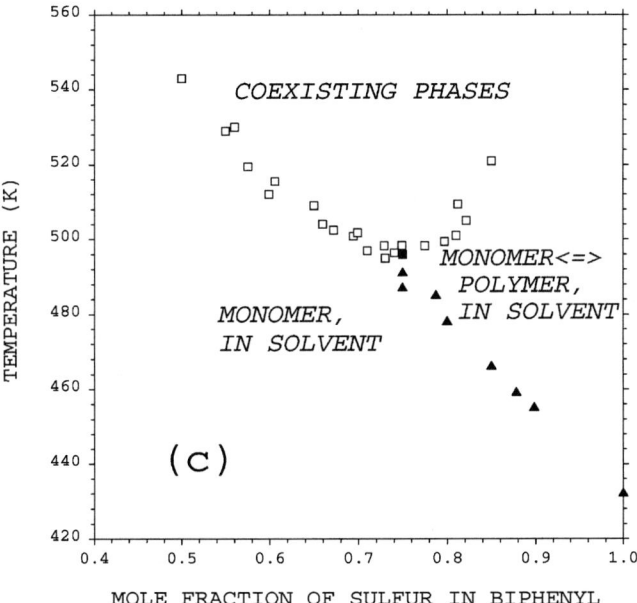

Figure 2 (*Continued*)

Transport Properties

The shear viscosity of sulfur has been measured (104, 110, 119) and shows the expected dramatic increase upon polymerization (120). However, there is reason to believe that the viscosity data are still not free of side reactions or impurities (104).

The absorption of sound in pure sulfur (109) does not show the expected (107) feature at T_p. Inelastic neutron-scattering experiments (121) are consistent with the polymerization transition.

Structure

Local ordering in sulfur has been studied by X-ray (122, 123) and neutron scattering (124). Of more interest here is the small-angle neutron-scattering study of sulfur in naphthalene and in biphenyl (125), where measurements of the osmotic compressibility are in qualitative agreement with mean field calculations.

Some (126–130) have used neutron-scattering determinations of the pair correlation function to argue that the underlying phenomenon in sulfur is a percolation transition rather than a polymerization transition, but no such predictive model has been put forward.

BIOMOLECULAR AGGREGATIONS

A number of proteins assemble reversibly by noncovalent forces and can therefore be considered with the same models used for organic and inorganic polymerizations. These include actin, tubulin, tobacco mosaic virus, and flagellin (20, 131, 132). The architectures of the resulting polymers vary from the semiflexible two-stranded helical polymer of actin, to the very rigid multifilamentary tubules of tubulin, to disc-like assemblies of *Tobacco mosaic virus*. We focus on the protein actin.

Model: Actin

Actin (133) occurs in all eukaryotic cells and does not vary much in amino acid sequence from species to species (131, 134). Globular actin (G-actin) consists of 375 amino acid residues with a total molecular weight of 42,000. X-ray studies (135) on the crystalline actin: DNase I complex indicate that the G-actin molecule has dimensions of 3.7 × 4.0 × 6.7 nm. Filamentary actin (F-actin), a polymeric form of G-actin (134, 136), is a double-stranded right-hand helix, with a diameter of about 8 nm and a turn about 72 nm long. Filament lengths can range up to 200 μm (20). Filamentary actin has major roles in cell structure (131) and motion (137, 138).

The aggregation or polymerization of G-actin to F-actin occurs under particular conditions of temperature, G-actin concentration, pH, and the concentrations of salts, actin-binding proteins and ATP and/or ADP. The transition from monomer to polymer is usually described as occurring above a "critical monomer concentration" for a given temperature. However, we describe the transition as occurring, for a given monomer concentration, above a floor temperature, T_p. There are no direct and reliable experimental determinations of ΔH_{prop} or ΔS_{prop} for actin (139).

The mechanism for the polymerization of actin is not understood, but can be considered as three steps (20, 134): (a) activation, whereby through an as yet undetermined interaction with a salt, the G-actin monomer changes to an activated form; (b) nucleation: although the nature of this step is controversial (140), one hypothesis is the formation of a trimer of G-actin and the conformational change of the trimer so as to form a helical nucleus for further polymerization (19); (c) propagation or elongation, the addition of G-actin monomers to the nucleus to form F-actin.

Actin polymerization differs from that of sulfur and organic monomers in several ways. (a) The actin filament is only semiflexible, whereas sulfur and organic monomers form very flexible polymer chains (141). (b) The nucleation step is rate determining for actin, whereas the propagation step is rate determining for organic polymers. The propagation step is diffusion-limited for actin (142), but not for organic polymers. (c) In the presence of adenosine triphosphate (ATP), actin continuously hydrolyzes the ATP to ADP, and a steady-state of this conversion is reached (143). (d) Under some conditions, the two ends of the F-actin have different affinities for G-actin, and a "treadmilling" of G-actin can result, in which case G-actin adds to one end of a chain, moves down the chain, and exits the other end of the chain (144). However, "the hydrolysis of the actin-bound ATP is not tightly coupled to polymer formation . . ." (145). In the absence of ATP and the presence of ADP, actin polymerizes to a true equilibrium with no treadmilling but slowly and with a tendency toward denaturation (145, 146). (e) The bonding of the G-actin monomers to form F-actin is not covalent but rather the result of hydrophobic and hydrophilic interactions, electrostatic interactions, and hydrogen bonding. (f) There is evidence of the formation of dimers of actin over a broad temperature range (147, 148).

These aspects of the polymerization of actin form the basis for a mechanism implemented in the mean field lattice model of DFD (38). The mechanism has three steps:

1. Monomer activation, where A_1 is the monomeric G-actin
    $A_1 \leftrightarrow A_1^*$, with an equilibrium constant $K_{init}(T)$;

2. Dimerization of two activated monomers,
    $A_1^* + A_1^* \leftrightarrow A_2$, with an equilibrium constant $K_{dimer}(T)$;

3. Trimer formation and propagation,
    $A_2 + A_1^* \leftrightarrow A_3^*$, with equilibrium constant $K_{prop}(T)$;
    $A_1 + A_i^* \leftrightarrow A_{i+1}^*$, where i $\geq$ 3, with equilibrium constant $K_{prop}(T)$,

where the subscript denotes the DP. This mechanism is treated via the Flory-Huggins incompressible lattice model (35–38) with 6 adjusted parameters (ΔH_{init}, ΔS_{init}, ΔH_{dimer}, ΔS_{dimer}, ΔH_{prop}, ΔS_{prop}). This mean field calculation is more complicated than that for sulfur or the organic systems, and requires numerically solving an equation while fitting the free parameters.

Experimental Considerations

The study of protein aggregations requires the careful isolation and purification of the protein. The protein must be freshly prepared, and its purity verified before and after an experiment by size exclusion chromatography. Because proteins adhere to glass, which can give off ions that initiate actin polymerization (149), plastic or quartz vessels should be used where possible. Actin can be prepared in ADP solution or in ATP solution, but it is more stable in the presence of ATP. The most widely studied actin is rabbit muscle actin, which requires access to fresh rabbit tissue. The work discussed here is in the presence of ATP, in aqueous buffer.

Thermodynamics

The extent of polymerization for actin is being studied (38, 150) by means of fluorescence spectroscopy. For the experiments discussed here, the salt is KCl and the salt concentrations are low (5 mM–15 mM) in order to have T_p fall in a convenient range near room temperature. Figure 3a shows data at an initial actin concentration, $[G_0]$, of 2 mg/mL; curves at lower/higher values of $[G_0]$ show higher/lower transition temperatures. Note that T_p decreases as [KCl] increases. Note also that $\Phi(T)$ shows a maximum above T_p, a feature not seen in either α-methylstyrene or sulfur. This maximum is a result of the behavior of $K_{init}(T)$: The fitted values of ΔH_{init} and ΔS_{init} are both positive, so the activation step itself has a ceiling temperature very near T_p, and the competition between initiation and propagation leads to the maximum in $\Phi(T)$ (38). The ceiling temperature for activation may be related to the transition temperature for helix formation that was predicted by computer simulations (151). The DVD mean field calculation is shown as a dotted line for [KCl] = 9 mM and gives rather good agreement with the data, capturing the maximum in $\Phi(T)$. The maximum in $\Phi(T)$ means that the system reaches a maximum in the extent of polymerization and, upon further heating, begins to depolymerize rather than continuing to polymerize. If data were available at still higher T than in Figure 3a, we might find a second inflection point, reflecting a depolymerization line (see below) in the phase diagram.

We now consider the phase diagram at fixed salt concentration. Figure 3b shows experimental values of T_p as a function of $[G_0]$. The triangles are determined from the inflection points of the $\Phi(T)$ measurements on actin, such as in Figure 3a (38). The filled circles are determined by inverting the $\Phi(T)$ data at 15 mM (as discussed above for α-methylstyrene) and applying Equation 4; we discount the "tail-off" at 2.6 mg/mL because the sample had reached its maximum of $[G_0]$. Clearly, $T_p([G_0])$ at 15 mM KCl as taken from the inflection points does not agree with the values

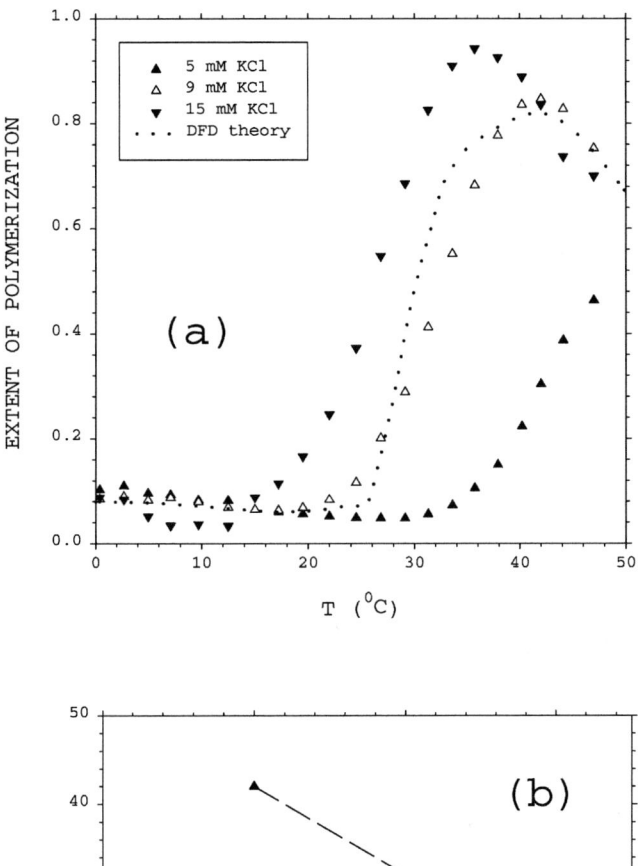
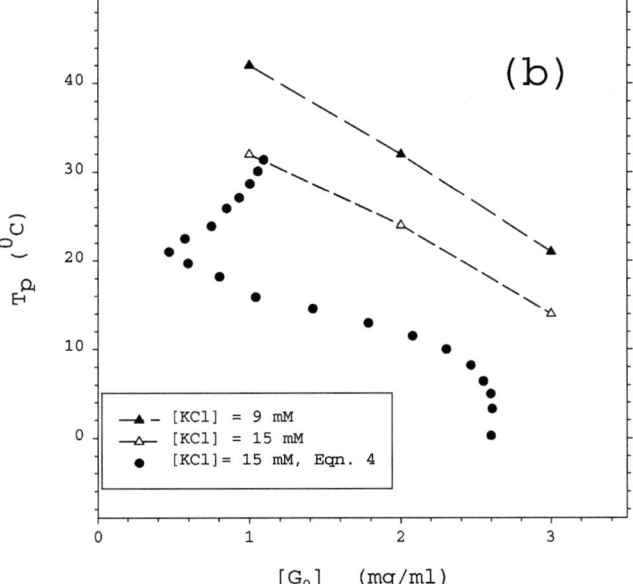

Figure 3 (*Continued*)

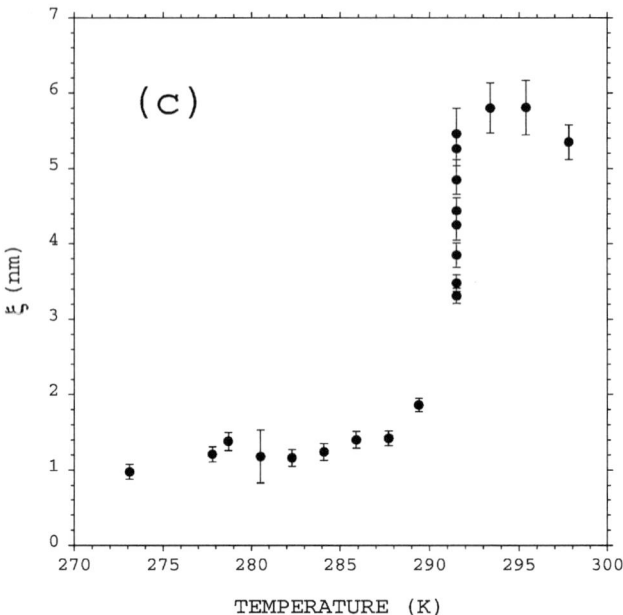

Figure 3 The equilibrium polymerization of actin: (*a*) the extent of polymerization, $\Phi(T)$, for $[G_0] = 2$ mg/mL, at various salt concentrations (38); the symbols are the data and the dotted lines represent the mean field model. (*b*) The phase diagram in aqueous buffer at different salt concentrations; the triangles are experimental determinations of T_p from the inflection points in $\Phi(T)$ (38), and the circles are determined from Equation 4 (see text). (*c*) The correlation length from small-angle neutron scattering for a sample with $[G_0] = 3.00$ mg/mL and at 9 mM KCl (139). Part (*c*) is from Reference (6).

obtained from Equation 4. This indicates that these solutions are highly nonideal, invalidating the use of Equation 4 and therefore the Van't Hoff plots often used for these systems. However, the low $[G_0]$ "turn-back," though not quantitatively accurate, is related to the maximum in $\Phi(T)$ and may indicate the presence at low $[G_0]$ and at high T, of a re-entrant depolymerization, a region in which the polymer returns to monomer. This is an interesting implication for the phase diagram of actin. A phase diagram has also been constructed for tubulin, but this diagram does not show re-entrant depolymerization over its range of T (132).

The heat capacity of polymerizing actin has not been measured. Using the values of ΔH_{prop} obtained from fits to the DFD theory (38) and assuming an actin concentration of 3 mg/mL, we expect a total heat signal over the T range of the polymerization of about 1 μJ, which is a challenging measurement.

The concentration susceptibility of polymerizing actin has been measured by small-angle neutron scattering (139) and increases as expected when polymerization occurs, but theoretical calculations have not yet been made for comparison with these data, in part because the required enthalpies and entropies are not well known.

For actin, there has been considerable study of the MWD because the F-actin molecules are large enough to be directly observed by fluorescence or electron microscopy, and because the time scales are reasonable. The stage 1 takes less than an hour, and full equilibration takes days (20, 142). The transition from Poisson to exponential MWDs has actually been observed (152, 153). Indeed, exponential MWDs have been observed frequently in equilibrated actin (154). The MWD of tubulin has also been found to be exponential (132).

Chemical Kinetics and Molecular Weight Distribution

There is an enormous literature on the chemical kinetics of actin which we cannot attempt to review here. The chemical kinetics has not yet been discussed in the context of the polymerization transition. The metastability of the polymeric species, discussed above for α-methylstyrene, has also been observed for actin (139).

Transport Properties

Although the shear viscosity has been used as a signal for polymerization in actin (20), neither the viscosity nor the diffusion coefficient has been studied near T_p.

Structure

The correlation length, ξ, has been measured by small-angle neutron scattering (139) and is shown in Figure 3c. Compare this plot to the $\xi(T)$ for α-methylstyrene in Figure 1e. The expected increase in ξ is present, and there is some indication of the maximum due to overlap. However, the magnitude of ξ for actin does not make physical sense because the size of the monomer is about 5 nm; an interpretation of the size of ξ remains to be made.

CONCLUSIONS

Summary

For organic equilibrium polymerizations, the $n \rightarrow 0$ model is consistent with all available data sets and is slightly better in describing the data than are mean field models, but the mean field models give a very good description of the data. For the polymerization of sulfur, the accumulated evidence is that the mean field model works better than the $n \rightarrow 0$ model, probably because the formation of nonlinear species in sulfur changes its universality class from $n \rightarrow 0$ to $n = 1$. For the polymerization of actin, the evidence to date is that a mean field model can describe the behavior within the reproducibility of the data; nonmean field theories have not yet been applied to actin. The study of the thermodynamics of actin polymerization is producing valuable clues about the microscopic reaction mechanism.

We have limited our discussion here to the three cases: organic polymerizations, inorganic polymerizations, and biomolecular aggregations. Many other systems show reversible polymerization and could be profitably considered in this framework (37). In particular, micelles (45, 155, 156), gelations (157), polymer aggregations (158), and microemulsions (159) warrant attention from this point of view.

Questions Remaining

Many questions remain. For equilibrium polymerizations in general, what causes the metastable persistence of living polymers in regions where they should revert to monomer? What are the effects of pressure (160–162)? For example, tetrahydrofuran shows a peak in the extent of polymerization under pressure, a peak not predicted by theories (163). Can this peak be explained in a model that allows for compressibility? What can we say about equilibrium copolymerizations (164–170)?

For organic equilibrium polymerizations, what is the nature of the Coulombic aggregations of organic living polymers to make meta-living polymers, and on what variables do these aggregations depend? How do these aggregations affect other properties? How do changes of initiator concentration affect the behavior of the system? What can we learn from the osmotic pressure about these systems? What if we change the nature of the solvent? Can we quantify their nonideality? How does the tacticity of the polymer change as the polymer grows? What factors affect the tacticity? Does a solution of a living polymer in a poor solvent show liquid-liquid critical behavior like that in terminated polymer solutions? That is, do the thermodynamic properties have Ising behavior, or do the polydispersity and the Coulombic interactions affect the nature of the critical anomalies? What are the equilibrium molecular weight distributions in the coexisting phases of a living polymer in a poor solvent, and how is that equilibrium approached? Does the living polymer show a collapse transition? How does this collapse depend on such factors as the nature of the counterion, the average degree of polymerization, and the nature of the solvent?

For inorganic polymerizations, is the polymerization of sulfur described by an $n = 1$ model? Can we develop a full theory for this model? Can we make more precise measurements of $\Phi(T)$ or some other telling quantity in order to test the model? Many predictions for phase equilibria in sulfur solutions have not been tested experimentally. The time development of the MWD has not been studied.

For biopolymer polymerizations, explorations of the physical chemistry of the aggregation of actin and of tubulin have only just begun. The study of other aggregating proteins will be of great interest.

ACKNOWLEDGMENTS

This work was supported by the Chemistry Division of the NSF and by the National Institute of Arthritis and Musculoskeletal and Skin Diseases of the NIH.

Visit the Annual Reviews home page at www.annualreviews.org

LITERATURE CITED

1. Wheeler JC, Kennedy SJ, Pfeuty P. 1980. *Phys. Rev. Lett.* 45:1748–52
2. Szwarc M. 1956. *Nature* 178:1168–69
3. Szwarc M. 1968. *Carbanions, Living Polymers, and Electron Transfer Processes.* New York: Wiley. 695 pp.
4. Webster OW. 1991. *Science* 251:887–93
5. Cates ME. 1987. *Macromolecules* 20:2289–96
6. Greer SC. 1998. *J. Phys. Chem.* 102:5413–22
7. Greer SC. 1996. In *Advances in Chemical Physics*, ed. I Prigogine, SA Rice, pp. 261–96. New York: Wiley
8. Greer SC. 1995. *Comp. Mater. Sci.* 4:334–38
9. Sawada H. 1976. *Thermodynamics of Polymerization.* New York: Dekker. 403 pp.
10. Dainton FS, Ivin KJ. 1948. *Nature* 162:705–7
11. McGlashan ML. 1979. *Chemical Thermodynamics.* New York: Academic. 345 pp.
12. Tobolsky AV, Eisenberg A. 1962. *J. Coll. Sci.* 17:49–65
13. Tobolsky AV, Eisenberg A. 1960. *J. Am. Chem. Soc.* 82:289–93
14. Tobolsky AV, Rembaum A, Eisenberg A. 1960. *J. Polym. Sci.* 45:345–66
15. Tobolsky AV, Eisenberg A. 1959. *J. Am. Chem. Soc.* 81:780–82
16. Tobolsky AV, Eisenberg A. 1959. *J. Am. Chem. Soc.* 81:2302–5
17. Tobolsky AV. 1957. *J. Polym. Sci.* 25:220–21
18. Oosawa F, Asakura K, Hotta K, Imai N, Ooi T. 1959. *J. Polym. Sci.* 37:323
19. Oosawa F, Kasai M. 1962. *J. Mol. Biol.* 4:10–21
20. Oosawa F, Asakura S. 1975. *Thermodynamics of the Polymerization of Protein.* New York: Academic. 204 pp.
21. Grosberg AY, Khokhlov AR. 1994. *Statistical Physics of Macromolecules.* New York: Am. Inst. Physics. 350 pp.
22. de Gennes PG. 1972. *Phys. Lett. A* 38:339–40
23. de Gennes PG. 1979. *Scaling Concepts in Polymer Physics.* Ithaca: Cornell Univ. Press. 324 pp.
24. des Cloizeaux J. 1975. *J. Phys. (Paris)* 36:281–91
25. Schofield P, Litster JD, Ho JT. 1969. *Phys. Rev. Lett.* 23:1098
26. Kennedy SJ, Wheeler JC. 1983. *J. Chem. Phys.* 78:1523–27
27. Wheeler JC, Pfeuty P. 1981. *Phys. Rev. Lett.* 46:1409–13
28. Wheeler JC, Pfeuty P. 1981. *J. Chem. Phys.* 74:6415–30
29. Wheeler JC. 1984. *Phys. Rev. Lett.* 53:174–77
30. Wheeler JC. 1984. *J. Chem. Phys.* 81:3635–40
31. Kennedy SJ, Wheeler JC. 1983. *J. Chem. Phys.* 78:953–62
32. Wheeler JC, Pfeuty PM. 1993. *Phys. Rev. Lett.* 71:1653–56
32a. van der Schoot P. 2001. *Macromolecules.* In press
33. Flory PJ. 1953. *Principles of Polymer Chemistry.* Ithaca, NY: Cornell Univ. Press. 672 pp.
34. Scott RL. 1965. *J. Phys. Chem.* 69:261–70
35. Dudowicz J, Freed KF, Douglas JF. 2000. *J. Chem. Phys.* 113:434–46
36. Dudowicz J, Freed KF, Douglas JF. 2000. *J. Chem. Phys.* 112:1002–10
37. Dudowicz J, Freed KF, Douglas JF. 1999. *J. Chem. Phys.* 111:7116–30
38. Niranjan PS, Forbes JG, Greer SC, Dudowicz J, Freed KF, Douglas JF. 2001. *J. Chem. Phys.* 114:10,573–76
39. Cordery R. 1981. *Phys. Rev. Lett.* 47:457–59
40. Helfrich W, Muller W. 1980. In *Continuous Models of Discrete Systems*, pp. 753. Waterloo, Ont.: Univ. Waterloo Press
41. Helfrich W. 1983. *J. Phys. (Paris)* 44:13

42. Gujrati PD. 1983. *Phys. Rev. B* 27:4507
43. Wheeler JC, Petchek RG, Pfeuty P. 1983. *Phys. Rev. Lett.* 50:1633–36
44. Petschek RG, Pfeuty P, Wheeler JC. 1986. *Phys. Rev. A* 34:2391–421
45. Cates ME, Candau SJ. 1990. *J. Phys. Condens. Matter* 2:6869–92
46. Milchev A, Landau DP. 1995. *Phys. Rev. E* 52:643–41
47. Milchev A, Rouault Y, Landau DP. 1997. *Phys. Rev. E* 56:1946–53
48. Wittmer JP, Milchev A, Cates ME. 1998. *Europhys. Lett.* 41:291–96
49. Leiderer P, Bosch W. 1980. *Phys. Rev. Lett.* 45:727–29
50. Corrales LR, Wheeler JC. 1989. *J. Chem. Phys.* 90:5030–55
51. Corrales LR, Wheeler JC. 1992. *J. Phys. Chem.* 96:9479–87
52. Knobler CM, Scott RL. 1984. In *Phase Transitions and Critical Phenomena*, ed. C Domb, JL Lebowitz, p. 318. New York: Academic
53. Brown WB, Szwarc M. 1958. *Trans. Faraday Soc.* 54:416–19
54. Das SS, Andrews AP, Greer SC, Guttman CM, Blair W. 1999. *J. Chem. Phys.* 111:9406–17
55. Miyake A, Stockmayer WH. 1965. *Makromol. Chem.* 88:90–116
56. Taganov NG. 1984. *Sov. J. Chem. Phys.* 1:2389–440
57. Ballone P, Jones RO. 2001. *J. Chem. Phys.* 115:3895–905
58. Tobolsky AV. 1944. *J. Chem. Phys.* 12:402–4
59. Peebles LH Jr. 1971. *Molecular Weight Distributions in Polymers*. New York: Wiley. 331 pp.
60. Whitten TA, Schaeffer L. 1978. *J. Phys. A.* 11:1843–54
61. van der Schoot P. 1997. *Europhys. Lett.* 39:25–30
62. Le Guillou JC, Zinn-Justin J. 1985. *J. Phys. Lett.* 46:L137
63. Bouchaud JP, Ott A, Langevin D, Urbach W. 1991. *J. Phys. II France* 1:1465–82
64. Rouault Y. 1998. *Phys. Rev. E* 58:6155–57
65. Schäfer L. 1992. *Phys. Rev. B* 46:6061–70
66. Das SS, Andrews AP, Greer SC. 1995. *J. Chem. Phys.* 102:2951–59
67. Szwarc M, Levy M, Milkovich R. 1956. *J. Am. Chem. Soc.* 78:2656–57
68. Szwarc M. 1996. *Ionic Polymerization Fundamentals*. New York: Hanser. 211 pp.
69. Szwarc M, Beylen MV. 1993. *Ionic Polymerization and Living Polymers*. New York: Chapman & Hall. 380 pp.
70. Fetters LJ. 1966. *J. Res. Natl. Bur. Stand.* 70A:421–33
71. Ndoni S, Papadakis CM, Bates FS, Almdal K. 1995. *Rev. Sci. Instrum.* 66:1090–95
72. Zheng KM, Greer SC, Corrales LR, Ruiz-Garcia J. 1993. *J. Chem. Phys.* 98:9873–80
73. Roberts DE, Jessup RS. 1951. *J. Res. Natl. Bur. Stand.* 46:11–17
74. Pendyala K, Gu X, Andrews KP, Gruner K, Jacobs DT, Greer SC. 2001. *J. Chem. Phys.* 114:4312–22
75. Dreyfuss P, Dreyfuss MP. 1967. *Adv. Polym. Sci.* 4:528–90
76. Dreyfuss MP, Dreyfuss P. 1966. *J. Polym. Sci.* 4:2179–200
77. Sims D. 1964. *J. Chem. Soc.* 1964:864–65
78. Zhuang J, Andrews AP, Greer SC. 1997. *J. Chem. Phys.* 107:4705–10
79. Ruiz-Garcia J, Greer SC. 1990. *Phys. Rev. Lett.* 64:3204
80. Ruiz-Garcia J, Greer SC. 1990. *Phys. Rev. Lett.* 64:1983–85
81. Andrews AP, Andrews KP, Greer SC, Boué F, Pfeuty P. 1994. *Macromolecules* 27:3902–11
82. Zheng KM, Greer SC. 1992. *Macromolecules* 25:6128–36
83. Hui KM, Ng TL. 1969. *J. Polym. Sci. A-1* 7:3101–9
84. Zhuang J, Das SS, Nowakowski M, Greer SC. 1997. *Physica A* 244:522–35
85. Ruiz-Garcia J, Greer SC. 1997. *J. Mol. Liq.* 71:209–24
86. Ruiz-Garcia J, Castillo R. 1999. *J. Chem. Phys.* 110:10,657–59

87. Rubio MA, Conde L, Riande E. 1988. *Rev. Sci. Instrum.* 59:2041–44
88. Fetters LJ, Balsara NP, Huang JS, Jeon HS, Almdal K, Lin MY. 1995. *Macromolecules* 28:4996–5005
89. Stellbrink J, Willner L, Jucknischke O, Richter D, Lindner P, Fetters LJ, et al. 1998. *Macromolecules* 31:4189–97
90. Balsara NP, Fetters LJ. 1999. *Macromolecules* 32:5147–48
91. Stellbrink J, Willner L, Richter D, Lindner P, Fetters LJ, Huang JS. 1999. *Macromolecules* 32:5321–29
92. Tobolsky AV, MacKnight WJ. 1965. *Polymeric Sulfur and Related Polymers*. New York: Interscience. 140 pp.
93. Eisenberg A, Tobolsky AV. 1960. *J. Poly. Sci.* 46:19–28
94. Faivre G, Gardissat J-L. 1986. *Macromolecules* 19:1988–96
95. Zingaro RA, Cooper WC, eds. 1974. *Selenium*. New York: Van Nostrand Reinhold. 835 pp.
96. Lumbroso H. 1977. *Selenium*. Paris: Masson
97. Meyer B. 1976. *Chem. Rev.* 76:376–88
98. Gee G. 1952. *Trans. Faraday Soc.* 48:515–26
99. Fairbrother F, Gee G, Merrall GT. 1955. *J. Polym. Sci.* 16:459–69
100. Zheng KM, Greer SC. 1992. *J. Chem. Phys.* 96:2175–82
101. Harris RE. 1970. *J. Phys. Chem.* 74:3102–11
102. Steudel R, Strauss R, Koch L. 1985. *Angew. Chem. Int. Ed. Engl.* 24:59–60
103. Stillinger FH, Weber TA, LaViolette RA. 1986. *J. Chem. Phys.* 85:6460–69
104. Ruiz-Garcia J, Anderson EM, Greer SC. 1989. *J. Phys. Chem.* 93:6980–83
105. Anderson EM, Greer SC. 1988. *J. Chem. Phys.* 88:2666–71
106. Powell RE, Eyring H. 1943. *J. Am. Chem. Soc.* 65:648–54
107. Anisimov MA, Kugel KI, Lisovskaya TY. 1987. *High Temp.* 25:165–73
108. Koh JC, Klement W. 1970. *J. Phys. Chem.* 74:4280
109. Kozhevnikov VF, Viner JM, Taylor PC. 2001. *Phys. Rev. B*. In press
110. Bacon RF, Fanelli R. 1943. *J. Am. Chem. Soc.* 65:639–54
111. Alvarenga AD, Grimsditch M, Susman S, Rowland SC. 1996. *J. Phys. Chem.* 100:11,456–59
112. Hosokawa S, Matsuoka T, Tamura K. 1994. *J. Phys. Condens. Matter* 6:5273–82
113. West ED. 1959. *J. Am. Chem. Soc.* 81:29–37
114. Feher F, Goerler GP, Lutz HD. 1971. *Z. Anorg. Allg. Chem.* 382:135–47
115. Hammick DL, Holt WE. 1926. *J. Chem. Soc.* 2:1995–2003
116. Larkin JA, Katz J, Scott RL. 1967. *J. Phys. Chem.* 71:352–58
117. Greer SC. 1986. *J. Chem. Phys.* 84:6984–88
118. Wiewiorowski TK, Parthasathy A, Slaten BL. 1968. *J. Phys. Chem.* 72:1890–92
119. Tanford C. 1961. *Physical Chemistry of Macromolecules*. New York: Wiley. 710 pp.
120. Touro FJ, Wiewiorowski TK. 1966. *J. Phys. Chem.* 70:239–41
121. Descotes L, Bellissent R, Pfeuty P, Dianoux AJ. 1993. *Physica A* 201:381–85
122. Poltavtsev YG, Titenko YV. 1975. *Russ. J. Phys. Chem.* 49:178–82
123. Tompson CW, Gingrich NS. 1959. *J. Chem. Phys.* 31:1598–604
124. Bellissent R, Descotes L, Boué F, Pfeuty P. 1990. *Phys. Rev. B* 41:2135–38
125. Boué F, Ambroise JP, Bellissent R, Pfeuty P. 1992. *J. Phys. I France* 2:969–80
126. Winter R, Bodensteiner T, Szornel C, Egelstaff PA. 1988. *J. Non-Cryst. Solids* 106:100–3
127. Winter R, Szornel C, Pilgrim W-C, Howells WS, Egelstaff PA, et al. 1990. *J. Phys. Condens. Matter* 2:8427–37
128. Winter R, Egelstaff PA, Pilgrim W-C, Howells WS. 1990. *J. Phys. Condens. Matter:* SA215–18
129. Winter R, Pilgrim W-C, Egelstaff PA,

Chieux P, Anlauf S, Hensel F. 1990. *Europhys. Lett.* 11:225–28
130. Stolz M, Winter R, Howells WS, McGreevey RL, Egelstaff PA. 1994. *J. Phys. Condens. Matter* 6:3619–28
131. Amos LA, Amos WB. 1991. *Molecules of the Cytoskeleton*. New York: Guilford. 253 pp.
132. Fygenson DK, Braun E, Libchaber A. 1994. *Phys. Rev. B* 50:1579–88
133. Sheterline P, Clayton J, Sparrow JC. 1996. *Actins*. San Diego: Academic. 116 pp.
134. Sheterline P, Sparrow JC. 1994. *Protein Profile* 1:1–121
135. Kabsch W, Mannherz HG, Suck D, Pai EF, Holmes KC. 1990. *Nature* 347:37–44
136. Holmes KC, Popp D, Gebhard W, Kabsch W. 1990. *Nature* 347:44–49
137. Stossel TP. 1994. *Sci. Am.* 27:54–63
138. Roush W. 1995. *Science* 269:30–31
139. Ivkov R, Forbes JG, Greer SC. 1998. *J. Chem. Phys.* 108:5599–607
140. Matsudaira P, Bordas J, Koch MHJ. 1987. *Proc. Natl. Acad. Sci. USA* 84:3151–55
141. Kas J, Strey H, Tang JX, Finger D, Ezzell R, et al. 1996. *Biophys. J.* 70:609–25
142. Carlier M-F, Pantaloni D, Korn ED. 1984. *J. Biol. Chem.* 259:9987–91
143. Carlier M-F. 1990. *Adv. Biophys.* 26:51–73
144. Bonder EM, Fishkind DJ, Mooseker MS. 1983. *Cell* 34:491–501
145. Kinosian HJ, Selden LA, Estes JE, Gershman LC. 1991. *Biochim. Biophys. Acta* 1077:151–58
146. Kasai M, Nakano E, Oosawa F. 1965. *Biochim. Biophys. Acta* 94:494–503
147. Selden LA, Kinosian HJ, Estes JE, Gershman LC. 2000. *Biochemistry* 39:64–74
148. Zimmerle CT, Frieden C. 1986. *Biochemistry* 25:6432–38
149. Niranjan PS, Forbes JG, Greer SC. 2000. *Biomacromolecules* 1:506–8
150. Niranjan PS. 2000. *Thermodynamics of actin polymerization: extent of polymerization studies*. PhD thesis. Univ. Maryland, College Park, MD. 157 pp.
151. van Gestel J, van der Schoot P, Michels MAJ. 2001. *J. Phys. Chem. B*. In press
152. Kawamura M, Muruyama K. 1972. *Biochim. Biophys. Acta* 267:422–34
153. Arisaka F, Kawamura M, Murayama K. 1973. *J. Biochem.* 73:1211–15
154. Burlacu S, Janmey PA, Borejdo J. 1992. *Am. J. Physiol. Cell Physiol.* 262:C569–C77
155. Wang Z-G, Costas ME, Gelbart WM. 1993. *J. Phys. Chem.* 97:1237–42
156. Gelbart WM, Ben-Shaul A. 1996. *J. Phys. Chem.* 100:13,169–89
157. Kumar SK, Panagiotopoulos AZ. 1999. *Phys. Rev. Lett.* 82:5060–63
158. Sintes T, Toral R, Chakrabarti A. 1994. *Phys. Rev. E* 50:2967–76
159. Chen SH, Rouch J, Sciortino F, Tartaglia P. 1994. *J. Phys. Condens. Matter* 6:10,844–83
160. Weale KE. 1974. In *Reactivity, Mechanism, and Structure in Polymer Chemistry*, ed. AD Jenkins, A Ledwith, pp. 158–74. New York: Wiley
161. Weale KE. 1967. *Chemical Reactions at High Pressures*. London: E & FN Spon. 349 pp.
162. Weale KE. 1962. *Q. Rev.* 16:267–81
163. Rahman M, Weale KE. 1969. *Polymer A–1* 7:122–24
164. Tobolsky AV, Owen GDT. 1962. *J. Polym. Sci.* 59:329–37
165. O'Driscoll KF. 1969. In *Advances in Chemistry*, ed. RF Gould. Washington, DC: Am. Chem. Soc.
166. Theil MH. 1969. *Macromolecules* 2:137–42
167. Kennedy SJ, Wheeler JC. 1984. *J. Phys. Chem.* 88:6595–605
168. Jensen PJ, Bennemann K-H. 1985. *J. Chem. Phys.* 83:6457–66
169. Szwarc M, Perrin CL. 1985. *Macromolecules* 18:528–33
170. Szymanski R. 1986. *Macromolecules* 19:3003–4

SCANNING TUNNELING MICROSCOPY STUDIES OF THE ONE-DIMENSIONAL ELECTRONIC PROPERTIES OF SINGLE-WALLED CARBON NANOTUBES

Min Ouyang,[1] Jin-Lin Huang,[1] and Charles M. Lieber[1,2]

[1]Department of Chemistry and Chemical Biology and [2]Division of Engineering and Applied Science, Harvard University, Cambridge, Massachusetts 02138; e-mail: cml@cmliris.harvard.edu

Key Words SWNTs, STM, band structure, energy gaps, resonant scattering, Kondo

■ **Abstract** Recent developments in scanning tunneling microscopy studies of the electronic properties of single-walled carbon nanotubes are reviewed. A broad range of topics focused on the unique electronic properties of nanotubes are discussed, including (a) the underlying theoretical description of the electronic properties of nanotubes; (b) the roles of finite curvature and broken symmetries in perturbing electronic properties; (c) the unique one-dimensional energy dispersion in nanotubes; (d) the nature of end states; (e) quantum size effects in short tubes; (f) the interactions between local spins and carriers in metallic systems (the Kondo effect); and (g) the atomic structure and electronic properties of intramolecular junctions. The implications of these studies for understanding fundamental one-dimensional physics and future nanotube device applications are discussed.

INTRODUCTION

Single-walled carbon nanotubes (SWNTs) have aroused great excitement as unique systems for understanding one-dimensional (1D) physics and for building carbon-based nanoelectronics (1–4). The inherent high aspect ratio of SWNTs makes these materials ideal scanning probe microscope tips (5, 6), and mechanical measurements have demonstrated that carbon nanotubes have the largest Young's modulus of any known material (7, 8). However, it is the electronic properties of SWNTs, which depend uniquely on geometric structure, that are arguably the most significant characteristic of this material. A single SWNT can be either metallic or semiconducting (9–15), depending only on diameter and chirality, although the local carbon-carbon bonding remains constant. The ability to yield both metallic and semiconducting forms without doping is unique among solid-state materials, and has led to speculation that SWNTs might thus serve as a building block for carbon-based electronics (4).

A large number of measurements have been carried out in attempts to verify the remarkable electronic properties predicted for SWNTs (13, 14, 16–18), although the first clear elucidation of metallic and semiconducting SWNTs was obtained from scanning tunneling microscopy (STM) studies (13, 14). STM has proven to be exceptionally well suited for the elucidation of the fundamental electronic properties of SWNTs because it can determine simultaneously both the atomic structure and local electronic density of states (DOS). Initial STM studies of SWNTs focused on the interrogation of the basic relationship between atomic structure and electronic properties (3, 13, 14), and subsequently investigated the structures of SWNT ends and the consequences of finite size (3, 19, 20). Most recently, STM investigations have begun to address more fundamental issues about the electronic properties of SWNTs (15, 21–23), including the determination of unique 1D energy dispersion predicted for this material.

Herein, we review the fascinating electronic properties of SWNTs and discuss the implications of these properties for future electronics. The review is organized as follows: First, a π-only tight-binding scheme is presented to explain the basic electronic properties of SWNTs and to provide a framework for interpreting STM studies; second, experimental and theoretical studies addressing the roles of finite curvature and broken symmetries in perturbing the electronic properties of SWNTs are discussed; third, studies of energy-dependent resonant scattering, which enable the unique 1D energy dispersion to be characterized, are presented; fourth, the structure and electronic states of SWNT ends and quantum effects in finite size SWNTs are discussed; fifth, the interactions between local spins and carriers in metallic and finite size SWNTs are described; and sixth, the atomic structure and electronic properties of intramolecular semiconductor-metal and metal-metal junctions are presented. In concluding, implications of these studies for understanding fundamental 1D physics and future nanotube device applications are discussed.

THEORETICAL BACKGROUND

A SWNT can be viewed as a seamless cylinder obtained by rolling-up a single layer of a two-dimensional (2D) graphene sheet. The SWNT is uniquely characterized by the roll-up vector, $\mathbf{C_h} = n\mathbf{a_1} + m\mathbf{a_2} \equiv (n, m)$, where $\mathbf{a_1}$ and $\mathbf{a_2}$ are the graphene primitive vectors and n, m are integers (Figure 1, see color insert). The translation vector, $\mathbf{T}$, is directed along the SWNT axis and perpendicular to $\mathbf{C_h}$; the magnitude of $\mathbf{T}$ corresponds to the length of the (n, m) SWNT unit cell. Once (n, m) is specified, other structural properties, such as diameter (d_t) and chiral angle (θ), can be determined: $d_t = (3^{1/2}/\pi)a_{cc}(m^2 + mn + n^2)^{1/2}$; and $\theta = \tan^{-1}[3^{1/2}m/(2n+m)]$, where a_{cc} is the nearest-neighbor carbon atom distance of 0.142 nm. Among the large number of possible $\mathbf{C_h}$ vectors, there are two inequivalent high symmetry directions. The two high symmetry tubes are called zigzag and armchair (Figure 1, color insert), and are designated by (n, 0) and (n, n), respectively. These tubes are achiral; however, when $\mathbf{C_h}$ lies along a lower symmetry direction, the corresponding SWNTs will be chiral.

The electronic band structure of SWNTs can be analyzed using a tight-binding model, in which hybridization effects due to the finite curvature of the tube structure are ignored. With this approximation the tight-binding Hamiltonian H(k) is (24):

$$H(k) = \sum \gamma_0 e^{ik \cdot r_i}, \qquad 1.$$

where γ_0 is the amplitude of the nearest-neighbor overlap integral, and $\mathbf{r}_i$ (i = 1, 2, 3) are the bond vectors of the graphene sheet. The π and π^* band energy dispersions obtained using this model are:

$$E_{cv}(k_x, k_y) = \pm \gamma_0 \{1 + 4\cos(3k_x a_{cc}/2)\cos(3^{1/2} k_y a_{cc}/2)$$
$$+ 4\cos^2(3^{1/2} k_y a_{cc}/2)\}^{1/2}, \qquad 2.$$

where $+(-)$ corresponds to the π^* (π) bands. Figure 2a (see color insert) shows the 2D graphene sheet band structure in the first Brillouin zone obtained from Equation 2. The Fermi surface of graphene sheet determined from H(k_F) = 0 is reduced to the six corners ($\mathbf{K_B}$) of the hexagonal first Brillouin zone (Figure 2b, color insert). Because there are two atoms per unit cell, the valence band of graphene is completely filled.

The 1D band structure of a SWNT can be derived from this 2D model by imposing periodic boundary conditions. All allowed wavevectors k are quantized in the direction perpendicular to the rolled-up vector $\mathbf{C_h}$: $k \cdot \mathbf{C_h} = 2\pi q$, where q is an integer. Therefore, only a particular set of lines, which are parallel to the corresponding tube axis with a spacing of $2/d_t$, are allowed (Figures 2c, 2d, color insert). If the allowed wavevector passes through Fermi points $\mathbf{K_B}$ of the graphene sheet, the SWNT is predicted to be metallic, and otherwise semiconducting. From the criteria $\mathbf{K_B} \cdot \mathbf{C_h} = 2\pi q$, we thus expect:

$$\begin{cases} n - m = 3q: \text{metallic} \\ n - m \neq 3q: \text{semiconducting}. \end{cases} \qquad 3.$$

The 1D band structure of SWNTs can be constructed by zone-folding the 2D graphene band structure into the 1D Brillouin zone of a (n, m) SWNT (1, 19). The allowed states for the (n, m) SWNT lie on a set of parallel lines in the reciprocal space along the tube axis. The first Brillouin zone of a (n, m) SWNT is determined from the translation vector $\mathbf{T}$ with the length of $2\pi/|\mathbf{T}|$; that is, the segment between $\pm\pi/\mathbf{T}$ (Figure 3, color insert). The π and π^* bands of the SWNT are obtained by assuming the values of Equation 2 along this line segment. The remaining bands are generated by zone folding using:

$$E_\mu(k) = E_{cv}\left(k\mathbf{T}/|\mathbf{T}| + \mu(2\pi)\mathbf{C_h}/|\mathbf{C_h}|^2\right)$$
$$(\mu = 0, \ldots, N-1, \text{ and } -\pi/|\mathbf{T}| < k < \pi/|\mathbf{T}|), \qquad 4.$$

where N is the number of hexagons in the nanotube unit cell, which is determined by $N = |\mathbf{C_h} \times \mathbf{T}|/|\mathbf{a_1} \times \mathbf{a_2}|$.

The DOS of SWNTs can be computed from the band structure by summing the number of states at every energy level (Figure 3, color insert). The calculated DOS of SWNTs exhibit a distinct signature of 1D systems; that is, divergent singularities that are commonly referred to as van Hove singularities (VHS) (25). In three dimensions (3D), VHS should appear as kinks, although in 2D they appear as stepwise discontinuities. Hence, the VHS in 1D SWNTs represent well-defined (spike-like) features that should enable direct clear comparisons between experiment and theory.

Lastly, there are several other important characteristics of the electronic properties of SWNTs suggested by π-only tight-binding models (3, 26–28). First, the DOS at E_F has a zero value for semiconducting SWNTs $(n - m \neq 3q)$ but is nonzero (and small) for metallic SWNTs $(n - m = 3q)$. Second, the VHS spacing has a characteristic 1–2–4 pattern from E_F (with spacing 1ξ–2ξ–4ξ) for semiconducting SWNTs, and 1–2–3 from E_F (with spacing 3ξ–6ξ–9ξ) for metallic SWNTs, where $\xi = 2\pi/3|\mathbf{C_h}|$. Third, the first VHS band gaps for semiconducting and metallic SWNTs are $E_g^S = 2\gamma_0 a_{cc}/d_t$ and $E_g^M = 6\gamma_0 a_{cc}/d_t$, respectively, and are independent of chiral angle θ to first order.

INITIAL SCANNING TUNNELING MICROSCOPY STUDIES

The first studies that directly addressed these theoretical predictions were carried out by Odom et al. (13) and Wildöer et al. (14) using low-temperature STM. These initial STM studies characterized both the atomic structures and the DOS of individual SWNTs and SWNTs packed in bundles, and showed the existence of both semiconducting and metallic SWNTs for a wide range of structures as predicted by theory. Subsequently, Kim et al. (19) and Odom et al. (3) reported the first detailed comparisons of experimentally determined SWNT VHS with tight-binding calculations for metallic and semiconducting tubes. The good agreement between theory and these experiments showed that much of the physics of SWNT band structure is captured by the simple π-only model. However, other important issues, such as the effects of finite curvature and broken rotational symmetry, which are essential to a complete understanding of the electronic properties and potential device applications of SWNTs, were not addressed. We examine these and other fascinating questions below.

FINITE CURVATURE EFFECT OF SINGLE-WALLED CARBON NANOTUBES

Theoretical Models

As the diameter of a SWNT decreases, the hybridization between orthogonal molecular orbitals cannot be neglected. For a SWNT with sufficiently small diameter, the hybridization of σ, σ^*, π and π^* orbitals can be quite large (29). Full-valence tight-binding calculations (30) and analytical calculations for a

Hamiltonian on a curved surface have suggested that the finite curvature of SWNTs (31) will strongly modify the electronic behavior of SWNTs and open up small energy gaps at E_F.

To understand the effect of finite curvature more directly, Ouyang et al. (15) and Kleiner & Eggert (32) have independently developed a Fermi-point shifting model in which finite curvature or tube deformation (33) produces shifts of the Fermi points from $\mathbf{K_B}$ (Figure 4, color insert). In general, when the bond symmetry of neighboring carbon atoms is broken, the corresponding nearest-neighbor overlap integral, γ_0, is no longer constant and can be expressed as $\gamma_i = \gamma_0 + \Delta\gamma_i$ (i = 1, 2, 3). Here, the $\Delta\gamma_i$ are estimated from misalignment angle α, which describes the misorientation of neighboring atoms (Figure 4a, color insert) and is determined by the local curvature. The corresponding Hamiltonian can be written as

$$H'(k) = \sum \gamma_i e^{ik \cdot r_i}, \qquad 5.$$

where the Fermi points with curvature are given by $H'(k_F) = 0$. For small changes of γ_i, we expect that the shift of k_F is small and can be given as $k_F = \mathbf{K_B} + \Delta k_F$. The magnitude and direction of Δk_F enable the role of curvature to be quantified. For a metallic zigzag (n, 0) SWNT, the magnitude is:

$$|\Delta k_F| = \{3^{1/2}\pi^2 / [a_{cc}(12n^2 + \pi^2)]\}$$
$$- \{\pi^3(8n^2 + \pi^2)^{1/2} / [3^{1/2}a_{cc}n^2(48n^2 + 4\pi^2)]\} \qquad 6.$$

and the direction of Δk_F is perpendicular to the tube axis along the circumferential direction **c** (Figure 4b, color insert). From this shift the energy gap width can be defined as

$$E_g^c = [3\gamma_0 a_{cc}^2/(16R^2)], \qquad 7.$$

where R is the radius of the SWNT.

In addition, the energy dispersion near E_F under finite curvature effect remains linear,

$$|E(k)| \propto (3^{1/2}\gamma_0 a_{cc})[1 + \pi^2/(8n^2)]|k - k_F|, \qquad 8.$$

and thus the DOS near E_F can be expressed as a universal function (28, 34):

$$\begin{cases} \rho(E) \propto |E|/[|E|^2 - E_g^{c2}/4]^{1/2} & \text{for } |E| > E_g^c \\ 0 & \text{for } |E| < E_g^c. \end{cases} \qquad 9.$$

In summary, our Fermi-point shifting model predicts that metallic zigzag tubes are in fact small gap semiconductors.

We have also applied this model to armchair (n, n) SWNTs with finite curvature (Figure 4c, color insert). In contrast to zigzag SWNTs, the symmetry of armchair

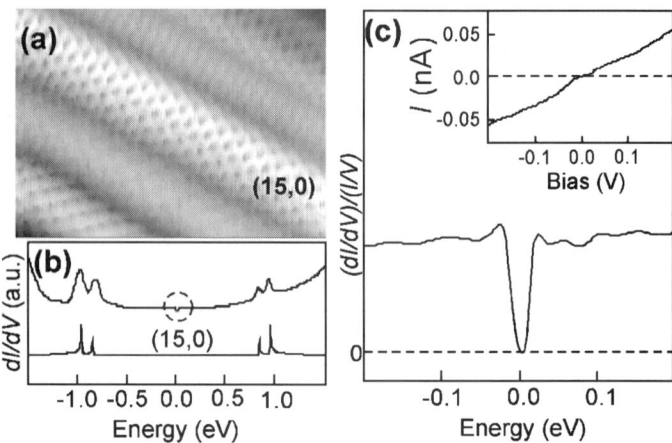

Figure 5 (*a*) Atomically resolved constant current STM image of a (15, 0) SWNT. (*b*) Tunneling conductance, dI/dV, for (15, 0) zigzag SWNTs, with corresponding calculated DOS shown below. The new feature near E_F is highlighted with a dashed circle. (*c*) High energy resolution normalized conductance, $(dI/dV)/(I/V)$, and measured *I–V* curves (inset) for the (15, 0) tube.

SWNTs requires that Δk_F is perpendicular to **c** and along the tube axis. Hence, even with finite curvature, the new Fermi points of an armchair SWNT still lie on the allowed wavevectors and suggest that an isolated armchair SWNT will remain truly metallic.

Experimental Studies of Curvature-Induced Gaps

Ouyang et al. carried out the first direct studies of curvature-induced gaps in SWNTs using low-temperature STM (15). Figure 5*a* shows a typical atomically resolved image of a (15, 0) SWNT. The VHS determined experimentally and calculated using a π-only tight-binding model show excellent agreement (Figure 5*b*), and suggest that finite curvature does not perturb the larger energy features of the SWNT electronic structure. However, these data also show a reduction in the DOS or gap-like feature near E_F. High-resolution normalized tunneling spectra, which are proportional to the local DOS, clearly show that the local DOS at E_F are reduced to zero (i.e., a true energy gap) with sharp increases at energies that depend inversely on the zigzag tube radius. Values of the energy gaps for different zigzag SWNT radii were determined using Equation 9 broadened with a Gaussian convolution to account for instrumentation and temperature effects (Figure 6) (15), and are well fit by the $1/R^2$ dependence predicted by Equation 7.

Lastly, these studies have enabled γ_0 to be determined independently by studying the radius dependence of gap widths. The value obtained from a fit to Equation 7, 2.60 eV, is in good agreement with the range of 2.5–2.7 eV determined in

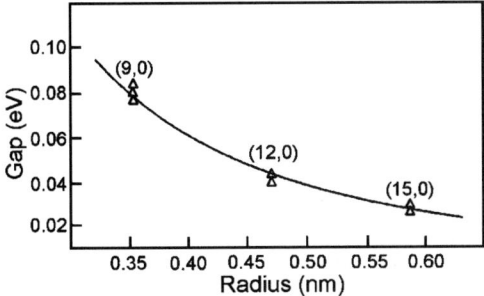

Figure 6 Curvature-induced energy gaps vs. tube radius. Every data point (triangle) represents the averaged gap value for one distinct (n, 0) SWNT. The solid line corresponds to a fit of $3\gamma_0 a_{cc}^2/(16R^2)$ with γ_0 value of 2.60 eV. Adapted from Reference 15.

earlier STM studies (13, 14). In summary, these experimental studies have shown conclusively that predicted metallic (n, 0) SWNTs are actually narrow gap semiconductors.

BROKEN SYMMETRY: ARMCHAIR SWNTs

The Fermi-point shifting model suggests that finite curvature will not affect the metallic nature of isolated armchair SWNTs. This suggestion is consistent with atomically resolved, low-temperature STM investigations (15) of isolated (8, 8) SWNTs (Figure 7a). These studies showed the local DOS was nonzero and constant at E_F, and thus the SWNT was metallic (Figure 7b). In contrast, studies of armchair SWNT bundles by Odom et al. (3) and Ouyang et al. (15) first reported the existence of a small energy gap feature near the E_F. For example, the local DOS measured for a (8, 8) SWNT packed in the bundle (Figures 7b, 7d) show a clear suppression around E_F in contrast to the data for the isolated (8, 8) SWNT. Comparison of these data with the calculated and experimental DOS of an isolated (8, 8) SWNT shows good agreement for the position of the VHS peaks, which suggests that the isolated and bundle tubes have very similar electronic band structure for greater energies, >0.1 eV. This comparison thus suggests that tube-tube interactions do not perturb the electronic band structure on larger energy scales.

High-resolution data recorded near E_F on the (8, 8) SWNT in the bundle show that the gap has a magnitude of ~100 meV (Figure 7d). The gap structure observed in armchair tubes in bundles differs from the finite curvature-induced gaps in zigzag SWNTs discussed above. Specifically, the local DOS are suppressed but not reduced completely to zero, and for this reason, these gaps are termed pseudogaps (15). Similar gap features were observed in other armchair SWNT bundles (Figure 8a, color insert), with gap values ranging from 80 to 100 meV in the (10,10) through (7, 7) SWNTs, respectively (Figure 8b, color insert).

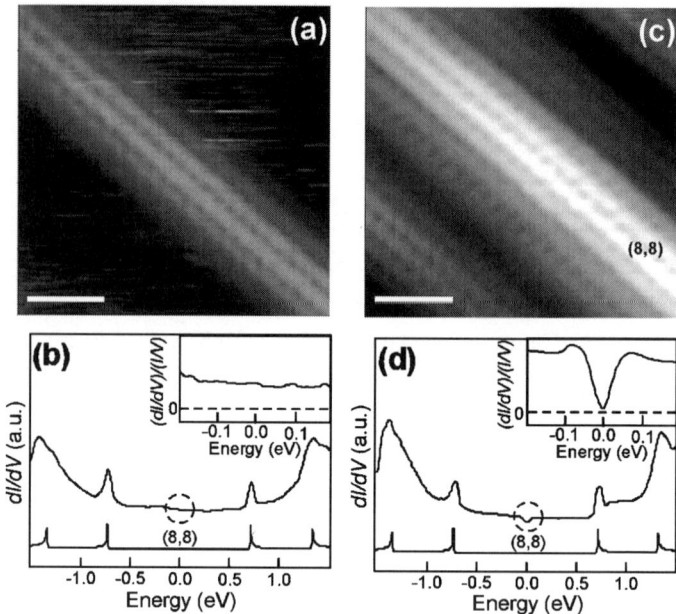

Figure 7 (*a*) STM image of an isolated (8, 8) SWNT on Au(111) substrate. Scale bar is 1 nm. (*b*) (*dI/dV*) recorded on the isolated (8, 8) tube. The calculated DOS for an isolated (8, 8) tube is displayed below the experimental data. The dashed circles highlight the energy near E_F. (*Inset*) High-resolution (*dI/dV*)/(*I/V*) data near E_F. (*c*) STM image of an (8, 8) SWNT exposed at the top of a bundle. Scale bar is 1 nm. (*d*) Same as (*b*) but data recorded on (8, 8) tube in the bundle of (*c*).

The appearance of pseudogaps in the armchair SWNTs is at first glance surprising, but can be understood in terms of the broken n-fold rotational symmetry of armchair SWNTs due to interactions in a bundle (15) (Figure 8*c*, color insert). Briefly, in an isolated armchair SWNT, π and π^* bands have different parity under reflection and cross at the E_F without opening a gap (27). However, when the armchair tubes are packed into a bundle, strong tube-tube interactions (26, 35–38) will break rotational symmetry and enable mixing of π and π^* bands and the formation of a pseudogap at E_F. Theoretical studies of armchair SWNT bundles (26, 35–38) have predicted the existence of pseudogaps at E_F, and our measured pseudogaps have similar magnitudes (Figure 8*b*, color insert). In addition, our experimental data exhibit a weak inverse dependence on SWNT radius, which is qualitatively consistent with theoretical predictions (37). However, deviations between experiment and theory are also clear when making detailed comparisons. One possible origin for these differences is that STM experiments were carried out on SWNTs at the surfaces of bundles, whereas theoretical calculations have been based on periodic lattices with higher local coordination. Future calculations carried out on structures accurately modeling STM experiments should help to address such differences.

Lastly, our experimental studies enable several important points to be addressed. First, experimental evidence for true 1D metallic behavior in isolated armchair tubes on Au(111) substrates implies that the tube-substrate interaction does not perturb strongly the band structures of nanotubes (3). Second, the presence of sizeable pseudogaps in armchair tube bundles will modify electrical transport, and the very low DOS at E_F will make extended states in such tubes susceptible to localization. From a positive perspective, the existence of pseudogaps in armchair SWNTs should make these samples sensitive to doping, which may enable their use in sensor applications.

1D ENERGY DISPERSION

SWNTs are predicted to exhibit a unique linear energy dispersion at low energies ($E/\gamma_0 \ll 1$), which contrasts with the parabolic dependence expected from a conventional free-electron picture, and will impact significantly electronic behavior near E_F. In the case of isolated armchair (n, n) SWNTs, their metallic nature arises from the linear crossing of π and π^* energy bands at E_F. However, experimental determination of this important characteristic of the band structure has been lacking. Use of conventional momentum analysis methods, which average over substantial area, is difficult because SWNT samples consist of a wide range of structures each with different energy dispersions (3, 13–15, 39). STM, which can be used to interrogate individual nanotubes and has been used previously to determine the energy dispersions of 2D surface states on metals (40, 41) and semiconductors (42, 43), can, however, address this critical point.

Modulations in the local DOS of 1D metallic SWNTs near defects (3, 23, 44) and at the tube ends (45) have been observed in STM studies, and qualitatively attributed to the interference between incident and scattered electron waves in the nanotubes. However, these studies did not investigate the energy dependence of the DOS modulations, which would be needed to probe the 1D energy dispersion. Ouyang et al. first applied the STM to elucidate the 1D energy dispersion of SWNTs by characterizing in detail energy-dependent quantum interference of electrons scattered by defects in metallic nanotubes (22).

High-resolution STM images (Figure 9a) of an armchair SWNT with a defect show clearly a strong modulation in the DOS around the defect with a period larger than that of the armchair atomic lattice. Tunneling spectra (Figure 9b) recorded away from the defect (~8 nm) exhibit the VHS features characteristic of an isolated (13, 13) SWNT. Significantly, tunneling spectra recorded near the defect region show nine additional low-energy peaks (Figure 10). The most striking aspect of these peaks in the DOS is that their amplitude varies along the SWNT axis; these oscillations can be seen clearly in plots of the DOS vs. position for fixed energies

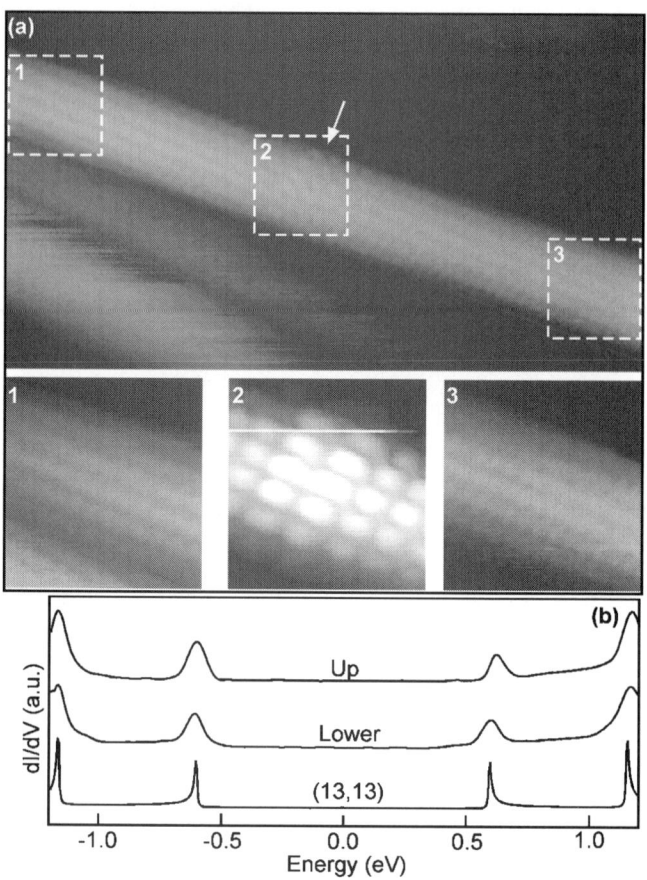

Figure 9 (*a*) (*Upper*) Atomically resolved STM image of an isolated armchair SWNT containing one defect; three numbered dashed squares highlight different regions along the tube axis. The white arrow highlights the defect position. (*Lower*) High-resolution images corresponding to the three distinct areas in upper panel. (*b*) dI/dV data recorded on the upper and lower portions of the tube ~8 nm away from the central defect area, and the calculated DOS for an isolated (13, 13) armchair SWNT.

(Figure 11, color insert). Qualitatively, these data also show that the oscillation wavelength decreases as the energy increases.

The observation of DOS oscillations at specific energy values is consistent with resonant electron scattering. Resonant scattering from defect-related quasibound states has been reported in recent SWNT theoretical studies (46) and transport measurements (47). We have developed this idea to understand quantitatively the DOS oscillations observed in STM studies (22). First, a specific energy-dependent oscillation is modeled using a 1D plane wave e^{ikx} (where x is position) with an energy equal to that of a defect quasibound state and incident on the defect at

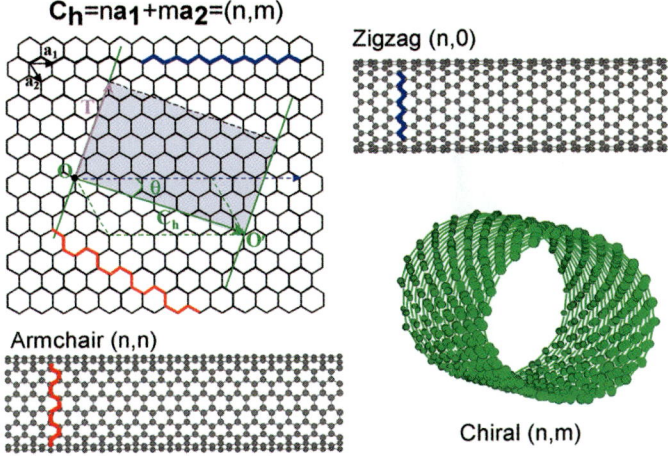

Figure 1 Schematics of graphene sheet and typical SWNTs structures. OO′ defines the chiral vector $\mathbf{C}_h = n\mathbf{a}_1 + m\mathbf{a}_2 \equiv (n,m)$. Translation vector, **T**, is along the nanotube axis and perpendicular to the $\mathbf{C}_h$. The shaded, area represents the unrolled unit cell formed by **T** and $\mathbf{C}_h$. Chiral angle, θ, is defined as the angle between the $\mathbf{C}_h$ and (n,0) zigzag direction. Two limiting achiral cases of (n,0) zigzag and (n,n) armchair SWNTs are indicated in *blue* and *red*, respectively. Molecular models of typical armchair, zigzag and chiral SWNTs are also shown.

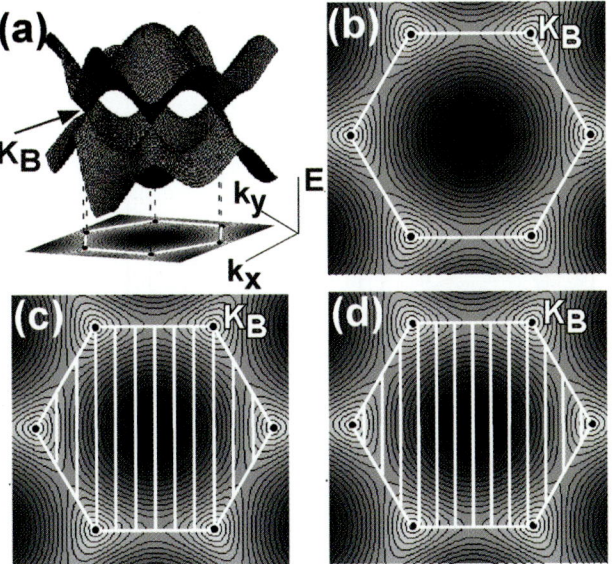

Figure 2 (*a*) Three-dimensional plot of the π and π* graphene energy bands and (*b*) a 2D projection. (*c*) Allowed 1D wavevectors for a metallic (9,0) SWNT. (*d*) Allowed 1D wavevectors for a semiconducting (10,0) tube. The *white hexagons* define the first Brillouin zone of a graphene sheet, and the *black dots* in the corners are the $\mathbf{K}_B$ points.

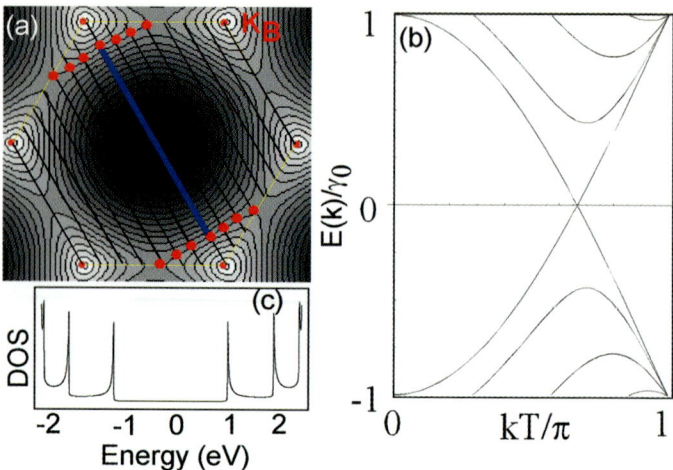

Figure 3 (*a*) Zone-folded representation of the 1D subbands for a (7,7) SWNT. The distance between the *red dots* segmenting these *parallel black lines* represents the size of the Brillouin zone, with the first Brillouin zone indicated by the *blue bold line*. The *small red dots* at the corners of graphene's hexagonal Brillouin zone are the graphene K_B points. (*b*) Energy dispersion for the (7,7) SWNT described in the zone-folding representation. (*c*) DOS calculated from a graphene based π-only tight-binding scheme.

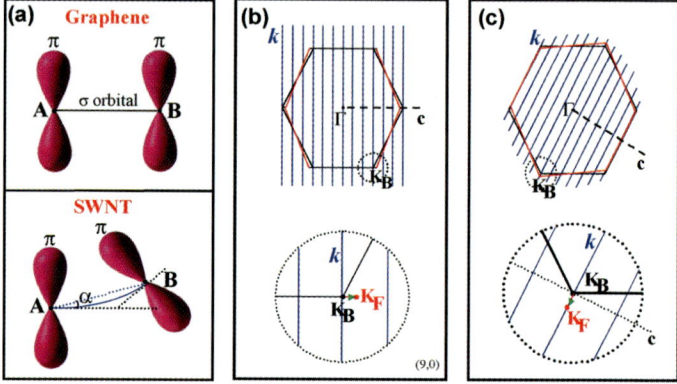

Figure 4 (*a*) (*top*) A and B denote two nearest neighbor parallel π orbitals on a flat graphene sheet. (*bottom*) Corresponding π orbitals on a curved surface, where α is the angle between the π orbitals. (*b*) (*top*) Hexagonal Brillouin zone of graphene sheet is defined by *black lines*. *Blue lines* are the allowed wavevectors *k* of (9,0) zigzag SWNTs, which are parallel to the tube axis. *Red lines* represent the effect of curvature. **c** is the circumferential direction perpendicular to the tube axis. (*bottom*) Fermi point shift is indicated in *red* for a "metallic" zigzag SWNTs; k_F moves along the **c** direction away from the graphene's K_B point. (*c*) Same as (*b*) but for case of armchair SWNT. For armchair SWNTs, k_F moves along the tube axis direction away from the K_B point.

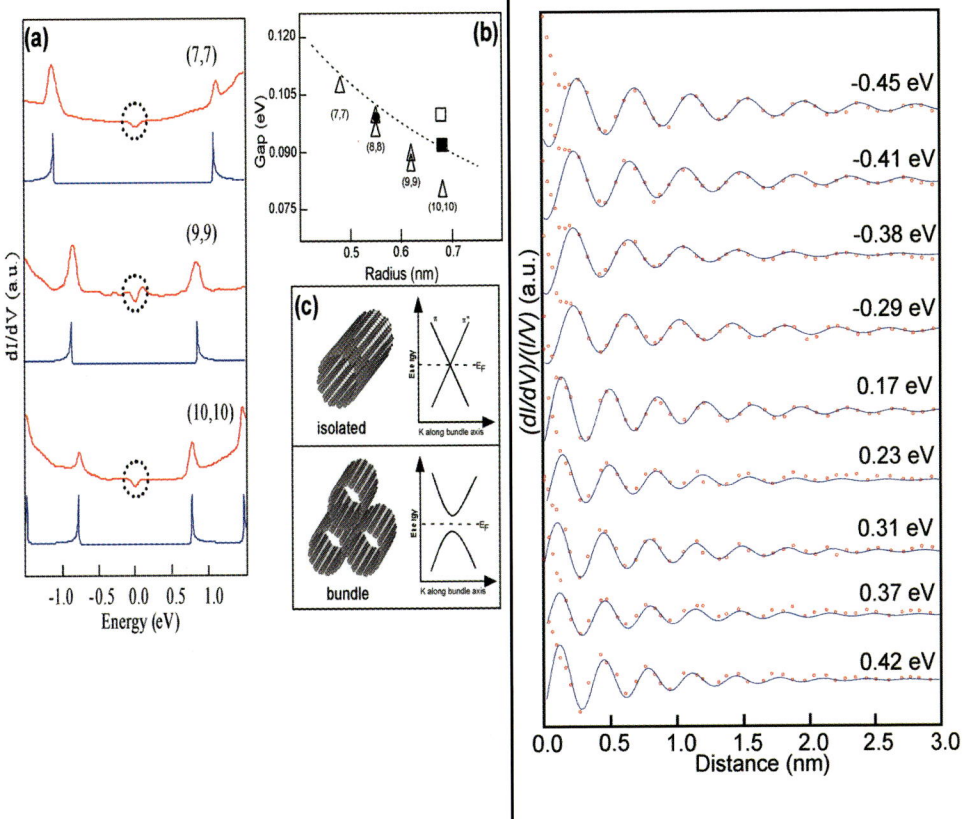

Figure 8 (left) (a) Tunneling conductance data, dI/dV, for different armchair SWNTs in bundles, with corresponding calculated DOS shown below each experimental curve. The new features in the low energy region of the (7,7), (9,9), and (10,10) tubes are highlighted by *dashed circles*. (b) Summary of the observed pseudogaps vs. tube radius. Each experimental data point (*triangle*) represents an average gap value measured on a distinct (n,n) tube. Theoretical results are also shown for comparison: the *solid square* and *dashed line* correspond, respectively, to the gap value of a (10,10) tube and radius dependence of the gap widths (37); the *open square* is for a (10,10) tube from (35, 36). And the *solid circle* shows the value calculated for an (8,8) tube (26). The (10,10) result from (38), ~200meV, is off the scale. (c) Models of an isolated (8,8) SWNT and bundle of three (8,8) tubes, and corresponding schematics of the π and π* bands near E_F.

Figure 11 (right) Spatial oscillation of *(dI/dV)/(I/V)* at all observed nine energies. The origin represents the defect position, which corresponds to the white arrow in Figure 9. *Red open circles*: experimental data; *blue solid lines*: theoretical fits of Eq.11 to the data. The fitting parameters [k (nm^{-1}), δ, l_φ (nm)] for data recorded at -0.45, -0.41, -0.38, -0.29, 0.17, 0.23, 0.31, 0.37 and 0.42 eV are [7.45, -3.82, 1.82], [7.59, -3.48, 1.82], [7.72, -3.49, 2.84], [7.97, -3.48, 2.08], [8.69, -2.32, 1.82], [8.85, -2.22, 1.80], [8.99, -1.95, 1.55], [9.16, -1.84, 1.74], [9.32, -1.99, 1.37], respectively.

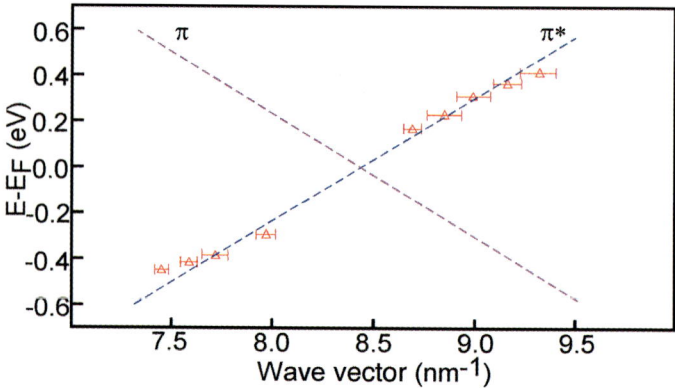

Figure 12 Energy dispersion near E_F for the (13,13) armchair SWNT. The experimental data points are *red triangles*, and the calculated bands are *dashed lines*.

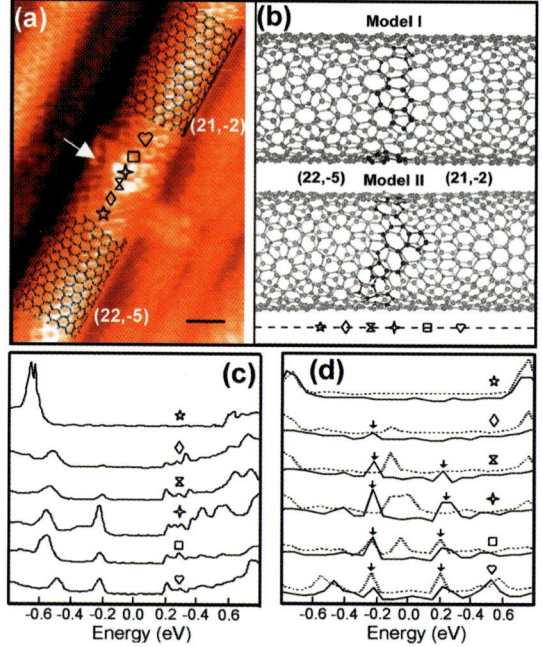

Figure 15 (*a*) Atomically resolved STM image of a SWNT containing an IMJ, which is highlighted with a *white arrow*. *Black honeycomb meshes* corresponding to (21, -2) and (22, -5) indices are overlaid on the *upper* and *lower* portions, respectively, to highlight the distinct atomic structures of the different regions of the SWNT. Scale bar, 1 nm. (*b*) Model I has three separated 5/7 pairs, and Model II has two isolated 5/7 pairs, and one 5/7-7/5 pair. The *solid black spheres* highlight the atoms forming the 5/7 defects. (*c*) Spatially resolved dI/dV data acquired across the M-S IMJ at the positions indicated by six symbols in (*a*). (*d*) Calculated local DOS for Model I (*solid line*) and for Model II (*dashed line*). The *arrows* highlight the same features as in (*c*).

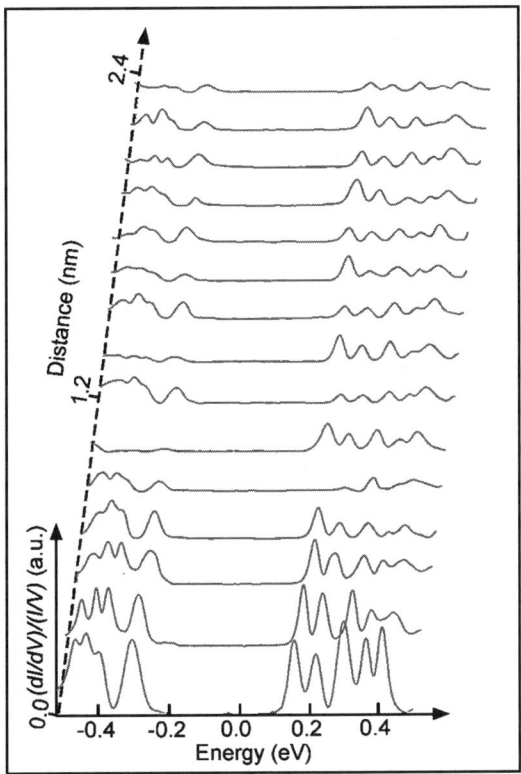

Figure 10 $(dI/dV)/(I/V)$ vs. energy recorded as a function of distance from the defect in Figure 9. Curves start at the central defect position (white arrow position in Figure 9) and progress along the upper left portion of the nanotube.

$x = 0$. Second, the incident plane wave can resonantly backscatter from the quasibound state with a reflectivity $|R|^2$ ($R = |R|e^{-i\delta}$, where δ is the phase shift). The corresponding standing wavefunction is written as $\psi(k, x) = e^{ikx} + |R|e^{-i(kx+\delta)}$, which corresponds to a spatial oscillation in the DOS, $\rho(k, x)$:

$$\rho(k, x) = |\psi(k, x)|^2 = 1 + |R|^2 + 2|R|\cos(2kx + \delta). \qquad 10.$$

In addition, processes such as electron-electron (e-e) and electron-phonon (e-ph) interactions can dephase the wave and lead to a decay of the interference (48). Including a phenomenological dephasing coherence length l_φ, we rewrite $\rho(k, x)$ as:

$$\rho(k, x) = |\psi(k, x)|^2 = 1 + |R|^2 + 2|R|\cos(2kx + \delta)e^{-2x/l_\varphi}. \qquad 11.$$

We have found this model provides excellent fits to all energy-dependent oscillations (Figure 11, color insert) (22). Significantly, these fits provide values for $E(k)$ vs. k, and demonstrate that this dispersion is indeed linear near E_F (Figure 12, color insert). The experimental energy dispersion agrees well with the linear dispersion predicted for armchair SWNTs (27), $|E(k)| = (3/2)a_{cc}\gamma_0|k - k_F|$. Such fitting also yields values of $\gamma_0 = 2.51$ eV and $k_F = 8.48 \pm 0.05$ nm^{-1}. This γ_0 value is in good agreement with previous experimental values from STM studies (13–15), and the value of k_F is consistent with the predicted value $k_F = 2\pi/(3^{3/2}a_{cc}) = 8.52$ nm^{-1} for an armchair SWNT (1). In addition, deviations in the fits close to the defect exist, although these are likely due to the simplified model used to treat defects; more detailed theoretical treatments that explicitly account for defect structure could provide further insight into this issue in the future.

Lemay et al. recently also reported an STM study of 1D energy dispersions of metallic SWNTs by imaging the wavefunctions in finite size (~40 nm) nanotubes (49). The physical principle behind this study is actually the same as our quantum interference method: the ~40-nm-long tubes can be taken as the system with two defects (both ends), which act as the total or partial reflection boundary for those two pairs of Bloch waves, and therefore the defect scattering picture still holds. However, this study focused on chiral tubes packed in a bundle and neglected the significant effects of finite curvature and tube-tube interactions reviewed above. These important issues will need to be evaluated to draw more meaningful conclusions in the future.

In the case of the data of Ouyang et al. (22), comparison of the experimental energy dispersion with the band structure calculated for a (13, 13) SWNT reveals an excellent agreement between reported experimental data and theoretical π^* band, but no data overlap with the π band. The absence of the π band can be understood in terms of parity matching (50), and further reveals an interesting but important relationship between the defect parity and the symmetry of the energy bands in armchair SWNTs. Specifically, the π and π^* bands of armchair SWNTs have opposite parity because they are constructed from rotationally invariant combinations of symmetric and antisymmetric sums of p_z orbitals. Therefore, if the defect in Figure 9 has odd parity, electrons in the π band (even parity) will not be scattered by the defect, whereas electrons in the π^* band (odd parity) can be scattered when their energies match a defect quasi-bound state. This interesting parity effect was supported by several observations: (*a*) The defect does not break the structural symmetry of the SWNT [i.e., both sides of the defect are consistent with (13, 13) armchair nanotubes], which prevents mixing of π and π^* bands; (*b*) analysis of the position- and energy-dependent local DOS data obtained from the other side of the SWNT yields the same π^* energy dispersion as plotted in Figure 12 (see color insert). Hence, we believe resonant scattering represents an exciting approach for investigating fundamental issues of parity and symmetry breaking in 1D SWNTs.

SWNT END STATES

Similar to the surface of a 3D structure and the edge of a 2D system, the ends of SWNTs can be treated as the surface of the 1D system, which reduces the symmetry of the tube and produces corresponding end states. Therefore, to understand thoroughly the electronic properties of SWNTs, knowledge of end states is necessary. In general, there are two different types of end states (51): localized and delocalized (i.e., resonant) states. If the energy of an end state falls within the energy gap of semiconducting SWNTs, it will be localized with the wavefunction decaying exponentially into the tube. Alternatively, if the energy of the end state is coincident with a bulk band, this state can connect to the extended bulk state and is delocalized.

Kim et al. (19) reported the first detailed investigations of atomic structure and electronic properties of SWNT ends. In these studies, tunneling spectra data recorded on a metallic (13, −2) SWNT showed two new peaks at 250 and 500 meV near the SWNT end. To investigate the origin of the SWNT end state and identify the specific end structure, Kim et al. further developed different cap models of SWNTs based on the Euler's rule and isolated pentagon rule (52), and carried out the numerical π-only tight-binding calculations for all possible models to compare with experimental data. Significantly, these studies demonstrated clearly that a specific topological arrangement of pentagons could be attributed to low energy peaks in the DOS at the SWNT end.

The decay of the end states was also quantified in this study (19), by examining the tip-nanotube separation, h(x), as a function of position, x, which is proportional to the DOS

$$\exp[k_d h(x)] \propto \int_0^{eV} \rho(x, E) dE, \qquad 12.$$

where k_d is the characteristic inverse decay length. Comparison of experimental $\exp[k_d h(x)]$ at the peak energies of 250 and 500 meV with integration of the calculated DOS showed good agreement with experimental data, with a characteristic decay length scale ∼10.2 nm (19). These functions also retain a finite magnitude in the bulk, which suggests that they correspond to resonant states.

In addition, there are many open and interesting questions about the properties of SWNT ends. For example, how do the electronic characteristics of end states differ for capped vs. open ends, and how do these differences affect the coupling to metals and/or the emission of electrons? Field emission properties of nanotubes, which have been recently demonstrated (53, 54), could be strongly influenced by the presence of localized or resonant end states (55), and hence such fundamental knowledge could impact significantly this application area.

FINITE SIZE EFFECTS

As the length (L) of a SWNT is reduced, one ultimately will reach the limit of a molecular cluster. Changes in the electronic properties as this finite size regime is approached are of importance to our fundamental understanding and to potential applications. At first order, the effect of finite size or length can be analyzed with a 1D "particle-in-a-box" model, where a tube of length L has a set of discrete wavevectors given by $k = n\pi/L$ (n is integer), and corresponding energy level spacing of $\Delta E = \hbar \Delta k = hv_F/2L = 1.67$ eV/L (nm), where $v_F = 8.1 \times 10^5$ m/s is the Fermi velocity of graphene (1).

Finite size SWNTs can be prepared by using a voltage pulse between an STM tip and SWNTs to cut a long nanotube into short pieces (3). With this technique, Venema et al. (20, 56) reported discrete energy levels due to the quantum confinement in a L $\sim$30 nm armchair SWNT. Odom et al. carried out more detailed and comprehensive studies of both semiconducting and metallic SWNTs, with lengths down to and below 10 nm (3). Studies of different length metallic SWNTs showed well-defined steps in I/V curves, which were attributed to the resonant tunneling through discrete eigenstates of the finite size SWNTs (Figure 13). The energy level spacing for 6 and 5 nm tubes, 300 and 370 meV, respectively, agreed quantitatively with the particle-in-a-box model. However, data recorded on an ultra-short 3 nm SWNT showed distinct deviation from this simple model. Both ab initio (57) and semi-empirical calculations (58, 59) have predicted that SWNTs <4 nm may deviate owing to asymmetry and shifting of the linear bands crossing at E_F, which could open a gap around E_F. Therefore, the unusual energy level structure observed for the 3 nm SWNT (3) could signify the transition to a system truly molecular in nature and represents an important area to address in the future.

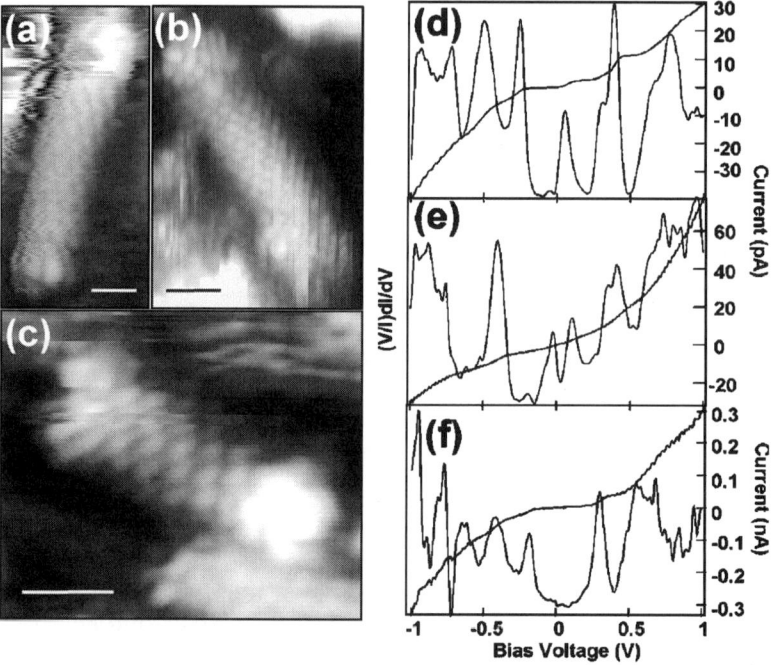

Figure 13 (*a–c*) STM images of SWNTs cut to lengths of 6 nm, 5 nm, and 3 nm, respectively. (*d–f*) I–V and normalized conductance data recorded on the SWNTs in (*a–c*), respectively. Adapted from Reference 3.

On the other hand, semiconducting SWNTs (T.W. Odom, J.L. Huang, & C.M. Lieber, unpublished manuscript) have not shown significant finite size effect as observed in metallic SWNTs (60). Even for a 3 nm ultra-short semiconducting SWNTs, the tunneling spectra measured in the middle of the tube shows almost the same VHS features as the bulk system. Qualitatively, the observation of semiconducting SWNTs in our studies is consistent with theoretical predictions (62) that semiconducting SWNTs have electron mean free paths of only several nanometers, and thus the distinction between bulk and finite size may be more difficult to make (63). Nevertheless, we believe this represents an important issue, which lies at the intersection of molecular and extended systems, to address in the future.

KONDO PHENOMENA IN 1D

A fundamental issue that can now be addressed owing to the availability SWNTs is how 1D electron systems respond to a local spin; that is, the Kondo effect in 1D. The Kondo effect, which describes the interaction between a magnetic moment and conduction electron spins of a nonmagnetic host, is a well-known phenomenon that leads to anomalous transport behavior in bulk systems of dilute magnetic alloys (64). Below the Kondo temperature, T_K, electrons of the host tend to screen the local spin of the magnetic impurity, resulting in the emergence of a Kondo resonance. The Kondo resonance for 2D systems was recently observed in STM investigations of magnetic atoms on noble metal surfaces (65–67).

Odom et al. first took advantage of the metallic SWNTs as 1D metallic hosts to study the interaction between the local spin on Co clusters and extended as well as finite size SWNTs (21). STM images showed that Co clusters could be readily observed on atomically resolved metallic SWNTs (Figure 14a), and spectroscopy data recorded directly above a cluster exhibited a strong resonance peak near E_F. Spatially resolved measurements further showed that these peak features systematically decrease in amplitude and ultimately disappear after several nanometers (Figure 14b). The new resonance peaks were not observed with nonmagnetic Ag clusters or with Co clusters on semiconducting SWNTs (21), which demonstrates that the observed peaks are due to the interaction of magnetic spins with the SWNT conduction electrons. In addition, analyses of the resonances revealed new information about the effect of dimensionality compared with 2D systems (65–67). In the 2D systems, the Kondo resonance is usually evident as a dip or antiresonance in the tunneling conductance, although in the measurement of Co on SWNTs the Kondo resonance always appears as a sharp peak. This difference can be qualitatively understood in terms of the small number of continuum final states available for tunneling in a SWNT vs. a noble metal (21, 65–67).

Finite size SWNTs have also provided an opportunity to study the Kondo effect under conditions where the energy level spacing of the conduction electrons is larger than $k_B T_K$ (vs. normally $\ll k_B T_K$ in extended systems) (68). Odom et al. tackled this question clearly by characterizing the DOS on and near Co clusters before and after cutting a SWNT host to ~11 nm (21). The spectra recorded after

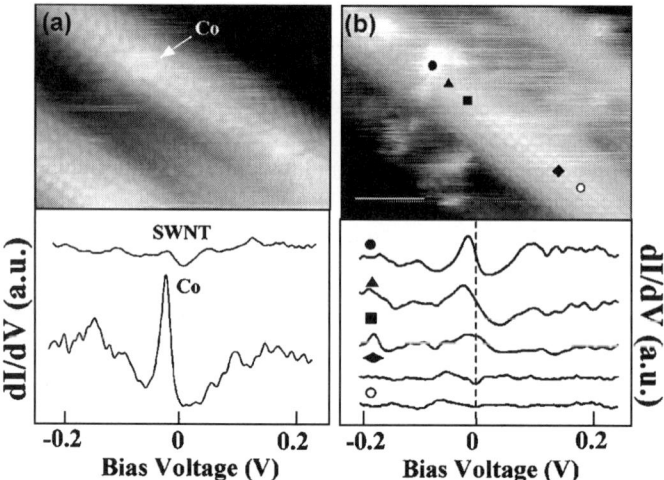

Figure 14 (*a*) Atomically resolved image of a 0.5 nm Co cluster on a SWNT, and dI/dV vs. V recorded at the position of the Co cluster and ∼7 nm away from cluster. (*b*) Constant current image of a larger Co cluster (∼1 nm) on a SWNT, and corresponding dI/dV vs. V measured at points indicated in the image. Adapted from Reference (21).

cutting showed the expected finite size eigenstates separated by ∼0.15 eV (3, 21), and also showed that the peak amplitude at E_F was markedly enhanced relative of the other level peaks. The enhanced conductance at E_F provides evidence of how sensitive the electronic properties of metallic nanotubes are to magnetic impurities, even in finite size structures where the level spacing is much larger than $k_B T_K$.

Lastly, these experiments suggest that metallic SWNTs are ideal hosts for studying basic features of the Kondo effect in 1D systems. However, the magnetic clusters (<1 nm) explored in this first report (21) also complicate analysis compared to the ideal of a single atom spin center. In this regard, future studies of the 1D Kondo effect could benefit substantially by using single magnetic atoms or molecules that have well-defined (and controllable) spin and can be registered precisely with respect to the underlying SWNT atomic lattice.

INTRAMOLECULAR JUNCTIONS

An exciting idea in the area of nanoscale electronic devices is to create intramolecular SWNT junctions. Theoretical work suggested early on that two different tubes could be connected seamlessly by interposing one or more pentagon-heptagon

(5/7) topologic defects between two nanotube segments of different helicity (69–74). Although some transport measurements (75) provided indirect evidence for the existence of proposed intramolecular junctions (IMJs), direct characterization of these potentially important structures was lacking until the recent work of Ouyang et al. (23).

Analysis of atomically resolved images has enabled the identification of IMJs (23). For example, the data in Figure 15a (see color insert) shows that the upper and lower portions of the nanotube have similar diameters (1.57 ± 0.07 nm) but significantly different chiral angles, where θ is $-3.9° \pm 0.8°$ and $-10.5° \pm 0.8°$ for the upper and lower portions, respectively. These data showed that the SWNT indices of the upper and lower regions were $(21,-2)$ and $(22,-5)$, respectively (23), and moreover, enabled the interface between SWNTs to be modeled (Figure 15b, color insert). Further strong evidence for this specific IMJ assignment was obtained from spectroscopy studies (Figure 15c, color insert), which showed that the upper and lower portions corresponded to a semiconductor and metal, respectively. Hence, these results show that this IMJ is a metal-semiconductor (M-S) junction—a potentially basic building block for nanoelectronics.

The detailed interface properties of this IMJ were also investigated by spatially resolved tunneling spectroscopy (Figure 15c, color insert). These measurements showed that this junction has a very sharp metal-semiconductor interface: the semiconducting VHS gap decays across the IMJ into the metallic segment within <1 nm, whereas the distinct spectroscopic features of the metallic tube appear to decay more quickly across the junction interface. Ouyang et al. used atomic models of the interface to compare calculated and experimentally measured DOS of this IMJ. Two low-energy structural models (Figure 15b, color insert), which contained three and four 5/7 pairs, were analyzed. Significantly, the local DOS (Figure 15d, color insert) obtained from tight-binding calculations for Model I agreed well with the measured local DOS across the IMJ interface, whereas Model II exhibited low-energy states around -0.10 eV not observed in experiment. Hence, it was possible to assign the structure of Model I as a reasonable representation for the observed $(21,-2) - (22,-5)$ IMJ.

Ouyang et al. also characterized a metallic-metallic (M–M) IMJ using similar methods (23). These studies exhibited several new features compared to the M–S intramolecular junctions. First, atomic resolution images and subsequent analysis showed that the IMJ formed between SWNTs with substantial differences in diameter: 1.23 ± 0.05 nm and $d_t = 1.06 \pm 0.05$ nm, respectively. Second, this M–M IMJ exhibited localized states at -0.55 and -0.27 eV in the junction area, and both of these states decay on a nanometer length scale into the SWNTs on either side of the IMJ. These features contrast with the abrupt, clean M–S junction described above.

Direct atomically resolved characterization of IMJs by STM has important implications for applications because it provides a clear means for assessing synthetic efforts designed to prepare specific IMJs. We believe that such work could open significant opportunity both for fundamental investigations of resonant scattering and for the controlled growth of intramolecular SWNT nanoelectronics devices.

CONCLUSIONS AND PROSPECTS

Scanning tunneling microscopy has proved an ideal tool to study the structural and electronic properties of individual and bundle SWNTs. Studies of defect-free SWNTs have shown that the electronic properties of nanotubes depend not only on radius and chirality but also on their detailed curvature and local environment. From a graphene based π-only tight-binding scheme, SWNTs with indices (n, n + 3q) should behave as metallic quantum wires. However, tunneling spectroscopy measurements made on metallic zigzag tubes showed energy gaps with magnitudes depending inversely on the square of the tube radius, whereas isolated armchair tubes do not have energy gaps and are truly metallic. In addition, armchair SWNTs in bundles exhibit pseudogaps at E_F due to tube-tube interactions that break the n-fold rotational symmetry of this system. STM was also used to determine directly the unique linear 1D energy dispersion of SWNTs through the analysis of resonant scattering and quantum interference. In addition, STM was exploited as a tool to create and investigate finite size effects in SWNTs. These studies show that it is possible to access readily a regime of "0D" behavior where finite length produces quantization along the tube axis. These results suggest future opportunities to probe, for example, connections between extended and molecular structures. As one of the applications of nanotubes as 1D systems, nanotubes were found to be ideal 1D hosts for the investigation of fundamental interactions between magnetic clusters and 1D electron systems. Lastly, direct atomically resolved characterization of intramolecular junctions in as grown SWNT samples by STM demonstrated the existence of IMJs and could have important implications for the controlled introduction of defects in nanoelectronic device applications.

We believe that this review has shown that many of the overall structural and electronic properties of SWNTs are now in hand, although this has really only opened the window to this fascinating 1D system. Future efforts promise to be rewarded with answers to very fundamental scientific questions, and importantly, these results should push from a firm intellectual footing the application of SWNTs in future nanotechnologies.

ACKNOWLEDGMENTS

We are indebted to our colleagues and collaborators. In particular, we thank Prof. H. Park, L.J. Lauhon, J.F. Wang, M.S. Gudiksen, and K. Kim for stimulating discussions, and thank T.W. Odom, P. Kim, and C.L. Cheung for their contributions to the work summarized herein. C.M.L. gratefully acknowledges support of this work by the Solid State Chemistry and Chemistry Divisions of the NSF.

Visit the Annual Reviews home page at www.annualreviews.org

LITERATURE CITED

1. Saito R, Dresselhaus G, Dresselhaus M. 1998. *Physical Properties of Carbon Nanotubes*. London: Imperial College
2. Dekker C. 1999. *Phys. Today* 52:22–28
3. Odom TW, Huang JL, Kim P, Lieber CM. 2000. *J. Phys. Chem. B* 104:2794–809
4. McEuen PL. 1998. *Nature* 393:15–16
5. Dai HJ, Hafner JH, Rinzler AG, Colbert DT, Smalley RE. 1996. *Nature* 384:147–50
6. Hafner JH, Cheung CL, Lieber CM. 1999. *Nature* 398:761–62
7. Treacy MMJ, Ebbesen TW, Gibson JM. 1996. *Nature* 381:671–74
8. Wong EW, Sheehan PE, Lieber CM. 1997. *Science* 277:1071–75
9. Mintmire JW, Dunlap BI, White CT. 1992. *Phys. Rev. Lett.* 68:631–34
10. Hamada N, Sawada S, Oshiyama A. 1992. *Phys. Rev. Lett.* 68:1579–82
11. Saito R, Fujita M, Dresselhaus G, Dresselhaus M. 1992. *Appl. Phys. Lett.* 60:2204–6
12. Saito R, Fujita M, Dresselhaus G, Dresselhaus M. 1992. *Phys. Rev. B* 46:1804–11
13. Odom TW, Huang JL, Kim P, Lieber CM. 1998. *Nature* 391:62–64
14. Wildöer JWG, Venema LC, Rinzler AG, Smalley RE, Dekker C. 1998. *Nature* 391:59–62
15. Ouyang M, Huang JL, Cheung CL, Lieber CM. 2001. *Science* 292:702–5
16. Tans SJ, Devoret MH, Dai H, Thess A, Smalley RE, et al. 1997. *Nature* 386:474–77
17. Journet C, Maser WK, Bernier P, Loiseau A, Lamy de la Chapelle M, et al. 1997. *Nature* 388:756–59
18. Rao AM, Bandow S, Chase B, Eklund PC, Williams KW, et al. 1997. *Science* 275:187–91
19. Kim P, Odom TW, Huang JL, Lieber CM. 1999. *Phys. Rev. Lett.* 82:1225–28
20. Venema LC, Wildöer JWG, Janssen JW, Tans SJ, Tuinstra HLJT, et al. 1999. *Science* 283:52–55
21. Odom TW, Huang JL, Cheung CL, Lieber CM. 2000. *Science* 290:1549–52
22. Ouyang M, Huang JL, Lieber CM. 2001. *Phys. Rev. Lett.* 88:066804-1-4
23. Ouyang M, Huang JL, Cheung CL, Lieber CM. 2001. *Science* 291:97–100
24. Wallace PR. 1947. *Phys. Rev.* 71:622–34
25. Ashcroft NW, Mermin ND. 1976. *Solid State Physics*. New York: Holt, Rinehart & Winston
26. Rubio A. 1999. *Appl. Phys. A* 68: 275–82
27. Mintmire JW, Robertson DH, White CT. 1993. *J. Phys. Chem. Solids* 54:1835–40
28. Mintmire JW, White CT. 1998. *Phys. Rev. Lett.* 81:2506–9
29. Blasé X, Benedict LX, Shirley EL, Louie SG. 1994. *Phys. Rev. Lett.* 72:1878–81
30. White CT, Robertson DH, Mintmire JW. 1996. In *Clusters and Nanostructures Materials*, ed. P Jena, S Behera, pp. 231–37. New York: Nova
31. Kane CL, Mele EJ. 1997. *Phys. Rev. Lett.* 78:1932–35
32. Kleiner A, Eggert S. 2001. *Phys. Rev. B* 63: 734081–84
33. Yang L, Han J. 2000. *Phys. Rev. Lett.* 85: 154–57
34. White CT, Mintmire JW. 1998. *Nature* 394:29–30
35. Delaney P, Choi HJ, Ihm J, Louie SG, Cohen ML. 1998. *Nature* 391:466–69
36. Delaney P, Choi HJ, Ihm J, Louie SG, Cohen ML. 1999. *Phys. Rev. B* 60:7899–904
37. Maarouf AA, Kane CL, Mele EJ. 2000. *Phys. Rev. B* 61:11156–65
38. Kwon YK, Saito S, Tomanek D. 1998. *Phys. Rev. B* 58:13314–17
39. Odom TW, Huang JL, Kim P, Ouyang M, Lieber CM. 1998. *J. Mater. Res.* 13:2380–87
40. Crommie MF, Lutz CP, Eigler DM. 1993. *Nature* 363:524–27

41. Hasegawa Y, Avouris Ph. 1993. *Phys. Rev. Lett.* 71:1071–74
42. Kanisawa K, Butcher MJ, Yamaguchi H, Hirayama Y. 2001. *Phys. Rev. Lett.* 81: 3384–87
43. Yokoyama T, Okamoto M, Takayanagi K. 1998. *Phys. Rev. Lett.* 81:3423–26
44. Clauss W, Begeron DJ, Freitag M, Kane CL, Mele EJ, Johnson AT. 1999. *Europhys. Lett.* 47:601–7
45. Hassanien A, Tokumoto M, Umek P, Mihailovic D, Mrzel A. 2001. *Appl. Phys. Lett.* 78:808–10
46. Choi HJ, Ihm J, Louie SG, Cohen ML. 2000. *Phys. Rev. Lett.* 84:2917–20
47. Bockrath M, Liang W, Bozovic D, Hafner JH, Lieber CM, et al. 2001. *Science* 291: 283–85
48. Burgi L, Jeandupeux O, Brune H, Kern K. 1999. *Phys. Rev. Lett.* 82:4516–19
49. Lemay SG, Janssen JW, van den Hout M, Mooij M, Bronikowski MJ, et al. 2001. *Nature* 412:617–20
50. Mizes HA, Foster JS. 1989. *Science* 244: 559–62
51. Lannoo M, Friedel P. 1991. *Atomic and Electronic Structure of Surfaces*. Berlin: Springer-Verlag
52. Dresselhaus MS, Dresselhaus G, Eklund PC. 1996. *Science of Fullerene and Carbon Nanotubes*. San Diego: Academic
53. de Heer WA, Chatelain A, Ugarte D. 1995. *Science* 270:1179–82
54. Collins PG, Zettle A. 1996. *Appl. Phys. Lett.* 69:1969–72
55. de Heer WA, Bonad J, Fauth K, Chatelain A, Forro L, Ugarte D. 1997. *Adv. Mater.* 9: 87–92
56. Venema LC, Wildöer JWG, Tuinstra HLJT, Dekker C, Rinzler A, Smalley RE. 1997. *Appl. Phys. Lett.* 71:2629–32
57. Rubio A, Sanchez-Portal D, Attach E, Ordoejon P, Soler JM. 1999. *Phys. Rev. Lett.* 82:3520–23
58. Bulusheva LG, Okotrub AV, Romanov DA, Tomanek D. 1998. *J. Phys. Chem. A* 102:975–81
59. Rochefort A, Salahub DR, Avouris PH. 1999. *J. Phys. Chem. B* 103:641–46
60. Odom TW, Hafner JH, Lieber CM. 2001. *Top. Appl. Phys.* 80:173–211
61. Deleted in proof
62. McEuen PL, Bockrath M, Cobden DH, Yoon YG, Louie SG. 1999. *Phys. Rev. Lett.* 83:5098–101
63. Jishi RA, Bragin J, Lou L. 1999. *Phys. Rev. B* 59:9862–65
64. Kondo J. 1964. *Prog. Theor. Phys.* 32:37–87
65. Li J, Schneider WD, Berndt R, Delley B. 1998. *Phys. Rev. Lett.* 80:2893–96
66. Madhavan V, Chen W, Jamneala T, Crommie MF, Wingreen NS. 1998. *Science* 280:567–70
67. Manoharan HC, Lutz CP, Eigler DM. 2000. *Nature* 403:512–15
68. Wolfgang B, Thimm JK, von Delft J. 1999. *Phys. Rev. Lett.* 82:2143–46
69. Dunlap BI. 1994. *Phys. Rev. B* 49:5643–51
70. Charlier JC, Ebbesen TW, Lambin PH. 1996. *Phys. Rev. B* 53:11108–13
71. Chico L, Benedict LX, Louie SG, Cohen ML. 1996. *Phys. Rev. B* 54:2600–6
72. Saito R, Dresselhaus G, Dresselhaus MS. 1996. *Phys. Rev. B* 53:2044–50
73. Lambin PH, Fonseca A, Vigneron JP, Nagy JB, Lucas AA. 1995. *Chem. Phys. Lett.* 76:971–74
74. Chico L, Benedict LX, Louie SG, Cohen ML. 1996. *Phys. Rev. Lett.* 76:971–74
75. Yao Z, Postma HWCh, Balents L, Dekker C. 1999. *Nature* 402:273–76

ELECTRON TRANSFER AT MOLECULE-METAL INTERFACES: A Two-Photon Photoemission Study

X.-Y. Zhu
Department of Chemistry, University of Minnesota, Minneapolis, Minnesota 55455; e-mail: zhu@chem.umn.edu

Key Words anionic resonance, molecular electronics, interfacial electron transfer, organic semiconductor, unoccupied states

■ **Abstract** Electron transfer between a molecular resonance and a metal surface is a ubiquitous process in many chemical disciplines, ranging from molecular electronics to surface photochemistry. This problem has been probed recently by two-photon photoemission spectroscopy. The first photon excites an electron from an occupied metal state to an unoccupied molecular resonance. Subsequent evolution of the excited electronic wavefunction is probed in energy, momentum, and time domains by the absorption of a second photon, which ionizes the electron for detection. These experiments reveal the important roles of molecule-metal wavefunction mixing, intermolecular band formation, polarization, and localization in interfacial electron transfer.

INTRODUCTION

What is common to the following interfacial processes? Consider an organic light emitting device or field effect transistor (FET), a molecular switch contacted by two metal electrodes, substrate-mediated surface photochemistry, and solar energy conversion on dye-sensitized semiconductors. All these processes involve electron transfer (ET) between a molecule and a metal or semiconductor surface. In organic light-emitting devices electron injection from a metallic electrode into the lowest unoccupied molecular orbital (LUMO) is a critical step, and the efficiency of the device is intimately related to interfacial ET rate (1, 2). Current understanding of this issue remains at the level of energetics, i.e., the relative position of the LUMO to the metal Fermi level, but the interfacial ET rate is also related to the electronic coupling matrix element as well as the dynamics of electron relaxation and localization. Similar issues are found in organic FETs (3). Interfacial ET is key to the emerging field of molecular electronics. Building a successful molecular electronic device often requires making electronic contacts to one or a group of molecules (4). In this regard electronic coupling between a metal electrode and a molecule determines not only contact resistance but also the nature of the molecular device. In surface photochemistry on metals the transfer of photo-excited substrate electrons to molecular resonances is believed to be a dominant mechanism, but few experiments have provided direct evidence for such a transient anionic resonance (5). In dye-sensitized solar energy conversion the injection of a photo-excited electron from the LUMO of the dye molecule to the conduction band of the semiconductor, as well as hole injection from the highest occupied molecular orbital to valence band, is the central issue (6, 7). Strong electronic coupling between the two leads to ultrafast ET rates, but the lack of precise control over molecular adsorption in a solution phase experiment makes this a particularly difficult problem. Similar issues are well known in the more mature field of electrochemistry (8).

This account summarizes recent experiments using two-photon photoemission (2PPE) spectroscopy to characterize the electronic structure and ET dynamics at molecule-metal interfaces. Two features of this approach are noteworthy: (*a*) The molecule-surface interaction is well characterized and controlled for a single crystal surface/adsorbate system in an ultrahigh vacuum environment, and (*b*) the 2PPE technique is sensitive to the electronic interaction, i.e., wavefunction overlap, between unoccupied molecular orbitals and substrate band structures and can be applied in a time-resolved manner to directly measure the ultrafast ET rate. The principle of this technique is illustrated in Figure 1.

In Figure 1*a* the molecule interacts strongly with a metal surface, as illustrated by a molecular wavefunction (e.g., LUMO) mixed with the substrate band

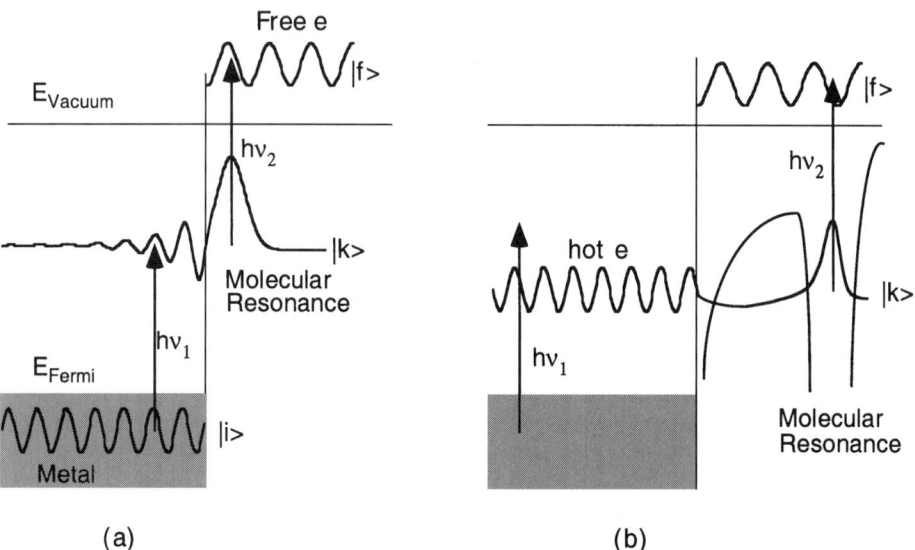

Figure 1 Schematic illustration of two-photon photoemission via an intermediate molecular resonance, |k⟩, in (*a*) a direct photo-induced electron transfer mechanism and (*b*) an indirect hot-electron transfer mechanism.

structure. Such mixing is represented schematically as an oscillating tail of the wavefunction in the periodic substrate lattice. The first photon excites an electron from an occupied metal state to the mixed molecular state; this is ET from the metal to the molecular resonance. The second photon ionizes the transient molecular anion for detection. The kinetic energy of the photo-emitted electron provides the energetic position of the molecular resonance. For excitation with one-color laser light under the continuous wave (CW) approximation, the rate (w_{if}) of such a resonant two-photon ionization process involving an initial metal state $|i\rangle$, an intermediate molecular resonance $|k\rangle$, and a final free-electron state $|f\rangle$ is given in (9) by

$$w_{if} \propto \left|(\vec{\mu}_{ik} \cdot \vec{E})(\vec{\mu}_{kf} \cdot \vec{E})\right|^2, \qquad 1.$$

where μ_{ik} and μ_{kf} are the transition dipole moments for the excitation step and the photo-ionization step, respectively. E is the electric field vector of the laser light. For the molecular resonance to be observable in 2PPE, both μ_{ik} and μ_{kf} must be nonzero. This leads to a few requirements on the molecular wavefuction, $|k\rangle$: (*a*) The molecular wavefunction must possess the right symmetry with respect to the initial metal wavefunction; (*b*) because $|i\rangle$ is delocalized parallel to the surface plane, the molecular wavefunction must possess nonvanishing dispersion in the surface plane; and (*c*) because the final state is a free-electron wave with certain periodicity (depending on kinetic energy), the spatial modulation or confinement of the molecular wavefunction in the surface normal direction must match that of the free-electron wave to give a nonvanishing μ_{kf} (10). Quantitatively the rate of photo-induced ET from metal to molecule, as well as back ET from molecule to the metal substrate, can be taken as proportional to the square amplitude of the oscillating tail of the mixed wavefunction $|k\rangle$ inside the metal. This tail is a schematic representation of the electronic coupling matrix element between the molecular resonance and the metal substrate. Note that Equation 1 applies to a one-color experiment under the CW approximation. In a pump-probe experiment with variable time delay, one needs to include the rate of decay from the transient molecular anionic resonance in the set of rate equations.

In Figure 1*b* the electronic interaction between the molecule and the metal substrate is weak. The coherent two-photon mechanism is not important here owing to the negligible matrix element for the excitation step. Instead, formation of the transient molecular anion is a result of scattering of the excited substrate electron into the molecular orbital. In the figure the wavefunction for the hot electron may represent different k values, not limited to $k_{||} = 0$. The tunneling barrier is determined by the nature of molecular spacers between the electron acceptor and the metal substrate. Such an indirect process can be distinguished from the direct mechanism in Figure 1*a* by the dependence of 2PPE yield on light polarization (9). The weak electronic coupling in Figure 1*b* should lead to a relatively long lifetime, and the electronic coupling matrix element should have a predominantly tunneling contribution. The overall rate of this process is given by

$$w_{if} \propto w_{ik} \cdot |(\vec{\mu}_{kk} \cdot \vec{E})|^2, \qquad 2.$$

where w_{ik} is the metal-to-molecule hot-electron tunneling rate, which has been treated in other contexts, such as hot-electron-induced resonant desorption, inelastic resonant electron-molecule scattering, and resonant tunneling. Interested readers are referred to this literature for details [(11) and references therein]. Pump-probe experiments are also applicable to this mechanism to establish the rate of ET.

In addition to the two major mechanisms illustrated in Figure 1, a transient molecular anion may also form indirectly from the decay of higher-lying electronic states, e.g., image states. However, the limited number of examples to date points to the dominance of the direct photo-excitation channel in 2PPE.

The two scenarios in Figure 1 illustrate the richness and power of the 2PPE technique in probing interfacial electronic structure and electron transfer dynamics. 2PPE is a particularly useful addition to more traditional surface spectroscopic techniques. For example, ultraviolet photoemission spectroscopy is limited to occupied electronic states (12). Inverse photoemission probes unoccupied electronic states [(13) and references therein] but has technical shortcomings, including limited energy resolution and the requirement for high electron fluxes that can easily damage molecular layers. Electron stimulated desorption has been successfully used to probe anionic resonances in a large number of adsorbate/metal systems that give desorption products from electron attachment (14). Other techniques such as X-ray absorption fine structure and electron energy loss spectroscopy provide limited probing of unoccupied electronic states. Compared to these traditional techniques, 2PPE is the only one capable of time-resolved measurements, particularly with time resolution compatible with ultrafast electron dynamics at interfaces.

Two-photon photoemission has been applied to the study of electron dynamics on metal and semiconductor surfaces, at metal-metal, metal-insulator, and metal-dielectric interfaces (10, 15–19). Recently, in this series, Harris and coworkers provided an authoritative review on 2PPE studies of image potential states at metal-adsorbate interfaces (20). These authors demonstrated the success in treating the adsorbate film as a dielectric layer and in elucidating the effects of such a dielectric layer on the energetics and dynamics of image states, on quantum well structures, and on polaron formation. In another chapter in this volume Petek & Ogawa probe an antibonding surface state formed from the adsorption of Cs on Cu(111) and demonstrate the use of time-resolved 2PPE in following the dynamics of Cs desorption (21).

In this account I focus on 2PPE studies of transient anionic resonances at molecule-metal interfaces and demonstrate this approach in addressing the central problem of interfacial ET. The review starts with a summary on experimental techniques, followed by a brief introduction to the dielectric continuum model. I then present experimental evidence for an anionic molecular resonance in a model system: hexafluorobenzene/Cu(111). The relationship of 2PPE results to interfacial ET dynamics is discussed. The next section addresses the important role of polarization and localization. Substrate mediated surface photochemistry is discussed within the context of localization. Finally I present some perspectives on the role of 2PPE studies in addressing key issues in molecular and organic electronics before concluding with some comments.

EXPERIMENTAL TECHNIQUES IN TWO-PHOTON PHOTOEMISSION

A 2PPE experiment can be carried out to determine the energetics, parallel dispersion, and lifetimes of interfacial electronic states. These experiments use laser light (one color or two color for pump-probe) with photon energies below the surface work function to avoid one-photon photoemission. Energetics are obtained in the spectroscopic mode, i.e., the kinetic energy of photo-emitted electrons are analyzed. The origins of photo-electrons from occupied, unoccupied, and final states can be easily distinguished based on the dependence of electron kinetic energy on photon energy. This is illustrated schematically for a one-color experiment in Figure 2, which shows three possible scenarios in 2PPE. In Figure 2A the absorption

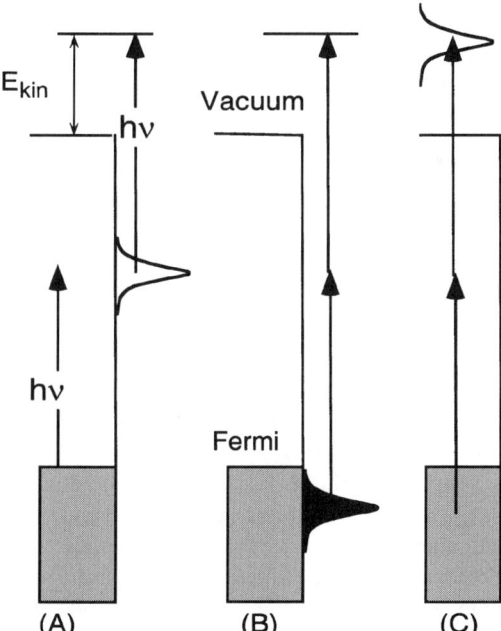

Figure 2 Schematic illustration of three possibilities in two-photon photoemission involving unoccupied states below (*A*) or above (*C*) the vacuum level and an occupied state below the Fermi level (*B*).

of the first photon resonantly excites an electron from occupied metal bands to an unoccupied intermediate state; the absorption of a second photon excites this transient electron above the vacuum level. In this case the change in electron kinetic energy scales with that in photon energy, i.e., $\Delta E_{kin} = 1 \cdot \Delta h\nu$. In Figure 2B two-photon excitation from an occupied state leads to the ejection of an electron and $\Delta E_{kin} = 2 \cdot \Delta h\nu$. Finally, if the state is above the vacuum level, (Figure 2C) E_{kin} is independent of photon energy. This can be viewed as a resonant scattering event in which the photo-excited electron resides transiently in the molecular orbital, followed by detachment and detection. Note the above simple arguments are only valid for interfacial states for which dispersion in the surface normal (z) direction is not important. They do not apply to bulk bands with dispersions in the z-direction.

Parallel dispersions of interfacial states are obtained in angle-resolved measurements. For an interface with long range order, the momentum parallel to the surface is conserved in the photoemission process. At an angle of detection of θ from the surface normal, the momentum parallel to the surface is obtained from the projection of the total linear momentum:

$$k_{//} = \sqrt{\frac{2m_e E_k}{\hbar^2}} \sin\theta, \qquad 3.$$

where m_e is the electron mass, E_k is the electron kinetic energy, and $k_{//}$ is the parallel momentum vector. The dispersion curve (E_k vs. $k_{||}$) can be represented by an effective electron mass, m_{eff}, derived by fitting to the following free-electron–like parabolic function:

$$E_k = E_o + \frac{(\hbar k_{//})^2}{2m_{eff}}. \qquad 4.$$

The effective electron mass reflects the extent of localization/delocalization of the wavefunction in the surface plane.

Lifetimes of unoccupied electronic resonances (between the vacuum level and the Fermi level; Figure 2A) are derived from pump-probe measurements. In this case formation and ionization of the transient state are induced by two different laser pulses (often with different color) at a variable time delay. Lifetimes can be obtained from fitting the time-dependent signal with simple rate equations, provided the shape of the laser pulses in the time domain are well accounted for and calibrated. This kind of time-resolved measurement can be carried out with energy and angular resolution to provide a complete picture of the dynamics.

Early 2PPE experiments employed nanosecond dye lasers (for energetics and dispersion), but more recent experiments have used femtosecond lasers. Figure 3 shows a typical femtosecond 2PPE spectrometer available in my laboratory. It consists of a mode-locked Ti:sapphire oscillator pumped by an 8-W solid-state laser. The output from the oscillator (700–1000 nm, 76 MHz) is frequency doubled or tripled. The UV output can be split into two pulses for pump and probe.

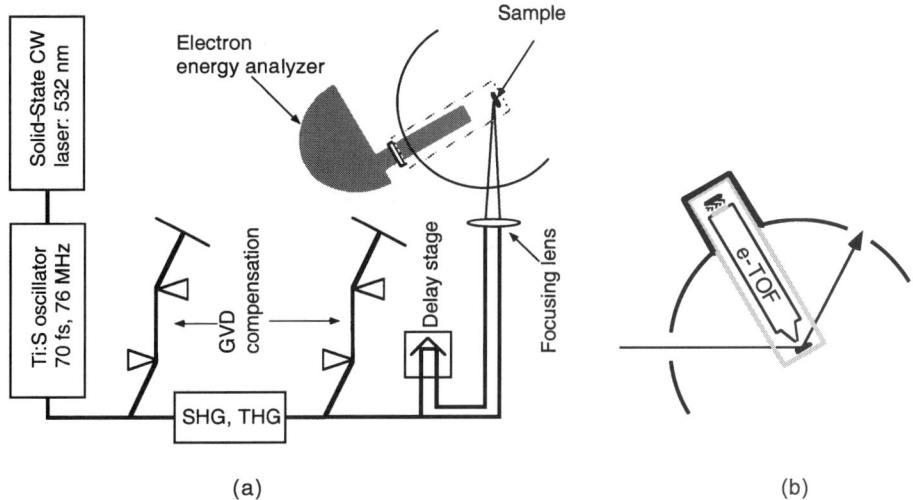

Figure 3 (*a*) Schematics of femtosecond two-photon photoemission spectrometer with a hemispherical electron energy analyzer housed in an ultrahigh vacuum chamber. The laser system consists of a solid state pump laser and a tunable mode-locked Ti:sapphire oscillator. GVD, group velocity dispersion; SHG, second harmonic generation; THG, third harmonic generation. (*b*) Electron detection by a time-of-flight detector.

Alternatively the UV and residual fundamental can be used as pump and probe, respectively. Time delay between the two laser pulses is controlled by a translation stage. The pump and probe beams are focused slightly by an f = 50-cm lens and directed onto the surface of the sample located in an ultrahigh vacuum chamber. Photoelectrons are detected by a hemispherical analyzer. Variations of this setup include the use of an amplifier for better tunability, a time-of-flight detector for electron detection (20), and an interferometric setup for improved time resolution (18).

IMAGE STATES AND THE DIELECTRIC CONTINUUM MODEL

To gain a better understanding of molecular anionic resonances, I first review a class of electronic states specific to interfaces: image states. This subject has been studied most extensively by 2PPE; interested readers are referred to several excellent review articles (15, 16, 20). The break of bulk symmetry at the crystalline surface gives rise to surface states that are confined to the interface in the surface normal direction but are of free electron Bloch wave character parallel to the surface. Among these, a series of unoccupied states at 1 eV or less below the vacuum level are the image states. Image states are of pure electrostatic origin. Consider an electron in front of and at a distance z from a metal surface. The electron induces polarization in the metal and the resulting positive charge cloud attracts the electron toward the surface. For a perfect metal the net result is equivalent to that of a Coulomb potential between the electron and a fictitious positive image charge (hence the name "image states") located at $-z$ inside the metal.

$$V_{im} = -\frac{e^2}{4\varepsilon z},\qquad 5.$$

where ε is the dielectric constant ($\varepsilon = 1$ for vacuum). If a project band gap exists in the surface normal direction, the electron is confined by the image potential on the vacuum side and the energy barrier on the metal side. A solution to this one-dimensional Hamiltonian gives a series of Rydberg states converging to the vacuum level. Note that image states are referenced to the vacuum level; the Fermi level does not come into the picture here. The consequence of this is made clear below.

What are the effects of an adsorbate layer on image states? The simplest picture involves treating the adsorbate layer as a dielectric film that screens the image potential ($\varepsilon > 1$). However, molecules are not mere dielectrics. They possess internal electronic structures and, in the case of an ordered film, energy bands. Consider the three scenarios in Figure 4.

In Figure 4a the adsorbate layer possesses negative electron affinity (EA). In other words, the bottom of the conduction band is above the vacuum level. The effective vacuum level or reference level (V_{eff}) an electron experiences when going from the vacuum side (*right*) through the molecular layer (*middle*) to the metal

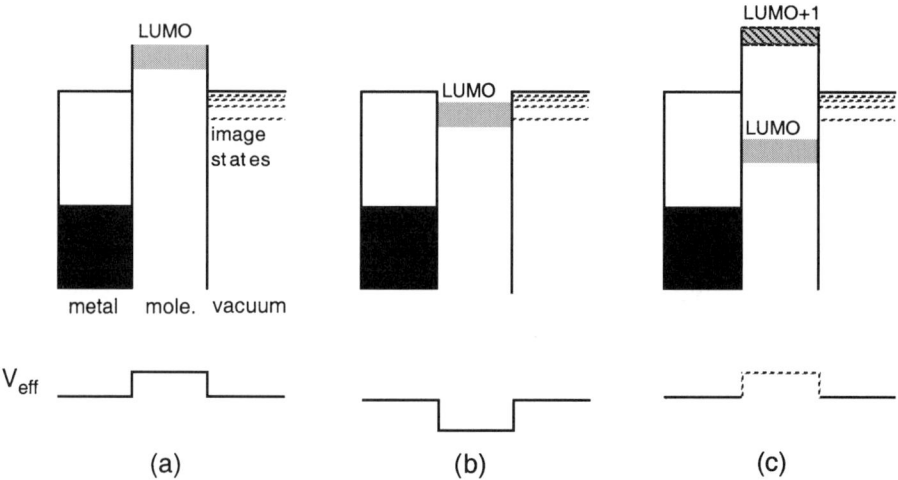

Figure 4 Energy level diagrams showing the relative positions of the lowest unoccupied molecular orbital band with respect to the vacuum level in a metal-molecule-vacuum system. An effective vacuum level is shown below each energy level diagram. (*a*) Negative-electron affinity, (*b*) small positive-electron affinity, (*c*) large positive-electron affinity. Image states are shown as *dashed lines*.

side (*left*) is shown below the energy level diagram. The adsorbate layer serves as an effective energy barrier.

In Figure 4*b* the adsorbate layer is characterized by a small but positive EA, with the bottom of the conduction band (LUMO) below the image states. The effective potential is shown as having an attractive well within the adsorbate layer. However, this picture is only valid for image states located within the LUMO band. For image states above the top of the conduction band, the situation will be the same as in Figure 4*c*.

In Figure 4*c*, with a large, positive EA, the adsorbate LUMO band is below the image states. When an image-state electron tunnels through the adsorbate layer, the wavefunction contains a superposition of the LUMO band and some higher-lying bands, such as LUMO + 1. An accurate modeling of the effect of the adsorbate layer on image states requires a multiband coupling treatment. Intuitively, one expects the adsorbate layer to serve as a tunneling barrier.

When the effective potential picture does apply, the modified image states can be obtained from solutions to a Hamiltonian with potential equal to the sum of V_{im} and V_{eff} ($=-EA$) within the adsorbate layer:

$$V = -\frac{e^2}{4\varepsilon z} - EA \quad (0 < z < d), \qquad 6.$$

where d is the thickness and ε is the dielectric constant of the adsorbate layer. On the vacuum side (outside the adsorbate layer) the potential is given by the polarization of the composite interface:

$$V = -\frac{\beta e^2}{4(z-d)} + \frac{(1-\beta^2) \cdot e^2}{4\beta} \cdot \sum_{n=1}^{\infty} \frac{(-\beta)^n}{z-d+nd} \quad (z > d), \qquad 7.$$

where $\beta = (\varepsilon - 1)/(\varepsilon + 1)$. Inside the metal with a projected band gap, a two-band nearly free electron model is satisfactory.

The above is the dielectric continuum model (DCM) that Harris and coworkers (20) have applied successfully to a number of metal-dielectric interfaces with both positive and negative EA. As an illustration of the DCM, Figure 5 shows DCM simulation for the n = 1 image state on the naphthalene/Cu(111) surface (22). Owing to the attractive V_{eff}, the image state wavefunction becomes increasingly confined within the adsorbate layer with thickness. Note that for image states located within the adsorbate conduction band, a more accurate treatment requires the consideration of quantum well formation, i.e., image wavefunction confined by the thickness of the adsorbate layer. This requires treating the potential inside the adsorbate layer as a flat band, with an effective electron mass obtained from quantum well analysis, as has been done for Xe/Ag(111) (23).

A word of caution: The application of the DCM model to Figure 4*c*, in which the image states are located within the adsorbate band gap, is ill-considered. One may adjust parameters in the model to fit some aspects of experimental observation. The results are nevertheless unphysical.

Image states serve as excellent model systems for investigating the physics of electron dynamics at surfaces. They may couple strongly to molecular bands, as in

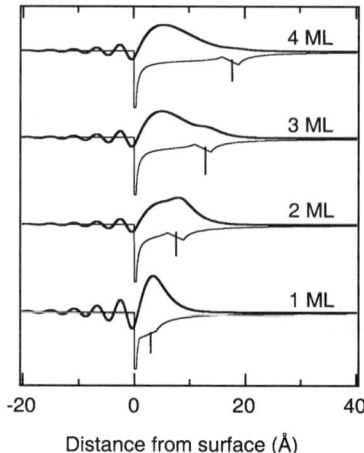

Figure 5 Calculated image-state (n = 1) wavefunctions (*thick curves*) for 1–4 ML of naphthalene/Cu(111). The *thin dotted line* is the potential used. The position of z = 0 is the Cu(111)/adsorbate interface and the vertical line on each potential is the adsorbate/vacuum interface.

Figure 2b. However, of more interest to interfacial ET and molecular/organic electronics is probing molecular resonances directly. There have been a few examples in which transient anionic resonances have been observed in 2PPE spectra. These include the observation of the π^* resonance in CO adsorbed on Cu(111) (24) and the σ^* antibonding state in Cs adsorbed on Cu(111) and other metal surfaces [(25) and references therein]. Of a more controversial nature is the system of benzene on coinage metal surfaces. Whereas Wolf and coworkers reported the observation of the π^* LUMO above the vacuum level (i.e., final state) for benzene adsorbed on Cu(111) (26), Harris and coworkers did not observe such a resonance on Ag(111) (27). An earlier claim of seeing the LUMO with vibrational fine structure for benzene/Cu(111) was not reproduced (28). Vondrak & Zhu first reported more definitive evidence for a transient anionic resonance for the system of hexafluorobenzene (C_6F_6) on Cu(111) (29–32). Below, I focus on this model system and discuss a few key issues in interfacial ET.

MOLECULAR RESONANCES AND INTERFACIAL ELECTRON TRANSFER: A CASE STUDY OF C_6F_6/Cu(111)

The C_6F_6/Cu(111) system falls into the category of Figure 2c, in which the LUMO of the adsorbate layer is below the image states. Figure 6 compares 2PPE spectra for clean Cu(111), 1 ML C_6F_6/Cu(111), and 2 ML C_6F_6/Cu(111). The clean surface spectrum shows peaks owing to the d-band (~2 eV below E_{Fermi}), the occupied

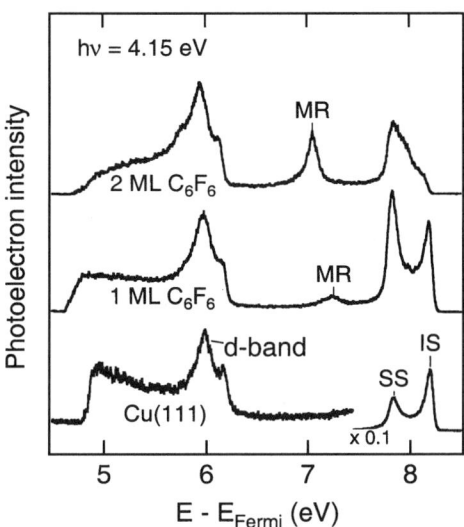

Figure 6 Two-photon photoemission spectra for (bottom to top) Cu(111), 1 ML C_6F_6/Cu(111), and 2 ML C_6F_6/Cu(111). $h\nu = 4.15$ eV (29, 32).

surface state (0.4 eV below E_{Fermi}), and the unoccupied image state (n = 1, 0.84 eV below E_{vac}). The strong intensity of the surface and image state result in part from the fact that they are located within the projected bandgap from -0.85 eV (below E_{Fermi}) to 4.2 eV (above E_{Fermi}) for Cu in the $\langle 111 \rangle$ direction. After the adsorption of 1 and 2 ML of C_6F_6 both the surface state and the image state are attenuated and a new peak (labeled MR for molecular resonance) develops. The unoccupied nature of this resonance is established by the one-photon dependence of electron kinetic energy (Figure 7). This state is located at 3.18 and 2.99 eV above the Fermi level at 1 or 2 ML coverages, respectively. The coverage dependence in the position of the resonance is discussed below (see Figure 10). The observed resonance (MR) is assigned to the LUMO of C_6F_6 because it is well separated from the image states and because it is energetically located at a position expected from gas phase EA (29, 32). The vertical EA of gas phase C_6F_6 is ~ 0 eV (adiabatic EA = 0.8 eV). The anionic resonance is stabilized by 0.8 eV in the condensed phase owing to polarization. Further stabilization of up to 1 eV is expected upon adsorption on the metal surface, owing to the image potential as well as wavefunction mixing between the LUMO and metal bands.

Excitation Mechanisms

In principle a transient molecular anionic resonance can be formed from direct photo-excitation, hot-electron transfer, and indirect channels owing to the relaxation of higher-lying electronic states, such as image states. The limited number of examples to date have shown the dominance of the direct photo-excitation channel.

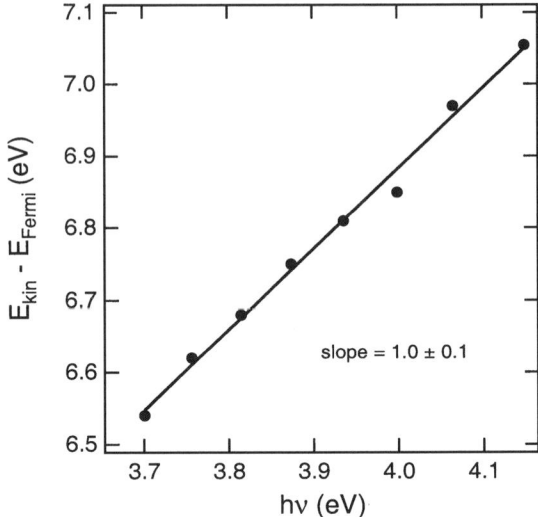

Figure 7 Photo-electron kinetic energy for the molecular resonance as a function of photon energy for bilayer C_6F_6 on Cu(111) (29, 32).

Measurements of photoemission yields as a function of light polarization have established that populations of the σ^* LUMO of C_6F_6 (33) and the π^* LUMO in CO on Cu(111) (9) are both due to direct photo-excitation. The same is true for the Cs/Cu(111) system in which a resonance enhancement was observed for the surface state to σ^* antibonding state excitation [(25) and references therein]. The direct photo-excitation mechanism is similar to the metal-to-ligand electron transfer mechanism in organometallic photochemistry.

The rate of photoemission in such a direct photo-excitation mechanism, given by Equation 1, is proportional to the square amplitude of the transition dipole moment, μ_{ik}, for metal-to-molecule ET. A weakening in electronic coupling between the molecule and the metal should result in a decrease in μ_{ik} and, thus, a decrease in photoemission yield. This effect is observed experimentally in a recent 2PPE study of C_6F_6 on hydrogen-passivated Cu(111) (Figure 8) (32). With increasing coverage ($\theta = 0$–0.34 ML) of preadsorbed atomic H, which systematically weakens the bonding between C_6F_6 and Cu(111), the 2PPE yield from the molecular resonance decreases by more than one order of magnitude. This is accompanied by an upward shift in energetic position by as much as 0.2 eV when θ_H increases from 0 to 0.34 ML. As the electronic interaction weakens, we expect the effect of stabilization of the transient anionic resonance by molecule-metal wavefunction mixing to diminish.

The requirement of nonvanishing μ_{ik} in the direct photo-excitation mechanism puts restrictions on the symmetry of the LUMO. Whereas the LUMO is readily observed in C_6F_6, it is invisible in pentafluorobenzene (C_6F_5H) on Cu(111) (31). For

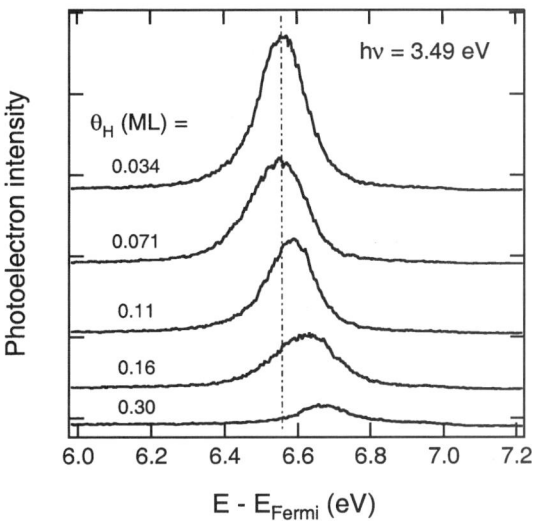

Figure 8 Two-photon photoemission spectra of bilayer C_6F_6 on H/Cu(111) at the indicated surface hydrogen coverages (0.034–0.30 ML). Only the energy region for the molecular resonance is shown (32).

naphthalene and anthracene on Cu(111) (22, 34) and Ag(111) (35), the expected LUMOs based on gas phase EA were not seen in 2PPE, but they were readily observed in inverse photoemission [(13) and references therein]. The absence of a molecular resonance in 2PPE spectra may be attributed to vanishing μ_{ik} owing to symmetry. This issue deserves further investigation.

In the case of Cs/Cu(111) direct photo-excitation into the σ^* antibonding state was observed at very low coverages [(25) and references therein]. In this case the requirement of nonvanishing μ_{ik} may be satisfied by the fact the σ^* state results from a combination of localized Cs atomic orbital and delocalized surface wavefunctions.

It is interesting to note that the hot-electron transfer mechanism (Figure 1b) has not been observed in the limited number of existing 2PPE studies. This mechanism must exist, as suggested by extensive evidence from surface photochemistry (5). I believe the direct observation of a transient anionic resonance from hot-electron transfer is only a matter of time, as more chemical systems are investigated by 2PPE. The same may be true for other indirect channels for transient anion formation. In this regard Ishioka et al. attributed the unusually short lifetime of image states on C_6F_6/Cu(111) to their efficient decay into the σ^* LUMO (36).

Intermolecular Interaction and Band Formation

Intermolecular interaction in molecular crystals results in the formation of electronic bands (37). This effect is more significant for unoccupied orbitals, such as LUMO and LUMO + n, than for occupied orbitals, owing to the fact that electronic

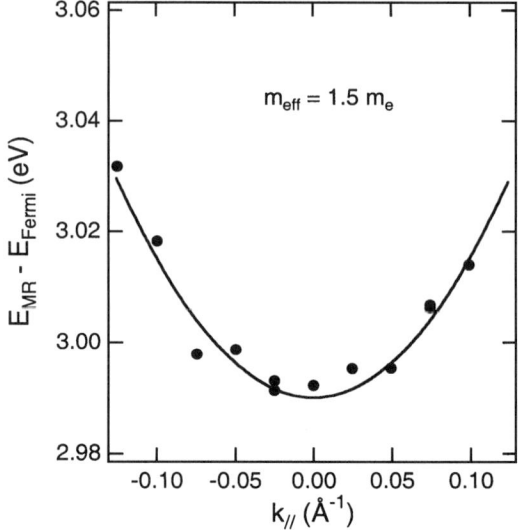

Figure 9 Parallel dispersion curve for bilayer C_6F_6/Cu(111). The solid line is a fit to the free-electron–like equation that yields the indicated effective electron mass (32).

wavefunctions are more spatially diffuse at higher energies. The widths of conduction bands in van der Waals crystals are typically of the order of 0.1–0.2 eV (37) and, in some cases, can be as high as 0.5–0.6 eV (38). When adsorbed in an ordered structure on metal surface, the molecular anionic resonance can disperse into a conduction band owing to intermolecular electronic interaction and/or mixing with highly dispersed metal bands. In the direct photo-excitation mechanism discussed above, the presence of dispersion is required for nonvanishing μ_{ik}.

The molecular anionic resonance in C_6F_6/Cu(111) displays significant dispersion. Figure 9 shows parallel dispersion of the molecular resonance for bilayer C_6F_6/Cu(111) obtained from angle-resolved 2PPE. The data can be fit to the free-electron-like parabolic dispersion in Equation 4 with an effective electron mass of 1.5 m_e. Experimental evidence points to the dominance of intermolecular interaction, not molecule-metal wavefunction mixing, as the source of such a highly dispersed molecular conduction band. Measurements of the dispersion of the molecular resonance in bilayer C_6F_6 on H/Cu(111) showed that m_{eff} decreased from 1.5 m_e on clean Cu(111) to 0.6 m_e on 0.34 ML H covered surface. With the weakening of the molecule-surface interaction by preadsorbed H, intermolecular interaction energy plays a more important role in determining the overlayer structure. As a result, intermolecular packing is improved. This may explain the decrease in effective electron mass within the C_6F_6 LUMO band as the coverage of preadsorbed H increases. Gahl et al. showed that the effective electron mass decreases with increasing adsorbate layer thickness from 1.9 m_e at 1 ML to 1.0 m_e at 5 ML coverage (31). With increasing coverage, the role of the surface in weakening intermolecular interaction is relaxed. As a result, we expect improved packing within the molecular layer to give better wavefunction overlap.

In view of the parallel dispersion and band formation, we expect that the molecular resonance should also disperse in the surface normal (z) direction for a film with finite thickness. Supporting this, one finds that the energetic position of the molecular resonance decreases with increasing thickness of the adsorbate film. A plausible interpretation of this observation is that the anionic resonance is stabilized owing to further delocalization (in the z direction) as film thickness increases, or in other words, vertical dispersion. Such a vertical dispersion may be analyzed using quantum well or particle in a one-dimensional box analysis, as for Xe/Ag(111) (23). In the direction perpendicular to the quantum well, the delocalized LUMO wavefunction extends over but is also confined by the total thickness of the film. The boundary condition dictates that the quantum well wavefunction should contain a component of $\sin(k_z z)$, where the perpendicular wave vector is given by

$$k_z = \frac{n\pi}{d}, \qquad 8.$$

where n (1, 2, ...) is the quantum number of the quantum well state, d (Nd_o) is the thickness of the molecular film, N is the total number of molecular layers, and d_o is the interlayer spacing. Taking $n = 1$ and using an average interlayer spacing of 5 Å for the solid C_6F_6 film (39), one can present the coverage dependence (31) of the molecular resonance position in the form of a perpendicular dispersion curve (Figure 10). The dashed curve is a fit of low k_z data points to a free-electron-like

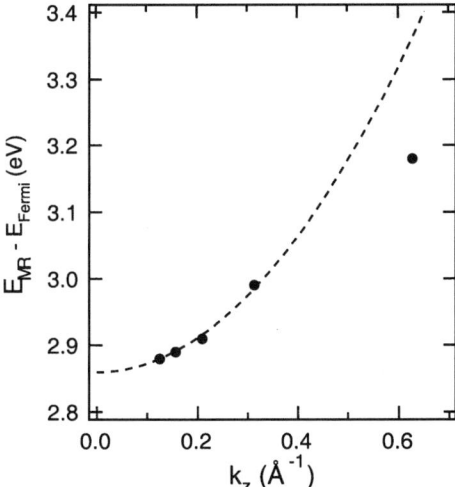

Figure 10 Perpendicular dispersion for 1–5 ML C_6F_6/Cu(111) within the quantum well approximation. The *dashed line* is a fit to the free-electron–like equation for low k_z.

parabolic relationship:

$$E_k = E_o + \frac{(\hbar k_z)^2}{2m_{eff}}. \qquad 9.$$

The fit yields an effective electron mass of 3 m_e in the surface normal direction and puts the bottom of the molecular band at 2.86 eV. Note that such a quantum well picture predicts states at higher quantum numbers ($n > 1$). For 1(2) ML coverage, the $n = 2$ state is located at ~2 (0.5) eV higher than the $n = 1$ state and is not supported by the limited width of the molecular conduction band. However, for C_6F_6 coverages ≥ 3 ML, a weak and broad molecular resonance at ≤ 0.5 eV higher than the LUMO was indeed observed. Whether this broad peak is due to states at $n \geq 2$ remains an open question.

Lifetimes and Molecule-to-Metal Electron Transfer

A transient anionic resonance, once formed, is subject to a number of dynamic processes, including polarization, localization, and decay to substrate states. The last is perhaps most important on a metal surface. Excited electronic states on metal surfaces are characterized by ultrashort lifetimes. This is due to the ease with which an electron in an excited molecular orbital can elastically transfer to the vast number of resonant electronic states (band) in the metal or can inelastically scatter with the large population of cold electrons at the Fermi sea. In the case of elastic molecule-to-metal ET, the lifetime, τ_{decay}, of an excited molecular anion can often be correlated with the electronic coupling matrix element between the molecular orbital and the metal band structure: $1/\tau_{decay}$ is proportional to the square amplitude of the oscillating tail of the mixed wavefunction inside the metal (see Figure 1a). Thus, lifetime measurement provides a quantitative measure of the electronic coupling strength. Experimentally, this is done by varying the pump-probe delay time, with the pump laser pulse populating the anionic molecular resonance and the probe pulse ionizing the transient ion.

Figure 11a shows such a pump-probe experiment for 1–5 ML of C_6F_6 on Cu(111). Fitting to experimental data yields τ_{decay} shown in Figure 11b. The decay time increases from ~7 fs at 1 ML to ~32 fs at 5 ML coverage. The coverage dependence of the lifetime suggests that the electronic coupling between the LUMO and the metal surface weakens as film thickness increases. This is expected from dispersion in the surface normal direction. With increasing film thickness, the delocalized wavefunction (in the z direction) within the adsorbate layer is characterized by an increasing distance of the center-of-gravity of the wavefunction from the surface and a diminishing amplitude of the tail inside the metal.

Although thermal desorption measurement indicates that C_6F_6 is weakly adsorbed on the Cu(111) surface (29), the lifetime of ~7 fs for the LUMO at monolayer coverage is of the same order as those in strong chemisorption systems determined by both experiment and theory. For example, the lifetime of the π^* anionic resonance in CO chemisorbed on Cu(111) was measured to be $1 < \tau < 5$ fs

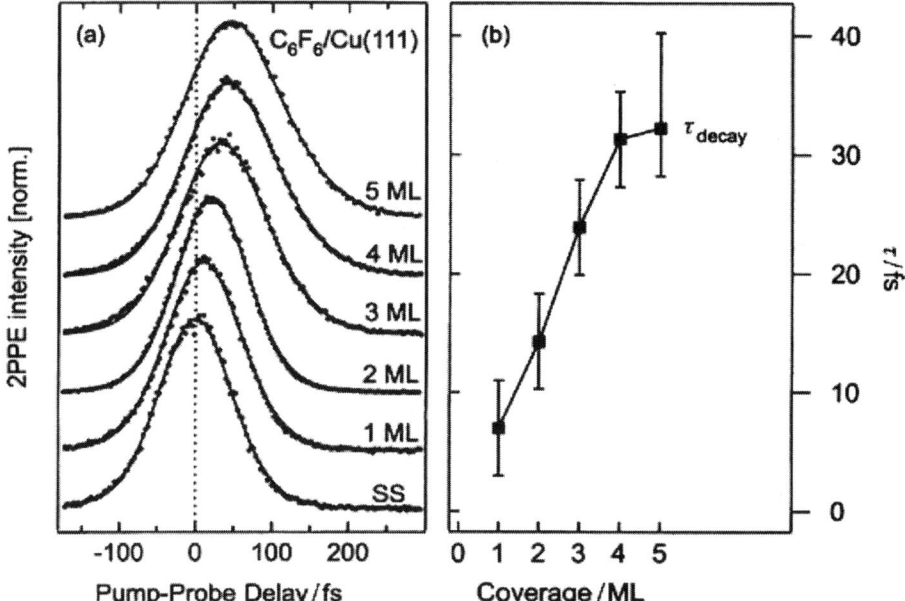

Figure 11 Time-resolved two-photon photoemission spectroscopy. (*a*) Cross-correlation curves of the molecular resonance of C_6F_6/Cu(111) at the indicated coverages, recorded with $h\nu_1 = 4.26$ eV pump and $h\nu_2 = 2.13$ eV probe pulses. The *dotted line* indicates zero delay time determined from the cross-correlation signal of the occupied surface state (SS) on clean Cu(111). (*b*) Lifetimes (population decay) for the molecular resonance as a function of coverage. Note that a finite rise time was included to obtain a satisfactory fit. After Gahl et al. (31).

by time-resolved 2PPE and spectral line width (40), whereas those for the σ^* antibonding states in alkali atoms on metal surfaces range from a few femtoseconds to a few tens of femtoseconds [(25) and references therein]. The electronic coupling strength between an unoccupied molecular orbital and the metal substrate is not immediately obvious from bonding strength because LUMOs are often not involved to a great extent in bonding.

Whereas the measured lifetime for the σ^* anionic resonance in C_6F_6/Cu(111) indicates efficient molecule-to-metal resonant (elastic) electron transfer and/or inelastic electron scattering, the actual ET mechanism is not clear. The σ^* LUMO of C_6F_6 is located within the projected bandgap of Cu(111). Elastic ET from σ^* LUMO of C_6F_6 to resonant metal states at or near $k_\parallel = 0$ is forbidden. In such a band picture the barrier for elastic ET increases as $k_\parallel$ increases [(25) and references therein]. However, as discussed in more detail below, polarization effects (electronic, intramolecular, and intermolecular) can relax the momentum requirement in resonant ET. A localized state essentially possesses all k components, and the band picture becomes invalid. As a result, elastic ET becomes allowed. On the other hand, decay of the σ^* resonance owing to inelastic scattering with the vast number of Fermi electrons is not restricted by the presence of a projected bandgap. The relative importance of each channel is not known; this is an excellent problem for further experimental and theoretical studies.

POLARIZATION AND LOCALIZATION

Molecule-to-metal electron transfer is not the only process for the molecular resonance. Once formed, the transient molecular resonance is subject to a number of dynamic processes, such as electronic polarization, nuclear polarization, localization, chemical changes, and other inelastic scattering events. These dynamic processes must be considered because we are employing concepts of single-electron excitation in describing an inherently multibody problem. I now present a qualitative analysis of this complex picture. Interested readers may find relevant concepts in the fascinating book on organic molecular crystals by Silinsh & Capek (37) and in several excellent review articles on surface electron dynamics (17, 18, 20).

Competition Between Delocalization and Localization

The complex dynamics following the excitation of a molecular anionic resonance on a metal surface is illustrated schematically in Figure 12 based on two axes: the spatial localization of excitation and the relaxation of excitation energy. This process is a result of competition between delocalization and localization, the conditions for which can be intuitively understood from energetic considerations based on two inequalities introduced below.

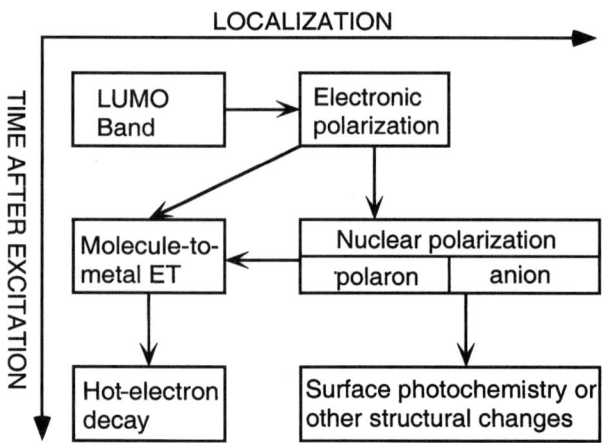

Figure 12 Schematic illustration of dynamic processes for a transient molecular resonance.

For an excited molecular anionic resonance, delocalization is determined by the resonant interactions between neighboring molecular orbitals and between a molecular orbital and metal band structure. Both processes are characterized by electronic coupling matrix elements or transfer integrals (37): J_{MM} and J_{MS} for intermolecular and molecule-substrate electron transfer, respectively. The values of these transfer integrals are of the same order as bandwidths. For molecular crystals, J_{MM} is of the order of a few tenths of eV. For molecule-metal interactions, J_{MS} for an excited molecular orbital can be as high as 1 eV [(41) and references therein]. For large values of J_{MM}, the wavefunction in the molecular layer is a Bloch wave with definitive wave vector, k. This picture, similar to that in inorganic semiconductors, is valid when the mean-free-path of the excited electron is significantly larger than the lattice parameter or intermolecular distance. For large values of J_{MS}, the molecular wavefunction is strongly mixed with Bloch-type waves of the metal substrate. A consequence of large values of J_{MS} is the ultrafast rates of ET between the metal and the molecule. In the limit of strong coupling, the rate approaches the Bohr frequency, and ET is essentially a charge redistribution process within the mixed wavefunction. The two delocalization effects result in energy gains (stabilization) represented by $\delta E_{del\text{-}MM}$ and $\delta E_{del\text{-}MS}$ for intermolecular and molecular-substrate delocalization, respectively. Both are negative values of the order of transfer integrals.

The opposite trend is localization, which is determined by polarization in electronic and nuclear coordinates of the surrounding environment. An excess electron in an adsorbate film induces charge redistribution in the valence wavefunctions of neighboring molecules as well as in the metal substrate. Such an electronic polarization process occurs on the timescale of the inverse of the Bohr frequency, i.e., 10^{-16}–10^{-15} s. One may also estimate this timescale from the electronic excitation energy based on the uncertainty principle. Because this timescale is substantially shorter than that for the movement (hopping) of a localized electron from one molecule to another, we can simply consider the electronic polarization effect as the total electrostatic attraction between a fixed electron and induced dipoles in surrounding neutral molecules and in the metal substrate; the latter is the image potential discussed earlier. These electrostatic attraction potentials create a localized trap for the electron. Additional localization comes from polarization of nuclear subsystems on a longer timescale. There are two main flavors: intramolecular polarization, i.e., vibronic coupling and molecular polaron formation (10^{-14} s), and intermolecular polarization, i.e., lattice polaron formation ($\sim 10^{-13}$ s). The timescale in each case is the inverse of the frequency involved, intramolecular vibrations or lattice phonons. I discuss polarization and localization of the nuclear subsystem in more detail below.

Both electronic and nuclear polarizations lead to energy gain, δE_{loc}, for the localization of the electron. The magnitude of $|\delta E_{loc}|$ is of the order of eV, depending on the polarizability of molecules and dipole moments for vibrational excitation. In addition to polarization energy, the localized electron can also form direct bonding with a molecule, resulting in an anion. In other words, the electron resides in a particular molecular orbital. This is accompanied by an energy gain of δE_b.

If the following inequality is satisfied,

$$|\delta E_{del-MM} + \delta E_{del-MS}| > |\delta E_{loc} + \delta E_B|, \qquad 10.$$

the delocalization trend wins, and the dynamic process is dominated by the left side of Figure 12. Excitation is quenched owing to hot-electron decay on a 10^{-15}–10^{-14} timescale. This is the fate for molecular layers with relatively broad bandwidth and/or with strong molecule-metal electronic coupling.

If the opposite is true, i.e.,

$$|\delta E_{del-MM} + \delta E_{del-MS}| < |\delta E_{loc} + \delta E_B|, \qquad 11.$$

localization dominates, and this leads to the right side of the dynamic pathway in Figure 12: the excitation of molecular and lattice vibrations, as well as chemical or structural changes. This applies to narrow bandwidth molecular films and weak molecule-metal coupling; instead of a Bloch-type wave, the wavefunction is better described by a localized quasi-particle.

There are two important issues one must remember in applying the inequalities. First, these inequalities should only serve as qualitative guidelines, not quantitative criteria dictating dynamics. Time is of the essence. For example, energetic conditions favoring localization, such as polaron formation and surface photochemistry, may not be observed experimentally because the timescale for localization could be too slow to compete with delocalization with the metal (resonant molecule-to-metal ET). On the other hand, inelastic scattering events may also lead to localization under conditions favoring delocalization. Second, the picture in Figure 12 starts with direct photo-excitation of a resonance with nonvanishing dispersion. However, inelastic scattering may lead to the formation of a localized molecular anion directly from hot electron transfer (Figure 1b), thus bypassing the dynamic steps leading to localization. In this scenario the dynamic process is a competition among the following possibilities: resonant molecule-to-metal ET, inelastic scattering with metal electrons at the Fermi sea, and chemical/structural changes.

The case study of C_6F_6/Cu(111) may fall into the category of Equation 10. Angle-resolved measurement showed no sign of localization; there were no features of nondispersive states within the lifetime of the molecular resonance. A minor photo-dissociation channel may be attributed to the direct formation of localized anions from hot-electron transfer/scattering events (31, 32). Whereas the molecular resonance was observed in 2PPE spectra at photon energies above the energetic threshold, the photo-dissociation channel was only observed at high photon energies; this can be attributed to the low cross sections for hot-electron transfer or inelastic scattering at low photon energies (e.g., $h\nu = 3.49$ eV) (Figure 8).

The only example of polaron formation at adsorbate/metal interfaces came from the work of Harris and coworkers (42). These authors showed that image-state electrons in the alkane/Ag(111) system can be trapped at the surface of the alkane film, leading to the formation of small polarons (mainly two-dimensional). The decay dynamics of delocalized image states to localized small polarons was directly followed in time- and angle-resolved 2PPE. Similar experiments for molecular resonances are warranted in the near future.

Molecular Polarons and Surface Photochemistry

Charge localization due to polarization of nuclear subsystems is of particular interest not only because it is an integral part of the dynamic process but also because it can lead to chemical and structural consequences. The following nuclear coordinates need to be considered: intramolecular and molecule-surface vibrations. The timescale (Δt) for each interaction can be estimated from the characteristic vibrational frequency based on the uncertainty principle: $\Delta E \Delta t = h\nu \Delta t \approx h$; thus, $\Delta t \approx 1/\nu$.

Polarization of intramolecular coordinates results in a molecular polaron or a transient anion. If dipole active vibrational modes in neighboring molecules are also involved in trapping the molecular polaron, the net result is a nearly small molecular polaron. Typical intramolecular vibrational frequencies of 3000–300 cm^{-1} correspond to interaction times of 2×10^{-15}–2×10^{-14} s. For the molecule-surface coordinate, the interaction time is of the order of 10^{-14} s. These timescales compete with that for molecule-to-metal ET on a similar or shorter timescale. Polaron hopping (to neighboring molecules) occurs on a longer timescale and is not competitive. During the lifetime of the molecular polaron, i.e., the residence time of the electron on the molecule, the molecule evolves from the equilibrium configuration for the neutral to that for the negatively charged state. Thus, following the decay of the molecular polaron, intramolecular and molecule-surface vibrations are excited. Whereas vibrational excitations within the adsorbate layer may be eventually dissipated to the phonon bath of the substrate metal, a competitive channel is that of chemical change, i.e., surface photochemistry. In fact, the above description for polaron formation and decay is identical to the widely accepted model of hot electron–mediated surface photochemistry, within the frameworks of the Menzel-Gomer-Redhead (MGR) and Antoniewicz models for photodissociation and desorption (5).

In an account entitled "Surface Photochemistry" in this series eight years ago, I concluded that the majority of photochemical processes on metal surfaces could be attributed to hot-electron mechanisms (5). Thus, metal surface photochemistry is part of the big picture for polarization and localization of an excited electron; in other words, surface photochemistry results from the decay of molecular polarons. However, most experimental evidence has shown that the rate of hot electron–mediated surface photochemistry scales with light absorbance in the metal, not with electric field strength (5). One must then conclude that the direct photo-excitation mechanism (Figure 1a) is not responsible in these systems. Rather, localized molecular polarons or anions form directly from the hot-electron scattering or tunneling mechanism in Figure 1b. In most surface photochemical systems investigated to date, molecular anionic resonances should be of highly localized character. The small intermolecular transfer integral, J_{MM}, points to the invalidity of the electronic band picture and excludes the direct photo-excitation mechanism.

In fact, the localized nature of these anionic resonances is the reason for the competitiveness of photochemical pathways. One can also predict that adsorbate systems featuring large J_{MM} values and, hence, delocalized electronic bands, are unlikely candidates for surface photochemistry.

In another chapter in this volume Petek & Ogawa (21) present an elegant 2PPE study of the photodesorption dynamics of Cs from Cu(111) at low adsorbate coverages. The σ^* antibonding resonance is populated by direct photo-excitation. In this case the requirement for delocalization was satisfied by the large adsorbate-substrate transfer integral, J_{MS}, not the interadsorbate interaction, J_{MM}. Photodesorption dynamics is essentially localization dynamics. As the Cs-surface bond stretches, J_{MS} decreases and the excited electron is increasingly localized to the adsorbate.

IMPLICATIONS FOR MOLECULE ELECTRONIC DEVICES

The use of molecules in electronic devices is attractive for a number of reasons, among them, the great flexibility for tuning electronic properties by chemical modification and the natural scalability to the nanometer or single molecule level. One can divide molecule-based electronic devices into two categories: conventional devices using molecular semiconductors and molecular electronics using one or a small group of molecules. The issues of electronic structure and electron transfer dynamics addressed by 2PPE are of fundamental importance to the understanding and design of both types of devices. I use three examples below to illustrate my point.

Energetics and Dynamics

Recent demonstration of high mobility field effect transistors (FETs) including superconducting FETs based on molecular crystals (43–46) such as pentacene, tetracene, and C_{60} have attracted considerable interest in these molecular materials. In such devices, charge transport is often determined by a single or a few molecular layers at the interface. Thus, understanding the band structure of these molecular thin films is critical. 2PPE is perhaps the most powerful technique in establishing the structure of conduction bands. For example, Figure 13 shows a set of 2PPE spectra taken for bilayer C_{60} epitaxially grown on Cu(111). The spectral dependence on $h\nu$ establishes three unoccupied states, LUMO, LUMO + 1, and LUMO + 2, and the highest occupied molecular orbital. Time-resolved experiments are necessary to characterize the dynamics of electrons in each unoccupied state and the dependence of dynamics on momentum and temperature. These dynamics measurements can establish quantitatively the electron-phonon coupling, which is responsible for the transition from band conduction at low temperatures to polaron-hopping conduction at high temperatures (47). For example, a recent 2PPE study determined, for the first time, the electron-phonon coupling constant in carbon nanotubes (48).

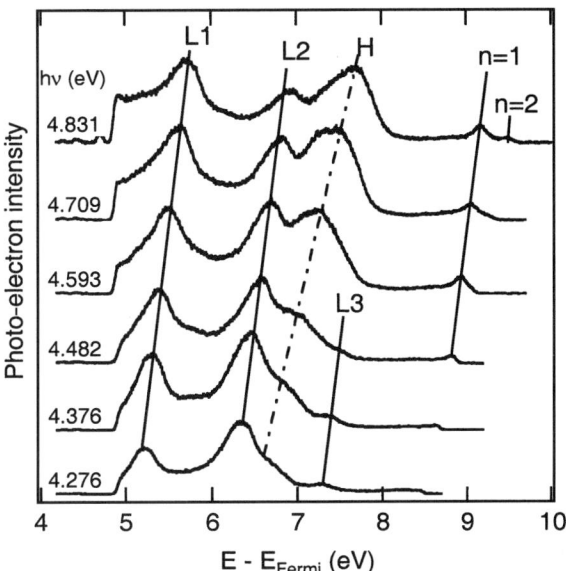

Figure 13 Two-photon photoemission spectra of 2 ML C60/Cu(111) taken with one-color laser light at the indicated photon energies. The *solid lines* correspond to one-photon dependence and the *dot-dashed line* is a two-photon dependence in peak positions. H, HOMO; L1, LUMO; L2, LUMO+1; L3, LUMO+2; image states (n = 1, 2). (G. Dutton & X.-Y. Zhu, unpublished results).

Delocalization or Localization?

In FETs based on molecular crystals the figure of merit is carrier mobility. Band formation owing to delocalization or a large transfer integral is most desirable. Localization and polaron formation can only decrease carrier mobility and should be minimized in molecular design and crystal engineering.

Is the same true for molecular electronics? The answer is probably "no." For a molecular device to function, one would like to see changes in electronic properties, such as conductivity, to occur locally. In this sense delocalization owing to large transfer integrals between molecules or between a molecule and a metal electrode is undesirable. Instead, the figure of merit should be the strong tendency to localize owing to electronic and nuclear polarization. Recently, negative differential resistance and switching effects have been observed in two-terminal devices based on single or a group of conjugating molecules assembled on metal electrodes (Figure 14) (4, 49). A possible interpretation is that when an electron passes through the molecule, the strong polarization effect from electron-vibration coupling may lead to the formation of a molecular polaron, which shuts off further electron transport. Even after the decay of the molecular polaron, the resulting excitation in nuclear coordinates may lead to meta-stable structural or conformational

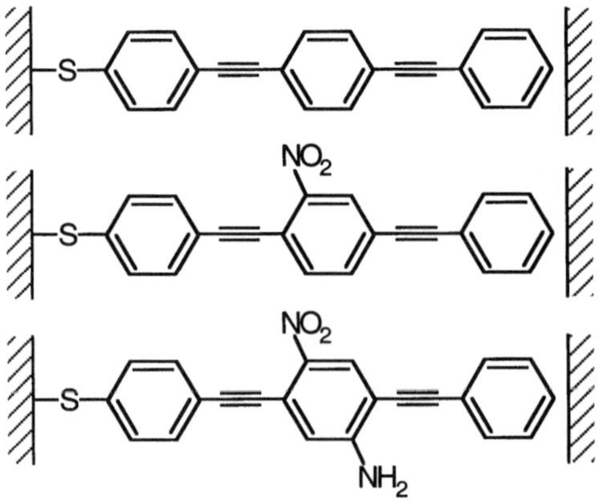

Figure 14 Molecules for which negative differential resistance or switching effects were observed (4, 49).

change that gives rise to the switching behavior. The fact that negative differential resistance or switching in molecules in Figure 14*B*, *C* is much more significant than that in Figure 14*A* supports the argument of localization. The addition of the $-NO_2$ group to the middle phenyl ring not only enhances electron-vibration coupling owing to the increased vibrational dipole moment but also creates a more efficient trap for the molecular polaron.

The Contact Problem

Making electronic contact to molecules is critical in both conventional and single-molecule devices. In organic FETs and light-emitting devices this is the charge injection issue. Energetics is not the only factor to consider in charge injection. Both electronic coupling and interfacial relaxation dynamics, as discussed extensively in this account, are important to charge injection at interfaces. These issues are not easy to probe in device measurements. Frisbie and coworkers recently carried out transport measurements using conduction-probe atomic-force microcopy to make nano-contacts to organic crystals or self-assembled monolayers (SAMs) (50, 51). These experiments clearly point to the critical importance of contacts. A recent experiment by Houston and coworkers established the role of contact pressure on conductance through alkanethiol SAMs (52).

The importance of contacts only increases as device dimensions decrease, particularly to the molecular level. For example, SAMs of thiols on metal surfaces have been popular choices in the construction and testing of molecular electronic devices. The easy formation of the thiolate-metal contact is an attractive method to connect molecular components to metal electrodes. The questions are: What is the electronic structure of the thiolate-metal contact? How does this contact affect electron transport through the molecular wire? Recent 2PPE measurements and ab initio calculations on model SAM/Cu(111) systems (53) showed the presence of two σ^* states localized to the C-S-Cu linker. For symmetry reasons these localized σ^* states introduced by the anchoring bond cannot couple to the delocalized π^* states within a conjugated molecular framework. Thus, the thiolate contact can be considered insulating for electron transport through a self-assembled monolayer of molecular wires. On the other hand, ab initio calculation showed that the highest occupied molecular orbital (π) is delocalized between the molecular framework and the metal surface via the –S-bridge. Therefore, unlike for electron transport, the thiolate contact is conducting for hole transport.

COMMENTS

To recycle a sentence from eight years ago, I hope the key issues addressed in this account have raised more questions than provided answers (5). The application of 2PPE to interfacial electronic structure and electron transfer dynamics is only beginning to show its exceptional attractiveness. These studies allow us to probe the central issue of interfacial ET and to provide a unified view on many exciting research areas. In the current rush toward demonstrating molecular electronic devices, it may be helpful to pause for a moment and consider some of the physical chemistry principles we can learn from the kind of fundamental studies demonstrated here. Whereas experimental work using time- and angle-resolved 2PPE on more model systems is needed, theoretical work will also be essential to guarantee the success of this research field.

ACKNOWLEDGMENTS

I thank my students and coworkers, particularly Gregory Dutton, for most of the experimental results included in this account and for a critical reading of the manuscript. This work was supported by the National Science Foundation (DMR-9982109) and the University of Minnesota. Acknowledgement is made to the donors of the Petroleum Research Fund, administered by the American Chemical Society, for partial funding of this work.

Visit the Annual Reviews home page at www.annualreviews.org

LITERATURE CITED

1. Salaneck WR, Brédas JL. 1997. *MRS Bull.* 22:46–51
2. Brutting W, Berleb S, Muckl AG. 2001. *Organ. Electron.* 2:1–36
3. Garnier F, Kouki F, Hajlaoui R, Horowitz G. 1997. *MRS Bull.* 22:52–56
4. Chen J, Reed MA, Rawlett AM, Tour JM. 2000. *Science* 286:1550–52
5. Zhu X-Y. 1994. *Annu. Rev. Phys. Chem.* 45:113–44
6. Hagfeldt A, Gratzel M. 1995. *Chem. Rev.* 95:49–68

7. Miller RJD, McLendon GL, Nozik AJ, Schmickler W, Willig F. 1995. *Surface Electron Tansfer Processes*. New York: VCH
8. Bard AJ, Faulkner LR. 2001. *Electrochemical Methods: Fundamentals and Applications*. New York: Wiley. 2nd ed.
9. Wolf M, Hotzel A, Knoesel E, Velic D. 1999. *Phys. Rev. B* 59:5926–35
10. Hofer U, Shumay IL, Reuss Ch, Thomann U, Wallauer W, Fauster Th. 1997. *Science* 277:1480–82
11. Gadzuk JW. 1995. *Surf. Sci.* 342:345–58
12. Bonzel HP, Kleint C. 1995. *Prog. Surf. Sci.* 49:107–53
13. Frank KH, Yannoulis P, Dudde R, Koch EE. 1988. *J. Chem. Phys.* 89:7569–77
14. Sanche L. 2000. *Surf. Sci.* 451:82–90
15. Fauster Th, Steinmann W. 1995. In *Photonic Probes of Surfaces*, ed. P Halevi. Amsterdam: Elsevier
16. Osgood RM Jr, Wang X. 1998. *Solid State Phys.* 51:1–80
17. Haight R. 1995. *Surf. Sci. Rep.* 21:275–325
18. Petek H, Ogawa S. 1998. *Prog. Surf. Sci.* 56:239–311
19. Aeschlimann M, Bauer M, Pawlik S. 1996. *Chem. Phys.* 205:127–41
20. Harris CB, Ge N-H, Lingle RL Jr, McNeill JD, Wong CM. 1997. *Annu. Rev. Phys. Chem.* 48:711–44
21. Petek H, Ogawa S. 2002. *Annu. Rev. Phys. Chem.* 53:507–31
22. Wang H, Dutton G, Zhu XY. 2000. *J. Phys. Chem. B* 104:10332–38
23. McNeill JD, Lingle RL, Jordan RE, Padowitz DF, Harris CB. 1996. *J. Chem. Phys.* 105:3883–91
24. Hertel T, Knoesel E, Hasselbrink E, Wolf M, Ertl G. 1994. *Surf. Sci.* 317:L1147–51
25. Petek H, Nagano H, Weida MJ, Ogawa S. 2001. *J. Phys. Chem. B* 105:6767–79
26. Velic D, Hotzel A, Wolf M, Ertl G. 1998. *J. Chem. Phys.* 109:9155–65
27. Gaffney KJ, Wong CM, Liu SH, Miller AD, McNeill JD, Harris CB. 2000. *Chem. Phys.* 251:99–110
28. Munakata T, Sakashita T, Tsukakoshi M, Nakamura J. 1997. *Chem. Phys. Lett.* 271:377–80
29. Vondrak T, Zhu X-Y. 1999. *J. Phys. Chem. B* 103:3449–56
30. Zhu X-Y, Vondrak T, Wang H, Gahl C, Ishioka K, Wolf M. 2000. *Surf. Sci.* 451:244–49
31. Gahl C, Ishioka K, Zhong Q, Hotzel A, Wolf M. 2000. *Faraday Discuss.* 117:191–202
32. Dutton G, Zhu X-Y. 2001. *J. Phys. Chem. B* 105:10912–17
33. Gahl C. 1999. Diploma thesis. Fritz Haber Inst., Berlin
34. Wang H. 2001. PhD thesis. Univ. Minnesota, Minneapolis
35. Gaffney KJ, Liu SH, Miller AD, Szymanski P, Harris CB. 2000. *J. Chin. Chem. Soc.* 47:759–63
36. Ishioka K, Gahl C, Wolf M. 2000. *Surf. Sci.* 454:73–77
37. Silinsh EA, Capek V. 1994. *Organic Molecular Crystals: Interaction, Localization, and Transport Phenomena*. Woodbury, CT: AIP
38. Cornil J, Calbert JPh, Bredas JL. 2001. *J. Am. Chem. Soc.* 123:1250–51
39. Boden N, Davis PP, Stam CH, Wesselink GA. 1973. *Mol. Phys.* 25:81–88
40. Bartels L, Meyer G, Rieder KH, Velic D, Knoesel E, et al. 1998. *Phys. Rev. Lett.* 80:2004–7
41. Braun J, Nordlander P. 2000. *Surf. Sci.* 448:L193–99
42. Ge NH, Wong CM, Linge RL Jr, McNeill JD, Gaffney KJ, Harris CB. 1998. *Science* 279:202–5
43. Schon JH, Kloc Ch, Batlogg B. 2000. *Nature* 406:702–4
44. Schon JH, Kloc Ch, Haddon RC, Batlogg B. 2000. *Science* 288:656–58
45. Schon JH, Kloc Ch, Batlogg B. 2000. *Science* 288:2338–40
46. Schon JH. 2001. *Synth. Metals* 122:157–60
47. Schon JH, Kloc Ch, Batlogg B. 2001. *Phys. Rev. Lett.* 86:3843–46

48. Hertel T, Moos G. 2000. *Phys. Rev. Lett.* 84:5002–5
49. Donhauser ZJ, Mantooth BA, Kelly KF, Bumm LA, Monnell JD, et al. 2001. *Science* 292:2303–7
50. Kelley TW, Granstrom EL, Frisbie CD. 1999. *Adv. Mater.* 11:261–64
51. Wold DJ, Frisbie CD. 2000. *J. Am. Chem. Soc.* 122:2970–71
52. Son KA, Kim HI, Houston JE. 2001. *Phys. Rev. Lett.* 86:5357–60
53. Vondrak T, Wang H, Winget P, Cramer CJ, Zhu X-Y. 2000. *J. Am. Chem. Soc.* 122:4700–7

AB INITIO MOLECULAR DYNAMICS WITH DENSITY FUNCTIONAL THEORY

John S. Tse
Steacie Institute for Molecular Sciences, National Research Council of Canada, Ottawa, Ontario, K1A 0R6; e-mail: John.Tse@nrc.ca

Key Words Molecular structures, reaction mechanisms, spectroscopic properties

■ **Abstract** Recent applications of density functional theory base ab initio molecular dynamics in chemical relevant systems are reviewed. The emphasis is on the dynamical aspect in the study of structures, reaction mechanisms, and electronic properties in both the molecular and condensed phases. Examples were chosen from fluxional molecules, solution reactions, and biological systems to illustrate the broad potential applications and unique information that can be obtained from ab initio molecular dynamics calculations. Recent advances in the development of efficient numerical algorithms for the prediction of spectroscopic properties are highlighted.

INTRODUCTION

In the early 1960s, Hohenberg and Kohn (1, 2) showed that the ground state properties of an interacting electron gas could be calculated from the electron density, independent of the nature of the external potential. These ideas were soon exploited for numerical electronic structure calculations and since then considerable progress has been achieved. In the 1980s density functional theory (DFT) was established as a convenient alternative to solve the Schrödinger equation, which provided comparable electronic structure information to the well-known Hartree-Fock theory (3). Despite a few unresolved shortcomings, in view of the relative computational simplicity, favorable scaling with the number of electrons, and significant improvements in the accuracy of density functionals for the estimation of exchange-correlation effects, DFT methods are now standard tools in quantum chemistry and are particularly widespread in computational materials science and condensed matter physics. A full account of recent progress on the development of DFT methods and their wide-ranging applications is beyond the scope of this review. There are several excellent books and monographs covering these topics (4–6). This article focuses on the applications of DFT in the study of the dynamics of chemical systems, such as vibrational properties of nonclassical or floppy molecules and the exploration of mechanistic pathways in chemical reactions.

The genesis of the application of DFT in molecular dynamics was the seminal paper by Car & Parrinello (CP) in 1985 (7), in which they proposed a revolutionary approach to the solution of the electronic structure problem of very large solid state systems. Realizing that the variational principle is simply a minimization procedure, they proposed that parameters in the electronic wave function may be treated as dynamical variables (electron dynamics) and the electronic structure problem can be solved by the application of the steepest descent method to the classical Newtonian equations of motions. It was further recognized that the fictitious electron dynamics can be coupled with the classical motions of the atom nuclei. Thus, the evolution of the electronic wave function and the forces acting on the atoms can be computed simultaneously, provided that the electronic state of the system is conserved (i.e., remains on the Born-Oppenheimer surface). This is achieved by the exploitation of the quantum mechanical adiabatic timescale separation of fast electronic and slow nuclear motion by transforming the electronic problem into classical mechanical adiabatic energy scale separation in the framework of dynamical systems theory. Hence, both the electronic structure problem and the dynamics of the atoms are solved concurrently by a set of Newton's equations. In contrast to the traditional approach, this theoretical breakthrough allows calculations of the fully dynamic time evolution of a structure (molecular dynamics) without resorting to a predefined potential energy surface. This ab initio or first-principles "on-the-fly" calculation of electronic potential energy surface and nuclear dynamics, ab initio molecular dynamics (AIMD) has extended the reign of traditional methods and opened new opportunities for the study of the nuclear dynamics in fluxional molecules and the determination of pathways of chemical reactions.

A number of articles reviewing the principles of AIMD and technical details on its implementation have already appeared (8–11). A particularly useful reference is a recent monograph by Marx & Hutter (12). To help readers better appreciate the theoretical origin of some of the applications in the ensuing discussion, a brief outline of the CP method and related techniques is given below.

Within the basis-set representation (localized or plane wave) (13) of the molecular electronic wave function, the self-consistent solution to the Schrödinger equation in the independent electron approximation reduces to a pseudo-eigenvalue problem that is commonly solved by iterative diagonalization of the secular determinant. Treating the coefficients of the basis functions as dynamical variables, Car & Parrinello postulated a classical Lagrangian for this system,

$$L_{CP} = \sum_i \frac{1}{2}\mu_i \langle \dot{\psi}_i | \dot{\psi}_i \rangle + \frac{1}{2}\sum_I M_I \dot{R}_I^2 - \langle \Psi_0 | H | \Psi_0 \rangle + constr. \qquad 1.$$

$$E = \langle \Psi_0 | H | \Psi_0 \rangle = \int V_{ext}(\vec{r})\rho(\vec{r})d\vec{r} + \langle \Psi_0 | \hat{T} + \hat{V}_{ee} | \Psi_0 \rangle, \qquad 2.$$

where the total wave function $|\Psi_0\rangle = det\{\psi_i\}$, M_I is the nuclear mass, and μ_i is the fictitious electron mass. T is the kinetic energy of noninteracting electrons and V_{ee} is the electron exchange and correlation potential. The constraints might be a function of both the set of orbitals $\{\psi_i\}$, e.g., orthonormality, and the nuclear positions $\{R_I\}$, e.g.,

$$constr. = \sum_{i,j} \Lambda_{i,j} \left(\int \psi_i^* \psi_j d^3\vec{r} - \delta_{ij} \right), \qquad 3.$$

where Λ_{ij} are the Lagrangian multipliers.

In principle there is no restriction to the form of the energy Hamiltonian, H. Very often in AIMD calculations, the Hamiltonian operator corresponding to the DFT is employed (3–6):

$$E = \langle \Psi_0 | H | \Psi_0 \rangle = \int V_{ext}(\vec{r}) \rho(\vec{r}) d\vec{r} + \langle \Psi_0 | \hat{T} + \hat{V}_{ee} | \Psi_0 \rangle$$

$$= \int V_{ext}(\vec{r}) \rho(\vec{r}) d\vec{r} + F[\rho(\vec{r})], \qquad 4.$$

where $F[\rho(r)]$ is a universal functional independent of the external potential and is identical for every system. The exact form of $F[\rho(r)]$ is unknown, and almost all of the current effort in the development of DFT is focused on the construct of this functional.

The corresponding Newtonian equations of motions are obtained from the Euler-Largange equations:

$$\frac{d}{dt}\left(\frac{\partial L_{CP}}{\partial \dot{R}_I}\right) = \frac{\partial L_{CP}}{\partial R_I}$$
$$\frac{d}{dt}\left(\frac{\partial L_{CP}}{\partial \dot{\psi}_i^*}\right) = \frac{\partial L_{CP}}{\partial \psi_i^*}. \qquad 5.$$

Thus, the CP equations of motions for the coupled electron-ion molecular dynamics become

$$L_{CP} = \sum_i \frac{1}{2}\mu_i \langle \dot{\psi}_i | \dot{\psi}_i \rangle + \frac{1}{2}\sum_I M_I \dot{R}_I^2 - E(\Psi_0, \vec{R}) + constr. \qquad 6.$$

The classical forces corresponding to the atomic and electron degree of freedoms are

$$M_I \ddot{R}_I(t) = -\frac{\partial}{\partial R_I}\langle \Psi_0 | H | \Psi_0 \rangle + \frac{\partial}{\partial R_I}\{constr.\}$$
$$\mu_i \ddot{\psi}_i(t) = -\frac{\partial}{\partial \psi_i^*}\langle \Psi_0 | H | \Psi_0 \rangle + \frac{\partial}{\partial \psi_i^*}\{constr.\}. \qquad 7a,b.$$

For a given nuclear geometry, the ground state electronic wave function can be obtained by standard techniques such as steepest descent or other higher-order methods:

$$\dot{\psi}_i(t) = -\frac{1}{2}\frac{\delta E}{\delta \psi} + \sum_j \Lambda_{ij}\psi(t) = -\frac{1}{2}H\psi_i(t) + \sum_j \Lambda_{ij}\psi(t). \qquad 8.$$

Once the ground state wave function has been determined, the dynamical behavior of a system can be studied at a given time; the force acting on the nuclei is given by Equation 7a. In the dynamical simulation both the electronic and nuclear (ionic) degrees of freedom can propagate concurrently. There is no need to obtain an iterative solution to the electronic structure as long as the nuclei-electron couplings are small, therefore the electrons will stay close to their instantaneous ground state on the Born-Oppenheimer surface and adiabatically follow the nuclear motion (CP dynamics). The atomic trajectory can be generated from the time integration of the equations of motion. The energy obtained at a given instantaneous structure $\{R_I\}$ generally differs slightly from the exact Kohn-Sham energy. However, if the energy exchange between the two subsystems is small, the trajectory will be very close to the correct one (14). Dynamical information can then be extracted from the atomic trajectories in the usual manner. Alternatively, self-consistent energy can be obtained at each time step using efficient eigenvalue solvers. Many numerical techniques, such as conjugate gradient, preconditioning, and the residual minimization method in direct inversion of iterative space, have been developed to accelerate the self-consistent solution (10). If such methods are used, the electrons are guaranteed to be on the Born-Oppenheimer surface, and larger time steps can be used in dynamical simulations at the expense of the more demanding electronic structure solution.

For light elements, full quantum dynamic effects can be studied using the Feynman path-integral molecular dynamics method (15–19). The basic idea is to employ the isomorphism between the quantum mechanical system of particles (mass m) and a purely classical system consisting of a cyclic chain of "beads" (P beads) coupled by harmonic springs with a spring constant of $mP/\hbar^2\beta^2$ ($\beta = 1/k_BT$). This implies that at low temperature, the spring constant is small and the chains become relatively extended as would be suggested by the uncertainty principle. At high temperature, the spring constant becomes stiff and the chains shrink to a small size and behave like classical particles. Quantum fluctuations are described by motions of the beads. These additional calculations can be conveniently incorporated into a generalized AIMD CP molecular dynamics scheme (20–23). Static properties including quantum effects are evaluated from the time-averaged values. Quantum dynamical information may be obtained from the time evolution of the center of mass or centroid of the beads (22).

$$R_I^c(t) = \frac{1}{P}\sum_{s=1}^{P} R_I^s(t) \qquad 9.$$

The quasiclassical power spectrum is simply the Fourier transform of the centroid position-position time correlation function.

As mentioned above, there are no restrictions on the Hamiltonian or the functional form of basis sets in the implementation of AIMD. The most common implementation is based on DFT employing plane wave basis sets within the pseudopotential approximation (13). An appealing feature of this formulation is the ease of numerical calculations of various relevant integrals. The derivatives in real space are simply multiplications in reciprocal space and can be efficiently evaluated via fast Fourier transform (12). Moreover, plane waves are not atom-centered. This implies that the Pulay force vanishes exactly even within a finite basis set. This greatly facilitates the calculation of Hellman-Feynman forces. The delocalized nature of plane waves also eliminates basis-set superposition errors. One disadvantage, however, is the large number of plane waves required to represent localized orbitals, such as transition metal d orbitals or inner core levels. Other schemes, such as ultrasoft potential (24), curvilinear adaptive coordinate transformation (25), and the projected augmented-wave (PAW) method (26), have been developed to alleviate some of the potential problems. The PAW method developed by Blöchl (26, 27) goes beyond the pseudopotential approach and retains information about the all-electron calculation. AIMD has also been implemented within the Hartree-Fock formalism using conventional localized basis sets with and without explicit treatment of electron correlation effects (28–30). In view of the very demanding computation requirements, only a few demonstrative calculations were performed with the Hartree-Fock scheme. The majority of AIMD applications reported in the literature were performed with either the pseudopotential approximation or PAW method.

An advantage of the CP extended-variable Lagrangian is that new constraints, such as temperature and strain, can be incorporated into the energy expression in a straightforward manner. This feature is particularly useful in the calculation of the free energy of a chemical reaction. In a canonical (NVT) ensemble the free energy difference ΔG can be calculated from the work done by the mean force $\langle F^c(R, \lambda) \rangle$ by thermodynamic integration (31, 32):

$$\Delta G = \int_0^1 \frac{\partial E(\lambda)}{\partial \lambda} d\lambda = \int_0^1 \left\langle \frac{\partial E(\vec{R}, \lambda)}{\partial \lambda} \right\rangle_\lambda d\lambda = -\int_0^1 \langle F^c(\vec{R}, \lambda) \rangle_\lambda d\lambda. \quad 10.$$

In principle these statistical averages can be evaluated by any sampling technique such as Monte Carlo or molecular dynamics (MD). A convenient method for MD applications is the blue-moon ensemble method (33), in which the ensemble average is estimated by a time average over a constrained trajectory fixed at a specified value of λ. The reaction coordinate, λ, can be treated as a geometric constraint. Once it is defined, a reaction is allowed to evolve adiabatically from the initial ($\lambda = 0$) to the final state ($\lambda = 1$) by a smooth continual increment along the reaction coordinate. The conditional and constrained averages of the force are related by (33)

$$\langle F^c(\vec{R}, \lambda) \rangle = \frac{\langle Z^{-1/2} F^c(\vec{R}) \rangle_{\lambda'}}{\langle Z^{-1/2} \rangle_{\lambda'}}, \quad 11.$$

where

$$Z = \sum_i \frac{1}{m_i} \left(\frac{\partial \lambda}{\partial \vec{R}_i} \right)^2 \quad 12.$$

compensates for the bias introduced by the mechanical constraint on the reaction coordinate. Consequently the calculation of the free energy of a reaction is evaluation of the force of constraint countering the total force in the constrained dynamics and a geometric correction term. A full description of the theory and the relevant equations has been given in a recent paper (33).

The AIMD free-energy calculation has a significant advantage over ordinary quantum chemical approaches because entropic effects and anharmonicity are explicitly included. In contrast, a normal static DFT procedure requires that the vibrational entropy be added via harmonic frequency calculations. The computational demands may be quite substantial for large molecular systems. Furthermore, the harmonic approximation may not even be valid in situations where weak interactions are dominant.

The unified scheme of combining molecular dynamics and density functional theory proposed by Car & Parrinello (7) has had profound impacts on a broad spectrum of research disciplines. It is evident from the vast number of recent publications that this method has revolutionized computational applications in materials science, condensed matter physics, quantum chemistry, and biology. As already noted in the seminal 1985 paper (7), "the main advantage of the present approach lie[s] in its ability to perform a global minimization of the energy DF, and, more importantly, in offering a convenient and, in principle, exact tool for studying finite temperature effects and dynamical properties." The last comments of this statement are particularly significant for chemical applications. Temperature and dynamics are two very important parameters in chemical reactions. Although traditional quantum chemistry approaches can provide very accurate predictions of energetics and geometry, the temperature effects can only be included in an ad hoc manner. In view of the near exponential increase (12) of papers dealing with the applications of AIMD methods, a comprehensive survey, even within the past 3–5 years, would be an enormous undertaking. A fairly complete tabulation of earlier work has already been presented in the excellent monograph by Marx & Hutter (12). This review focuses on more recent applications, with emphasis on dynamical studies of chemical significance.

ISOLATED SYSTEMS

Among the first applications of CP molecular dynamics is the determination of metal cluster structures (34). For a given size, a cluster may adopt several energy minima. It is difficult, if not impossible, to make an educated guess about all the possible structures and explore their relative stability using conventional quantum chemistry electronic structure calculations. In combination with the stimulated annealing technique, the CP method is well-suited for this purpose. Finite temperature CP molecular dynamics has been used to explore the potential energy surface to search for possible candidate structures and act as an efficient geometry optimizer.

A few selected classes of molecules, owing to the presence of multiple minima with very low activation barriers, are highly fluxional, and the concept of molecular structure becomes problematic. The lack of well-defined molecular structures makes theoretical prediction of observables (e.g., vibrational and tunneling spectra) for direct comparison with experiments difficult. One such system is the protonated alkanes. The first member of this series is protonated methane, or methonium ion CH_5^+ (35). Converged highly correlated Hartee-Fock calculations suggest at least three low-energy structures with nearly identical energies: (*a*) $C_s(1)$ and (*b*) $C_s(2)$ structures with the H_2 moiety eclipsed or staggered with respect to the C–H bond of the CH_3 radical and (*c*) a C_{2v} structure with three coplanar C–H bonds (see Figure 1) (35, 36). If zero-point vibrations are taken into account, the energies of these three structures become almost identical, and in practice CH_5^+ does not have a definite equilibrium structure. The near degeneracy and dynamical exchange of protons complicated the vibrational spectra and hampered spectroscopic identification of this molecule. The theoretical work also indicates that the five protons in CH_5^+ are well bound to the central carbon and exchange rapidly among the minima. Therefore, CH_5^+ may not be described as a $CH_3^+ + H_2$ complex. More significantly, the theoretical results imply that the usual concept of three-center two-electron (3c-2e) bonding for hypercoordination carbocations may no longer be valid (37).

Owing to extensive proton-tunneling and zero-point motions, a theoretical characterization of those molecules requires the consideration of quantum dynamics. To this end, classical (38) and path-integral AIMD calculations (39, 40) have been performed. Classical simulations at low temperature (<500°K) show that clearly two types of C–H bonds are presented. As expected, the proton dynamics are highly temperature dependent. The protons execute large-amplitude motions indicative of rapid exchange. On time average, there is a tendency for two protons to pair up. In this regard, the classical view of a $CH_3^+ \ldots H_2$ complex still prevails. Simulations performed at temperatures above 500°K show that the intermolecular dynamics can be broadly described as protons moving on a sphere around the carbon. Path-integral simulations at 5°K show substantial quantum effects on the structure and proton exchanges (41) (Figure 2, see color insert). This effect, as expected, is most pronounced in the H-H radial distribution function (40). The peaks in the H-H distribution are broadened and overlapped significantly by quantum fluctuations, indicating tunneling between sites. On a configuration average, the chemical structure of CH_5^+ is still characterized as 3c-2e bonding between CH_3^+ and H_2 (Figure 2).

Although these theoretical results clarify the structure and dynamics of CH_5^+, it is still difficult to make direct comparisons with the observed infrared spectrum of CH_5^+, and the true identity of CH_5^+ is far from being settled (42–45). The theoretical conclusion seems to be in conflict with the mass spectrometry study of the isotopmers, which shows CH_4D^+ and CD_4H^+ to be exceptionally stable in the absence of intermolecular collisions. Recently Bunker and colleagues developed a quantitative theory by using a mechanical model in which CH_3^+ and H_2 are treated

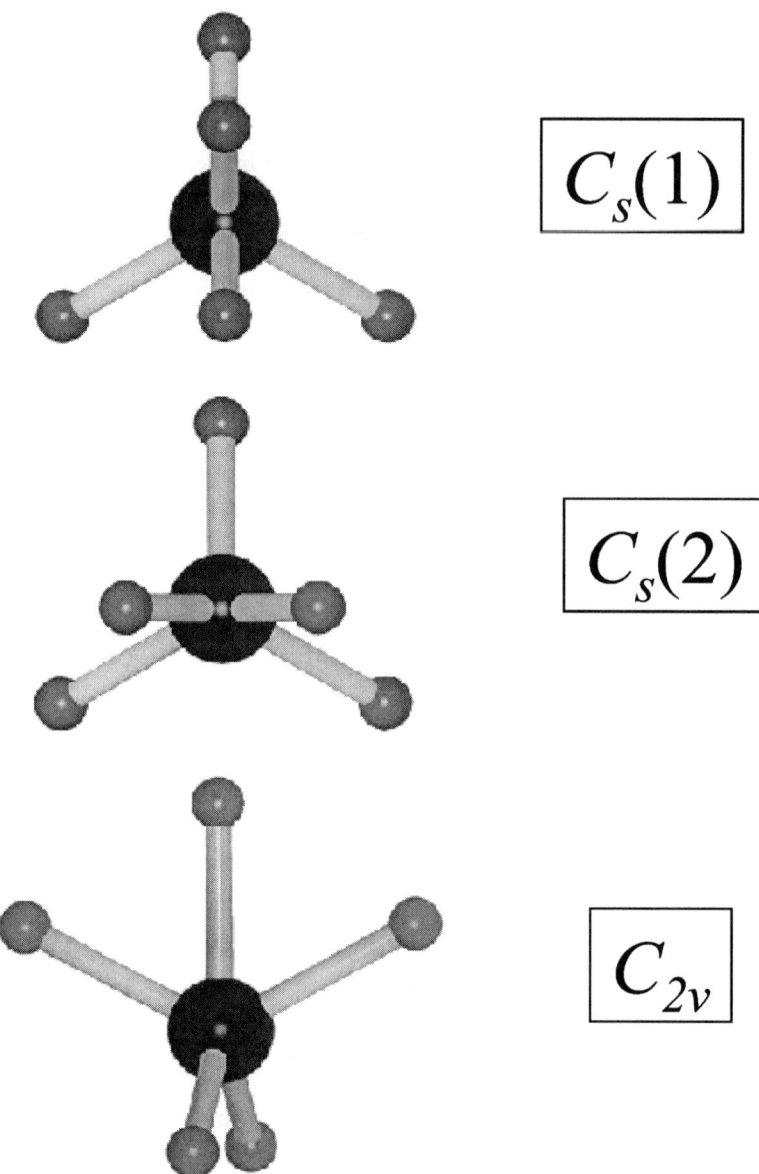

Figure 1 Three lowest energy conformations of CH_5^+ predicted by ab initio quantum chemistry calculations.

as separate moieties with full consideration of the symmetry of the scrambling protons (46, 47). It remains to be seen if such elaborate treatment is useful for assigning the complex high-resolution infrared spectrum (48) consisting of over a hundred lines.

The dynamics of CH_5^+ solvated by H_2 ($CH_5^+(H_2)_n$, n = 1, 3) has also been studied by classical AIMD (49). The results indicate an impediment of the proton scrambling motions in the core CH_5^+ as the amount of solvated H_2 increases. This suggestion is consistent with the experimental infrared spectra that show that as the amount of solvent H_2 increases, the C–H stretch vibrational bands become better resolved (50).

The structure and dynamics of protonated acetylene ($C_2H_3^+$) have been subjects of special interest. Quantum chemical calculations have established that the classical vinyl cation (Y-structure) and the nonclassical H-bridged structures are nearly equal in stability (51). The experimental determination of the equilibrium structure proves to be difficult. Although a bridge structure has been inferred from the analysis of the infrared spectrum (52), there is also strong evidence suggesting that the protons migrate between the apex and the two end equilibrium positions (53). A recent Coulomb explosion imaging study (54) suggests that the bridging H is delocalized and orbits about the C=C bond. This conclusion contradicts both proposed structural models. An initial classical AIMD calculation at 300°K showed that the three protons in $C_2H_3^+$ are very mobile and permute cyclically among the end carbon atoms and the bridging position (38). Subsequent AIMD calculations using different pseudopotentials, with and without quantum effects, showed different results (55). In these calculations the nonclassical H-bridged structure was still found to be more stable, but there is no thermally activated proton exchange (Figure 3, see color insert). In the quantum ground state at 5°K, the protons are mainly localized around the bridged configuration and exchange with the terminal protons through tunneling (Figure 3). The discrepancy between the two AIMD calculations can be attributed to a much lower calculated energy difference between the classical vinyl cation and the nonclassical H-bridged structure (1.2 kcal/mol versus 3.6 kcal/mol) by the former calculation. At 3000°K the basic description of the structure is somewhat similar to the classical proton exchange (38), but the anisotropic spatial distribution of the protons becomes even more significant. Interestingly, at quite high temperature the proton density profile shares some resemblance to the Coulomb explosion imaging (54, 55). This strongly suggests that Coulomb explosion imaging (54) did not probe the quantum ground state of $C_2H_3^+$, but did probe an ensemble of the ro-vibrationally excited molecule.

An interesting combination of AIMD and quantum chemical calculations was used in the characterization of the structure and stability of protonated ethane ($C_2H_7^+$) (56). $C_2H_7^+$ is akin to CH_5^+ but with a methyl group substituted for one of the H atoms. An exhaustive search of possible stationary point geometry by ab initio quantum chemical calculations has revealed three stable isomeric forms at low temperature: the nonclassical H-bridged isomer that features a 3c-2e CCH bond, an open CH_3-CH_4^+ isomer with a 3c-2e CHH bond, and a solvated-ion isomer with an H_2 group roughly 3 Å from the CC bond of ethane. All three isomers have many flexible rotational degrees of rotational freedom. Furthermore the open isomer is expected to exhibit fast H-exchange as in the case of CH_5^+. Dynamical AIMD calculations were performed and the infrared spectra of the three floppy isomers were computed (56). With the help of the simulation, a qualitative understanding of the experimental spectrum of Yeh et al. (57) is now possible, although the nature of all the observed features still defy explanation. The simulation showed that the solvated-ion isomer is stable under experimental conditions (rotational temperature $<40°K$) and the features that appeared in the calculated spectrum are in qualitative agreement with experiment. Therefore, the species observed in the Yeh et al. spectrum of $C_2H_7^+$ is likely to be that of the high-energy solvated-ion isomer. A cautionary note emerged for the application of DFT to floppy systems. Although the DFT potential surface has the essential features, a significant number of minimum or stationary structures located by high-level correlated quantum chemical calculations did not exist in the DFT calculations. The suitability of current DFT to these fluxional molecules needs to be critically reexamined.

AIMD has been used to study the influence of a knot on the strength of a polymer strand and the evolution of fragments formed at the rupture of a knotted polymer (58, 59). The presence of a knot is known to have a significant influence on the mechanical properties of polymers under tension. Relatively little is known about knots in polymer strands at the atomic level. In this study linear n-alkanes were used as model polymers. A "trefoil" or "overhand knot," the simplest kind of knot, was introduced by adding an appropriate number of gauche defects. The chain was subjected to uniaxial strain generated by pulling the two end atoms. During the elongation period the individual bonds underwent large-amplitude thermal fluctuation. The dissociation breaking occurred at a bond just outside the entrance to the knot. The breaking process is analogous to macroscopic knotted rope, where breaks under tension always begin near the entrance to the knot. The dynamical evolution after the rupture of the chain reveals some novel chemistry (59). The dissociation produces two radicals and a biradical. Instead of recombination forming an unknotted chain, the diradical forms a cyclic alkane. The remaining two radical fragments eventually interact and transfer a H from the longer fragment to the shorter one (see Figure 4). Subsequently the two terminal atoms of the longer fragment disproportionate to form a double bond.

Proton Sharing in Water Complexes

The quantum nature of the shared proton in the H-bonded complexes $H_5O_2^+$ and $H_3O_2^-$ at $300°K$ have been examined with classical and path-integral CPMD (60). High-level correlated quantum mechanical calculations show that the shared proton in $H_5O_2^+$ occupies a symmetric position. In contrast, the $H_3O_2^-$ was found to be an asymmetric complex with one covalent O–H bond $-[O–H \ldots H]^-$. Path-integral CPMD calculations reinforced the classical single potential symmetric distribution of the shared proton in $H_5O_2^+$. There is little difference between the

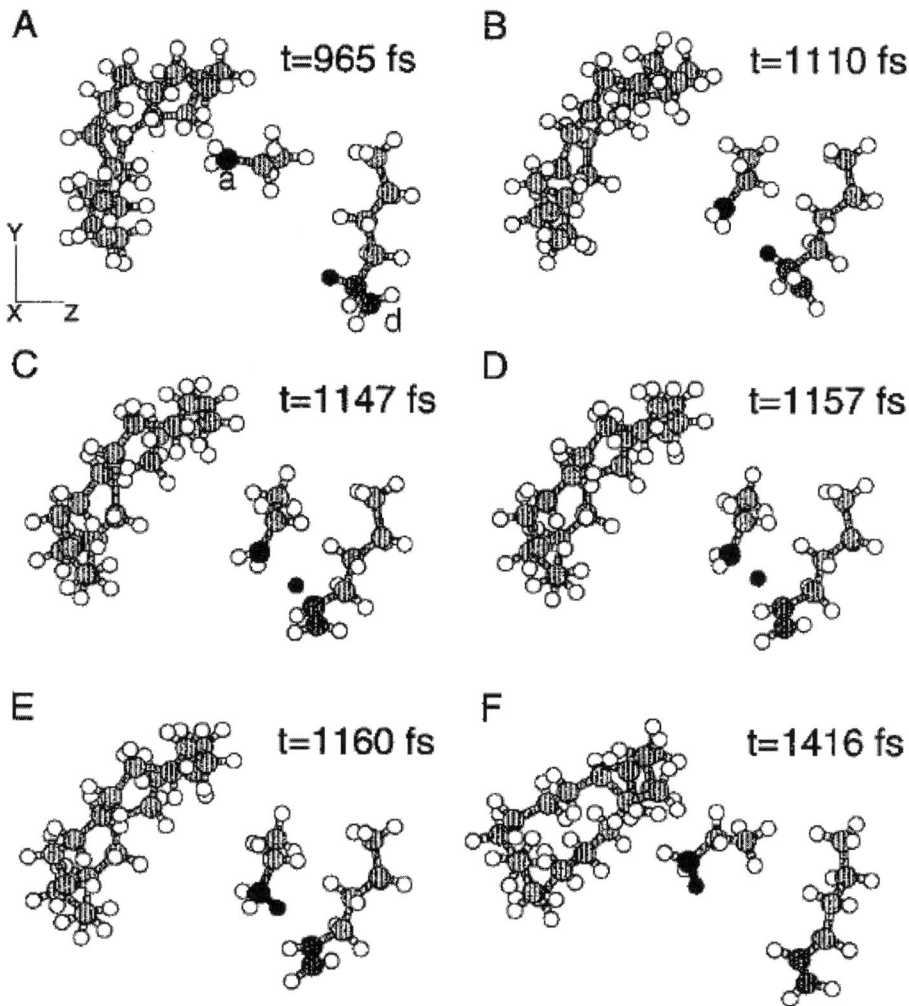

Figure 4 Temporal evolution of the polymer chains after the rupture of a knotted alkane molecule. (*A*) The formation of biradical pairs and a cyclic alkane. (*B, C, D*) Proton transfer between the two shorter chains. (*E, F*) Disproportional reaction and formation of C=C bonds in the two fragments.

structures obtained from classical and path-integral calculations (Figure 5*a*, see color insert). However, the shared proton in $H_3O_2^-$ shows pronounced quantum effects. Classical CPMD calculations show a double-peak structure with clustering of the shared proton at the positions within the standard O–H distance of the two oxygen end atoms (Figure 5*b*). This suggests that the proton moves on a potential with a double-well structure. The double-peak structure is completely washed out after the consideration of the quantum effects. The calculated effective free-energy profile shows a small classical barrier to proton transfer at 300°K that is removed in the quantum free-energy profile.

Proton Transfer in Malonaldehyde

Malonaldehyde has been the prototypical molecule for the study of proton transfer mechanisms with intermediate barriers (a few kcal/mol). Because the C=C and C–C=O π-conjugation patterns change during proton transfer, the effects of these heavy atoms may have a nonvanishing effect on the transfer rate. This problem was recently studied theoretically (61). Tunneling was shown to be dominant even at room temperature. The most significant result is that treating the migrating proton alone as a quantum particle is not sufficiently accurate. The calculated free-energy barrier is significantly higher than full quantum treatment of all the heavy atoms, leading to a discrepancy of a factor of two in the calculated transfer rate. An intricate interplay between the quantum effects in the the C–O bond rearrangement affects the tunneling process. Quantum and thermal fluctuation of double proton transfer in the formic acid dimer has been investigated wtth CPMD simulation (62).

The studies featured above clearly illustrated (*a*) that quantum effects are very important in protonated species and cannot be neglected and (*b*) that the quantum effects can be incorporated in dynamic DFT simulations, such as the CPMD methods, that yield sufficiently accurate structural and energetic information.

LIQUID AND SOLUTION

AIMD within the plane wave implementation is well-suited for the study of the electronic structure in the condensed phase. Periodic boundary conditions, which are explicitly imposed in the plane wave method, are most suitable to mimic an extended system. The DFT AIMD method therefore covers an important area that is not yet thoroughly studied using conventional quantum chemical techniques. There are many good examples of the application of the study of pure liquids and solutions, but the studies of chemical reactions in the liquid phase are perhaps the best for illustrating the potential of this method.

Liquid Sulfur

Molten sulfur is a very complex liquid. The structure of liquid sulfur is strongly dependent on temperature. Several liquid → liquid transitions have been characterized and many experiments have been performed to interpret the results. In spite of these efforts, the structure and dynamics of liquid sulfur is still speculative, and a complete understanding at the molecular level is lacking. Recently, AIMD has been used to predict the structure, vibrations, and dynamic behavior of liquid sulfur over a wide temperature range (63). The effect of photo-excitation was also studied. To avoid bias toward the final structure, two sets of calculations, one starting from random S_8 rings and the other from randomly positioned S atoms, were performed. After simulated annealing, the temperature of the system was raised gradually. The calculated radial distribution function at 500°K is in very good agreement with that observed, and the theoretical predicted structure is consistent with that obtained from reverse Monte Carlo analysis of the experimental data showing the S_8 rings are still intact. Upon heating and/or photo-excitation, the S_8 rings broke open, leading to entanglement of the ring fragments (Figure 6). At the highest temperature (~1400°K) there were no S_8 rings remaining, the system was composed of sulfur chains of various lengths ranging from S_2 to S_{11}, and there

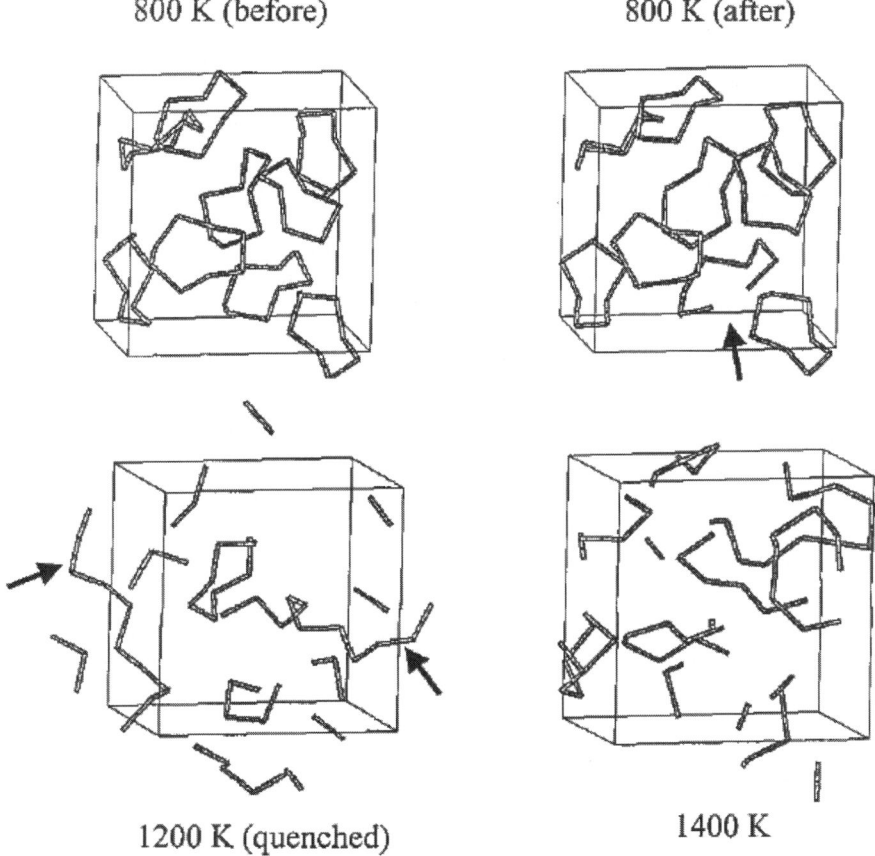

Figure 6 Instantaneous structure of liquid sulfur at different temperatures. The upper two frames indicate the cleavage (*arrow*) of a S_8 ring as a result of photo-excitation. The lower two frames show the fragmentation of the S_8 into longer-chain polymers and then into shorter-chain molecules at high temperature.

were signs of cross-linkages (onset of polymerization). The calculated trends in the vibrational spectra were also in good accord with those observed. AIMD calculations offer significant new insights into the molecular structure of liquid sulfur. The predicted sequences of structural transformations from molecular S_8 rings, to long and cross-linked polymeric chains, and finally to shorter S_n fragments are in good agreement with the proposals in the literature. The theoretical study helps resolve a long-standing suggestion that liquid sulfur starts to polymerize at a temperature above the λ-transition.

Liquid Water

Liquid water is arguably the most important solvent. It is the archetype of hydrogen bonding, an essential medium of all biology, a major component in geology, and a key reagent in the physics and chemistry of the atmosphere. There have been many intermolecular potential models proposed for water dimer interactions and numerous attempts to make simulations based on these models, but none of them are fully satisfactory. This is particularly true when many-body and polarization effects are neglected because they are the largest changes of the electronic property passing from the gas to the condensed phase. AIMD is a powerful tool to circumvent these limitations because all electronic effects are explicitly included.

In a series of papers Parrinello and colleagues (64–66) demonstrated that the structure, dynamics, and transport properties of liquid water can be simulated successfully from first-principles. In the latest calculation (66), employing 64 water molecules and a Becke, Lee, Yang, Parr (BLYP) gradient-corrected functional at 300°K and normal density (1 gm/cm^3), the agreement between observed and predicted atom radial distribution functions has improved significantly. More pleasingly, the calculated molecular diffusion coefficient of $2.8 \pm 0.5 \times 10^{-5}$ cm^2/sec is in substantial agreement with the experimental values of 2.4×10^{-5} cm^2/sec (67). One major issue in liquid water is the effective dipole moment of the water molecules. Owing to dipole-induced dipole interactions, the effective dipole moment of a water molecule in liquid water is larger than the free molecule value. The most commonly accepted value is 2.6 D obtained by Coulson & Eisenberg (68) on ice I_h. For the first time, this value can be estimated from first-principles theoretical calculation via the calculation of maximally localized Wannier functions—a generalization of the Boys localized orbitals (69, 70). Wannier functions represent very useful tools to study, at a quantitative level, the electronic charge distribution and the bond properties of water molecules in the liquid. To this end, a Wannier function center (WFC) is defined and the dipole of single water can be estimated, assuming that the electronic charge is concentrated in point charges located on the WFCs (70). The calculated average value is ∼3.0 D, somewhat higher than the previous accepted value. However, Coulson & Eisenberg's analysis has been refined recently using more accurate input, and a value of 3.09 D was obtained (71). The infrared absorption spectrum for liquid water has also been computed (72) from the time correlation of the total dipole moment that was evaluated from the electronic polarization using the Berry phase method (73, 74). AIMD circumvents the potential difficulty of modeling the dynamic electron polarization using effective potentials. All major features of the experimental spectrum are well reproduced by the simulation and related to specific motions of the water molecules (72).

Calculations of the structure and dipole moment of supercritical water at 647°K and 22.1 MPa and a density of 0.32 gm/cm^3, and at 635°K and 0.76 gm/cm^3 have been reported (75, 76). The computed atom pair–distribution function for a D_2O system at T = 635°K and 0.76 gm/cm^3 is in reasonably good agreement with experiment (75). Little is known about the structure of water in the supercritical region in a microscopic level. The simulation can provide unique insight into the structure. The salient feature of supercritical water is the presence of nonlinear H-bonds, especially cyclic H-bonds, bifurcated H-bonds, and twofold H-bonds, as in normal water. At very low density a large number of waters are unbounded and the structure is mainly composed of linear chains of dimers and trimers. The significant changes in the structure and H-bond pattern will have important consequences for the screening properties and reactivity of supercritical water. The calculated averaged dipole moment is 2.02 ± 0.14 D (75), which is only slightly larger than the free molecule value of 1.86 D. Once again the estimated diffusion constants of $46.2 \pm 0.6 \times 10^{-5}$ cm^2/sec at 0.73 gm/cm^{-3} and $103.5 \pm 2.1 \times 10^{-5}$ cm^2/sec at 0.32 gm/cm^{-3} are in agreement with experimental findings of 47.4×10^{-5} cm^2/sec at 0.73 gm/cm^{-3} and 112×10^{-5} cm^2/sec at 0.3 gm/cm^{-3} (77).

Water molecules in the liquid state are not static entities. Instead they dissociate spontaneously to produce hydroxium (H_3O^+) and hydroxyl (OH^-) ions. An understanding of the origin and thermodynamics of this reaction is fundamental to aqueous chemistry. Because the pH of liquid water at standard temperature and pressure is 7, the autoionization process is very rare: For a single water molecule, such an event occurs about every 11 hours. Atomistic dynamical simulation of this process in this timescale is not feasible. In an initial attempt to study the energy profile of this process, CPMD simulations with geometric constraint were performed (78, 79). The H–O distance of a randomly chosen water molecule in a model liquid system with 32 water molecules was chosen as the reaction coordinate. The O–H distance was allowed to evolve incrementally during the simulation from 1.1 to 1.8 Å. At each step the ensemble-averaged force owing to the constraint along the reaction coordinate was evaluated. The dissociation can be characterized by four stages: (*a*) predissociation intact molecule, (*b*) formation of constrained H_3O^+ and OH^- ion pair, (*c*) migration of H from H_3O^+ to adjacent water, and, (*d*) independently solvated H_3O^+ and OH^-. The calculated energy of the separated solvated ions is not stable and is about 17.5 kcal/mol higher than the liquid. Obviously, the mechanism for autoionization is more complicated than a simple proton transfer and therefore cannot be adequately described by a simple reaction coordinate.

According to a theory of Eigen & De Maeyer (80), the microscopic mechanism of autoionization is driven by rearrangement of solvating water molecules. Therefore, solvent fluctuation should play an important role. They also proposed the formation of a transition-state solvated ion pair separated by ~6 Å (80, 81). To explore this possibility and to overcome the timescale, CPMD calculations combined with transition path sampling were performed (82). Unlike conventional methods for barrier crossing, transition path sampling does not require the specification of a reaction coordinate. The sampling was based on a Monte Carlo scheme (83) requiring that the trajectory leads from neutral water to a separated ion pair. Analysis of the trajectories reveals details of the mechanism for autoionization (Figure 7, see color insert). In the first step, the cleavage of the O–H bond of a water molecule was due to fluctuation of the electric field of the solvent. The unbound proton migrated to an adjacent water molecule, forming a transient H_3O ion. Within 30 fs, the proton on the hydroxium ion jumped to the next hydrogen-bonded water such that the ion pair was no longer the nearest neighbor. These consecutive proton shifts, or structural diffusion, was first suggested by Grotthuss (84). The fluctuating electric field of the solvent continued to push the ions apart through rapid successive proton transfer along a path of connected H-bonds. At the transition state, the hydrogen-bonded path connecting the OH^- and H_3O^+ ions was broken through the reorganization of H-bonds. At this point the maximum separation between the ions was estimated to be ~ 10 Å. The energy of the dissociation of water in the liquid was calculated with the potential of protons in the connected H-bond path for a chosen trajectory with other degrees of freedom being held fixed.

The results show that the fluctuating electric field mediated and lowered the activation energy and stabilized the formation of the ions. For the first time, theoretical calculations have provided an atomistic mechanism for the autoionization of water. The calculations show the importance of the collaborative nature and the solvent polarization effects. This work clearly demonstrated the power and potential of the CPMD when combined with the appropriate statistical mechanics methods. Microscopic insights cannot be otherwise achieved with current conventional quantum chemistry approaches.

Acid and Superacid

The proton mobility in acidic solutions is anomalously high compared with common ions. Studies of this novel phenomenon have a very history. Grotthuss (84) postulated more than 200 years ago that the proton diffusion might occur through continual hopping between water molecules. More recently, it was recognized that protons may be chemically bonded to water, forming a structural defect. Thus, proton migration in fact involves the diffusion of the structure defect. Two structural models have been proposed: the Eigen ion $H_9O_4^+$ (85), in which the H_3O^+ core donates three H-bonds to neighboring water molecules, and the Zundel ion $H_5O_2^+$ (86), in which the proton is shared by two water molecules. The true nature of hydrated excess protons in water has recently been uncovered by CPMD calculations with and without quantum corrections (87–89). The study showed that proton migration in liquid water is a highly cooperative process with the solvent involving both the $H_9O_4^+$ and $H_5O_2^+$ structure. This process can be envisaged as follows. Starting from the $H_9O_4^+$ structure defect, one of the three protons of the H_3O^+ core migrates along its H-bond and forms a $H_5O_2^+$ defect when it reaches the midpoint. When the proton transfer is completed another $H_9O_4^+$ is formed. Interestingly, this mechanism does not require substantial displacement of the participating atoms even though the acidic site has shifted by twice the water-water distance, or ~5 Å. A cautious note is warranted here. Because the proton shift is a very fast process, in the timescale of 30 ps, the $H_9O_4^+$ and $H_5O_2^+$ ions are very mobile and short-lived and may only be regarded as transient species. It is arguable whether the conventional concept of a static structure can be applied to these fluxional entities. An important observation that emerged from the quantum path integral calculation is that the proton transfer need not be facilitated by tunneling through the barrier and/or soliton-mediated coherent transfer. The proton transfer barrier is mediated and lowered by the electric field of the solvent. The zero-point vibration of the active proton already has sufficient energy to shift the lowest vibrational state toward the top the classical barrier.

Transfer of excess proton in linear water chains has also been studied with classical and path integral CPMD (90). The most interesting observation is that when quantum effects are included, the instantaneous path integral configurations reveal an extended $H_7O_3^+$ complex in a pentamer chain. This is in contrast to the prediction from classical CPMD, which favored the formation of H_3O^+. This result serves as a warning about the application convention approach to the study of proton transfer, in which delocalization of protons owing to quantum effects are not included.

The structure and dynamics of protons in aqueous hydrochloric acid (91), anhydrous liquid (92), and other superacids have been studied using the CPMD method. The salient feature of the structure in hydrochloric acid is that the HCl dissociates readily into solvated Cl^- and hydroxonium ions (91). However, at high concentration, with an acid:water ratio of 1:3.1, the solvated ions break down with the formation of a hydrogen-bonded Cl–H ... Cl^-. This differs from the classical textbook description of the acid. In HF short zigzag chains of hydrogen-bonded HF were found (92). The calculated diffusion constant of 8×10^{-5} cm^2/sec is in excellent agreement with experiment (93). The shortening of the H–F bond from the vapor to the condensed phase was confirmed. Kim & Klein (94) studied the structure of an excess proton in liquid HF, a prototype for superacids. Protonated HF chains with H_2F^+ and $H_3F_2^+$ structures were identified. Furthermore, proton transport in HF follows a similar structure diffusion mechanism as observed in liquid water. Other related systems, such as KF/HF(95) and SbF_5/HF (96) solutions have also been studied.

Other Systems

CPMD methods are most useful in studying solution structure when there is strong coupling between solute and solvent. One important area is the solvation of cation in aqueous solution. Molecular dynamics (MD) calculations using empirical potentials have met with varying degree of success. Recently, CPMD has been applied to investigate the solvation of Li^+ (97), K^+ (98), and Mg^{2+} (99) in water where there is strong coupling between solute and solvent. In general the structures predicted by the calculations are in good agreement with experiment. In the case of solvated Li^+ the ab initio results were used to derive a Li^+-water interaction potential using an inverse Monte Carlo technique (97). The fitted potential reproduces the liquid structure very well. It is well known that available empirical interaction potentials often fail to accurately describe the structure of carbohydrates. CPMD has been performed to investigate the structural aspects of anomeric equilibrium of glucose in aqueous solution (100). The structural behavior of an azide anion in water has also been studied (101). The role of H-bonds in the surprisingly fast rotational dynamics of NH_4^+ ions in liquid water has been reported (102). AIMD calculations reproduced the measured nuclear magnetic resonance (NMR) relaxation time and provided a microscopic dynamical model to account for the restricted rotational motions of NH_4^+ in water.

CHEMICAL REACTIONS

Direct Approach

Quantum mechanical study of the mechanism of a chemical reaction with or without prior knowledge of the products often starts with an initial guess of a reaction coordinate. In polyatomic or complex systems sometimes the choice of a simple reaction coordinate, such as bond length or bond angles, may not be obvious or even feasible. A straightforward but perhaps somewhat naive approach is to use the CPMD to search for the potential energy surface and to probe for possible reaction channels. This approach has been used to study the thermal dissociation of acetic acid (103). Interestingly, even with a rather limited number of trial trajectories, the three dominant dissociation channels observed in experiments were identified and are in qualitative agreement with transition state theory. These results, however, may not be statistically meaningful. Nevertheless, they illustrate that the AIMD-based method can be a powerful tool in identifying possible reaction pathways in priori, on a first-principle basis. A subsequent more elaborate study reported the channels for unimolecular thermal decomposition of a more complicated NTO (5-nitro-2,4-dihydro-3H-1,2,4,triazol-3-one) molecule (104):

A larger number (up to 200) of trajectories were investigated. All the major channels suggested in previous works were identified. In particular, the C-NO$_2$ dissociation was found to be most favorable at high temperature. The frequency of occurrence of this mechanism from randomly chosen initial conditions is significantly higher than the other competitive routes even though the sampling size is still not of statistical significance. One of the most important results from this study is that several low-energy reaction channels were revealed from the energy search. These reaction channels were neither intuitively obvious nor discovered from careful extensive systematic search of the reaction using a standard quantum chemistry approach (104, 105). For instance, at low temperature the most probable dissociation pathway is through a two-step process involving a ring opening at the C3–N2 bond followed by the cleavage of the C5–N6 bond and the loss of HONO.

A lesson learned here is that the straightforward application of AIMD can be very useful in exploring reaction pathways that otherwise may not be apparent. This strategy perhaps is best for the study of unimolecular decomposition of complex molecules where the lack of product information hindered the initial guess of possible reaction coordinates.

Dynamic Simulation of Chemical Reactions

One of the early applications of AIMD was the study of transition metal catalysis reactions. Ziegler and his co-workers, applying the projected augmented-wave–CPMD method have studied the reactions of olefin polymerization catalyst X-Cp-Zr-R$^+$ with R=H and SiH$_2$NH, R=CH$_3$, C$_2$H$_5$, or Cl, to chain termination and long-chain branching processes (106–109). They studied several possible reaction mechanisms with well-defined reaction coordinates. The free energy of a reaction was computed with a "slow-growth" procedure (106). In essence, the free energy difference ΔG between two stationary points defined by $\lambda = 0$ and $\lambda = 1$, where λ is a linear parameter representing any path connecting the precursor and the product along a specific reaction coordinate, is evaluated by thermodynamic integration of the constraint force (Lagrangian multiplier) with λ changing continuously. With this slow-growth technique, the mean value may be an approximate for the ensemble average as long as λ is scanned slowly. This method differs slightly from the blue-moon strategy (33) that the true ensemble average is not evaluated and the geometric correction to the free energy is neglected (31). Nevertheless, the free-energy profile calculated in this manner includes the enthaply and the vibrational entropy. This leads to better agreement with experiment in the calculated activation barrier. Because the instantaneous atomic dynamics along the reaction path are sampled, these quantities can be examined a posterori, which leads to a better understanding of the cooperative effects of "spectator" atoms and provides unique insight into the details of the reaction mechanism.

These results clearly demonstrate that AIMD calculations complement and corroborate results obtained from the conventional static approach. Moreover, the MD method provides an efficient means to explore important regions of phase space for processes that possess many reaction channels. It should be noted that the reaction barrier obtained from thermodynamic integration is dependent on the choice of the reaction coordinate. The selection of the appropriate reaction coordinate for a chemical reaction is not often trivial or unambiguous. This is particularly true in complex systems with many degrees of freedom, such as chemical reactions in solution. Recently the slow-growth AIMD method has been adapted to perform simulation along a predetermined intrinsic reaction path (110). The intrinsic reaction path method of Fukui (111), by definition, connects the reactants with the products through a transition state along the bottom of the valley along the reaction path.

Hybrid Quantum Mechanics–Molecular Mechanics Methods

Recently Ziegler and colleagues implemented a combined quantum mechanics–molecular mechanics (QM/MM) methodology in conjunction with Car-Parinello (CP) MD to study the effects of bulky ligands on the activation barrier of an extended system (106). In this approach the QM, via the PAW method, was applied to the atoms directly involved in the chemical reaction; the environmental effects owing to surrounding ligands are described by the AMBER95 force field. The initial application to a Brookhart polymerization reaction of $(ArN=C(R)-C(R)=NAr)Ni(II)-R'^+$, where $R = Me$ and $Ar = 2,6-C_6H_3(i-Pr)_2$ catalyzed by a Ni(II) dimine at ambient temperature. It is shown that the energetics and free-energy profiles obtained from QM/MM and pure QM without bulky ligands are qualitatively and quantitatively different. The inclusion of nonbonded environmental effects is crucial to the characterization of the transition state and reaction mechanism. Moreover, the combined QM/MM simulation provides a free-energy barrier that is in excellent agreement with experiment.

More recently the combined QM/MM method has been extended to simulation of solution (112). In solvation reaction, there is a natural separation of the QM/MM regions. The solute is described by QM and the interactions between solvent, by MM. To account for the polarization effects, the coupling of the QM/MM electrostatic interactions must be treated properly. To the first approximation, only the long-range Coulomb interactions are considered. In this scheme, the charges on the solvent are fixed and are chosen either from the tabulation of well-known force fields or derived from fitting to reproduce results of reference calculations. The charges on the solute, however, were determined using the charge isolation scheme from the PAW wave function (113). In this method the charge density in the periodic simulation cell is approximated by a linear combination of Gaussian charges entered on each QM atom. Usually 3–4 Gaussians are needed. The parameters determining the Gaussian functions are determined by a fitting procedure such that the electrostatic multiple moments of the true density in the simulation cell are reproduced. This fitting procedure is very efficient for a plane wave basis set (113). The polarization effects of the MM solvent are incorporated into the CP calculation as an additional force term owing to electrostatic solute-solvent interactions. Recall the CP equation of motion,

$$\mu \ddot{c}_{l,k} = -\frac{\partial E}{\partial c_{l,k}} - \sum_j \lambda_{i,j} c_{l,k} \langle \Psi_i | \Psi_k \rangle, \qquad 13.$$

and that the force owing to electrostatic interactions with a MM point charge, q_j, for each Gaussian on QM atoms is

$$\frac{\partial E_{l,n}}{\partial Q_{l,n}} = \frac{1}{4\pi\varepsilon} \sum_j^{N_{MM}} \frac{q_j}{|\vec{r}_j - \vec{R}_I|} \, erf\left(\frac{|\vec{r}_j - \vec{R}_I|}{\varsigma_n}\right). \qquad 14.$$

This interaction can be substituted into the equation of motion,

$$\frac{\partial E}{\partial c_{i,k}} = \frac{\partial E}{\partial Q_I} \frac{\partial Q_I}{\partial c_{i,k}}. \qquad 15.$$

Because the Gaussian charge is atom-centered, a Pulay correction term must be applied to ensure the total energy is consistent with the gradients. This coupling scheme for QM/MM electrostatic interactions can be conveniently incorporated into the dynamical equations of CP dynamics. It has also been shown that the PAW QM/MM coupling scheme is consistent with energy conserving MD simulations.

To evaluate the quality of the PAW QM/MM coupling scheme, test calculations were performed on the interaction potential between a Li^+ cation with a water molecule (112). The water is treated as the QM region and the Li^+ is treated as a classical MM atom with unit charge ($+1\ e$). In the initial calculations the Li^+ ... water interactions were computed with the AMBER95 van der Waal parameter for Li and the TIP3P potential for water. This interaction is added to the quantum mechanical energy of the water owing to the presence of the Li^+. The results are compared with the reference pure density functional theory (DFT) calculation in Figure 8. At a long distance the PAW/QM/MM potential compares well with the reference potential. At close range the deviation becomes more significant. In particular both the shape of the repulsive region and the energy minimum of the reference potential are not very well reproduced. Because the long-range potential, dominated by electrostatic interactions, is consistent with the reference calculation, only the short-range interactions need to be modified. To this end, the

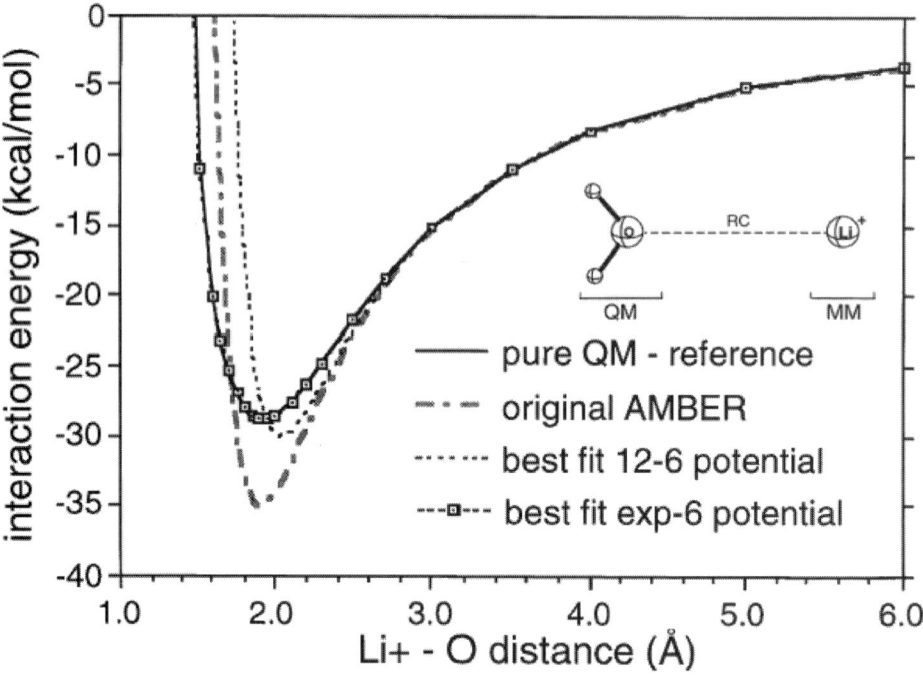

Figure 8 Comparison of the Li$^+$-water interaction potential for various hybrid projected augmented-wave/quantum mechanics (QM)/molecular mechanics (MM) potentials. The reference QM potential is obtained from density functional theory calculations with Born-Oppenheimer 86 functional. RC, reaction coordinates.

van der Waals interaction parameters and the functional form of the potential for Li$^+$ are optimized until a reasonable fit to the reference potential is obtained. As shown in Figure 8, the best agreement is achieved with an exp-6 potential. This potential model in combination with PAW QM/MM simulation may be useful to study the solvated Li$^+$ in liquid water. It is interesting to compare the application of this method with ab initio CPMD calculations in which all the atoms are treated with QM. Through use of a similar procedure, a set of optimal van der Waals parameters were derived by fitting the PAW QM/MM potential to the reference calculations on 27 H-bonded bimolecular interactions. The quality of the fit is reflected in the very small root-mean-square deviation in the 27 interaction energies of only 0.98 kcal/mol (112). To further improve the model, an empirical scaling factor reducing the electrostatic energy between the QM wave function and the MM charges by 95% was introduced. The final fit further reduced the root-mean-square error to a remarkable 0.70 kcal/mol. The PAW QM/MM method provides an alternate computational scheme for the study of solvent effects or chemical reactions in the condensed phase.

The method described above for coupling the QM/MM interface region is not well suited for systems in which the QM and MM fragments are linked by covalent bonds. The usual linked-atom method, that is, the replacement of a covalently bonded atom in the MM fragment by an artificial terminating atom for the purpose of saturating the chemical valence of the connecting atom of the QM fragment, is often not appropriate. This approximation is particularly problematic in dynamical simulation. The replacement atom perturbs the wave function of the QM fragment near the vicinity of the replaced atom and correspondingly changes the forces acting on the nearby nuclei. These will have a drastic effect on the dynamics of the system. A remedy to eliminate these artifacts within the CPMD scheme has been suggested (114). Instead of fixing the valence angle and bond length of the linked-atom pair, they are allowed to vary according to the local bonding environment. This is achieved by constraining the linked atom along the original bond direction. Instead of setting the artificial linked-atom bond length arbitrary to some preselected standard values according to the force field, it is scaled to the ratio of the original, and the linked-atom force constants. For instance, the bond length of an artificial r_{C-H} bond, in which a C–C bond of bond length r_{CC} that has been replaced by a linked H atom can be approximated as

$$r_{CH} = r_{CH}^0 + \frac{k_{CC}}{k_{CH}}(r_{CC} - r_{CC}^0), \qquad 16.$$

where k_{CH} and k_{CC} are the C–H and C–C force constants, respectively. r_{CH}^0 and r_{CC}^0 are the standard equilibrium bond lengths and r_{CC} is the C–C bond distance in the system. Furthermore, correction terms associated with the change in stretching and bending energies and dipole to the external potential and also with the partial charges of nonbonded atoms need to be added. This "scaled position linked-atom method" (114) compensates for the changes in the nature of chemical bonding between the original and the linkage atom. Few results have been reported that used this improved QM/MM scheme. Their applications to realistic systems are keenly anticipated.

Reactions in Solution

The ideal scheme for the study of reactions in solution is to treat the electronic structure of both the solvent and the solute with quantum mechanics. At present it is only computationally feasible for small solute molecules and a limited number of solvent molecules enclosed in a periodic replicated simulation box to mimic the extended system effects. The first application that explicitly includes the solvent molecules in the CPMD scheme is the investigation of the kinetics and thermodynamics of acid-catalyzed polymerization in liquid 1,3,5-trioxane (115). This is also the first calculation of free-energy profile for a chemical reaction in a condensed phase system using the blue-moon ensemble methodology mentioned above. The initial reaction involves a very rapid acid-catalyzed opening of 1,3,5-trioxane followed by decomposition into formaldehyde and methyenic carbocation (reactions 1–3).

$$R^+ + \underset{\text{1,3,5-trioxane}}{\text{(trioxane ring)}} \rightleftharpoons R-O^+\text{(trioxane ring)} \quad (1)$$

$$R-O^+\text{(trioxane ring)} \rightleftharpoons ROCH_2OCH_2OCH_2^+ \quad (2)$$

$$H^+ + \text{(trioxane)} \rightleftharpoons H-O^+\text{(trioxane)} \rightleftharpoons HOCH_2^+ + 2O{=}CH_2 \quad (3)$$

CPMD simulation of this reaction started from a model system consisting of seven well-equilibrated 1,3,5-trioxanes with a H^+ placed near the acetal oxygen of one of the molecules. The fragmentation of the 1,3,5-trioxane ring into two formaldehydes and a methylenic carbocation coordinated to another 1,3,5-trioxane occurred almost instantaneously. This observation concurs with the very fast kinetics determined from experiment. Through rapid succession of proton transfers and further cleavage of CO bonds, three other 1,3,5-trioxanes decomposed into formaldehydes. As the concentration of formaldehyde increases, the second phase of the reaction proceeds with the insertion of formaldehyde to protonated 1,3,5-trioxane followed by breaking of the ring structure and initiation of the growth of the polymer.

$$O{=}CH_2 + H-O^+\text{(trioxane)} \rightleftharpoons H^+ + \text{(8-membered ring)} \quad (4)$$

The 1,3,5-trioxane-formaldehyde solution was found to be inhomogeneous. The formaldehyde clustered around the $HOCH_2^+$. Subsequently, excess formaldehyde may interact forming dimers or oligermize into $H(OCH_2)_nOCH_2^+$ polymers by further reaction with the methylenic carbocation. An important reaction governing the kinetics of the polymerization process is the competing reverse degradation reaction,

$$H(OCH_2)_nOCH_2^+ \leftrightarrow H(OCH_2)_{n-1}OCH_2^+ + O=CH_2. \qquad 5.$$

The free energies of dissociation of a $HOCH_2OCH_2^+$ dimer into OCH_2 and $HOCH_2^+$ monomers in the gas phase and in the liquid were determined with constrained dynamics. In both cases it was found that the dissociation is without barriers. The major difference is that the free energy of reaction ($\sim$10 kJ/mole) in the liquid state is much smaller than in the gas phase ($\sim$30 kJ/mol). Examination of the temporal structures indicated that the dissociation process in the liquid is much more complex and the solvent plays an important role in the mechanism and kinetics of the reaction. The reduction of the free energy of formation is due to a more efficient solvation effect for the $HOCH_2^+$ and the screening of the effective attraction between the two fragments by the intervening formaldehyde. The acid-catalyzed polymerization in 1,3,5-trioxane is a complicated multi-step process, and it illustrates the importance of explicitly treating the solvent quantum mechanically and the potential advantages and problems associated with the application of AIMD condensed phase reactions.

The importance of the cooperative effect of the solvent is also demonstrated in a study of the reaction of water and formaldehyde in sulphric acid—a three-component system (116). The protonation of formaldehyde to the carbonyl oxygen, either by a H_3O^+ ion or HSO_4^-, led to simultaneous nucleophilic attachment by a water yielding the diol product after deprotonation.

$$H_2O + \text{>C=O} + H^+ \xrightleftharpoons{a} H_2O + {}^+\text{C--OH} \quad \mathbf{3}$$

$$\xrightleftharpoons{b} H_2O^+\text{--C--OH} \xrightleftharpoons{c} H^+ + HO\text{--C--OH} \quad \mathbf{4}$$

Ab initio calculations of the dissociation of Cl_2 in water also demonstrate the importance of the solvent cooperative effect. In contrast to gas phase cluster calculations, the dissociation has no energy barrier and only involves one water molecule. The ability to realistically treat the changes in electronic structures associated with bond breaking and formation processes at finite temperature help advance our understanding of the microscopic details of reaction mechanisms.

Reaction in the Excited State

The extended variables Lagrangian method of CPMD has recently been extended to the approximated calculation of low-spin excited states (117). This attempt may open up a new way to investigate the excited state molecular dynamics of photo-induced chemical reactions. This approach is based on Ziegler's sum method (118). It was shown that the energy for the lowest singlet excited state, S, can be computed from the energies of the pure triplet (T) and mixed-spin (M) wave functions.

$$E(S) = 2E(M) - E(T) \qquad 17.$$

However, instead of calculating the energy for the different spin states, if one assumes a common set of spin-restricted orbital $\{\psi(r)\}$ for both the triplet and mixed-spin wave functions, the energy expression can be approximated as

$$E(S) = 2\langle \psi_M(\vec{r})|H|\psi_M(\vec{r})\rangle - \langle \psi_T(\vec{r})|H|\psi T(\vec{r})\rangle. \qquad 18.$$

According to this ansatz, the triplet and mixed-spin wave functions only differ by the spin of one orbital and therefore, the total density for both the mixed and triplet states must be identical. The energy of the singlet state, S, can then be obtained with variational density functional theory subjected to orthogonalization of the orbital wave functions.

$$H(\psi(\vec{r})) = 2\langle \psi_M(\vec{r})|H|\psi_M(\vec{r})\rangle - \langle \psi_T(\vec{r})|H|\psi_T(\vec{r})\rangle$$
$$- \sum_{i,j=1}^{n+1} \varepsilon_{ij} \left\{ \int \psi_i^*(\vec{r})\psi_j(\vec{r})dr - \delta_{ij} \right\} \qquad 19.$$

To simplify the calculation, the Hamiltonian H is approximated by

$$\langle \Gamma|H|\Gamma\rangle = T[\rho] + J[\rho] + \int v(r)\rho r(r)dr - E_{xc}[\rho_\Gamma^\alpha, \rho_\Gamma^\beta], \qquad 20.$$

where Γ is the spin label $= M$, or T, $T[\rho]$ is the kinetic energy function, $J[\rho]$ is the Coulomb energy, $\rho = \rho_M^\alpha + \rho_M^\beta = \rho_T^\alpha + \rho_T^\beta$ is the total density, $v(\rho)$ is the external potential, and E_{xc} is the exchange-correlation energy.

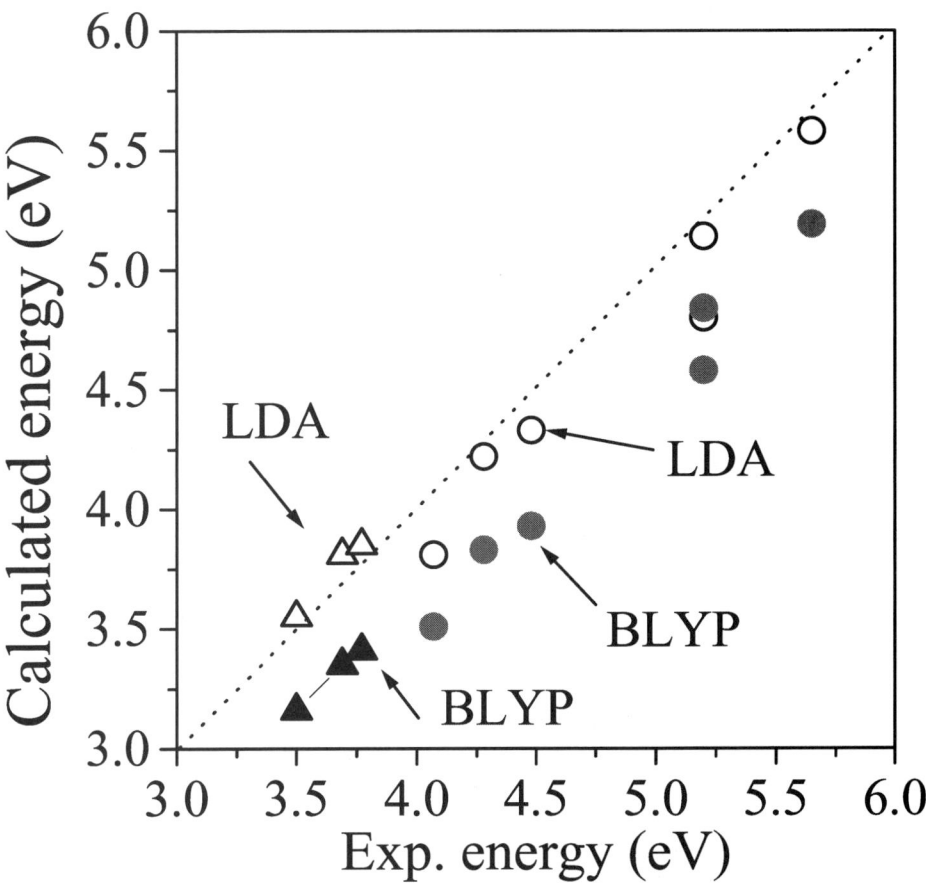

Figure 9 Comparison of calculated and observed low-spin singlet excitation energy for a selected set of organic molecules. *Open symbols*, LDA results; *filled symbols*, BLYP results; *circles*, vertical excitations; *triangles*, adiabatic transitions.

A test of this procedure on a number of organic molecules has yielded reasonable geometries and $S_0 \to S_1$ excitation energies (see Figure 9). The calculated excitation energies using the BLYP or local density approximation (LDA) functional are consistently lower than the observed values. The calculated adiabatic excitation energies are closer to the experiment than the vertical excitations. It is apparent that the LDA results are somewhat better than the BLYP results. A systematic underestimation of the excitation energy as compared with experiment has also been reported with the time-dependent density functional theory (TDDFT) method (119).

The excited-state CPMD has been applied to the photo-isomeriztion reaction of formaldine (117). The calculated reaction mechanism and energetics are in general agreement with more accurate multi-reference doubles configuration interactions (MRDCI) calculations. An additional insight from the study of the molecular dynamics is that the isomerization process involves the participation of all atoms in the molecule and not simply the proton being transferred.

The kinetic, mechanism, molecular dynamics and energetics for the photoisomerization of *cis*-11 to *trans* in the chromophore of rhodopsin have been subjects of intense theoretical and experimental investigations. The rhodopsin chromophore consists of a retinal covalently bound to the apoprotein opsin via a protonated Schiff base. The origin of the high selectivity, high quantum yield, and very low energy barrier are still unsolved problems. CPMD has been applied to the study of the molecular dynamics and energetics in both the ground state and lowest-singlet excited state for the isomerization in neutral and protonated retinal models (120). The calculation showed that the activation energy in protonated retinal is much higher than the experimental measurement. Furthermore, details of the mechanism also differ from a complete active space self-consistent field (sa-CASSCF) calculation (121, 122). It was then suggested that the protein environment effect would play a unique role in catalyzing the photo-isomerizaton of the retinal chemomphore. It should be noted that the nonadiabatic process was not incorporated into the present CPMD scheme. This effect may be included via the state-hopping method within the framework of molecular dynamics (123).

BIOLOGICAL SYSTEMS

In recent years, there has been an increased use of quantum mechanical calculations to elucidate the properties and functions of biological systems. Given the typical size of biological systems, a majority of studies were performed on model systems and, very often, with density functional theory (DFT) methods. In view of the subtlety of ligands and environment to the geometry and electronic properties, the use of hybrid QM/MM methods is becoming more important. Although a great deal of progress has been made in the past years, the development of a seamless scheme for linking QM and MM remains a challenge. There are a number of fundamental problems that need to be resolved (124). A full ab initio approach to biomolecular systems is still elusive, at least for the foreseeable future. The application of DFT to biological-relevant metal centers has been reviewed (125). An excellent overview of the state-of-the-art on the application of DFT-based molecular dynamics in computational biology has been published recently (126).

Early applications of dynamical calculations using DFT were mainly on the geometry optimization of key biomolecules, such as the structure of DNA base pairs *in vaccuo* (127), the ground and excited ($S_z = 0$) crystal structure of methyl bacteriophorbide (128), the structure of hydrated RNA duplex (129), platinum modified nucleotase pairs (130), and self-assembled peptide nanotubes (131). These calculations help benchmark and establish the performance of density function theory, in particular the plane wave–based method, to large bio-systems. A noteworthy example illustrating the usefulness of ab initio studies is the recent full ab initio calculation of the structure of a small protein 1PNH (Figure 10, see color insert). 1PNH is a scorpion toxin consists of 31 residues (approximately 500 atoms) (126). The structure of the protein in its standard +6 charge state was first optimized *in vaccuo* starting from the geometry determined from NMR measurements. The calculated root-mean-square displacement (RMSD) of the computed structure from the experimental structure was only 0.181 Å/atom. Analysis of the structure showed that the RMSD for the C, N, O, and H atoms was very reasonable (~0.1–0.2 Å/atom) except for the S atoms, for which the RMSD of 0.567 Å/atom is unexpectedly high. In fact, this discrepancy was not due to inadequacy of the theory, but was ascribed to the lack of accuracy in the parameterization of the force field for $-$S-S-linkage, which was used in the structural refinement from the NMR results. DFT calculations can be used to complement the structural refinement of protein structures.

The plane wave–based DFT method becomes computationally more efficient than traditional localized basis set methods when the number of atoms in the system is very large. Static calculations have been performed to study the structure and electronic properties of iron [Fe(II) and Fe(III)]–sulfur bonding in heme protein cyctochrome c (132), *cis*-Platin, and platinum-modified nucleobase pairs in the solid state (133), and the interactions between *cis*-Platin and DNA (134) and steric effects in Co-corrin macrocyclic Co-corrole, Co-corrin, and Co-prophyrin (135). Early dynamical studies of biomolecules were mainly performed on systems of limited size to model the active sites. AIMD simulations have been carried out for model systems representing the reverse transciptase triphosphate binding site in HIV-1 (136). The conformation transformation of an alanine dipeptide analog was found to occur within the picosecond timescale (137). These results contradicted the classical force field calculations, which fail to yield any observable transformation even for nanosecond simulations. This serious discrepancy certainly creates a challenge to develop new force fields. The geometry, electronic structure, and vibrational spectra of the chromophore of green fluorescent protein in the four protonated states have been reported (138). Dynamical studies reveal an intramolecular mode-coupling mechanism, coupling of the electronic excitation to the low frequency vibrational modes around 500–600 cm^{-1} that helps to rationalize the coherent dynamics in ultrafast pump-and-probe optical spectroscopy (138). The proton dynamics in benzoylacetone, a model substrate for short H-bond low-barrier H-bond (LBHB) enzymatic reactions has been investigated (139). Along this theme, calculations have been performed on serine protease and Ser133 phosphate-PIX (140). These calculations provide new insight to the crucial role of LBHB in the stabilization of the structure. More recently, hybrid QM/MM CPMD methods have been applied to the study of the environmental influence on conformation, electronic structure, and vibrational dynamics of the Fe–CO bond in myoglobin (141, 142).

The potential and impact of AIMD on the study of biological systems have not been fully explored. However, it is clear that in the near future the hybrid QM/MM approaches will be most promising. Together with advances in the development of order N methods, plane wave–based DFT methodologies are very attractive techniques, particularly for very large systems, i.e., systems that consist of a few hundreds to a few thousand atoms. The ability to treat a large region with QM may help alleviate some of the problems associated with the QM/MM interface. Because the MM region will be farther away from the QM region, the artifacts owing to the artificial boundary may be less severe.

CALCULATION OF SPECTROSCOPIC PROPERTIES

The rapid advancement of the plane wave–based CPMD method has benefited from numerical algorithms, computational strategies, and chemical concepts developed over the years for conventional quantum chemistry calculations. The development of new methodologies for the prediction of common spectroscopic properties, such as infrared, Raman, ESR, and NMR spectra, is highly desirable.

Analysis of Valence Charge Distribution

MULLIKEN POPULATION ANALYSIS AND ELECTRON LOCALIZATION FUNCTION There have been attempts to calculate atomic charges and bond populations by means of Mulliken analysis (143) through the projection of non-atom-centered plane wave functions onto linear combination of localized atomic basis sets (144). The projection wave functions only reproduce the qualitative trend and are of limited use for detailed analysis of electron charge distribution. The results obtained from the projection procedure are very sensitive to the choice of localized basis set, and there are inherent ambiguities in the handling of the overlap region. A more useful scheme for the elucidation of valence electronic structure is the use of electron localization function (ELF) proposed originally by Becke (145) and further developed by Savin (146, 147). Although ELF is not a replacement for the popular Mulliken analysis, this method is basis set–independent and has proven to be very convenient for the analysis of electronic properties for molecules consisting mainly of s, p elements.

MAXIMALLY LOCALIZED WANNIER FUNCTIONS In a periodic system, such as in a solid or a supercell model for molecular calculation, the electronic wave functions are described by delocalized Blöchl orbitals. The unitary transformation from delocalized orbitals into localized orbitals or Wannier function is nontrivial. Recently Marzari & Vanderbilt (69) proposed a practical scheme for the determination of an optimally localized set of Wannier functions from a set of Blöchl orbitals. This procedure is essentially the generalization of the well-known Boys localized orbital method (148, 149).

As shown by Resta (73), the fundamental parameter for the localization of the electronic wave function is the dimensionless complex quantity, z, and in a one-dimensional periodic system of length L is defined by

$$z = \int_0^L dx e^{i(2\pi/L)x}|\varphi(x)|^2 = \langle\varphi(x)|e^{i(2\pi/L)x}|\varphi(x)\rangle. \qquad 21.$$

For a completely delocalized system, the normalization condition requires that $|\varphi(x)|^2 = 1/L$ and $z = 0$, whereas for extreme localization, $z = \exp(i(2\pi/L)x_0)$, where x_0 is the center of the distribution of $|\varphi(x)|^2$. In the most general case the center of a periodic distribution $|\varphi(x)|^2$ (Wannier function center, WFC) is given by the expectation value

$$x_0 = \langle x \rangle = \frac{L}{2\pi} \text{Im} \ln z \qquad 22.$$

and the spread of the electronic distribution

$$\lambda^2 = \langle x^2 \rangle - \langle x \rangle^2 = -\left(\frac{L}{2\pi}\right)^2 \ln |z|^2. \qquad 23.$$

The three-dimensional generalization of the spread of the Wannier function Ω is given by

$$\Omega = \sum_n \left(\langle r^2 \rangle_n - \langle r \rangle_n^2\right) = \sum_n \left(|x_{nn}|^2 + |y_{nn}|^2 + |z_{nn}|^2\right). \qquad 24.$$

The maximization of the Wannier function is equivalent to the minimization of the spread function Ω and can be achieved with a steepest descent algorithm (150).

An interesting property of the Wannier function is that the dipole moment of a molecule can be calculated from the ion and the WFC positions by assuming that the electronic charge is concentrated in the point charges located on the WFCs (151). The calculated WFCs for an isolated water molecule are shown in Figure 11 (151). There are four sites of charge concentration corresponding to the two O–H bonds and the two lone pairs. The WFC from the oxygen atom for the covalent O–H bond is 0.53 Å and 0.30 Å for the lone pair. The molecular dipole moment computed from the positions of the WFC is 1.87 D, which is in excellent agreement with the experimental value of 1.86 D. There has been an ambiguity in assigning effective dipole moment of a water molecule in solution. The optimally localized Wannier function overcomes this difficulty. It has been employed to estimate the effective dipole moment of a water molecule under different thermodynamic states: in liquid water, in supercritical liquid, and in crystalline ice. Recently the localization procedure has been extended to wave functions obtained with Vanderbilt ultrasoft pseudo-potential (149).

Infrared Spectrum

Infrared (IR) activity is due to the change of the dipole moment of a molecule with vibration. In a molecular dynamics scheme employing periodic boundary conditions, the IR absorption coefficient $\alpha(\omega)$ can be computed from the formula (72)

$$\alpha(\omega) = \frac{4\pi \omega \tanh(\hbar\omega/2kT)}{3\hbar n(\omega)cV} \int_{-\infty}^{\infty} dt e^{-i\omega t} \langle M(t) \cdot M(0) \rangle, \qquad 25.$$

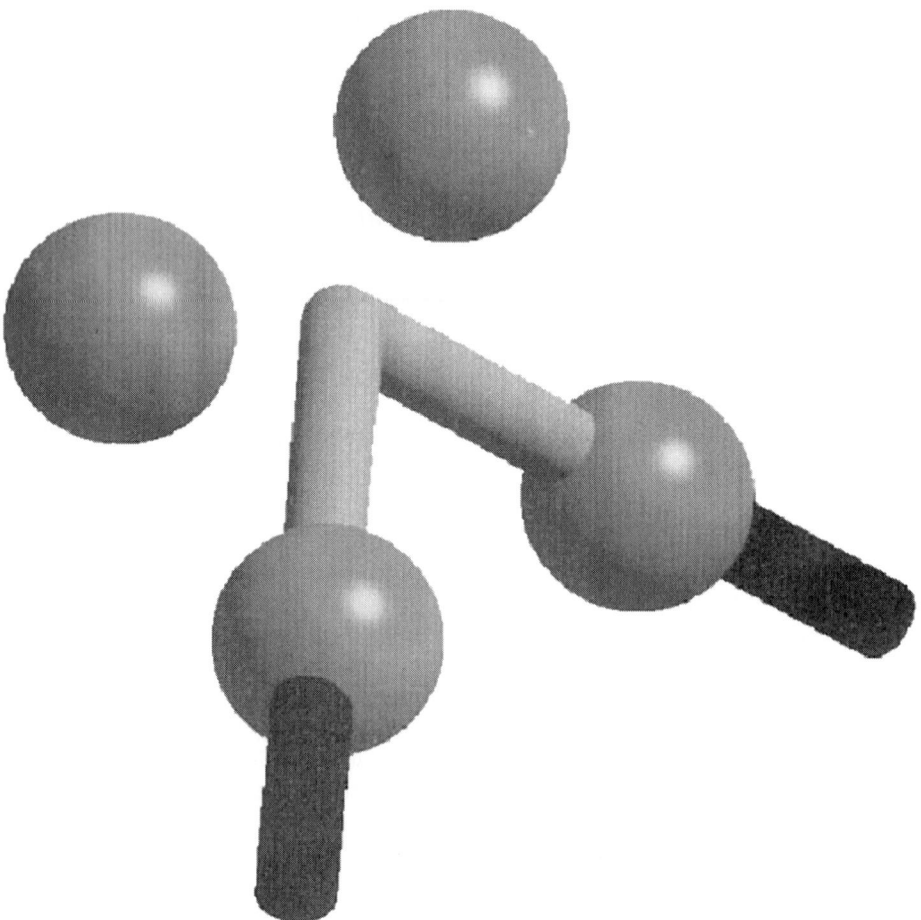

Figure 11 Wannier function centers (WFCs) (*spheres*) for an isolated water molecule.

where V is the volume, k is the Boltzmann constant, $n(\omega)$ is the refractive index, c is the speed of light, and M is the total electronic dipole moment of the simulation cell, which can be calculated from an MD simulation. For systems with periodic boundary conditions, the total dipole moment of the supercell is an ill-defined quantity. The problem associated with the calculation of $M(t)$ was resolved only recently using the Berry phase scheme for electronic polarization. In the limit only the zone-center point (Γ) is used; the total electronic dipole moment projection onto the unit cell vector is given as

$$M_\alpha^{el}(t) = \frac{2|e|}{|G_\alpha(t)|}\varphi(G_\alpha(t)), \qquad 26.$$

where $\alpha = 1, 2, 3$, the three components of the unit cell vector, $\{G_\alpha\}$, are the reciprocal lattice basis vectors and $\varphi(G_\alpha)$ the berry phase, which is defined as

$$\varphi(G_\alpha(t)) = \text{Im} \ln \det S(t), \qquad 27.$$

where S is defined as

$$S_{m,n}(t) = \langle u_m(r,t)|e^{-iG_0 \cdot r}|u_n(r,t)\rangle. \qquad 28.$$

$u_n(t)$ is the electronic wave function at the Γ point at the time step. The self-correlation function $[M(t) \cdot M(0)]$ can be evaluated from the trajectory of the simulation.

The Berry phase method has been applied with success to the interpretation of the IR spectra of amorphous Si (70), amorphous SiO_2 (152), liquid water (72), and high pressure ice VII, VIII to centro-symmetric ice transformation (153). In the latter case, the subtle changes in the infrared absorption features with pressure are well reproduced (see Figure 12) indicating the accuracy of the method.

Vibration Modes, Raman Scattering, and NMR Chemical Shift Using Generalized Density Functional Perturbation Theory

A straightforward way to calculate spectroscopic properties is to ultilize the response functions appropriate to the external perturbations. Denisty functional perturbation theory (DFPT), originally formulated by Baroni et al. (154) and refined by Gonze (155), is a very powerful technique for the evaluation of these properties. The use of the $2n + 1$ theorem (156) allows the calculation of second order properties, for example, the second energy derivative, with only first-order variation in the Kohn-Sham orbitals. Recently the DFPT has been generalized in a functional formulation to include perturbation that cannot be represented in a Hamiltonian form and is suitable for a pseudo-potential plane wave basis set (157).

The vibrational spectrum of a system is the result of the perturbation of the self-consistent ground-state potential by the displacement of the ions. The perturbation function is given by

$$\varepsilon^{pert} = \int d^3 r n(r) \sum_{\beta=1,N_l} \frac{\partial V_{l-e}(\vec{r}-\vec{R})}{\partial R_{\alpha i}} \delta_{\alpha\beta}. \qquad 29.$$

The second-order energy derivative $(\partial^2 E_{KS}/\partial R_{\alpha i} \partial R_{\beta j})$ to the Kohn-Sham orbitals owing to the perturbation can be computed from DFPT. The normal modes can be obtained by diagonalization of the dynamical matrix,

$$D_{\alpha i, \beta j} = \frac{1}{\sqrt{m_\alpha m_\beta}} \left(\frac{\partial^2 E_{I-l}}{\partial R_{\alpha i} \partial R_{\beta j}} + \frac{\partial^2 E_{KS}}{\partial R_{\alpha i} \partial R_{\beta j}} \right). \qquad 30.$$

Test calculations show that the results obtained from DFPT using a plane wave basis set are in good agreement with the finite difference method and also with the localized basis set quantum chemistry method (157).

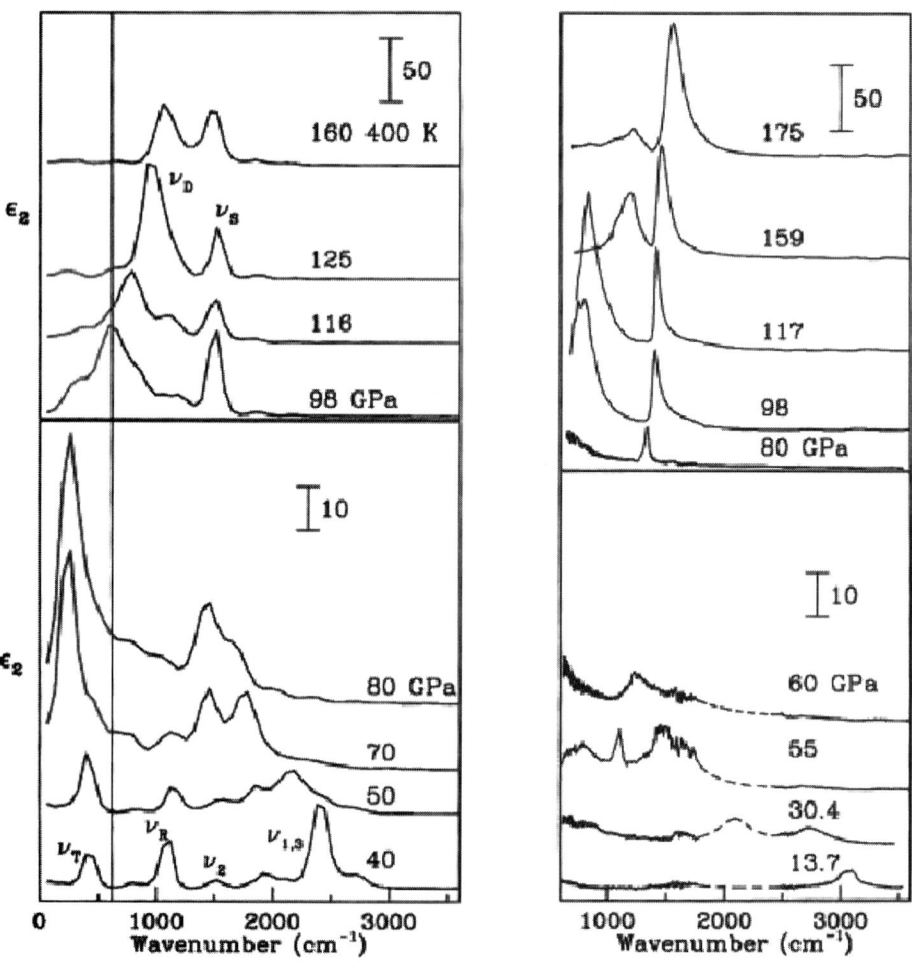

Figure 12 Comparison of theoretical (*left panel*) and experimental (*right panel*) infrared spectra of high pressure ice. Note the change of spectral features at ~80 GPa (theoretical) and ~60 GPa (experiment).

The calculation of the Raman spectra cannot be treated with DFPT in the Hamiltonian formulation. The perturbing potential is a coupling of the external electric field E_{ext} with the electronic polarization P^{ele}. However, in the functional formulation, it can be expressed in the following form:

$$\varepsilon^{pert} = -\vec{E}_{ext} \cdot \vec{P}^{ele} \qquad 31.$$

Recall that according to the Berry phase electronic polarization theory, the

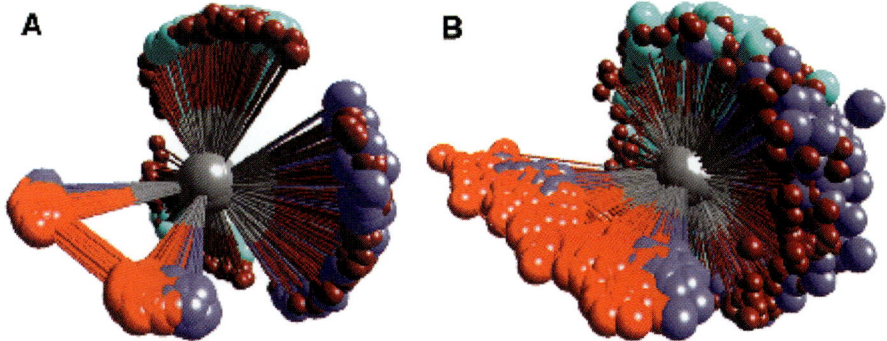

Figure 2 Superimposed configurations of CH_5^+ obtained from AIMD simulations. (*A*) Low temperature results show the persistence of the CH_3^+-H_2 species and (*B*) high temperature results show large fluctuation of protons around the C atom.

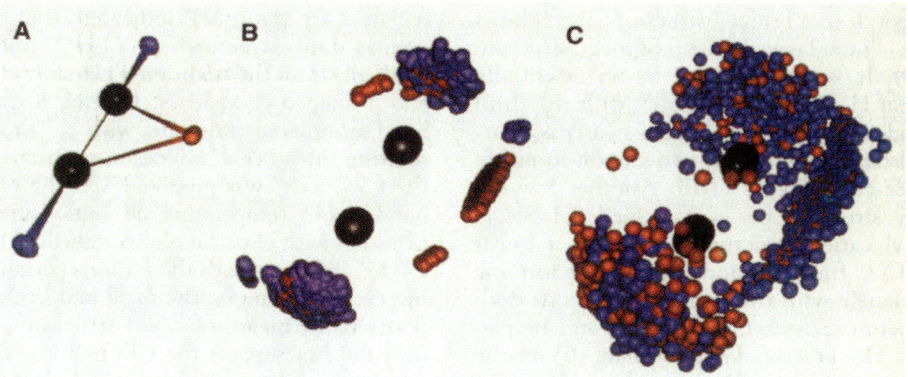

Figure 3 Representative superimposed configurations from AIMD runs for $C_2H_3^+$. (*A*) 5 K classical MD, (*B*) 5 K path-integral MD and (*C*) classical MD at 3000 K.

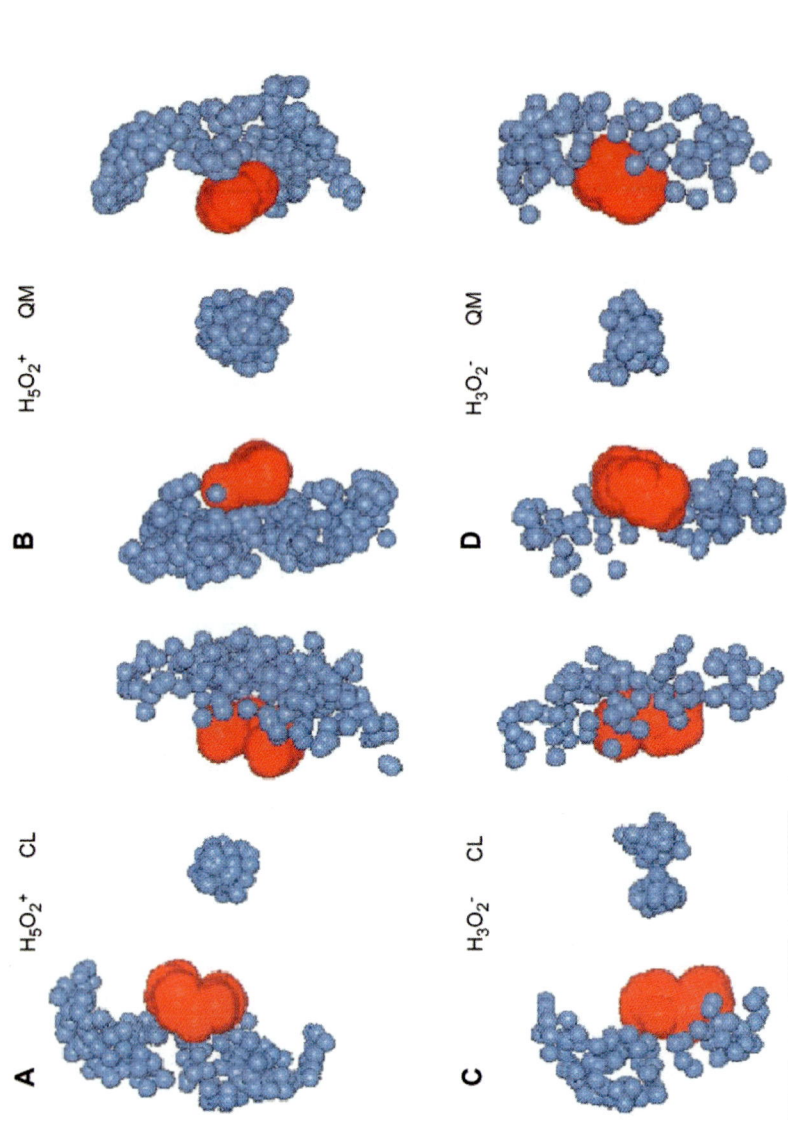

Figure 5 Representative superimposed configurations from AIMD calculations at 300 K, (*A*) Classical (CL) $H_5O_2^+$; (*B*) quantum (QM) $H_5O_2^+$; (*C*) Classical (CL) $H_3O_2^-$; and (*D*) quantum (QM) $H_3O_2^-$.

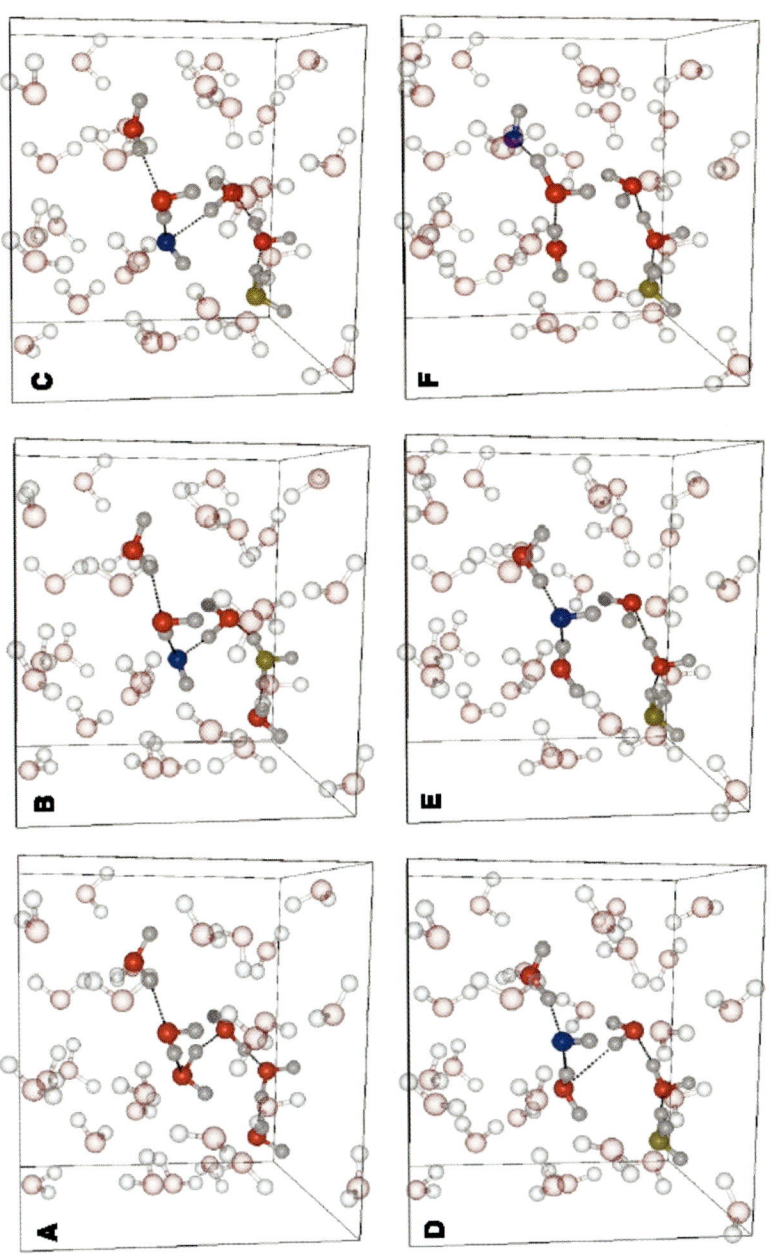

Figure 7 Snapshots from an autoionization trajectory of liquid water. (*A*) Initial stage of proton transfer between two neighbouring water. (*B*) separation of the hydoxide and hydroxonium ions by one water. (*C*) and (*D*) proton transfer along connected hydrogen-bonded paths. (*E*) and (*F*) disconnection of the continuous hydrogen-bond path and the eventual separation of the two ions.

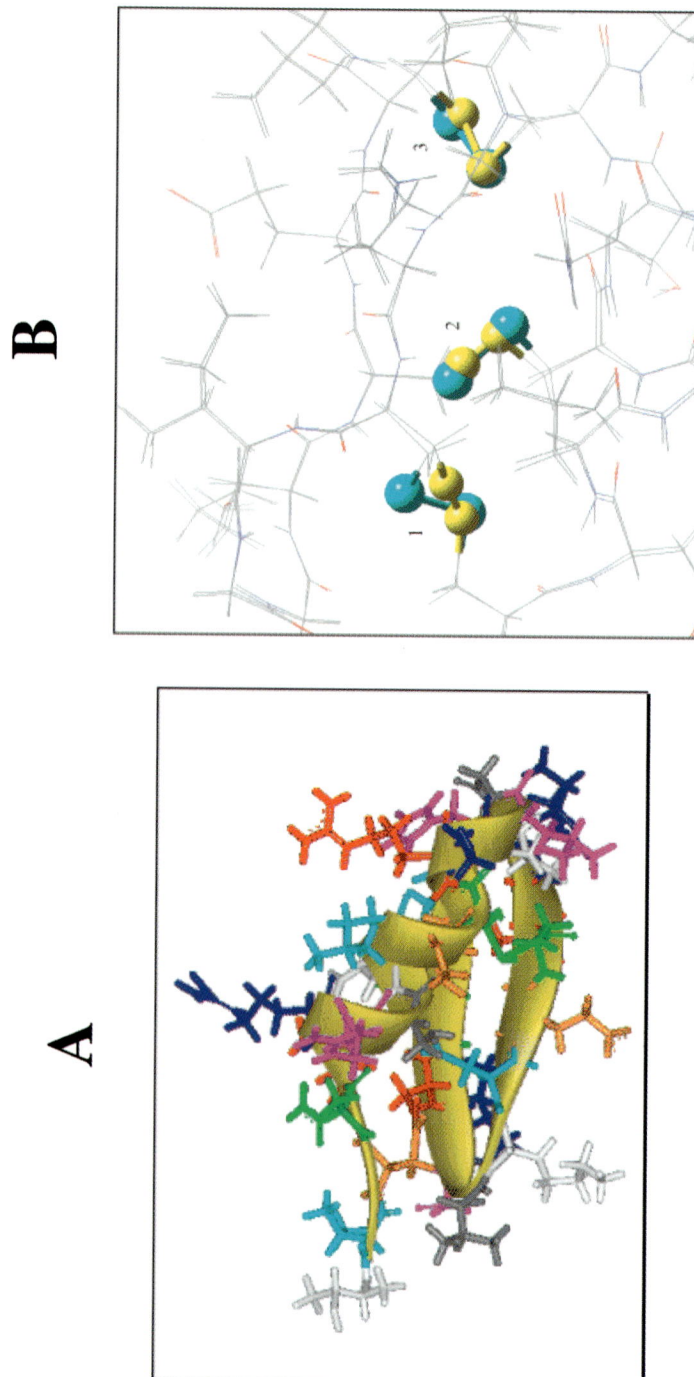

Figure 10 Structure of 1PNH protein. (*B*) The positions of –S-S- bridges in 1PNH protein.

macroscopic polarization is given by

$$P_\mu^{ele} = \frac{2|e|}{|G_\alpha|} \operatorname{Im} \ln S_{mn}, \qquad 32.$$

where $S_{mn} = \langle \varphi_m(r)|e^{iG\cdot r}|\varphi_n(r)\rangle$, and therefore, the perturbation function becomes

$$\varepsilon^{pert} = -\sum_v Ev \frac{2|e|}{|G_\alpha|} \operatorname{Im} \ln S_{mn}. \qquad 33.$$

The polarizability, $\alpha_{\mu v} = -\partial P_\mu / \partial E_v$, can then be computed. The polarizability of an isolated water molecule calculated by this method using a plane wave basis set is in good agreement with experiment (157). In constrast, the convergence of the calculated dielectric constant [$\varepsilon = 1 + 4\pi(\partial P/\partial E)$] with respect to the system size for crystalline Si, an extended system, is rather poor (157). The Raman spectrum, however, is related to the derivative of the polarizability with respect to atomic displacements from equilibrium that can be calculated by the finite difference method. Surprisingly, the derivative of the polarizability of crystalline Si shows very good convergence with the system size (157).

The chemical shift of a nucleus observed in an NMR experiment is a second-order property that is proportional to the ratio between the induced magnetic field, B_{ind} and an external magnetic field B_{ext} (158),

$$\vec{B}_{ind}(\vec{r}) = -\vec{\sigma}(\vec{r})\vec{B}_{ext}. \qquad 34.$$

The perturbation Hamiltonian to the first and second order is given by

$$H^{pert} = \frac{e}{m}\vec{p}\cdot\vec{A}(\vec{r}) + \frac{e^2}{2m}\vec{A}(\vec{r})\cdot\vec{A}(\vec{r}), \qquad 35.$$

where A is the vector potential. The vector potential for a homogeneous magnetic field B_{ext} is

$$\vec{A}(r) = -\frac{1}{2}(\vec{r} - \vec{R}) \times \vec{B}_{ext}, \qquad 36.$$

R being the gauge origin. This leads to a second-order perturbation functional,

$$\varepsilon^{pert} = \sum_k \frac{e}{m}(\langle\varphi_k^{(1)}|\vec{p}\cdot\vec{A}(\vec{r})|\varphi_k^{(0)}\rangle + c.c.) + \sum_k \frac{e^2}{2m}\langle\varphi_k^{(0)}|\vec{A}(\vec{r})\cdot\vec{A}(\vec{r})|\varphi_k^{(0)}\rangle \qquad 37.$$

$\varphi_k^{(0)}$ and $\varphi_k^{(1)}$ are the ground-state and perturbed orbitals and c.c. is the complex conjugate.

Several methods have been proposed in quantum chemistry calculations to handle problems associated with the choice of the origin. Some of the most well-known techniques are the gauge invariant atomic orbitals (159), independent gauges for localized orbitals (160), continuous set of gauge transformation (161), and independent gauge for atom in molecule (162). To minimize a numerical problem owing to the cancellation effect of two large terms, the most appropriate choice is the continuous set of gauge transformation method with the so-called $R = r$ (163). In this method for each point r' in space, the current is calculated with the gauge origin R located at r'. This ansatz reduces the diamagnetic contribution to the induced current to identical zero and the cancellation problem no longer exists.

Unlike isolated systems, there is a technical problem in the evaluation of the expectation values of the perturbative Hamiltonian for a uniform field between extended (delocalized) orbitals in a periodic system. A possible solution to overcome this problem was first suggested and implemented by Mauri, Pfrommer, & Louie (MPL) (158). The MPL method replaces the homogeneous external magnetic field with a modulated one with a finite wave vector q, thus making the perturbation Hamiltonian also periodic in q. The physically relevant case of the magnetic field at $q = 0$ can be obtained via differentiation with a small but nonzero wave vector between q and $-q$. First calculations of the NMR chemical shift on a variety of isolated molecules, surfaces, and bulk solids using the pseudo-potential plane wave method yield very promising results as compared with the quantum chemistry methods. One drawback of the pseudo-potential approximation is that the core contribution to the NMR chemical shift needs to be added in an ad hoc manner. A very successful application of the MPL method is the calculation of the chemical shift of liquid water and ice Ih (164). In the former case chemical shift calculations were performed on a model liquid water system consisting of 32 molecules. Chemical shifts for both the H and O nuclei were obtained by averaging over a number of instantaneous liquid water structures obtained from separate AIMD calculations. The calculated gas to solid isotropic shift of ice Ih with respect to the gas phase is about -8.0 ppm for proton and ~ 48.4 ppm for oxygen at $0°$K. These values are to be compared with the measured proton shift of -7.4 ppm at $77°$K (165). The agreement between the calculated and observed gas to liquid isotropic shift is equally impressive. The calculated proton shift is -5.8 ppm and -36.6 ppm for oxygen as compared with the experimental value of -4.3 ppm and -36.1 ppm, respectively (166).

The MPL method requires the wave functions to be complex even with the Γ point approximation and increases the computational effort. An alternative scheme to resolve the problem on the evaluation of matrix elements of the perturbation Hamiltonian has been proposed by Sebastiani & Parrinello (167). Instead of transforming the uniform magnetic field into a periodic one, a new position operator is defined such that the spatial extension of the extended wave function is minimized by maximally localized Wannier wave function. An individual virtual cell is then assigned to each Wannier wave function. These virtual cells are chosen such that the cell walls are located where the orbital density is almost zero. In this way, the position operator has a saw-tooth shape that matches the periodic boundary conditions. This method is only efficient when the wave functions are sufficiently localized and the decay lengths are small compared to the size of the simulation cell. Tests of this method on small organic molecules show good convergence properties with plane wave cutoff and the size of the simulation cell (167). The calculated chemical shifts are in good agreement with the MPL method and also with localized basis set quantum chemistry methods using the continuous set of gauge transformation scheme.

The new method was applied to elucidate the ^{13}C NMR chemical shifts and isotopic shifts of the protonated state of Asp 25-Asp 25' dyad in pepstatin A/HIV-1 protease (168). Experimentally, two resonance lines were detected and interpreted as a singly protonated dyad structure. CPMD calculations on a suggested structural model for a mono-protonated Asp 25' (A1) were found to be unstable on the picosecond timescale. Moreover, the calculated ^{13}C NMR chemical shifts do not agree with experiment. Analysis of the theoretical results led to the suggestion of an alternative structure in which both aspartic (Asp25 and Asp25') groups are protonated (B1).

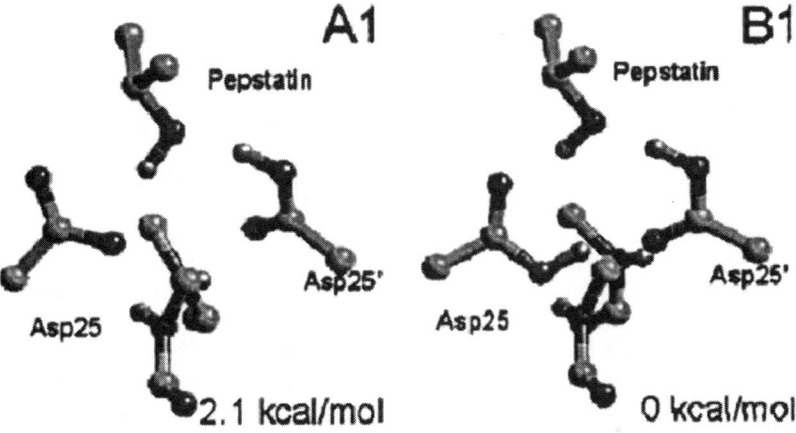

The calculated chemical shifts and isotopic shifts are found to be in substantial agreement with experiment despite the apparent symmetric proton states. This study clearly demonstrates the power of the new combined quantum mechanical and molecular dynamic approach together with property calculations over static quantum chemistry calculations. In addition to providing static structure and energetic information, AIMD provides unique insight into the dynamical structure and processes and predicts spectroscopic properties of large macromolecular systems at finite temperature.

CONCLUSION

Since the inception of the concept of combined dynamical and quantum mechanical calculations in 1985, the CPMD method, initially intended to alleviate the problem in the diagonalization of large matrices in condensed-matter physics calculations, has developed into a very useful tool in quantum chemistry. The extended variable Lagrangian (Hamiltonian) formalism lends itself to efficient implementation of other dynamical variables. The pseudo-potential plane wave approach with periodic boundary conditions is computationally highly efficient for the calculation of large systems, although in principle these conditions need not be imposed (169–171). In a short 16 years, AIMD methods have found applications to a variety of systems of chemical importance—solution chemistry, surface chemistry, and chemical biology—and under a broad range of thermodynamic conditions, such as at finite temperature and finite pressure, in the gas and condensed phases. This method is becoming more acceptable to the quantum chemistry community as witnessed by its growing use by computational chemists. It is anticipated that the AIMD will be further developed to more functions relevant to the elucidation of chemical phenomena.

Visit the Annual Reviews home page at www.annualreviews.org

LITERATURE CITED

1. Hohenberg P, Kohn W. 1964. *Phys. Rev. B* 136:864–71
2. Kohn W, Sham LJ. 1965. *Phys. Rev. A* 140:1133–38
3. Chong DP. 1997. *Recent Advances in Computational Chemistry*, Vol. I. Singapore: World Sci.
4. Koch W, Holthausen M. 2001. *A Chemist's Guide to Density Functional Theory*. Weinheim: Wiley-VCH
5. Parr RG, Yang W. 1996. *Density-Functional Theory of Atoms and Molecules*. Berlin/New York: Springer-Verlag
6. Nalewajski RF, ed. 1996. *Density Function Theory*, Vols. I–IV. Berlin/New York: Springer-Verlag
7. Car R, Parrinello M. 1985. *Phys. Rev. Lett.* 55:2471–74
8. Andreoni WA, Curioni A. 2000. *Parallel Comput.* 26:819–42
9. Remler DK, Madden PA. 1990. *Mol. Phys.* 70:921–66
10. Payne MC, Teter MP, Allan DC, Arias TA, Joannopoulos JD. 1992. *Rev. Mod. Phys.* 64:1045–97
11. Gillan MJ. 1997. *Contemp. Phys.* 38:115–30
12. Marx D, Hutter J. 2000. In *Modern Methods and Algorithms of Quantum Chemistry*, ed. J Grotendorst, 1:301–449. Forschungszentrum Jülich, NIC Ser. http://www.fz-juelich.de/wsqc/proceedings/
13. Ihm J, Zunger A, Cohen ML. 1979. *J. Phys. C* 12:4409–22
14. Patore G, Smargriass E, Buda F. 1991. *Phys. Rev. A* 44:6334–47
15. Gillan MJ. 1987. *J. Phys. C* 20:3621–41
16. Marx D, Parrinello M. 1994. *Z. Phys. B* 95:143–44
17. Tuckerman ME, Marx D, Klein ML, Parrinello M. 1996. *J. Chem. Phys.* 104:5579–58
18. Gillan MJ. 1990. In *Computer Modelling of Fluids, Polymers and Solids*, ed. CRA Catlow, SC Parker, MP Allen. Dordrecht, The Netherlands: Kluwer
19. Cheng HP, Barnett RN, Landman U. 1995. *Chem. Phys. Lett.* 237:161–70
20. Cao J, Voth GA. 1993. *J. Chem. Phys.* 99:10070–73
21. Cao J, Martyna GJ. 1996. *J. Chem. Phys.* 104:2028–35
22. Marx D, Tuckerman ME, Martyna GJ. 1999. *Comput. Phys. Commun.* 118:166–84
23. Pavese M, Jang S, Voth GA. 2000. *Parallel Comput.* 26:1025–41
24. Vanderbilt D. 1990. *Phys. Rev. B* 41:7892–95

25. Gygi F. 1995. *Phys. Rev. B* 51:11190–93
26. Blöchl PE. 1994. *Phys. Rev. B* 50:17953–79
27. Kresse G, Joubert D. 1999. *Phys. Rev.* 59:1758–75
28. Hartke B, Carter EA. 1992. *J. Chem. Phys.* 97:6569–78
29. Field MJ. 1991. *J. Phys. Chem.* 95:5104–8
30. Tachikawa H. 1998. *J. Phys. Chem. A* 102:7065–69
31. Sprik M, Ciccotti G. 1998. *J. Chem. Phys.* 109:7737–44
32. Margl P, Lohrenz JCW, Ziegler T, Blöchl PE. 1996. *J. Am. Chem. Soc.* 118:4434–41
33. Carter EA, Ciccotti G, Hynes JT, Kapral R. 1989. *Chem. Phys. Lett.* 156:472–77
34. Andreoni W. 1987. In *Elemental and Molecular Clusters*, ed. G Benedek, TP Martin, G Pacchiono. Berlin: Springer-Verlag
35. Scuseria G. 1993. *Nature* 362:512–13
36. Müller H, Kutzelnigg W, Noga J, Klopper W. 1997. *J. Chem. Phys.* 106:1863–69
37. Schreiner PR, Kim SJ, Schaefer HF III, Schleyer PvR. 1993. *J. Chem. Phys.* 99:3716–20
38. Tse JS, Klug DD, Laasonen K. 1995. *Phys. Rev. Lett.* 74:876–79
39. Marx D, Parrinello M. 1995. *Nature* 375:216–18
40. Marx D, Parrinello M. 1997. *Z. Phys. D* 41:253–260
41. Marx D, Parrinello M. 1999. *Science* 284:59–61
42. White ET, Tang J, Oka T. 1999. *Science* 284:135–47
43. Kramer GM. 1999. *Science* 286:1051
44. Oka T, White ET. 1999. *Science* 286:1051
45. Marx D, Parrinello M. 1999. *Science* 286:1051
46. East ALL, Kolbuszewski M, Bunker PR. 1997. *J. Phys. Chem. A* 101:6746–52
47. East ALL, Bunker PR. 1997. *J. Mol. Spectrosc.* 183:157–62
48. Oka T, White ET. 1999. *Science* 284:35–137
49. Boo DW, Liu ZF, Suits AG, Tse JS, Lee YT. 1995. *Science* 269:57–59
50. Tse JS, Rousseau R. 2000. In *Computational Molecular Spectroscopy*, ed. P Jensen, PR Bunker. Chichester, England: Wiley
51. Lindh R, Rice JE, Lee TJ. 1991. *J. Chem. Phys.* 94:8008–14
52. Crofton MW, Jagod M, Rehfus BD, Oka T. 1986. *J. Chem. Phys.* 85:3437–43
53. Escribano R, Bunker PR, Gomez PC. 1988. *Chem. Phys. Lett.* 150:60–62
54. Vager Z, Zajfman D, Graber T, Kanter EP. 1993. *Phys. Rev. Lett.* 71:4319–22
55. Marx D, Parrinello M. 1996. *Science* 271:179–81
56. East ALL, Liu ZF, McCague C, Cheng K, Tse JS. 1998. *J. Phys. Chem. A* 102:10903–11
57. Yeh LI, Price JM, Lee YT. 1989. *J. Am. Chem. Soc.* 111:5597–604
58. Saitta AM, Soper PD, Wasserman E, Klein ML. 1999. *Nature* 399:46–48
59. Saitta AM, Klein ML. 1999. *J. Am. Chem. Soc.* 121:11827–30
60. Tuckerman ME, Marx D, Klein ML, Parrinello M. 1997. *Science* 275:817–20
61. Tuckerman ME, Marx D. 2001. *Phys. Rev. Lett.* 86:4946–49
62. Miura S, Tuckerman ME, Klein ML. 1998. *J. Chem. Phys.* 109:5290–99
63. Tse JS, Klug DD. 2000. *Phys. Rev. B* 59:34–37
64. Laasonen K, Sprik M, Parrinello M, Car R. 1993. *J. Chem. Phys.* 99:9080–89
65. Sprik M, Hutter J, Parrinello M. 1996. *J. Chem. Phys.* 105:1142–52
66. Silvestrelli PL, Parrinello M. 1999. *J. Chem. Phys.* 111:3572–80
67. Krynicki K, Green CD, Sawyer DW. 1978. *Faraday Discuss. Chem. Soc.* 66:199–208
68. Coulson CA, Eisenberg D. 1966. *Proc. R. Soc. London Ser. A* 291:445–53
69. Marzari N, Vanderbilt D. 1997. *Phys. Rev. B* 56:12847–65
70. Silvestrelli PL, Marzari N, Vanderbilt D, Parrinello M. 1998. *Solid State Commun.* 107:7–11

71. Batista ER, Xantheas SS, Jonsson H. 1998. *J. Chem. Phys.* 109:4546–51
72. Silvestrelli PL, Bernasconi M, Parrinello M. 1997. *Chem. Phys. Lett.* 277:478–82
73. Resta R. 1994. *Rev. Mod. Phys.* 66:899–915
74. King-Smith RD, Vanderbilt D. 1993. *Phys. Rev. B* 47:651–54
75. Boero M, Terakura K, Ikeshoji T, Liew CC, Parrinello M. 2000. *Phys. Rev. Lett.* 85:3245–48
76. Boero M, Terakura K, Ikeshoji T, Liew CC, Parrinello M. 2001. *J. Chem. Phys.* 113:2219–27
77. Lamb WJ, Hoffman GA, Jonas J. 1980. *J. Chem. Phys.* 74:6875–50
78. Trout BI, Parrinello M. 1998. *Chem. Phys. Lett.* 288:343–47
79. Trout BI, Parrinello M. 1999. *J. Phys. Chem. B* 103:7340–45
80. Eigen M, De Maeyer L. 1955. *Z. Elektrochem.* 59:986–93
81. Natzle WC, Moore CB. 1985. *J. Phys. Chem.* 89:2605–12
82. Geissler PL, Dellago C, Chandler D, Hutter J, Parrinello M. 2001. *Science* 291:2121–24
83. Dellago C, Bolhuis PG, Chandler D. 1999. *J. Chem. Phys.* 110:6617–25
84. Atkins PW. 1998. *Physical Chemistry*. London/New York: Oxford Univ. Press. 6th ed.
85. Eigen M. 1964. *Angew. Chem. Int. Ed. Engl.* 3:1–19
86. Zundel G. 1976. In *The Hydrogen Bond— Recent Developments in Theory and Experiments. II. Structure and Spectroscopy*, ed. P Schuster, G Zundel, CA Sandrofy, pp. 683–766. Amsterdam: North-Holland
87. Tuckerman M, Laasonen K, Sprik M, Parrinello M. 1995. *J. Phys. Chem.* 99:5749–52
88. Tuckerman M, Laasonen K, Sprik M, Parrinello M. 1995. *J. Chem. Phys.* 103:150–60
89. Marx D, Tuckerman ME, Hutter J, Parrinello M. 1999. *Nature* 397:601–4
90. Mei HS, Tuckerman ME, Sagnella DE, Klein ML. 1998. *J. Phys. Chem. B* 102:10446–58
91. Laasonen K, Klein ML. 1997. *J. Phys. Chem.* 101:98–102
92. Rothlisberger U, Parrinello M. 1997. *J. Chem. Phys.* 106:4658–64
93. Cole RH. 1973. *J. Chem. Phys.* 59:1545–46
94. Kim D, Klein ML. 1999. *J. Am. Chem. Soc.* 121:11251–52
95. von Rosenvinge T, Parrinello M, Klein ML. 1997. *J. Chem. Phys.* 107:8012–19
96. Kim D, Klein ML. 2000. *J. Phys. Chem. B* 104:10074–100
97. Lyubartsev AP, Laasonen K, Lasksonen A. 2001. *J. Chem. Phys.* 114:3120–26
98. Ramaniah LM, Parrinello M, Bernasconi M. 1999. *J. Chem. Phys.* 111:1587–91
99. Lightstone FC, Schegler E, Hood RQ, Gygi F, Galli G. 2001. *Chem. Phys. Lett.* 343:549–55
100. Molteni C, Parrinello M. 1998. *J. Am. Chem. Soc.* 120:2168–71
101. Yarne DA, Tuckerman ME, Klein ML. 2000. *Chem. Phys.* 238:163–69
102. Bruge F, Bernasconi M, Parrinello M. 1999. *J. Am. Chem. Soc.* 121:10883–88
103. Liu ZF, Siu CK, Tse JS. 1999. *Chem. Phys. Lett.* 314:317–25
104. Yim W-L, Liu ZF. 2001. *J. Am. Chem. Soc.* 123:2343–50
105. Wang YM, Chen C, Lin ST. 1999. *J. Mol. Struct.-THEOCHEM* 460:79–102
106. Woo TK, Margl P, Blöchl PE, Ziegler T. 1997. *J. Phys. Chem. B* 101:7877–80
107. Woo TK, Margl P, Ziegler T, Blöchl PE. 1997. *Organometallics* 16:3454–68
108. Woo TK, Margl P, Lohrenz JCW, Blöchl PE, Ziegler T. 1996. *J. Am. Chem Soc.* 118:13021–30
109. Woo TK, Margl P, Blöchl PE, Ziegler T. 1998. *J. Am. Chem. Soc.* 120:2174–75
110. Michalak A, Ziegler T. 2001. *J. Phys. Chem. A* 105:4333–43
111. Fukui K. 1981. *Acc. Chem. Res.* 14:363–68

112. Woo TK, Blöchl PE, Ziegler T. 2000. *J. Mol. Struct.-THEOCHEM* 506:313–34
113. Blöchl PE. 1995. *J. Chem. Phys.* 103:7422–28
114. Eichinger M, Tavan P, Hutter J, Parrinello M. 1999. *J. Chem. Phys.* 110:10452–67
115. Curioni A, Sprik M, Andreoni W, Schiffer H, Hutter J, Parrinello M. 1997. *J. Am. Chem. Soc.* 119:7210–29
116. Meijer EJ, Sprik M. 1998. *J. Am. Chem. Soc.* 120:6345–55
117. Frank I, Hutter J, Marx D, Parrinello M. 1998. *J. Chem. Phys.* 108:4060–69
118. Ziegler T, Rauk A, Baerends EJ. 1977. *Theor. Chim. Acta* 43:261–71
119. Bauernschmitt R, Ahlrichs R. 1996. *Chem. Phys. Lett.* 256:454–64
120. Molteni C, Frank I, Parrinello M. 1999. *J. Am. Chem. Soc.* 121:12177–83
121. Yamamoto S, Waseda H, Kakitani K. 1998. *J. Mol. Struct.-THEOCHEM* 451:151–62
122. Yamamoto S, Waseda H, Kakitani K, Yamato T. 1999. *J. Mol. Struct.-THEOCHEM* 462:463–71
123. Coker DF. 1993. In *Computer Simulation in Chemical Physics*, ed. MP Allen, DJ Tildesley, pp. 315–78. Dordrecht: Kluwer Acad.
124. Monard G, Merz KM Jr. 1999. *Acc. Chem. Res.* 32:904–11
125. Seigbahn PEM, Blomberg MRA. 1999. *Annu. Rev. Phys. Chem.* 50:221–49
126. Andreoni W, Curioni A, Mordasini T. 2001. *IBM J. Res. Dev.* 45:397–407
127. Guerra CF, Bickelhaupt FM, Snijders JG, Baerends EJ. 2000. *J. Am. Chem. Soc.* 122:4117–28
128. Marchi M, Hutter J, Parrinello M. 1996. *J. Am. Chem. Soc.* 118:7847–48
129. Hutter J, Carloni P, Parrinello M. 1996. *J. Am. Chem. Soc.* 118:8710–12
130. Carloni P, Andreoni W. 1996. *J. Phys. Chem.* 100:17797–800
131. Carloni P, Andreoni W, Parrinello M. 1997. *Phys. Rev. Lett.* 79:761–64
132. Rovira C, Carloni P, Parrinello M. 1999. *J. Phys. Chem. B* 103:7031–35
133. Carloni P, Andreoni W, Hutter J, Curioni A, Giannozzi P, Parrinello M. 1995. *Chem. Phys. Lett.* 234:50–56
134. Tornaghi E, Andreoni W, Carloni P, Hutter J, Parrinello M. 1995. *Chem. Phys. Lett.* 246:469–74
135. Rovira C, Kunc K, Hutter J, Parrinello M. 2001. *Inorg. Chem.* 40:11–17
136. Alber F, Carolini P. 2000. *Protein Sci.* 9:2535–46
137. Wei D, Guo H, Salahub DR. 2001. *Phys. Rev. E* 64:011907-1–4
138. Tozzini V, Nifosi R. 2001. *J. Phys. Chem. B* 105:5797–803
139. Pantano S, Alber F, Carloni P. 2000. *J. Mol. Struct.-THEOCHEM* 530:177–81
140. Dal Peraro M, Alber F, Carloni P. 2001. *Eur. Biophys. J.* 20:75–81
141. Rovira C, Parrinello M. 2000. *Biophys. J.* 78:93–100
142. Rovira C, Schulze B, Eichinger M, Evanseck JD, Parrinello M. 2001. *Biophys. J.* 81:435–45
143. Mulliken RS. 1995. *J. Chem. Phys.* 23:1833–40
144. Segall MD, Shah R, Pickard CJ, Payne MC. 1996. *Phys. Rev. B* 54:6317–20
145. Becke AD, Edgecombe KE. 1990. *J. Chem. Phys.* 92:5397–403
146. Silvi B, Savin A. 1994. *Nature* 371:683–86
147. Marx D, Savin A. 1997. *Angew. Chem. Int. Ed. Engl.* 136:2077–79
148. Boys SF. 1960. *Rev. Mod. Phys.* 32:296–99
149. Bernasconi L, Madden PA. 2001. *J. Mol. Struct.-THEOCHEM* 544:49–60
150. Berghold G, Mundy CI, Romero AH, Hutter J, Parrinello M. 2000. *Phys. Rev. B* 61:10040–48
151. Silvestrelli PL, Parrinello M. 1999. *Phys. Rev. Lett.* 82:3308–11
152. Pasquarello A, Car R. 1998. *Phys. Rev. Lett.* 80:5145–57
153. Bernasconi M, Silvestrelli PL, Parrinello M. 1998. *Phys. Rev. Lett.* 81:1235–38
154. Baroni S, Giannozzi P, Testa A. 1987. *Phys. Rev. Lett.* 58:1861–63

155. Gonze X. 1995. *Phys. Rev. A* 52:1096–114
156. Hirschfelder JO, Byers Brown W, Epstein ST. 1964. *Adv. Quant. Chem.* 1:255–374
157. Putrino A, Sebastiani D, Parrinello M. 2000. *J. Chem. Phys.* 113:7102–9
158. Mauri F, Pfrommer BG, Louie SG. 1996. *Phys. Rev. Lett.* 77:5300–3
159. Ditchfield R. 1972. *J. Chem. Phys.* 56:5688–91
160. Kutzelnigg W. 1980. *Isr. J. Chem.* 19:193–280
161. Keith TA, Bader RFW. 1993. *Chem. Phys. Lett.* 210:223–31
162. Keith TA, Bader RFW. 1992. *Chem. Phys. Lett.* 194:1–8
163. Gregor T, Mauri F, Car R. 1999. *J. Chem. Phys.* 111:1815–22
164. Pfrommer BG, Mauri F, Louie SG. 2000. *J. Am. Chem. Soc.* 122:123–29
165. Burum DP, Rhim WK. 1979. *J. Chem. Phys.* 70:3553–54
166. Hindman JC. 1966. *J. Chem. Phys.* 44:4582–92
167. Sebastiani D, Parrinello M. 2001. *J. Phys. Chem.* 105:1951–58
168. Piana S, Sebastiani D, Carloni P, Parrinello M. 2001. *J. Am. Chem. Soc.* 123:8730–37
169. Barnett RN, Landman U. 1993. *Phys. Rev. B* 48:2081–97
170. Martyna GJ, Tuckerman ME. 1999. *J. Chem. Phys.* 110:2810–21
171. Krack M, Parrinello M. 2000. *Phys. Chem. Chem. Phys.* 2:2105–12

TRANSITION PATH SAMPLING: Throwing Ropes Over Rough Mountain Passes, in the Dark

Peter G. Bolhuis
Department of Chemical Engineering, Nieuwe Achtergracht 166, 1018 WV Amsterdam, The Netherlands; e-mail: bolhuis@science.uva.nl

David Chandler
Department of Chemistry, University of California, Berkeley, California 94720; e-mail: chandler@cchem.berkeley.edu

Christoph Dellago
Department of Chemistry, University of Rochester, Rochester, New York 14627; e-mail: dellago@chem.rochester.edu

Phillip L. Geissler
Department of Chemistry and Chemical Biology, Harvard University, Cambridge, Massachusetts 02138; e-mail: geissler@chemistry.harvard.edu

Key Words potential surfaces, kinetics, transition states, complex systems, trajectories, basins of attraction, rare events

■ **Abstract** This article reviews the concepts and methods of transition path sampling. These methods allow computational studies of rare events without requiring prior knowledge of mechanisms, reaction coordinates, and transition states. Based upon a statistical mechanics of trajectory space, they provide a perspective with which time dependent phenomena, even for systems driven far from equilibrium, can be examined with the same types of importance sampling tools that in the past have been applied so successfully to static equilibrium properties.

INTRODUCTION

During the past several years, we and our coworkers have developed a general computational method for finding the transition pathways for infrequent events in both equilibrium and nonequilibrium systems (1–14). The method requires no preconceived notion of mechanism or transition state. Called "transition path sampling," it is metaphorically akin to throwing ropes over rough mountain passes, in the dark. "Throwing ropes" in the sense that one shoots short trajectories, attempting to reach one stable state from another. "In the dark" because high-dimensional systems are so complex that it is generally impossible to literally visualize the topography of relevant energy surfaces. In such cases, it is unlikely that the first throw of the rope will be successful, but one can learn from failures; and there should be an optimum procedure, i.e., sequence of throws, with which success is obtained efficiently. We have discovered and demonstrated this type of sequence, opening the way for many heretofore impossible computational studies of the dynamic pathways of chemical and physical transformations in clusters and in condensed materials.

RARE BUT IMPORTANT EVENTS IN COMPLEX SYSTEMS

DISPARITY OF TIMESCALES   Often, dynamical processes of interest occur on timescales that are very long compared to the shortest significant timescale. For example, the dissociation of a weak acid in water might occur with a half-life of, say, 1 ms, while elementary steps of molecular motions in water occur in 1 fs. Similarly, timescales for folding the smallest of proteins are in the range of microseconds to milliseconds, whereas that for small-amplitude motions of amino acid side chains and water solvent is again 1 fs.

This wide disparity of timescales can present serious computational challenges. For instance, consider a computed trajectory for a system containing a weak acid molecule and a bath of a few hundred water molecules. Within one or two orders of magnitude—depending on computing equipment and algorithm—1 s of computation time is required to advance the system for what would correspond to 1 fs of physical time. As such, typically 10^{12} s of computing time seems to be required to find one example of an event leading to acid dissociation. A representative sampling of pathways to dissociation would therefore seem to be an impractical computational task.

TRANSITION STATE THEORY   One way to get around this problem is to focus on the dynamical bottleneck for the rare event—the transition state surface. In a rare event, it is this surface or threshold that is rarely visited and thus rarely crossed. If its location is known, however, one may construct a scheme where the system is first moved reversibly to the transition state surface and then many fleeting trajectories are initiated from that surface. The first step determines the reversible work and thus the probability for reaching the transition state, and the subsequent trajectories determine the probability for successfully crossing the threshold. Together, they give the rate for the rare event. This approach was pioneered by Anderson (15), Bennett (16), and Chandler (17). It has been recently reviewed by Anderson (18), and a tutorial on it has been written by Chandler (19). Elementary discussions are found in textbooks [e.g., References (20, 21)]. Although theoretically sound, this two-step procedure is limited in applicability because it presupposes knowledge of the transition state. In most interesting cases, transition state surfaces are not known and not easily characterized.

DIFFICULTY OF IDENTIFYING TRANSITION STATE SURFACES   For low-dimensional systems involving only a few atoms, transition state surfaces usually intersect saddle points in the potential energy surface. In those cases, transition state surfaces can be identified with various algorithms that examine gradients of the potential energy surface and systematically search for saddle points on that surface (22, 23). For higher-dimensional systems, however, the potential energy surface will typically contain many saddle points, most if not all of which are irrelevant to the dynamics that carries the system from one stable (or metastable) state to another. Figure 1 (see color insert) illustrates this point. Explicit enumeration of saddle points is feasible for a cluster of the order of ten or fewer atoms, but this enumeration provides no means to distinguish saddle points that are dynamically irrelevant from those that are dynamically relevant. For complex chaotic systems—large polyatomic molecules, large clusters, condensed phases, and so forth—potential energy surfaces are rough on the scale of thermal energies, $k_B T$, and dense in saddle points. Effectively, therefore, there is generally an uncountable number of saddle points. Searching for a few such points is therefore insufficient and likely irrelevant. Instead, one wants to locate and sample the ensemble of true dynamical bottlenecks. This task can be accomplished with transition path sampling.

TRANSITION PATH SAMPLING

IMPORTANCE SAMPLING   The basic idea is a generalization of standard Monte Carlo procedures (20, 21, 24, 25) that focuses upon chains of states constituting dynamical trajectories (26) rather than upon individual states. In its standard form, a Monte Carlo calculation performs a random walk in configuration space. The walk is biased to ensure that the most important regions of configuration space are adequately sampled. Specifically, in a Monte Carlo random walk, configuration x is visited in proportion to its probability $p(x)$. The walk may be initiated far from a typical configuration [i.e., x far from values of x where the weight from $p(x)$ is largest], but after some equilibration period, the bias drives the system to those important regions of configuration space. This feature is crucial to the success of Monte Carlo sampling. It is called importance sampling and is illustrated in Figure 2.

Importance sampling can be generalized to trajectory space, as we have done to create the methods of transition path sampling. Consider the ensemble of all trajectories that are, say, 1 ps long. Most of these trajectories will be localized near some basin of attraction—a long-lived collection of neighboring microstates. Rare transition state crossings will comprise a small subset of these 1-ps trajectories. For example, if the process of interest occurs roughly once every millisecond, then only one out of a billion 1-ps trajectories will exemplify that process. Transition path sampling provides an efficient means to sample such rare subensembles.

IMPORTANCE SAMPLING OF TRAJECTORY SPACE   Let us suppose the rare processes of interest are transitions between states or regions A and B. These regions are characterized by their respective population operators, $h_A(\chi)$ and $h_B(\chi)$. Here, χ denotes a point in phase space—configuration space and momentum space combined. (The applications of transition path sampling discussed in the following

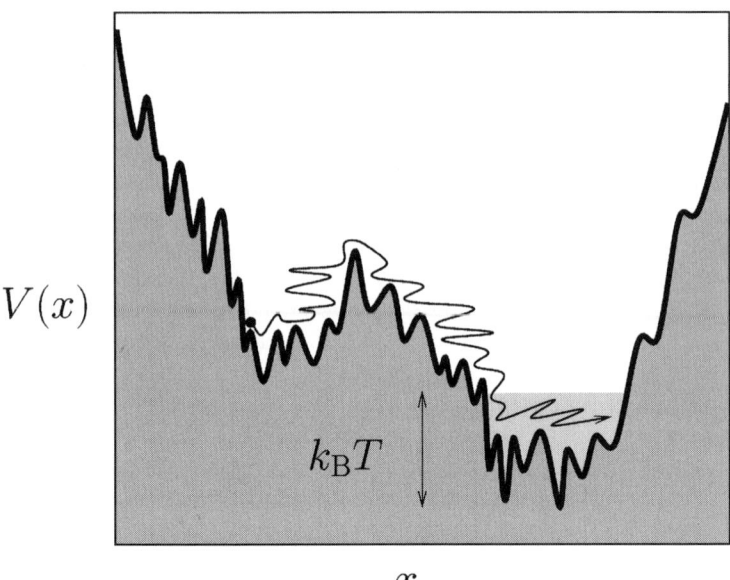

Figure 2 In a Metropolis Monte Carlo simulation, one generates a random walk in configuration space according to the probability distribution $p(x) \propto \exp[-V(x)/k_B T]$. If the distribution were that of a canonical ensemble, $V(x)$ would denote the potential energy for configuration x. Along this walk, a new configuration x' is generated by displacing the old configuration x by a randomly chosen small step, Δ. Then x' is accepted or rejected. If the step goes downhill in energy, i.e., if the new configuration has a higher probability than the old one, x' is always accepted. Uphill moves, on the other hand, are only accepted with a probability $w(x, \Delta)\, p(x')/p(x)\, w(x', -\Delta)$, where $w(x, \Delta)$ is the distribution for the random step, Δ, given the configuration x. In this way, barriers of the order of $k_B T$ or smaller do not hinder the random walk, and a system will move quickly to configurations of high probability (the lightly shaded region) even when initiated far away from that important region in configuration space.

sections of this review use characteristic functions of configuration space, x, only, but this limitation is not required.) When χ is within region A, $h_A(\chi) = 1$, otherwise, $h_A(\chi) = 0$. The corresponding population operator for region B, $h_B(\chi)$, is similarly defined. Transitions between regions A and B coincide with trajectories connecting these regions. A trajectory of time duration t, $\chi(t) = (\chi_0, \chi_1, \ldots, \chi_t)$, is a chronological sequence of phase space points generated by repeated application of a dynamical propagation rule. Trajectories we imagine are consistent with Liouville's equation or one of its analogues (27, 28). Namely, they must be reversible, must preserve the norm of the distribution of states, and must preserve an equilibrium distribution. For simplicity, but not for necessity, we might be considering deterministic dynamics, in which case χ_t is entirely determined by the initial phase space point, χ_0. The statistical weight for the rare trajectories connecting A and B is $h_A(\chi_0)\rho[\chi(t)]h_B(\chi_t)$, where $\rho[\chi(t)]$ is the unconstrained distribution functional for trajectories. For deterministic trajectories,

$$\rho[\chi(t)] = \rho(\chi_0) \prod_{0 < t' \leq t} \delta[\chi_{t'} - \chi_{t'}(\chi_0)], \qquad 1.$$

where $\rho(\chi_0)$ is the unconstrained distribution of initial phase space points, χ_0. Transition path sampling is done by carrying out a random walk in trajectory space, biased to be the importance sampling for the distribution $h_A(\chi_0)\rho[\chi(t)]h_B(\chi_t)$. Figure 3 illustrates how it is done in a practical and simple fashion.

In this perspective, stable or long-lived states A and B must be well characterized at the outset. This characterization can be difficult, as we discuss below. Nevertheless, we see that nothing need be presupposed about the dynamical pathways

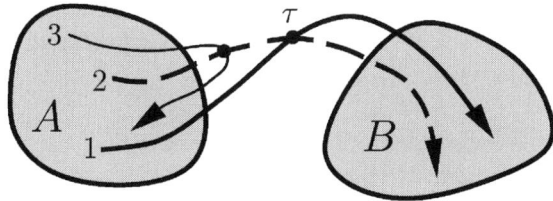

Figure 3 Illustration of "shooting moves," generating a random walk in trajectory space for Newtonian trajectories connecting regions A and B. For example, trajectory 2 is generated by changing trajectory 1 by a small amount. This change can be accomplished, for example, by first choosing a time slice point τ lying between 0 and t. At this time slice, the momentum of trajectory 1 can be altered by some small randomly chosen amount. The resulting new momentum can be used along with the configuration of trajectory 1 at time τ as the initial conditions for a new trajectory created by propagating forward from that phase space point for $t - \tau$ steps and backward from that phase space point for τ steps. Because regions A and B remain connected, this second path will be accepted as the new trajectory, provided the value of $\rho(\chi_0)$ for the new trajectory compares favorably with that for the first trajectory. Specifically, the probability to attempt a step from a trajectory $\chi(t) = (\chi_0, \chi_1, \ldots, \chi_t)$ to $\chi'(t) = (\chi'_0, \chi'_1, \ldots, \chi'_t)$ is the joint probability for choosing time slice τ and assigning a momentum change δ at that time slice, $w(\chi, \tau, \delta)$. The acceptance probability for that trial step is min[1, $w(\chi, \tau, \delta)$ $h_A(\chi'_0) \rho(\chi'_0) h_B(\chi'_t)/h_A(\chi_0)\rho(\chi_0)h_B(\chi_t) w(\chi', \tau, -\delta)$]. By the same type of procedure, trajectory 3 is generated from trajectory 2. This time, however, the new path does not connect A and B, and it is rejected. This sequence of acceptances and rejections ensures that the correct path ensemble is sampled—namely, the ensemble that is weighted by the distribution $h_A(\chi_0)\rho(\chi_0)h_B(\chi_t)$. There is great flexibility in the choice of random walk steps. This flexibility can be exploited in efforts to improve the efficiency of transition path sampling. In practice, shooting moves are only one of several moves employed in transition path sampling. References (2, 10, 62) describe other useful moves.

(i.e., trajectories) that join these states. This feature is the major strength of the method. Transition path sampling is a random walk through the ensemble of all paths connecting A and B. From studying the trajectories visited during this walk, the nature of the dynamical pathways is discovered.

COMPUTATIONAL COST   The computational effort in carrying out a transition path sampling calculation scales linearly with the number of trajectories harvested. This scaling is optimum. In particular, to harvest N statistically independent transition pathways of length t requires the same order of effort as that required to perform a single trajectory of length N. In practice, random moves like those illustrated in Figure 3 are accepted with probabilities between 0.1 and 0.5. In addition, the correlations in that random walk persist typically for only two or three accepted moves. Thus, for instance, 1000 statistically independent 1 ps trajectories are obtained with roughly the same computational resources required for a single straightforward trajectory of length 1–10 ns. The straightforward trajectory, however, will almost certainly not show an example of a rare event occurring on the timescale of, say, 1 ms, while each of 1000 transition path trajectories will exhibit an independent example of the event.

INITIAL TRAJECTORY   Before the sampling of typical transition paths begins, one requires a representative member of the ensemble of trajectories with distribution $h_A(\chi_0)\rho[\chi(t)]h_B(\chi_t)$. This member, i.e., this first example of a typical trajectory linking regions A and B, can be obtained in a variety of ways. All of these ways coincide with some sort of equilibration run. The situation is much like that encountered in standard Monte Carlo. In that case, the Monte Carlo walk is initiated at some chosen configuration. The configuration may be far from a typical equilibrium configuration, as illustrated in Figure 2. Nevertheless, after repeated steps in the random walk, each one satisfying detailed balance, the system eventually reaches the region of typical equilibrium configurations. It is at this point where equilibrium sampling is initiated. Similarly, in transition path sampling, one may begin with literally no concept of a reasonable dynamical trajectory. Any initial path can be drawn to initiate an equilibration run. After equilibration, i.e., after the walk through trajectory space begins to visit trajectories typical of the weight functional $h_A(\chi_0)\,\rho[\chi(t)]\,h_B(\chi_t)$, sampling can begin.

For example, suppose trajectories connecting regions A and B are easily found in a dynamical simulation run at a temperature T', but the actual temperature of interest, T, is much smaller than T'. In other words, suppose one has examples of trajectories taken from the distribution $h_A(\chi_0)\,\rho[\chi(t); T']\,h_B(\chi_t)$, but one wants to sample the distribution $h_A(\chi_0)\,\rho[\chi(t); T]\,h_B(\chi_t)$. One may use the high-temperature trajectory taken from the former and initiate an equilibration run with the latter. If there is poor overlap between the distributions $\rho[\chi(t); T']$ and $\rho[\chi(t); T]$, this run may be done in stages, lowering the temperature by only a fraction of $T' - T$ at each stage.

Some initial paths may be farther from the desired ensemble than others, and some equilibration walks may be slower than others. Nevertheless, this illustrations shows that there is great flexibility as to how one may proceed. We discuss this point further below.

REVERSIBLE WORK

Standard Monte Carlo sampling of microstates follows from the principles of equilibrium statistical mechanics, and quantities computed from it are thermodynamic properties. Similarly, transition path sampling follows from a statistical mechanics of trajectory space, and quantities computed from it are dynamical properties, like rate constants. The two techniques share an important similarity—namely, they both move through their respective spaces (configuration space and trajectory space) in fashions that preserve their prescribed distributions. In other words, they both obey conditions of detailed balance. This similarity can be used to establish an isomorphism between thermodynamical quantities and dynamical properties. The isomorphism is of practical importance because it makes accessible to the study of dynamics all the computational advantages of methods used to determine the statistics of rare configurations in an equilibrium system.

REVERSIBLE WORK IN EQUILIBRIUM STATISTICAL MECHANICS   To illustrate the isomorphism, consider first the traditional connection between thermodynamics and equilibrium statistical mechanics. The partition function associated with a thermodynamic state A, Z_A, is the sum over the configurations that characterize state A weighted by the distribution $p(x)$, i.e., $Z_A = \sum_x h_A(x) p(x)$. (In the context of equilibrium statistical mechanics, we define states in terms of configurational variables, x, rather than phase space variables, χ.) The reversible work to move from thermodynamic state A to thermodynamic state B, W_{AB}, is the free energy difference between those states. Namely,

$$\exp(-W_{AB}/k_B T) = \frac{\sum_x h_B(x) p(x)}{\sum_x h_A(x) p(x)}, \qquad 2.$$

or $W_{AB} = -k_B T \ln(Z_A/Z_B)$. In addition, for a system with distribution $p(x)$, $\exp(-W_{AB}/k_B T)$ is the probability that the system is found in state B relative to that of being found in state A. As such, one may efficiently compute the relative probability for being in state B, even when this probability is extremely small, i.e., even when $W_{AB} \gg k_B T$. In particular, because reversible work is independent of path, it can be evaluated by moving the system reversibly through an arbitrarily chosen series of intermediate states. A specific reversible path is created by a specific series of steps for converting $h_A(x)$ to $h_B(x)$. For instance, one can introduce a class of functions, $h^{(\lambda)}(x)$, that smoothly interpolate between $h_A(x)$ at $\lambda = 0$ to $h_B(x)$ at $\lambda = 1$. For a given λ, the partition function is $Z^{(\lambda)} = \sum_x h^{(\lambda)}(x) p(x)$, and provided that $h^{(\lambda)}(x)$ has a reasonable overlap with $h^{(\lambda-\Delta\lambda)}(x)$, we can also write

$$Z^{(\lambda)} = \sum_x \left[h^{(\lambda)}(x)/h^{(\lambda-\Delta\lambda)}(x)\right] h^{(\lambda-\Delta\lambda)}(x) p(x)$$
$$= Z^{(\lambda-\Delta\lambda)} \langle h^{(\lambda)}(x)/h^{(\lambda-\Delta\lambda)}(x) \rangle_{\lambda-\Delta\lambda}, \qquad 3.$$

where $\langle \ldots \rangle_\lambda$ denotes the average with distribution $h^{(\lambda)}(x)p(x)$. By applying this result over and over again, with $\lambda = \Delta\lambda, 2\Delta\lambda, \ldots, 1$, the quantity Z_B/Z_A is determined. In order to ensure reasonable overlap of adjacent distributions, the number of steps required in this procedure is of the order of W_{AB}/k_BT, i.e., $\Delta\lambda \simeq k_BT/W_{AB}$. In contrast, a straightforward Monte Carlo sampling of $p(x)$ will provide a reasonable estimate of the probability ratio, Z_A/Z_B, in a computational timescale of $\bar{t} \exp(W_{AB}/k_BT)$, where $\bar{t}$ is a typical sampling time, such as that to arrive at reasonable statistics for just state A. This juxtaposition of linear vs. exponential computational cost shows that whenever $W_{AB} \gg k_BT$, the stepwise procedure makes feasible estimates that would be impossible to perform in a straightforward simulation.

REVERSIBLE WORK FOR CHANGING ENSEMBLES OF TRAJECTORIES With these ideas in mind, we now consider the "partition function" for trajectories of length t connecting regions A and B. Namely,

$$Z_{AB}(t) = \sum_{\chi(t)} \rho[\chi(t)] h_A(\chi_0) h_B(\chi_t). \qquad 4.$$

The sum over $\chi(t)$ denotes the sum over all trajectories $(\chi_0, \chi_1, \ldots, \chi_t)$. For deterministic trajectories, Equation 4 reduces to

$$Z_{AB}(t) = \sum_{\chi_0} \rho(\chi_0) h_A(\chi_0) h_B(\chi_t), \qquad 5.$$

where χ_t is determined solely by χ_0. This partition function counts the number of trajectories connecting A and B, weighted by the distribution functional, $\rho[\chi(t)]$. In contrast to this quantity, consider the similarly weighted number of trajectories that begin in A and end anywhere,

$$\sum_{\chi(t)} \rho[\chi(t)] h_A(\chi_0) = \sum_{\chi_0} \rho(\chi_0) h_A(\chi_0) = Z_A, \qquad 6.$$

where the first equality follows from the normalization of the distribution functional, and the second is true when $\rho(\chi)$ is an equilibrium distribution—microcanonical, or canonical, or so forth. In that case, $Z_{AB}(t)$ is the time correlation function, $\langle h_A(0) h_B(t) \rangle$. Here, $\langle \ldots \rangle$ denotes the equilibrium ensemble average over initial conditions, and $h_B(t)$ is the population of state B at time t. Thus, the ratio of partition functions,

$$Z_{AB}(t)/Z_A = \frac{\langle h_A(0) h_B(t) \rangle}{\langle h_A \rangle}, \qquad 7.$$

is the probability of finding the system in state B a time t after it was in state A. If A and B are separated by a single dynamical bottleneck, this probability will increase from 0 to $\langle h_B \rangle$ with a time dependence that is exponential after a transient time, τ_{mol}. The transient time is the typical time for a trajectory to cross the bottleneck and commit to one of the two basins of attraction. It is a relatively short time, far shorter than the exponential relaxation time, $\tau_{\text{rxn}} = 1/(k_{AB} + k_{BA})$, where k_{AB} is the rate constant for transitions from A to B, and k_{BA} is that for reverse transitions (17). As such, the rate constant for transitions from A to B can be computed as a ratio of partition functions,

$$Z_{AB}(t)/Z_A = k_{AB} t, \quad \tau_{\text{mol}} < t \ll \tau_{\text{rxn}}. \qquad 8.$$

The first inequality, $\tau_{\text{mol}} < t$, establishes the appropriate length for the trajectories harvested by transition path sampling for the crossing of a single bottleneck. Trajectories should be long enough to show that $Z_{AB}(t)/Z_A$ grows linearly in time. Trajectories of shorter length will be atypical of the transition path ensemble. In cases where B is not reached from A in typically one step, but through one or more intermediate long-lived states, $Z_{AB}(t)/Z_A$ will not exhibit linear behavior after a short period of time. This fact provides a criterion that can be used to discover the existence of intermediate states (2).

The partition function $Z_{AB}(t)$ converts to Z_A when the population operator $h_B(\chi)$ is converted to unity. Hence, $-k_B T \ln[Z_{AB}(t)/Z_A]$ can be viewed as the reversible work to change from the ensemble of trajectories initiated in A to the ensemble of trajectories connecting regions A and B. Furthermore, this work is independent of the specific path, provided the steps are taken reversibly. In other words, with a slight change in notation, the second equality in Equation 3 can apply to the calculation of $Z_{AB}(t)/Z_A$—namely,

$$Z^{(\lambda)}(t) = Z^{(\lambda - \Delta\lambda)}(t) \left\langle h^{(\lambda)}(\chi_t) / h^{(\lambda - \Delta\lambda)}(\chi_t) \right\rangle_{\lambda - \Delta\lambda} \qquad 9.$$

where $h^{(\lambda)}(\chi)$ interpolates between 1 at $\lambda = 0$ and $h_B(\chi)$ at $\lambda = 1$, and $\langle \ldots \rangle_\lambda$ denotes the average over trajectories of length t weighted by the distribution proportional to $h_A(\chi_0) \rho(\chi_0) h^{(\lambda)}(\chi_t)$. As such, $Z^{(\lambda)}(t)$ changes from Z_A when $\lambda = 0$ to $Z_{AB}(t)$ when $\lambda = 1$. As in the standard equilibrium case, Equation 3, the dynamical formula (9) is to be applied with a choice of $h^{(\lambda)}(\chi)$ that allows for reasonable overlap between adjacent ensembles. In addition, as in the equilibrium case, the dynamical formula (9) provides the basis for computing the desired partition function ratio with linear rather than exponential computational effort.

STEPWISE ROUTE TO THE INITIAL TRAJECTORY Finally, by converting from the ensemble where trajectories begin in A to the ensemble where trajectories link A and B, the stepwise procedure provides a method for preparing the initial trajectory for transition path sampling. It is a laborious preparation, moving from one ensemble to the next. For specific situations, more efficient preparation schemes can be devised, as we discuss below. Nevertheless, this example, illustrated in Figures 4 and 5,

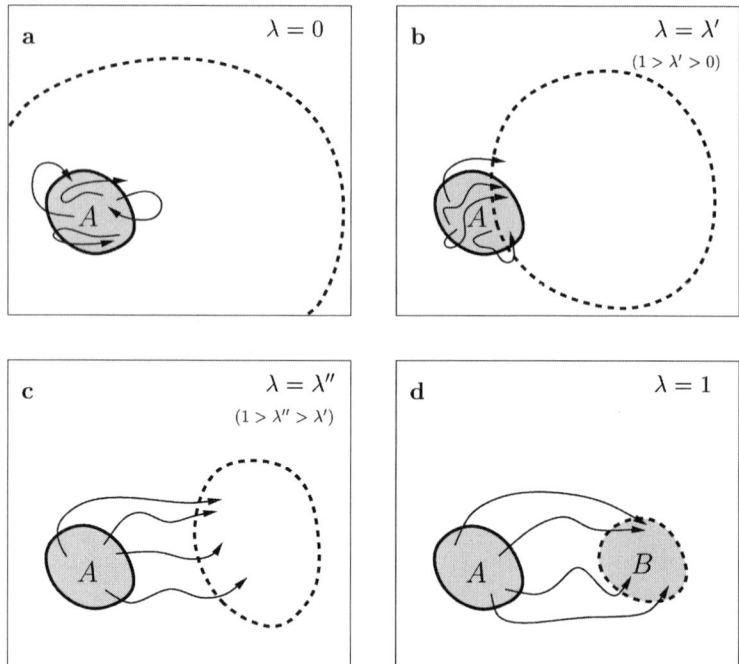

Figure 4 Schematic sequence of trajectory ensembles, with distribution functionals $h_A(\chi_0)\rho[\chi(t)]h^{(\lambda)}(\chi_t)$, changing from $\lambda = 0$ to $\lambda = 1$. Dashed lines surround regions where $h^{(\lambda)}(\chi)$ is nonzero. For $\lambda = 0$, trajectories may end anywhere in the accessible phase space. For $\lambda = 1$, trajectories must end in state B. In the initial stages of the sequence, the transition state surface lies within the region defined by $h^{(\lambda)}(\chi)$, and typical trajectories remain in the basin of attraction of state A. In the latter stages, the transition state surface lies outside the region defined by $h^{(\lambda)}(\chi)$, so that trajectories must cross the separatrix and typically continue deep into the basin of attraction of state B. This scheme will generally succeed at creating the desired final ensemble of trajectories passing from A to B, but the scheme is not satisfactory for computing rate constants. A satisfactory scheme must reach the final ensemble reversibly. The latter stages of the sequence illustrated in this figure will usually fail to be reversible, because they do not efficiently sample trajectories that end near the transition state surface on the side of state B. To ensure reversibility, this scheme can be modified to use a sequence of more confined "window" ensembles (4), much as is done with umbrella sampling in equilibrium statistical mechanics (20, 21, 24, 25). The ith such window includes only trajectories that end in the region defined by $h^{(\lambda_i - \Delta)}(\chi)[1 - h^{(\lambda_{i+1} + \Delta)}(\chi)]$. Here, Δ is a small, positive number that allows for reasonable overlap between adjacent ensembles. With appropriately chosen values for λ_i, the reversible work is comparable to $k_B T$ for each step in this modified scheme.

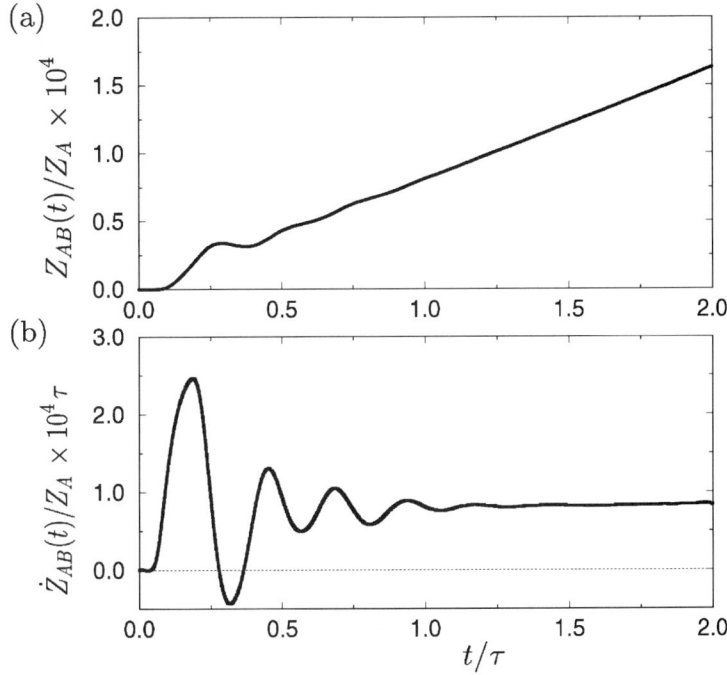

Figure 5 (*a*) The ratio of partition functions $Z_{AB}(t)$ and Z_A as a function of time t, calculated using the scheme illustrated in Figure 4, for the process described in the section "Isomerization of a Solvated Model Dimer," below. For times longer than that required to commit to a basin of attraction ($\approx \tau$ in this example) but short compared to the characteristic time of spontaneous transitions, this ratio is a linear function of time. (See Equation 8.) The corresponding slope, i.e., the plateau value of $d[Z_{AB}(t)/Z_A]/dt$ in (*b*), is the rate constant for transitions from A to B.

indicates that without any prior knowledge of dynamical pathways, manageable procedures exist for using transition path sampling to harvest transition paths and to compute rate constants.

UNBAISED DYNAMICS An important feature of transition path sampling is that harvested paths are true dynamical pathways, unhindered by arbitrarily imposed forces. Ensembles of paths are prepared by manipulating the constraints that define the ensembles. These manipulations do not affect the equations of motion governing the dynamics of the system. Other methods used to build ensembles of paths connecting basins of attraction introduce unphysical forces that affect the paths directly and do not preserve the intended equilibrium distribution (29–38). As such, these other methods harvest paths that are not the actual trajectories of the system.

DISTINGUISHING BASINS OF ATTRACTION Often, a significant challenge in transition path sampling work is the characterization of the stable states. It requires a choice of discriminating order parameters—variables that uniquely distinguish states A and B. Establishing that a variable, say q, has mostly one range of values in one state and a nearly distinct range of values in the other is not sufficient. There must be no overlap between the region spanned by $h_A(\chi)$ and the basin of attraction of state B, and vice versa. Otherwise, sampling with the weight functional $\rho[\chi(t)]\,h_A(\chi_0)h_B(\chi_t)$ will fail to harvest trajectories crossing from one basin to the other. This point is illustrated in Figure 6, and is exemplified by the difficulty

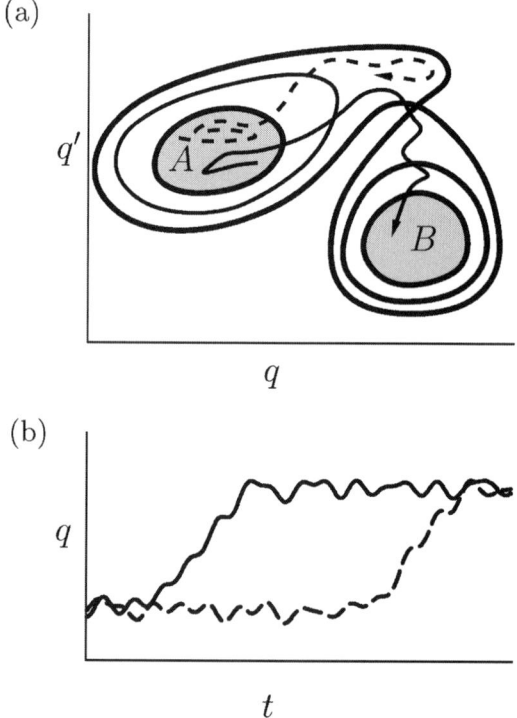

Figure 6 (*a*) Contours of a free energy surface, $F(q, q')$, for which the coordinate q does not successfully discriminate between the basins of attraction of states A and B. Although the distributions of q within A and B do not overlap, some microstates belonging to the basin of attraction of A have values of q characteristic of B. (*b*) q as a function of time for the two trajectories sketched in (*a*). The trajectory depicted as a solid line makes a transition from A to B, passing through the transition state surface. The trajectory depicted as a dashed line remains within the basin of attraction of state A, but, when projected onto q, appears to visit state B. Transition path sampling with q as an order parameter would yield primarily trajectories of the latter type, which do not pass through the transition state surface.

of sampling pathways for excess proton transport in liquid water. In this application, basins of attraction for hydronium ion structures are poorly characterized by molecular geometries (39) and weights of empirical valence bond states (40). Day et al. have attempted to circumvent this problem by studying proton transfer, from a hydronium ion to a nearby water molecule, through an intervening water molecule (41). Because the order parameter Day et al. use may not distinguish among the pertinent states, however, it is possible that the trajectories they have harvested do not represent true proton transfer events. These pathways may comprise instead large fluctuations within the basin of attraction of the intermediate state. In our experience, identifying discriminating order parameters can involve a significant degree of experimentation, performing transition path sampling with various choices of order parameters until a satisfactory discriminating choice is determined.

COMMITTORS, THE SEPARATRIX, AND THE TRANSITION STATE ENSEMBLE

Harvested transition paths can be examined to determine examples of configurations lying on the transition state surface. This examination is done with the concepts of committors and the separatrix. The committor, $p_A(x, t_s)$, is the probability (or fraction) of fleeting trajectories $\overline{\chi}(t_s)$ initiated from configuration x to end in state A a short time t_s later—namely,

$$p_A(x, t_s) = \sum_{\overline{\chi}(t_s)} \rho[\overline{\chi}(t_s)] \, \delta(\overline{x}_0 - x) \, h_A(\overline{\chi}_{t_s})/p(x), \qquad 10.$$

where the δ-function has unit weight when the initial configuration of the fleeting trajectory, $\overline{x}_0$, coincides with x and is zero otherwise. We often use the abbreviated symbol p_A for the committor, leaving the dependence upon x and t_s to be understood implicitly. In the context of protein folding, this object has been called p_F—for "p-fold"—or $(1 - p_F)$, depending on whether the protein ends in a folded or unfolded state (42, 43). For the sequence of configurations visited in a specific trajectory connecting A and B, $(x_0, x_1, \ldots, x_\tau, \ldots, x_t)$, p_A can be viewed as a function of τ. For physical situations where transitions between A and B exhibit the typical timescale separation $\tau_{\text{mol}} \ll \tau_{\text{rxn}}$, $p_A(\tau)$ will be either 1 or 0, except for one or a few short periods of time where the function changes between these two values. As illustrated in Figure 7, the short period(s) coincide with crossing(s) of the dynamical bottleneck. Thus, meaningful examination of the bottleneck is obtained from a committor if the short time, t_s, is of the order of the commitment time, τ_{mol}. A time slice on a trajectory connecting A and B is committed to state A if $p_B \ll p_A \simeq 1$ for the configuration at that time slice. Here, p_B is defined in the same way as p_A. Similarly, a time slice is committed to state B if $p_A \ll p_B \simeq 1$. On the other hand, a time slice where $p_A \simeq p_B \simeq 1/2$ coincides with the location of the bottleneck. It is a configuration on a separatrix—a surface in configuration

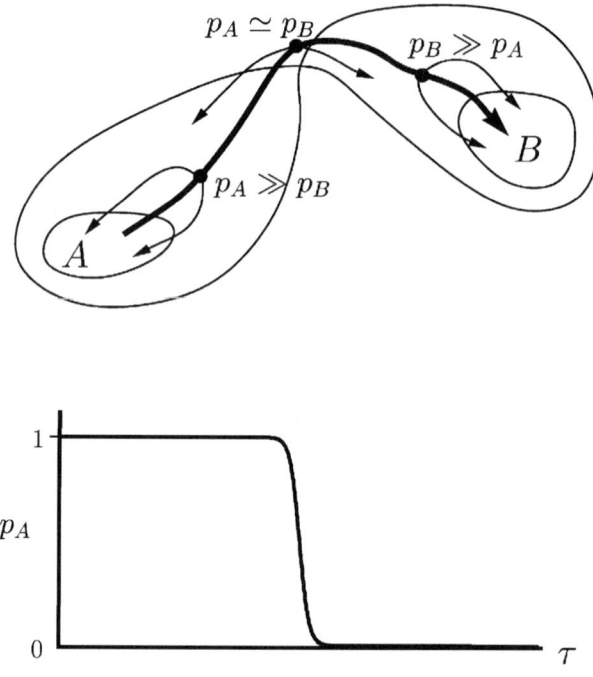

Figure 7 The committor, p_A, is computed along a single path in the transition path ensemble (thick solid line, top panel) by determining the percentage of fleeting trial trajectories starting from the configuration at time slice τ (with random momenta) that has reached region A in a time t. Typically 10–100 of these fleeting trajectories are needed to obtain p_A accurately. For instance, $p_A \approx 1$ for the left time slice in the top panel, because nearly all trajectories started from that time slice end in A. The configurations for which $p_A \simeq p_B$ are considered transition states.

space where initiated trajectories have equal likelihood of ending in either state A or state B (44).

For a system of few enough dimensions, the separatrix simply locates saddle points on the potential energy surface—the simplest conception of transition states. For complex high-dimensional systems, however, saddle points are not necessarily signatures of dynamical bottlenecks. For such systems, the separatrix provides the generally applicable definition of a transition state surface (42–44). The definition is particularly useful in connection with transition path sampling. Suppose the sampling has been employed to harvest N trajectories connecting A and B. Configurations along each of these trajectories can be examined statistically to determine which configurations have $p_A \simeq p_B \simeq 1/2$, as illustrated in Figure 7. Each trajectory will pass through one or more such configurations. A given trajectory may pass through the surface more than once. Those that pass through once exhibit one barrier or bottleneck crossing. Those that pass through more than once exhibit multiple crossings. As such, this analysis will yield N or more examples of the transition state surface. Each example is a member of the transition state ensemble.

Access to an ensemble of typical transition state configurations proves useful for understanding the mechanism of a rare event in a complex system. Relevant dynamic variables are usually collective coordinates, and identifying these variables through explicit visualization of specific dynamic pathways is usually impossible. In addition, for a many-particle system, there is generally a huge variety of atomistic pathways that accomplish the transformation from A to B. Viewing just one or a few examples is unlikely to reveal what is typical. Rather, statistical analysis of the process is needed. An ensemble of transition states provides data for carrying out such an analysis. In particular, averaging a dynamical variable over this ensemble can be compared with averaging the variable over configurations typical to states A and B. Substantial differences between the transition state average and the stable state averages would suggest that the variable is significant to the mechanism of the transitions between A and B. Ascertaining the degree to which the variable describes the dynamical mechanism in its entirety requires additional analysis, of the sort we turn to in the next section.

ORDER PARAMETERS VS. REACTION COORDINATES AND COMMITTOR DISTRIBUTIONS

There is an important distinction between variables that characterize basins of attraction and variables that characterize dynamical mechanisms. We refer to the former as "order parameters" and the latter as "reaction coordinates." Order parameters are used to construct the population functions $h_A(\chi)$ and $h_B(\chi)$. Reaction coordinates can be used to define the transition state ensemble. For example, suppose that a configurational variable q is presumed to be the reaction coordinate. Its free energy $W(q)$—the reversible work function for controlling q—is determined by the partition function for the system when constrained to that value of q—namely,

$$\exp[-W(q)/k_B T] \propto \sum_x p(x)\, \delta[q(x) - q]. \qquad 11.$$

Viewing the δ-function in Equation 11 as requiring $q(x)$ to lie in a small but finite interval, $q \pm \Delta q/2$, $W(q)$ can be evaluated in steps, as with the method illustrated by Equation 3. To the extent that q is truly relevant to the dynamical mechanism, $W(q)$ will have a maximum at some intermediate value, q^*, and that value of q coincides with the location of the transition state surface. Figure 8 illustrates this behavior. Of course, if q is particularly irrelevant, it could exhibit no maximum. Figure 8 also illustrates the important distinction between order parameters and reaction coordinates. Even when a variable q serves well to distinguish equilibrium states A and B, the location of q^* and the value of $W(q^*)$ may have nothing to do with the dynamical bottleneck for $A \rightarrow B$ transitions. Indeed, the transmission

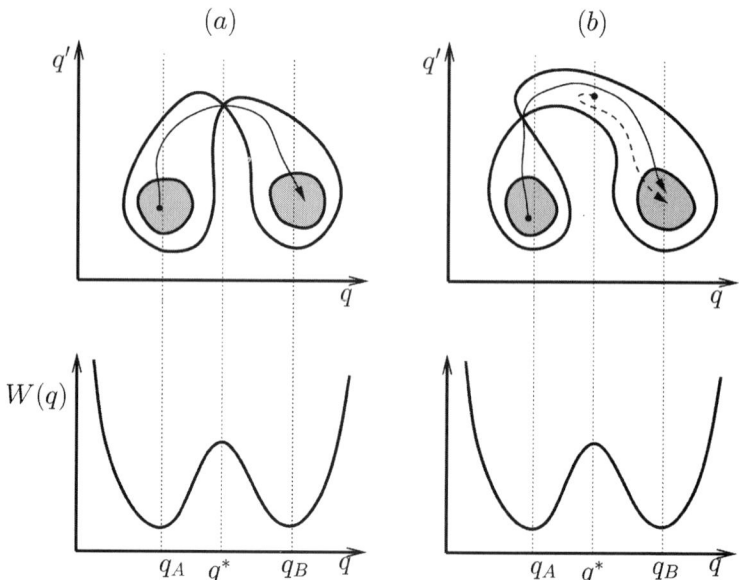

Figure 8 Two illustrative potential energies $V(q, q')$ and their corresponding free energy functions $W(q) = -k_B T \ln \sum_{q'} \exp[-V(q,q')/k_B T]$. (a) The coordinate q serves as a reasonable order parameter, distinguishing basin A from basin B. It is also a reasonable reaction coordinate, because the transition state surface coincides with $q = q^*$. (b) Here, q might appear to be a reasonable order parameter because its typical values in state A are indeed different than those for state B, but it is not a discriminating order parameter. Further, it is not a reasonable reaction coordinate. The orthogonal variable q' plays an important role in $A \to B$ transitions, and the maximum in $W(q)$ at $q = q^*$ does not coincide with the transition state surface. The dashed trajectory beginning at q^* and ending in B illustrates this point.

probability for trajectories launched from the $q = q^*$ surface of Figure 8b (i.e., the fraction of these trajectories that reach A or B without recrossing the $q = q^*$ surface) will be close to zero.

The illustration in Figure 8b is not far-fetched. Consider the kinetics of a liquid-vapor phase transition in circumstances where the liquid, for example, is metastable, and its density, ρ_l, is much greater than that of the vapor, ρ_v. The bulk density of the fluid, ρ, serves as a reasonable order parameter because microstates with $\rho \approx \rho_l$ or $\rho \approx \rho_v$ will coincide with the liquid or vapor phase, respectively. In contrast, the kinetics of forming one from the other will involve the formation of an interface and critical nucleus—a vapor bubble in the liquid. An illustration of this dynamic is found in a transition path sampling study of a surface-induced evaporation (8). The dynamically relevant variables describe the size and shape of the bubble. These variables are virtually orthogonal to the bulk density. Thus, the picture in Figure 8b is a reasonable caricature in this case. Similarly, consider the dissociation of an ion pair, say Na$^+$ and Cl$^-$, in liquid water. The distance between the ions, r, can serve as an order parameter, distinguishing the state where the ions are in contact from the state where they are separately solvated. The free energy or reversible work function in this case, $W(r)$, is the potential of mean force (21). It shows a deep minimum at small r, corresponding to ions in contact, and a barrier to a stable state at larger r in which the ions are separately solvated (45, 46). The barrier at $r = r^*$ corresponds to a least likely separation of the ions, where no water can fit between them. But r^* is not a good indicator of the transition state ensemble as suggested by the low transmission probability for trajectories initiated at states with $r = r^*$ (47, 48). In fact, microstates prepared with $r = r^*$ most likely coincide with one or the other of the stable states as shown in Reference (6). The kinetic mechanism for the ion dissociation involves a fluctuation in the water density surrounding the ion pair, creating space for the ions to move apart and inserting water molecules between them (6). The variables describing this solvent rearrangement are virtually orthogonal to r. Figure 8b is thus close to a reasonable caricature in this case. Indeed, given the complexity of a high-dimensional system, the coincidence of order parameter and reaction coordinate would seem unlikely. Something like Figure 8b would seem to be more like the rule than the exception.

Committor distributions provide a statistical diagnostic for the correctness of a presumed reaction coordinate, q. Specifically, one may compute the committor $p_A(x, t_s)$ for configurations in the ensemble with $q(x) = q^*$. This ensemble is sampled at the stage where $q \approx q^*$ in the stepwise calculation of $W(q)$ (see Equation 11). The distributions of these computed committors is $P(p_A) = \langle \delta[p_A(x, t_s) - p_A] \rangle_{q^*}$, where $\langle \ldots \rangle_{q^*}$ denotes the average over the ensemble with $q(x) = q^*$. To the extent that q is indeed a good reaction coordinate, $P(p_A)$ will be sharply peaked at $p_A \approx 1/2$. Different behaviors suggest different involvements of other coordinates. Various behaviors are illustrated in Figure 9.

The idea of considering the committor distribution was introduced in Reference (6), where the kinetics of ion pair dissociation was studied. For that situation, using the interionic separation, r, as the presumed reaction coordinate, $P(p_A)$ was found to be bimodal, with peaks at 0 and 1. This sort of behavior is illustrated in panel (b) of Figure 9. It indicates that a barrier must be crossed moving in a direction other than that of r. Truhlar & Garrett have noted that the bimodal character of $P(p_A)$ can be captured analytically with a two-dimensional parabolic barrier model, where the presumed reaction coordinate is essentially orthogonal to the actual saddle point surface (49). It remains unknown how to apply the simple model to ion dissociation (where the orthogonal variable is a collective coordinate describing density fluctuations near the ions) or to any other kinetic process in a complex system.

The utility of computing committor distributions is not specific to transition path sampling. This diagnostic alone indicates whether a postulated reaction coordinate indeed drives a transition or is instead simply correlated with its progress.

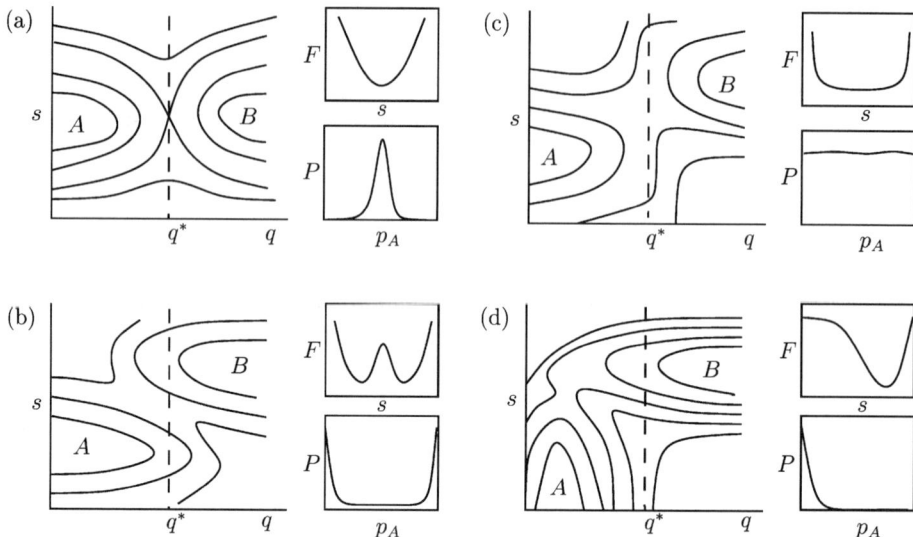

Figure 9 Four different potential or free-energy landscapes $V(q, s)$. Alongside each are plotted the corresponding free energy, $F(q^*, s)$, and committor distribution, $P(p_A)$, for the ensemble of microstates with $q = q^*$. For landscape (a), the reaction coordinate is adequately described by q, and $P(p_A)$ is peaked at $p_A = 1/2$. For landscape (b), the reaction coordinate has a significant component along s, as indicated by the barrier in $F(q^*, s)$ and the bimodal shape of $P(p_A)$. In (c), s is again an important dynamical variable. In this case $P(p_A)$ is nearly constant, suggesting that motion along s is diffusive when q is near q^*. Finally, for landscape (d), the reaction coordinate is orthogonal to q, reflected by the single peak of $P(p_A)$ near $p_A = 0$. In this case, almost none of the configurations belonging to the constrained ensemble with $q = q^*$ lie on the transition state surface.

Averaging variables over many examples of a transition does not provide equivalent information. Day et al., for example, have demonstrated that certain hydrogen bond angles change on average during transfer of an excess proton in liquid water (41). But in order to establish that the proton transfer mechanism can be described using only these coordinates, it will be necessary to compute the appropriate distribution of committors. Similarly, determining only the mean of a committor distribution does not provide information about the possible importance of orthogonal coordinates. In remarkable experimental studies of colloidal crystallization, Gasser et al. have, in effect, determined $\langle p_A \rangle_R$ for various crystallite sizes R (50). The monotonic decrease of $\langle p_A \rangle_R$ with increasing R, passing through $\langle p_A \rangle_{R_c} \simeq 1/2$ for a critical size R_c, indicates that cluster size is indeed correlated with the progress of nucleation. But it does not guarantee that the ensemble of configurations with $R = R_c$ coincides with the transition state surface for nucleation.

APPLICATIONS

The previous sections have outlined the essential concepts of transition path sampling. What remains to be discussed are technical issues that a practitioner will encounter when actually attempting a transition path sampling calculation. Several of these issues are mentioned below in the context of different applications of transition path sampling. The reader can find detailed discussions in the papers presenting these applications. In addition, computer programs with simple illustrative examples are found at http://gold.cchem.berkeley.edu/TPS_code.html.

Heptamer of Cold Lennard-Jones Disks

This model system was first investigated in Reference (2) with transition path sampling of stochastic trajectories. The lowest-energy state of the cluster has one disk at the center with the remaining six packed in a circle around it. There are, of course, many such states, each one a particle label permutation of the first. Transitions from one such ground state to another involve transitions between intermediate states. Minimization (or quenching) of the path action was used to discover the intermediate states and the possible chronologies with which they are visited. Rate constants were computed from transition path sampling of the trajectories connecting adjacent intermediate states.

An earlier paper (1) set down many of the principles of transition path sampling, but with move sets and rules that are more complicated and less efficient than those introduced in Reference (2). It is here that shooting moves like those illustrated in Figure 3 were introduced. In a shooting move, a new path is created from an old one by slightly changing the momenta at a randomly selected time slice. Then, the equations of motion are integrated forward and backward in time starting from this modified phase space point. If the new trajectory is reactive, i.e., it starts in A and ends in B, it can be accepted. Otherwise, it is rejected. The average acceptance probability can be adjusted by varying the magnitude of random momentum displacements. These shooting moves together with shifting moves, in which the path is simply translated in time (2), prove indispensable for efficient transition path sampling. A second study of the Lennard-Jones heptamer introduced the use of transition path sampling for deterministic trajectories (3). We recommend Reference (3) as the simplest place to start learning about transition path sampling.

Isomerization of a Solvated Model Dimer

Equations 7 and 8 relate the time correlation function $\langle h_A(0)h_B(t)\rangle$ to a free energy difference between path ensembles and provide the theoretical basis for the calculation of transition rate constants. An efficient way to exploit this relationship was developed in Reference (4). In this approach the reversible work required to confine the endpoint of the transition pathways into region B at time t is decomposed into two terms. The first term is the free energy of confinement for a particular time t', and the second term is the free energy required to change the length of the path from t' to t. Whereas calculation of the first term requires a computationally expensive thermodynamic integration, the second term can be readily evaluated in a single transition path sampling calculation.

This method was demonstrated by calculating isomerization rate constants for a diatomic molecule immersed in a solvent of soft spheres. Because isomers of the diatomic differ in bond length, interconversion is mediated by the solvent. For a sufficiently high internal energy barrier, isomerization events are rare.

In these simulations, shooting and shifting moves were supplemented with path reversal moves in which new initial conditions are obtained by exchanging the final and the initial point of the path and inverting the momenta. Because no new integration of the equations of motion is necessary for path reversals, the computational cost for this path move is negligible. Path reversal moves can facilitate ergodic sampling if qualitatively different transition pathways exist.

The efficiency of transition path sampling depends on the degree of correlation between successive steps in the random walk through trajectory space. On one hand, these correlations hinder rapid sampling because subsequent pathways bear certain similarities. On the other hand, it is exactly this similarity that guarantees a nonvanishing acceptance probability for trial steps. As in any Monte Carlo procedure, these two aspects should be balanced. A systematic study of sampling efficiency for the solvated diatomic indicated that optimum sampling is obtained for acceptance probabilities ranging from 30% to 60%. This range of values should be used as a rule of thumb when an efficiency analysis is computationally impractical.

Water Clusters

A cluster consisting of three water molecules and an excess proton may be viewed as the simplest aqueous system in which activated proton transfer occurs. This transfer results in a permutation of atomic labels in the cluster's stable state—a distinct hydronium ion solvated by two neutral water molecules. Transition path sampling was used in References (5, 11, 51) to determine rate constants and transition states for proton transfer in several empirical models of $(H_2O)_3H^+$. Harvesting transition paths for an ab initio model of the cluster, accomplished with Car-Parrinello molecular dynamics (CPMD) (52, 53), required selective storage of trajectory data. The temporal locations of future Monte Carlo moves were chosen (at random) before computing trial pathways. In this way, the massive amount of data detailing the electronic wave function was stored only at a few, predetermined times as each trajectory was integrated. This scheme is useful in general for applications in which data storage is burdensome.

Two structurally distinct classes of transition states control proton transfer in this system. Although a large energetic barrier lies between them on the separatrix, the temporally nonlocal nature of shooting moves allows for both transition state regions to be visited in a single walk through trajectory space. The kinetics computed by path sampling thus correctly deviate from predictions of Rice, Ramsperger, Kassel, and Marcus (RRKM) theory (54–57), when assuming a single harmonic transition state region.

Sampling proton transfer pathways with vanishing total linear momentum, **P**, and angular momentum, **L**, requires that trial moves be performed carefully. Proper construction of shooting moves consistent with the microcanonical ensemble and with constraints on linear functions of particle momenta (such as **P** and **L**) is discussed in the Appendix of Reference (5). This construction has also been used in other applications to properly incorporate constraints on interparticle distances (6). Improper treatment of such constraints incorrectly biases the walk through trajectory space, and we have found it to generate qualitatively erroneous results in the case of ion pair dissociation in liquid water.

At low temperatures, neutral clusters of a few water molecules exist in a manifold of solid-like stable states which interconvert infrequently. At higher temperatures, these crystalline structures are replaced by amorphous liquid-like ones. Transition path sampling was used in Reference (58) to collect pathways for both low-temperature isomerizations and the melting transition in the water octamer, $(H_2O)_8$. Because the liquid state is stabilized by entropy, transition states for the melting transition do not correspond to saddle points on the potential energy surface. While conventional methods for exploring potential energy surfaces can locate only the stationary points of the energy landscape, the statistically defined separatrix allows identification of energetic as well as entropic bottlenecks.

Diffusion of Isobutane in Silicalite

In silicalite, a zeolite of great importance in petrochemical applications, branched alkanes are preferentially adsorbed at channel intersections. Through hops from one intersection to the next these adsorbates can diffuse through the three-dimensional channel network. By analyzing transition states for diffusion, Vlugt et al. showed that the hopping mechanism involves both translation and rotation of the isobutane molecule (10).

Slightly modified shooting moves were used to improve sampling efficiency in this study. In these shooting moves, a random displacement was applied not only to the momenta but also to the position of the butane molecule. Both the momentum and the position displacements were chosen from a uniform distribution in a certain interval. In fact, a variety of configurational trial moves can be used in conjunction with shooting moves. For instance, the orientation of long-branched molecules could be modified efficiently with configurational Monte Carlo methods or rotations around a randomly selected axis (10). It is necessary to employ acceptance probabilities for such moves that correctly guide the random walk through trajectory space without imposing artificial biases.

Parallel Tempering

In some systems, transitions between stable states occur through many qualitatively different mechanisms. As a result, the corresponding pathways may reside in disconnected parts of trajectory space. The two distinct mechanisms for proton transfer in $(H_2O)_3H^+$ described above in the section "Water Clusters" exemplify this diversity. In such systems, ergodic sampling of trajectory space may be difficult to achieve. This situation is analogous to sampling problems encountered in Monte Carlo simulations of glassy systems, in which ergodic sampling is hindered by high free-energy barriers separating adjacent metastable states. Various methodologies developed to overcome these problems, including J-walking (59), multicanonical sampling (60), and parallel tempering (61), can, in principle, also be utilized in transition path sampling. Vlugt & Smit have demonstrated that parallel tempering is particularly simple to combine with transition path sampling, dramatically increasing the rate at which trajectory space is explored (62).

The basic idea of parallel tempering is to perform several transition path sampling simulations simultaneously at different temperatures. At each temperature level, individual trial moves, such as shooting and shifting, are performed. In addition to these moves, exchange of pathways between adjacent temperature levels is periodically attempted. While low-temperature pathways cannot easily cross barriers in trajectory space, high-temperature pathways can. Through path exchange between different temperature levels, ergodic sampling is achieved simultaneously at all temperature levels.

Biomolecule Isomerization

The folding of a protein molecule from a denatured state to its native conformation is a rare event of central biological importance. Although the denatured and native states of several model proteins have been reasonably well characterized, the dynamical variables that drive folding remain elusive (42, 43, 63, 64). The results of importance sampling for alanine dipeptide isomerization (9) (among the simplest of biomolecular rearrangements) suggest that these variables are indeed complex, involving collective intramolecular fluctuations as well as solvent degrees of freedom. Analysis of transition states revealed that, even in the absence of solvent, a coincidence of dihedral and torsional motions is required to cross the separatrix. With solvent molecules explicitly included, a nearly uniform committor distribution (as in Figure 9c) indicated that intramolecular variables are insufficient to describe the isomerization mechanism. The solvent variables needed to distinguish between the isomers' basins of attraction involve more than simply numbers of hydrogen bonds and density of coordinating molecules.

A significant difficulty arises in harvesting folding pathways for molecules larger than the alanine dipeptide. Owing to the low frequency of backbone motions and buffeting by solvent molecules, relatively long times are required for trajectories initiated on the separatrix to commit to either the folded or unfolded state (say, nanoseconds rather than picoseconds). As a result, the appropriate length of harvested paths is much greater than the characteristic timescale of chaos, i.e., the inverse rate of divergence of a displacement in phase space. The efficiency of shooting moves is greatly diminished in this case. Even the smallest obtainable momentum displacements (limited by the finite precision of numerical simulations) lead to large trial steps in trajectory space, few of which are accepted. Only shooting moves initiated near the separatrix have a reasonable acceptance probability. The alanine dipeptide is sufficiently small that this problem is not yet severe. In Reference (9), the magnitude of momentum displacements was adjusted to obtain an average acceptance probability of 30%. But a relatively long commitment time is apparent from the fact that most isomerization trajectories cross the transition state surface several times before settling into the basin of attraction of the final state.

Water Autoionization

Trajectories of simulated liquid water exhibiting dissociation of a water molecule to form hydronium (H_3O^+) and hydroxide (OH^-) ions have been harvested using transition path sampling in conjunction with CPMD (13). As Eigen imagined (65, 66), the product ions of this process are separated on a nanometer scale and are metastable. From this intermediate state in liquid water, the ions may become stable by diffusing away from each other via the Grotthuss mechanism (39, 67, 68). Alternatively, they may recombine on a picosecond timescale by returning through the transition state surface for dissociation.

Characterizing basins of attraction is a significant aspect of sampling autodissociation trajectories: An order parameter describing only the separation of charges does not successfully discriminate between the intermediate dissociated ion state and neutral water molecules. Indeed, ions artificially separated by 1 nm most often recombine within 100 fs along a wire of hydrogen bonds. The majority of trajectories leading to charge separation thus constitute fluctuations within the neutral basin of attraction and do not cross the transition state for dissociation. A successfully discriminating order parameter instead describes the existence and length of hydrogen bond wires connecting the two ions.

Owing to the considerable computational expense of performing ab initio molecular dynamics, and the extreme rarity of autoionization events, preparing an initial pathway for sampling was also an important step in this application. By artificially separating ions, while simultaneously ensuring the absence of short hydrogen bond wires, trajectories evincing recombination on a picosecond timescale were generated. Time reversal of such a trajectory, which passes through the dynamical bottleneck, produced a suitable starting point for transition path sampling.

Solvation Dynamics

Importance sampling of trajectories is useful not only for harvesting rare events at equilibrium but also for studying the dynamics of systems out of equilibrium. In

Reference (12), the methods of path sampling were extended to efficiently sample the wings of a nonequilibrium distribution function. Umbrella sampling (20, 21) has been used to show that the energy gap, ΔE, between ground and excited states of a solute in a polar solvent obeys Gaussian statistics at equilibrium, even for values of ΔE that are several standard deviations from the mean. The dynamical linear response suggested by these statistics (69) has been observed in many, but not all, simulations of solvent relaxation following instantaneous excitation of the solute. Because these straightforward simulations rarely encounter values of ΔE far from the mean, however, they cannot determine whether deviations from linear response behavior are accompanied by non-Gaussian statistics as the system relaxes. For this purpose, it is necessary to bias the sampling of trajectories according to their solvation dynamics, $\Delta E(t)$. Using this generalization of equilibrium umbrella sampling, the statistics of $\Delta E(t)$ were shown to remain Gaussian even in states far from equilibrium (12). But the variance of these statistics changes in time, i.e., solute excitation breaks time-translational symmetry of the linear susceptibility. This nonstationarity is in fact the source of apparently nonlinear response.

The efficient sampling of nonequilibrium trajectories in this application also required careful construction of appropriate trial moves. Because paths of interest are very short (10–100 fs), correlations in the random walk through trajectory space decay quickly only for large-amplitude shooting moves. Such moves, however, tend to heat the system considerably, so that trial paths are accepted with low probability. Controlling the distribution of kinetic energies for large-amplitude shooting moves, as described in the Appendix of Reference (12), is sufficient to restore a reasonable acceptance probability.

FOR THE FUTURE

The preceding sections describe applications from several branches of chemical and biological physics. It would seem that any rare event whose underlying dynamics can be simulated for times as long as the commitment time, τ_{mol}, is amenable to transition path sampling. Indeed, we expect to see the general methodology of importance sampling in trajectory space widely applied. Many applications are possible without significant changes to the methods we have presented, including phenomena quite different from those we have studied so far. For example, the techniques outlined in the sections "Transition Path Sampling" and "Reversible Work," could be used to sample the dynamical structures of highly chaotic systems out of equilibrium. Other applications will require improvements and generalizations of our methods. In this section, we point to three issues that are truly problematic for the specific methodology we have developed. It is our hope that others' experience and fresh perspectives will lead to advances in these areas.

HARVESTING LONG TRAJECTORIES   As discussed in the section "Biomolecule Isomerization," very long commitment times pose a serious difficulty for path

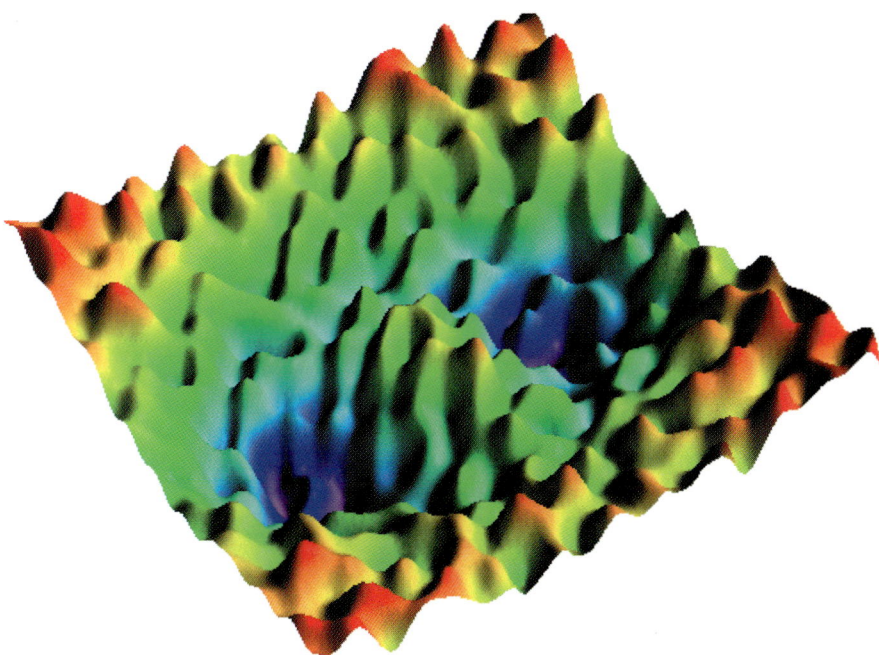

Figure 1 Schematic depiction of the potential energy surface of a complex system. Even though such an energy landscape is dense in saddle points, only a few of them are relevant for transitions between different basins of attraction. At finite temperature all details of the surface smaller than $k_B T$ are of minor importance. Because the transition path sampling method does not rely on identifying saddle points in the potential energy surface, it is the tool of choice to study transitions in complex systems.

sampling. Shooting, our basic technique for generating trial steps in trajectory space, is ineffective when τ_{mol} greatly exceeds the timescales characterizing chaos. For this reason, processes such as protein folding, structural rearrangement of deeply supercooled liquids, and condensation of a supersaturated vapor are frontier applications. Harvesting pathways in these cases will likely require invention of a random walk step whose magnitude can be tuned even for very long trajectories.

RECOGNIZING PATTERNS IN STABLE STATES, METASTABLE STATES, AND TRANSITION STATES We have described a systematic method for generating correctly weighted examples of transition pathways and transition states, given an order parameter that discriminates between stable states. We have also shown how distributions of the committor may be used to test an interpretation of the reaction mechanism. Characterizing stable states and generating mechanistic interpretations, however, remain subjective endeavors. When the relevant fluctuations involve only a few atomic coordinates, or are linear combinations of preconceived variables, they can usually be discerned through visual inspection or techniques such as principal component analysis (70). But in complex systems, the pertinent coordinates, such as electric fields and density fields, are more often nonlinear functions of very many atomic coordinates. Identifying the few important variables is a significant challenge in these cases, even when many examples of stable states and transition states are known. Generalizations of principal component analysis for nonlinear systems (71, 72) may be helpful in systematically approaching this problem of pattern recognition.

Recognizing patterns that characterize long-lived intermediate states poses a similar challenge. In the section "Reversible Work," we described a criterion for detecting the presence of metastable regions between reactants and products. But identifying the segments of harvested pathways that belong to these regions, and subsequently characterizing each region, is not straightforward. For this purpose it may be necessary to generalize the concept of a committor, because a significant fraction of fleeting trajectories initiated near metastable states will reach neither reactants nor products.

COMPUTING QUANTUM DYNAMICS The nuclear dynamics we have considered in this review, and moreover the very notion of distinct trajectories in phase space, are entirely classical. Quantum mechanical phenomena arise from fluctuations about, and interference between, such classical trajectories. These effects are in many cases captured accurately within the semiclassical initial value representation (SC-IVR) (73), which expresses quantum mechanical correlation functions as superpositions of classical trajectories. It is thus tempting to use the ensemble of trajectories generated by transition path sampling in conjunction with SC-IVR to compute the dynamical effects of quantization on high-dimensional systems. The "weight" of a pair of trajectories in SC-IVR, however, is a highly oscillatory function. Summation over trajectories, therefore, results in significant cancellation, and numerical convergence is extremely slow. It remains to be seen whether an importance sampling of trajectories can be appropriately biased to generate groups of strongly interfering pathways, so as to overcome this problem.

ACKNOWLEDGMENTS

The work reviewed in this article was supported by the NSF and DOE through a variety of grants. Several members and visitors of the Chandler Group and other colleagues have contributed significantly to the development of the ideas described in this review: Christian Bartels, Gavin Crooks, Felix Csajka, Daniel Laria, Jin Lee, Ka Lum, Jordi Marti, Vijay Pande, Daniel Rokhsar, and Udo Schmitt.

Visit the Annual Reviews home page at www.annualreviews.org

LITERATURE CITED

1. Dellago C, Bolhuis PG, Csajka FS, Chandler D. 1998. *J. Chem. Phys.* 108:1964–77
2. Dellago C, Bolhuis PG, Chandler D. 1998. *J. Chem. Phys.* 108:9236–45
3. Bolhuis PG, Dellago C, Chandler D. 1998. *Faraday Discuss.* 110:421–36
4. Dellago C, Bolhuis PG, Chandler D. 1999. *J. Chem. Phys.* 110:6617–25
5. Geissler PL, Dellago C, Chandler D. 1999. *Phys. Chem. Chem. Phys.* 1:1317–22
6. Geissler PL, Dellago C, Chandler D. 1999. *J. Phys. Chem. B* 103:3706–10
7. Crooks GE. 1999. *Excursions in statistical dynamics.* PhD thesis. Univ. Calif., Berkeley. 107 pp.
8. Bolhuis PG, Chandler D. 2000. *J. Chem. Phys.* 113:8154–60
9. Bolhuis PG, Dellago C, Chandler D. 2000. *Proc. Natl. Acad. Sci. USA* 97:5877–82
10. Vlugt TJH, Dellago C, Smit B. 2000. *J. Chem. Phys.* 113:8791–99
11. Geissler PL, Dellago C, Chandler D, Hutter J, Parrinello M. 2000. *Chem. Phys. Lett.* 321:225–30
12. Geissler PL, Chandler D. 2000. *J. Chem. Phys.* 113:9759–65
13. Geissler PL, Dellago C, Chandler D, Hutter J, Parrinello M. 2001. *Science* 291:2121–24
14. Crooks GE, Chandler D. 2001. *Phys. Rev. E* 64:026109
15. Anderson JB. 1973. *J. Chem. Phys.* 58:4684–92
16. Bennett CH. 1977. In *Algorithms for Chemical Computations*, ed. RE Christoffersen, pp. 63–97. Washington, DC: Am. Chem. Soc.
17. Chandler D. 1978. *J. Chem. Phys.* 68:2959–70
18. Anderson JB. 1995. *Adv. Chem. Phys.* 91:381–431
19. Chandler D. 1998. In *Classical and Quantum Dynamics in Condensed Phase Simulations*, ed. BJ Berne, G Ciccotti, DF Coker, pp. 3–23. Singapore: World Sci. In this book version of this article, the numbering of equations does not match that of the text. A corrected version can be obtained as a PDF file at http://gold.cchem.berkeley.edu/bibliography.html.
20. Frenkel D, Smit B. 1996. *Understanding Molecular Simulation: From Algorithms to Applications*, San Diego, CA: Academic
21. Chandler D. 1987. *Introduction to Modern Statistical Mechanics.* New York: Oxford Univ. Press
22. Cerjan CJ, Miller WH. 1981. *J. Chem. Phys.* 75:2800–6

23. Wales D, Miller MA, Walsh TR. 1998. *Nature* 394:758–60
24. Binder K, Heermann DW. 1992. *Monte Carlo Methods in Statistical Physics: An Introduction.* New York: Springer-Verlag
25. Newman MEJ, Barkema GT. 1999. *Monte Carlo Methods in Statistical Physics.* New York: Oxford Univ. Press
26. Pratt LR. 1986. *J. Chem. Phys.* 85:5045–48
27. Toda M, Kubo R, Saito N. 1995. *Statistical Physics I.* New York: Springer. 2nd. ed.
28. Zwanzig R. 2001. *Nonequilibrium Statistical Mechanics.* New York: Oxford Univ. Press
29. Elber R, Karplus M. 1987. *Chem. Phys. Lett.* 139:375–80
30. Elber R, Meller J, Olender R. 1999. *J. Phys. Chem. B* 103:899–911
31. Henkelman G, Johannesson G, Jonsson H. 2000. In *Progress on Theoretical Chemistry and Physics*, ed. SD Schwartz. Dordrecht, Netherlands: Kluwer
32. Voter AF. 1997. *Phys. Rev. Lett.* 78:3908–11
33. Grubmüller H. 1995. *Phys. Rev. E* 52:2893–906
34. Gillilan RE, Wilson KR. 1996. *J. Chem. Phys.* 105:9299–315
35. Gillilan RE. 1992. *J. Chem. Phys.* 97:1757–72
36. Sevick EM, Bell AT, Theodorou DN. 1993. *J. Chem. Phys.* 98:3196–212
37. Eastman P, Grönbech-Jensen N, Doniach S. 2001. *J. Chem. Phys.* 114:3823–41
38. Zuckerman DM, Woolf TB. 1999 *J. Chem. Phys.* 111:9475–84
39. Marx D, Tuckerman ME, Hutter J, Parrinello M. 1999. *Nature* 397:601–4
40. Schmitt UW, Voth GA. 1999. *J. Chem. Phys.* 111:9361–81
41. Day TJF, Schmitt UW, Voth GA. 2000. *J. Am. Chem. Soc.* 122:12027–28
42. Du R, Pande VS, Grosberg AY, Tanaka T, Shakhnovich EI. 1998. *J. Chem. Phys.* 108:334–50
43. Bryant Z, Pande VS, Rokhsar DS. 2000. *Biophys. J.* 108:584–89
44. Klosek MM, Matkowsky BJ, Schuss Z. 1991. *Ber. Bunsenges. Phys. Chem.* 95:331–37
45. Belch AC, Berkowitz M, McCammon JA. 1986. *J. Am. Chem. Soc.* 108:1755–61
46. Guardia E, Rey R, Padró JA. 1991. *Chem Phys.* 155:187–95
47. Karim OA, McCammon JA. 1986. *J. Am. Chem. Soc.* 108:1762–66
48. Rey R, Guardia E. 1992. *J. Phys. Chem.* 96:4712–18
49. Truhlar DG, Garrett BC. 2000. *J. Phys. Chem. B* 104:1069–72
50. Gasser U, Weeks ER, Schofield A, Pusey PN, Weitz DA. 2000. *Science* 292:258–62
51. Geissler PL, Van Voorhis T, Dellago C. 2000. *Chem. Phys. Lett.* 324:149–55
52. Car R, Parrinello M. 1985. *Phys. Rev. Lett.* 55:2471–74
53. Parrinello M. 1997. *Solid State Commun.* 102:107–20
54. Rice OK, Ramsperger HC. 1927. *J. Am. Chem. Soc.* 49:1617–29
55. Rice OK, Ramsperger HC. 1928. *J. Am. Chem. Soc.* 50:617–20
56. Kassel LS. 1928. *J. Phys. Chem.* 32:225–42
57. Marcus R, Rice OK. 1951. *J. Phys. Colloid Chem.* 55:894–908
58. Laria D, Rodriguez J, Dellago C, Chandler D. 2001. *J. Phys. Chem. A* 105:2646–51
59. Frantz DD, Freeman DL, Doll JD. 1990. *J. Chem. Phys.* 93:2769–84
60. Berg BA, Neuhaus T. 1992. *Phys. Rev. Lett.* 68:9–12
61. Geyer CJ. 1995. *J. Am. Stat. Assoc.* 80:909–20
62. Vlugt TJH, Smit B. 2001. *Phys. Chem. Commun.* 2:1–7
63. Pande VS, Grosberg AYu, Rokhsar DS, Tanaka T. 1998. *Curr. Opin. Struct. Biol.* 8:68–79
64. Karplus M. 2000. *J. Phys. Chem. B* 104:11–27
65. Eigen M, de Maeyer L. 1955. *Z. Elektrochem.* 59:986–93

66. Eigen M. 1964. *Angew. Chem. Int. Edit.* 3:1–19
67. Tuckerman M, Laasonen K, Sprik M, Parrinello M. 1995. *J. Chem. Phys.* 103:150–61
68. Agmon N. 1999. *Isr. J. Chem.* 39:493–502
69. Bagchi B, Oxtoby DW, Fleming GR. 1984. *Chem. Phys.* 86:257–67
70. Mardia KV, Kent JT, Bibby JM. 1979. *Multivariate Analysis*. London: Academic
71. Tenenbaum JB, de Silva V, Langford JC. 2000. *Science* 290:2319–23
72. Roweis ST, Saul LK. 2000. *Science* 290:2323–26
73. Miller WH. 1998. *Faraday Discuss.* 110:1–21

ELECTRONIC STRUCTURE AND CATALYSIS ON METAL SURFACES

Jeff Greeley,[1] Jens K. Nørskov,[2] and Manos Mavrikakis[1]*

[1]*Department of Chemical Engineering, University of Wisconsin, Madison, Wisconsin 53706; e-mail: jgreeley@cae.wisc.edu, manos@engr.wisc.edu*
[2]*Center for Atomic-Scale Materials Physics, Technical University of Denmark, DK-2800, Lyngby, Denmark; e-mail: norskov@fysik.dtu.dk*

Key Words first principles, theory, metals, catalysis, surface chemistry

■ **Abstract** The powerful computational resources available to scientists today, together with recent improvements in electronic structure calculation algorithms, are providing important new tools for researchers in the fields of surface science and catalysis. In this review, we discuss first principles calculations that are now capable of providing qualitative and, in many cases, quantitative insights into surface chemistry. The calculations can aid in the establishment of chemisorption trends across the transition metals, in the characterization of reaction pathways on individual metals, and in the design of novel catalysts. First principles studies provide an excellent fundamental complement to experimental investigations of the above phenomena and can often allow the elucidation of important mechanistic details that would be difficult, if not impossible, to determine from experiments alone.

INTRODUCTION

First principles quantum mechanical calculations are fast becoming an indispensable tool in the fields of surface science and heterogeneous catalysis. Qualitative and oftentimes quantitative insights into surface chemistries can be obtained with first principles techniques.

In this review, we give a brief overview of the state-of-the-art computational techniques that are making the above investigations possible, and we discuss the application of these techniques to several classes of problems. The review is not intended as exhaustive, and readers interested in additional information should consult previous reviews on these subjects (1–8). First, we describe several areas in which first principles calculations have been used to successfully explain unusual experimental phenomena. Next, we examine the successful elucidation of chemisorption trends across the transition metals for specific adsorbates. Then, we consider the detailed exploration of certain classes of reactions on individual

---

*Corresponding Author.

metal surfaces. We conclude with two examples of how fundamental computational methods have allowed researchers to progress beyond the information-gathering stage of research to the design of improved catalytic materials.

OVERVIEW OF FIRST PRINCIPLES TECHNIQUES

The Time-Independent Schrödinger Equation

The fundamental equation upon which electronic structure theories are based is the Time-Independent Schrödinger Equation (TISE),

$$H\psi = E\psi, \qquad 1.$$

where H is the Hamiltonian (total energy operator), E is the total energy of the system, and ψ is the wavefunction, a function of space thought to contain all knowable information about the system. Solution of this equation yields fundamental information about the system, including probability distributions for all particles within it and energetic information about particular particle configurations.

For systems of interest to surface physicists and chemists, the full TISE is simplified by the Born-Oppenheimer Approximation. This approximation relies on the fact that atomic nuclei move thousands of times more slowly than electrons. In effect, this approximation allows the TISE to be split into nuclear and electronic structure calculations that can be performed separately. The electronic problem, which is of primary concern here, may be visualized as fixed configurations of nuclei surrounded by an "electron gas." The electronic structure of this gas is determined by solution of the electronic TISE, and the resulting total energy is interpreted as a potential energy for the nuclei. Solution of the TISE for many different nuclear arrangements permits the construction of potential energy surfaces (PES's) for the nuclei. In practice, it is often only the ground-state PES that is of interest in chemical analyses; this PES can be used to analyze the nuclear dynamics either in a classical (molecular dynamics, classical Monte Carlo, etc.) approximation or in a full quantum formalism. The former approximation is by far the most commonly used in the surface science and catalysis communities.

Electronic Structure Methods

The solution of the electronic structure problem for given configurations of nuclei is an extremely difficult problem. Analytical solutions are not possible for systems of chemical interest, and approximate numerical schemes must be used. A large fraction of these schemes is based on the solution of the electronic TISE in terms of one-electron orbitals. In the following paragraphs, the simplest technique for the calculation of these orbitals, the Hartree Fock Self-Consistent Field (HFSCF) approach, is described (9). More elaborate approaches are also discussed.

In the HFSCF approach, the full, many-electron wavefunction for the system is written as a product of one-electron wavefunctions that contain adjustable parameters. The number of one-electron orbitals is equal to the number of electrons in the system. The full wavefunction is constructed so as to be antisymmetric with respect to electron exchange. Using this approximate wavefunction, the full TISE is separated into many one-electron equations,

$$-\frac{\hbar^2}{2m}\nabla^2\psi + v\psi = \varepsilon\psi, \qquad 2.$$

where $\hbar$ is Planck's constant, m is the rest mass of an electron, ∇^2 is the Laplacian, v is an effective one-electron potential energy function, and ε is a one-electron eigenvalue. The one-electron potential energy function (v) is calculated from the exact electrostatic attraction of the nucleus for the electrons and from an average electron-electron electrostatic interaction energy. An iterative scheme is used to solve Equation 2. It involves the choice of initial guesses for the one-electron wavefunctions (usually written as linear combinations of basis functions), the calculation of an average electron-electron interaction energy (which depends on all of the one-electron orbitals) from these wavefunctions, the insertion of this interaction energy term as part of v in Equation 2, and the solution of Equation 2 for improved one-electron orbitals. This process is repeated until orbital convergence (self-consistency) is obtained. Full self-consistency in the HFSCF method requires $\sim N^4$ calculations, where N is the number of basis functions used.

Although this solution technique is appealing in its simplicity, the HFSCF approach gives very inaccurate molecular energies. The problem stems primarily from the lack of explicit electron correlation effects in the technique. Only correlation effects imposed by the antisymmetric wavefunction are accounted for; the neglect of other correlation energies can give very poor results (6). Good methods of correcting for this deficiency do exist, but they are generally extremely computationally expensive. Configuration Interaction (CI) methods make use of unoccupied (virtual) states to account for correlation effects. The mechanics of the technique are very similar to the HFSCF technique, but additional (unoccupied) one-electron wavefunctions are used to construct the total wavefunction. This procedure essentially allows the incorporation of excited electron configurations into the wavefunction (10). The result of CI calculations is an accurate solution of the TISE but at extreme computational cost (the HFSCF method requires $\sim N^4$ calculations per basis function, while CI methods can require more than N^7 calculations per basis function) (6). Typically, excitations beyond the triple excitation level are infeasible, and the method is truncated at that point.

Quantum Monte Carlo (QMC) techniques (11–14) present an alternative to CI methods for the incorporation of electron correlation effects into electronic structure calculations. These techniques use random sampling approaches to calculate energies for many-electron systems. In the Variational Monte Carlo method, for example, random sampling is used to evaluate integrals that arise naturally during CI calculations (15). In the Diffusion Monte Carlo method, on the other hand, the Schrödinger equation is recast into an integral Green's function form, and the Green's function is approximated by successive Monte Carlo sampling (15). QMC techniques are computationally expensive, but they are beginning to be applied to realistic solid-state systems (16–18). These techniques have many attractive features; they offer the possibility of obtaining exact solutions to the TISE, and they are well suited to describe systems where electronic interactions are particularly important, such as superconductors, materials with large low-temperature-specific heat coefficients, and systems where van der Waals forces play a significant role (14).

The computational burden of the exact calculation of correlation effects makes it highly desirable to use an approximate scheme for evaluating such effects. This scheme, known as Density Functional Theory (DFT), is widely used in chemical computations. The methodology of DFT is substantially similar to that described for the HFSCF method, but an additional term is added to the effective one-electron potential energy functions v (see Equation 2) that are used in HFSCF (7, 19–21). This extra term is called the exchange-correlation (XC) energy and represents a heuristic, but highly efficient, way to account for these effects. By incorporation of these extra exchange-correlation terms, DFT can yield accurate total energies with comparable computational effort, as the number of calculations scales with N^3 (6), which is a substantial improvement over CI methods. Although many XC energy expressions have been developed, all share the property of being explicit functionals of the electron density. The LDA (Local Density Approximation) uses an XC functional that depends only on the electron density itself and takes the XC energy to be the exact XC energy for a homogeneous electron gas (22, 23). The GGA (Generalized Gradient Approximation) incorporates density gradient terms into the XC functional. LDA calculations often produce poor estimates of binding energies and molecular structures (24, 25). State-of-the-art GGA calculations (together with suitable models for the surface structure, discussed below), however, can give much better values. Bond lengths and solid lattice constants, for example, are reproduced to within several hundredths of an angstrom (7, 26), and vibrational frequencies are calculated with an accuracy of ∼5% (4, 6, 7). Comparison of calculated adsorption energies with experimental values (which are obtained most accurately from Single-Crystal Adsorption Calorimetry) (27) demonstrates that these energies can be found to within ∼0.15 eV (7, 28–30).

Core Electron Representations

When implementing one of the above electronic structure calculation techniques, it is important to choose an appropriate method to represent core electrons. These electrons could certainly be treated with the same quantum techniques as valence electrons, but because the core electrons are located in the innermost shells of atoms, they do not generally play an active role in chemical bonding. Hence, it is often not necessary to explicitly include the core electrons in the quantum calculations, and their behavior can be treated approximately to reduce computational burdens. One approximation that retains almost all of the important features of the core electrons but reduces computational time (31, 32) is the Frozen Core Approximation (FCA). In the FCA, the one-electron wavefunctions for the core electrons are not recalculated during every self-consistent iteration; they are simply fixed in the functional form that the isolated atomic orbitals would have. Another core-level approximation is the pseudopotential method in which the core electrons are completely removed from the problem. Their effect is felt only through an effective core potential energy function (33, 34). Although this technique represents a rough approximation of the core states, the current generation of ultrasoft pseudopotentials (35) yields accurate energies with great computational efficiency.

Models of Surface Structure

Other factors not directly related to the electronic structure determination problem can have significant effects on the accuracy of theoretical models in surface science and heterogeneous catalysis. One of the most critical issues is the representation of the surface structure. Cluster calculations use finite ensembles of metal atoms to model surfaces; these calculations are computationally convenient because they employ atomic or molecular orbital basis sets to satisfy the boundary condition of zero electron density at infinite distance from the cluster. Such basis sets have long been used by theoretical chemists to model isolated molecules, and hence no significant modifications are required to apply theoretical chemistry codes to surface cluster calculations. Unfortunately, the electronic structure of clusters can be quite different from the corresponding structure of semi-infinite surfaces (36), and thus the clusters may not in all cases be suitable models for heterogeneous catalyst surfaces. Slab calculations, on the other hand, use periodic boundary conditions to model extended surfaces. These models avoid the electronic structure artifacts that sometimes trouble cluster calculations. The slab models require the use of a periodic basis set to match the boundary conditions; plane waves are often used for this purpose. This basis works well for periodic systems, although convergence can be slow where there are sharp electron density gradients. The inclusion of some atomic orbitals (exactly represented with plane waves) has been shown to speed convergence for some systems (37).

SYNERGY BETWEEN THEORY AND EXPERIMENT

Experimental investigations in surface science and catalysis often raise intriguing questions that cannot be easily answered by the experiments themselves. In such situations, it is sometimes possible to explain the resulting puzzles by attacking the problem with ab initio simulation methodologies. First principles calculations have a number of features that allow them to make important contributions to these types of investigations. First, the simulations can be performed at almost any degree of spatial resolution, thus making it possible to accurately determine the geometries, energies, electronic structures, and site preferences of adsorbates on well-defined solid surfaces. Such data are not easily accessible in experiments. Second, the inputs to simulations can be easily controlled, eliminating concerns about the effects of contaminants or other unknown variables. Finally, the results of simulations are generally easier to interpret than are the results of experiments; although wavefunctions and other outputs from theoretical simulations can be quite complex, it is usually possible to create software to analyze these numerical data quickly and efficiently.

Below we discuss several areas of current interest to the surface science and catalysis communities to which substantial contributions have been made by theoretical methods. Our goal is to describe the fruitful and fascinating interplay between experimental and theoretical analyses that has allowed recent progress to be made in these areas.

Site Preferences and Vibrational Spectroscopies

On single-crystal transition metal surfaces, vibrational spectroscopies (either EELS, HREELS, or RAIRS) can provide important qualitative information about the chemical structure of adsorbed species. These techniques are used both for the identification and classification of surface species and for the determination of site preferences and adsorption geometries of these species. Often, however, it is difficult to conclusively assign a specific structure to a measured vibrational spectrum, and without a reliable means of doing this, the vibrational spectra are of limited use.

Lately, theoretical techniques have been increasingly used to aid in the determination of physical and chemical structures from measured vibrational spectra. An illustration of this procedure comes from the surface oxametallacycle literature. These cyclic reaction intermediates incorporate metal surface atoms into their rings (Figure 1a) and are likely intermediates in epoxidation reactions and in other reactions involving oxygenate molecules (38). Unfortunately, the oxametallacycles are difficult to isolate on metal surfaces, and prior to the work described here, spectroscopic standards for the molecules did not exist. Jones et al. (38) overcame this difficulty by comparing a theoretical vibrational spectrum for an iodoethanol-derived oxametallacycle attached to a silver cluster with an experimental HREELS spectrum taken on Ag{110}. The remarkable agreement between the theoretical and experimental spectra allowed the authors to conclusively identify the oxametallacycle and its exact structure (involving two metal atoms in the ring) on the indicated surface (Figure 1b). The same approach has also been used to identify oxametallacycles derived from 1-epoxy-3-butene on Ag{110} (39, 40) and from a tert-butoxy species on Pt{111} (41).

Another instance in which experimental surface spectra have been analyzed with the help of ab initio techniques involves the adsorption geometry of NO. The adsorption and dissociation of NO have been heavily studied on a wide variety of transition metal surfaces (42), but for the sake of brevity, the present review details NO adsorption geometry on only two such surfaces, Pd{111} and Rh{111}. On Pd{111}, site assignments have been made using a combination of vibrational frequency information from gas-phase nitrosyl compounds and LEED patterns (43, 44). The conclusions from these studies are that NO prefers either top-site or bridge-site binding (or a combination of the two), depending on coverage. These

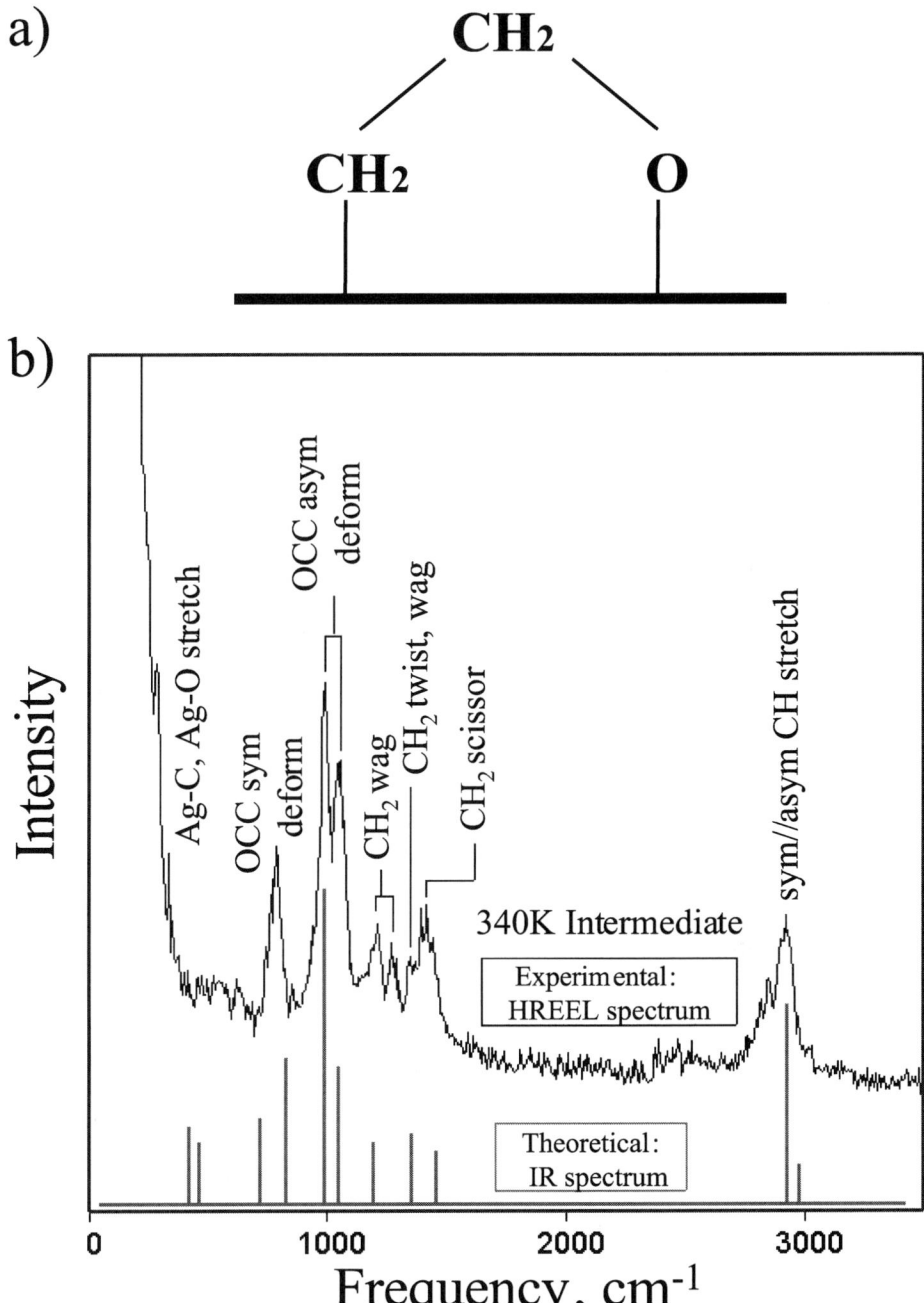

Figure 1 Surface oxametallacycles: (*a*) an oxametallacycle derived from iodoethanol and (*b*) comparison of theoretical and experimental oxametallacycle vibrational spectra on Ag{110} with the corresponding peak assignments. [Reprinted with permission from (38); Copyright 1998. Am. Chem. Soc.]

conclusions are contradicted, however, by a periodic DFT study of NO chemisorption on Pd{111} (45). In that work, it is shown that theoretical frequencies from hollow sites provide the best match with the experimental data at low coverages, and a combination top/hollow site occupation gives good agreement with experiments at high coverages. On Rh{111}, HREELS and LEED studies (46) are again taken to indicate that NO adsorption occurs on bridge and top sites. This interpretation is contradicted by DFT calculations that favor hollow sites at moderate coverages and a combination of top and hollow sites at high coverages (45). These theoretical corrections to the experimental vibrational results are confirmed by Temperature Programmed Static Secondary Ion Mass Spectroscopy (TPSSIMS) (47) and X-ray Photoelectron Diffraction (XPD) (48) studies. Brown & King (42) conclude from the above theoretical considerations that re-evaluation of the site-vibrational frequency relationship for NO is necessary to obtain accurate site assignments from experimental vibrational spectra.

Effect of Surface Defects

Another area in which there has been fruitful interaction between first principles calculations and experimental investigations involves the effect of defects on surface chemistry. It has long been speculated that defects and, more specifically, steps are more reactive than terraces, but it has not been until relatively recently that quantitative descriptions of the relationship between these defects and surface reactivity have been developed. Periodic DFT calculations indicated that N_2 dissociation at steps on Ru{0001} is activated by 0.4 eV, while dissociation on terraces is activated by $\sim$1.9 eV (Figure 2). These predictions helped in the interpretation of detailed experiments by Dahl et al. (49). The authors measured the activation barrier for dissociative adsorption of N_2 on both clean and gold-passivated Ru{0001}. The clean surface barrier was $\sim$0.4 eV, whereas the barrier on the passivated surface was $\sim$1.3 eV. The experimental findings, together with the theoretical calculations, demonstrated convincingly that nitrogen dissociation on clean Ru{0001} is completely dominated by steps, while the same reaction on gold-passivated Ru{0001} (where steps and other defects have been saturated by gold) occurs on terraces. Even the gold-passivated sample may not show the true terrace barrier but rather the sum of the step barrier and the energy required to remove gold from the step.

Reactions with Subsurface Species

In recent years, the effect of subsurface hydrogen on hydrogenation reactions has received considerable attention. In particular, the case of the hydrogenation of methyl to methane over nickel has been thoroughly studied. Johnson et al. (50) performed a pioneering study of this system in which they showed unambiguously that subsurface hydrogen is the active species in this reaction. Although they speculated that the reaction was C_{3V} symmetric, occurring via the combination of a hydrogen atom in an octahedral subsurface site with methyl adsorbed directly

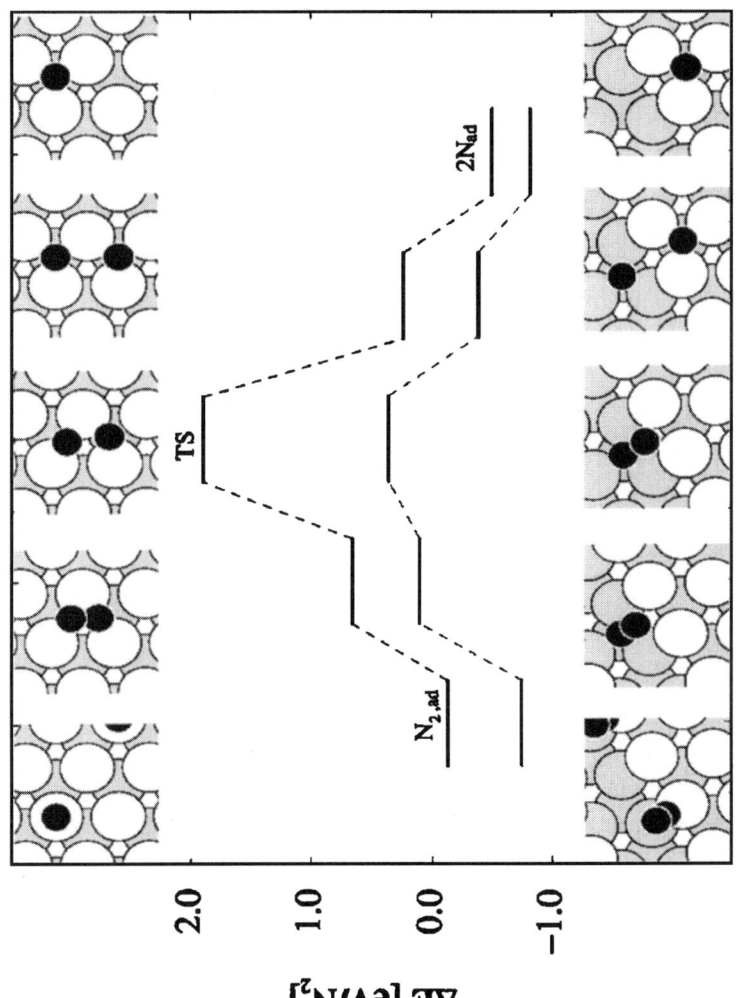

Figure 2 N_2 dissociation on flat (*top*) and stepped (*bottom*) Ru{0001}. Black circles indicate nitrogen, white circles indicate first layer Ru atoms, and gray circles indicate second layer Ru atoms. The leftmost images correspond to stable chemisorbed states for dinitrogen (the reactant) on both the flat and stepped surfaces, TS indicates the transition state on both surfaces, and the rightmost images correspond to two nitrogen atoms at infinite separation (the products). [Reprinted with permission from (49); Copyright 1999. Am. Phys. Soc.]

above it, they could not provide definitive descriptions of the microscopic pathways by which the reaction proceeded. More recently, two periodic DFT studies (51, 52) have been performed to elucidate these microscopic pathways. Both studies confirmed that surface hydrogen has a high barrier to methyl hydrogenation, and both showed that the symmetric C_{3V} pathway proposed by Johnson et al. (50) is highly energetically unfavorable, presumably due to strong Pauli repulsive forces along this pathway (52). Both studies found that the reaction mechanism involves a composite pathway consisting of hydrogen resurfacing from the bulk (accompanied by perturbations to the surface methyl structure), followed by recombination of the hydrogen with the surface methyl. It appears that the reaction barrier is highly sensitive to the exact location of the resurfacing event. Thus, small differences in the chemical environment (e.g., the subsurface hydrogen coverage) that change the preferred resurfacing site may also change the reaction barrier. The calculations also show that while subsurface hydrogen is more reactive than surface hydrogen, the difference in the reaction barrier is simply a result of the higher potential energy of the subsurface hydrogen. This means that under steady-state reaction conditions subsurface hydrogen is no more important than surface hydrogen for the overall reaction rate.

High- and Low-Pressure Reactivity

The pressure gap is an important issue at the interface between surface science and heterogeneous catalysis. It is sometimes found that surface science experiments on well-defined surfaces in ultra-high vacuum (UHV) conditions yield results that are different from results obtained in high-pressure catalytic reactors. An illustration of this effect can be found in the case of CO oxidation on ruthenium. In this system, at a total pressure of ~10 torr, the rate of CO oxidation on Ru{0001} is higher than on any other transition metal surface (53, 54). The oxidation rate is highest for a combination of high surface coverages of oxygen and extremely low coverages of CO. In stark contrast to these results, Ru{0001} is among the poorest of CO oxidation catalysts under UHV conditions (55).

This interesting puzzle was recently analyzed by Stampfl & Scheffler (56, 57) using periodic DFT calculations. Motivated by the fact that, under high-pressure conditions, high-coverage (1 × 1)-O overlayers can be created by O_2 dissociation on Ru{0001} (58, 59), the authors examined CO oxidation under these conditions. They describe an Eley-Rideal (E-R) mechanism for the reaction in which fast CO molecules approach the oxygen-covered surface directly above adsorbed oxygen atoms; after reaction, CO_2 desorbs. The thermochemistry of this reaction is highly favorable ($E_{rxn} \sim -1.95$ eV), although the kinetic barrier is quite high [~1.1 eV with respect to CO(g) and O(a)] . The oxidation rate calculated from this barrier (with a pre-exponential estimated from gas-phase kinetic theory) is much lower than the experimentally measured rates, so it is unlikely that the E-R mechanism alone could be responsible for the measured high-pressure oxidation rates. However, the E-R reactions (in conjunction with thermal fluctuations) open up vacancies in the (1 × 1)-O adlayer. These vacancies can be filled by CO molecules (the equilibrium CO coverage is estimated to be 0.0003) that are then oxidized in a Langmuir-Hinshelwood-type reaction. The estimated reaction barrier for this process is ~1.5 eV. Although this barrier is quite large, the authors estimate that the close proximity of the CO and O adspecies will lead to a very high attempt frequency (pre-exponential) for the reaction, thereby giving a high rate of CO oxidation via this mechanism.

Polymerization Catalysis

Theoretical techniques have made many contributions to the fields of heterogeneous and homogeneous polymerization chemistry over the past several years. The most well-known of the heterogeneous polymerization catalysts are the Ziegler-Natta (ZN) catalysts. Since their discovery in the 1950s (60), the ZN catalysts have been in widespread industrial use. In spite of their importance, however, the detailed mechanisms governing their operation are still not well understood, owing in large part to the difficulty of experimental studies of the extremely rapid surface polymerization reactions (61, 62). Hence, the use of first principles calculations to probe in detail the structure and energetics of reactants, catalyst, and products can provide valuable insights into the ZN polymerization chemistries. Boero et al. (62–65) have conducted extensive calculations to analyze olefin polymerization on $MgCl_2$-supported $TiCl_4$. Analyzing the dynamics on the $MgCl_2\{110\}$ facet, they determine that a Ti atom coordinated to three surface Cl atoms and two free Cl atoms is an excellent candidate for an active site precursor. To fully activate the active site, one free Cl atom must be replaced with a chain-terminating group (e.g., a methyl group). Then, an ethylene molecule spontaneously complexes to the Ti center and is later inserted between the Ti and the methyl chain termination group (Figure 3). This last process is found to have a barrier of ~0.28 eV, in excellent agreement with experimental results. The insertion process is apparently facilitated by so-called agostic interactions where hydrogen atoms, bonded either to the methyl chain-termination group or, later, to the growing polymer chain, interact with the Ti center, thereby destabilizing the Ti-C bond.

Boero et al. (65) considered the possibility that other $MgCl_2$ facets may be active in polymerization. They analyzed polymerization on the $\{100\}$, $\{104\}$, and $\{110\}$ facets and concluded that the $\{110\}$ facet described above is, indeed, the most suitable facet for the ZN reactions. They also extended their polymerization analyses to include propylene polymerization and determined that their proposed Ti active complex on $MgCl_2\{110\}$ can produce isotactic polypropylene with a high degree of stereoselectivity, in agreement with experimental observations of ZN catalyst activity.

Single-site organometallic catalysts for homogeneous polymerization have emerged as interesting alternatives to the ZN catalysts since their discovery 25 years ago (66). Many of these catalysts consist of a metal ion sandwiched between

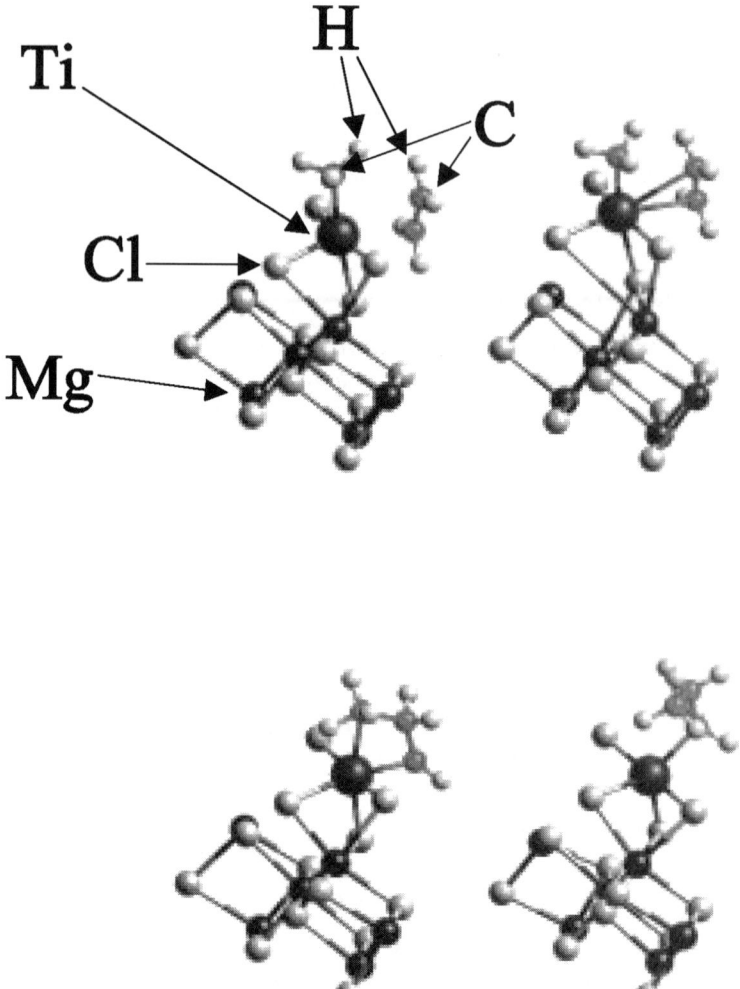

Figure 3 Ethylene insertion into a TiCl$_4$ ZN catalyst. Ethylene is seen to approach the Ti center from the right in the first and second images. One of the ethylene carbon atoms begins to bond with the Ti-coordinated methyl group, leading to a weakening of the methyl-Ti bond and eventually to complete insertion of the ethylene between methyl and Ti. [Reprinted with permission from (62); Copyright 1998. Am. Chem. Soc.]

two or more ligands (67, 68). Polymer chains are formed at the metallic ion site in an activation/complexation/insertion mechanism similar to the ZN mechanism (69). Metallocene catalysts, consisting of metal ions bonded to two ligands, are a particular class of organometallic polymerization catalysts that have attracted the attention of theoretical investigators. Ziegler and co-workers have performed extensive DFT investigations on these catalysts (69–72) and found, for example, that a wide variety of first-row transition metals can be active in metallocene chemistry. Complexation and insertion energies for ethylene vary depending on the d-orbital occupancy of the metallic ion, but all high-spin first-row transition metals with up to four d-electrons can be expected to have some catalytic activity (69). They have also considered the effect of various spectator species on the polymerization reactions. Co-catalysts (used to activate the metallocenes by converting them to cations), unactivated metallocenes, and solvent molecules could form ion pairs or so-called dormant complexes with the activated cation. The presence of ion pairs or dormant complexes would be expected to have an effect on the polymerization reactions. The authors predict that such species could well be present in typical polymerization conditions and that they lead to substantial changes in the polymerization chemistry (70, 71). Finally, they have examined the interaction of polar co-monomers with Ni and Pd-based catalysts (72). Ni, experimentally observed to be inactive for incorporation of polar co-monomers into polymer chains, is found to bind the co-monomers through their carbonyl oxygen atoms, thus precluding polymerization. In contrast, Pd binds the co-monomers through their olefinic functionalities, thereby facilitating polymerization in agreement with experimental results.

Catalysis with Zeolites

Zeolite chemistry is extremely important in the petrochemical industries where these catalysts are used for alkane aromatization, methanol-to-gas conversion, reactive separations, and other processes. Ion-exchanged zeolites, in which Si atoms in the silica framework have been replaced with Al atoms, have an excess of negative charge. This charge can be neutralized by a combination of protonation (resulting in solid acid sites) and metal cationic insertion. Zeolites resulting from Zn insertion in this manner are active for alkane dehydrogenation and aromatization (73, 74). Unfortunately, in spite of the industrial importance of these catalysts, little is known about the nature of the catalytically active Zn sites. Theoretical analysis has proved useful in providing more information about these catalytic centers. For instance, Barbosa et al. (75) have used cluster DFT calculations to simulate the interaction of methane probe molecules with Zn sites. They determined equilibrium structures and vibrational frequencies for methane interacting with bare Zn(II) cations, ZnO solid, and $[Zn-O-Zn]^{2+}$ in 4T and 5T rings. None of these models yielded methane vibrational frequency shifts (with respect to gas-phase methanol) as large as experiments had suggested. Shubin et al. (76) studied Zn^{2+} active sites in 4, 5, and 6-membered rings. They found that Zn^{2+} is least stable in 4-membered rings, indicating that these configurations should be the most reactive. This result was confirmed by calculations of heterolytic ethane dissociation over Zn^{2+} active sites; in these calculations, the reaction barrier was smallest in 4-membered rings. Barbosa et al. (77) performed a similar study using periodic DFT calculations in chabazite with a Si/Al ratio of 5.0. They, too, found that Zn^{2+} cations are less stable (and hence more reactive) in 4-membered rings. Probe molecules will generally bind to these sites with a chemisorption energy proportional to the stability of the site if the probes interact weakly with the zeolite lattice, as is found for methane. However, if the probes interact strongly with the lattice, they can cause significant deformations that may obscure the stability trends of the cationic sites.

In addition to active site characterization, theoretical studies are useful to elucidate the chemistry of particular reactions in zeolites. For example, the hydrolysis of acetonitrile has been examined with DFT cluster calculations on Zn-substituted 4T model zeolites (78). The reaction was found to proceed via activation of either water or acetonitrile, depending upon conditions. Interestingly, the active Zn(II) sites in the model zeolite behaved in a remarkably similar way to biological enzymatic Zn(II) sites that catalyze CO_2 hydrolysis. Another theoretical study, employing periodic DFT methods, has considered the isomerization of toluene and xylenes in an acidic mordenite zeolite (79). This reaction was found to be highly structure sensitive, with steric constraints from the zeolite cage structure exerting a strong influence on the reaction. Periodic DFT calculations have also been employed to study the conversion of methanol to dimethyl ether (a candidate for the first step in the methanol-to-gas reaction) in chabazite (80). Two possible pathways are found for this conversion: One pathway involves direct condensation of methanol molecules in the zeolite pores to form dimethyl ether; the other proceeds through a zeolite-bound methyl intermediate. Carbene and ylide species are found to be unstable in the zeolite matrix and hence are not likely reaction intermediates.

CHEMISORPTION TRENDS

One of the principal objectives of surface science is to develop a detailed, fundamental understanding of the factors that influence the chemical reactivity of surfaces. To this end, the determination of chemisorption trends across the periodic table is crucial. First principles calculations provide a convenient, accurate, and efficient means of studying these trends. It is straightforward to perform tests on a wide variety of metal surfaces because these tests do not require the purchase and preparation of a large number of samples (as would be the case for experimental periodic studies). In recent years, a body of literature on theoretical calculations has accrued that permits the analysis of chemisorption trends for many adsorbates and across a variety of transition metal surfaces. Here, we briefly summarize some of those trends and the methods used to determine them.

A Simple Model for Periodic Trends

Hammer & Nørskov (7, 81) have developed a simplified theory of adsorbate bonding on transition metal surfaces. The theory, based on the assumption that the interaction of a relatively few adsorbate orbitals with surface sp- and d-bands will determine periodic trends in the chemisorption energy of the system, has great explanatory and predictive power. For atomic adsorbates (where only a single type of adsorbate orbital interacts to an appreciable extent with the surface), the model is particularly simple:

$$E_{\text{d-hyb}} = -2(1-f)\frac{V^2}{|\varepsilon_d - \varepsilon_a|} + 2(1+f)\alpha V^2, \qquad 3.$$

where $E_{\text{d-hyb}}$ is the energy gained from hybridization of the adsorbate orbital with the metal d-bands, f is the fractional metal d-band filling, V is a Hamiltonian matrix element that describes the coupling between the metal d-band states and the adsorbate orbital, ε_d is the first moment of the metal d-band density of states (the d-band center), ε_a is the adsorbate orbital energy (renormalized by the metal sp-bands), and α is a constant that is independent of the metal and depends weakly on the identity of the adsorbate (7). Interactions of adsorbates with metal sp-bands generally depend only on the nature of the adsorbate, and so ε_a is essentially constant from metal to metal. Also, constant sp-band interaction energies imply that changes in $E_{\text{d-hyb}}$ for given adsorbates on various metals are nearly identical to corresponding changes in the full chemisorption energies (E_{chem}) of these adsorbates. This fact, in combination with Equation 3, leads to the major conclusion of the Hammer-Nørskov model; namely, that changes in adsorbate chemisorption energies (E_{chem}) over different metals are simply related to changes in the metal d-band centers.

This simple model accounts well for trends in atomic chemisorption energies on transition metals. It can, for example, predict trends in atomic oxygen and sulfur chemisorption energies to a high degree of accuracy (7) (Figure 4a). A similar model developed for molecular adsorbates works equally well, as evidenced by the remarkable parity plot of the model's predicted CO chemisorption energies versus full DFT-GGA chemisorption energies in Figure 4b (82).

In addition to describing periodic trends in atomic and molecular adsorption on pure metal surfaces, the model captures changes in chemisorption energies on pseudomorphic overlayers as well. Pallassana et al. (83) showed that this model could be used to describe trends in hydrogen chemisorption energies over Pd{111}, Re{0001}, and pseudomorphic overlayers of these two elements. For these systems, they demonstrated that the E_{chem} versus ε_d relationship is linear. In a subsequent paper, the same authors (84) developed an extension of the Hammer-Nørskov model to Pd and Re systems with alloyed surface layers. The extended model, using a weighted surface d-band center, again showed a linear relationship between the hydrogen chemisorption energy and the weighted d-bound center.

The Hammer-Nørskov model is also useful for describing periodic trends in dissociation energies. For example, this model accurately predicts changes in the energy of dissociative adsorption of dihydrogen on metals with a wide variety of sp- and d-state properties (81).

A compilation of a wide variety of applications of the Hammer-Nørskov model to systems of interest in surface science and heterogeneous catalysis is shown in Figure 5 (see color insert). This figure, from Mavrikakis et al. (85), includes

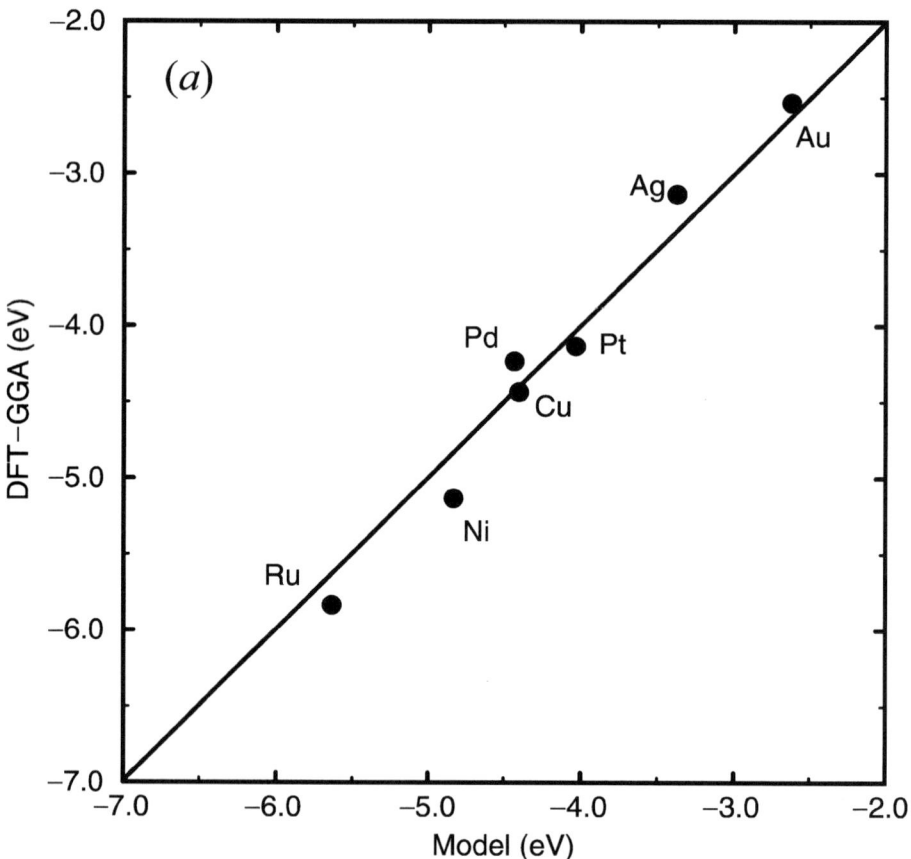

Figure 4 Hammer-Nørskov model analyses. Full DFT-GGA predictions versus model predictions. (*a*) Atomic oxygen chemisorption. [With kind permission from Kluwer Academic Publishers (7).] (*b*) CO chemisorption. [Reprinted with permission from (82); Copyright 1996. Am. Phys. Soc.]

correlations of chemisorption and dissociation energies with d-band centers (ε_d) for various metals, single-crystal metal facets, defect structures, bulk alloys, pseudomorphic overlayers, and strain levels. The remarkable quality of these correlations underlines the potential use of ε_d as an important catalyst design parameter.

Trends in Segregation Energies and Surface Alloy Formation

It is well known that the elemental chemical composition at the surface of an alloy may differ from the composition in the bulk; one of the alloy components is effectively enriched in the surface region. This enrichment can be caused, for example, by differences in surface segregation energies between the alloy components. Such

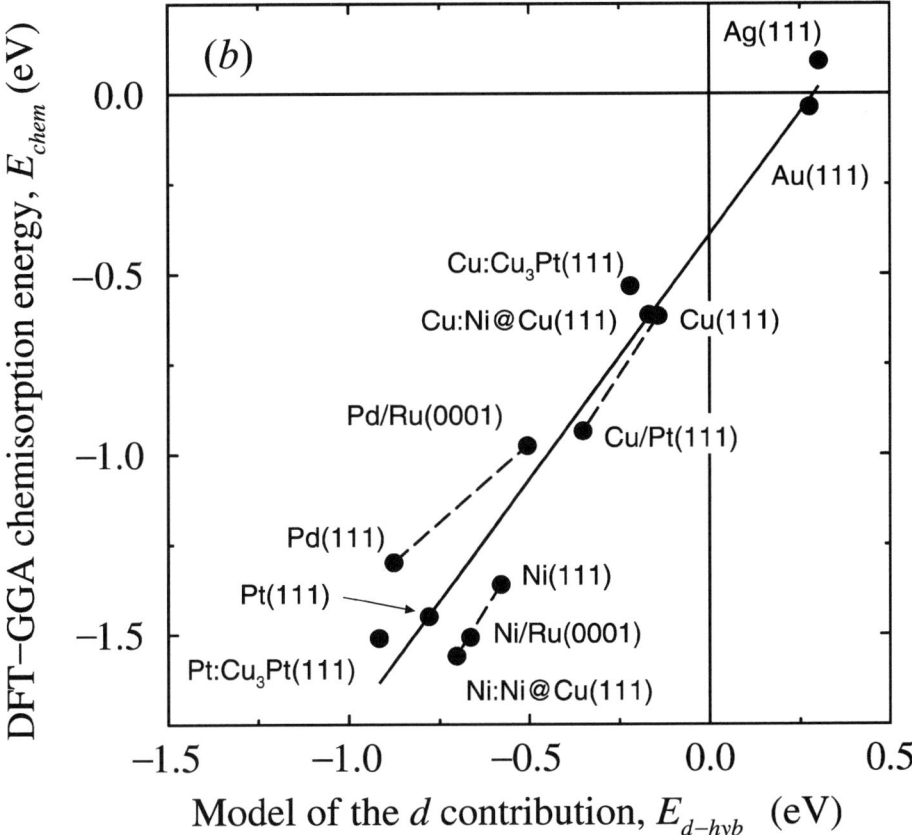

Figure 4 (*Continued*)

compositional changes can have a dramatic effect on the reactivity of the surface and are therefore of first order importance in heterogeneous catalysis. In an effort to develop a more fundamental understanding of the properties of surface alloys, Ruban et al. (86) have performed an extensive analysis of the electronic properties of pseudomorphic metal overlayers and of surface metal impurities. Following the Hammer-Nørskov reactivity model, they assume that reactivity trends will follow trends in d-band center values of overlayer and impurity atoms. They present an extensive database of d-band center shifts for these systems, and they conclude that, when metals with small lattice constants are overlayed or alloyed on metals with larger lattice constants, the d-band center shifts up. The reverse is true when metals with larger lattice constants are alloyed on metals with smaller lattice constants. These trends appear to result from a combination of d-band bandwidth changes upon overlaying or alloying, followed by d-band shifts to maintain a constant local d-band filling.

An important question in connection with alloy catalysis is determining the composition of the surface of the alloy. Christensen et al. (87) combine data from periodic DFT calculations with a detailed thermodynamic formalism describing surface compositions to produce phase diagrams for surface alloys. They find that this analysis provides a good description of many types of surface alloys, including random pseudomorphic overlayers, ordered overlayer structures, non-pseudomorphic structures, and systems where subsurface segregation occurs.

Finally, Ruban et al. (88) have created an extensive database of surface segregation energies for bimetallic systems (Figure 6, see color insert). The database points to two important periodic trends in surface segregation energies, and the authors interpret this information using a combination of several theories of surface structure. First, in agreement with the observations of Aldén et al. (89, 90), a substantial portion of the surface segregation energy is determined by the difference in surface energies of the host and the impurity. This conclusion is reached by interpreting the surface segregation data with the theory of Friedel (91), wherein cohesive energies of the transition metals are related to the d-occupation numbers of the metals. This theory, in combination with the observations of Aldén, implies that no impurities in the middle of a transition series are expected to segregate to the surface of early or late transition metals, an implication that is strongly supported by the calculated segregation energies. Second, crystal structure differences between the host and the impurity can have a significant effect on segregation energies. Such differences alter the local d-state character around the impurities and can lead to changes in the surface segregation energy of up to 1 eV. These changes are particularly significant if either the host or the impurity comes from the beginning of a transition metal series.

Oxygen on Transition Metals

Oxygen chemistry on transition metal surfaces is of fundamental importance in numerous surface processes, from oxidation reactions on heterogeneous catalysts to corrosion processes. In an effort to gain a fundamental understanding of such processes, a number of first principles studies of oxygen adsorption and dissociation on transition metal surfaces have been undertaken. Calculations of binding energies for atomic oxygen on a variety of surfaces have been performed; they generally indicate that the absolute binding energy increases from right to left on the periodic table (Figure 4a). Other studies have focused primarily on the overlayer structures of chemisorbed atomic oxygen (58, 59, 92). On both Ru{0001} and Rh{111}, high-coverage states (up to 1 ML) exist; there are kinetic barriers to forming these states, and hence they are only observed under high-pressure conditions or under conditions where highly energetic atomic oxygen has been prepared. The binding energy of the O atoms on Ru{0001} is higher than that on Rh{111}, a result explained in terms of the higher d-band center of Ru{0001} surface metal atoms. Additionally, chemisorbed oxygen leads to an increased separation between the first and second metal layers for both Ru{0001} and Rh{111}. However, the expansion is much smaller for Ru{0001}, an observation that is explained by the higher cohesive energy of bulk Ru.

Other studies of oxygen on transition metals have focused on molecular surface states and dissociation pathways. Hafner and coworkers (93–95) have examined O_2 adsorption and dissociation on Ni{111}, Pd{111}, and Pt{111}. Three molecular O_2 precursors were identified in each of these cases, and a monotonic increase of the absolute value of the binding energy (with respect to gas-phase O_2) with the metal d-band center energy was found [Ni{111} had both the largest binding energies and the highest d-band center, followed by Pd{111} and then Pt{111}]. This trend appears to extend to the work of Xu & Mavrikakis (96) on Cu{111}. The authors find an additional precursor state not present on Ni{111}, Pd{111}, or Pt{111}, but their results for precursor binding energies still fit nicely into the monotonic relationship of binding energy to d-band center mentioned previously.

CO on Transition Metals

CO chemistry on transition metal surfaces is of great importance in heterogeneous catalysis. Describing the full range of theoretical work that has been done in this area would be impossible in a brief review, so we focus on only four prominent contributions that first principles techniques have made to the understanding of CO chemistry on transition metals. First, we note that DFT calculations indicate that absolute binding energies for CO increase as one moves from right to left across the periodic table (Figure 4b). Second, we discuss a study of CO diffusion along the close-packed [1$\bar{1}$0] directions of Pt{110}-(1 × 1) (97). In these directions, CO chemisorbed on top sites has a binding energy about 0.02 eV stronger than does CO on bridge sites. A simple model of diffusion (in which the bridge site is arbitrarily assumed to be a transition state for the diffusion process) would suggest a diffusion barrier of this order, but calculations with a higher spatial resolution along the reaction coordinate show that the true diffusion barrier is ~0.13 eV (Figure 7). A similar bridge-top diffusion barrier has also been computed on Rh{100} (98). These calculations demonstrate the important role that first principles techniques can play in the detailed description of diffusion on transition metal surfaces.

Third, we examine a periodic DFT study of the interaction of CO with potassium promoters on Co{10$\bar{1}$0} (99). In that analysis, it is confirmed that CO adsorbs through carbon in an orientation roughly perpendicular to the surface. Potassium increases the reactivity of this configuration by donating charge to the CO $2\pi^*$ antibonding orbital. Furthermore, the potassium charge donation leads to a polarization of this orbital, with charge accumulating around the oxygen end of the molecule. These effects could facilitate either the dissociation of CO or the hydrogenation of the molecule to COH; either of these elementary pathways could be the first step in Fischer-Tropsch reactions on cobalt.

Lastly, we discuss a study of the investigation of localized demagnetization effects of CO on Ni{110} (100). In this study, Ge et al. find that CO adsorption on the surface leads to a strong, highly localized surface demagnetization. The key

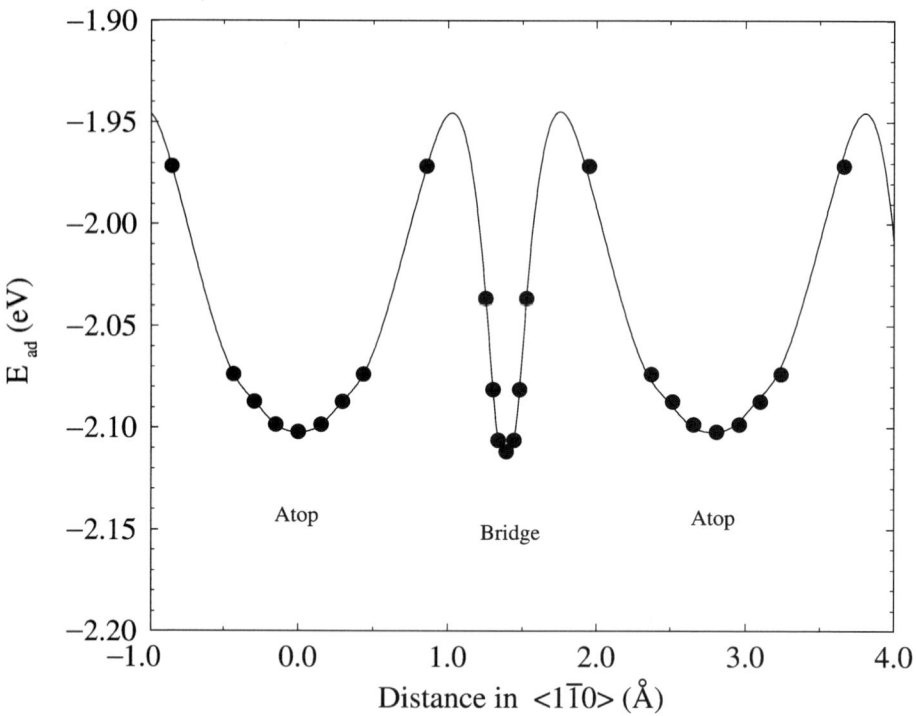

Figure 7 Potential energy surface for diffusion of CO on Pt{110}-(1 × 2) along ⟨1̄10⟩ directions. The reference energy state corresponds to the total energy of a gas-phase CO molecule at infinite separation from the platinum surface. Distances correspond to lateral displacements of the CO center of mass in angstroms. Atop and bridge refer to specific, stable surface sites for CO adsorption. The coverage is 0.25 ML. [Reprinted with permission from (97); Copyright 1999. Am. Inst. Phys.]

electronic contribution to this effect is a transfer of residual spin from the surface nickel atoms to the $2\pi^*$ or 1π orbitals of CO.

REACTIVITY TRENDS

First principles calculations are powerful tools for the analysis of reactivity trends on well-defined transition metal surfaces. Because of the ability of ab initio techniques to study transition states and reactive intermediates, the techniques can be used to generate large databases of chemisorption and activation energies for bond-breaking/bond-making events on given surfaces. From these energies, potential energy surfaces (PES's) for various reaction pathways can be developed that can be used to determine kinetically significant steps in the reaction pathways

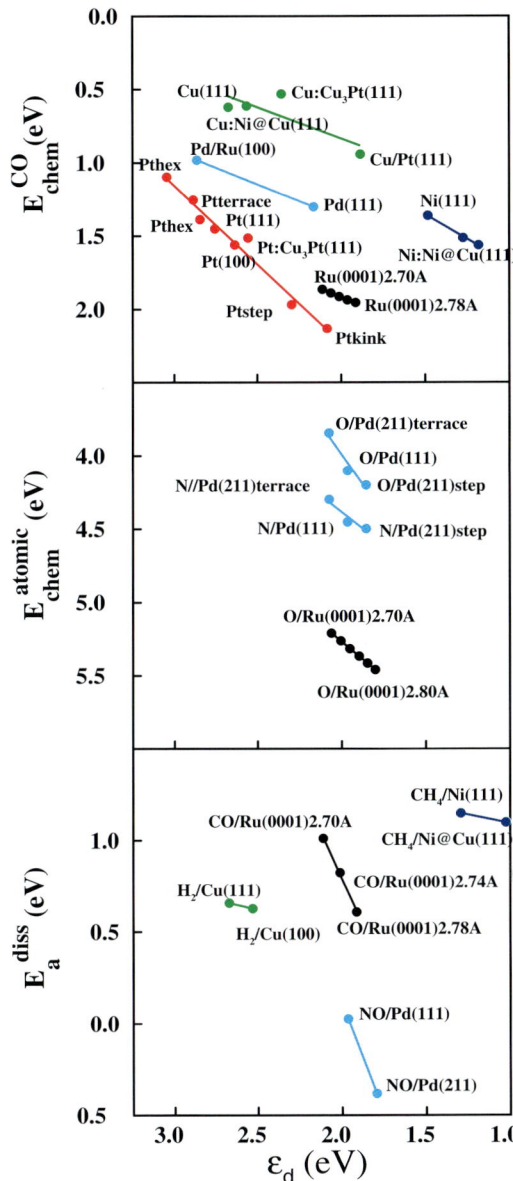

Figure 5 Chemisorption energies and dissociation barriers for various atomic and molecular species on a variety of transition metal surfaces versus the clean metal d-band centers (ε_d). The *top* panel summarizes trends in CO chemisorption energies; the *middle* panel provides similar information for atomic chemisorption energies; and the *bottom* panel shows trends in dissociation energies for various molecules. [Reprinted with permission from (85); Copyright 1998, Am. Phys. Soc.]

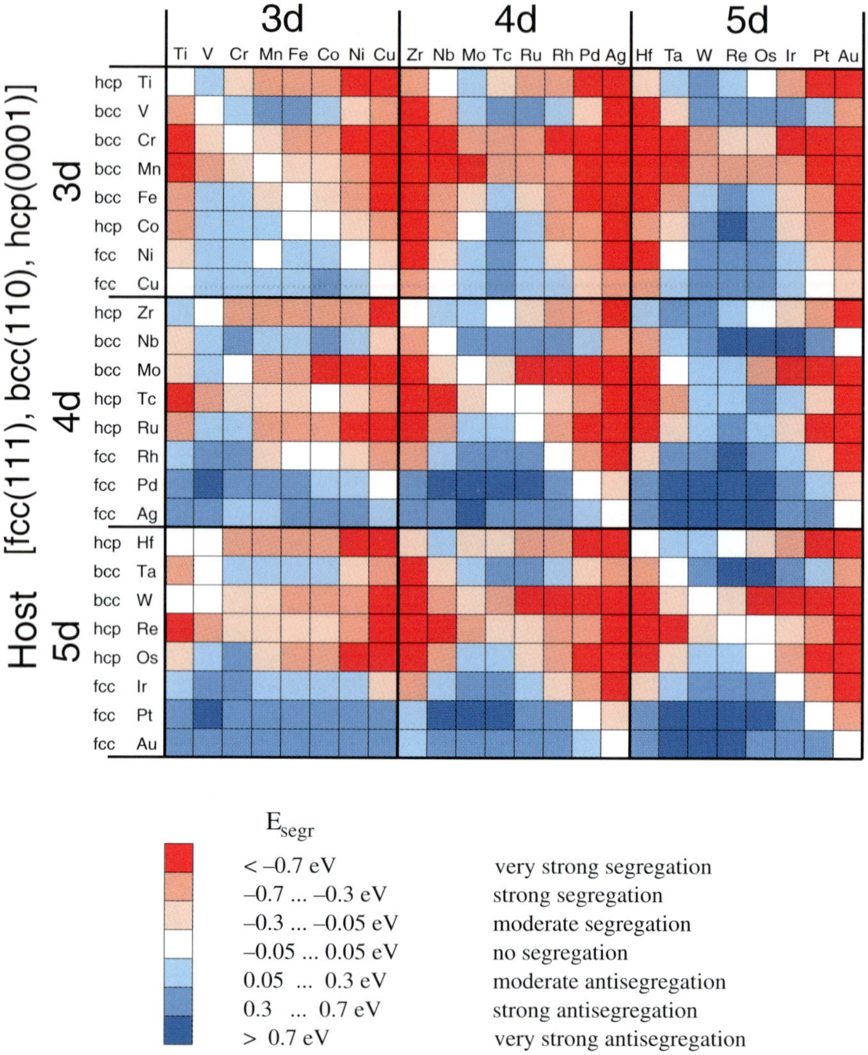

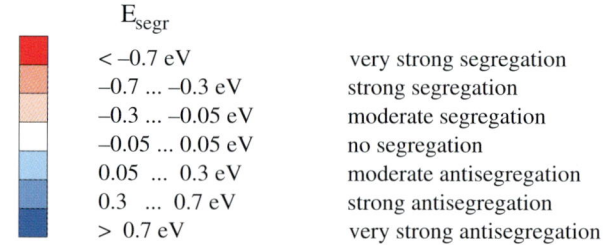

Figure 6 Surface segregation energies of transition metal impurities (solutes) for the close-packed surfaces of transition metal hosts. [Reprinted with permission from (88); Copyright 1999, Am. Phys. Soc.]

and to identify abundant surface species on the metal surfaces. In this section, we present several examples of systems in which theoretical techniques have been used to establish such trends on well-defined transition metal surfaces. Although these calculations are generally performed for specific surfaces, a comparison of reactivity trends on different metals can be extremely valuable, and we present some of these results below.

Hydrogenation-Dehydrogenation Reactions

Hydrocarbon hydrogenation and dehydrogenation reactions are crucially important to the chemical industry. They form the basis of isomerization reactions of alkanes and alkenes, syngas synthesis, and the Fischer-Tropsch synthesis, among other industrially important processes. A fundamental understanding of these processes is therefore extremely valuable, and many theoretical efforts have been made to determine hydrogenation/dehydrogenation trends on transition metal surfaces. Kua & Goddard (101), for example, have used DFT cluster calculations to study methane and ethylene fragments on Pt{111}. They find that CH_x species adsorb so as to maintain a tetravalent coordination for carbon. The binding energy of these fragments scales almost linearly with the number of Pt-C bonds. The CH radical is found to be the thermodynamic sink for the methane decomposition reaction on this surface.

The same authors have also performed a more general study of $CH_{n-m}(CH_3)_m$ ($n = 1, 2, 3$ with $m \leq n$) species on Pt, Ir, Os, Pd, Rh, and Ru (102). For methane fragments, they find that a tetravalent carbon configuration is preferred for all fragments on all metals. Furthermore, CH is the thermodynamic sink on all metals, and fcc metals appear to bind methyl fragments more than hcp metals. For fragments of higher carbon numbers, steric interactions of methyl groups with the surface lead to lower binding energies. The authors synthesize all of their data for Pt{111} surfaces to design an elegant group contribution approach for the estimation of hydrocarbon binding energies to that surface.

Papoian et al. (103) have also studied the behavior of small hydrocarbon fragments on Pt{111}. Using periodic DFT calculations, they show that atomic hydrogen moves freely about the Pt{111} surface, whereas methyl and ethyl fragments are strongly bound to top sites through their carbon atoms. These results, consistent with the findings described above, are analyzed in detail using the Crystal Orbital Hamilton Population Theory (104, 105).

Additional methane hydrogenation/dehydrogenation periodic DFT studies have been performed on Ru{0001} (106) and Ni{111} (107). On Ru{0001}, it is found that all methane fragments prefer threefold binding configurations; this result apparently contradicts the results of the cluster calculations described above (102). In the case of CH_3, at least, the authors attribute this result to reduced Pauli repulsion from the unfilled Ru d-bands leading to a preference for higher-coordination sites. CH is found to be the thermodynamic sink for the methane dehydrogenation reaction, and all reaction steps (with the exception of C–H scission in CH) are found to be exothermic. On Ni{111}, it is found that all methane fragments prefer threefold sites, and CH is again the thermodynamic sink of the reaction. In this study, kinetic barriers were also calculated, and it was determined that, with the exception of CH_2 conversion to CH, these barriers are significant ($\sim$0.85 eV for most steps). C–H bond scission occurs exclusively over top sites.

Michaelides & Hu (108) performed a detailed thermochemical and kinetic study of C, N, and O hydrogenation on Pt{111}, starting with the bare atoms and ending with CH_4, NH_3, and H_2O. Using periodic DFT calculations, they demonstrated that the transition states for hydrogenation reactions of all of these atoms possessed the preferred valency of the atoms themselves. Furthermore, the direction of approach of hydrogen atoms to the CH_x, NH_x, and OH_x species during hydrogenation coincided with the directions of the molecular orbitals of these species (Figure 8). The same authors extended this analysis to more general diatomic recombination reactions of C, O, N, and H atoms on Pt{111} and Cu{111} (109), reactions in which all species involved could have well-defined valencies and site preferences. In these reactions, transition states were always of the hollow-bridge or hollow-top form. The higher valency reactant was always found to occupy the hollow position. The lower valency reactant occupied the bridge position if its PES was highly corrugated and occupied the top position if its PES was smooth.

Partial Oxidation Reactions

Partial oxidation reaction pathways on transition metal surfaces can also be well described by theoretical techniques. For example, Greeley & Mavrikakis have studied the reaction pathway for methanol decomposition (partial oxidation) on both Cu{111} and Pt{111} surfaces using periodic DFT calculations (Figure 9). On Cu{111}, the authors find that the rate-limiting step (RLS) for this decomposition process is the conversion of methoxy to formaldehyde. The RLS for the reverse reaction, methanol synthesis from CO, is found to be the hydrogenation of CO to yield a surface formyl intermediate. For either the forward or reverse reactions, a high-pressure gas phase environment appears to be necessary to obtain a reasonable reaction rate. Three of the species in the pathway (methanol, formaldehyde, and CO) have higher reaction barriers than desorption barriers, making it likely that they will desorb before reacting in a UHV environment.

On Pt{111}, the reaction pathway is completely different. The RLS for the methanol decomposition reaction is the abstraction of hydrogen from methanol to produce methoxy; the RLS of the reverse reaction is the hydrogenation of CO to formyl. CO is the thermochemical sink of this reaction pathway (in fact, the CO binding energy is so strong that this species is a poison for the catalyst surface). Finally, because of the weak binding of methanol to Pt{111}, a high-pressure environment is necessary to observe this reaction experimentally.

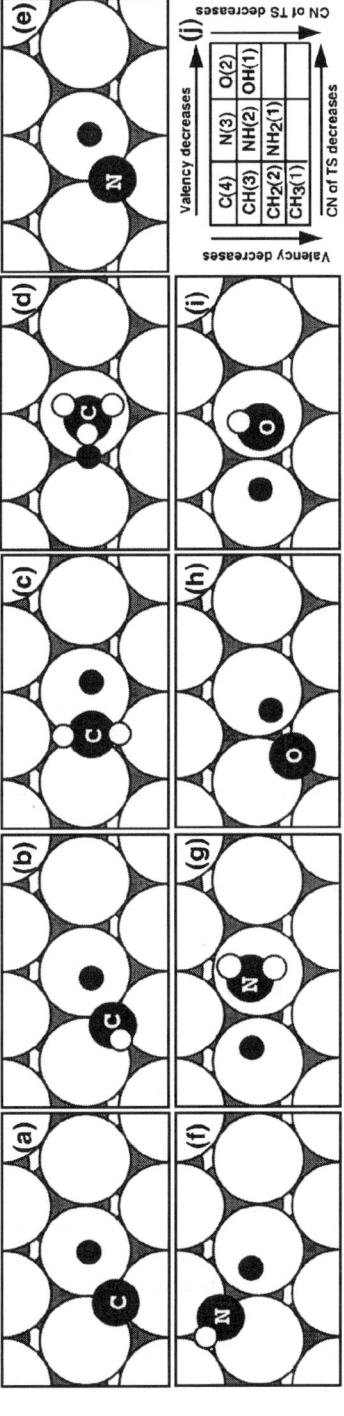

Figure 8 Transition state structures for intermediates of C, N, and O hydrogenation on Pt{111}. Small black and white circles indicate hydrogen. Large white circles indicate surface platinum atoms. Large gray circles indicate subsurface platinum atoms. [Reprinted with permission from (108); Copyright 2000. Am. Chem. Soc.]

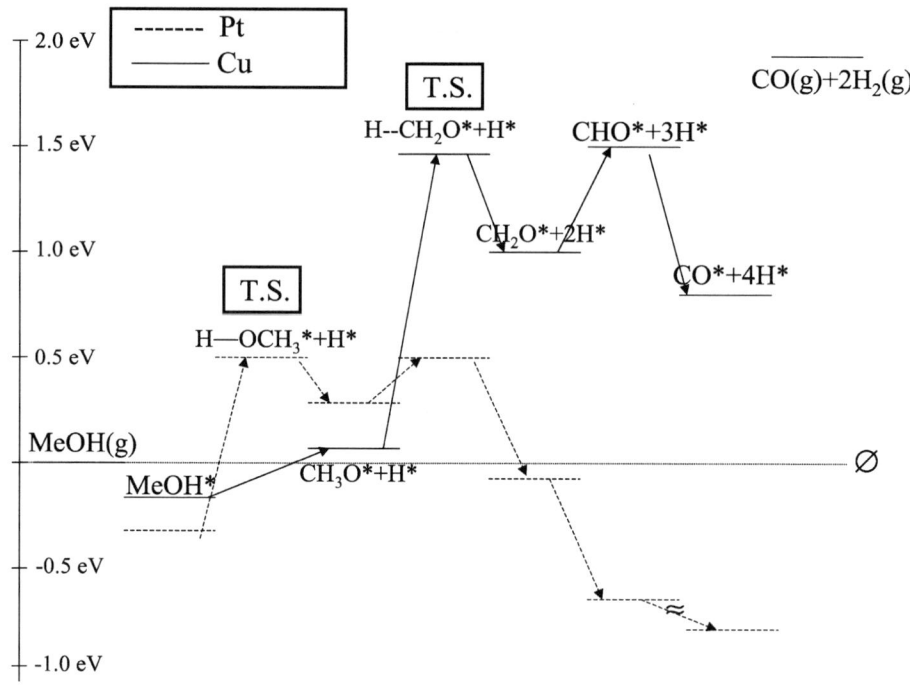

Figure 9 Calculated potential energy surfaces for MeOH decomposition on Cu{111} and Pt{111} surfaces (J. Greeley, M. Mavrikakis, unpublished data). The zero of the vertical axis corresponds to the total energy of a methanol molecule at infinite separation from the corresponding metal surface. The * indicates adsorbed species, whereas (g) is used to indicate species in the gas phase. T. S. represents local maxima in the reaction path, corresponding to the respective transition states. The broken arrow used for the adsorbed CO on Pt{111} indicates that the energy for that state is off with respect to the scale of the vertical axis.

CONCLUSION: FROM THEORY TO INDUSTRIAL APPLICATIONS

In this review, we have attempted to both highlight some of the recent contributions that theoretical techniques have made to the fields of surface science and catalysis and to show the great potential of theory for advancing these fields in the future. We have described how first principles methods can provide significant insight into puzzles that cannot be explained by purely experimental means, how these methods are well-suited for identifying trends in reactivity across the periodic table, and how insight into detailed reaction pathways and mechanisms on particular surfaces can be obtained with these techniques. We would be remiss, however, if we failed to mention two case studies in which theoretical analyses played a major role in the development of new catalysts, either directly through suggestions of new catalyst compositions based on explicit calculations, or indirectly through the development of concepts that have been used in catalyst development work. These examples epitomize the potential of theoretical methods in surface science and catalysis.

New Steam-Reforming Catalysts

In steam-reforming processes, hydrocarbons such as methane are converted to CO and H_2 over nickel catalysts. Graphite formation is a problem over nickel, however, and can lead to catalyst deactivation. As a possible solution to this problem, Besenbacher et al. (112) considered a gold-doped nickel catalyst. They performed two sets of theoretical calculations for methane steam reforming. First, they calculated the change in the reaction barrier for hydrogen abstraction from methane (the RLS for steam reforming of methane) (113), when 0.25 ML of Au was substitutionally alloyed into the Ni surface layer. The barrier increased by $\sim$0.31 eV (compared with pure Ni) for methane dissociation over Ni atoms with two Au neighbors. Second, they determined the effect of the same Au doping on the binding of atomic carbon to the surface. 0.25 ML of gold completely destabilized carbon on three quarters of the threefold sites (the preferred binding sites of C on Ni) and led to a $\sim$0.26 eV decrease in the binding energy of C on the remaining two sites. The authors concluded that graphite formation would be substantially inhibited by Au doping and that methane dissociation would be only marginally affected. Extensive surface characterization confirmed that a stable Au/Ni surface alloy could exist under in situ reaction conditions and that the catalyst could effectively catalyze the reforming reaction without substantial graphite formation. The extra cost of this new catalyst is reasonable because Au stays on the surface of Ni; Ni and Au do not produce a bulk alloy. Thus, only a small amount of Au is needed. As a result, the Halder Topsøe corporation is presently replacing its traditional steam-reforming catalyst with a Au/Ni alloy catalyst, which has substantially improved lifetime characteristics.

New Ammonia Synthesis Catalysts

The RLS in the heterogeneously catalyzed synthesis of ammonia from N_2 and H_2 is known to be the dissociative adsorption of N_2 (114). The transition state in this elementary step resembles adsorbed atomic nitrogen (115), and thus trends in the nitrogen dissociative chemisorption energy are strongly correlated with the reaction barrier and with the overall rate of reaction. For metals on which nitrogen is relatively weakly bound, decreases in the endothermicity of nitrogen chemisorption (corresponding to more strongly bound nitrogen) imply a lowering of the transition state energy and thus a higher rate of reaction. For metals on which nitrogen is strongly bound, however, increases in the strength of the metal-nitrogen interaction can lead to increased nitrogen surface coverage and thus to decreased catalytic activity. The net result of these two effects is a volcano plot in which a maximum in the ammonia synthesis rate versus nitrogen chemisorption energy relationship is observed. Jacobsen et al. (115) showed that alloying can be used to make new structures with nitrogen chemisorption energies close to the optimum. They identified CoMo as a potential high-activity alloy catalyst (i.e., an alloy with a suitable nitrogen chemisorption energy) by simple interpolation between the corresponding pure-metal components on the volcano curve (Figure 10). Then, they confirmed with first principles DFT calculations that nitrogen does, in fact, have the required intermediate chemisorption energy on CoMo surfaces. They also calculated that the N_2 dissociation energy on this alloy is intermediate between the dissociation energies on the pure metal components. Finally, they used a Co_3Mo_3N catalyst (synthesized by Jacobsen; 116) to demonstrate experimentally that this alloy has an ammonia synthesis activity comparable to that of the best industrial catalysts.

Theory-to-catalyst successes such as those described above are expected to become more common in the future. With continued improvement of electronic structure calculation techniques and concurrent increases in available computer resources, theoretical tools will play an increasingly important role in rational catalyst design in the coming years.

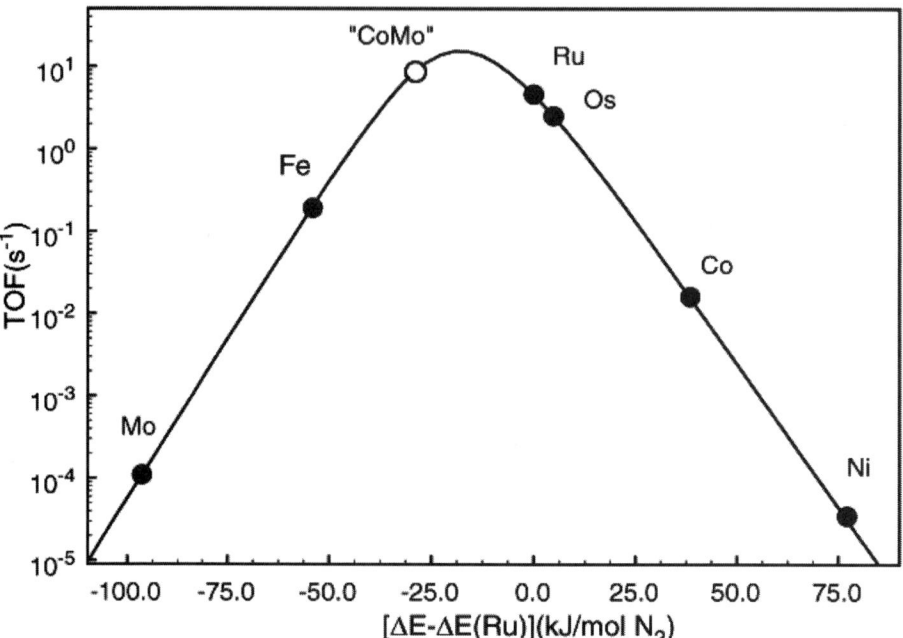

Figure 10 Calculated turnover frequencies for ammonia synthesis as a function of the adsorption energy of nitrogen for various transition metals and alloys. [Reprinted with permission from (115); Copyright 2001. Am. Chem. Soc.]

ACKNOWLEDGMENTS

The authors thank Mark Barteau, Bjork Hammer, Peijun Hu, David King, Michele Parrinello, and Andrei Ruban for permission to reprint their figures in this review. J. G. acknowledges partial financial support from a National Science Foundation predoctoral fellowship. M. M. acknowledges partial financial support from an NSF CAREER award (CTS-0134561) and a Shell Oil Company Foundation Faculty Career Initiation Award. The Center for Atomic-Scale Materials Physics (CAMP) is supported by the Danish National Research Foundation. J. G. and M. M. acknowledge partial support from NSF cooperative agreement ACI-9619020 through computing resources provided by the National Partnership for Advanced Computational Infrastructure (NPACI). Additional computational resources for J. G. and M. M. were provided by the DOEBES office, National Energy Research Scientific Computing Center (NERSC).

Visit the Annual Reviews home page at www.annualreviews.org

LITERATURE CITED

1. Lundqvist BI. 1983. *Vacuum* 33:639–49
2. van Santen RA. 1987. *Prog. Surf. Sci.* 25:253–74
3. Somorjai GA. 1991. *Catal. Lett.* 9:311–28
4. Ziegler T. 1991. *Chem. Rev.* 91:651–67
5. Payne MC, Teter MP, Allan DC, Arias TA, Joannopoulos JD. 1992. *Rev. Mod. Phys.* 64:1045–97
6. van Santen RA, Neurock M. 1995. *Catal. Rev. Sci. Eng.* 37:557–698
7. Hammer B, Nørskov JK. 1997. In *Chemisorption and Reactivity on Supported Clusters and Thin Films*, ed. RM Lambert, G Pacchioni, pp. 285–351. Netherlands: Kluwer
8. Hammer B, Nørskov JK. 2000. *Adv. Catal.* 45:71–129
9. McQuarrie DA. 1983. *Quantum Chemistry*. Mill Valley, CA: University Science
10. McWeeny R, Sutcliffe BT. 1969. *Methods of Molecular Quantum Mechanics*. London: Academic
11. Ceperley DM, Mitas L. 1996. *Adv. Chem. Phys.* 93:1–38
12. Mitas L. 1997. *Curr. Opin. Solid State Mater. Sci.* 2:696–700
13. Mitas L. 1997. *Physica B* 237:318–20
14. Foulkes WMC, Mitas L, Needs RJ, Rajagopal G. 2001. *Rev. Mod. Phys.* 73:33–83
15. Hammond BL, Lester WA, Reynolds PJ. 1994. *Monte Carlo Methods in Ab Initio Quantum Chemistry*. Singapore: World Scientific
16. Mitas L, Martin RM. 1994. *Phys. Rev. Lett.* 72:2438–41
17. Rajagopal G, Needs RJ, Kenny S, Foulkes WMC, James A. 1994. *Phys. Rev. Lett.* 73:1959–62
18. Williamson AJ, Hood RQ, Needs RJ, Rajagopal G. 1998. *Phys. Rev. B* 57:12140–44
19. Hohenberg P, Kohn W. 1964. *Phys. Rev.* 136:B864–71
20. Kohn W, Sham LJ. 1965. *Phys. Rev.* 140:A1133–38
21. Mortensen JJ. 1998. *A theoretical study of adsorption and dissociation on metal surfaces*. PhD Thesis. Tech. Univ. Denmark, Lyngby, Denmark. 174 pp.
22. Perdew JP, Zunger A. 1981. *Phys. Rev. B* 23:5048–79
23. Vosko SH, Wilk L, Nusair M. 1980. *Can. J. Phys.* 58:1200–211
24. Tschinke V, Ziegler T. 1991. *Theor. Chim. Acta* 81:65–78

25. White JA, Bird DM, Payne MC, Stich I. 1994. *Phys. Rev. Lett.* 73:1404–7
26. Fan L, Ziegler T. 1991. *J. Chem. Phys.* 95:7401–8
27. Brown WA, Kose R, King DA. 1998. *Chem. Rev.* 98:797–831
28. Hammer B, Hansen LB, Nørskov JK. 1999. *Phys. Rev. B* 59:7413–21
29. Perdew JP, Burke K, Ernzerhof M. 1996. *Phys. Rev. Lett.* 77:3865–68
30. Zhang YK, Yang WT. 1998. *Phys. Rev. Lett.* 80:890
31. Cohen M, Kelly PS. 1966. *Can. J. Phys.* 44:3227–40
32. McEachran RP, Tull CE, Cohen M. 1968. *Can. J. Phys.* 46:2675–78
33. Bachelet GB, Hamann DR, Schlüter M. 1982. *Phys. Rev. B* 26:4199–228
34. Troullier N, Martins JL. 1991. *Phys. Rev. B* 43:1993–2006
35. Vanderbilt D. 1990. *Phys. Rev. B* 41:7892–95
36. Mavrikakis M, Stoltze P, Nørskov JK. 2000. *Catal. Lett.* 64:101–6
37. Gülseren O, Bird DM, Humphreys SE. 1998. *Surf. Sci.* 404:827–30
38. Jones GS, Mavrikakis M, Barteau MA, Vohs JM. 1998. *J. Am. Chem. Soc.* 120:3196–204
39. Medlin JW, Barteau MA, Vohs JM. 2000. *J. Mol. Catal. A: Chem.* 163:129–45
40. Medlin JW, Sherrill AB, Chen JG, Barteau MA. 2001. *J. Phys. Chem. B* 105:3769–75
41. Ihm H, Medlin JW, Barteau MA, White JM. 2001. *Langmuir* 17:798–806
42. Brown WA, King DA. 2000. *J. Phys. Chem. B* 104:2578–95
43. Bertolo M, Jacobi K. 1990. *Surf. Sci.* 226:207–20
44. Chen PJ, Goodman DW. 1993. *Surf. Sci. Lett.* 297:L93–99
45. Loffreda D, Simon D, Sautet P. 1998. *Chem. Phys. Lett.* 291:15–23
46. Kao CT, Blackman GS, van Hove MA, Somorjai GA, Chan CM. 1989. *Surf. Sci.* 224:77–96
47. Borg HJ, Reijerse JFCJM, van Santen RA, Niemantsverdriet JW. 1994. *J. Chem. Phys.* 101:10052–63
48. Kim YJ, Thevuthasan S, Herman GS, Peden CHF, Chambers SA, et al. 1996. *Surf. Sci.* 359:269–79
49. Dahl S, Logadottir A, Egeberg RC, Larsen JH, Chorkendorff I, et al. 1999. *Phys. Rev. Lett.* 83:1814–17
50. Johnson AD, Daley SP, Utz AL, Ceyer ST. 1992. *Science* 257:223–25
51. Michaelides A, Hu P, Alavi A. 1999. *J. Chem. Phys.* 111:1343–45
52. Ledentu V, Dong W, Sautet P. 2000. *J. Am. Chem. Soc.* 122:1796–801
53. Peden CHF, Goodman DW. 1986. *J. Phys. Chem.* 90:4839–43
54. Peden CHF, Goodman DW, Weisel MD, Hoffmann FM. 1991. *Surf. Sci.* 253:44–58
55. Engel T, Ertl G. 1979. *Adv. Catal.* 28:1–78
56. Stampfl C, Scheffler M. 1997. *Phys. Rev. Lett.* 78:1500–3
57. Stampfl C, Scheffler M. 1999. *Surf. Sci.* 433–35:119–26
58. Stampfl C, Scheffler M. 1996. *Phys. Rev. B* 54:2868–72
59. Stampfl C, Schwegmann S, Over H, Scheffler M, Ertl G. 1996. *Phys. Rev. Lett.* 77:3371–74
60. Natta G. 1955. *Macromol. Chem.* 16:213
61. Somorjai GA. 1994. *Introduction to Surface Chemistry and Catalysis.* New York: Wiley
62. Boero M, Parrinello M, Terakura K. 1998. *J. Am. Chem. Soc.* 120:2746–52
63. Boero M, Parrinello M, Terakura K. 1999. *Surf. Sci.* 438:1–8
64. Boero M, Parrinello M, Hüffer S, Weiss H. 2000. *J. Am. Chem. Soc.* 122:501–9
65. Boero M, Parrinello M, Weiss H, Hüffer S. 2001. *J. Phys. Chem. A.* 105:5096–105
66. Andreson A, Cordes HG, Herwig J, Kaminsky W, Merck A, et al. 1976. *Inst. Anorg. Angew. Chem.* 88:689–90
67. Small BL, Brookhart M. 1999. *Macromolecules* 32:2120–30
68. Small BL, Brookhart M, Bennett AMA. 1998. *J. Am. Chem. Soc.* 120:4049–50

69. Schmid R, Ziegler T. 2000. *Organometallics* 19:2756–65
70. Chan MSW, Ziegler T. 2000. *Organometallics* 19:5182–89
71. Vanka K, Ziegler T. 2001. *Organometallics* 20:905–13
72. Michalak A, Ziegler T. 2001. *Organometallics* 20:1521–32
73. Mole T, Anderson JR, Creer G. 1985. *Appl. Catal.* 17:141–54
74. Ono Y. 1992. *Catal. Rev. Sci. Eng.* 34:179–226
75. Barbosa LAMM, Zhidomirov GM, van Santen RA. 2000. *Phys. Chem. Chem. Phys.* 2:3909–18
76. Shubin AA, Zhidomirov GM, Yakovlev AL, van Santen RA. 2001. *J. Phys. Chem. B* 105:4928–35
77. Barbosa LAMM, van Santen RA, Hafner J. 2001. *J. Am. Chem. Soc.* 123:4530–40
78. Barbosa LAMM, van Santen RA. 2001. *J. Mol. Catal. A: Chem.* 166:101–21
79. Rozanska X, van Santen RA, Hutschka F, Hafner J. 2001. *J. Am. Chem. Soc.* 123:7655–67
80. Shah R, Gale JD, Payne MC. 1997. *J. Phys. Chem. B* 101:4787–97
81. Hammer B, Nørskov JK. 1995. *Surf. Sci.* 343:211–20
82. Hammer B, Morikawa Y, Nørskov JK. 1996. *Phys. Rev. Lett.* 76:2141–44
83. Pallassana V, Neurock M, Hansen LB, Hammer B, Nørskov JK. 1999. *Phys. Rev. B* 60:6146–54
84. Pallassana V, Neurock M, Hansen LB, Nørskov JK. 2000. *J. Chem. Phys.* 112:5435–39
85. Mavrikakis M, Hammer B, Nørskov JK. 1998. *Phys. Rev. Lett.* 81:2819–22
86. Ruban A, Hammer B, Stoltze P, Skriver HL, Nørskov JK. 1997. *J. Mol. Catal. A: Chem.* 115:421–29
87. Christensen A, Ruban AV, Stoltze P, Jacobsen KW, Skriver HL, Nørskov JK. 1997. *Phys. Rev. B* 56:5822–34
88. Ruban AV, Skriver HL, Nørskov JK. 1999. *Phys. Rev. B* 59:15990–6000
89. Aldén M, Abrikosov IA, Johansson B, Rosengaard NM, Skriver HL. 1994. *Phys. Rev. B* 50:5131–46
90. Aldén M, Skriver HL, Johansson B. 1994. *Phys. Rev. B* 50:12118–30
91. Friedel J. 1976. *Ann. Phys. (Paris)* 1:257–307
92. Ganduglia-Pirovano MV, Scheffler M. 1999. *Phys. Rev. B* 59:15533–43
93. Eichler A, Hafner J. 1997. *Phys. Rev. Lett.* 79:4481–84
94. Eichler A, Mittendorfer F, Hafner J. 2000. *Phys. Rev. B* 62:4744–55
95. Mittendorfer F, Eichler A, Hafner J. 1999. *Surf. Sci.* 433–35:756–60
96. Xu Y, Mavrikakis M. 2001. *Surf. Sci.* 494:131–44
97. Ge Q, King DA. 1999. *J. Chem. Phys.* 111:9461–64
98. Eichler A, Hafner J. 1998. *J. Chem. Phys.* 109:5585–95
99. Jenkins SJ, King DA. 2000. *J. Am. Chem. Soc.* 122:10610–14
100. Ge Q, Jenkins SJ, King DA. 2000. *Chem. Phys. Lett.* 327:125–30
101. Kua J, Goddard WA. 1998. *J. Phys. Chem. B* 102:9492–500
102. Kua J, Faglioni F, Goddard WA. 2000. *J. Am. Chem. Soc.* 122:2309–21
103. Papoian G, Nørskov JK, Hoffmann R. 2000. *J. Am. Chem. Soc.* 122:4129–44
104. Dronskowski R, Blochl PE. 1993. *J. Phys. Chem.* 97:8617–24
105. Glassey WV, Papoian GA, Hoffmann R. 1999. *J. Chem. Phys.* 111:893–910
106. Ciobîca IM, Frechard F, van Santen RA, Kleyn AW, Hafner J. 1999. *Chem. Phys. Lett.* 311:185–92
107. Watwe RM, Bengaard HS, Rostrup-Nielsen JR, Dumesic JA, Nørskov JK. 2000. *J. Catal.* 189:16–30
108. Michaelides A, Hu P. 2000. *J. Am. Chem. Soc.* 122:9866–67
109. Michaelides A, Hu P. 2001. *J. Chem. Phys.* 114:5792–95
110. Deleted in proof

111. Deleted in proof
112. Besenbacher F, Chorkendorff I, Clausen BS, Hammer B, Molenbroek AM, et al. 1998. *Science* 279:1913–15
113. Lee MB, Yang QY, Ceyer ST. 1987. *J. Chem. Phys.* 87:2724–41
114. Ertl G, Huber M, Lee SB, Paal Z, Weiss M. 1981. *Appl. Surf. Sci.* 8:373–86
115. Jacobsen CJH, Dahl S, Clausen BS, Bahn S, Logadottir A, Nørskov JK. 2001. *J. Am. Chem. Soc.* 123:8404–5
116. Jacobsen CJH. 2000. *Chem. Commun.* 12:1057–58

CHEMICAL SHIFTS IN AMINO ACIDS, PEPTIDES, AND PROTEINS: From Quantum Chemistry to Drug Design

Eric Oldfield

Department of Chemistry and Center for Biophysics and Computational Biology, University of Illinois at Urbana-Champaign, 600 South Mathews Avenue, Urbana, Illinois 61801; e-mail: eo@chad.scs.uiuc.edu

Key Words electrostatics, porphyrins

■ **Abstract** This chapter discusses recent progress in the investigation and use of ^{13}C, ^{15}N, and ^{19}F nuclear magnetic resonance (NMR) chemical shifts and chemical shift tensors in proteins and model systems primarily using quantum chemical (ab initio Hartree-Fock and density functional theory) techniques. Correlations between spectra and structure are made and the techniques applied to other spectroscopic and electrostatic properties as well, including hydrogen bonding, ligand binding to heme proteins, J-couplings, electric field gradients, and atoms-in-molecules theory, together with a brief review of the use of NMR chemical shifts in drug design.

INTRODUCTION

It has been known for many years that folding a protein into its native conformation causes a large range of chemical shift nonequivalence. For reasons of sensitivity, the earliest protein nuclear magnetic resonance (NMR) investigations utilized the ^{1}H nucleus, but there were few resolved single atom resonances, typically just those of $H^{\varepsilon 2}$ of histidine residues, which fall outside the main protein spectral region (1). With the advent of Fourier transform NMR techniques, there was considerable interest in using ^{13}C NMR spectroscopy to investigate chemical shifts in proteins because it was thought that there would be far more resolved, single atom resonances owing to the well-known large chemical shift range of ^{13}C. Unfortunately, the earliest investigations were unsuccessful, because of low signal-to-noise ratios. However, in 1972 Allerhand developed a 20-mm NMR probe that enabled the first observation of single carbon atom resonances in proteins (2). The results were of interest because they showed a very large range of chemical shift nonequivalence due to folding, for example, up to 6 ppm in the case of C^{γ} of tryptophan residues (2). In addition, it was demonstrated that there were significant chemical shift differences between C^α chemical shifts in random coil versus helical polypeptides (3). These early observations suggested that ^{13}C chemical shifts might be useful probes of protein structure, but it took another decade of development in the area of heteronuclear NMR, together with isotopic labeling, to enable the routine measurement of ^{13}C (and ^{15}N) chemical shifts in proteins. Once this had been accomplished, it became clear that ^{13}C chemical shifts in proteins were sensitive to both backbone ϕ and ψ torsion angles, and Spera & Bax (4) showed that helical and sheet segments could be clearly differentiated based upon differences in their C^α and C^β chemical shifts. Because of the clearly defined dependence of the observed chemical shifts on these backbone torsion angles, their work strongly suggested that it should be possible to compute such experimental chemical shifts by using quantum chemical techniques.

Exactly how such calculations should be carried out was not clear, however. In particular, we had made a number of observations of the ^{13}C and ^{17}O NMR chemical shifts of the CO ligands in a variety of carbonmonoxyheme proteins. We had found that the ^{13}C and ^{17}O NMR chemical shifts were highly (anti) correlated and were also proportional to the CO vibrational stretch frequencies and the ^{17}O NMR-determined electric field gradients, as well as the Fe-C vibrational stretch frequencies determined from Raman spectroscopy (5). Augspurger et al. suggested that the origin of all these correlations was due to electrical polarization of the FeCO fragment in the protein, an electrostatic field effect, which appeared to be the case (6). The question then arose as to whether the ^{13}C NMR chemical shifts seen by Spera & Bax might also have a primarily electrostatic origin. It was also well-known that the ^{19}F NMR chemical shifts in proteins covered a very wide range, and in this case purely electronic structural differences caused by changes in backbone torsion angles would not reasonably be expected to have any direct effect on the fluorine chemical shifts.

Fortunately, it is now possible to compute essentially all of these chemical shifts in proteins by using quantum chemical techniques, suitably tailored to the nucleus of interest. In the case of ^{13}C and ^{15}N nuclei in amino-acid residues in peptides and proteins, the chemical shifts report on local geometric or electronic structure, in particular, the backbone and side-chain torsion angles, ϕ, ψ, and χ. However, in the case of ^{19}F nuclei in aromatic amino acids in proteins as well as in the case of ^{13}C and ^{17}O NMR chemical shifts in carbonmonoxyheme proteins, it is the electrostatic field generated by the protein that is responsible for the observed chemical shift nonequivalencies. As a result, it is necessary to use two rather different approaches to computing experimental spectra: one focusing on torsional angles and electronic effects, the second on electrostatics.

I begin this review by discussing the computation of the ^{13}C and ^{15}N NMR chemical shifts in amino acids, peptides, and proteins using both ab initio Hartree-Fock and density functional theory (DFT) techniques. This is followed by a brief description of recent progress in computing chemical shifts in paramagnetic systems. I then describe approaches to the use of chemical shifts and chemical shift tensors in predicting and refining aspects of peptide and protein structure. Next, I consider the computation of ^{19}F NMR chemical shifts in proteins and model systems, with emphasis on the effects of electrostatic fields. I then consider the case of metalloproteins such as carbonmonoxymyoglobin, in which questions about metal-ligand geometries and electrostatic field effects can be probed by using a combination of chemical shift, chemical shift tensor, Mössbauer, infrared, and Raman spectroscopy, together with quantum chemical calculations. This leads naturally into a more detailed discussion of electrostatics, including a consideration of hydrogen bonding in proteins, the covalence of the hydrogen bond, and the observation of through-space hydrogen bond J-couplings. Finally, I briefly describe results of the use of purely experimental NMR chemical shifts in the area of drug design, an important recent development in the use of chemical shift information.

With modern codes and high-speed computer workstations or workstation clusters and accessibility to supercomputer centers, it is now possible to evaluate essentially all of the spectroscopic observables in amino acids, peptides, or proteins with good accuracy. Whereas I focus on the chemical shift or chemical shielding and its associated tensor element magnitudes (and orientations), I also consider topics such as spin densities (which control hyperfine shifts in proteins), together with J-couplings and electric field gradients, the latter being of interest both in NMR and Mössbauer spectroscopy. Consideration of the electric field gradient tensor leads into a more general consideration of electrostatics, in which the results of calculations can be validated against charge densities, electrostatic potentials, and Laplacians/Hessians of the charge densities obtained from high-resolution X-ray diffraction experiments. In every instance I make a comparison between the experimental observable and the computed property, obtained by using quantum chemistry. Moreover, where possible, I consider complete tensor representations of the observable and the computed property, including tensor orientations, which for the chemical shielding, electric field gradient, and Hessian-of-$\rho(\mathbf{r})$ tensors may be conveniently combined and described in an icosahedral representation.

CARBON-13 AND NITROGEN-15 NMR CHEMICAL SHIFTS IN AMINO ACIDS, PEPTIDES, AND PROTEINS

How does one calculate the chemical shifts of an amino acid, a peptide, or a protein? In early work we attempted to find empirical correlations between experimental ^{13}C chemical shifts and obvious aspects of protein secondary structure, such as the backbone ϕ and ψ torsion angles, with electrostatic field effects and with ring current effects. However, this met with little success, even though the combination of electrostatic field and ring current effects had found some success in reproducing ^{1}H chemical shifts in proteins (7). The reason for this lack of success can be attributed primarily to the dominance of the paramagnetic term in determining shielding of the heavier elements ^{13}C and ^{15}N. Consequently, we decided to attempt the use of ab initio Hartree-Fock quantum chemical techniques to evaluate shielding in amino acids, peptides, and proteins. For amino acids, the use of Hartree Fock methods (or density functional theory techniques) might be expected to be successful because the molecules are quite small, although one might expect significant contributions from the electrostatic fields arising from neighboring zwitterionic amino acid molecules. However, in the case of proteins, it was unclear to what extent it would be possible to evaluate chemical shifts, because protein NMR (and crystal) structures have considerable structural uncertainties compared with the small molecule structures more typically investigated using quantum chemistry. In addition, the effects of solvation/dielectric constant/electrostatic field effects and internal motion might all make major contributions to shielding. It was also unclear just how large a structure would have to be investigated in order to adequately reproduce the experimental NMR spectra. Would ab initio or density functional theory computations of protein NMR chemical shifts be feasible? We thought so.

In early work we therefore decided to try to predict the ^{13}C NMR chemical shielding tensors in two amino acids: zwitterionic L-threonine and L-tyrosine. These molecules were chosen because they both had high-resolution X-ray structures. Additionally, we had carried out a single crystal ^{13}C NMR shielding tensor determination of each of the carbons in L-threonine (8), and Frydman and coworkers had carried out a determination of the shielding tensor elements of each of the carbons in L-tyrosine (9). When taken together, the experimental data provided 39 individual shielding tensor elements covering a 250-ppm chemical shift range, in addition to the orientations of the chemical shielding tensor elements in the molecular frame in L-threonine. This represented a sizeable body of data for investigation using quantum chemical techniques.

To carry out these calculations we used the TEXAS90 program of Pulay, Wolinski, and Hinton, which makes use of an efficient implementation of the gauge-including-atomic-orbital (GIAO) methods introduced by them in earlier work (10). Because essentially all atomic sites were of interest, we used a uniform 6-31G** basis in all calculations. The combined results for the ^{13}C shielding tensors in L-threonine and L-tyrosine were very promising, as shown in Figure 1A (11). The slope of the fitted line is -1.03 and the intercept is about 206 ppm [$R^2 = 0.987$, root mean squared deviation (RMSD) from the fitted line = 12.3 ppm]. However, several points (I-IV in Figure 1A) significantly deviated from the fitted line, and these points arose from the tensor components σ_{11} and σ_{22} of the carboxyl ^{13}C sites in both molecules. The σ_{11} component lies in the sp^2 plane and is perpendicular to the C$^\alpha$–C$^\beta$ bond, whereas σ_{22} lies perpendicular to the sp^2 plane. Upon comparison with experiment, σ_{11} is underestimated, whereas σ_{22} is overestimated, so the discrepancies are less dramatic when only the isotropic values are considered. That is to say, evaluation of the shielding tensor elements is a more rigorous test of the quality of a calculation than is a determination of solely the isotropic chemical shift.

The origins of these discrepancies were tentatively attributed to the effects of the electrostatic fields originating from nearby charge centers. We therefore

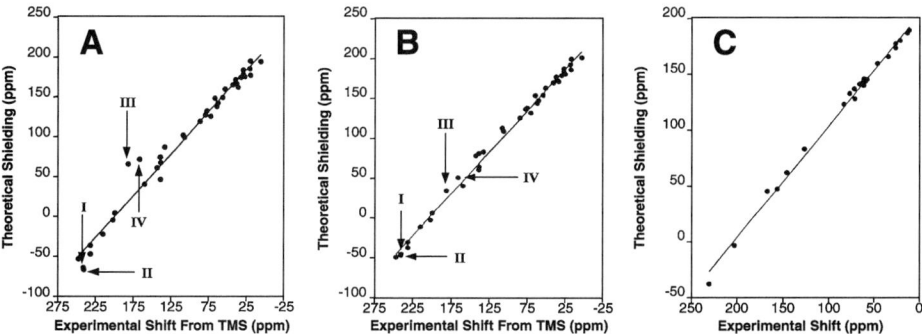

Figure 1 Graphs showing experimental shifts [ppm from tetramethylsilane (TMS)] versus theoretical shieldings (in ppm) for ^{13}C NMR shielding tensor elements in L-threonine and L-tyrosine. (*A*) Single-residue calculation; (*B*) charge-field perturbation–gauge-including-atomic-orbital (CFP-GIAO) calculation; (*C*) CFP-GIAO calculation for L-threonine only, icosahedral tensor representation. I = σ_{11} Tyr; II = σ_{11} Thr; III = σ_{22} Tyr; IV = σ_{22} Thr. Reproduced with permission from *J. Am. Chem. Soc.* 1994 116:7784–7786. Copyright 1994 Am. Chem. Soc.

carried out a second series of calculations in which unit cell translations were used to generate 28 additional molecules in the vicinity of the molecule of interest. Only the molecule of interest had basis sets assigned to its atoms, neighboring molecules being represented by point charges obtained from an AMBER force-field. Considerable improvements were achieved when the calculations were performed in the presence of point charges, as shown in Figure 1*B* (slope = −1.04, intercept = 207 ppm, R^2 = 0.996, RMSD = 6.4 ppm), where the previous outlying points for σ_{11} and σ_{22} are now much closer to the fitted line. The electrostatic contributions led to a significant increase in the value of σ_{11} for the carboxyl groups [16 ppm for L-tyrosine (I) and 11 ppm for L-threonine (II)] and a major decrease in σ_{22} [37 ppm for L-tyrosine (III) and 27 ppm for L-threonine (IV)]. These results demonstrated the ability to evaluate the magnitudes of the ^{13}C shielding tensors in amino acids (11).

Analysis of a shielding tensor involves examination not only of the principal components of the tensor but also the orientation of these components with respect to the molecular frame. Fortunately, in our previous NMR work on L-threonine we utilized a single crystal in our study, and consequently the orientations of the ^{13}C shielding tensors were well known for each site, enabling a comparison of tensor orientations. If one compares the experimental and calculated direction cosines with respect to the crystallographic axes of the ^{13}C shielding tensors, one obtains good agreement, although comparing tensor orientations in terms of direction cosines does not properly gauge the calculations because in addition to being difficult to visualize, errors in direction cosines do not translate directly into errors in terms of radians or degrees. We therefore used the icosahedral representation described previously by Alderman and coworkers (12), in which the three principal components of the shielding tensor together with the three direction cosines are transformed into six icosahedral shielding tensor elements. For a 220 ppm overall chemical shielding tensor element range, the so-called charge-field perturbation–gauge-including-atomic-orbitals (CFP-GIAO) method yielded excellent results, as shown in Figure 1C, in which there is an R^2 value of 0.997 and an RMSD of 4.4 ppm. Thus, the results of these early calculations clearly indicated, at least with amino acids, that it was possible to compute the shielding tensor elements, as well as their orientations, with good accuracy.

The next question that arises is, to what extent can the chemical shifts of peptides and in particular the much larger proteins of principal interest be predicted by using quantum chemical techniques? To investigate this question we first began to investigate the C^α and C^β chemical shifts for the helical and beta sheet residues in the database of Spera & Bax (4). These workers had shown that a clear distinction could be made between the chemical shifts of residues with these two distinct structure types, which suggested (but did not prove) a major, direct effect of the ϕ and ψ torsion angles on electronic structure and thence on shielding. However, other effects such as hydrogen bonding/electric field effects might also be important, as might solvation and dynamics. We therefore began to investigate the C^α and C^β chemical shieldings in a series of alanine molecular fragments, as shown in the following structure:

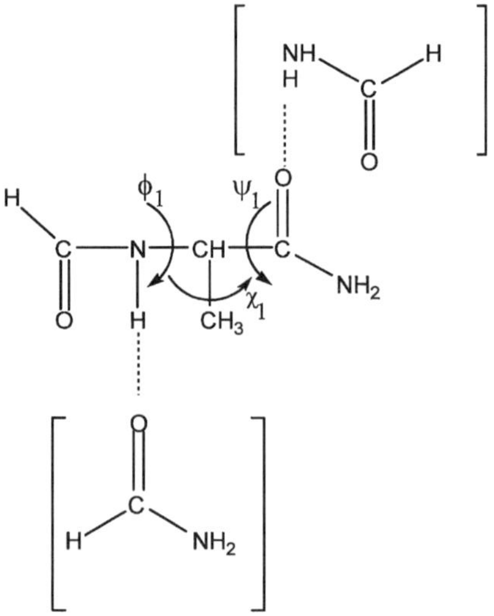

To begin, we used a relatively small and uniformly dense (6-311G**) basis for all atoms but neglected the formamide hydrogen bond partners shown above. Figure 2A shows histograms of the experimental C^α and C^β chemical shifts reported in the database of Spera & Bax, and Figure 2B shows the theoretical chemical shielding results computed by using ab initio Hartree-Fock GIAO calculations

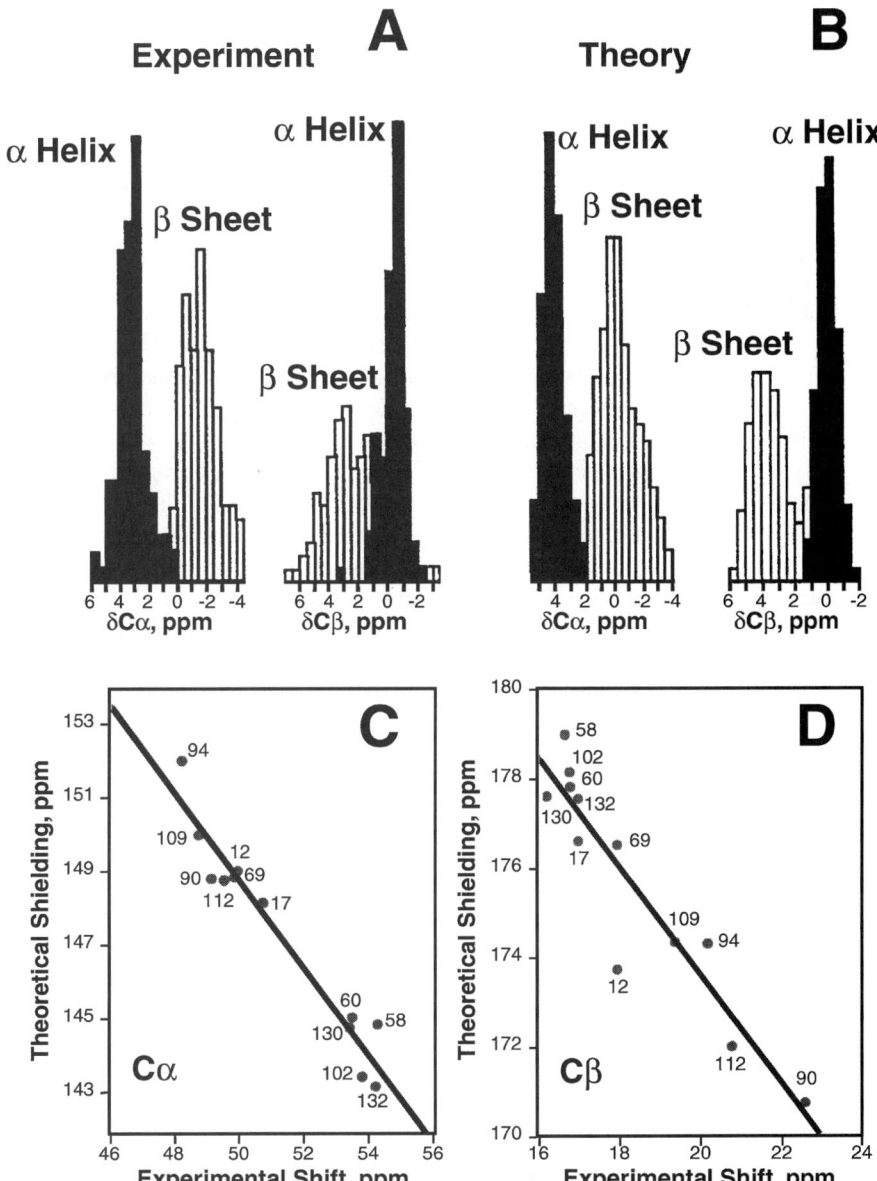

Figure 2 ^{13}C NMR shielding results for C^α, C^β sites in protein. (A) C^α/C^β shift separation histograms from Spera & Bax (4); (B) computed C^α/C^β shift histogram for model alanine fragments; (C) experimental shift/theoretical shielding results for 12 Ala C^α sites in SNase; and (D) same as (C) but with C^β sites. Reprinted with permission from *Science* 260:1491–96 (1993). Copyright 1993 Am. Assoc. Adv. Sci.

(13). Clearly, the general ~5 ppm chemical shift separation seen experimentally between the helical and sheet residues for both C^α and the C^β were reproduced in the calculations, but we did not have a one-to-one comparison between specific protein chemical shifts and experimentally computed shieldings, only a general trend. We therefore extended our calculations to investigate 12 specific alanine residues in a nuclease from *Staphylococcus aureus* [Staphylococcal nuclease (SNase)]. To get the best agreement between theory and experiment, we used a locally dense basis set approach proposed originally by Chesnut & Moore (14), in which a larger number of basis functions is given to the nucleus whose chemical shielding is to be calculated. In this case we used a 6-311++G(2d,2p) basis for C^α, C^β, C^o, N^H, and H^α, but only a 6-31G basis for the other atoms in the fragment.

Given a general concern about the possible effects of solvation, electrostatic field effects, hydrogen bonding, dynamics, and magnetic susceptibility effects, the experimentally observed slope of 0.85 and regression coefficient $r = 0.94$ represented a rather promising initial agreement between theory and experiment (13). We then investigated the effects of hydrogen bonding on shielding using the formamide hydrogen bond partners shown above, together with the effects of the protein's charge field, using the CFP-GIAO approach used with the amino acids. For both C^α and C^β, we found excellent agreement between theory and experiment, as shown in Figures 2C,D. Interestingly, however, there was only moderate accord between theory and experiment when fragment geometries based directly on the crystallographic structures were utilized, and we had to carry out an energy minimization of the protein (crystal) structure (suggested by C. Jameson) by using molecular mechanics techniques, in order to regularize all of the bond lengths and three-atom bond angles to those of an AMBER forcefield. The fact that the ^{13}C NMR chemical shifts are highly sensitive to bond lengths then led to an investigation of the bond-length and bond-angle shielding derivatives in glycine, alanine, and valine residues in which chemical shieldings were calculated for both C^α and C^β at various geometries by systematically varying one structural parameter (a bond length or a bond angle) while simultaneously fixing all other geometric parameters (15). The shielding traces were expressed in terms of Taylor series expansions of the shielding as a function of bond-length (r) or bond-angle (θ):

$$\sigma(r) = \sigma(r_0) + \left(\frac{\partial \sigma}{\partial r}\right)_{r_0} (r - r_0)$$

$$+ \frac{1}{2}\left(\frac{\partial^2 \sigma}{\partial r^2}\right)_{r_0} (r - r_0)^2 + \frac{1}{6}\left(\frac{\partial^3 \sigma}{\partial r^3}\right)_{r_0} (r - r_0)^3 + \cdots \qquad 1.$$

$$\sigma(\theta) = \sigma(\theta_0) + \left(\frac{\partial \sigma}{\partial r}\right)_{r_0} (\theta - \theta_0) + \frac{1}{2}\left(\frac{\partial^2 \sigma}{\partial \theta^2}\right)_{\theta_0} (\theta - \theta_0)^2$$

$$+ \frac{1}{6}\left(\frac{\partial^3 \sigma}{\partial \theta^3}\right)_{\theta_0} (\theta - \theta_0)^3 + \cdots \qquad 2.$$

with first, second, and third-order derivatives being reported. The results of these calculations clearly showed that the effects of individual bond-length and bond-angle changes were additive, making it possible to predict chemical shifts for a given set of structural parameters without having to perform further ab initio calculations, once a basic Ramachandran shielding surface [defined here as $\sigma(\phi, \psi)$] has been evaluated. Ramachandran shielding surfaces for all of the naturally occurring amino acids in their most popular side-chain conformations have now been evaluated and are available at http://feh.scs.uiuc.edu.

Of course, as one increases the complexity of the amino acids, from glycine to alanine to valine to isoleucine, for example, the number of structures that need to be computed increases. In the case of alanine there is an additional torsional angle, χ_1, but in this case the methyl group can simply be fixed at its fully staggered conformation. However, in the case of valine residues, there are three clearly defined χ_1 conformers, two of which occur frequently in proteins. In the case of isoleucine both χ_1 and χ_2 conformers exist, as shown below:

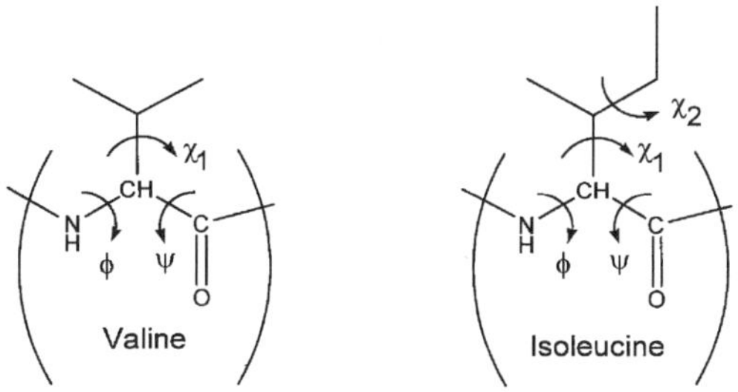

leading in principle to nine side-chain conformers, although once again only two of these occur with high frequency in proteins. For example, in calculations of valine shifts in proteins, we used an N-formyl-L-valine amide structure to evaluate complete Ramachandran shielding surfaces for all three χ_1 conformers, and shielding results for C^α and C^β are shown in Figure 3 (see color insert) (16). If the ϕ, ψ, and χ values in a protein are known, then it should be possible to use such shielding surfaces to compute the experimental C^α and C^β chemical shifts from the theoretical shielding surfaces. However, when we attempted to use the theoretical shielding surfaces to predict the ^{13}C NMR chemical shifts for valine residues in calmodulin using torsion angles determined from X-ray crystallography, we found only a modest agreement between theory and experiment. The results were particularly puzzling for helical C^β shifts because five out of seven valine residues in calmodulin have (solution NMR) C^β shifts in the 31.3–31.7 ppm range, even though the X-ray structures showed vastly different χ_1 angles, which would be expected to result in a much wider chemical shift distribution. However, assuming that all of the valine residues in calmodulin in solution were present as the *trans* conformer ($\chi_1 = 180°$) resulted in excellent agreement between experimental and predicted C^α and C^β chemical shifts (16). This result was consistent with J-coupling constant information obtained from Ikura, which showed that six out of seven valine residues existed in the $\chi_1 = 180°$ conformation in calmodulin in solution.

This success then led to the desire to calculate a complete valine chemical shielding hypersurface, the chemical shift as a function of ϕ, ψ, and χ, which involved the evaluation of 1728 points uniformly distributed in ϕ, ψ, and χ (17). The chemical shifts predicted using this shielding hypersurface were in good agreement with the chemical shift values determined experimentally. For example, the C^α chemical shifts in a Staphylococcal nuclease and an invertebrate calmodulin were predicted with an RMSD = 1 ppm and an R^2 = 0.842 (17).

But what about ^{13}C shielding tensors and their orientations? In order to answer this question, we evaluated the ^{13}C NMR shielding tensors in glycine, valine, isoleucine, serine, and threonine, comparing the tensor magnitudes and orientations for helical and sheet residues, including the effects of χ_1 on shielding (18). Results for glycine and alanine showed the expected ~4–5 ppm increase in the isotropic chemical shielding of sheet over helical residue geometries and generally similar overall breadths of shielding tensors for both helical and sheet fragments. However, for each of the C^β-branched amino acids (valine, isoleucine, and threonine) we found not only an increase in shielding of sheet over helical fragments, but we also noted a large increase in the overall tensor breadths for the sheet residues. On average, C^α sheet tensor breadths were ~50% larger than the overall tensor breadths found for helical residues. In addition, the C^α shielding tensor elements were all sensitive to χ_1, and the orientations of the shielding tensors were quite different in helical and sheet geometries (18).

We then began to investigate experimentally the C^α and C^β shielding tensor magnitudes in alanine containing tripeptides with a wide range of geometries, encompassing both helical and sheet regions of Ramachandran space. Using three alanine $^{13}C^\alpha$–labeled tripeptides and slow magic-angle sample-spinning, the individual components of the C^α shielding tensor were evaluated and compared with predictions made by using the theoretical shielding surface (19). For the nine tensor elements, an R^2 value of 0.99 and RMSD of 2.15 ppm was found, although the slope of the correlation was only -0.67, suggesting a small systematic error in the theoretical calculations. For a $^{13}C^\beta$-alanine labeled tripeptide, an R^2 value of 0.99 was found (20), together with a slope of -1.07 (compare with an ideal value of -1.0). When taken together, these results indicate that there is good to excellent agreement between the experimental and theoretical shielding tensor element magnitudes, with the helical/sheet tensor differences seen in the calculations being reproduced experimentally.

In other recent experimental work, Tjandra & Bax have shown that it is possible to determine the C^α chemical shift anisotropy (defined as the value of the C^α shielding along the C^α–H^α bond minus the average perpendicular to this bond) for proteins in solution by using dipole-dipole/chemical shift anisotropy interference techniques (21). Sitkoff & Case have compared the results of these experimental C^α "chemical shift anisotropies" with their calculations for ubiquitin and a calmodulin/M13 complex (22). They computed shielding surfaces for an alanine dipeptide using structures that were optimized from an all-atom CHARMM forcefield, using a 30° grid for their Ramachandran shielding surface. The anisotropies were approximated from the surface by using a Fourier series. The correlation between computed and experimental values was quite good, although the computed results appeared to be systematically somewhat larger than the experimental results, possibly owing to motional averaging about the mean structure (22). We obtained similar results using a density functional theory (DFT)–computed alanine C^α shielding surface, with a slight improvement when using an ab initio Hartree-Fock surface, for all residue types (20). Even better agreement was obtained when specific glycine, alanine, valine, serine, threonine, isoleucine, and phenylalanine surfaces were used, with the best correlations being found with ubiquitin, presumably because it is a smaller protein and has a higher resolution (X-ray crystallographic) structure.

At present, there are no direct determinations of ^{13}C shielding tensor element orientations in proteins or peptides, only the amino acid results discussed previously and what may be inferred from the relaxation interference experiments. Nevertheless, the fact that C^α chemical shift tensor breadths or chemical shift anisotropies in sheet structures are often considerably larger than those found in helical structures has led to the development of a novel chemical shift anisotropy "filtering" experiment by Hong and coworkers (23). This experiment enables the acquisition of solid-state NMR spectra of primarily helical spectra (in this case in the protein ubiquitin), together with a determination of chemical shift anisotropy (CSA) values, which were found to be in good agreement with the results of the quantum chemical calculations. Spectra of sheet residues can be obtained by using a second filtering experiment (24), and are of use in spectral assignments.

The ability to obtain resolved, high-resolution solid-state NMR spectra of proteins in the crystalline solid state is also of interest because it enables a rather direct comparison between crystal and solution NMR structures, based on an analysis of NMR chemical shifts. In early work, Cole and coworkers utilized one-dimensional magic-angle sample-spinning NMR to investigate selectively ^{13}C- and ^{15}N-labeled SNase and concluded that the chemical shift patterns seen in the solid state were similar to those observed in solution (25). More recently, using high-field, high-speed magic-angle sample-spinning and high-power proton decoupling, two-dimensional NMR spectra of microcrystalline bovine pancreatic trypsin inhibitor (BPTI) have been obtained and a number of specific spectral assignments have been made (26). There was once again good agreement between solution and crystal NMR shifts. In the case of BPTI, we have computed the chemical shifts for C^α, C^β, $C^{\gamma 1}$, $C^{\gamma 2}$, and C^δ for two isoleucine residues using the X-ray crystal structure and found good agreement between theory and experimental chemical shifts (27). Thus, it is possible to compute more than just C^α and C^β shieldings in proteins, and it seems likely that quantum chemical calculations may have a role to play in identifying side-chain conformations in proteins, both in solution and in the solid state.

In addition to investigating the chemical shifts of aliphatic carbons in proteins, there is considerable interest in understanding the chemical shifts of the carbon and nitrogen atoms in the peptide bond. As had been found for C^α chemical shifts, carbonyl carbons (C^o) in helical residues were known to resonate ~5 ppm downfield of carbonyl carbons in sheet residues in proteins. It would therefore seem reasonable that it should be possible to compute this chemical shift separation by using quantum chemical techniques, but early calculations showed the opposite trend to that observed experimentally, with the helical carbonyl sites being more shielded. Because it seemed likely that hydrogen bonding (as opposed to backbone torsion angles) might be responsible for the deshielding of helical residues, this was also investigated, but agreement with experiment could only be obtained at artificially short hydrogen bond distances. Based on single molecule geometry optimizations, it also appeared possible that the peptide nitrogen might be somewhat pyramidal, but this is contrary to recent experimental observations. However, when shielding computations were performed on SCF-optimized helical and β-turn N-formylpentaalanine amide structures, in which the hydrogen bond structure is geometry optimized (28), a helix-sheet chemical shift difference of 4.9 ppm was found (29), with the helical site being deshielded—as is found experimentally, where carbonyl nuclei in helical alanine sites are typically deshielded by ~4.6 ppm when compared with sheet-like residues.

Based on quantum chemical calculations, the ϕ, ψ contributions are about -1.3 ppm (29), so the overall hydrogen bond contribution to deshielding is about 6.2 ppm (29). The chemical shift separation seen for carbonyl groups appears to be due to the changes in C–O bond length that occur on hydrogen bonding, consistent with a very large first-order shielding derivative in the carbonyl group. More recently, Walling and coworkers have made a detailed investigation of the carbonyl shielding tensor in peptides and concluded that the principal axis system of the C^o shielding tensor follows the directions given by the amide plane and the C=O bond and that hydrogen bonding does not have a major effect on the orientation of this principal axis system (30).

Within the peptide group, the other hetero-atom for which there exists an extensive database of structural and chemical shift information is nitrogen. In the case of ^{15}N, however, it is not solely the torsion angles ϕ and ψ that dominate shielding, as is the case with C^α residues, owing of course to the fact that the peptide group bridges two different amino acid residues. As a result, it is the torsion angles ϕ and ψ_{i-1}, where i refers to the residue whose ^{15}N chemical shift is of interest, that dominate shielding. In early work, it was found that two-dimensional chemical shielding surfaces $\delta(\phi_i, \psi_{i-1})$ could be constructed that gave moderate predictions of the ^{15}N chemical shifts for a given structure (31). On average, the rms error between experiment and prediction was ~3.5 ppm for a chemical shift range of ~15–20 ppm. For β-branched residues, such as valine, the experimental chemical shift range is larger (up to ~35 ppm) and variations in χ_1 values (not reflected in the empirical shielding surfaces) resulted in somewhat larger rms errors of ~4.8 ppm. The results of ab initio quantum chemical calculations confirmed that indeed it is the two backbone dihedral angles closest to the peptide group, ϕ_{i-1} and ϕ_i, that contribute most (~20 ppm) to the experimental shielding, whereas the adjacent torsion angles ϕ_{i-1} and ψ_i make a contribution of up to ~8 ppm (32). Variations in side-chain conformation produced up to ~4 ppm chemical shift effects. In addition, as expected from earlier work, hydrogen bonding effects are also important contributors to ^{15}N shielding in proteins. Both ab initio Hartree-Fock and density functional theory calculations of ^{15}N shieldings gave a reasonably good description of experimental chemical shifts in a small series of proteins. For example, in the case of 13 alanine residues in a cytochrome c_{551}, DFT shieldings were found to have a correlation line of slope -1.04 and an R^2 value of 0.97, whereas the Hartree-Fock results had a slope of -1.11 and an R^2 value all 0.96, with the DFT results giving a closer absolute shielding intercept (251.3 ppm) to the experimental value of 244.6 ppm (32).

As for the chemical shift anisotropy of ^{15}N in proteins, there have been extensive investigations into the measurement and use of the ^{15}N CSA in the determination of peptide/protein structures in membranes using oriented samples (33, 34), together with numerous investigations of the ^{15}N CSA for proteins in solution using dipole-dipole/CSA cross-correlation techniques (35, 36). Results with ubiquitin show a larger number of statistically significant values far from the mean CSA, a result not observed in ribonuclease H. It may be possible to better understand the shielding tensor results by using DFT techniques, as described above, although such calculations have not yet been reported.

PARAMAGNETIC SYSTEMS

All of the systems discussed above have been diamagnetic. However, there are many proteins of interest that contain paramagnetic centers, and it is perhaps surprising that there has been very little work done on the computation of spectroscopic properties in these systems. Evaluating the chemical or purely orbital shifts in paramagnetic systems is generally not feasible because of the breakdown of the Ramsey formulation. Fortunately, however, we are not concerned with computing purely orbital shifts in paramagnetic systems, because the observed overall shift can to a good first approximation be broken down into the purely orbital (or diamagnetic) shift of a reference (diamagnetic) model compound and a hyperfine shift, made up of a Fermi contact contribution together with a dipolar or pseudocontact shift (37). The purely orbital or diamagnetic shift can be computed using standard quantum chemical methods or can be estimated from the chemical shifts of model compounds, and in some cases from temperature dependence studies. The pseudo-contact shift has already been used to carry out protein structure refinements (38). There has, however, been much less use of the Fermi contact shift. Nevertheless, in recent work, Wilkens et al. used DFT to compute ^{2}H, ^{13}C, and ^{15}N (protein) Fermi contact shifts for an iron-sulfur protein (39) and found very good correlation between theory and experiment. Their results clearly open the way to using Fermi contact shifts without any extensive semi-empirical parameterizations, as well as dipolar or pseudo-contact shifts, as structure refinement tools.

DFT methods appear to be particularly attractive for use in computing the Fermi contact shift, as well as Mössbauer quadrupole splittings (even in paramagnetic systems) because all are ground-state properties and, as is well known, the Hohenberg/Kohn formulation of DFT states that all ground-state properties are functions of the charge density, $\rho(r)$, where:

$$\rho(\mathbf{r}) = \sum_i [\phi_i(\mathbf{r})]^2 \qquad 3.$$

and ϕ_i represents a molecular orbital. The molecular orbitals can be occupied by either spin up (α) or spin down (β) electrons, so it is possible to form two different charge densities. Putting both α and β spins into the same molecular orbital (MO) is called a spin-restricted calculation, whereas putting α and β spins into different MOs is called a spin-unrestricted calculation. In the unrestricted case, it is possible to form two sets of charge densities, one for the α and one for the β MOs. Their sums give the overall charge density $\rho(r)$, whereas their differences give the spin density, $\rho_{\alpha\beta}$:

$$\rho_{\alpha\beta} = \sum_i \left[|\Psi_i^\beta(0)|^2 - |\Psi_i^\alpha(0)|^2 \right] \qquad 4.$$

and the Fermi contact shift, δ^{FC}, is directly proportional to the spin density, $\rho_{\alpha\beta}$:

$$\delta^{FC} = \frac{\mu_0 \mu_B^2 g_e^2 (S+1)}{9kT} \bullet \rho_{\alpha\beta}. \qquad 5.$$

For open shell molecules (radicals and other paramagnetic systems), DFT methods appear to be the most promising ones for handling spin polarization without significant spin contamination (the appearance of electronic states of higher multiplicity), as demonstrated by the remarkable success of Wilkens et al. in computing ^{2}H, ^{13}C, and ^{15}N Fermi contact shifts in rubredoxins from *Clostridium pasteurianum* (39, 40).

Wilkens et al. utilized a 104-atom model of the iron-sulphur cluster and neighboring amino acid residues in a rubredoxin from *C. pastuerium* as a basis for DFT calculations of the spin densities on the ^{1}H, ^{2}H, ^{13}C, and ^{15}N nuclei of threonine, valine, glycine, leucine, and the metal-cluster-coordinated cysteine residues in oxidized and reduced protein forms. They used the Gaussian-94 program with a B3LYP hybrid exchange-correlation functional and a uniform 6-311G** basis set to evaluate the Fermi contact spin densities and clearly showed that the experimental hyperfine shifts were dominated by the Fermi contact interaction. The computed spin densities were all found to correlate linearly with the experimental isotropic hyperfine shifts for Fe(III) rubredoxin, with R^2 values of 0.93–0.96 for 12 experimental ^{2}H NMR resonances (depending on the exact geometric model used) and 0.85–0.96 for 12 experimental ^{15}N resonances. Using the same hybrid exchange correlation functional, but with a smaller basis set (3-21G*), these workers carried out a geometry optimization of the H^N atom positions in their models,

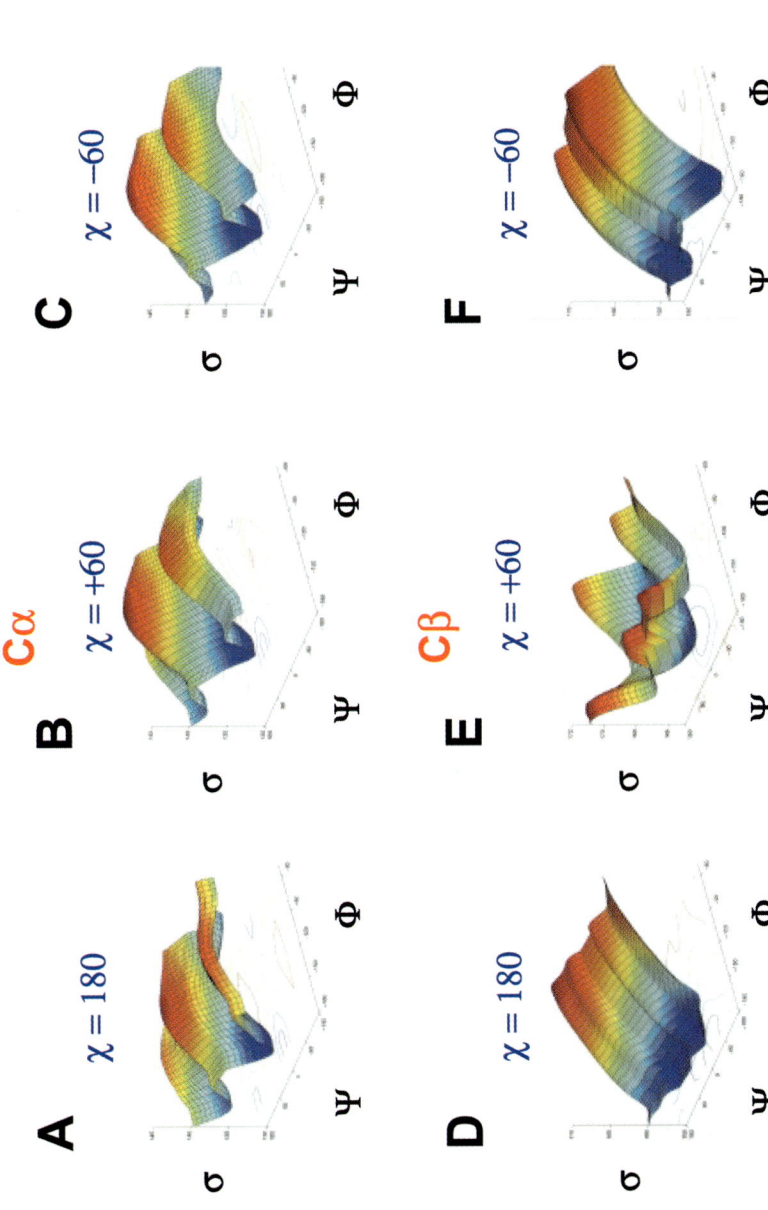

Figure 3 Calculated $^{13}C^\alpha$ and $^{13}C^\beta$ shieldings in N-formylvaline amide as a function of ϕ and ψ (at various χ^1 values). (*A*) C^α, $\chi^1 = 180°$; (*B*) C^α, $\chi^1 = 60°$; (*C*) C^α, $\chi^1 = -60°$; (*D*) C^β, $\chi^1 = 180°$; (*E*) C^β, $\chi^1 = 60°$; (*F*) C^β, $\chi^1 = -60°$. Reproduced with permission from *J. Am. Chem. Soc.* 1995. 117:9542–46. Copyright 1995 Am. Chem. Soc.

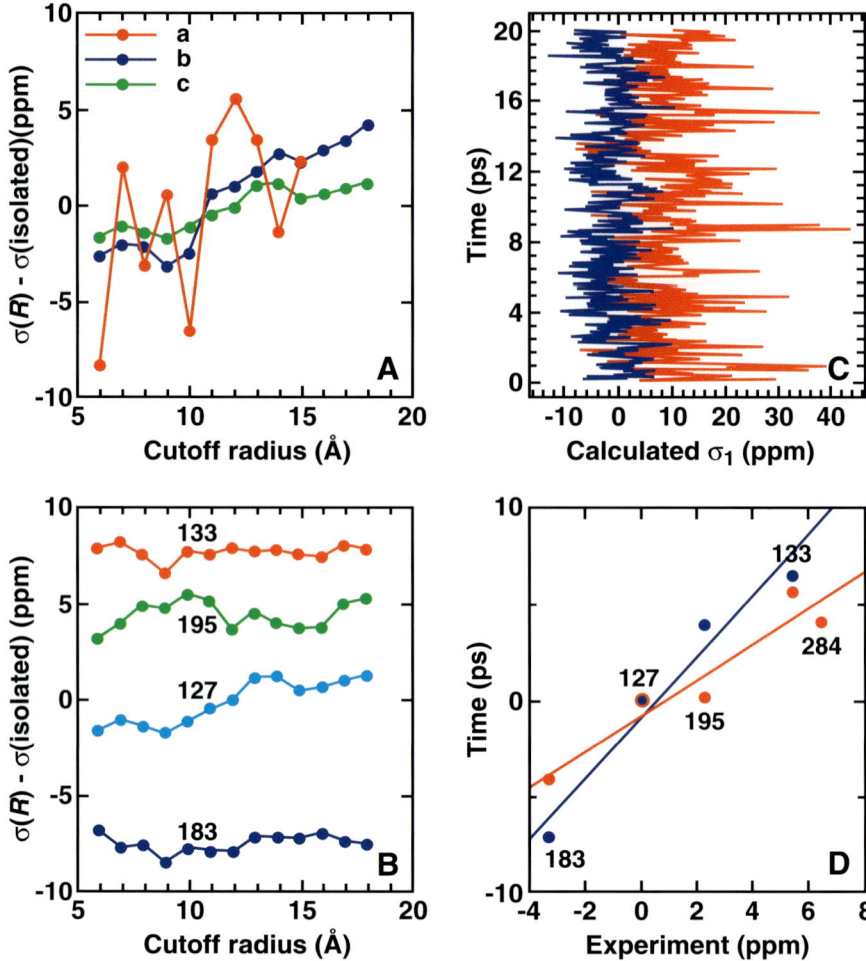

Figure 4 Fluorine shielding results for the Trp residues in the *Escherichia coli* galactose binding protein (GBP). (*A*) Ab initio [gauge-including-atomic-orbital (GIAO)] results with the use of a radial cutoff with atomic resolution (*a, red*), a radial cutoff with whole residue resolution (*b, blue*), and a radial cutoff with whole, electroneutral residue resolution (*c, green*). (*B*) GIAO shielding results for Trp residues 127, 133, 183, and 195 with the use of whole electroneutral residue resolution as a function of radial cutoff. (*C*) Results for 20-ps shielding trajectories for [5-F]Trp[183] (*blue*) and Trp[284] (*red*) residues in GBP. (*D*) Graph showing experimental versus theoretical shielding results in GBP: charge-field perturbation—GIAO (*blue*) and *red*, multipole shielding polarizability—local reaction field method. Reprinted with permission from *Science*, 260:1491–96 (1993). Copyright 1993 Am. Assoc. Adv. Sci.

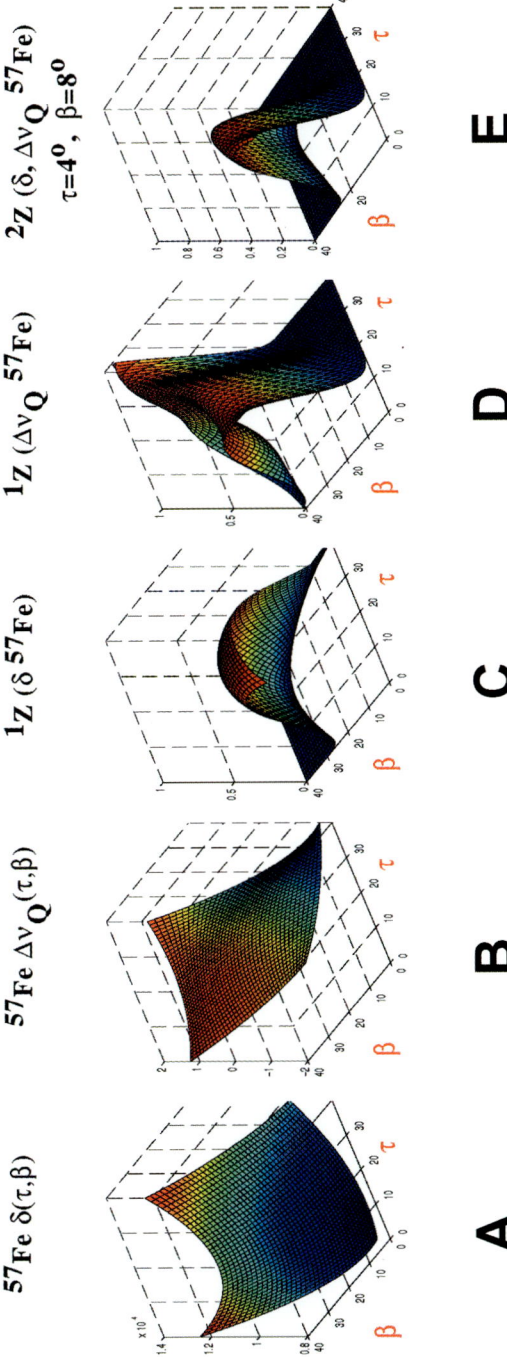

Figure 5 Property and τ,β probability surfaces for CO heme model system. (*A*) ^{57}Fe NMR chemical shift τ,β surface; (*B*) ^{57}Fe Mössbauer quadrupole splitting surface; (*C*) ^1Z chemical shift probability surface; (*D*) ^1Z quadrupole splitting probability surface; (*E*) ^2Z probability surface for tilt and bend, the product of (*C*) and (*D*) surfaces. Based on data in Reference 65.

which resulted in an improvement in R^2 (from 0.85 to 0.94) for the ^{15}N shifts for the ferric protein.

There was much worse agreement, however, for Fe(II) shifts using a hypothetical model for the ferrous protein structure derived from X-ray crystallographic structures of the ferric protein, suggesting that the ferrous protein undergoes structural changes upon reduction. Nevertheless, given a ^{2}H shift range of ~800 ppm and an ^{15}N chemical shift range of >600 ppm in the ferric protein, these results are very promising. Our group has been investigating the hyperfine shifts in paramagnetic heme proteins, and as with the work described above on rubredoxin, we have found good agreement between the experimentally observed hyperfine shifts (for ^{13}C) and those computed with density functional theory, a hybrid exchange-correlation functional, and moderately sized basis sets (J. Mao, Y. Zhang, E. Oldfield, unpublished data). It seems clear that in future work DFT techniques will prove to be very helpful in predicting hyperfine shifts in paramagnetic proteins, in addition to being used in structure prediction and refinement, as are semi-empirical correlations currently.

PREDICTING/REFINING ASPECTS OF PEPTIDE AND PROTEIN STRUCTURE

Simply being able to use quantum chemical techniques to predict chemical shifts and chemical shift tensors in amino acids, peptides, and proteins is, in and of itself, of some interest. However, it might be more useful to use experimental chemical shifts or chemical shift tensor information to determine or at least to refine aspects of peptide and protein structure. Such a capability might be particularly useful for noncrystalline materials, such as membrane-bound proteins, in addition of course to playing a role in structure determination for proteins in solution.

In a recent study, Van Alsenoy et al. used ab initio Hartree-Fock methods to carry out the first quantum chemical geometry optimization of a small protein (41), and with the availability of larger computational resources and refined codes, it is entirely possible that geometry optimizations of small proteins may become more generally accessible over the next few years. We found that energy minimization/geometry optimization appears to be quite useful in evaluating ^{13}C NMR chemical shifts, at least for valine residues in proteins (42). We first carried out both molecular mechanics (an AMBER forcefield in Discover, combined with a conjugate gradient minimization) and Hartree-Fock geometry optimization of 19 valine residues in 3 proteins. We then carried out chemical shift calculations on these structures; the results showed that the most successful prediction of both C^α and C^β chemical shifts was obtained for all 19 residues by utilizing the lowest energy χ_1 conformer obtained from these geometry optimizations, together with the X-ray ϕ, ψ values (42). This strongly suggests that the X-ray–determined backbone torsion angles are accurate and reflect the solution values, whereas the side-chain conformations appear to be less so, or at least they are more likely to differ between proteins in solution and in the crystalline solid state. This suggests that geometry optimization using quantum chemical techniques can play a role in structure determination, but faster computers are needed for more routine application.

We have also begun to more directly predict backbone torsion angles for amino acids in proteins. We developed a Bayesian probability technique that utilizes C^α, C^β, and H^α chemical shifts to predict the likelihood that a particular ϕ, ψ torsion angle combination represents the experimental backbone torsion angles. In its simplest form, one would have a parameter, P, which is a function of a single angle, α, such that

$$P = f(\alpha). \qquad 6.$$

We define the probability that the experimental value of P, P_{expt}, corresponds to a given angle, α, given by the unnormalized probability, Z, as (43):

$$Z = e^{-\frac{(P_{expt} - f(\alpha))^2}{W}}, \qquad 7.$$

where W represents a search width (to take into account any computational inadequacies or experimental uncertainties). The approach can be readily applied to a two-dimensional chemical shift surface in which ϕ and ψ are variables:

$$Z(\phi, \psi) = \exp\left[-(\delta_{expt} - \delta(\phi, \psi)^2/W\right]. \qquad 8.$$

As additional properties, $P_1, P_2 \ldots P_n$, are investigated, the overall conditional or Bayesian probability simply becomes the cumulative probability $^nZ = Z_1^\times Z_2^\times \ldots Z_n$. Using empirically determined parameter (chemical shift and J-coupling) surfaces, the backbone ϕ, ψ error between prediction and experiment was about 17°, but for 22 out of 24 ϕ, ψ torsion angles for alanine residues in Staphylococcal nuclease, this reduced to $\sim 10°$ when quantum mechanically computed ^{13}C shielding surfaces were utilized (43).

Using C^α and C^β chemical shift restraints for 12 alanine and 8 valine residues in the nuclease from *S. aureus*, we also used a direct approach to refining structure (43) and found that the ϕ, ψ, and χ values obtained from structures that included chemical shift restraints were much closer to the X-ray structure than were the solution NMR structures obtained without chemical shift restraints, yielding back-calculated chemical shifts in good accord with experiment and with only very minor nuclear Overhauser effect (NOE) violations (43). It should be noted that dipolar coupling information (44) may be even more useful in structure refinement (45) and is currently being employed in conjunction with chemical shift information by some groups.

Cornilescu and coworkers have reported a somewhat different approach to using multiple chemical shifts in structure determination. In their TALOS (torsion angle likelihood obtained from shifts) program (46), these workers used a database of C^α, C^β, C^o, H^α, and N^H chemical shifts for 20 proteins with highly resolved X-ray crystal structures and known solution NMR chemical shifts. The idea is to search the database for sequences of adjacent residues, having sequence and chemical shift similarity, that give the closest matches to a (triplet) sequence of interest. This work is a useful advance over earlier work by Wishart and coworkers (47), which utilizes solely (secondary) chemical shift information, because both chemical shift and sequence homology information is used to predict the most likely backbone torsion angles for a residue in a sequence of interest. The results are impressive and indicate a ~15° RMSD between the output of the TALOS program and the X-ray–derived backbone torsion angles (46). It may be possible to extend this approach even further, and progress is being made in using solely chemical shift information in generating three-dimensional protein structures (48).

FLUORINE-19 NMR AND ELECTROSTATIC FIELD EFFECTS

Why might one be interested in fluorine-19 NMR spectroscopic studies of fluorine-labeled proteins? Historically, fluorine-19 NMR studies of protein structure, or at least spectroscopy began in the early 1970s. In these early days high-field, NMR spectrometers were not available and as noted above, ^{1}H NMR spectra were poorly resolved because of the small chemical shift range of the ^{1}H nucleus. However, ^{19}F was known to have a much larger chemical shift range. Consequently, a number of groups began to chemically modify proteins with ^{19}F probes (49) or to incorporate fluorine-labeled amino acids biosynthetically into proteins (50) in order to effect large spectral simplifications. The observed chemical shift range can indeed be very large. For example, we found a 16.8 ppm shift range for a [4-F]-Trp-labeled hen egg-white lysozyme (51) using a *Gallus domesticus* expression system. The ^{19}F chemical shifts in this enzyme are very sensitive to inhibitor binding, and such shift effects may in the future be of use in, for example, drug screening. However, the question arises: What is the origin of the large ^{19}F chemical shift nonequivalencies seen in native proteins? In earlier work we had estimated that the electric field–induced shifts in proteins would be very large, on the order of 10 ppm for fluoroaromatic groups, owing to the low dielectric constants expected in protein interiors and the large polarizability of the C–F bond, and consequently they might dominate the ^{19}F shift nonequivalencies seen experimentally. However, before we discuss electrostatic field effects in ^{19}F NMR spectra of proteins, we need to be confident that theoretical techniques can adequately describe ^{19}F NMR spectra of small molecules. We therefore investigated the isotropic chemical shifts for a series of 13 fluorobenzenes whose isotropic chemical shifts covered a 63-ppm chemical shift range, as well as the shielding tensor elements (derived from solid-state NMR), which covered a 237-ppm range. We used geometry optimization and a locally dense basis gauge-including-atomic-orbital (GIAO) approach to calculate shieldings and found that for the isotropic, liquid state chemical shifts, the slope was -0.94, $R^2 = 0.975$, and the RMSD was 3.1 ppm over the 63 ppm chemical shift range (52). For the solid-state shielding tensor elements, the slope was 0.94, $R^2 = 0.989$, and the RMSD was 6.5 ppm over the 237 ppm chemical shielding range, essentially the uncertainty in the experimental measurements (52). This gives considerable confidence in the quality of ^{19}F NMR shielding calculations, at least in the absence of an electrical perturbation.

We then investigated the effects of electrostatic field perturbations on ^{19}F NMR chemical shifts. Unfortunately, there were no "model compounds" that had experimental chemical shifts with which the calculations could be validated. Consequently we decided to validate the calculations in a different way, by carrying out a series of different computations of the electrostatic contributions to shielding, using fluorobenzene as the model target compound. In the first series of calculations (53) we investigated how the addition of hydrogen fluoride molecules influenced the ^{19}F NMR shielding in fluorobenzene, using "super-molecule" cluster calculations. These clusters contained one fluorobenzene molecule and up to five hydrogen fluoride molecules, oriented in different ways (HF or FH) along the molecular x, y, and z coordinates. The results of these calculations, on $C_6H_5F\text{-}(HF)_n$ where $n = 1\text{--}5$, clearly demonstrated the additivity of the intramolecular contribution to shielding, as expected for a purely electrical perturbation. We also investigated the effects of using point charges (no basis functions) to model the long-range purely electrostatic field effects on shielding. The results showed good accord between the super-molecule calculations and those using point charges. This is consistent with a purely electrostatic (or polarization) effect of the hydrogen fluoride molecules on the shielding in fluorobenzene, and is also consistent with the additivity of the intramolecular shift.

We then used a second approach, which involved determining the multipole shielding polarizabilities (MSP) of fluorobenzene: how the tensor elements change with electrical polarization (54). Dykstra used derivative Hartree-Fock methods to calculate the MSP tensor elements for the ^{19}F nucleus in fluorobenzene with respect to a uniform field, a field gradient, and a field-hypergradient. These shielding polarizability tensor elements, when multiplied by the appropriate field, field gradient, or hypergradient terms (which can be readily calculated from the known charge distributions, at least in the H-F molecule), yield the electrical contributions to shielding. A comparison of the multipole shielding polarizability expansion truncated at the point of linear field, field gradient, and hypergradient terms was found to be in excellent accord with benchmark ab initio results, both for a dipole and for a point-charge perturbation (54). On this basis, we concluded that super-molecule, point charge, and low-order MSP expansions permitted the accurate determination of the electrical contributions to shielding in model systems and should therefore be applicable to proteins as well.

To test these methods, we first investigated the ^{19}F NMR spectrum of a [5-F]-Trp-labeled galactose binding protein (in the presence of Ca^{2+} and galactose) from *Escherichia coli*, because specific assignments had previously been reported by Luck & Falke (55). Galactose binding protein contains five tryptophan residues at positions 127, 133, 183, 195, and 284. To compute the ^{19}F NMR chemical shifts in galactose binding protein, we used two techniques: the charge field perturbation-gauge including atomic orbital (CFP-GIAO) method (13) and the MSP-local reaction field approach (56). In the first approach, we used ab initio Hartree-Fock GIAO methods to compute shieldings using (*a*) a radial cut-off with atomic resolution, (*b*) a radial cut-off with all residue resolution, and (*c*) a radial cut-off with all electroneutral residue resolution.

We first investigated shielding results for Trp-127. Using a cut-off radius with atomic resolution (only atoms within the cut-off being included in the calculation), there were large oscillations as a function of cut-off distance, owing to formal charge imbalances, as shown in Figure 4*A* (see color insert). We therefore next investigated how shielding varied with radial cut-off, using whole residue resolution, which avoided artificially truncating molecules and allowed the presence of charged sidechains. The oscillations were much less apparent (Figure 4*A*). However, based on experimental results in which we had chemically converted the charged lysine sidechains in lysozyme to electroneutral N-acetyl groups and had found that there were essentially no chemical shift changes (51), it appeared that inclusion of full formal charges was unnecessary and might indeed contribute to the small residual oscillations seen experimentally. We therefore carried out a third series of calculations, using whole electroneutral residue resolution, and as shown in Figure 4*A* the oscillation was completely damped. We then used this approach to compute the chemical shifts of the four buried fluorotryptophan residues in galactose binding protein, as a function of the cut-off radius. These theoretical results successfully reproduced the full 10-ppm chemical shift nonequivalence seen experimentally (Figure 4*B*) together with the correct ordering and position of the four buried fluorotryptophan residues. However, the CFP-GIAO technique appeared incapable of successfully reproducing the chemical shift of the exposed tryptophan residue, Trp-284.

We therefore turned our attention to the use of the MSP–local reaction field technique, which includes the effects of molecular dynamics and solvation. We calculated a series of molecular dynamics trajectories and sampled the fields and field gradients at the fluorine nuclei at selected points along each trajectory. Multiplying the dipole and quadrupole shielding polarizabilities (computed previously) by the uniform field and electric field gradient tensor components yielded the purely electrical contributions to shielding, as a function of the time along the molecular dynamics trajectory. Selected results are shown in Figure 4*C*. The average electrical contributions to shielding (the theoretical chemical shifts or shieldings) are plotted as a function of the experimental chemical shifts in Figure 4*D*. This comparison reveals an excellent correlation and slope between predicted and experimental ^{19}F NMR chemical shifts for each of the five fluorotryptophan residues in galactose binding protein, including the solvent-exposed Trp-284. The CFP-GIAO calculations also show a good correlation for the four buried fluorotryptophan residues. However, the slope of the theory-versus-experiment correlation in that case is about 1.6, which we attributed to a dielectric constant effect for these buried residues (13).

To confirm that these results were not purely fortuitous, we investigated the ^{19}F NMR chemical shifts in a series of [4-F]-Trp-labeled myoglobins and hemoglobins. In these systems the maximum observed chemical shift range induced by folding was smaller than in galactose binding protein, only 6.4 ppm. Using the MSP–local reaction field approach once again, we predicted the ^{19}F NMR chemical shifts and found that for fluorine atoms that did not have close contacts with neighboring groups there was a ∼1 ppm mean square deviation between experimental and theoretical chemical shifts (57): good agreement once again.

In the case of ^{19}F-labeled aliphatic amino acids in proteins the situation is more complex. Feeney and coworkers (58) found that a (2S,4S)-5-fluoroleucine-labeled dihydrofolate reductase from *Lactobacillus casei* exhibited a well resolved spectrum containing 12 peaks for the 13 leucine residues, covering a 15-ppm chemical shift range. For fluorine-labeled aliphatic carbons, the dipole and quadrupole shielding polarizabilities are expected to be smaller than in the more highly polarizable aromatic residues. The large range of chemical shifts seen experimentally could therefore not be explained solely in terms of electrical contributions to shielding. However, in the case of fluoroleucine it is clear that side-chain conformational differences (such as γ–gauche effects) would be expected to make major contributions to shielding. We therefore performed ^{19}F NMR chemical shift calculations of simple model systems and found that side-chain conformational differences can contribute up to 10 ppm to fluorine shielding. Similar calculations on a [5-F]-leucine-amide model indicated a 12-ppm difference in chemical shift between the two most likely leucine rotamers. For such aliphatic amino acids, the theoretical calculations predict about a 5–8 ppm maximal shift range owing to electrostatics, but up to a 12-ppm contribution to the chemical shift range owing to conformational effects. Both effects can be expected to contribute to fluorine NMR chemical shifts in fluoro-aliphatic amino acids in proteins, a somewhat more complex situation than is observed with fluoro-aromatic amino acid residues.

Interestingly, the same dihydrofolate reductase enzyme had previously been investigated by Feeney and coworkers, who used fluorotryptophan-labeled amino acids and found the very unusual result that a well resolved J-coupling was observable between two of the fluorotryptophan sites (59). These workers postulated, based on the magnitude of the J-coupling, that this must indicate that two of the Trp residues were close to each other in space, even though they were not close in the primary sequence. Such observations of long-range or through-space fluorine-fluorine J-couplings have been observed in the literature for many years in small organic molecules, and in recent work we have used DFT to try to predict them. We first investigated several of these small molecule systems and found excellent agreement between long-range J-couplings computed with DFT and those determined experimentally (60). In addition, we found that DFT techniques could also be used to estimate the J-coupling determined experimentally in dihydrofolate reductase (60). These results are of interest in the context of hydrogen bond J-couplings, because both are through-space interactions, as discussed in more detail below.

HEMEPROTEINS AND MODEL SYSTEMS

As described in the Introduction, in early experimental work we found strong correlations between the ^{13}C and ^{17}O NMR chemical shifts of the CO ligands in a wide range of carbonmonoxyheme proteins; these chemical shifts were also correlated with the ^{17}O electric field gradients and with the C–O and Fe–C vibrational stretch frequencies determined by infrared and Raman spectroscopy (5). These correlations are what would be expected if electrostatic polarization effects dominated the frequencies and electric field gradients seen experimentally, and in early work using only CO as a model it was found to be possible to reproduce all the experimental trends seen in proteins, with only electrical polarization (6). These correlations would almost certainly be expected to break down if there were large geometric distortions in Fe–C–O bonding geometry between the different proteins, but an examination of the literature revealed a that a wide range of such bonding geometries had actually been proposed, over the years, for different hemeproteins. Of particular interest, the bonding in carbonmonoxymyoglobin had been reported, based on X-ray crystallographic determinations, to be highly distorted, and this distortion had been proposed to be of particular importance in controlling the relative affinities of CO and O_2 molecule-binding to hemoglobin and myoglobin. Indeed, two modern biochemistry textbooks discuss this topic in some depth (61, 62). But is this view correct?

To investigate the question of Fe–C–O binding geometry in more detail, we undertook the synthesis of a broad range of heme model compounds and determined their structures using high-resolution small-molecule X-ray crystallography (63, 64). In particular, we synthesized carbonyl complexes of iron, ruthenium, and osmium tetraphenylporphyrinates and octaethylporphyrinates, complexed with pyridine and 1-methylimidazole. In all cases the metal-C–O geometry was very close to linear and was untilted. We carried out solid-state NMR determinations of the ^{13}C and ^{17}O NMR chemical shift tensors, together with an estimation of the ^{17}O electric field gradient (using nutation spectroscopy) and then used DFT to evaluate these properties (63, 64). As might now be expected, we found good agreement between theory and experiment. In addition, we used Mössbauer spectroscopy to determine the quadrupole splittings, and hence the electric field gradient, at the iron nuclei. Remarkably, the ^{13}C NMR shifts and shift anisotropy, the ^{17}O NMR chemical shift, the ^{17}O NMR electric field gradient, and the ^{57}Fe Mössbauer quadrupole splitting were all virtually identical to those found in the protein carbonmonoxymyoglobin. This strongly suggested the possibility that the ligand binding geometry in the carbonmonoxyheme proteins was also linear and untilted, just as it is in the model compound.

However, to investigate this question further it was necessary to determine just how all the spectroscopic observables would be influenced by changes in metal-ligand geometry. Perhaps they would not be sensitive to ligand distortions such as tilt and bend. We therefore computed ^{13}C and ^{17}O NMR chemical shielding surfaces as a function of ligand tilt (τ) and bend (β) (65), in much the same way that we had previously computed Ramachandran shielding surfaces for amino acids as a function of the backbone torsion angles ϕ and ψ. These results clearly indicated that the experimentally observed correlations between carbon and oxygen spectra could not be generated by variations in ligand tilt and bend (65).

We next began to investigate the possible utility of ^{57}Fe NMR chemical shifts (66) and Mössbauer electric field gradients (67) as probes of hemeprotein structure, using DFT techniques. In earlier work, we had labeled myoglobin with ^{57}Fe and had obtained ^{57}Fe NMR chemical shifts for carbonmonoxymyoglobin and a series of alkylisocyanide-labeled myoglobins (68). Because the ^{57}Fe NMR chemical shifts and electric field gradients in ferrocytochrome c and in bis-pyridine and bis-trimethylphosphine Fe(II) porphyrins were also known, this provided a moderate database for chemical shift and Mössbauer quadrupole splitting/electric field gradient tensor calculations, especially when supplemented with additional Mössbauer and chemical shift results on much smaller molecule model compounds (67). We used DFT and a locally dense basis approach to evaluate the ^{57}Fe NMR shielding and electric field gradients at iron and compared these results with the experimentally measured chemical shifts and quadrupole splittings. We found good accord between the experimental and theoretical iron-57 chemical shifts ($R^2 = 0.992$, slope $= -0.984$), as well as for the Mössbauer quadrupole splittings ($R^2 = 0.975$, slope $= 1.04$, RMSD $= 0.18$ mm sec^{-1}), using a Wachters' all electron representation for iron, a 6-311++ G (2d) basis for all atoms directly attached to iron, 6-31G* for the second shell of atoms, and a 3-21G* basis for more distant atoms, together with a B3LYP hybrid exchange-correlation functional. However, we found extremely poor accord between theory and experiment for the ^{57}Fe NMR chemical shifts and electric field gradients when using carbonmonoxymyoglobin models with the highly distorted structures proposed in earlier X-ray crystallographic investigations (66, 67), clearly suggesting that the Fe–C–O bond geometry is very close to linear and untilted, both in solution and in the solid state.

We then investigated to what extent we could actually predict the experimental tilt and bend angles, using the Bayesian probability or Z-surface approach described above for the amino acid residues in peptides and proteins. In the case of carbonmonoxyhemeproteins, we had six observables: the ^{13}C NMR chemical shift, the ^{13}C NMR chemical shift anisotropy, the ^{17}O NMR chemical shift, the ^{17}O electric field gradient, the ^{57}Fe NMR chemical shift, and the ^{57}Fe Mössbauer quadrupole splitting (or electric field gradient). We therefore computed six property surfaces as a function of tilt and bend, $P(\tau, \beta)$, and used them together with the experimentally determined property values to predict a ^6Z probability surface. For clarity, we show only two such property and probability results in Figure 5 (see color insert). The results of adding additional property surfaces had little effect on τ, β predictions and confirmed that the ligand tilt and bend angles are very close to zero. In the case of the so-called A_o substate of carbonmonoxymyoglobin, we determined most probable tilt and bend angles of 0° and 1°, and in the A_1 substate of carbonmonoxymyoglobin we determined $\tau = 4°$ and $\beta = 7°$ (65).

A year after our publication, Kachalova and coworkers (69) reported a reinvestigation of the X-ray crystallographic structure of carbonmonoxymyoglobin, using a synchrotron source. They reported results only 0.6° different from those we had determined by using the combination of NMR spectroscopy, Mössbauer spectroscopy, and quantum chemical calculations, confirming the correctness of our approach. In addition, deDios & Earle (70) have recently shown by using DFT methods on a charge balanced Fe(II)CO fragment that the ^{13}C NMR chemical shift/^{17}O NMR chemical shift/^{17}O electric field gradient/IR vibrational stretch frequency correlations observed previously can now all be very accurately reproduced by using the charge field perturbation technique, supporting the electrostatic origin of these correlations. In further recent work, Phillips and coworkers (71) have used Delphi (a finite-difference Poisson-Boltzmann algorithm) to compute the electrostatic potentials on CO in a variety of carbonmonoxymyoglobin mutants. They found a linear relationship between the infrared vibrational frequency and the potential, confirming once again the electrostatic origins of the vibrational frequency differences seen between different carbonmonoxyhemeproteins.

How then, are CO and O_2 differentiated in their binding to hemeproteins? Whereas there are undoubtedly a number of factors that contribute to the stabilization of oxygen binding (see e.g., Ref. 72 for detailed discussion), the results of density functional theory calculations on carbonmonoxy- and oxy-heme model compounds clearly indicate a much more negative electrostatic potential on the terminal oxygen in oxyheme when compared with carbonmonoxy-heme model compounds (73). That is to say, there is a much larger negative charge build-up on the terminal oxygen in oxyheme model compounds or oxyheme proteins than in the corresponding CO proteins, which can reasonably be thought to facilitate an electrostatic/hydrogen bonding interaction with the distal histidine residues found in hemoglobin and myoglobin. But how much confidence can be placed in such computations of charge density or electrostatic potential? That is my next topic.

HYDROGEN BONDING AND ELECTROSTATICS

The question of hydrogen bonding and electrostatics in proteins is of great importance from the standpoint of how the three-dimensional structure of a protein is held together, how enzymes function, how ligands such as CO and O_2 bind to proteins, how drugs bind, and so forth, and is clearly a topic worthy of a multivolume series. Nevertheless, when carrying out either ab initio Hartree-Fock or density functional theory (DFT) calculations of NMR chemical shifts, one can obtain, essentially as a by-product of the calculations, extensive details about electrostatics. The question then arises as to how much confidence should be placed in the accuracy of these results, because unlike the spectroscopic observables, electrostatic properties are almost always much more difficult to determine experimentally. We therefore wish to briefly review some recent results of the investigation of electrostatic properties in amino acids, including hydrogen bonding, that are of interest and relevant in the context of the observation of hydrogen bond J-couplings in proteins, as well as the observation of highly deshielded proton resonances in the active sites of many enzymes.

We have recently investigated the charge density, $\rho(\mathbf{r})$, its curvatures, $\partial^2\rho/\partial\mathbf{r}_{ij}$, the dipole moments, μ, the electrostatic potential, $\Phi(\mathbf{r})$, and the electric field gradients at the nitrogen sites, ∇E, in L-asparagine monohydrate, using ultra-high-resolution X-ray crystallography (a synchrotron source, area detector, 20°K sample temperature) and quantum chemical techniques (74). We used both ab initio Hartree-Fock and DFT methods with a variety of basis sets to calculate the electrostatic properties and used Bader's atoms in molecules (AIM) theory (75) to help analyze some of the experimental results. We found that the charge density, the dipole moment, and the molecular electrostatic potential, as well as the curvatures of the charge density at the bond critical points and at the hydrogen bond critical points, extracted from the high resolution X-ray diffraction results were in extremely good agreement with the values computed theoretically, using both Hartree-Fock and DFT methods (74). This gives us some confidence that the electrostatic properties, which can be readily obtained from the calculation, are accurate, a conclusion also reached by a number of workers investigating other amino acids (76, 77). For the electric field gradient, the results obtained from the crystallographic study were highly correlated with the theoretical results but had a poor slope. However, when using electric field gradient tensor elements determined from NMR spectroscopy, we found excellent agreement between calculation and experiment. We also used the icosahedral representation (mentioned above in the context of the chemical shielding tensor) to analyze the electric field gradient tensor as well as the Hessian-of-$\rho(\mathbf{r})$ tensor, which provided a very rigorous evaluation of experimental versus theoretical-electrostatic property comparisons. Once again, we found excellent agreement between theory and experiment (74).

This initial investigation into hydrogen bonding and electrostatics clearly indicated that hydrogen bond charge densities at hydrogen bond critical points, hydrogen bond principal curvatures, and hydrogen bond Hessian tensors in the icosahedral representation could all be extracted from high resolution crystallographic data and that the values of these properties could all be accurately predicted by using either Hartree-Fock or DFT methods. This is an important observation because of recent debate of the question of the chemical nature of biological hydrogen bonds that has followed the observation of NMR scalar couplings across protein backbone hydrogen bonds. New ways of looking at hydrogen bonds are of interest because the existence of such $^{3h}J_{NC'}$ couplings has led some workers to conclude that even weak biological hydrogen bonds must be partially covalent. This assertion might appear surprising, given that through-space scalar couplings in simple organic molecules have long been recognized. Furthermore, as outlined above, we have been able to show via DFT calculations that through-space couplings between fluorine nuclei, J_{FF}, do indeed occur in the absence of a bond network, covalent or otherwise. We therefore began an investigation of protein hydrogen bonds in terms of both calculated Hessian-of-$\rho(\mathbf{r})$ tensors and experimental hydrogen bond scalar couplings and ^1H chemical shifts.

By calculating $\rho(\mathbf{r})$ and $\partial^2\rho/\partial r_i r_j$ for protein backbone hydrogen bonds, we showed that the magnitude of $^{3h}J_{NC'}$ is an exponential function of the mutual penetration of the nonbonding van der Waals shells of the isolated hydrogen bond donor and acceptor fragments, and not the partial covalence invoked previously (78). Using nonbonded fluoromethane dimer models for the calculation of J_{FF}, our results also showed that the magnitude of these through-space couplings exhibits the same exponential dependence upon the penetration of nonbonding monomer charge densities (Figure 6A). These results suggested that both the hydrogen bond ^{19}F–^{19}F J-coupling and the through-space J-coupling are subject to the same inductive mechanism.

Atoms-in-molecules theory is also useful in classifying (hydrogen-) bonding. It is straightforward to extract the relative contributions of both kinetic and potential energies to a bond from the calculation of $\partial^2\rho/\partial r_i r_j$ (75, 78). A local expression of the virial theorem relates the trace of the bond critical point Hessian tensor to the electronic kinetic energy density, G(r), and the electronic potential energy, V(r) (75):

$$\frac{1}{4}\text{Tr(Hessian)} = 2G(r) + V(r) \qquad 9.$$

This means one can determine which energy density is in excess of the 2:1 kinetic:potential average and thus determine the character of the bond. Stabilization resulting from a concentration of electronic charge in the bonding region is reflected by an excess of potential energy. Thus, closed-shell (electrostatic) bonds have a virial excess of kinetic energy, $|2G(r)| > |V(r)|$, whereas shared-electron (covalent) bonds have a virial excess of potential energy, $|V(r)| > |2G(r)|$. Furthermore, one can characterize a closed-shell situation in which $|2G(r)| > |V(r)|$, but $|V(r)| > |G(r)|$, as partially covalent. Using these definitions, we found that protein backbone hydrogen bonds are purely closed-shell, electrostatic interactions: $|2G(r)| > |V(r)|$ and $|G(r)| > |V(r)|$.

The chemical shift is also a very useful probe of the hydrogen bond, especially when combined with a quantum chemical/AIM analysis (78). Peptide backbone hydrogen bonds appear at high-field and correlate with a purely closed shell or electrostatic interaction (no covalence). Between 12 and 14 ppm partial covalent character begins to develop in the hydrogen bonds as $|V(r)| > |G(r)|$ and is clearly seen in enzyme short-strong hydrogen bonds and in some acids (78) (Figure 6B). This covalent character increases exponentially as the ^1H nucleus becomes more deshielded, and at $\sim$21 ppm, the chemical shift region of experimentally observed low barrier hydrogen bonds, the hydrogen bond becomes a fully covalent, shared-electron interaction, with $|V(r)| > |2G(r)|$, as shown in Figure 6B (78).

TOWARDS DRUG DESIGN

Chemical shifts have also recently been used with great success as a tool with which to probe ligand binding to proteins in the drug discovery process. The idea is to use chemical shift changes on ligand binding to carry out high throughput

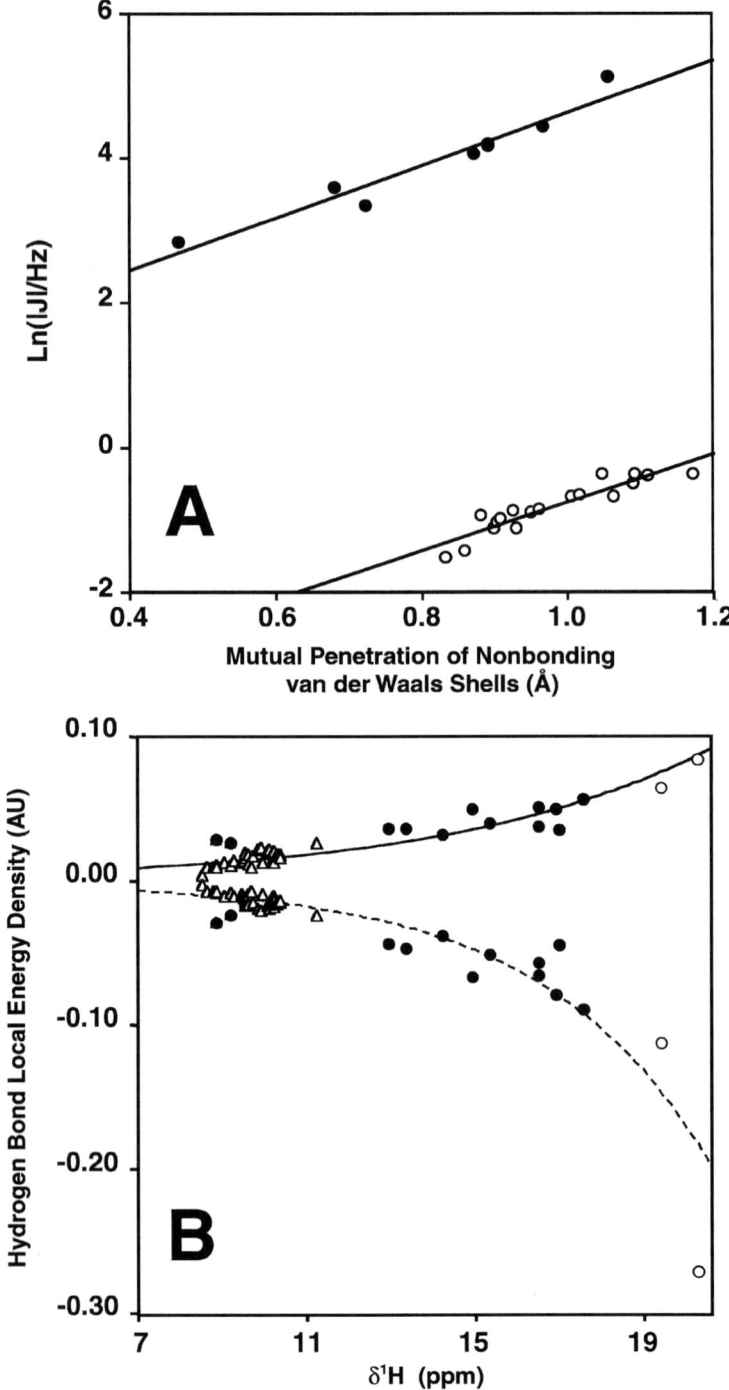

screening of small molecule libraries to discover lead compounds (typically with micromolar affinity constants) and then to use NMR or X-ray crystallographic structural information to help appropriately link two (or more) such compounds to provide inhibitors with nanomolar binding constants.

In its earliest embodiment (79), ^{15}N and ^{1}H amide chemical shifts obtained from two-dimensional spectra were used to find an initial lead compound, based on chemical shift changes observed on ligand binding. Ten small ligand molecules were added at a time to a protein solution, which enabled the screening of ~1000 compounds in a reasonable time period. Once a promising lead compound was identified, this compound could be optimized. A second series of screenings was carried out (possibly in the presence of the first, optimized lead compound) to detect a second ligand binding site, again based solely on changes in chemical shift owing to ligand binding. Optimization of the second lead compound, using high-resolution solution NMR or X-ray crystallographic studies, was then carried out to determine the location of the two binding ligands, a strategy that facilitates the relatively precise engineering of a suitable tether to chemically link the two individual ligands. When two micromolar binding ligands are tethered together a typically nanomolar binding inhibitor is obtained. For example, a tight binding inhibitor ($K_d = 19$ nM) of the FK506 binding protein was discovered by linking two weak binding inhibitors ($K_d = 2$ μM, 100 μM) that had been identified and optimized in less than two months, and it was only necessary to synthesize five tethered molecules to obtain the nanomolar inhibitor. In another example, a nanomolar nonpeptide inhibitor of stomelysin, a matrix metalloproteinase, was produced by linking together 17-mM and 20-μM inhibitors, and it has been pointed out that conventional high throughput screening of more than 100,000 compounds against stomelysin failed to produce a single inhibitor with better than a 10-μM K_d (80).

In early reports, this approach was limited to screening about 1000 compounds per day, a rather low throughput by high throughput screening standards, because relatively high protein concentrations were required to obtain NMR spectra in reasonable time periods. In addition, high protein concentrations limited the total number of compounds that could be screened in a given spectrum, based on solubility considerations. More recently, the use of cryoprobe technology has decreased typical sample concentrations required for data acquisition (81) and it has been reported that mixtures containing 100 compounds can be screened in a single experiment

---

Figure 6 (A) Dependence of the magnitude of through-space J-couplings upon the mutual penetration of nonbonding van der Waals shells, $\Sigma \Delta r$. Filled circles, J_{FF}; open circles, $^{3h}J_{NC}$. (B) Dependence of the hydrogen bond local energy densities upon the proton chemical shift. Open circles, low barrier hydrogen bonds; filled circles, short-strong hydrogen bonds; triangles, peptide backbone hydrogen bonds; solid line, $G(r) = 0.0066 \exp[0.12]$, $R^2 = 0.87$; dashed line, $V(r) = -0.0012 \exp[0.24]$, $R^2 = 0.82$. Based on data in Ref. 78.

at protein concentrations of 50 μM, enabling libraries of more than 200,000 compounds to be investigated in less than one month. This chemical shift–based approach to drug design also has considerable advantages over more standard combinatorial chemistry approaches because far fewer compounds need to be synthesized, because linker compounds can be specifically tailored based on known NMR or X-ray crystallographic structural information. Moreover, the technique has the advantage that no specific high throughput screening assays need to be developed.

This approach has recently been extended to the use of ^{13}C-labeled proteins. The method involves ^{13}C-labeling of the methyl groups of valine, leucine, and isoleucine (δ_1) residues from inexpensively produced ^{13}C α-ketoisovalerate and α-ketobutyrate and results in a sensitivity increase of almost a factor of three when compared with the ^{15}N/^{1}H chemical shift screening approach (82), owing to the favorable relaxation properties of methyl groups and the fact that three protons are attached to each ^{13}C-labeled methyl group. Whereas fewer resonances are monitored in the ^{1}H/^{13}C-experiment, an analysis of 191 crystal structures of proteins with bound ligands showed that 92% of the ligands were within 6 Å of at least one methyl carbon in valine, leucine, or isoleucine-δ_1, but only 82% of the ligands were within 6 Å of a backbone nitrogen. Moreover, even high molecular weight protein targets, such as dihydroneopterin aldolase (MW = 110 kDa), could be investigated using this chemical shift approach (82).

ACKNOWLEDGMENTS

I thank A.C. de Dios, C.J. Jameson, and P. Pulay for their contributions to the work reported here, and W. Arnold for help with preparing this article. This work was supported by the U.S. National Institutes of Health (Grants GM-50694, HL-19481) and by the National Computational Science Alliance (Grant MCB000018N).

Visit the Annual Reviews home page at www.annualreviews.org

LITERATURE CITED

1. Meadows DH, Markley JL, Cohen JS, Jardetzky O. 1967. *Proc. Natl. Acad. Sci. USA* 58:1307–13
2. Allerhand A, Childers RF, Oldfield E. 1973. *Biochemistry* 12:1335–41
3. Allerhand A, Oldfield E. 1973. *Biochemistry* 12:3428–33
4. Spera S, Bax A. 1991. *J. Am. Chem. Soc.* 113:5490–92
5. Park KD, Guo K, Adebodun F, Chiu ML, Sligar SG, Oldfield E. 1991. *Biochemistry* 30:2333–47
6. Augspurger JD, Dykstra CE, Oldfield E. 1991. *J. Am. Chem. Soc.* 113:2447–51
7. Osapay K, Case DA. 1991. *J. Am. Chem. Soc.* 113:9436–44
8. Janes N, Ganapathy S, Oldfield E. 1983. *J. Magn. Reson.* 54:111–21
9. Frydman L, Chingas GC, Lee YK, Grandinetti PJ, Eastman MA, et al. 1992. *Isr. J. Chem.* 32:161–64
10. Wolinski K, Hinton JF, Pulay P. 1990. *J. Am. Chem. Soc.* 112:8251–60
11. deDios AC, Laws DD, Oldfield E. 1994. *J. Am. Chem. Soc.* 116:7784–86

12. Alderman DW, Sherwood MH, Grant DM. 1993. *J. Magn. Reson.* 101:188–97
13. deDios AC, Pearson JG, Oldfield E. 1993. *Science* 260:1491–96
14. Chesnut DB, Moore KD. 1989. *J. Comput. Chem.* 10:648–59
15. deDios AC, Pearson JG, Oldfield E. 1993. *J. Am. Chem. Soc.* 115:9768–73
16. deDios AC, Oldfield E. 1994. *J. Am. Chem. Soc.* 116:5307–14
17. Laws DD, Le H, deDios AC, Havlin RH, Oldfield E. 1995. *J. Am. Chem. Soc.* 117:9542–46
18. Havlin RH, Le H, Laws DD, deDios AC, Oldfield E. 1997. *J. Am. Chem. Soc.* 119:11951–58
19. Heller J, Laws DD, King DS, Wemmer DE, Pines A, et al. 1997. *J. Am. Chem. Soc.* 119:7827–31
20. Szabo CM, Sanders LK, Arnold W, Grimley JS, Godbout N, et al. 1999. In *ACS Monograph on Chemical Shielding*, ed. A deDios, J Facelli, pp. 40–62
21. Tjandra N, Bax A. 1997. *J. Am. Chem. Soc.* 119:9576–77
22. Sitkoff D, Case DA. 1998. *Prog. NMR Spectrosc.* 32:165–90
23. Hong M. 2000. *J. Am. Chem. Soc.* 122:3762–70
24. Huster D, Yamaguchi S, Hong M. 2000. *J. Am. Chem. Soc.* 122:11320–27
25. Cole HB, Sparks SW, Torchia DA. 1988. *Proc. Natl. Acad. Sci. USA* 85:6362–65
26. McDermott A, Polenova T, Bockmann A, Zilm KW, Paulsen EK, et al. 2000. *J. Biomol. NMR* 16:209–19
27. Sanders L, Oldfield E. Unpublished results
28. Schafer L, Newton SQ, Cao M, Peeters A, Van Alsenoy C, et al. 1993. *J. Am. Chem. Soc.* 115:272–80
29. deDios AC, Oldfield E. 1994. *J. Am. Chem. Soc.* 116:11485–88
30. Walling AE, Pargas RE, deDios AC. 1997. *J. Phys. Chem. A* 101:7299–303
31. Le H, Oldfield E. 1994. *J. Biomol. NMR* 4:341–48
32. Le H, Oldfield E. 1996. *J. Phys. Chem.* 100:16423–28
33. Marassi FM, Opella SJ. 2000. *J. Magn. Reson.* 144:150–55
34. Wang J, Denny J, Tian C, Kim S, Mo Y, et al. 2000. *J. Magn. Reson.* 144:162–67
35. Fushman D, Tjandra N, Cowburn D. 1998. *J. Am. Chem. Soc.* 120:10947–52
36. Kroenke CD, Rance M, Palmer AG III. 1999. *J. Am. Chem. Soc.* 121:10119–25
37. Kurland RJ, McGarvey BR. 1970. *J. Magn. Reson.* 2:286–301
38. Qi PX, Beckman RA, Wand AJ. 1996. *Biochemistry* 35:12275–86
39. Wilkens SJ, Xia B, Weinhold F, Markley JL, Westler WM. 1998. *J. Am. Chem. Soc.* 120:4806–14
40. Wilkins SJ, Westler WH, Markley JL, Weinhold F. 1999. *J. Inorg. Biochem.* 74:338
41. Van Alsenoy C, Yu C-H, Peeters A, Martin JML, Schäfer L. 1998. *J. Phys. Chem. A* 102:2246–51
42. Pearson JG, Le H, Sanders L, Godbout N, Havlin RH, Oldfield E. 1997. *J. Am. Chem. Soc.* 119:11,941–50
43. Pearson JG, Wang J-F, Markley JL, Le H, Oldfield E. 1995. *J. Am. Chem. Soc.* 117:8823–29
44. Tjandra N, Bax A. 1997. *Science* 278:1111–14
45. Chou JJ, Li S, Bax A. 2000. *J. Biomol. NMR* 18:217–27
46. Cornilescu G, Delaglio F, Bax A. 1999. *J. Biomol. NMR* 13:289–302
47. Wishart DS, Sykes BD, Richards FM. 1992. *Biochemistry* 31:1647–51
48. Wishart DS, Nip AM. 1998. *Biochem. Cell Biol.* 76:153–63
49. Bode J, Blumenstein M, Raftery MA. 1975. *Biochemistry* 14:1153–60
50. Sykes BD, Weingarten HI, Schlesinger MJ. 1974. *Proc. Natl. Acad. Sci. USA* 71:469–73
51. Lian C, Le H, Montez B, Patterson J, Harrell S, et al. 1994. *Biochemistry* 33:5238–45
52. deDios AC, Oldfield E. 1994. *J. Am. Chem. Soc.* 116:7453–54

53. deDios AC, Oldfield E. 1993. *Chem. Phys. Lett.* 205:108–16
54. Augspurger JD, deDios AC, Oldfield E, Dykstra CE. 1993. *Chem. Phys. Lett.* 213:211–16
55. Luck LA, Falke JJ. 1991. *Biochemistry* 30:4248–56
56. Pearson JG, Oldfield E, Lee FS, Warshel A. 1993. *J. Am. Chem. Soc.* 115:6851–62
57. Pearson JG, Montez B, Le H, Oldfield E, Chien EYT, Sligar SG. 1997. *Biochemistry* 36:3590–99
58. Feeney J, McCormick JE, Bauer CJ, Birdsall B, Moody CM, et al. 1996. *J. Am. Chem. Soc.* 118:8700–6
59. Kimber BJ, Feeney J, Roberts GCK, Birdsall B, Griffiths DV, et al. 1978. *Nature* 271:184–85
60. Arnold WD, Mao J, Sun H, Oldfield E. 2000. *J. Am. Chem. Soc.* 122:12164–68
61. Stryer L. 1995. *Biochemistry*. New York: Freeman. 4th ed.
62. Garrett RH, Grisham CM. 1999. *Biochemistry*. Orlando, FL: Harcourt Brace. 2nd ed.
63. Salzmann R, Ziegler CJ, Godbout N, McMahon MT, Suslick KS, Oldfield E. 1998. *J. Am. Chem. Soc.* 120:11323–34
64. Salzmann R, McMahon MT, Godbout N, Sanders LK, Wojdelski M, Oldfield E. 1999. *J. Am. Chem. Soc.* 121:3818–28
65. McMahon MT, deDios AC, Godbout N, Salzmann R, Laws DD, et al. 1998. *J. Am. Chem. Soc.* 120:4784–97
66. Godbout N, Havlin R, Salzmann R, Debrunner PG, Oldfield E. 1998. *J. Phys. Chem. A* 102:2342–50
67. Havlin RH, Godbout N, Salzmann R, Wojdelski M, Arnold W, et al. 1998. *J. Am. Chem. Soc.* 120:3144–51
68. Chung J, Lee HC, Oldfield E. 1990. *J. Magn. Reson.* 90:148–57
69. Kachalova GS, Popov AN, Bartunik HD. 1999. *Science* 284:473–76
70. deDios AC, Earle EM. 1997. *J. Phys. Chem. A* 101:8132–34
71. Phillips GN Jr, Teodoro M, Li TS, Smith B, Olson JS. 1999. *J. Phys. Chem. B* 103:8817–29
72. Springer BA, Sligar SG, Olson JS, Phillips GN Jr. 1994. *Chem. Rev.* 94:699–714
73. Godbout N, Sanders LK, Salzmann R, Havlin RH, Wojdelski M, Oldfield E. 1999. *J. Am. Chem. Soc.* 121:3829–44
74. Arnold WD, Sanders LK, McMahon MT, Volkov AV, Wu G, et al. 2000. *J. Am. Chem. Soc.* 122:4708–17
75. Bader RFW. 1990. *Atoms in Molecules—A Quantum Theory*. Oxford: Clarendon
76. Koritsanszky T, Flaig R, Zobel D, Krane H-G, Morgenroth W, Luger P. 1998. *Science* 279:356–58
77. Flaig R, Koritsanszky T, Zobel D, Luger P. 1998. *J. Am. Chem. Soc.* 120:2227–38
78. Arnold WD, Oldfield E. 2000. *J. Am. Chem. Soc.* 122:12835–41
79. Shuker SB, Hajduk PJ, Meadows RP, Fesik SW. 1996. *Science* 274:1531–34
80. Hajduk PJ, Meadows RP, Fesik SW. 1997. *Science* 278:497–99
81. Hajduk PJ, Gerfin T, Boehlen J-M, Häberli M, Marek D, Fesik SW. 1999. *J. Med. Chem.* 42:2315–17
82. Hajduk PJ, Augeri DJ, Mack J, Mendoza R, Yang J, et al. 2000. *J. Am. Chem. Soc.* 122:7898–904

REACTIVE COLLISIONS OF HYPERTHERMAL ENERGY MOLECULAR IONS WITH SOLID SURFACES

Dennis C. Jacobs
Department of Chemistry and Biochemistry, University of Notre Dame, Notre Dame, Indiana 46556; e-mail: jacobs.2@nd.edu

Key Words reaction dynamics, electron transfer, ion/surface scattering, self-assembled monolayers, abstraction

■ **Abstract** Recent experimental advances have uncovered many of the diverse reaction pathways following an energetic collision between a molecular ion and a solid surface. Hyperthermal translational energies (5–500 eV) are sufficient to activate a number of chemical transformations in the near-surface region, including charge transfer, dissociation, abstraction, and deposition. State-of-the-art scattering studies probe the consumption and disposal of energy and the effects of approach geometry and surface electronic structure on the operative reaction mechanisms. These fundamental investigations provide insight relevant to the fabrication of microelectronics devices, the interaction of space vehicles with the earth's atmosphere, and the development of analytical techniques in mass spectrometry.

INTRODUCTION

Surface science traces its roots back to the investigation of heterogeneous catalysts for the petrochemical industry. Single-crystal metals in ultra-high vacuum (UHV) were selected as well-characterized models for industrial catalysts. Gas-phase reagents were introduced into UHV chambers under thermal conditions while reaction products were studied with mass-spectrometry or surface-sensitive spectroscopies. In the last decade, more attention has been focused on studying thermal reactions at semiconductor and metal oxide substrates—the former is motivated by the booming microelectronics sector, and the latter is recognized to be a more realistic model for supported catalysts than are single-crystal metals (1).

When a molecule strikes a surface, the chemical nature of the incident particle and the electronic structure of the surface are critical in determining the outcome of the collision. Under thermal conditions, desorption and decomposition often compete with surface chemical reactions, leaving the experimentalist with limited control over the distribution of products. In contrast, the introduction of nonthermal energies into the molecule/surface system can activate chemical reactions that go unobserved under thermal conditions. Figure 1 identifies some of the more common events that can occur when a molecular projectile strikes a surface target. For a given

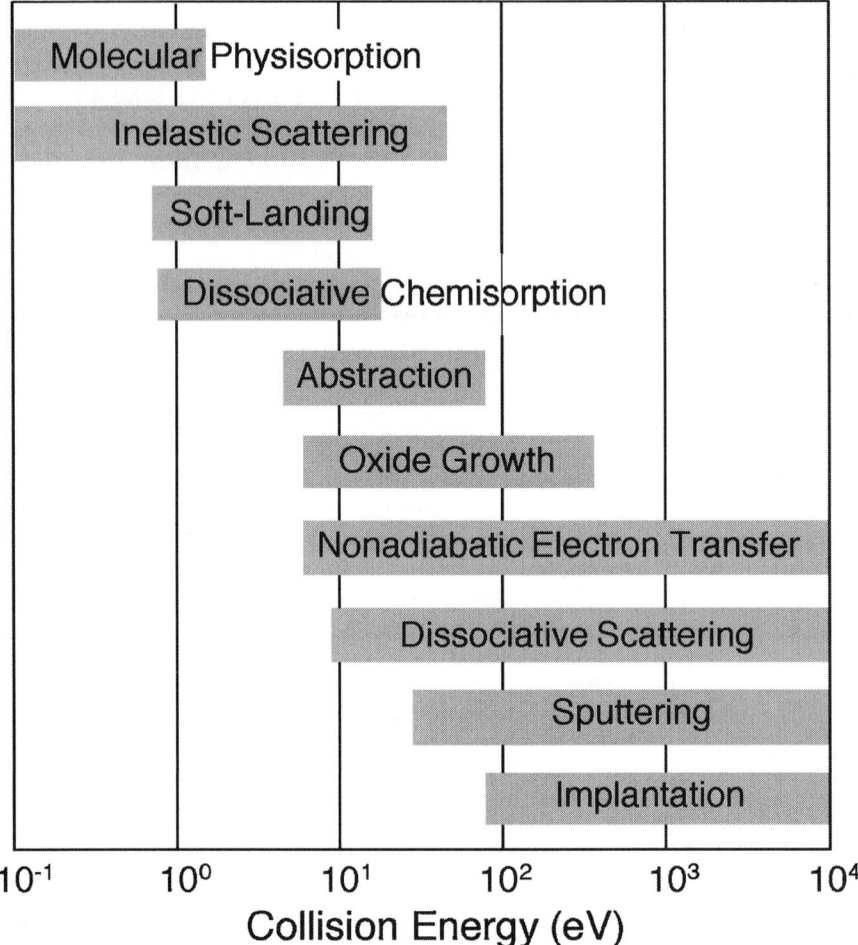

Figure 1 Fundamental processes associated with molecule/surface collisions. The shaded regions indicate the approximate energy ranges over which the processes occur or are typically studied.

molecule/surface system, the translational energy of the incident molecule largely determines the branching ratio between the various reaction pathways. The reaction mechanisms listed in Figure 1 that operate predominantly in the hyperthermal energy range are nonadiabatic electron transfer, dissociative scattering, abstraction, and processes related to ion beam deposition. These fundamental mechanisms will be discussed individually within this chapter.

The use of energetic reagents to modify surfaces has found its way into a number of industrial applications (2). Plasma processing, ion implantation, sputtering, and reactive ion etching are routinely employed in the fabrication of microelectronic devices (3). Protective thin films are deposited on optical materials using plasma sources (4). Ion/surface collisions can be exploited within a tandem mass spectrometer to help analytical chemists identify the chemical composition of unknown sample (5). The low-earth orbit (LEO) and geosynchronous-earth orbit (GEO) environments degrade spacecraft materials through energetic bombardment by atmospheric particles (neutrals; ions such as O^+, NO^+, and O_2^+; photons; electrons; and micrometeorites) (6, 7). Gaining a detailed understanding of the reaction mechanisms associated with molecular ion/surface collisions will better inform engineers in the optimization and control of these processing environments.

EXPERIMENTAL METHODOLOGY

Delivering hyperthermal energy molecules to a surface is technically challenging. Although supersonic molecular beams are capable of generating tunable, nonthermal energy distributions of neutral molecules, these sources are typically limited to translational energies less than a few electron volts. Alternatively, molecular ions can be accelerated to virtually any translational energy between $1-10^8$ eV. In practice, however, it is difficult to generate stable, focused ion beams below 100 eV without using nonstandard ion optics. Nevertheless, several research groups have interfaced a variety of hyperthermal energy ion sources to surface scattering chambers (5).

Ion Sources and Transport

Whereas atomic ion sources are usually designed to generate large ion currents through efficient ionization schemes, these are not always preferred for the generation of molecular ions. For example, plasma ion sources rely on numerous energetic electron-neutral and ion-neutral collisions to ionize the neutral gas within the hot plasma. Molecular gases readily fragment in this environment, making it difficult to generate sufficiently high beam currents of the desired molecular species. Alternatively, electron-impact ion sources involve only single electron-molecule collisions, reducing fragmentation significantly. Producing a beam of large molecular ions presents new challenges because of their many fragmentation pathways and the low volatility of the neutral precursor. Electrospray ionization techniques remarkably overcome both of these limitations by introducing the neutral macromolecules of interest within a solvent stream that evaporates during field ionization within the vacuum (8).

Unfortunately, all of the aforementioned ionization methods will form molecular ions in a distribution of vibronically excited states. As will be discussed later in this chapter, excited-state ions may react differently than ground-state ions when impacting a surface. A powerful tool for studying the influence of an ion's internal state on its reaction with a surface is state-selective ionization through a technique such as resonance enhanced multiphoton ionization (REMPI) (9). In REMPI, an integer number of laser photons is resonant with an intermediate state in the molecule, causing the overall rate of multiphoton ionization to be significantly enhanced. Because the intermediate state is bound with respect to both dissociation and electron loss, the resonant transitions are quite sharp, and the laser wavelength can selectively populate an individual rovibronic level within the intermediate state. Furthermore, this state-selectivity can be passed along to the ion if one has chosen a resonant intermediate state with strong Rydberg character (10). Ionization out of a Rydberg state largely preserves the vibrational and rotational quantum numbers prepared within the neutral intermediate state, because the final ionizing photon predominantly ejects the Rydberg electron while leaving the rotational and vibrational state of the ion core intact. REMPI has been demonstrated to achieve a high degree of state-selectivity for forming H_2^+, N_2^+, O_2^+, NO^+, CO^+, HBr^+ (11), Br_2^+ (12), OCS^+, NH_3^+, $C_2H_2^+$, and a variety of substituted aromatic cations (10). This is not an exhaustive list of ions capable of being state-selectively prepared through REMPI; instead, it merely reflects those molecules for which photoelectron spectroscopy or laser-induced fluorescence has been utilized to confirm that a preponderance of the molecular ions are indeed produced within a single vibronic state.

After the molecular ions are formed within the source, they must be extracted and transported to the surface. Cracking of the parent ion within the source can rarely be avoided; therefore, mass filtration of the ion beam through a magnetic sector, Wien filter, or quadrupole mass filter is often necessary to insure that fragment ions do not contribute to the molecular ion beam (5). Space-charge interactions, i.e., repulsion between like-charged ions, can severely limit the current of ions that can be transmitted at kinetic energies below 100 eV. To minimize space-charge effects, ions are typically transported at ~keV energies through much of their path length. Then, in close proximity to the surface, the ions are decelerated within a specially designed set of ion optics. Care must be taken to minimize beam divergence and current loss within this critical region. Whereas deceleration optics typically consist of a close-packed series of cylinders, Cowin utilized a pair of fine meshes immediately in front of the surface to approximate a parallel-plate geometry (13). A different strategy for minimizing space-charge broadening at hyperthermal energies is to inject the decelerated ions into an octapole radio-frequency (RF) field. Here, the ions drift down the longitudinal axis of the octapole rods while being trapped laterally in the guiding RF fields (14). Anderson has designed a novel but elaborate phase-space compressor for increasing the ion current density delivered to the surface by approximately one order-of-magnitude (15). Regardless of the method for ion transport, the common goal is a focused (collimated), monoenergetic beam of hyperthermal ions directed at the surface target (diameter $\leq$ 1 cm).

Surface Preparation

Experiments are more reproducible and easier to interpret when they involve surfaces that are uniformly prepared and well-characterized. Two basic approaches have been widely adopted. In the first, single-crystal substrates are cut and polished to expose a single crystallographic plane. They are transferred to an ultrahigh vacuum ($<10^{-9}$ torr) chamber where ion sputtering and annealing treatments are typically employed to generate an atomically clean surface. Surface-sensitive techniques such as Auger electron spectroscopy, X-ray photoelectron spectroscopy, and low-energy electron diffraction can verify that the surface is free of contaminants and well-ordered. The second type of well-characterized surface, a self-assembled monolayer (SAM), does not require a battery of diagnostic techniques. SAM films can be prepared, for example, by simply dipping a gold surface into an alkane thiol solution. The resulting monolayer presents a dense uniform array of tilted hydrocarbon chains at the vacuum interface (16). The relatively inert nature of the resulting hydrocarbon surface allows SAM films to remain free of contamination under less stringent vacuum conditions ($<10^{-7}$ torr).

Product Detection

The reaction of a molecular ion with a surface is evidenced by the changed nature of the scattered particle or by a surface modification near the point of impact. Because gas-phase particle detection remains more sensitive than surface-sensitive spectroscopies, the former approach is more widely adopted. Scattered ions can be detected with near-unit efficiency, because they are easily accelerated to keV energies where intensified detectors, e.g., channeltrons and microchannel plates (MCP), can operate in single-particle counting mode. Naturally, it is desirable to gain mass, energy, and exit-angle information about the scattered ions; the following four paragraphs describe various experimental approaches to accomplish this goal.

Time-of-flight methods employ a pulsed source of ions and a lengthy flight tube in which the scattered ions disperse according to their final velocity. This approach has the advantage of high throughput and multichannel processing efficiency where ions at all energies are collected from each ion pulse. However, energy-resolution is nonlinear, detection angles are usually fixed rather than rotatable, and mass separation relies on distinctly different velocity distributions for each product channel (a rarely met condition for more complex systems).

Electrostatic sectors provide calibrated linear, high-resolution energy dispersion within a rotatable unit. Unfortunately, single anode instruments (as opposed to multichannel anode arrays) must be repeatedly scanned in both pass energy and angle. Additionally, electrostatic sectors suffer from not having an independent measure of the product mass. Coupling time-of-flight methods with an electrostatic sector could give mass and energy information, but the resulting signal levels are often too low to be practical (17).

Quadrupole mass filters offer excellent mass resolution for low energy, ~5–15 eV, product ions. However, higher energy ions will suffer degraded mass resolution unless they are decelerated before entering the quadrupole. Quadrupole mass filters are usually employed at a fixed angle without a high-resolution energy

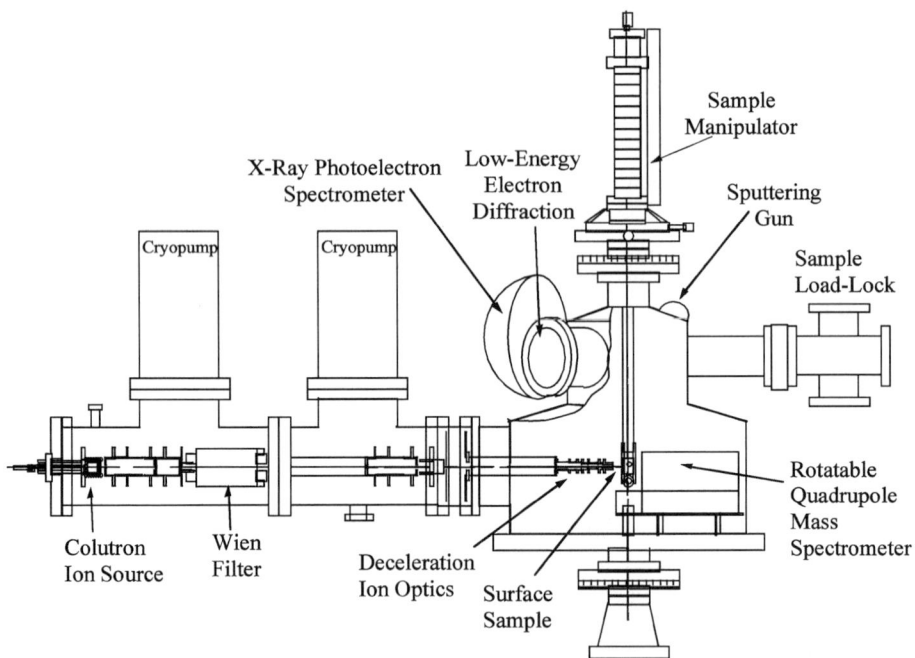

Figure 2 Schematic diagram of an ion/surface scattering chamber at the University of Notre Dame. The ions are formed in a Colutron ion source, accelerated to ~1 keV, mass-selected in a Wien filter, and decelerated just prior to striking the surface. Scattered products are detected in a rotatable energy-resolved, quadrupole mass spectrometer.

filter. An exception to this is shown in Figure 2, where a rotatable chamber within the UHV scattering chamber contains an electrostatic energy analyzer coupled to a quadrupole mass filter. Some designs have combined a time-of-flight approach with a quadrupole mass filter to gain velocity distributions for each distinct product mass (18).

Imaging detection allows one to gain multichannel information by storing the position where the ion strikes an area detector at a particular time of arrival. For example, a pulsed ion beam can be combined with a pulsed imaging detector to gain two-dimensional in-plane velocity images for products at a selected mass (19, 20). An alternative design uses the spatial resolution of the area detector to collect products as a function of both in-plane and out-of-plane angles at a given exit velocity (21).

Detection of scattered neutral products often requires that the products first be ionized and then collected. Ionization schemes are inherently inefficient and represent a limiting factor in the detector's sensitivity. Electron impact ionization is the most common ionization method because it can be universally applied to all molecules; however, one must carefully distinguish between the cracking that occurs within the ionizer and the fragments that are formed at the target surface. Laser based detection methods, e.g., REMPI, or laser induced fluorescence (LIF), can be used to sensitively measure the populations of individual quantum states within the scattered neutral products (21).

Surface modifications induced by hyperthermal energy ion bombardment are most directly observed by a change in the morphology or chemical composition of the surface. A battery of surface sensitive probes has been developed over the years. The more commonly used methods are Auger electron spectroscopy, X-ray photoelectron spectroscopy, temperature programmed desorption, Kelvin probe for work function measurements, infrared absorption spectroscopy, and scanning probe or electron microscopies (22).

REACTION PROCESSES

Hyperthermal molecular ions readily undergo chemical change upon collision with a surface (23). A reaction mechanism may involve one or more fundamental steps in which the molecule is fragmented and/or electrons or nuclei are exchanged between the projectile and the surface. Although the fundamental steps are often interrelated within a given mechanism, they will be discussed individually below.

Nonadiabatic Charge Transfer

Molecular ions are efficiently (>99%) neutralized within a few Angstroms of a metallic or semiconducting surface (24). In contrast, molecular ions are less likely to suffer charge exchange on an insulating surface (e.g., a metal oxide or SAM) (5). Much has been learned about electron transfer at surfaces through studies of atomic ion/surface interactions. Before considering the charge transfer dynamics of a molecular ion near a surface, it is useful to review the major concepts that have been developed for the case of atomic projectiles on metallic or semiconducting substrates. The predominant mechanisms for charge exchange at surfaces are resonant and Auger electron transfer. When the affinity level is resonant with the occupied electronic states of the solid, resonant neutralization is allowed. Although Auger electron transfer can compete, it represents a two-electron reorganization that is, in general, less probable than a one-electron reorganization. Notwithstanding, Auger electron transfer is more common when the affinity level lies well below the Fermi level. Molecules often have lower ionization potentials than atoms and have a higher density of excited states. Hence, molecular ions are usually well-suited for resonant neutralization.

The charge state of an atom near a surface is dependent on the atom's instantaneous position above the surface and the history of its trajectory. In the adiabatic limit, where nuclear motion is slow compared with the timescale for electron transfer, the charge state of an atom tracks the equilibrium charge state. For resonant charge exchange, the atomic charge state can be calculated by considering the overlap of the atom's lifetime-broadened affinity level with the occupied electronic

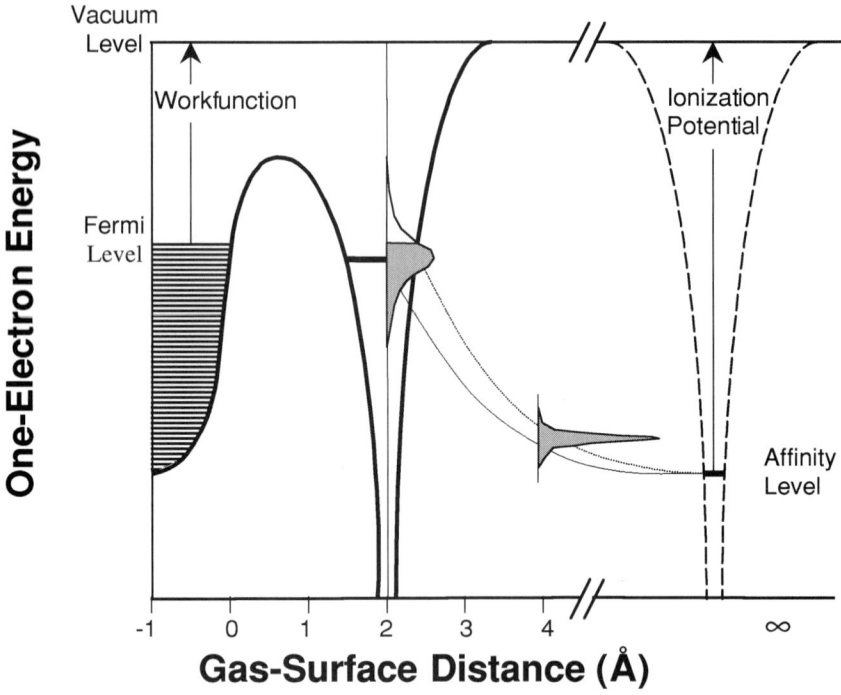

Figure 3 One-electron energy level diagram for neutralization of a positive ion near a metal surface. The metal conduction band is shown on the left. The dashed and solid curves represent the potential energy for an electron when the ion is at infinite and close distances to the surface, respectively. The affinity level for the ion shifts upward and broadens as the ion approaches the surface.

states in the solid (25). Figure 3 illustrates how the affinity level for a positive ion shifts as the particle approaches the surface and interacts with its image charge. The equilibrium charge state is the fraction of the Lorentzian affinity level that lies below the Fermi level. For example, if the atom were frozen at $z = 2$ Å above the surface, the shaded region of the leftmost Lorentzian profile in Figure 3 represents ~70% of the area under the curve; therefore, the particle would adopt an equilibrium charge of $(1.0 - 0.7 = +0.3)$. The charge state for a nonstationary particle will try to track the equilibrium charge state at every point along the trajectory. The tunneling rate that governs electron transfer decreases exponentially with particle-surface distance. Close to the surface, the tunneling rate is fast compared with the timescale for translation; thus, the particle maintains its equilibrium charge state in the near-surface region. Along the exit trajectory, the scattered particle will come to a point, commonly called the freezing distance, where the tunneling rate becomes too slow to allow for any additional charge exchange (25). Here, the particle's charge state becomes frozen at whatever value the equilibrium charge state was at the freezing distance. This simple picture provides great insight for understanding resonant electron transfer in the case of atomic projectiles.

The situation becomes more complicated with molecular projectiles because of the additional degrees of freedom introduced. For the simplest case of a diatomic molecule, the internuclear separation and axis orientation are integrally involved. For hyperthermal collision energies, rotational and vibrational motion are often slower than translational motion. When modeling a polyatomic molecule undergoing charge transfer at a surface, as many as $3N - 3$ internal coordinates may need to be considered along with z, the distance from the molecule's center-of-mass to the surface plane. Because nuclear motion is slow compared with electronic motion, it is appropriate to apply the Franck-Condon principle to electron transfer. In the classical limit, the energy of an electronic transition within the molecule equals the difference in energy between the two relevant potential energy surfaces evaluated at the point corresponding to the instantaneous nuclear coordinates of the molecule. Consequently, the molecular affinity level and tunneling rate is continuously shifting throughout the trajectory as the molecule translates, rotates, and vibrates. This will be discussed in further detail within the section on dissociative scattering.

Miller et al. conducted a systematic study of the total scattered ion yields of $Fe(C_5H_5)_2^+$ and Cs^+ ions colliding with Si(100) (26). They assumed that the survival probability for Cs^+ scattering with its charge intact was approximately unity. In comparing the scattered Cs^+ ion signal with the total ion intensities from 5–100 eV $Fe(C_5H_5)_2^+$ scattering on Si(100), the authors determined that 90–99.999%, respectively, of the incident ferrocene ions are neutralized.

Kempter and coworkers have used ion impact electron spectroscopy to study the charge transfer dynamics that accompany molecular ions scattering on single-crystal metal surfaces (27–30). Ion impact electron spectroscopy probes the electronic transitions involved in charge transfer by observing the kinetic energy distribution of the associated emitted electrons. For example, an Auger electron will emerge with a kinetic energy less than or equal to the energy of the molecular affinity level minus twice the workfunction. Figure 4a shows the kinetic energy spectra for electrons emitted when O_2^+ ions impact W(110) at 50 eV (29). The observed electron emission provides a signature of which charge transfer or deexcitation mechanisms listed in Figure 4b are operative in the O_2^+/W(110) system. On a clean surface, only Auger capture into the ground ($X^3\Sigma_g^-$) and first excited bound state (a $^1\Delta_g$) are observed. As the workfunction decreases below 4 eV with increasing potassium coverage, Auger deexcitations of the O_2^* (1, 3) Π_g core-excited states and the O_2^* (1, 3) Π_u repulsive valence states are observed, indicating that resonant neutralization populates these excited states as soon as it becomes energetically feasible.

Unfortunately, ion impact electron spectroscopy is blind to processes that do not eject electrons into the vacuum, e.g., resonant neutralization to the ground state or Auger neutralization/deexcitation processes that generate hot electrons at energies below the vacuum level. A less precise but nevertheless informative

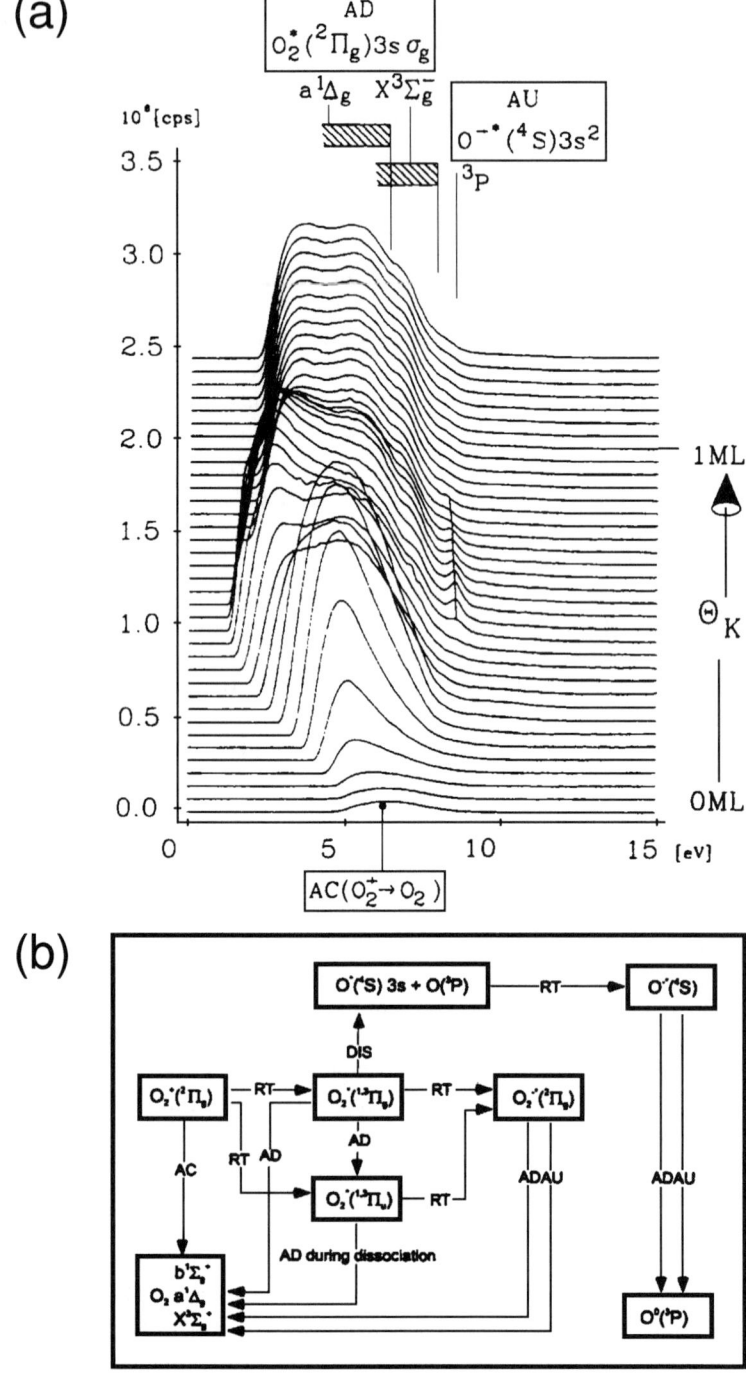

method for deducing the electronic state populated upon neutralization is to follow the dynamics of the nascent neutral molecule as it reacts with and/or captures a second electron from the surface.

Although the electron affinity of an isolated molecule is typically less than the workfunction of a metal surface, the negative ion state becomes more stable as the molecule approaches the surface and interacts with its image charge. The process of electron attachment proceeds similarly to resonant neutralization, except that the affinity level for attaching an electron to a neutral particle drops as the particle approaches the surface, as shown in Figure 5. The freezing distance model applies equally well to electron attachment as it does to neutralization. Consequently, an anion formed at the surface will often lose its electron on the outgoing trajectory, unless it is moving fast enough to escape nonadiabatically (9). Accordingly, negative ions are only observed as scattered products if the component of their translational energy directed along the surface normal exceeds a few electronvolts (25). The prevalence of electron loss along the outgoing trajectory implies that for every negative ion that scatters from the surface, many more neutral products are emerging—often undetected.

On metal and semiconductor surfaces, electron transfer is inextricably linked with a molecular ion's reactivity on a surface, because electron transfer usually precedes surface impact. As such, charge transfer prepares a distribution of molecular states that serve as precursors for subsequent reaction with the surface. The importance of charge transfer to dissociative scattering is illustrated in the following section.

Dissociative Scattering

Unlike dissociative chemisorption, where an incident molecule dissociates upon impact and the resulting fragments bind tightly to the surface, dissociative scattering produces only gas-phase fragments. The mechanisms governing dissociative scattering are often attributed to two types of excitation: mechanical and electronic (31). Mechanical dissociation refers to the impulsive transfer of incident translational energy into ro-vibrational energy that subsequently ruptures one or more molecular bonds. For small molecules, and necessarily for diatomics, mechanical dissociation is prompt, forming products on the timescale of half a vibrational period ($\sim 10^{-14}$ s). In this rapid dissociation event, the fragments are formed in

Figure 4 (*a*) Experimental electron kinetic-energy spectra for slow (50 eV) O_2^+ collisions with W(110), partially covered by K atoms, under grazing incidence (5° with respect to the surface). Spectra are presented as a function of the K coverage; bottom spectra are for clean W(110). (*b*) Schematic diagram of the possible electronic transitions: resonant neutralization (RN), Auger capture (AC), Auger de-excitation (AD), dissociation (DIS), and autodetachment of the atomic Feshbach resonance (AU). Reproduced with permission from Reference (30).

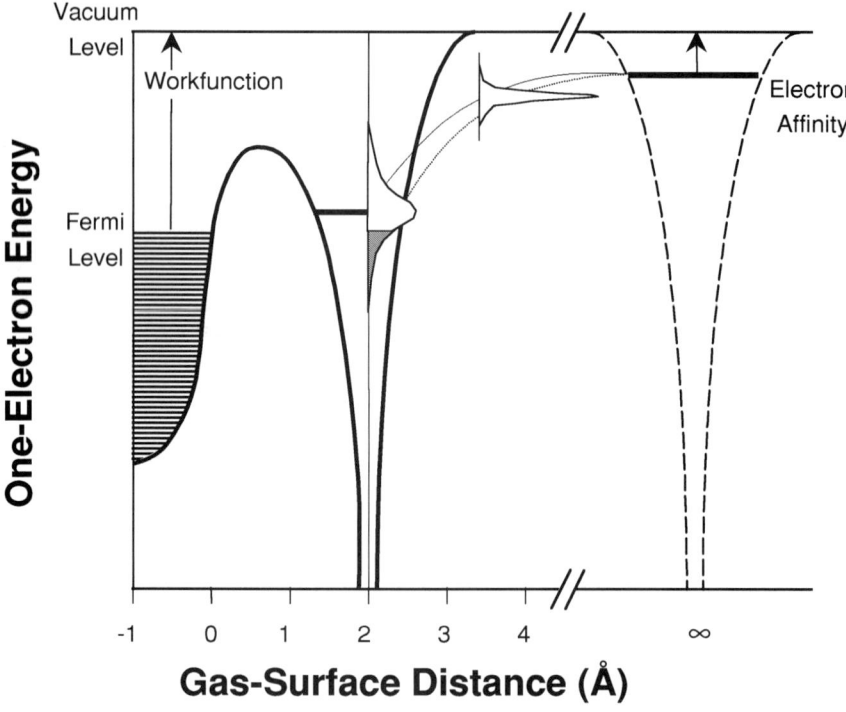

Figure 5 One-electron energy level diagram for negative ion formation near a metal surface. The dashed and solid curves represent the potential energy for an electron when the atom is at infinite and close distances to the surface, respectively. The affinity level for the neutral atom shifts downward and broadens as the atom approaches the surface.

close proximity with the surface where they may undergo additional electron transfer processes before scattering into the gas phase. Researchers have implicated a prompt mechanical dissociation mechanism in the dissociative scattering of O_2^+ on Ag(111) (32, 33), CO^+ and N_2^+ on Pt(100) (34), BF_2^+ on Au (35), SiF^+ and SiF_3^+ on Cu(100) (36), H_2^+ on Ag(111) (37), and NO^+ on GaAs(110) and Ag(111) (9, 38).

In the latter two studies, state-selective preparation of the NO^+ ion allowed investigators to measure the individual roles that collision energy, initial vibrational state, and alignment of the internuclear axis had on the dissociative scattering event. Figure 6 illustrates how the yield of O^-, in the scattering of NO^+ on GaAs(110), depends on both the translational and vibrational energies of the incident NO^+ molecular ions (9). The data indicate an O^- appearance threshold of 25 eV, approximately five times the reaction endoergicity (4.7 eV). This inefficiency for coupling collision energy into motion along the reaction coordinate

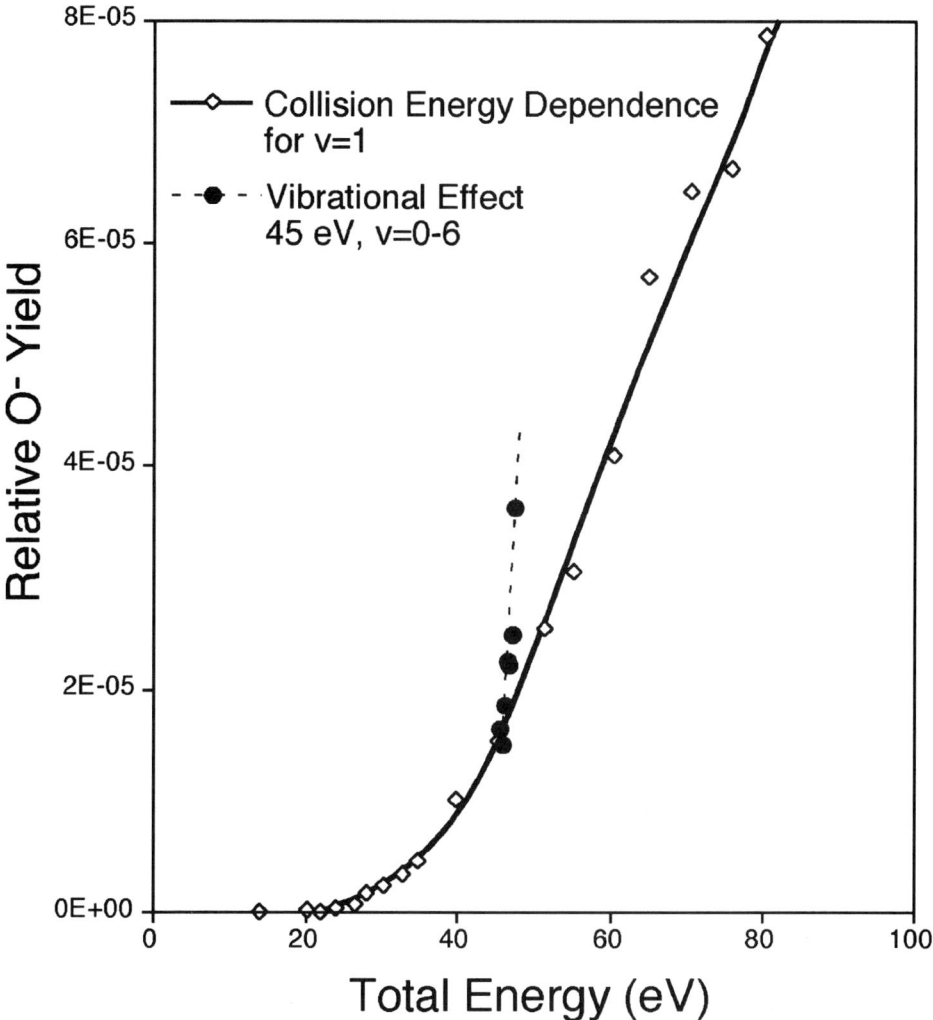

Figure 6 Relative yield of scattered O⁻ fragments as a function of the total energy of $NO^+(X^1\Sigma^+)$ incident on GaAs(110). The solid circles correspond to $NO^+(v=1)$ at various translational energies. The open squares represent vibrational levels, $v = 0$–6, for $NO^+(v)$ incident at a collision energy of 45 eV. Reproduced with permission from Reference (40).

(bond stretching) is underscored by the large transfer of energy into surface vibrations. Approximately 75% of the incident energy is channeled into the surface, although only a small fraction leads to rovibrational excitation of NO. Figure 6 also reveals that the efficacy of initial vibrational energy is one order of magnitude greater than the efficacy of incident translational energy at 45 eV collision energy. This appears intuitively reasonable; after all, vibrational motion is directed along the reaction coordinate, whereas translational energy must be converted into rovibrational energy to affect dissociation.

Surprisingly, classical trajectories failed to predict that translational and vibrational energies should exhibit remarkably different efficiencies for activating dissociation in the NO^+/GaAs(110) system (39). The shortcoming of the classical trajectory approach is that it ignores electron transfer—a process that is highly coupled to vibration as suggested by the dramatically different bond lengths for NO^+, NO, and NO^-. A time-dependent quantum-mechanical model, which explicitly treated both electron transfer and mechanical excitation, quantitatively reproduced the observed behavior (40). The quantum mechanical description reveals the sequence of events that leads to the emergence of O^- ions in the scattering of hyperthermal energy NO^+ on GaAs(110): (*a*) Efficient neutralization of the impinging molecule creates a coherent vibrational wavepacket along the incident trajectory. (*b*) Bond compression upon surface impact leads to a rapid separation of the N and O atoms. (*c*) Electron attachment to the separating diatom occurs while the oxygen atom is still proximate to the surface.

The importance of impulsive bond compression through collisional impact is supported by scattering experiments in which incident NO^+ molecular ions were aligned prior to impact with a Ag(111) surface (38, 41). In this way, the experiments directly determined whether "end-on" or "side-on" orientations of the internuclear axis were preferred for mechanical dissociation. Both the alignment experiments and classical trajectories showed that near end-on collisions had the greatest probability for fragmentation (38).

In contrast to the dynamics of impulsive dissociative scattering for diatomics, mechanical dissociation of large polyatomic molecules can occur over a much longer time scale. A large polyatomic molecule will often scatter from the surface intact, albeit with a significant amount of vibrational excitation. Because of the large number of modes available for accommodation of internal energy, fragmentation will not occur until a critical amount of internal energy becomes localized in a particular bond. Thus a polyatomic molecule may scatter from the surface and travel a significant distance in the gas-phase before unimolecular dissociation transpires. Hanley and coworkers compared the dissociation dynamics for 25 eV $Si(CH_3)_3^+$ scattering on clean Au(111) versus a C_6 self-assembled monolayer (SAM) (8, 42). The investigators measured the degree of dissociation and the kinetic energy distribution of the resulting fragments. Figure 7 shows the kinetic energy and velocity distributions for the dominant scattered ions when $Si(CH_3)_3^+$ scatters on clean Au(111). The high velocity peak in Figure 7*b* corresponds to delayed dissociation in the gas-phase, whereas the low energy peak in Figure 7*a* is consistent with fragmentation occurring while the ion is still in contact with the surface. Molecular dynamics simulations and Rice-Ramsperger-Kassel-Marcus modeling of the data predicted the residence time of the collision complex, the distribution of energy after scattering, and the time scale for dissociation. The brief contact period (20 fs) in which $Si(CH_3)_3^+$ interacts with clean

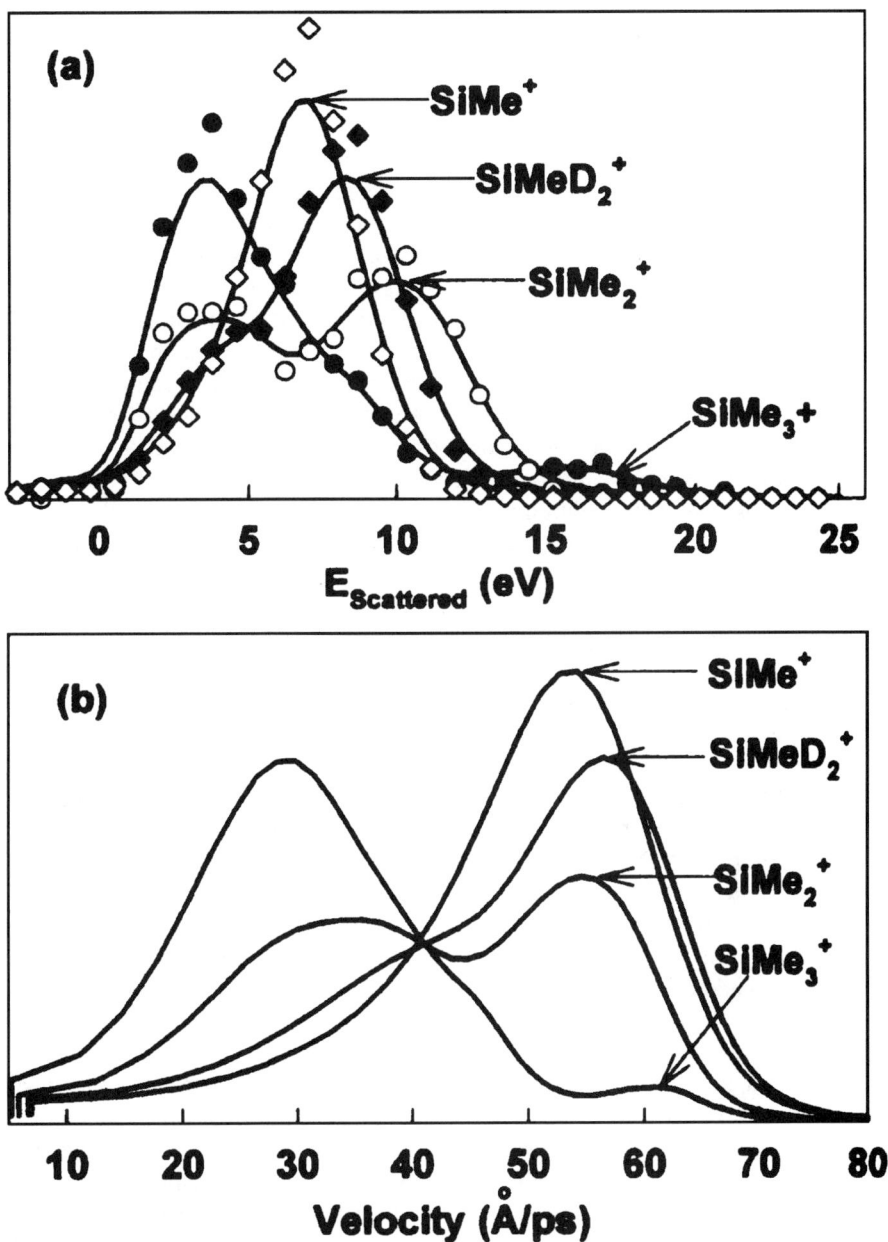

Figure 7 (*a*) Experimental kinetic energy distribution for the dominant scattered ions from 25 eV SiMe$_3^+$ incident on clean Au(111). Normalized data = points, FFT smoothed fit = curves. Arrows denote the features that scale with fragment mass. (*b*) Transformation of the kinetic energy distributions into velocity space. Reproduced with permission from Reference (42).

Au(111) doesn't permit much of the fragmentation to occur near the surface. In contrast, the long residence time (210 fs) of Si(CH$_3$)$_3^+$ on the SAM allows considerable fragmentation to occur while the molecular ion is still in contact with the collisionally deformed SAM.

The dissociation dynamics for metal cluster ions are similar to those for molecular ions. Murakami and coworkers scattered Ag$_x^+$ (x = 1–4) on highly ordered pyrolytic graphite (HOPG) and observed the distribution of fragments (43). At low impact energies, fragmentation arises from unimolecular dissociation of a scattered hot parent cluster after collisional heating. At high collision energies, shattering of the cluster at the point of impact is the dominant fragmentation process (44). Beck et al. observed similar behavior in the scattering of C$_{60}^+$ on a HOPG surface (45). Figure 8 shows the fragment distribution for fullerene cations striking the surface at six different collision energies. At collision energies below 300 eV, the mechanical transfer of translational to vibrational energy produces a hot parent ion that undergoes sequential evaporation of C$_2$ units. Above 300 eV collision energies, rapid disintegration of the parent fullerene ion produces a distribution of small odd- and even-numbered carbon cluster cations. This appearance threshold for shattering corresponds to the point where the cluster receives more than 70 eV of internal energy from surface impact.

Cooks and coworkers compared the extent to which 20 eV Si(C$_2$H$_5$)$_4^+$ ions fragment on a hydrocarbon SAM (H-SAM), a hydroxyl-terminated SAM (HO-SAM) and a fluorocarbon SAM (F-SAM) (46). Figure 9 demonstrates that the extent of fragmentation, which is related to the amount of internal energy transferred during the collision, was greatest for the F-SAM, followed by the HO-SAM, and the H-SAM. This ranking is consistent with a greater momentum transfer as a result of the larger effective mass of the CF$_3$ surface group, compared with the CH$_2$OH and CH$_3$ groups (47). Miller et al. systematically compared the fragmentation distributions of Fe(C$_5$H$_5$)$_2^+$ ions colliding with clean Si(100) versus an alkane thiol SAM film (26, 48). Although the overall yield of scattered cations was lower for the clean Si(100) surface, the dissociation thresholds and degree of ferrocene fragmentation was remarkably similar on the alkane thiol SAM and Si(100) surfaces, indicating that collisions on either surface result in a similar amount of translational to vibrational energy transfer.

Impulsive mechanical dissociation upon surface impact can occur only when the incident molecular ion has enough kinetic energy to transfer a sufficient portion into rovibrational energy. An alternative mechanism for dissociative scattering, one that does not suffer from the same collision energy constraint as impulsive dissociation, involves electronic excitations in the molecule. When a repulsive electronic state within the molecule is populated through neutralization, the molecule can fragment spontaneously. Dissociative neutralization has been implicated as the cause for fragmentation in the scattering of H$_2^+$, N$_2^+$, and O$_2^+$ on Ni(111) (49, 50), H$_2^+$ on Ag(111) (51), H$_2^+$ on Cu(111) (52, 53), and OCS$^+$ on Ag(111) (54).

In addition to dissociative neutralization, electron attachment to a neutral molecule can form an unstable negative ion, which will immediately fragment. Kleyn

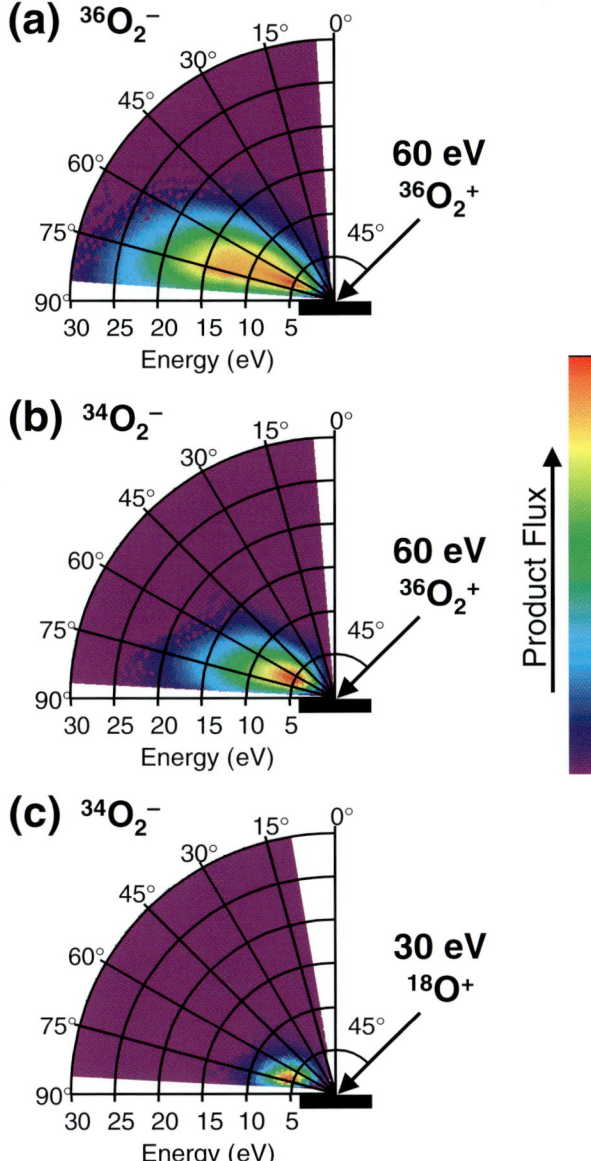

Figure 12 Scattered product distributions for 60 eV O_2^+ incident on silicon oxide, SiO_x. The ions approach at 45° from the horizontal surface. The flux of scattered ions, depicted in a false color scale, is plotted versus the exit energy and angle. Each plot is individually scaled such that the maximum flux is depicted as *red*. (*a*) Charge inversion of scattered $^{36}O_2^+$ yields $^{36}O_2^-$. (*b*) $^{36}O_2^+$ which undergoes a substitution reaction on $Si^{16}O_x$ produces $^{34}O_2^-$. (*c*) As a point of comparison, 30 eV $^{18}O^+$, travelling at the same incident velocity as 60 eV O_2^+, abstracts an oxygen atom from $Si^{16}O_x$ to also form $^{34}O_2^-$.

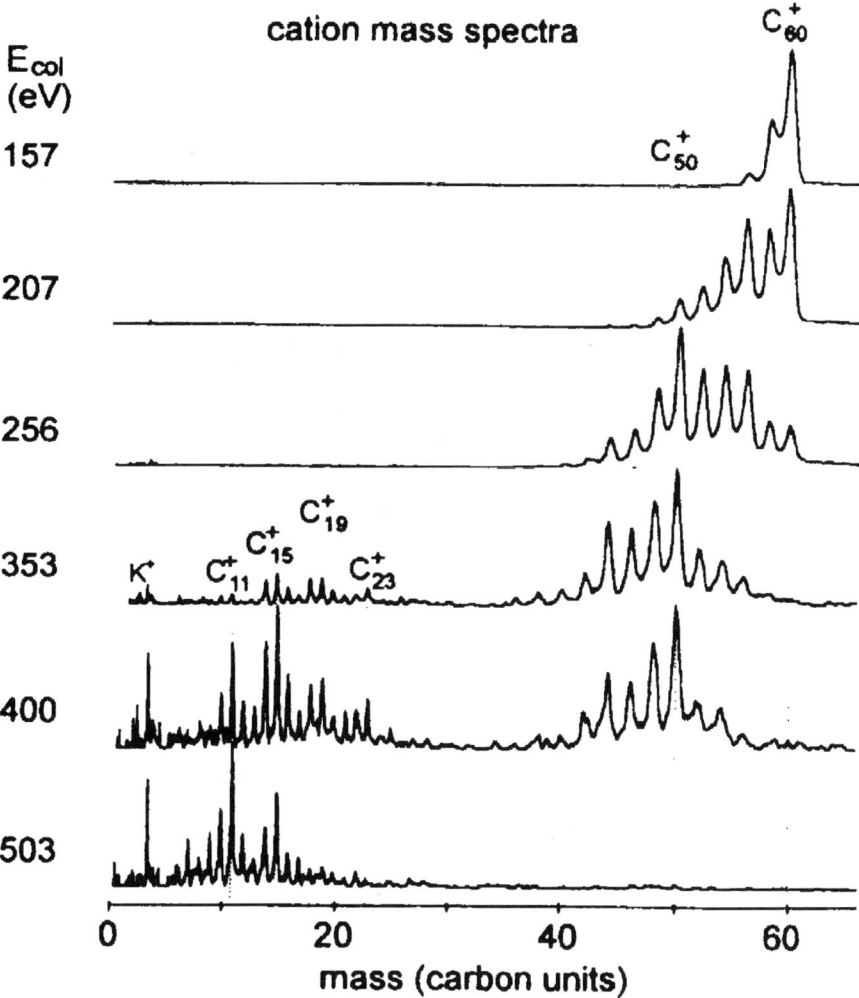

Figure 8 Mass distribution of scattered carbon cluster cations emerging from surface impact of C_{60}^+ on graphite. The fullerene ions were incident at 45° with the collision energies, E_{col}, listed, and the detector was centered at a 90° scattering angle. Reproduced with permission from Reference (45).

and coworkers compared the scattering of CF_3^+ ions on Ag(111) and HOPG graphite surfaces (55, 56). Whereas no parent negative ions survived the collision with Ag(111), an appreciable amount of CF_3^- emerged from the graphite surface. It is argued that Auger neutralization on the latter surface populates the ground state of CF_3, in which fragmentation is minimal. In contrast, resonant neutralization to an excited state of CF_3, followed by electron attachment, leads to efficient

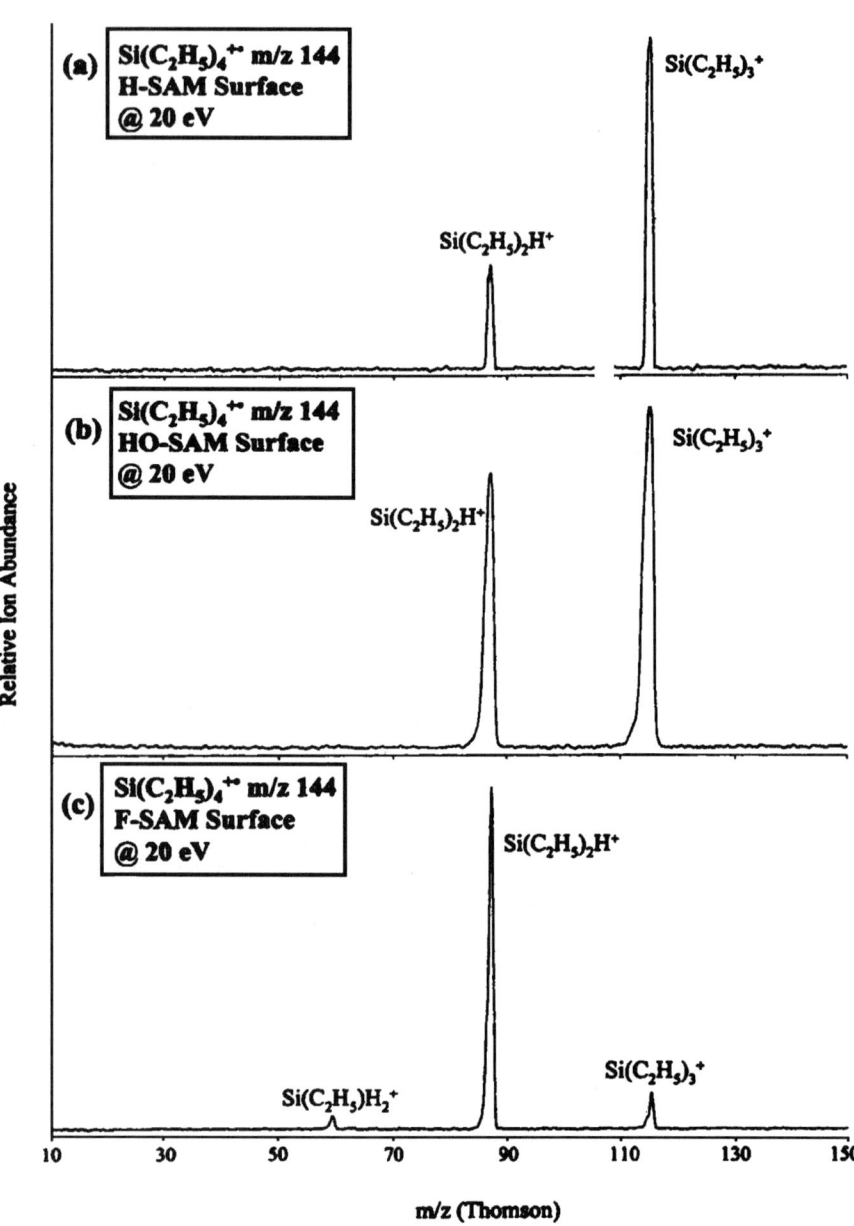

Figure 9 Scattered ion mass spectra recorded upon collision of Si-$(C_2H_5)_4^+$ at (*a*) a H-SAM surface, (*b*) a HO-SAM surface, and (*c*) a F-SAM surface at a collision energy of 20 eV. Reproduced with permission from Reference (46).

dissociation of the parent species on Ag(111). The excited CF_3^{*-} ion has a weakened C–F bond that ruptures upon impact with the Ag(111) surface. Heiland and coworkers argue that dissociative electron attachment is the primary fragmentation mechanism in the scattering of CO_2^+ and CO_2 on Ni(111) and Pd(110) (57–59).

In some systems, both mechanical and electronic dissociative scattering can occur. For example, OCS^+ incident on Ag(111) exhibits at least three different product channels. Figure 10 shows the relative yield of S^-, O^-, and SO^- as a

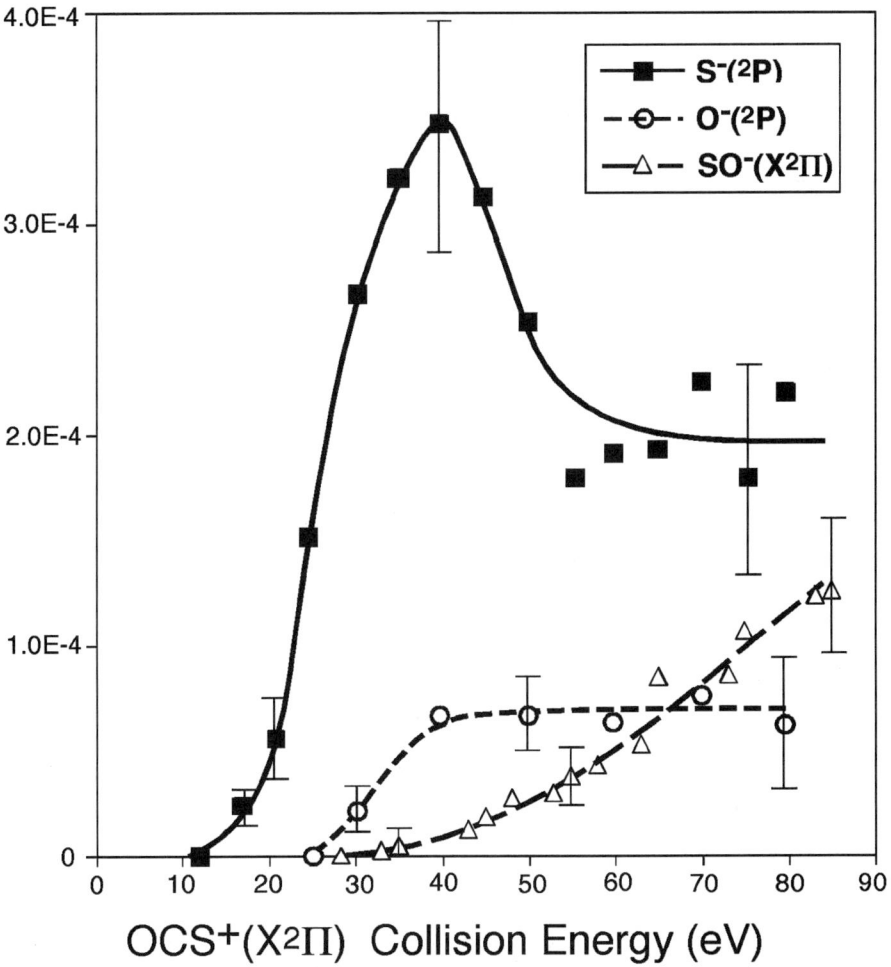

Figure 10 Relative product yields of $S^-(^2P)$, $O^-(^2P)$ and SO^- ($X\ ^2\Pi$) as a function of collision energy for $OCS^+(X\ ^2\Pi)$ incident on Ag(111). Representative error bars indicate 90% confidence limits in the data. Curves are drawn to guide the eye. Reproduced with permission from Reference (54).

function of the OCS$^+$ collision energy (54). Whereas O$^-$ and SO$^-$ appear only above collision energies of 25 eV, the S$^-$ product emerges at collision energies as low as 12 eV. Moreover, a detailed comparison of the scattered S$^-$ velocity distributions, resulting from incident OCS$^+$ versus incident S$^+$, reveals that the sulfur fragment is produced before OCS impacts the surface. The data strongly suggest that resonant neutralization of OCS$^+$ can partially populate the $^1\Delta$ dissociative state of OCS as the molecule approaches the surface. The nascent sulfur fragment goes on to attach an electron in the vicinity of the surface and scatters as S$^-$. This electron-transfer induced dissociation mechanism is in sharp contrast to the other two fragmentation channels for the OCS$^+$/Ag(111) system, where bond cleavage is induced by impulsive momentum transfer at the point of impact.

A signature for identifying the specific dissociation mechanism can be found in the velocity distributions of the scattered parent and daughter ions. When prompt mechanical dissociation occurs at the surface, each fragment emerges with approximately the same mean kinetic energy, whereas in delayed fragmentation, the mean velocities of the fragments are all comparable with the mean velocity of the scattered parent (42). In the case of dissociative neutralization along the inbound trajectory, the nascent fragments impact and scatter from the surface independent of one another. Consequently, the outgoing fragment's kinetic energy distribution is similar to that which would be observed if the fragment were to approach the surface, with an incident velocity equal to the parent molecule's incident velocity, and scatter intact from the surface (54). In summary, the reaction mechanism is most easily identified through a detailed study of the scattering dynamics (31).

Abstraction Reactions

Atom abstraction is an elementary process by which an atom is transferred to or from an incident molecule as the molecule impacts a surface. In the case of hyperthermal molecular ion/surface scattering, the incident molecule is more likely to abstract an atom from the surface. For example, Wu & Hanley established that 32 eV pyridine projectiles incident on pyridine-covered Ag(111) efficiently neutralize on the inbound trajectory; furthermore, a fraction goes on to abstract a proton from the surface and scatter with one additional mass unit (14). Cooks and coworkers demonstrated that polyatomic projectile ions, such as pyrazine and pyrene, can abstract H, F, CH$_3$, or C$_2$H$_3$ fragments from self-assembled monolayers (60).

Wysocki and coworkers reported a fascinating result in the dependence of the abstraction probability on the structure of the SAM (61). These authors observed that 20 eV pyrene ions more easily abstract H-atoms if the SAM alkanethiolates contain an odd number rather than an even number of alkyl carbons (see Figure 11). Pyrene ions exhibit the opposite preference in the abstraction of methyl groups. The authors argue that the odd-even effect can be attributed to the alternating cant angle of the terminal methyl group on the tilted hydrocarbon chains. When an even number of aklyl carbon atoms exist in each SAM molecule, the terminal methyl groups are directed towards the vacuum in an orientation favorable for

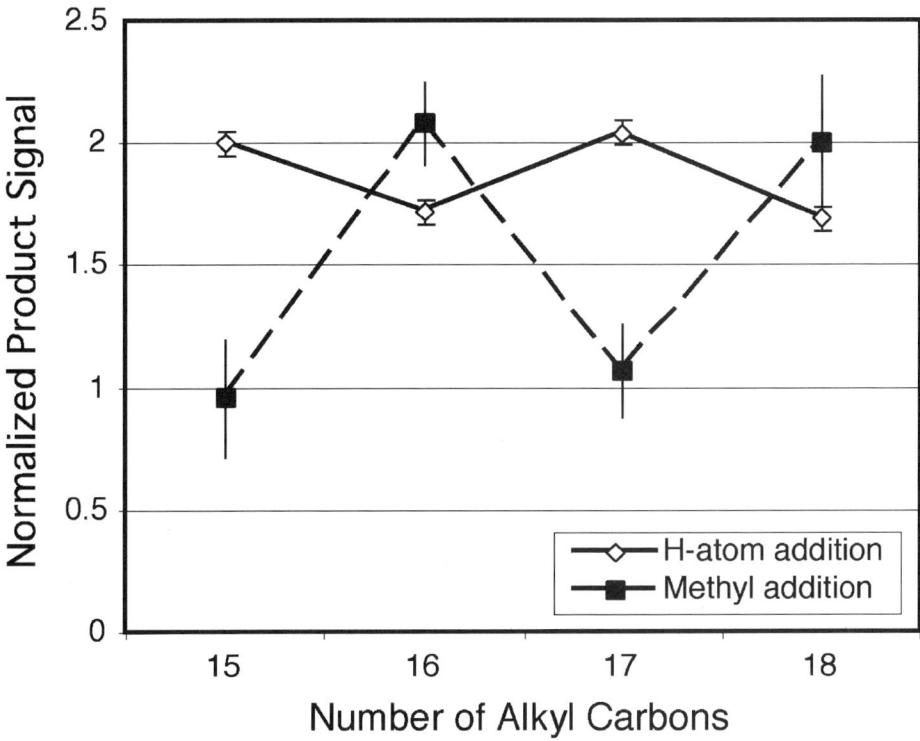

Figure 11 Modulation of scattered ion product signal with alkyl chain-length in the SAM layer. A beam of 20 eV pyrene cations is incident on a family of 4-(4-alkoxyphenyl) benzenethiol self-assembled monolayer films. Adapted from Reference (61).

methyl abstraction (62). In contrast, the terminal C–C bond lies nearly parallel to the surface, in a geometry favorable for hydrogen abstraction, when the SAM thiols contain an odd number of alkyl carbons.

The strong binding of oxygen on aluminum makes O-atom abstraction on aluminum oxide an improbable event. Notwithstanding, Maazouz et al. observed O-atom abstraction when NO^+ was incident on oxidized aluminum (63). The authors measured the product yield and translational energy distribution, as a function of the NO^+ collision energy. The experimental data are consistent with NO_2^- being formed by a three-step mechanism: Incident NO^+ is neutralized close to the surface, nascent NO directly impacts an adsorbed oxygen atom, and O^- is abstracted by NO to form NO_2^- via an Eley-Rideal (ER) mechanism. This represents the first demonstration of a molecular ion abstracting an oxygen atom from a metal oxide surface.

Tzvetkov et al. studied the abstraction of oxygen atoms from silicon oxide by incident oxygen ions (64). A Si(100) wafer, exposed to 60 eV O^+ or O_2^+ ions, develops a thin oxide film less than 30 Å thick (65). Prolonged exposure to the ion beam does not result in a thicker oxide film, because the surface is being eroded at the same rate that it is oxidized. Scattering experiments performed within this steady-state regime revealed the complex behavior associated with ion beam oxidation.

For incident O_2^+ ions on silicon oxide, both O^- and O_2^- product channels were detected (T. Tzvetkov, X. Qin, C. Quinteros, D.C. Jacobs, in preparation). Isotopic labeling techniques helped to define the origin of each oxygen atom in the scattered products. Initially, an isotopically pure $Si^{16}O_x$ film was grown using a mass-selected $^{16}O^+$ beam. Following the oxide growth phase, scattering commenced with a mass-selected $^{36}O_2^+$ beam. Both $^{16}O^-$ and $^{18}O^-$ product ions emerged from $^{36}O_2^+$ bombardment, consistent with sputtering and dissociative scattering pathways, respectively. The O_2^- products were comprised of $^{32}O_2^-$, $^{34}O_2^-$ and $^{36}O_2^-$ signals. Whereas the $^{32}O_2^-$ and $^{36}O_2^-$ species could be assigned to sputtering and scattering processes, respectively, the $^{34}O_2^-$ signal must correspond with a substitution reaction at the surface. Figure 12b (see color insert) shows the angle-resolved energy distribution of $^{34}O_2^-$ ions produced in the scattering of 60 eV $^{36}O_2^+$ on $Si^{16}O_x$. The $^{34}O_2^-$ products emerge in the specular direction with a mean energy of 7.8 eV. The forward scattering behavior is consistent with $^{34}O_2^-$ being formed in a direct collision with the surface. In contrast, if the projectile ion had encountered multiple bounces on the surface, memory of its momentum direction would be lost, and the $^{34}O_2^-$ product would emerge at an angle close to the surface normal.

It is interesting to compare the angle-resolved energy distribution for the $^{34}O_2^-$ substitution product (Figure 12b) with that for the directly scattered parent, $^{36}O_2^-$ (Figure 12a, see color insert). The shapes of the two angular distributions are similar, but the mean kinetic energy of $^{34}O_2^-$ is 70% of that for the directly scattered $^{36}O_2^-$ species. It is conceivable that $^{34}O_2^-$ could be formed in either a stepwise or concerted fashion. A stepwise approach would first involve dissociation of the incident $^{36}O_2^+$ projectile into ^{18}O fragments, which then go on to abstract ^{16}O from the surface. As a point of comparison, scattering 30 eV $^{18}O^+$ projectiles (same incident velocity as $^{36}O_2^+$ projectiles) on the identical $Si^{16}O_x$ surface yielded $^{34}O_2^-$ abstraction products (Figure 12c, see color insert) with significantly less kinetic energy than that observed in Figure 12b. Consequently, both of the O-atoms in the $^{36}O_2^+$ projectile must be involved in the transition state for the substitution reaction. The $^{34}O_2^-$ product is formed by a direct concerted substitution of ^{18}O, from the $^{36}O_2^+$ projectile ion, with ^{16}O from the $Si^{16}O_x$ surface. In addition to a substitution of atoms on the molecular projectile, isotopic analysis of the topmost layer of the surface oxide reveals facile replacement of ^{16}O-atoms in the lattice with ^{18}O as a result of $^{36}O_2^+$ bombardment.

Ion Deposition and Surface Modification

Selective control of the physical and chemical properties of interfaces is critical to a variety of technologies. The following examples demonstrate how mass-selected molecular and cluster ions can be deposited on or react with substrates as a means to grow thin films of novel materials.

The bombardment of Si(100) with hyperthermal N_2^+ ions produces a Si_3N_4 layer, approximately 5 Å thick (66, 67). The initial reaction probability for N_2^+ on Si(100) increases monotonically from 0 to 0.25 as the collision energy increases from 0 to 25 eV; thereafter, the reaction probability remains relatively constant up to collision energies of 100 eV. Park et al. (66) argue that neutralization and dissociation precedes nitridation. Furthermore, nitrogen removal via N-N recombination competes with nitridation, eventually leading to an overall steady-state behavior at saturation. Similarly, the deposition of 10–100 eV CO^+ ions on Si(111) proceeds through dissociation of the incident molecule subsequent to neutralization (68). Prolonged CO^+ exposure yields a stable mixed carbide and oxide layer on the silicon substrate. Rabalais and coworkers studied the growth of TiC films through energetic deposition of $TiCl_x^+$ (x = 1, 3) on HOPG graphite (69). They found that thicker films can be grown using higher beam energies. The experiments also established that $TiCl^+$ produces a more stoichiometric titanium carbide layer than does $TiCl_3^+$.

Lau & Kwok reacted hyperthermal energy, CH_3^+ ions with a 40 Å-thick film of InP (70). At 20 eV collision energies, CH_3^+ fragmented on the 350 K surface, damaging the top 5–10 Å of the InP film and forming a thick carbon overlayer. In contrast, 3 eV CH_3^+ ions did not dissociate and instead formed methyl indium that subsequently desorbed from the 350 K surface. Consequently, prolonged exposure to 3 eV CH_3^+ ions had the effect of etching the InP film. In a similar study, Bello et al. bombarded Si(100) and oxidized silicon with hyperthermal CF^+ ions (71). At 2 eV collision energies, molecular adsorption and fluorination of the clean silicon occurred; however, no reaction transpired on silicon oxide. As the collision energy increased to 20 eV, dissociation became more prevalent, with fluorocarbon accumulation on the silicon oxide surface exceeding that on the clean Si(100) surface. Under 100 eV CF^+ bombardment, a compound SiC and SiF layer formed on both the clean Si(100) and oxidized silicon surfaces. These two studies underscore the important role that collision energy plays in determining the type of surface modification induced by ion beam exposure.

Hanley and coworkers utilized 25–100 eV CF_3^+ and $C_3F_5^+$ ions to modify a polystyrene surface (72, 73). The ions were deposited on and reacted with the polystyrene surface to promote crosslinking and produce fluorocarbon functional groups, whose structure and composition depended on the ion, its collision energy and fluence. Consequently, the nature of the ion-bombarded surface could be continuously varied from organic to inorganic. Cooks and coworkers employed $ClC(CN)_2^+$ ions as chemical reagents to modify fluorocarbon SAM surfaces (74). Both the Cl and CN groups on the incident ion were capable of undergoing halogen (or pseudohalogen) exchange with F-atoms on the SAM.

When it is critical that the molecular projectile be deposited intact, a soft-landing approach is preferable. Here, the incident molecular ions are deposited at energies below their dissociation threshold. Cowin and coworkers soft-landed

D_3O^+ ions on crystalline D_2O ice films to study the transport of hydronium ions through water ice (75). Ion deposition at 1 eV accurately placed D_3O^+ on top of the ice multilayer. The authors were able to track the extent to which hydronium ions diffused into the ice layer by using a Kelvin probe. They demonstrated that hydronium ions have very little mobility in ice at temperatures below 160 K.

Cooks and coworkers utilize a soft-landing approach to deposit molecular ions intact onto self-assembled hydrocarbon and fluorocarbon monolayers (76). The trapped ions are quite stable within the hydrocarbon layer for a period of days, even at room temperature under atmospheric conditions. They find, for example, that 8 eV N,N-dimethyl-p-toluidine cations will embed into a hydrocarbon SAM with little or no fragmentation.

The catalytic properties of small metal clusters can be significantly enhanced relative to those for a macroscopic metal surface (77). The technological challenge lies in finding a reliable procedure for depositing size-selected metal clusters on a solid support. One successful approach has utilized ion beams to deliver mass-selected metal and semiconductor clusters to the surface target. If the cluster ions impact the surface at low collision energies, they can be deposited intact, producing size-selected metal islands with unique physical and chemical properties. Systems studied include Si_n^+ (n = 10, 13, 40–50) on amorphous carbon (78), C_n^+ (n = 1, 2, 4) on Ag (79), Si_{10}^+ on Au(001) (80), Sb_n^+ (n = 3, 4, 8) on HOPG graphite (81), Cu_n^+ (n = 1, 2) on Mo (82), Ag_n^+ (n = 1–10, 50–400) on HOPG (83–85), Si_n^+ (400 < n < 9000) on CaF_2 and LiF (86, 87), Ag_{19}^+ on Pd(100) (88), Ni_{11}^+ on MgO(100) (89, 90), and Pd_n^+ (1 ≤ n ≤ 30) on MgO(100) (77).

The efficiency at which metal clusters are deposited on a surface depends strongly on the collision energy and the cluster size. Figure 13 shows the total amount of silver atoms deposited on a HOPG graphite surface for five different Ag_n^+ cluster sizes (83). In Figure 13a, the data is plotted versus the collision energy of the cluster. Using an alternative energy scale, Figure 13b shows the data as a function of the energy per atom in the cluster, where the collision energy has been divided by n, the number of atoms in the cluster. Figure 13c uses a third energy scale in which the collision energy is divided by $n^{2/3}$. This latter scale corresponds to the energy density, where the collision energy is divided by the contact area between the cluster and the surface. The data for different cluster sizes coalesce onto virtually a single curve in Figure 13c, indicating that energy density is the important parameter in predicting deposition efficiency. Furthermore, the data in Figure 13c exhibit three different regimes. At low energy, the cluster lands softly

Figure 13 Integrated photoelectron intensity for the Ag $3d_{5/2}$ core level. Deposition of Ag_1 (*solid circles*), Ag_3 (*open circles*), Ag_5 (*solid triangles*), Ag_7 (*open triangles*), and Ag_9 (*solid squares*) clusters on graphite as a function of collision energy. (*a*) Data plotted on an "energy per cluster" scale. (*b*) Data replotted on an "energy per atom" scale. (*c*) Data replotted on a normalized energy scale where the scaling factor is $1/n^{2/3}$ for Ag^n. Reproduced with permission from Reference (83).

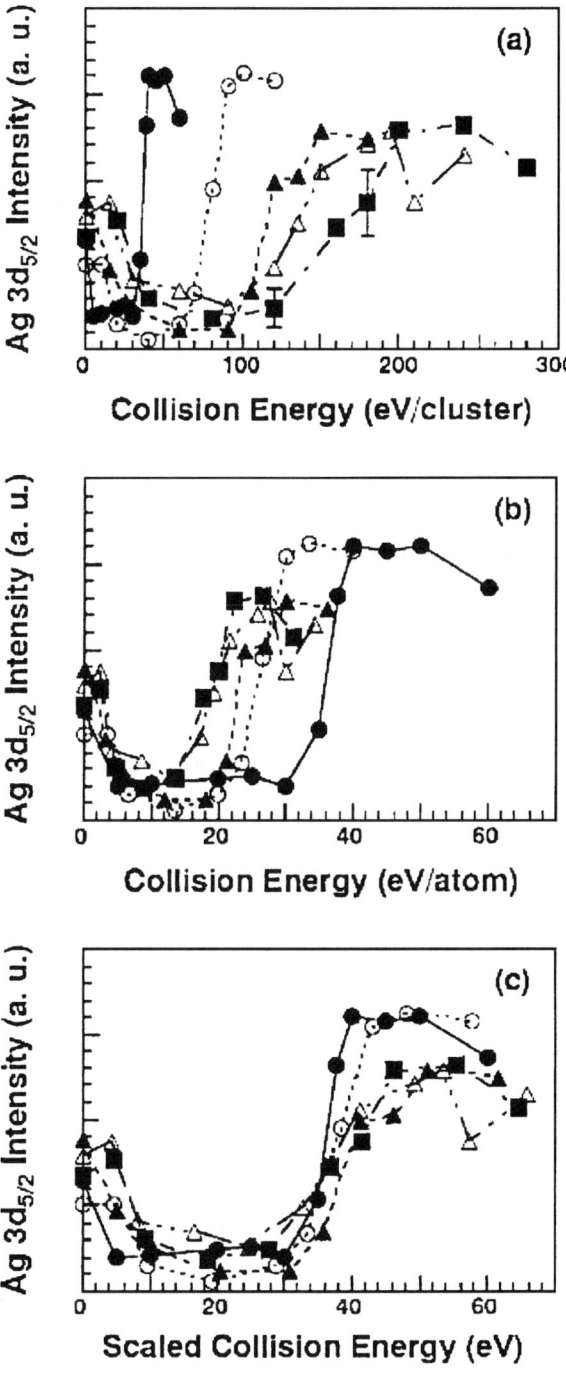

on the surface and deposits intact. At moderate energies, the cluster rebounds and escapes from the surface without sticking. At high energies, the cluster shatters and its atoms implant within the solid.

SUMMARY

This review has highlighted some of the many ways in which molecular ions can induce chemical change at the vacuum/solid interface. The hyperthermal energy of an incident ion is sufficient to overcome the chemical barriers to most surface reactions, leading to both a competition and a synergy between various fundamental processes/mechanisms. Electron transfer can populate the neutral or anionic charge states of the projectile ion just before it strikes the substrate, often making the nascent species more reactive with the surface target. Impulsive energy transfer during surface impact can promote prompt or delayed fragmentation of the molecular projectile. The incident molecule or fragment may then undergo an abstraction reaction, whereby an atom or chemical moiety from the surface adlayer is removed and attached to the scattered particle. Finally, the surface itself is often transformed through ion beam deposition and/or modification. The technological application of molecular ion/surface reactions shows great promise for impacting the microelectronics industry, optimizing spacecraft design, and developing new advances in mass spectrometry.

ACKNOWLEDGMENTS

The author gratefully acknowledges the support of the National Science Foundation and the Air Force Office of Scientific Research for funding his ongoing work in the field of ion/surface scattering. Tochko Tzvetkov, Mostafa Maazouz, Toni Barstis, Patricia Maazouz, Xiangdong Qin, and Jennifer Gagliardi are thanked for their critical reading of this manuscript.

Visit the Annual Reviews home page at www.annualreviews.org

LITERATURE CITED

1. Street SC, Xu C, Goodman DW. 1997. *Annu. Rev. Phys. Chem.* 48:43–68
2. Rabalais JW, ed. 1994. *Low-Energy Ion Surface Interactions*. New York: Wiley. 594 pp.
3. Winters HF, Coburn JW. 1992. *Surf. Sci. Rep.* 14:161–269
4. Martinu L, Poltras D. 2000. *J. Vac. Sci. Tech. A* 18:2619–45
5. Grill V, Shen J, Evans C, Cooks RG. 2001. *Rev. Sci. Instrum.* 72:3149–79
6. Murad E. 1998. *Annu. Rev. Phys. Chem.* 49:73–98
7. Jacobs DC. 2000. In *Chemical Dynamics in Extreme Environments*, ed. R Dressler, pp. 349–89. Singapore: World Sci.
8. Hanley L, Lim H, Schultz DG, Garbis S, Yu C, et al. 1999. *Nucl. Instrum. Methods B* 157:174–82
9. Martin JS, Greeley JN, Morris JR, Jacobs DC. 1992. *J. Chem. Phys.* 100:6791–12
10. Anderson SL. 1992. *Advances in Chemical*

Physics, Vol. 82, Part 1, ed. C-Y Ng, M Baer, pp. 177–212. New York: Wiley
11. Xie J, Zare RN. 1989. *Chem. Phys. Lett.* 159:399–405
12. Koenders BG, Kuik GJ, Drabe KE, DeLange CA. 1988. *Chem. Phys. Lett.* 147:310–14
13. Biesecker JP, Ellison GB, Wang H, Iedema MJ, Tsekouras AA, Cowin JP. 1998. *Rev. Sci. Instrum.* 69:485–95
14. Wu Q, Hanley L. 1993. *J. Phys. Chem.* 97:2677–85
15. Boyd KJ, Lapicki A, Aizawa M, Anderson SL. 1998. *Rev. Sci. Instrum.* 69:4106–15
16. Schwartz D. 2001. *Annu. Rev. Phys. Chem.* 52:107–37
17. Koppers WR, Tsumori K, Beijersbergen JHM, Weeding TL, Kistemaker PG, Kleyn AW. 1998. *Int. J. Mass Spect. Ion Proc.* 174:11–34
18. Beck RD, Weis P, Bräuchle G, Rockenberger J. 1995. *Rev. Sci. Instrum.* 66:4188–97
19. Corr D, Jacobs DC. 1992. *Rev. Sci. Instrum.* 63:1969–72
20. Maazouz M, Morris JR, Jacobs DC. 2000. In *Imaging in Chemical Dynamics*, ed. A Suits, R Continetti, pp. 139–50. Washington, DC: Am. Chem. Soc.
21. Vandeweert E, Meserole CA, Sostarecz A, Dou Y, Winograd N, Postawa Z. 2000. *Nucl. Instrum. Methods Phys. Res. B* 164:820–26
22. Woodruff DP, Delchar TA. 1990. *Modern Techniques of Surface Science*. Cambridge, UK: Cambridge Univ. Press. 453 pp.
23. Kasi SR, Kang H, Sass CS, Rabalais JW. 1989. *Surf. Sci. Rep.* 10:1–104
24. Lorente N, Teillet-Billy D, Gauyacq J-P. 1999. *Nucl. Instrum. Methods B* 157:1–10
25. Los J, Geerlings JJC. 1990. *Phys. Rep.* 190:133–90
26. Miller SA, Riederer DE, Cooks RG, Cho WR, Lee HW, Kang H. 1994. *J. Phys. Chem.* 98:245–51
27. Müller H, Gador D, Kempter V. 1994. *Surf. Sci.* 318:403–12
28. Müller H, Gador D, Kempter V. 1995. *Surf. Sci.* 358:313–21
29. Stracke P, Wiegershaus F, Krischok S, Kempter V. 1998. *Surf. Sci.* 396:212–20
30. Krischok S, Müller H, Kempter V. 1999. *Nucl. Instrum. Methods B* 157:198–207
31. Morris JR, Kim G, Barstis TLO, Mitra R, Quinteros CL, Jacobs DC. 1997. *Nucl. Instrum. Methods B* 125:185–93
32. Reijnen PHF, van den Hoek PJ, Kleyn AW, Imke U, Snowdon KJ. 1989. *Surf. Sci.* 221:427–53
33. Kleyn AW. 1992. *J. Phys. C* 4:8375–94
34. Akazawa H, Murata Y. 1990. *J. Chem. Phys.* 92:5560–68
35. Shen YG, Bello I, Lau WM. 1993. *Nucl. Instrum. Methods B* 73:35–40
36. Yamamoto H, Baba Y, Sasaki TA. 1996. *Appl. Surf. Sci.* 100/101:333–37
37. van Slooten U, Andersson DR, Kleyn AW, Gislason EA. 1992. *Surf. Sci.* 274:1–20
38. Greeley JN, Martin JS, Morris JR, Jacobs DC. 1995. *J. Chem. Phys.* 102:4996–5011
39. Martin JS, Feranchak BT, Morris JR, Greeley JN, Jacobs DC. 1996. *J. Phys. Chem.* 100:1689–97
40. Qian J, Jacobs DC, Tannor DJ. 1995. *J. Chem. Phys.* 103:10764–78
41. Greeley JN, Martin JS, Morris JR, Jacobs DC. 1994. *Surf. Sci.* 314:97–106
42. Schultz DG, Hanley L. 1998. *J. Chem. Phys.* 109:10976–83
43. Tai Y, Yamaguchi W, Maruyama Y, Yoshmura K, Murakami J. 2000. *J. Chem. Phys.* 113:3808–13
44. Raz T, Even U, Levine RD. 1995. *J. Chem. Phys.* 103:5394–409
45. Beck RD, Warth C, May K, Kappes MM. 1996. *Chem. Phys. Lett.* 257:557–62
46. Wade N, Evans C, Pepi F, Cooks RG. 2000. *J. Phys. Chem. B* 104:11230–37
47. Callahan J, Somogyi A, Wysocki VH. 1993. *Rapid Commun. Mass Spectrom.* 7:693–99
48. Kang H, Lee HW, Cho WR, Lee SM. 1998. *Chem. Phys. Lett.* 292:213–17
49. Willerding B, Heiland W, Snowdon KJ. 1984. *Phys. Rev. Lett.* 53:2031–34

50. Willerding B, Snowdon KJ, Imke U, Heiland W. 1986. *Nucl. Instrum. Methods B* 13:614–18
51. Reijnen PHF, van den Hoek PJ, Kleyn AW, Imke U, Snowdon KJ. 1989. *Surf. Sci.* 221:427–53
52. Rechtien JH, Harder R, Herrmann G, Röthig C, Snowdon KJ. 1992. *Surf. Sci.* 270:213–18
53. Rechtien JH, Harder R, Herrmann G, Snowdon KJ. 1992. *Surf. Sci.* 272:240–46
54. Morris JR, Kim G, Barstis TLO, Mitra R, Jacobs DC. 1997. *J. Chem. Phys.* 107:6448–59
55. Gleeson MA, Kropholler M, Kleyn AW. 2000. *Appl. Phys. Lett.* 77:1096–98
56. Koppers WR, Beijersbergen JHM, Tsumori K, Weeding TL, Kistemaker PG, Kleyn AW. 1996. *Phys. Rev. B* 53:11207–10
57. Schmidt K, Franke H, Schlathölter T, Narmann CHA, Heiland W. 1994. *Surf. Sci.* 301:326–36
58. Schlathölter T, Heiland W. 1995. *Surf. Sci.* 323:207–18
59. Schlathölter T, Heiland W. 1995. *Surf. Sci.* 333:311–16
60. Morris MR, Riederer DE, Winger BE, Cooks RG, Ast T, Chidsey CED. 1992. *Int. J. Mass Spectrom. Ion Proc.* 122:181–217
61. Angelico VJ, Mitchell SA, Wysocki VH. 2000. *Anal. Chem.* 72:2603–8
62. Bryant MA, Pemberton JE. 1991. *J. Am. Chem. Soc.* 113:8284–93
63. Maazouz M, Barstis TLO, Maazouz PL, Jacobs DC. 2000. *Phys. Rev. Lett.* 84:1331–34
64. Quinteros CL, Tzvetkov T, Qin X, Jacobs DC. 2001. *Nucl. Instrum. Methods Phys. Res. B* 182:187–92
65. Todorov SS, Fossum ER. 1998. *J. Vac. Sci. Technol. B* 6:466–69
66. Park KH, Kim BC, Kang H. 1992. *J. Chem. Phys.* 97:2742–49
67. Baek DH, Kang H, Chung JW. 1994. *Phys. Rev. B* 49:2651–57
68. Kim BC, Hahn JR, Kang H. 1995. *Nucl. Instrum. Methods Phys. Res. B* 106:137–41
69. Ada ET, Lee SM, Lee H, Rabalais JW. 2000. *J. Phys. Chem. B* 104:5132–38
70. Lau WM, Kwok RWM. 1998. *Int. J. Mass Spectrom. Ion Proc.* 174:245–52
71. Bello I, Chang WH, Lau WM. 1994. *J. Vac. Sci. Tech.* 12:1425–30
72. Ada ET, Kornienko O, Hanley L. 1998. *J. Phys. Chem. B* 102:3959–66
73. Wijesundara MB, Ji Y, Ni B, Sinnott SB, Hanley L. 2000. *J. App. Phys.* 88:5004–16
74. Shen J, Grill V, Evans C, Cooks RG. 1999. *J. Mass Spectrom.* 32:354–63
75. Tsekouras AA, Iedema MJ, Ellison GB, Cowin JP. 1998. *Int. J. Mass Spectrom. Ion Proc.* 174:219–30
76. Shen J, Yim YH, Feng B, Grill V, Evans C, Cooks RG. 1999. *Int. J. Mass Spectrom.* 182/183:423–35
77. Abbet S, Sanchez A, Heiz U, Schneider WD. 2001. *J. Catalysis* 198:122–27
78. Bower JE, Jarrold MF. 1992. *J. Chem. Phys.* 97:8312–21
79. Haslett TL, Fedrigo S, Moskovits M. 1995. *J. Chem. Phys.* 103:7815–19
80. Kuk Y, Jarrold MF, Silverman PJ, Bower JE, Brown WL. 1989. *Phys. Rev. B* 39:11168–70
81. Kaiser B, Bernhardt TM, Rademann K. 1998. *Appl. Phys. A* 66:S711–14
82. Boyd KJ, Lapicki A, Aizawa M, Anderson SL. 1999. *Nucl. Instrum. Methods B* 157:144–54
83. Yamaguchi W, Yoshimura K, Tai Y, Maruyama Y, Igarashi K, et al. 2000. *J. Chem. Phys.* 112:9961–66
84. Goldby IM, Kuipers L, von Issendorff B, Palmer RE. 1996. *Appl. Phys. Lett.* 69:2819–21
85. Carroll SJ, Weibel P, von Issendorff B, Kuipers L, Palmer RE. 1996. *J. Phys. Condens. Matter* 8:L617–24
86. Ehbrecht M, Kohn B, Huisken F, Laguna MA, Paillard V. 1997. *Phys. Rev. B* 56:6958–64
87. Laguna MA, Paillard V, Kohn B, Ehbrecht

M, Huisken F, et al. 1999. *J. Luminescence* 80:223–28
88. Félix C, Vandoni G, Massobrio C, Monot R, Buttet J, Harbich W. 1998. *Phys. Rev. B* 57:4048–52
89. Heiz U, Vanolli F, Trento L, Schneider WD. 1997. *Rev. Sci. Instrum.* 68:1986–94
90. Vanolli F, Heiz U, Schneider WD. 1997. *Chem. Phys. Lett.* 277:527–31

MOLECULAR THEORY OF HYDROPHOBIC EFFECTS:
"She is too mean to have her name repeated."*

Lawrence R. Pratt
Theoretical Division, Los Alamos National Laboratory, Los Alamos, New Mexico 87545; e-mail: lrp@lanl.gov

Key Words aqueous solutions, biomolecular structure, statistical thermodynamics, quasichemical theory, pressure denaturation

■ **Abstract** This paper reviews the molecular theory of hydrophobic effects relevant to biomolecular structure and assembly in aqueous solution. Recent progress has resulted in simple, validated molecular statistical thermodynamic theories and clarification of confusing theories of decades ago. Current work is resolving effects of wider variations of thermodynamic state, e.g., pressure denaturation of soluble proteins, and more exotic questions such as effects of surface chemistry in treating stability of macromolecular structures in aqueous solution.

INTRODUCTION

In the past several years, there has been a breakthrough, associated with the efforts of a theoretical collaboration at Los Alamos (1–11) but with important antecedents (12–16), on the problem of the molecular theory of hydrophobic effects. That breakthrough is the justification for this review.

One unanticipated consequence of that work has been the clarification of the Pratt-Chandler theory (17). Judged empirically, that theory was not less successful than is typical of molecular theories of liquids. But the supporting theoretical arguments had never been compelling and engendered significant confusion. That confusion was signaled already in 1979 by the view (18): "The reason for the success of their theory may well be profound, but could be accidental. We cannot be sure which." Today, the correct answer is both, though accidental first. In an amended form, it is a compelling theory. In reviewing these developments, a significant volume of intervening theoretical work must eventually be viewed again in this new light. In addition, the work that clarified the Pratt-Chandler theory suggested several improvements and extensions, and was deepened by the parallel development of the molecular quasichemical theory of solutions (11, 19–23). On

---
*W. Shakespeare: All's well that ends well.

this basis, I predict an extended period of consolidation of the theory of these systems and inclusion of more realistic, interesting, and exotic instances.

This review seeks scholarly patience in addressing the foundations without prejudging speculations ellicited by the fascinating biophysical motivations. Thus, physical chemists working at strengthening those foundations are the audience for this review.

Nevertheless, clarity of the biophysical goals is important. Thus, an uncluttered expression of the motivation and the basic problem is essential. The molecular theory of hydrophobic effects is an unsolved facet of a molecular problem foundational to biophysics and biochemistry: the quantitative molecular scale understanding of the forces responsible for structure, stability, and function of biomolecules and biomolecular aggregates. I estimate that the term hydrophobic appears in every biophysics and biochemistry textbook. An intuitive definition of hydrophobic effects is typically assumed at the outset. Hydrophobic effects are associated with demixing under standard conditions of oil-like materials from aqueous solutions. A more refined appreciation of hydrophobic effects acknowledges that they are a part of a subtle mixture of interactions that stabilize biomolecular structures in aqueous solution over a significant temperature range while permitting sufficient flexibility for the biological function of those structures.

Preeminent characteristics of these hydrophobic interactions are temperature dependences and concurrent entropies that can be exemplified by cold denaturation of soluble proteins (24). If hydrophobic effects stabilize folded protein structures, then folding upon heating suggests that hydrophobic interactions become stronger with increasing temperature through this low temperature regime. This is a counterintuitive observation.

A primitive correct step in relieving this contrary intuition is the recognition that water molecules of the solution participate in this folding process (25). Specific participation by small numbers of water molecules is not unexpected but also isn't the mark of hydrophobic effects. For hydrophobic effects, on the contrary, a large collection of water molecules are involved nonspecifically. It is the statistics of the configurations of these water molecules at the specified temperature that lead to the fascinating entropy issues. Because of the significance of these entropies, hydrophobic effects are naturally a topic for molecular statistical thermodynamics.

Given the acknowledged significance of this topic, it is understandable that the literature that appeals to them is vast. Many researchers from a wide range of backgrounds and with a wide variety of goals have worked on these problems. Thus, the assertion that they are unsolved will challenge those researchers. But the unsolved assertion also reflects a lack of integration of principles, tools, and results to form a generally accepted mechanism of hydrophobic effects.

For example, it is widely, but not universally, agreed that the hydrogen bonding interactions between water molecules are a key to understanding hydrophobic effects. Conventional molecular simulation calculations with widely accepted molecular interaction models for small hydrophobic species in water broadly agree with experimental results on such systems (26–134). In this sense, everything is known. Nevertheless, this complete knowledge hasn't achieved a consensus for a primitive mechanism of hydrophobic effects. By "mechanism" we mean here a simpler, physical description that ties together otherwise disparate observations. Though this concept of mechanism is less than the complete knowledge of simulation calculations, it is not the extremely "poetic 'explanation'" famously noted by Stillinger (135). The elementary simplifications that lead to a mechanism must be more than rationalizations; they must be verifiable and consistent at a more basic level of theory, calculation, and observation.

The breakthrough mentioned in the first paragraph is particularly exciting because it hints at such a mechanism for the most primitive hydrophobic effects. A great deal more research is called for, certainly. But discussion of that development is a principal feature of this review.

WHAT CHANGED

The breakthrough required a couple of steps. The first step was the realization that feasible statistical investigations of spontaneous formation of atomic sized cavities in liquid solvents should shed light on operating theories of hydrophobic hydration (12–15, 136). Those studies could be based upon the formal truth (3, 10),

$$\Delta \mu_A = -RT \ln p_0, \qquad 1.$$

where $\Delta \mu_A$ is the interaction contribution to the chemical potential of a hard core hydrophobic solute of type A, and p_0 the probability that an observation volume defined by the excluded volume interactions of A with water molecules would have zero (0) occupants. These cavity formation studies were not an attempt to calculate hydration-free energies for realistic hydrophobic solutes. The goal was just to examine simple theories and to learn how different solvents might be distinguished on this basis.

The Small Size Hypothesis

There was also a significant physical idea alive at the time those studies were undertaken (12): "The low solubility of nonpolar solutes in water arises not from the fact that water molecules can form hydrogen bonds, but rather from the fact that they are small in size." As a simple clear hypothesis, this view contributed to the breakthrough although the hypothesis was eventually disputed (13, 137).

Packing and molecular sizes are important concerns for liquids because they are dense. The idea was that since water molecules are smaller than, say, CCl_4 molecules, the interstitial spaces available in liquid water would be smaller than those in liquid CCl_4. The first disputed point was that this hypothesis was suggested by the approximate scaled particle model (138, 139) and that model was known to have flaws (135, 140) as applied to hydrophobic hydration. The second disputed point was that treatment of the coexisting organic phase, liquid CCl_4 in this discussion, was less convincing than the treatment of liquid water: the modeling of a

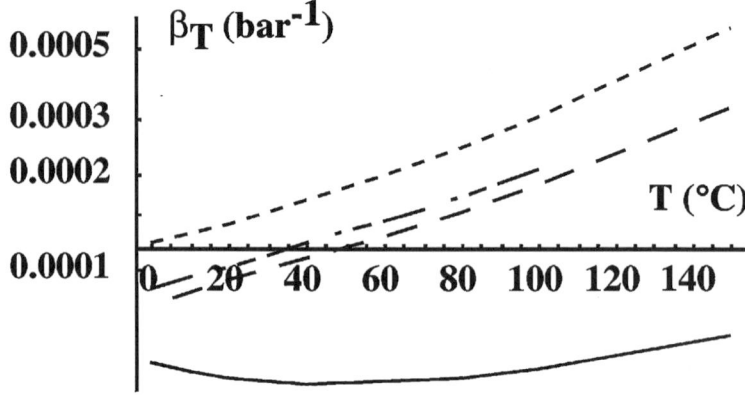

Figure 1 The isothermal compressibilities $\beta_T \equiv -\left(\frac{1}{\rho}\right)\left(\frac{\partial \rho}{\partial p}\right)_T$ along the liquid-vapor coexistence curve of several organic solvents compared to water (143) in common units. From top to bottom are n-heptane, carbon tetrachloride, benzene, and water. The compressibility is smaller for water than for these organic solvents, is less strongly temperature dependent, and has a minimum near 46°C. The critical temperatures of these organic solvents are all substantially less than the critical temperature of liquid water.

CCl_4 molecule as a simple ball is an oft-convenient canard but shouldn't be taken too literally. Another significant consideration is that liquid water is less dense on a packing fraction basis than most coexisting organic solvents.

Eventually (13–15, 136), the direct simulation investigations at the low pressures of first interest indicated that the most probable spontaneously occurring cavities might be smaller in typical organic solvents than in water, though the differences are small. What was decidedly different between water and organic solvents was the flexibility of the medium to open cavities larger than the most probable size. Water is less flexible in this regard, stiffer on a molecular scale, than typical organic liquids. Figure 1 gives a macroscopic experimental perspective on this relative stiffness. Furthermore, the results for the organic phase were not at all similar to what scaled particle models suggested (14, 15). So as an explanation of the distinction between water and common organic solvents, the small size mechanism must be discounted (91, 141, 142; T. Lazaridis, private communication). Nevertheless, we can anticipate subsequent discussion by noting that the equation of state of the solvent is important in establishing thermodynamic signatures of hydrophobic hydration.

Transient Cavities Probing Packing and Fluctuations

As an analysis tool for assessment of packing in disordered phases, studies of cavity statistics should be more widely helpful (23, 40, 47, 78, 86, 88, 99, 105, 112, 133,

144–154). The interesting work of Kocher et al. (147) is notable. Those calculations studied the cavity formation work in protein interiors and that cavity formation work was seen to be larger than for comparable organic solvents. In that respect, the packing of those protein interiors was tighter and less flexible than a simple oil droplet. This conclusion is significant for our pictures of protein structures and deserves further investigation. Structural and compositional heterogeneity are undoubtedly also important features of the cores of globular proteins. These observations should be helpful in distinguishing the interiors of micelles from the cores of folded globular proteins (155).

Modeling Occupancy Probabilities

The decisive second step in achieving the present breakthrough was the modeling of the distribution p_n of which p_0 is the $n = 0$ member (1). Several specific distributions p_n had been tried in analyzing the results of Pratt & Pohorille (14). But it was eventually recognized that a less specific approach, utilizing a maximum entropy procedure to incorporate successively more empirical moment information, was prudent and effective. Surprisingly, direct determination of the distribution p_n showed that a two-moment model,

$$-\ln p_n \approx \zeta_0 + \zeta_1 n + \zeta_2 n^2, \qquad 2.$$

was accurately borne out in circumstances of computer simulation of liquid water (1). The parameters ζ_j are evaluated by fitting the predicted moments $\langle n^j \rangle$, $j = 0, 1, \ldots$, to moment data. A practical virtue of this two-moment model is that the required moment data can be obtained from long-available experimental results. An additional surprise was that it had previously been shown, in different contexts and with additional assumptions (16, 156), that theories of a Percus-Yevick analog type had a structure derivable from a Gaussian or harmonic density field theory. The Pratt-Chandler theory was proposed as just this type of this Percus-Yevick analog. To the extent that the empirical observation Equation 2 suggests a normal distribution, the Pratt-Chandler theory is given a better foundation than was available at its genesis (10).

Nonequivalence with the Pratt-Chandler Theory

In fact, the two-moment model shown in Equation 2 is not precisely the same as the Pratt-Chandler theory. Several related direct observations can make that point clear. For example, the probability model shown in Equation 2 assigns probability weight only to nonnegative integer occupancies. That is not the case for the harmonic density field theory. The restriction that such density field theories should not permit the negative occupancy of any subvolume is an obvious but interesting requirement. Additionally, the Percus-Yevick theory for hard sphere mixtures can predict negative probabilities (157). These are technical issues, however, and the performance of the two-moment model Equation 2 gives strong and unexpected support for the Pratt-Chandler theory.

There is a different respect in which the correspondence of the two-moment model shown in Equation 2 with the Pratt-Chandler theory is imprecise. The kinship indicated above is based upon calculation of solvation-free energies when the solvent can be idealized as an harmonic density field. That idea can be straightforwardly carried over to consideration of nonspherical solutes. For example, classic potentials of mean force might be addressed by consideration of a diatomic solute of varying bond length. The Pratt-Chandler theory does not do that directly but utilizes the structure of the Ornstein-Zernike equations together with yet another Percus-Yevick style closure approximation. Those distinctions have not yet been discussed fully.

Good Theories are either Gaussian or Everything

A curious feature of the $\ln p_n$ moment modeling is that convergence of predictions for $\Delta \mu_A = -RT \ln p_0$ with increasing numbers of utilized moments is nonmonotonic (6, 8, 10). The predicted thermodynamic results are surprisingly accurate when two moments are used but become worse with three moments, before eventually returning to an accurate prediction with many more moments available. The two-moment model, and also the Pratt-Chandler theory, is fortuitous in this sense. But this does conform to the adage that good theories are either Gaussian or everything.

TECHNICAL OBSERVATIONS AND EXTENSIONS

The theory above has always been understood at a more basic level than the description above (10). The most important observation is that the Mayer-Montroll series (10, 158) can be made significantly constructive with the help of simulation data (10, 13, 136), and those approaches can be more physical than stock integral equation approximations. Simulation data can provide successive terms in a Mayer-Montroll series. In that case the binomial moments $\langle \binom{n}{j} \rangle_0$ are the stylistically preferred data (10). Then the maximum entropy modeling is a device for a resummation based upon a finite number of initial terms of that series (10). Several additional technical points are helpful at this level.

Default Models

The adage that good theories are either Gaussian or everything is serious but doesn't address the physical reasons why these distributions are the way they are. In fact, painstaking addition of successive moments is not only painful but often unsatisfying. And the two-moment model of Equation 2 has been less satisfactory for every additional case examined carefully beyond the initial one that was connected with this breakthrough (1); recent examples can be seen in References (23, 133). It is better to consider physical models for the distribution p_n and approximations:

$$-\ln\left[\frac{p_n}{\hat{p}_n}\right] \approx \zeta_0 + \zeta_1 n + \zeta_2 n^2 + \cdots, \qquad 3.$$

where $\hat{p}_n$ is a model distribution chosen on the basis of extraneous considerations. Because $p_n = \hat{p}_n$ in the absence of additional information, $\hat{p}_n$ is called the default model. Utilization of a default model in this way compromises the goal of predicting the distribution and instead relies on the moments to adapt the default model to the conditions of interest. The default model of first interest (1) is $\hat{p}_n \propto 1/n!$ This default model produces the uncorrelated result (the Poisson distribution) when the only moment used is $\langle n \rangle_0$. Another way to identify default models is to use probabilities obtained for some other system having something in common with the aqueous solution of interest (8).

An important practical point is that this approach works better when the default model is not too specific (22). This can be understood as follows: The moment information used to adapt the default model to the case of interest is not particularly specific. If the default model makes specific errors, a limited amount of that nonspecific data will not correct those errors adequately. This argument gives a partial rationalization for the accurate performance of the flat default model that leads to Equation 2.

In Equation 2, $\zeta_0 = \Delta\mu_A/RT$, but it is helpful to notice that this thermodynamic quantity can be alternatively expressed as (10)

$$\Delta\mu_A = RT \ln\left\{1 + \sum_{n=1}\left(\frac{\hat{p}_n}{\hat{p}_0}\right)\exp\left[-\sum_{k=1}^{k_{max}}\zeta_k\binom{n}{k}\right]\right\}, \qquad 4.$$

where binomial moments through order k_{max} are assumed and the default model is included. ζ_0 does not appear on the right because that normalization factor is being expressed through the thermodynamic property. The point is that this is a conventional form of a partition function sum. The interactions are n-function interactions, in contrast to density or ρ-functional theories, but with strength parameters adjusted to conform to the data available. The fact that $\Delta\mu_A$ is extracted from a fully considered probability distribution, and this consequent structure, is the substance behind our use of the adjective physical for these theories. These theories are still approximate, of course, and they will not have the internal consistency of statistical mechanical theories obtained by exact analysis of a mechanical Hamiltonian system.

Quasichemical Theory

The quasichemical theory (11, 19) adapted to treat hard core solutes (23) gives an explicit structure for the $\Delta\mu_A$ formula as in Equation 4. That result can be regarded as a formal theorem

$$\Delta \mu_A = RT \ln\left[1 + \sum_{m \geq 1} K_m \rho_W^m\right].\qquad 5.$$

The K_m are equilibrium ratios

$$K_n = \frac{\rho_{\bar{A}W_n}}{\rho_{\bar{A}W_{n=0}} \rho_W^n}\qquad 6.$$

for binding of solvent molecules to a cavity stencil associated with the AW excluded volume, understood according to the chemical view

$$\bar{A}W_{n=0} + nW \rightleftharpoons \bar{A}W_n.\qquad 7.$$

$\bar{A}$ is a precisely defined cavity species (23) corresponding to the AW excluded volume. Equation 4 should be compared to Equation 5; because of the structural similarity, it is most appropriate to consider Equation 4 as a quasichemical approximation. The K_m are well-defined theoretically (23) and observable from simulations. So again this approach can be significantly constructive when combined with simulations (11, 19, 20). But the first utility is that the low density limiting values of K_m, call them $K_m^{(0)}$, are computable few body quantities (23). The approximation

$$\Delta \mu_A \approx RT \ln\left[1 + \sum_{m \geq 1} K_m^{(0)} \rho_W^m e^{-m\zeta_1}\right]\qquad 8.$$

is then a simple physical theory, the primitive quasichemical approximation (23). The Lagrange multiplier ζ_1 serves as a mean field that adjusts the mean occupancy to the thermodynamic state of interest. For a hard-sphere solute A, in a hard sphere solvent, this theory produces sensible results though it does not achieve high accuracy in the dense fluid regime $\rho d^3 > 0.6$ (23), where d is the diameter of the solvent hard spheres. (The foremost questions for aqueous solutions are at the lower boundary of this conventional demarcation of dense fluids.) When the accuracy of this theory for hard-sphere systems degenerates, it is because the distribution

$$\hat{p}_n = \frac{K_n^{(0)} \rho_W^n e^{-n\zeta_1}}{1 + \sum_{m \geq 1} K_m^{(0)} \rho_W^m e^{-m\zeta_1}}\qquad 9.$$

is too broad in the low n extreme (23); see Figure 2. Because this theory thus directly treats short range molecular structure somewhat too broadly, it is a natural suggestion for generating default models. No direct experience along those lines is presently available.

The investigation of how such a simple theory (Equation 9) breaks down is a yet more interesting aspect of the development of the quasichemical theories for these problems (11, 23). For the hard-sphere fluid at higher densities, the primitive quasichemical theory shown in Equation 9 remains a faithful descriptor of the $n \geq 1$

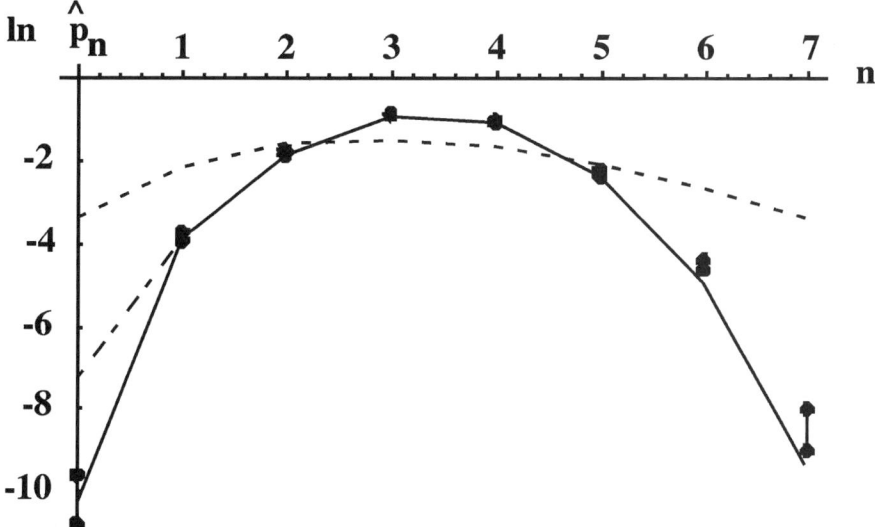

Figure 2 Distribution $\hat{p}_n$ for a isodiameter hard-sphere solute in a hard-sphere fluid at density $\rho d^3 = 0.8$ and various models. The dots are simulation results (23) and the spread indicates a 67% confidence interval. The upper dashed curve is the Poisson distribution with the required mean $\langle n \rangle_0 = 4\pi \rho d^3 / 3$. The long-dashed curve next down is the primitive quasichemical model, Equation 9 (23). The solid line is an iterated quasichemical theory that incorporates an empirical correlation correction (23) and provides a simple accurate description of these quantities for the hard-sphere fluid. Note that the primitive quasichemical approximation is a good description of this distribution for $n \geq 1$ but significantly overestimates p_0 at these densities.

features of the distribution. But the actual p_0 (see Figure 2) becomes depressed relative to the model; p_0 breaks away from the rest of the primitive quasichemical distribution. In fact, the suggested correlation correction can be effectively empiricized and provides an accurate description of these distributions for hard-sphere fluids. These exotic complexities with p_0 are not reflected in Equation 2. Tiny features like that were noticed, however, in the initial simulation studies of these probability models (1). Additionally, we anticipate discussion below by noting that proximity to a low pressure liquid-vapor transition point, associated with solvent-solvent attractive interactions and potential dewetting of hard surfaces, is expected to increase p_0 (1, 11).

Importance Sampling to Correct Occupancies

A more workman-like investigation of these theories can be based upon the potential distribution theorem (3, 23, 159)

$$e^{-\Delta \mu_A / RT} = \langle\!\langle e^{-\Delta U / RT} \rangle\!\rangle_0 . \qquad 10.$$

The brackets $\langle\langle \cdots \rangle\rangle_0$ indicate the average over the thermal motion of a distinguished A solute and the solvent under the condition of no interactions between these subsystems; the latter restriction is conveyed by the subscript 0. Distributions $\hat{p}_n$ should be helpful as importance functions. It would be natural to use this estimate to revise the calculation of all the probabilities p_n. But we have seen a case, Figure 2, where the distinction between $n = 0$ and $n \geq 1$ is most interesting. In addition, this quantity averaged here (Equation 10) takes the values zero (0) if $n \geq 1$ and one (1) for $n = 0$. Thus we consider the importance function,

$$W = \begin{Bmatrix} \hat{p}_0, & n = 0, \\ 1 - \hat{p}_0, & n > 0. \end{Bmatrix} \qquad 11.$$

The standard importance sampling ideas (160, 161) then produce

$$e^{-\Delta\mu_A/RT} = \frac{\langle\langle We^{-\Delta U/RT}\rangle\rangle_{1/W}}{\langle\langle W\rangle\rangle_{1/W}}. \qquad 12.$$

The sampling distribution is the Boltzmann weight in Equation 10 multiplicatively augmented by the configurational function $1/W$ and corresponds to a finite probability step $\frac{p_0}{1-p_0}$ as the first solvent molecule enters the observation volume. Typically, this leads to a diminished occupancy of the observation volume. After some rearrangement, Equation 12 is

$$\frac{p_0 - \hat{p}_0}{\hat{p}_0} = \frac{(2\pi_0 - 1)}{1 - \pi_0 \left(\frac{1 - 2\hat{p}_0}{1 - \hat{p}_0}\right)}. \qquad 13.$$

Here, π_0 is the probability of the observation volume being empty with the reweighted sampling,

$$\pi_0 = \langle\langle e^{-\Delta U/RT}\rangle\rangle_{1/W}. \qquad 14.$$

These formulae could be used directly with simulation calculations. The best available approximate p_0 (135, 163) could be used as $\hat{p}_0$ in order to achieve higher accuracy; that would be interesting but not easy because achieving $n \gg 0 \rightarrow n = 0$ transitions requires rare collective processes. On the other hand, $n = 0 \rightarrow n \geq 1$ transitions will have a low acceptance probability. But our argument here is directed toward understanding physical features left out of simple models such as the quasichemical model of Equation 4. The choice here of W in Equation 11 is transparently directed toward discussion of a two-state picture of the hydration.

With that goal, the conceptual perspective is the more interesting one. This is a discrete example of a procedure common in density functional arguments. Local particle occupancies are altered by a reweighting. Just as with p_0, π_0 can be studied with a Mayer-Montroll series and moment modeling (10). The moments involved now would be obtained from study of the designed nonuniform system. The average on the right side of Equation 14 is a functional of the density induced by the reweighting.

For example, if the reweighting can serve to nucleate a bubble because of proximity of the thermodynamic state to a liquid-vapor transition, then, physically viewed, π_0 is expected to be composed of two important cases: (*a*) vapor with a depletion region surrounding the observation volume; this gives contribution one (1) to π_0 for these cases, and (*b*) liquid; typically these cases contribute zero (0) to π_0. But occasional configurations, roughly with frequency $\hat{p}_0$, give $e^{-\Delta U/RT} = 1$. Thus as a rough estimate, we expect,

$$\pi_0 \approx \frac{e^{-\Delta F/RT}\left(\frac{1-\hat{p}_0}{\hat{p}_0}\right) + \hat{p}_0}{e^{-\Delta F/RT}\left(\frac{1-\hat{p}_0}{\hat{p}_0}\right) + 1}, \qquad 15.$$

where ΔF is the free energy for formation of a bubble from the liquid corresponding to boundary conditions $n = 0$ on the observation volume. That free energy might be approximated by a combination of van der Waals theories and molecular theories appropriate for the vapor phase (162). Let's consider a one-phase, dense liquid thermodynamic state, not far from coexistance with a vapor phase, so that $1 > e^{-\Delta F/RT} > \hat{p}_0$. With these estimates, Equations 13 and 15 evaluate to

$$p_0 \approx e^{-\Delta F/RT} > \hat{p}_0. \qquad 16.$$

The insertion probability is the probability of bubble formation, surely the only simple guess, and this estimated change raises the value $\hat{p}_0$.

van der Waals Picture

On the basis of the observations above, we can construct the following picture (23) by considering how these theories would work for an atomic-sized hard-sphere solute in a simple van der Waals fluid system (164). Consider packing effects first, then subsequently the effects of attractive interactions. For dense liquid cases with full-blown packing difficulties, models such as those shown in Equation 2 or 9 overestimate p_0 because those models aren't accurate for packing problems in the dense fluid regime. Next, consider attractive interactions and the possibility of dewetting. Those effects raise p_0. The model of Equation 2 does not reflect those phenomena. But these two errors can compensate, so Equation 2 can be empirically accurate for atomic solutes despite the naiveté of that model. This is another rationalization of the astonishing, fortuitous accuracy of the two-moment model and the Pratt-Chandler theory.

Two additional points may be noted. The first point is that in discussing errors in treating packing effects, we have been concerned about errors of the same type as those in the Percus-Yevick theory for the hard-sphere fluid. But the Percus-Yevick theory of the hard-sphere fluid is the most successful theory of a dense liquid, "gloriously accurate, considering its simplicity" (165). So this discussion brings a high-sensitivity view to the problem. This is necessary because of the high interest in these important problems. The second point to note is that attractive interactions in this argument involve solvent-solvent interactions, not solute-solvent attractive interactions. Part of the subtlety of these discussions is that the approach that offers models such as Equation 2 is sufficiently empirical that a unique identification of the source of a particular inaccuracy is nontrivial.

HYDROPHOBIC HYDRATION AND TEMPERATURE DEPENDENCES

The most astonishing result of the new theory shown in Equation 2 is its explanation of the hydrophobic temperature dependence known as entropy convergence (2, 9–11, 133), and those temperature dependences are discussed here.

Solubilities

The solubilities of simple gases in water have some interesting complexities (166). Many simple gases have a solubility minimum in water at moderate temperatures and pressures. Because the solubility is governed by $\Delta\mu_A/RT$, the temperature variation of the solubility at a fixed pressure requires information on

$$\left(\frac{\partial \Delta\mu_A/RT}{\partial T}\right)_p = -\frac{1}{T}\left(\frac{\Delta h_A}{RT}\right), \qquad 17.$$

with Δh_A the partial molar enthalpy change upon dissolution of species A. (We consider the low concentration limit here.) Thus, a solubility minimum leads us to anticipate a temperature of zero enthalpy change for the dissolution where $\Delta\mu_A/RT$ plotted as a function of T has a maximum.

At a higher temperature, the dissolution of many simple gases shows approximately zero partial molar entropy change:

$$\left(\frac{\partial \Delta\mu_A}{\partial T}\right)_p = -\Delta s_A. \qquad 18.$$

$\Delta\mu_A$ plotted as a function of T has a maximum. More puzzling is the fact that this temperature of zero entropy change is common to a number of different gases. This phenomenon is referred to as entropy convergence because the entropies of hydration of different solutes converge to approximately zero at a common temperature (167, 168).

Model Explanation

The puzzle "why?" was first answered on a molecular level in Reference (2) on the basis of the model shown in Equation 2; see also (9–11, 133). Establishing these two temperature behaviors should go a long way toward establishing the temperature variations of hydrophobic effects throughout an extended range relevant to biomolecular structure.

If we agree to be guided by an estimate of p_n based upon a continuous normal distribution (2, 9–11), then evaluation of the Lagrange multipliers of Equation 2 is not a problem and

$$\Delta\mu_A/RT \approx \frac{1}{2}\left\{\frac{\langle n\rangle_0^2}{\langle\delta n^2\rangle_0} + \ln\left[2\pi\langle\delta n^2\rangle_0\right]\right\} \qquad 19.$$

$$= \frac{1}{2}\left\{\frac{(\rho v)^2}{\langle\delta n^2\rangle_0} + \ln\left[2\pi\langle\delta n^2\rangle_0\right]\right\} \qquad 20.$$

with v the volume of the AW excluded volume, expected to be weakly temperature dependent. The surprise is that $\langle\delta n^2\rangle_0$ varies only slightly with temperature over the interesting temperature range. This is suggestive of the data shown in Figure 1. Furthermore, the second term of Equation 20 is smaller than the first. Therefore the plot of $\Delta\mu_A$ as a function of T experiences a maximum because the combination $T\rho^2$ has a maximum as $\rho = \rho_{sat}(T)$ decreases with increasing temperature along the coexistence curve. To the extent that the rightmost term of Equation 20 can be neglected and $v^2/\langle\delta n^2\rangle_0$ is independent of temperature, then entropy convergence will occur and the temperature at entropy convergence will be the same for all hydrophobic solutes.

This development applies to model hard core solutes, and the convergence temperature does appear to shift slowly but systematically to lower temperatures as the volume of the solute increases. But it was disturbingly noted, nearly forty years ago, that the success of the scaled particle model in evaluating hydration entropies "... suggests an almost thermodynamic independence of molecular structure" (140). In the entropy convergence phenomena, we see that this almost thermodynamic independence of molecular structure is a feature of the data (168) and that the current theory gives a simple molecular explanation that solves this puzzle.

The lower temperature iso-enthalpy solubility minimum is expected to be tied to a different aspect of the solution-water interactions, the van der Waals attractive interactions. Following a van der Waals view, these effects should be reasonably described by first order perturbation theory, so as a qualitative model we have (9, 11)

$$\Delta\mu_A \approx -A\rho + BT\rho^2 + CT \qquad 21.$$

with fitting parameters A, B, and C. This equation does indeed have the correct qualitative behaviors (9, 11). If Equation 21 is used as a fitting model and the parameters are unrestricted, it is essentially perfect. If the parameters are constrained by physical expectations for the temperature-independent fitted parameters v, $\langle\delta n^2\rangle_0$, and A, then this model is only qualitatively and crudely successful in describing experimental solubilities.

Several of these considerations have been reexamined recently with results consistent with this picture (169; H.S. Ashbaugh, T.M. Truskett, P.G. Debenedetti, private communication). Although these temperature behaviors were not always so clearly recognized as this, there remains specific solubility issues that aren't resolved including at the simulation level. An interesting case was provided by the important simulation calculations of Swope & Andersen (39) on the solubility of inert gas atoms in water. With regard to the Lennard-Jones solute water (oxygen) interaction models, they concluded: "For the potentials used in the present simulations, it is not possible to fit the value, slope, and curvature for helium and neon without choosing what we believe to be unreasonably large values of σ. We can, however, obtain fits to the data for argon and krypton with reasonable values of the diameters." Simulation calculations for such cases have been pursued several times since then; those activities up to 1998 are summarized by Arthur & Haymet (106), and the latter effort also concludes with some ambiguity about the case of the He solute.

It is also important to emphasize that this model is used here only over a limited temperature range and at low pressure. Lin & Wood (95) used molecular dynamics to model the thermodynamic properties of small hydrocarbons in water over a wide range of temperature and pressure, and Errington et al. (109) studied the phase equilibria of water-methane and water-ethane systems over wide ranges of temperature and pressure using Monte Carlo techniques.

RECENT EXPERIMENTAL STUDIES OF HYDRATION STRUCTURE

In recent years, efforts to measure directly the structure of water surrounding simple hydrophobic solutes have produced results that should be of quantitative relevance to the theories discussed here (100, 170–174). A more quantitative consideration is warranted, but the initial impression is that these results are in good agreement with the calculations that have been done. These data give the weight of evidence to important basic conclusions also.

Pressure Dependence of Hydrophobic Hydration

One such conclusion is that this structuring of water appears to be independent of variations of the pressure to 700 bar (70 MPa) (173). This may be important to the current issue of pressure denaturation of soluble proteins as is discussed below. But investigation in a higher pressure range would be necessary for that purpose (119, 175).

Clathrate is in the Eye of the Beholder

Another important conclusion follows from the direct comparison of the radial distribution of oxygen atoms surrounding Kr in liquid aqueous solution and in a solid clathrate phase. Those radial distributions are qualitatively different in the two different phases. This is important because a clathrate picture of hydration structure of nonpolar solutes in liquid water is a common view of hydrophobic hydration that has not been of quantitative relevance; it has been a pictorial theory (176). In contrast, several theoretical calculations that have had quantitative value assume roughly the antithesis of clathrate, that the conditional density of water surrounding a nonspherical nonpolar solute can be built-up by superposition of proximal radial information (84, 85, 94, 177–179). The clathrate language is widely used, hardly explicitly justified, and leads to misunderstandings. A number of studies have explicitly considered the issue of how valid the clathrate description is (54, 56, 82, 108, 116, 180, 181). The conclusion seems to be that if you look for clathrate-style hydration structures you probably see them, but if you ask whether they are necessary for a correct quantitative understanding, the answer is no. Clathrate is in the eye of the beholder. A reasonable recommendation is that when clathrate is used as a descriptor in these liquid solutions, it should be explicitly defined and justified. Attempts to formulate quantitative theories on the basis of chemical models of these hydration shells are known but ill-developed (11).

POTENTIALS OF THE MEAN FORCES AMONG PRIMITIVE HYDROPHOBIC SPECIES IN WATER

The theories discussed above are straightforwardly applicable to nonspherical solutes. For a solute composed of two atoms with varying interatomic separation, the comparison

$$\Delta\mu_{AA'}(r) - \Delta\mu_A - \Delta\mu_{A'} \equiv w_{AA'}(r) \qquad 22.$$

leads to the classic issue of the potential of mean force (pmf), an issue of long-standing interest (26, 51, 65, 68, 175, 182–190, 192–194, 196–208, 210–214). The model shown in Equation 2 for a simple case was tested against simulation results (1) and the comparison was close. That theory, including Equation 20, sheds new light on these properties.

Contact Hydrophobic Interactions

Consider $w_{AA}(r)$ for atomic solutes in contact. When the solute atoms are in van der Waals contact, we still anticipate that $\langle \delta n^2 \rangle_0$ will be only weakly temperature dependent. In addition, the volume excluded to the solvent by the pair is less than twice the volume of an atom alone because of the overlap of their excluded regions. Thus the dominant contribution to Equation 22 is negative, stabilizing the contact pair, and that stabilization increases with temperature below the entropy convergence temperature. [Note that the subtraction in Equation 22 requires some additional thought when the logarithmic term of Equation 20 is addressed (6). If the subtraction were naively carried out, that r-independent nonzero difference might imply pathologically long-ranged interactions. A more careful consideration of the statistical approaches satisfactorily resolves that pathology (6). This detail shows again that these theories are not naively equivalent to the Pratt-Chandler theory.]

Noncontact Hydrophobic Interactions

On the other hand, when the atomic solutes are separated enough that a water molecule may fit between them, the volume excluded to the solvent by the pair is more nearly twice the excluded volume of a separated atom. The theories following Equation 2 then are more sensitive to water molecule correlations of longer range because the information $\langle \delta n^2 \rangle_0$ depends on those correlations. These theories then produce more subtle effects. Noncontact hydrophobic interactions may be, nevertheless, significant because of the larger configurational volume corresponding to those solvent-separated configurations. In addition, as discussed by Pratt & Chandler (215, 216), free energies of these solvent-separated configurations may be more sensitive to details of van der Waals attractive interactions than are contact configurations.

Simulation Results

These views seem to be borne out by the available simulation results, although "the conclusions drawn from previous simulation calculations have been very contradictory" (203). Substantial stability for contact hydrophobic pairs is probably the least contradictory of the possible conclusions. The recent preponderance of simulation results indicates that these contact pairs are stabilized by favorable hydration entropies (68, 199–201, 203, 204, 207, 214, 217). This would agree with the view established from the simple model shown in Equation 20, but those model temperature variations have been checked mostly along the liquid-vapor coexistence curve. The interesting results of References (203, 204) show substantial temperature variations at fixed water density. It was noted (201) that the conditions of temperature increase at fixed density strengthen the apparent entropic stabilization of the contact pair. That is also how Equation 20 works; the density decrease serves to moderate the temperature increase and eventually, at the entropy convergence temperature, to dominate it. But the large effects seen by the Vienna group (203, 204) make it unclear that the simple model shown in Equation 20 will be accurate for those phenomena.

Simulation results for the solvent-separated configurations are less clear also. There are entropic and enthalpic temperature effects in opposite directions with small net results (200, 204, 207). Because of the larger configurational volume for the solute pair in this configuration, these smaller hydration-free energies are not negligible and "the puzzling finding that the marked hydrophobic behavior of methane-like solutes concluded from the free energy data is not reflected in a similarly clear manner by the second osmotic virial coefficients requires a closer inspection of the underlying phenomena" (204); see also (195). There is precedent, depending on a variety of additional details, for simulation results to exhibit either hydrophobic clustering or not, i.e., "hydrophobic repulsion" (191).

Polarizability?

The complications suggested for these solvent-separated configurations seem to have lead to other contradictory results. It was suggested long ago (188) that solute polarizability might change the character of hydrophobic interactions predicted by simple theories. A later simulation calculation that included explicit polarizability (65) in the water-water interactions also suggested that these more complicated descriptions might qualitatively change the hydration of noncontact atom pairs. Subsequent calculations again suggested that polarizability could lead to substantial changes but in a different direction from those seen earlier (202); treatment of long-ranged interactions was noted as a significant issue in these calculations. More recent studies of interaction models that include polarizability, however, have restored an original small change view for the moment (207).

Alkane Conformational Equilibrium in Water

The potentials of mean force just discussed are relevant to consideration of the conformational equilibrium of small flexible hydrophobic molecules in water. The first test case for theories has always been the trans-gauche isomerization of n-butane (84, 91, 177, 179, 218–224). For the case of n-butane, solvent separated possibilities are not available, so the contact hydrophobic interactions are relevant. The populations of the more compact gauche configurations are enhanced by an entropic hydration effect. Hummer et al. (1) applied the model shown in Equation 2 to the case of conformational equilibrium of n-butane in water and found close agreement with the latest simulation results. Hummer has effectively adapted the model of Equation 2 so that it can be simply applied to other alkanes (225). Much longer chain molecules that have been strictly hydrophobic are less well studied primarily because they would be so unusual as isolated components of aqueous solutions. Hydrophilic groups are necessary to solubilize large molecules. Perhaps the simplest such soluble molecules would be polyethylene oxide chains that are a specific interest (226). But Gallicchio et al. (129) has recently studied the hydration of slightly larger alkanes in additional detail.

PRESSURE DEPENDENCE OF HYDROPHOBIC INTERACTIONS

The intense current interest in pressure studies of protein structure is due to the alternative light this research can shed on protein conformational dynamics. A recent example can be found in (227), but we are unable to review adequately that body of interesting work here.

The complications of the noncontact hydrophobic interactions mentioned above appear to be involved in understanding pressure denaturation of proteins, however. We identify some of that work because views of hydrophobic effects have been paradoxical for these issues (228) and because this discussion gives additional perspective on the theories above. Wallqvist reported initial studies of pressure dependence of hydrophobic interactions (180, 229). Atomic hydrophobic solutes in water clustered at low pressure but dispersed at a substantially higher pressure that wasn't further quantified in that study. Later, the Rutgers group took an important step (205) in Monte Carlo calculations of the effects of pressure on the pmf between Lennard-Jones model hydrophobic atomic solutes in water. Hummer et al. (4)

then developed the theory shown in Equation 2 for these pmfs as a function of pressure. That theory suggests that noncontact configurations of hydrophobic pairs become progressively more stable relative to contact pairs, and this might be a feature of pressure denaturation known to produce less disrupted structures than does heat denaturation. Subsequent molecular dynamics calculations (175) confirm this picture of dispersal at higher pressures and the pressure variations of the pmfs: as pressure is increased, these pmfs become more structured, the contact minimum deepens, the desolvation barrier becomes higher; the solvent-separated well becomes better defined and deeper, and it appears to deepen faster than the contact well. This work also observed clustering of hydrophobic atomic solutes at low pressure (1 atm) but dispersal at high pressure (8000 atm, 810.6 MPa). This work also saw changes in the solute-water (oxygen) radial distribution that should be observable in experimental studies such as those of Reference (173). But the simulation results are for considerably higher pressures than the experimental work reported. These calculations also considered spherical hydrophobic solutes of a larger size, more comparable with valine, leucine, or isoleucine side chains. The responses to substantial pressure increases were similar but perhaps slightly more pronounced. A physical view is that as the pressure is increased, water molecules can be jammed between contact hydrophobic pairs; this evidently results in more efficient, lower-volume packing, and consequently a negative hydration-free energy change with increasing pressure for noncontact configurations.

SIZE DEPENDENCE OF HYDROPHOBIC HYDRATION FOR HARD SPHERE SOLUTES

For a hard sphere solute, the rate of increase of the hydration-free energy shown in Equation 1 with the distance of closest AO approach, denoted by λ, produces a particularly interesting quantity,

$$\rho_W G(\lambda) = \frac{1}{4\pi\lambda^2} \left(\frac{\partial \Delta\mu_A/RT}{\partial \lambda} \right). \qquad 23.$$

This $\rho_W G(\lambda)$ is the conditional density of the solvent water (oxygen) at contact with the spherical solute. Because of the involvement of $\partial \Delta\mu_A/\partial\lambda$, this relation describes the compressive force exerted by the solvent on the solute.

Contact Densities

Direct studies of these quantities have shown (14, 15) that in the range $2.0 \text{ Å} < \lambda < 3.0 \text{ Å}$ $G(\lambda)$ for liquid water is approximately two times larger than for n-hexane. Water exerts a higher compressive force on the surface of an inert solute than do typical organic liquids so that water squeezes out hydrophobic solutes (230). More pertinent for the present discussion is that the Pratt-Chandler theory overestimates this compressive force and the original scaled particle model underestimates it

(14, 15). The revised scaled particle model (135) lands in the middle and does a better job at describing this compressive force. Recent work (163) has studied these quantities over a much larger range of λ and confirmed the accuracy of the revised scaled particle model.

The reason for the differences between the scaled particle models and the Pratt-Chandler theory is associated with the known behavior $\rho_W G(\lambda) \sim p/RT$ for large λ. For the cases of first interest, $p/\rho_W RT \ll 1$. Because $G(\lambda)$ is initially one (1) and increases initially, $G(\lambda)$ decreases for large λ to achieve the small value $p/\rho_W RT$. In fact, this decreasing behavior obtains for $\lambda > 3$ Å, approximately. Thus, for hard sphere solutes with $\lambda \gg 3$ Å the contact density can be small. This low pressure for a dense liquid is due to attractive forces between the solvent molecules. If the conditions are adjusted for liquid-vapor coexistence, then p is also the pressure of the coexisting vapor and p/RT would be the density of the vapor under the assumption that it can be treated as ideal. Under these conditions we can say, therefore, that a sufficiently large hard sphere solute nucleates a bubble of the vapor. These behaviors are built into the approximate scaled particle models but not into the Pratt-Chandler theory. These issues have been directly investigated for spherical model solutes moderately larger than canonical for methane (5); weak effects were found for those cases and convincing models were developed (5).

PMF for Stacked Plates in Water

This issue has been of particular interest recently because a previous calculation (231) studied the pmf between modeled stacked plates with exclusively repulsive interactions with water molecules. That work suggested that contact hydrophobic interactions in that case could be dominated by a dewetting event: the last two layers of water molecules intervening between parallel plates evacuated together.

Benzene-Benzene PMF

The comparable results for more realistically modeled benzene, or toluene, or other small aromatic solute molecule pairs are also interesting but more complicated (52, 195–198, 209, 210, 213). These molecules are slightly smaller than the stacked plates that were studied. The available thermodynamic data (232) suggest that benzene, toluene, and ethyl benzene should display conventional entropy convergence behavior. T-shaped contact pairs are more probable for benzene than a stacked arrangement is. The opposite is true for toluene (210). This is also true for the gas-phase potential energy surface though the hydration seems to enhance this distinction slightly (210). Variations in these pmfs (52, 195, 197, 198, 210, 213) are much smaller than for the modeled stacked plates (231). A dewetting transition is not obvious for the calculations with higher molecular realism. It may be that the significance of any dewetting would be more obvious near transitional configurations such as the desolvation barrier region that separates contact from solvent-separated configurations. Though the hydration of neither of the high probability configurations discussed here (52, 197, 198, 210, 213) seemed remarkable in this way, the variation of the free energy in the desolvation barrier region might be unusual; this deserves further checking. How these complications are affected by the more complicated environment of an amino acid side chain, e.g., phenyl alanine, in a hydrated protein is not known; the peptide backbone is, of course, highly polar. The hydrogen bonding possibilities of tyrosine or tryptophan side chains complicate things yet again. An interesting study of pairing of tryptophan-histidene side chains (213) suggested that hydration of these side chains results in stacked pairing near protein surfaces but T-contacts in protein interiors. [Continuum dielectric models did not provide a rationalization of that observed tendency (213).] The variety of the results obtained suggested (213) "... the importance of the atomic details of the solvent in determining the free energy for the solute-solute interactions."

Theory of Interface Formation

The development of the theory (162) corresponding to the stacked plates data was initiated by Lum, Chandler, & Weeks; see Figure 1 there (233). Subsequently (234–237), a focus has been the study of how the entropy dominated hydration-free energies discussed above for inert gas solubilities change to the surface tension–dominated behavior expected for ideal mesoscopic hydrophobic species; see also (128). That behavior was explicitly built into the revised scaled-particle model for hard sphere solutes decades ago (135); that model is known to be accurate for water (163) and for a simple liquid (234, 236). A fundamental niche for that theoretical work (162) is the development of a molecular description of the interface formation mechanism built into the revised scaled-particle model at large sizes (162).

How these theoretical developments will accommodate heterogeneity of chemistry and structure typical of biomaterials, in contrast to the model stacked plates, is not yet established. A nice example of the issue of heterogeneity, absorption of water on activated carbon, was discussed recently by Müller & Gubbins (238). It is unreasonable to imagine that the biophysical applications will be simpler than this. It might be more appropriate at this stage of development to regard the Berkeley project as ambitiously directed toward an implicit hydration model (20) rather than an assertion of specific physical relevance of drying to biomolecular structure (P. Rein ten Wolde, S.X. Sun, D. Chandler, private communication). These problems require consideration of several distinct issues together. One such issue is the direct contributions of solute-solvent attractive interactions to hydration-free energies for a specific structure (D.M. Huang, D. Chandler, private communication). A second issue is the indirect effects of solute-solvent interactions in establishing structures and switching between structures as was initially anticipated (233). That switching can be sensitive to details of van der Waals attractive interactions (163, 239–241; G. Hummer, J.C. Rasaiah, J. Noworyta, private communication).

It is worthwhile attempting to articulate a down-to-earth view of the claims of Reference (233) specifically. For biomolecules, there likely are uncommon transitional structures and conditions for which localized water occupancies can change abruptly. As these transitional structures become identified, they will be interesting. Considering water-hydrocarbon liquid interfaces not compromised by hydrophilic contacts, it is likely that these interfacial regions will be looser than adjoining bulk phases and more accomodating to imposition of hydrophobic species (145). Surfactants probably change that conclusion qualitatively (145). This is likely to be relevant to protein hydration and function. Most solute configurations, except for a few transitional structures, won't require specific acknowledgment of drying. The claims to Reference (233) don't seem to require modification of the discussion above, "Potentials of the Mean Forces Among Primitive Hydrophobic Species in Water," that separated contact from noncontact configurations and entropy effects from the rest, and then suggested that the more poorly understood noncontact questions are likely to show the most variability. The specific claims of Reference (233) and the general issues remain questions for research.

CONCLUDING DISCUSSION

This review has adopted a narrow theoretical focus and a direct style with the goal of identifying primitive conclusions that might assist in next stages of molecular research on these problems. These theory and modeling topics haven't been reviewed with this goal recently, and a review of the bigger topic of hydrophobic effects would not be feasible in this setting. More comprehensive and formal reviews of these topics are in progress and that must be my excuse for considering such a small subset of the work in this area. Nevertheless, some historical perspective is necessary in identifying valuable primitive conclusions.

One such conclusion is the rectification of the antique Pratt-Chandler theory (1–16). This was an unexpected development because the advances reviewed above had bypassed stock integral equation theories. It may have been the stock aspect of the earlier approach (17) that caused the greatest confusion on this topic. There is no obvious point to doing that type of integral equation approximation again for the more complex solutes to which the theoretical interest has progressed. For the problems addressed, however, this amended Pratt-Chandler theory is now seen to be a compelling, approximate theory with empirical ingredients. The research noted above, "Theory of Interface Formation" and "Importance Sampling to Correct Occupanices," emphasizes that the treatment of those attractive force effects on the hydration problem by the Pratt-Chandler theory was less satisfactory than that of the scaled particle models.

Another primitive conclusion is that the scaled-particle models (135, 138) have been the most valuable theories for primitive hydrophobic effects. This is due to the quantitative focus of those models, which permitted more incisive analyses (12–15, 135, 139, 140) in contrast to pictorial theories. Those analyses have lead to significant advances in our understanding of these problems. The connection from scaled-particle models, to Mayer-Montroll series (10, 158), the potential distribution theorem (3, 159), and the quasichemical approach (11, 19–23) identify a promising line for further molecular theoretical progress on these problems. Comparing Equations 4 and 5, the amended Pratt-Chandler theory is most appropriately viewed as a quasichemical theory. The anticipated theoretical progress will treat more thoroughly the effects of changes in temperature, pressure, and composition of the solution, including salt effects, and will treat neglected context hydrophobicity (195) in detail. That work will study cold denaturation and chemical denaturation in molecular detail and will begin to discriminate hydrophobic effects in the cores of soluble proteins from hydrophobic effects in membranes and micelles. That work will begin to consider hydrophobic effects in nanotechnology with molecular specificity.

A natural explanation of thermodynamic signatures of hydrophobic hydration, particularly entropy convergence, emerges from these theoretical advances. How those temperature behaviors are involved in cold denaturation or the stability of thermophilic proteins will be a topic for future research.

Much has been made of the Gaussian character of results such as Figure 2. The observation (1) best supporting this view can be explained as a cancellation of approximation errors (23); slight inaccuracies of a Percus-Yevick (Gaussian) approximation are balanced by neglect of incipient interface formation for atomic sized solutes. In a number of other cases where these distributions have been investigated carefully, simple parabolic models of results such as Figure 2 are less accurate for thermodynamic properties and not only because of the influence of a second thermodynamic phase nearby. Nevertheless, quadratic models provide convenient, reasonable starting points for these analyses.

A final conclusion regards the better discrimination of contact and noncontact hydrophobic interactions. The contact hydrophobic interactions express the classic picture of entropy dominance at lower temperatures. The noncontact interactions have more variability and are likely to be involved in more unusual effects such as pressure denaturation where a historical picture of hydrophobicity had been paradoxical.

ACKNOWLEDGMENTS

The theoretical collaboration at Los Alamos mentioned in the Introduction has included G. Hummer (NIH), A.E. García (LANL), S. Garde (Rensselaer Polytechnic Institute), M.A. Gomez (Vassar College), R.A. LaViolette (INEEL), M.E. Paulaitis (Johns Hopkins University), and A. Pohorille (NASA Ames Research Center). I thank those collaborators for their numerous essential contributions that I have discussed in a personal way here. I thank H.S. Ashbaugh and M.E. Paulaitis for helpful discussions of this review. This work was supported by the U.S. Department of Energy under contract W-7405-ENG-36 and the LDRD program at Los Alamos, LA-UR-01-4900.

Visit the Annual Reviews home page at www.annualreviews.org

LITERATURE CITED

1. Hummer G, Garde S, García AE, Pohorille A, Pratt LR. 1996. *Proc. Natl. Acad. Sci. USA* 93:8951–55
2. Garde S, Hummer G, García AE, Paulaitis ME, Pratt LR. 1996. *Phys. Rev. Lett.* 77:4966–68
3. Pratt LR. 1998. *Encyclopedia of Computational Chemistry: "Hydrophobic Effects,"* pp. 1286–94. Chichester, UK: Wiley
4. Hummer G, Garde S, García AE, Paulaitis ME, Pratt LR. 1998. *Proc. Natl. Acad. Sci. USA* 95:1552–55
5. Hummer G, Garde S. 1998. *Phys. Rev. Lett.* 80:4193–96
6. Hummer G, Garde S, García AE, Paulaitis ME, Pratt LR. 1998. *J. Phys. Chem. B* 102:10469–82
7. Pohorille A. 1998. *Pol. J. Chem.* 72:1680–90
8. Gomez MA, Pratt LR, Hummer G, Garde S. 1999. *J. Phys. Chem. B* 103:3520–23
9. Garde S, García AE, Pratt LR, Hummer G. 1999. *Biophys. Chem.* 78:21–32
10. Pratt LR, Hummer G, Garde S. 1999. In *New Approaches to Problems in Liquid State Theory*, ed. C Caccamo, J-P Hansen, G Stell, 529:407–20. Netherlands: Kluwer
11. Hummer G, Garde S, García A, Pratt LR. 2000. *Chem. Phys.* 258:349–70
12. Lee B. 1985. *Biopolymers* 24:813–23
13. Pohorille A, Pratt LR. 1990. *J. Am. Chem. Soc.* 112:5066–74
14. Pratt LR, Pohorille A. 1992. *Proc. Natl. Acad. Sci. USA* 89:2995–99
15. Pratt LR, Pohorille A. 1993. In *Proc. EBSA Int. Workshop Water-Biomolecule Interactions*, ed. MU Palma, MB Palma-Vittorelli, F Parak, pp. 261–68. Bologna, Italy: Società Italiana di Fisica
16. Chandler D. 1993. *Phys. Rev. E* 48:2898–2905
17. Pratt LR, Chandler D. 1977. *J. Chem. Phys.* 67:3683–704
18. Chan DYC, Mitchell DJ, Ninham BW, Pailthorpe BA. 1979. In *Water: A Comprehensive Treatise*, ed. F Franks, 6:239–78. New York: Plenum
19. Pratt L, LaViolette RA. 1998. *Mol. Phys.* 94:909–15
20. Pratt LR, Rempe SB. 1999. In *Simulation and Theory of Electrostatic Interactions in Solution. Computational Chemistry, Biophysics, and Aqueous Solutions. AIP Conf. Proc.*, ed. LR Pratt, G Hummer, 492:172–201. Melville, NY: Am. Inst. Phys.
21. Rempe SB, Pratt LR, Hummer G, Kress JD, Martin RL, Redondo A. 2000. *J. Am. Chem. Soc.* 122:966–67
22. Rempe SB, Pratt LR. 2001. *Fluid Phase Equilibria* 183–184:121–32
23. Pratt LR, LaViolette RA, Gomez MA, Gentile ME. 2001. *J. Phys. Chem. B* 105:11662–68
24. Privalov PL. 1990. *Crit. Rev. Biochem. Mol. Biol.* 25:281–305
25. Parsegian VA, Rand RP, Rau DC. 2000. *Proc. Natl. Acad. Sci. USA* 97:3987–92
26. Dashevsky VG, Sarkisov GN. 1974. *Mol. Phys.* 27:1271–90
27. Owicki JC, Scheraga HA. 1977. *J. Am. Chem. Soc.* 99:7413–18
28. Swaminathan S, Harrison SW, Beveridge DL. 1978. *J. Am. Chem. Soc.* 100:5705–12
29. Geiger A, Rahman A, Stillinger FH. 1979. *J. Chem. Phys.* 70:263–76
30. Pangali C, Rao M, Berne BJ. 1979. *J. Chem. Phys.* 71:2982–90
31. Okazaki S, Nakanishi K, Tonhara H, Watanabe N. 1981. *J. Chem. Phys.* 74:5863–77
32. Bigot B, Jorgensen WL. 1981. *J. Chem. Phys.* 75:1944–52
33. Postma JPM, Berendsen HJC, Haak JR.

1982. *Faraday Symp. Chem. Soc.* pp. 55–67
34. Okazaki S, Touhara H, Nakanishi K, Watanabe N. 1982. *Bull. Chem. Soc. Jpn.* 55:2827–30
35. Kincaid RH, Scheraga HA. 1982. *J. Comp. Chem.* 3:525–47
36. Rossky PJ, Zichi DA. 1982. *Faraday Symp. Chem. Soc.* 69–78
37. Rapaport DC, Scheraga HA. 1982. *J. Phys. Chem.* 86:873–80
38. Tani A. 1983. *Mol. Phys.* 48:1229–40
39. Swope W, Andersen HC. 1984. *J. Phys. Chem.* 88:6548–56
40. Remerie K, van Gunsteren WF, Postma JPM, Berendsen HJC, Engberts JBFN. 1984. *Mol. Phys.* 53:1517–26
41. Linse P, Karlstrom G, Jonsson B. 1984. *J. Am. Chem. Soc.* 106:4096–102
42. Jorgensen WL, Gao J, Ravimohan C. 1985. *J. Phys. Chem.* 89:3470–73
43. Straatsma TP, Berendsen HJC, Postma JPM. 1986. *J. Chem. Phys.* 85:6720–27
44. Zichi DA, Rossky PJ. 1986. *J. Chem. Phys.* 84:2814–22
45. Zichi DA, Rossky PJ. 1986. *J. Chem. Phys.* 84:1712–23
46. Fois ES, Gamba A, Morosi G, Dementis P, Suffritti GB. 1986. *Mol. Phys.* 58:65–83
47. Tanaka H. 1987. *J. Chem. Phys.* 86:1512–20
48. Fleischman SH, Brooks CL. 1987. *J. Chem. Phys.* 87:3029–37
49. Koop OY, Perelygin IS. 1988. Transl. Z. *Fiz. Khim.* 62:2085–90 (From German)
50. Jorgensen WL, Blake JF, Buckner JK. 1989. *Chem. Phys.* 129:193–200
51. Rao BG, Singh UC. 1989. *J. Am. Chem. Soc.* 111:3125–33
52. Linse P. 1990. *J. Am. Chem. Soc.* 112:1744–50
53. Guillot B, Guissani Y, Bratos S. 1991. *J. Chem. Phys.* 95:3643–48
54. Laaksonen A, Stilbs P. 1991. *Mol. Phys.* 74:747–64
55. Cummings PT, Cochran HD, Simonson JM, Mesmer RE, Karaborni S. 1991. *J. Chem. Phys.* 94:5606–21
56. Tanaka H, Nakanishi K. 1991. *J. Chem. Phys.* 95:3719–27
57. Andaloro G, Sperandeo-Mineo RM. 1992. *Eur. J. Phys.* 11:275–82
58. Fleischman SH, Zichi DA. 1991. *J. Chim. Phys. Phys.-Chim. Biol.* 88:2617–22
59. Lazaridis T, Paulaitis ME. 1992. *J. Phys. Chem.* 96:3847–55
60. Wallqvist A. 1992. *J. Chem. Phys.* 96:1655–56
61. Sun Y, Spellmeyer D, Pearlman DA, Kollman P. 1992. *J. Am. Chem. Soc.* 114:6798–801
62. Guillot B, Guissani Y. 1993. *J. Chem. Phys.* 99:8075–94
63. Guillot B, Guissani Y, Bratos S. 1993. Transl. Z. *Fiz. Khim.* 67:25–31 (From German)
64. Smith D, Haymet ADJ. 1993. *J. Chem. Phys.* 98:6445–54
65. van Belle D, Wodak SJ. 1993. *J. Am. Chem. Soc.* 115:647–52
66. Guillot B, Guissani Y. 1993. *Mol. Phys.* 79:53–75
67. Zeng J, Hush NS, Reimers JR. 1993. *Chem. Phys. Lett.* 206:318–22
68. Skipper NT. 1993. *Chem. Phys. Lett.* 207:424–29
69. Beglov D, Roux B. 1994. *J. Chem. Phys.* 100:9050–63
70. Forsman J, Jonsson B. 1994. *J. Chem. Phys.* 101:5116–25
71. Madan B, Lee B. 1994. *Biophys. Chem.* 51:279–89
72. Matubayasi N. 1994. *J. Am. Chem. Soc.* 116:1450–56
73. Matubayasi N, Reed LH, Levy RM. 1994. *J. Phys. Chem.* 98:10640–49
74. Bushuev YG. 1994. *Zh. Obshch. Khim.* 64:1931–34
75. Lazaridis T, Paulaitis ME. 1994. *J. Phys. Chem.* 98:635–42
76. Kaminski G, Duffy EM, Matsui T, Jorgensen WL. 1994. *J. Phys. Chem.* 98:13077–82

77. Head-Gordon T. 1994. *Chem. Phys. Lett.* 227:215–20
78. Beutler TC, Beguelin DR, van Gunsteren WF. 1995. *J. Chem. Phys.* 102:3787–93
79. Head-Gordon T. 1995. *J. Am. Chem. Soc.* 117:501–7
80. Sun YX, Kollman PA. 1995. *J. Comp. Chem.* 16:1164–69
81. Wallqvist A, Berne BJ. 1995. *J. Phys. Chem.* 99:2885–92
82. Head-Gordon T. 1995. *Proc. Natl. Acad. Sci. USA* 92:8308–12
83. Mancera RL, Buckingham AD. 1995. *J. Phys. Chem.* 99:14632–40
84. Ashbaugh HS, Paulaitis ME. 1996. *J. Phys. Chem.* 100:1900–13
85. Garde S, Hummer G, García AE, Pratt LR, Paulaitis ME. 1996. *Phys. Rev. E* 53:R4310–13
86. Prevost M, Oliveira IT, Kocher JP, Wodak SJ. 1996. *J. Phys. Chem.* 100:2738–43
87. Chau P, Forester T, Smith W. 1996. *Mol. Phys.* 89:1033–55
88. Re M, Laria D, Fernández-Prini R. 1996. *Chem. Phys. Lett.* 250:25–30
89. Matubayasi N, Levy RM. 1996. *J. Phys. Chem.* 100:2681–88
90. Durell SR, Wallqvist A. 1996. *Biophys. J.* 71:1695–1706
91. Wallqvist A, Covell DG. 1996. *Biophys. J.* 71:600–8
92. Skipper NT, Bridgeman CH, Buckingham AD, Mancera RL. 1996. *Faraday Disc.* 103:141–50
93. Haymet ADJ, Silverstein KAT, Dill KA. 1996. *Faraday Disc.* 103:117–24
94. Garde S, Hummer G, Paulaitis ME. 1996. *Faraday Disc.* 103:125–39
95. Lin CL, Wood RH. 1996. *J. Phys. Chem.* 100:16399–409
96. Mancera RL. 1996. *J. Chem. Soc. Faraday Trans.* 92:2547–54
97. Meng EC, Kollman PA. 1996. *J. Phys. Chem.* 100:11460–70
98. Lynden-Bell R, Rasaiah J. 1997. *J. Chem. Phys.* 107:1981–91
99. Floris F, Selmi M, Tani A, Tomasi J. 1997. *J. Chem. Phys.* 107:6353–65
100. DeJong PHK, Wilson JE, Neilson GW, Buckingham AD. 1997. *Mol. Phys.* 91:99–103
101. Radmer RJ, Kollman PA. 1997. *J. Comp. Chem.* 18:902–19
102. Mancera RL, Buckingham AD, Skipper NT. 1997. *J. Chem. Soc. Faraday Trans.* 93:2263–67
103. Silverstein KAT, Dill KA, Haymet ADJ. 1998. *Fluid Phase Equilibria* 151:83–90
104. Silverstein KAT, Haymet ADJ, Dill KA. 1998. *J. Am. Chem. Soc.* 120:3166–75
105. Ikeguchi M, Shimizu S, Nakamura S, Shimizu K. 1998. *J. Phys. Chem. B* 102:5891–98
106. Arthur J, Haymet A. 1998. *J. Chem. Phys.* 109:7991–8002
107. Panhuis M, Patterson C, Lynden-Bell R. 1998. *Mol. Phys.* 94:963–72
108. Mountain RD, Thirumalai D. 1998. *Proc. Natl. Acad. Sci. USA* 95:8436–40
109. Errington JR, Boulougouris GC, Economou IG, Panagiotopoulos AZ, Theodorou DN. 1998. *J. Phys. Chem. B* 102:8865–73
110. Mancera RL. 1998. *J. Chem. Soc. Faraday Trans.* 94:3549–59
111. Silverstein KAT, Haymet ADJ, Dill KA. 1999. *J. Chem. Phys.* 111:8000–9
112. Tomas-Oliveira I, Wodak SJ. 1999. *J. Chem. Phys.* 111:8576–87
113. Arthur JW, Haymet ADJ. 1999. *J. Chem. Phys.* 110:5873–83
114. Pomes R, Eisenmesser E, Post CB, Roux B. 1999. *J. Chem. Phys.* 111:3387–95
115. Urahata S, Canuto S. 1999. *Chem. Phys. Lett.* 313:235–40
116. Fois E, Gamba A, Redaelli C. 1999. *J. Chem. Phys.* 110:1025–35
117. Slusher JT. 1999. *J. Phys. Chem. B* 103:6075–79
118. Somasundaram T, Lynden-Bell RM, Patterson CH. 1999. *Phys. Chem. Chem. Phys.* 1:143–48
119. Chau PL, Mancera RL. 1999. *Mol. Phys.* 96:109–22

120. Smith PE. 1999. *J. Phys. Chem. B* 103:525–34
121. Madan B, Sharp K. 1999. *Biophys. Chem.* 78:33–41
122. Guisoni N, Henriques VB. 2000. *Braz. J. Phys.* 30:736–40
123. Svishchev I, Zassetsky A, Kusalik P. 2000. *Chem. Phys.* 258:181–86
124. Rasaiah JC, Noworyta JP, Koneshan S. 2000. *J. Am. Chem. Soc.* 122:11182–93
125. Urbic T, Vlachy V, Kalyuzhnyi Y, Southall N, Dill K. 2000. *J. Chem. Phys.* 112:2843–48
126. Noworyta JP, Koneshan S, Rasaiah JC. 2000. *J. Am. Chem. Soc.* 122:11194–202
127. Schurhammer R, Wipff G. 2000. *J. Phys. Chem. A* 104:11159–68
128. Southall NT, Dill KA. 2000. *J. Phys. Chem. B* 104:1326–31
129. Gallicchio E, Kubo MM, Levy RM. 2000. *J. Phys. Chem. B* 104:6271–85
130. Hernández-Cobos J, Mackie AD, Vega LF. 2001. *J. Chem. Phys.* 114:7527–35
131. Bergman DL, Lynden-Bell RM. 2001. *Mol. Phys.* 99:1011–21
132. Raschke TM, Tsai J, Levitt M. 2001. *Proc. Natl. Acad. Sci. USA* 98:5965–69
133. Garde S, Ashbaugh HS. 2001. *J. Chem. Phys.* 115:977–82
134. Kaira A, Tugcu N, Cramer SM, Garde S. 2001. *J. Phys. Chem. B* 105:6380–86
135. Stillinger FH. 1973. *J. Solid Chem.* 2:141–58
136. Pratt LR. 1991. In *CLS Division 1991 Annual Review*, LA-UR-91-1783, Natl. Tech. Inform. Serv. US Dept. Comm., Springfield, VA
137. Tang KES, Bloomfield VA. 2000. *Biophys. J.* 79:2222–34
138. Pierotti RA. 1976. *Chem. Rev.* 76:717–26
139. Lucas M. 1976. *J. Phys. Chem.* 80:359–62
140. Ben-Naim A, Friedman HL. 1967. *J. Phys. Chem.* 71:448–49
141. Madan B, Lee B. 1994. *Biophys. Chem.* 51:279–89
142. Silverstein TP. 1998. *J. Chem. Ed.* 75:116–18
143. Rowlinson JS, Swinton FL. 1982. *Liquids and Liquid Mixtures*. London: Butterworths
144. Postma JPM, Berendsen HJC, Haak JR. 1982. *Faraday Symp. Chem. Soc.* 17:55–67
145. Pohorille A, Wilson MA. 1993. *J. Mol. Struct. Theochem.* 103:271–98
146. Wolfenden R, Radzicka A. 1994. *Science* 265:936–37
147. Kocher JP, Prevost M, Wodak SJ, Lee B. 1996. *Structure* 4:1517–29
148. Crooks GE, Chandler D. 1997. *Phys. Rev. E* 56:4217–21
149. Stamatopoulou A, Ben-Amotz D. 1998. *J. Chem. Phys.* 108:7294–7300
150. Tanaka H. 1998. *Chem. Phys. Lett.* 282:133–38
151. Mountain R. 1999. *J. Chem. Phys.* 110:2109–15
152. Garde S, Khare R, Hummer G. 2000. *J. Chem. Phys.* 112:1574–78
153. in'tVeld PJI, Stone MT, Truskett TM, Sanchez IC. 2000. *J. Phys. Chem. B* 104:12028–34
154. Kussell E, Shimada J, Shakhnovich E. 2001. *J. Mol. Biol.* 311:183–93
155. Lesemann M, Thirumoorthy K, Kim YJ, Jonas J, Paulaitis ME. 1998. *Langmuir* 14:5339–41
156. Percus JK. 1993. *J. Phys. IV* 3:49–57
157. Mitchell D, Ninham B, Pailthorpe B. 1977. *Chem. Phys. Lett.* 51:257–60
158. Stell G. 1985. In *The Wonderful World of Stochastics. A Tribute to Elliot W. Montroll*, ed. MF Schlesinger, GH Weiss, XII:127–56. New York: Elsevier
159. Widom B. 1963. *J. Chem. Phys.* 39:2808–12
160. Torrie GM, Valleau JP. 1977. *J. Comp. Phys.* 23:187–99
161. Valleau JP, Torrie GM. 1977. *Statistical Mechanics, Part A: Equilibrium Techniques*, ed. BJ Berne, 5:169–94. New York: Plenum
162. Weeks JD. 2002. *Annu. Rev. Phys. Chem.* 53:533–62
163. Ashbaugh HS, Paulaitis ME. 2001. *J. Am. Chem. Soc.* 123:10721–28

164. Chandler D, Weeks JD, Andersen HC. 1983. *Science* 220:787–94
165. Stell G. 1977. Modern theoretical chemistry. In *Statistical Mechanics, Part A: Equilibrium Techniques*, ed. BJ Berne, 5:47–84. New York: Plenum
166. Pollack GL. 1991. *Science* 251:1323–30
167. Baldwin RL. 1986. *Proc. Nail. Acad. Sci. USA* 83:8069–72
168. Lee B. 1991. *Proc. Natl. Acad. Sci. USA* 88:5154–58
169. Boulougouris GC, Voutsas EC, Economou IG, Theodorou DN, Tassios DP. 2001. *J. Phys. Chem. B* 105:7792–98
170. Filipponi A, Bowron DT, Lobban C, Finney JL. 1997. *Phys. Rev. Lett.* 79:1293–96
171. Bowron DT, Filipponi A, Lobban C, Finney JL. 1998. *Chem. Phys. Lett.* 293:33–37
172. Bowron DT, Filipponi A, Roberts MA, Finney JL. 1998. *Phys. Rev. Letts.* 81:4164–67
173. Bowron DT, Weigel R, Filipponi A, Roberts MA, Finney JL. 2001. *Mol. Phys.* 99:761–65
174. Sullivan DM, Neilson GW, Fischer HE. 2001. *J. Chem. Phys.* 115:339–43
175. Ghosh T, Garde S, García AE. 2001. *J. Am. Chem. Soc.* In press
176. Frank HS, Evans MW. 1945. *J. Chem. Phys.* 13:507–32
177. Pellegrini M, Doniach S. 1995. *J. Chem. Phys.* 103:2696–702
178. Pellegrini M, Gronbech-Jensen N, Doniach S. 1996. *J. Chem. Phys.* 104:8639–48
179. Ashbaugh HS, Garde S, Hummer G, Kaler EW, Paulaitis ME. 1999. *Biophys. J.* 77:645–54
180. Wallqvist A. 1991. *J. Phys. Chem.* 95:8921–27
181. Cheng Y, Rossky PJ. 1998. *Nature* 392:696–99
182. Pangali C, Rao M, Berne BJ. 1978. *ACS Symp. Ser.* 1978:32–34
183. Pangali C, Rao M, Berne BJ. 1979. *J. Chem. Phys.* 71:2975–81
184. Swaminathan S, Beveridge DL. 1979. *J. Am. Chem. Soc.* 101:5832–33
185. Ravishanker G, Mezei M, Beveridge DL. 1982. *Faraday Symp. Chem. Soc.* 17:79–91
186. Beveridge DL, Mezei M, Ravishanker G, Jayaram B. 1985. *J. Biosci.* 8:167–78
187. Ravishanker G, Beveridge DL. 1985. *J. Am. Chem. Soc.* 107:2565–66
188. Backx P, Goldman S. 1985. *Chem. Phys. Lett.* 113:578–81
189. Backx P, Goldman S. 1985. *Chem. Phys. Lett.* 119:144–48
190. Watanabe K, Andersen HC. 1986. *J. Phys. Chem.* 90:795–802
191. Watanabe K, Andersen HC. 1986. In *Molecular-Dynamics Simulation of Statistical-Mechanical Systems*, ed. G Ciccotti, WG Hoover, 97:418–23. Amsterdam: *ASI-NATO Int. Sch. Phys. "Enrico Fermi"*
192. Wallqvist A, Berne BJ. 1988. *Chem. Phys. Lett.* 145:26–32
193. Jorgensen WL, Buckner JK, Boudon S, Tiradorives J. 1988. *J. Chem. Phys.* 89:3742–46
194. Berne BJ, Wallqvist A. 1989. *Chem. Scr.* 29A:85–91
195. Rossky PJ, Friedman HL. 1980. *J. Phys. Chem.* 84:587–89
196. Jorgensen WL, Severance DL. 1990. *J. Am. Chem. Soc.* 112:4768–74
197. Linse P. 1992. *J. Am. Chem. Soc.* 114:4366–73
198. Linse P. 1993. *J. Am. Chem. Soc.* 115:8793–97
199. Smith DE, Zhang L, Haymet ADJ. 1992. *J. Am. Chem. Soc.* 114:5875–76
200. Smith DE, Haymet ADJ. 1993. *J. Chem. Phys.* 98:6445–54
201. Dang LX. 1994. *J. Chem. Phys.* 100:9032–34
202. New MH, Berne BJ. 1995. *J. Am. Chem. Soc.* 117:7172–79
203. Lüdemann S, Schreiber H, Abseher R, Steinhauser O. 1996. *J. Chem. Phys.* 104:286–95
204. Lüdemann S, Abseher R, Schreiber H,

Steinhauser O. 1997. *J. Am. Chem. Soc.* 119:4206–13
205. Payne VA, Matubayasi N, Murphy LR, Levy RM. 1997. *J. Phys. Chem. B* 101:2054–60
206. Young WS, Brooks CL. 1997. *J. Chem. Phys.* 106:9265–60
207. Rick SW, Berne BJ. 1997. *J. Phys. Chem. B* 101:10488–93
208. Rank JA. Baker D. 1998. *Biophys. Chem.* 71:199–204
209. Gao JL. 1993. *J. Am. Chem. Soc.* 115:6893–95
210. Chipot C, Jaffe R, Maigret B, Pearlman D, Kollman P. 1996. *J. Am. Chem. Soc.* 118:11217–24
211. Rick SW. 2000. *J. Phys. Chem. B* 104:6884–88
212. Shimizu S, Chan HS. 2000. *J. Chem. Phys.* 113:4683–4700
213. Gervazio FL, Chelli R, Marchi M, Procacci P, Scettino V. 2001. *J. Phys. Chem. B* 105:7835–46
214. Ghosh T, Garde S, García AE. 2001. *J. Chem. Phys.* In press
215. Pratt LR, Chandler D. 1980. *J. Chem. Phys.* 73:3434–41
216. Pratt LR. 1985. *Annu. Rev. Phys. Chem.* 36:433–49
217. Hummer G. 2001. *J. Chem. Phys.* 114:7330–37
218. Jorgensen WL. 1982. *J. Chem. Phys.* 77:5757–65
219. Rosenberg RO, Mikkilineni R, Berne BJ. 1982. *J. Am. Chem. Soc.* 104:7647–49
220. Jorgensen WL, Buckner JK. 1987. *J. Phys. Chem.* 91:6083–85
221. Tobias DJ, Brooks CL. 1990. *J. Chem. Phys.* 92:2582–92
222. Wallqvist A, Covell DG. 1995. *J. Phys. Chem.* 99:13118–25
223. Garde S, Hummer G, Paulaitis ME. 1996. *Faraday Disc.* 103:125–39
224. Ashbaugh HS, Kaler EW, Paulaitis ME. 1998. *Biophys. J.* 75:755–68
225. Hummer G. 1999. *J. Am. Chem. Soc.* 121:6299–305
226. Borodin O, Bedrov D, Smith GD. 2001. *Macromolecules* 34:5687–93
227. Akasaka K, Li H. 2001. *Biochemistry* 40:8665–71
228. Kauzmann W. 1987. *Nature* 325:763–64
229. Wallqvist A. 1992. *J. Chem. Phys.* 96:1655–56
230. Richards FM. 1991. *Sci. Am.* 264:545
231. Wallqvist A, Berne BJ. 1995. *J. Phys. Chem.* 99:2893–99
232. Privalov PL, Gill SJ. 1989. *Pure Appl. Chem.* 61:1097–1104
233. Lum K, Chandler D, Weeks JD. 1999. *J. Phys. Chem. B* 103:4570–77
234. Huang DM, Geissler PL, Chandler D. 2001. *J. Phys. Chem. B* 105:6704–9
235. Huang DM, Chandler D. 2000. *Proc. Natl. Acad. Sci. USA* 97:8324–27
236. Huang DM, Chandler D. 2000. *Phys. Rev. E* 61:1501–6
237. Sun SX. 2001. *Phys. Rev. E* 64:1512–20
238. Müller EA, Gubbins KE. 1998. *Carbon* 36:1433–38
239. Brovchenko I, Paschek D, Geiger A. 2000. *J. Chem. Phys.* 113:5026–36
240. Wallqvist A, Gallicchio E, Levy RM. 2001. *J. Phys. Chem. B* 105:6745–53
241. Brovchenko I, Geiger A, Oleinikova A. 2001. *Phys. Chem. Chem. Phys.* 3:1567–69

STUDIES OF POLYMER SURFACES BY SUM FREQUENCY GENERATION VIBRATIONAL SPECTROSCOPY

Zhan Chen
Department of Chemistry, University of Michigan, Ann Arbor, Michigan 48109;
e-mail: zhanc@umich.edu

Y. R. Shen[1] and Gabor A. Somorjai[2]
[1]Department of Physics and [2]Department of Chemistry, University of California at Berkeley and Materials Science Division, Lawrence Berkeley National Laboratory, Berkeley, California 94720; e-mail:shenyr@socrates.berkeley.edu; somorjai@socrates.berkeley.edu

Key Words nonlinear optics, laser, surface restructuring, surface segregation, polymer blend

■ **Abstract** Recently, sum frequency generation (SFG) vibrational spectroscopy has been developed into a powerful technique to study surfaces of polymer materials. This review summarizes the significant achievements in understanding surface molecular chemical structures of polymer materials obtained by SFG. It reviews in situ detection at the molecular level of surface structures of some common polymers in air, surface segregation of small end groups, polymer surface restructuring in water, and step-wise changed polymer blend surfaces. Studies of surface glass transition and surface structures modified by rubbing, plasma deposition, UV light irradiation, oxygen ion and radical irradiation, and wet etching are also discussed. SFG probing of polymer surfaces provides valuable insights into the relations between polymer surface structures and surface properties, which will assist in the design of polymer materials with desired surface properties.

INTRODUCTION

The many techniques of surface science that were developed over the past 35 years permitted atomic and molecular level investigations of the surface monolayer of solids (1, 2). Because of the dominant use of electrons, atoms, and ions as surface probes, low pressures or ultrahigh vacuum were necessary for most of these surface studies. In recent years, optical (photon in–photon out) and scanning probe techniques (atomic force and scanning tunneling microscopies) were developed that permitted molecular level studies at the solid-liquid and solid–high pressure gas interfaces (3–9). Among these new techniques, sum frequency generation vibrational spectroscopy (SFG-VS) emerged as a uniquely monolayer surface-sensitive and informative method to provide surface structure and surface composition via the vibrational spectrum (3–5, 10–25).

The performance of structural polymers such as polyethylene (PE), polypropylene (PP), polystyrene (PS), or poly(methyl methacrylate) (PMMA) is often dominated by interface properties such as wettability, friction, lubrication, adhesion, and compatibility with applied coatings, such as paint (26–31). Biopolymer-liquid

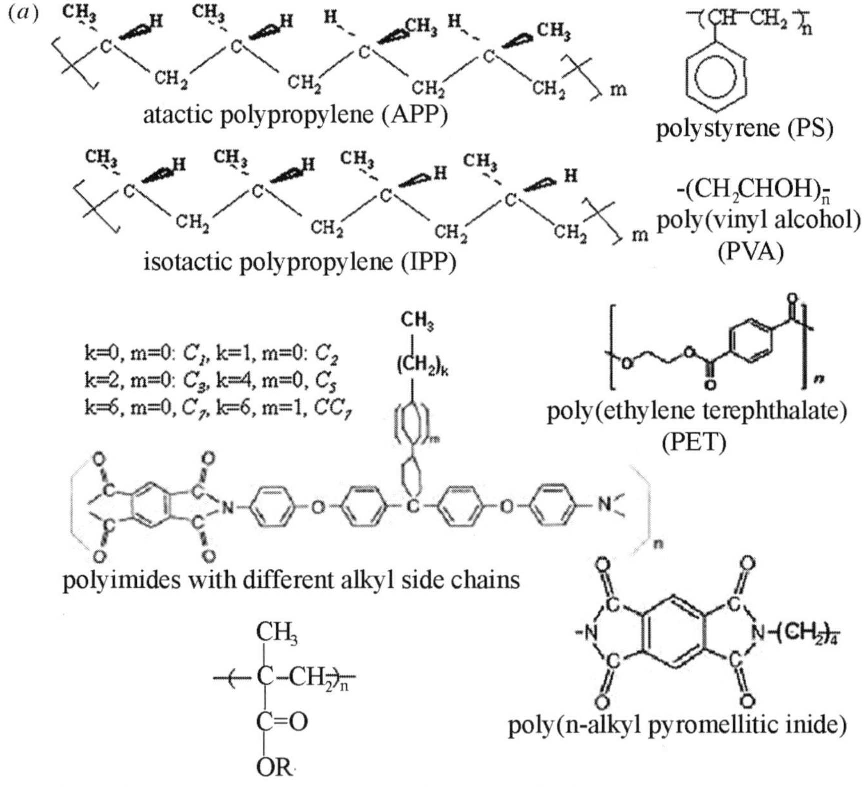

(a) atactic polypropylene (APP)

polystyrene (PS)

isotactic polypropylene (IPP)

$-(CH_2CHOH)_n$ poly(vinyl alcohol) (PVA)

$k=0, m=0: C_1, k=1, m=0: C_2$
$k=2, m=0: C_3, k=4, m=0: C_5$
$k=6, m=0: C_7, k=6, m=1: CC_7$

poly(ethylene terephthalate) (PET)

polyimides with different alkyl side chains

poly(n-alkyl pyromellitic inide)

R= $-CH_3$, poly(methyl methacrylate) (PMMA)
$-(CH_2)_3CH_3$, poly(n-butyl methacrylate) (PBMA)
$-(CH_2)_7CH_3$, poly(n-octyl methacrylate) (POMA)
$-(CH_2)_2OH$, poly (2-hydroxyethyl) methacrylate (PHEMA)

$HO(CH_2CH_2O)_nH$ poly(ethylene glycol) diol
$CH_3O(CH_2CH_2O)_nH$ poly(ethylene glycol) methyl ether
$CH_3O(CH_2CH_2O)_nCH_3$ poly(ethylene glycol) dimethyl ether

Figure 1 Molecular formulas of polymers studied in this paper.

interfaces are overwhelmingly important in human biology as our body is a polymer-water interface system. Polymeric implants gained increasing acceptance in recent years (contact lenses, knee-cap replacements, heart valves, and others), and their interactions with biological molecules in aqueous media influence their biocompatibility (28, 32, 33).

SFG has been successfully employed to study polymer, copolymer, and polymer blend surfaces to provide molecular level information at air and water interfaces (34–48). In this paper, we summarize recent progress in SFG studies of polymer and biopolymer surfaces. Important results obtained by SFG that are discussed here include: (*a*) surface structures of some common polymers in air, (*b*) change of polymer surface structures by varying end groups of polymer units, (*c*) modification of polymer surfaces by surface treatment, (*d*) surface structures of polymer blends, and (*e*) restructuring of polymer surfaces in aqueous solution. The molecular formulas for the polymers to be discussed are displayed in Figures 1*a* and 1*b*.

Other techniques have been used for studies of polymer interfaces. Among them contact angle measurements are macroscopic in nature but have a molecular origin as they respond to changes of interfacial energy (49, 50). Good correlation between changes of molecular surface structures as monitored by SFG and variations of

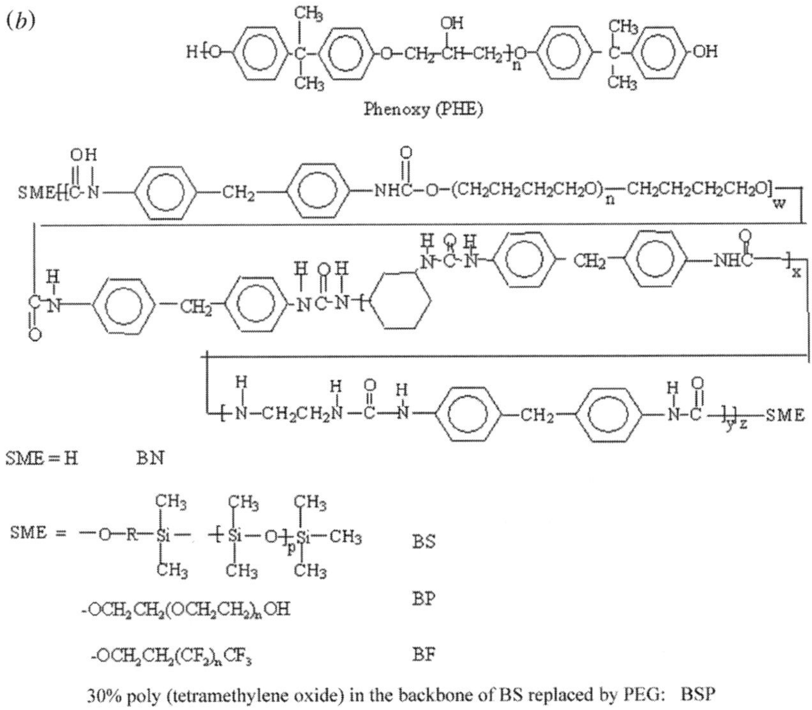

Figure 1 (*Continued*)

contact angles has been found (42, 43, 45). Atomic level information on polymer surfaces has been obtained by X-ray photoelectron spectroscopy (XPS) (51), secondary ion mass spectroscopy (SIMS) (52, 53), near-edge X-ray absorption fine structure spectroscopy (NEXAFS) (54), ellipsometry (55), neutron reflection (56), and infrared and Raman spectroscopies (57–59). Although all these techniques provide valuable information about polymer interfaces, we believe that SFG excels in these studies for two reasons: It is monolayer surface sensitive and it can be employed in situ, at the solid-liquid or solid–high pressure gas (air) interface without loss of molecular surface information.

SFG TECHNIQUE

SFG is a process in which two input beams at frequencies ω_1 and ω_2 mix in a medium and generate an output beam at the sum frequency $\omega = \omega_1 + \omega_2$. As a second-order nonlinear optical process, it is forbidden under the electric-dipole approximation in media with inversion symmetry. At surfaces or interfaces where inversion symmetry is broken, SFG is naturally allowed and can therefore be used as an effective surface probe. If either $\omega_1(\omega_2)$ or $\omega_1 + \omega_2$ or both are scanned over surface resonances, SFG is resonantly enhanced, thus producing a spectrum characteristic of the surface.

SFG can be used to probe electronic as well as vibrational transitions in a medium. For molecular systems, vibrational spectroscopy (VS) is often more informative. SFG-VS requires a tunable infrared input (ω_2) for scanning over resonances. In contrast to infrared or Raman spectroscopy, it often probes only vibrational modes that are both infrared and Raman active. Unlike infrared or Raman spectroscopy, SFG is surface-specific in media with inversion symmetry. For example, no SFG can be observed from the IR-inactive stretching modes of symmetric molecules such as O_2 or N_2 adsorbed on a surface, if the surface-molecule interaction is weak. However, if the surface-molecule interaction is strong enough to make the stretch modes IR active, then SFG will be observed. A plot of SFG intensities versus the IR input frequency yields the vibrational spectrum of the surface species. With different input/output polarization combinations with respect to beam and sample geometries, additional information on surface molecules and structures can be obtained (60–64).

A representative experimental arrangement for SFG-VS is depicted schematically in Figure 2. Both the visible and the infrared beams are generated by a picosecond pulsed Nd:YAG laser/optical parametric generator-amplifier system, with ~20 pico-second pulse width and ~20 Hz repetition rate (65). The visible input is fixed at frequency 18,800 cm^{-1} (532 nm), but the IR input is tunable from 15,000 to ~4,000 cm^{-1}. Via difference frequency generation in a nonlinear optical crystal, the IR tunability can be extended down to 1,100 cm^{-1}. SFG output can be detected and collected by a photomultiplier tube and a gated photon counting system. Variations of SFG setups are used in different research laboratories, depending on the types of lasers employed. For example, the laboratory at the National Institute

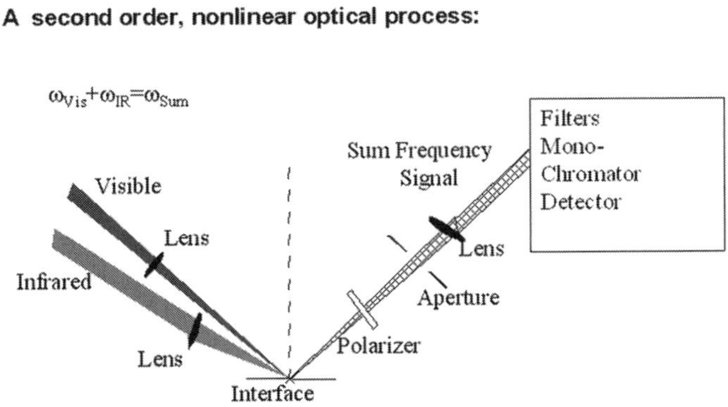

Figure 2 A second-order, nonlinear optical process. Schematic of the SFG experimental arrangement.

of Standards and Technology has developed a setup using femtosecond pulses as the light source (22).

Various experiments show that SFG is indeed surface specific and sensitive as a polymer surface probe. Comparing reflected and transmitted SFG spectra from the same sample (66), for example, suggests that the polymer bulk contribution in reflected SFG is usually negligible. Modifying a polymer surface by immersing it in water or other liquids or by exposing it to plasma treatment for a short time often dramatically changes the reflected SFG spectra, again indicating that the surface contribution dominates in reflected SFG (37, 38, 40, 41, 46). Independence of SFG spectral (18, 41) intensity from polymer film thickness is another way to prove that reflected SFG is dominated by surface contribution.

ANALYSIS OF SFG SPECTRA

The SFG output intensity in the reflected direction can be written as (64)

$$I(\omega) = \frac{8\pi^3 \omega^2 \sec^2 \beta}{c^3 n(\omega) n(\omega_1) n(\omega_2)} |\chi_{eff}^{(2)}|^2 I_1(\omega_1) I_2(\omega_2), \qquad 1.$$

where $n(\Omega)$ is the refractive index of the medium at frequency Ω, β is the reflection angle of the sum frequency field, $I_1(\omega_1)$ and $I_2(\omega_2)$ are the intensities of the two input fields, and $\chi_{eff}^{(2)}$ is the effective second-order nonlinear susceptibility tensor of the surface, which is proportional to the concentration of the surface functional groups. With IR-visible SFG, if the IR frequency (ω_2) is near vibrational resonances, we can write

$$\chi_{eff}^{(2)} = \chi_{nr} + \sum_q \frac{A_q}{\omega_2 - \omega_q + i\Gamma_q}, \qquad 2.$$

where χ_{nr} arises from the nonresonant background contribution, and A_q, ω_q, and Γ_q are the strength, resonant frequency, and damping coefficient of the qth vibrational mode. Values for these parameters can be deduced from fitting of the observed SFG spectra using Equations 1 and 2. Due to the interferences between nonresonant and resonant contributions, or between different resonant modes (12), resonant peaks in the SFG spectra may appear asymmetric. For most polymer materials covered in this review, however, the nonresonant contribution is small, making the spectra less complex.

The effective second-order nonlinear surface susceptibility tensor, $\chi_{eff}^{(2)}$, in relation to the second-order surface nonlinear susceptibility, is defined as

$$\chi_{eff}^{(2)} = \vec{e}(\omega) \cdot \ddot{\chi}^{(2)} : \vec{e}(\omega_1)\vec{e}(\omega_2), \qquad 3.$$

where $\vec{e}(\Omega) = \hat{e}(\Omega) \cdot \ddot{L}(\Omega)$, with $\hat{e}(\Omega)$ being the unit polarization vector and $\ddot{L}(\Omega)$ denoting the transmission Fresnel coefficient at Ω. In SFG with a selective input/output polarization combination, different components of $\ddot{\chi}^{(2)}$ can be accessed. For example, with ssp and sps polarization combination, we have

$$\chi_{eff,ssp}^{(2)} = L_{yy}(\omega)L_{yy}(\omega_1)L_{zz}(\omega_2) \sin\beta_2 \chi_{yyz}^{(2)}$$
$$\chi_{eff,sps}^{(2)} = L_{yy}(\omega)L_{zz}(\omega_1)L_{yy}(\omega_2) \sin\beta_1 \chi_{yzy}^{(2)}. \qquad 4.$$

Here, $\hat{z}$ is along the surface normal and $\hat{x} - \hat{z}$ is the plane of incidence.

The nonlinear susceptibility $\ddot{\chi}^{(2)}$ is directly connected to the molecular hyperpolarizability $\ddot{\alpha}^{(2)}$ by a coordinate transformation

$$\chi_{ijk}^{(2)} = \langle (\hat{i} \cdot \hat{\xi})(\hat{j} \cdot \hat{\eta})(\hat{k} \cdot \hat{\zeta})\rangle \alpha_{\xi\eta\zeta}^{(2)}, \qquad 5.$$

assuming the microscopic local field correction is negligible. The angular brackets here refer to an average over molecular orientations. In simple cases, only a few $\alpha_{\xi\eta\zeta}^{(2)}$ elements are nonvanishing by symmetry. Considering the symmetric stretch of methyl groups at an azimuthally isotropic surface as an example, we have

$$\chi_{xxz}^{(2)} = \chi_{yyz}^{(2)} = \frac{1}{2}N_s\alpha_{ccc}\langle\cos\theta(1+r) - \cos^3\theta(1-r)\rangle$$
$$\chi_{xzx}^{(2)} = \chi_{yzy}^{(2)} = \chi_{zxx}^{(2)} = \chi_{zyy}^{(2)} = \frac{1}{2}N_s\alpha_{ccc}\langle\cos\theta - \cos^3\theta\rangle(1-r)$$
$$\chi_{zzz}^{(2)} = N_s\alpha_{ccc}\langle r\cos\theta + \cos^3\theta(1-r)\rangle. \qquad 6.$$

Here, $\alpha_{aac} = r\alpha_{ccc}$, with $\hat{c}$ being along the symmetric axis of the methyl group, θ is the angle between $\hat{c}$ and the surface normal $\hat{z}$, and N_s is the surface density of methyl groups. Assuming that the molecular orientational distribution in θ is a δ-function, we can deduce θ and r from the ratios of $\ddot{\chi}^{(2)}$ components obtained from measurements.

However, the orientation distribution is generally not a δ distribution. In the above example, we can model it by a Gaussian distribution (40): $f(\theta) = C \exp[-\frac{(\theta-\theta_0)^2}{2\sigma^2}]$, with θ_0 and σ as adjustable parameters. If the value of $r = \alpha_{aac}/\alpha_{ccc}$

is known from other sources, then θ_0 and σ can also be deduced from measured ratios of $\ddot{\chi}^{(2)}$ components. Examples to deduce the orientation and orientational distributions of functional groups on polymer surfaces are shown in References (23, 39, 40).

The above discussion assumes the surface or interface is smooth. Surface roughness clearly would broaden the molecular orientational distribution, although the effect has not yet been studied systematically.

POLYMER SURFACES IN AIR

SFG has been used to study surface structures of a number of polymers, namely, PE, PP, PS, PMMA, and polyimide (22, 23, 34, 35, 37, 39). The molecular formulas for these polymers are shown in Figure 1a. PE and PP represent chemically simple polymer systems, as they are composed of only singly bonded carbon and hydrogen. The surface tension of PP is lower than that of PE (68), indicating that the methyl side groups of PP have significant effect on the surface structure and properties. Surface properties of PS have been widely studied (69–71). The side groups (phenyl groups) of PS also influence its surface structure, resulting in a much higher surface tension than that of PE (68). The surface tension of PMMA is higher than that of PE and PP, but similar to that of PS (68). Comparison of the water contact angles on these polymer surfaces (68) indicates that PP has the most hydrophobic surface, whereas PMMA has the most hydrophilic surface. In this section, we summarize SFG studies on molecular surface structures of PE, PP, PS, and PMMA. We also discuss surface structures of polyimides with different alkyl side groups.

Polymer film samples for SFG studies were prepared by either solvent casting or spin-coating polymer solutions on different substrates, such as fused silica, silicon, or sapphire. The polymer film thicknesses were varied from solvent casting samples ($\sim$100 μm) to spin-coating samples ($\sim$100 nm). Samples were annealed above the glass transition temperature before the SFG measurements. Particularly, polyimide samples were prepared by spin-coating polyamic acid solutions on fused silica substrates and then baked at high temperature for imidization reaction (36, 37).

PE and PP

PE and PP have enormously wide applications. Examples include plastic bags, tubing, electrical insulation, and orthopedic implants (33, 72). The polymers studied here, in increasing order of percentage crystallinity, were atactic PP (APP) ($\sim$2%), low-density PE (LDPE) (20–35%), isotactic PP (IPP) ($\sim$63%), and ultrahigh molecular weight PE (UHMWPE) (70–75%). APP has a lower density compared to IPP because of the random appearance of methyl groups on two sides of the polymer backbone (Figure 1a).

IR and Raman spectra for various PE samples are identical (34), but their SFG spectra (Figure 3) are markedly different, indicating that although their bulk

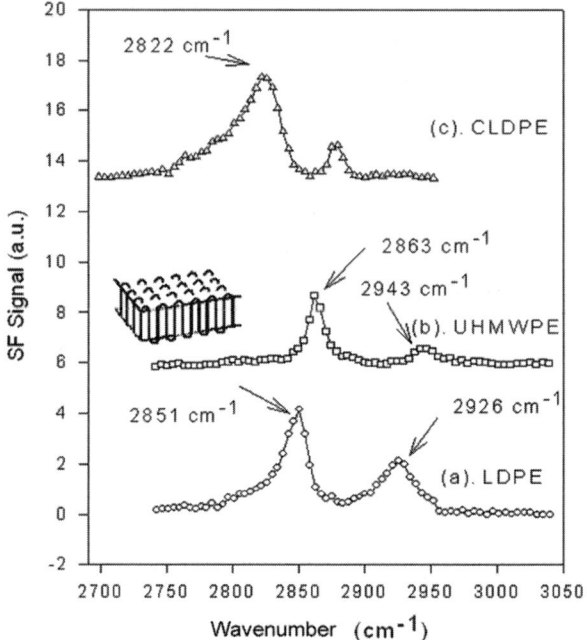

Figure 3 SFG spectra of various polyethylenes: (*a*) LDPE; (*b*) UHMWPE; (*c*) CLDPE.

compositions are basically the same, their surface structures are not. For LDPE (Figure 3*a*), the bands at 2851 and 2926 cm^{-1} can be attributed to the symmetric and asymmetric stretches of CH$_2$ trans conformers. The average orientation of surface CH$_2$ groups is 55°. The random packing of the polymer chains and the disorder of the polymer surface are evidenced by the larger bandwidths of the peaks in the SFG spectrum. However, for UHMWPE (Figure 3*b*) the symmetric and antisymmetric CH$_2$ stretching peaks are narrower and shift to higher frequencies. This discrepancy can be explained by the presence of more gauche conformers on the polymer surface (34). The crystalline phase of PE is composed of thin lamellae of about 10 nm thick, extending up to 1∼10 mm (73). At the PE surface, the lamellar structure has the molecular chains fold back on themselves repetitively (inset of Figure 3*b*), creating a high density of gauche conformers. The average orientation of the methylene group at such a surface is ∼42°. We have also obtained the SFG spectrum of a commercial LDPE (CLDPE) (Figure 3*c*). It is very different from those of LDPE and UHMWPE. The peaks are characteristic of methoxy derivatives, known to be present in PE additives. The spectrum shows that the additives in CLDPE must have preferentially segregated to the surface. This example demonstrates that SFG can differentiate surface structures of three similar polymers at the molecular level.

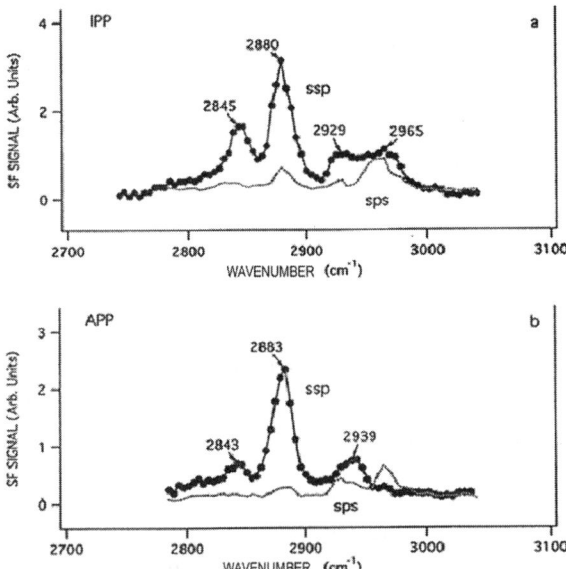

Figure 4 SFG spectra of polypropylenes: (*a*) IPP; (*b*) APP.

The IR and Raman spectra of IPP and APP are again very similar, but their SFG spectra (Figure 4) are quite different (34). From the SFG spectra of APP (Figure 4*b*), the average surface orientation was found to be ~30° for the methyl group, and ~59° for the methylene group. This suggests that the hydrocarbon backbone (methylene groups) tends to lie, on average, parallel to the surface (to optimize its interaction with the underlying chains) whereas the methyl groups project out of the surface. For IPP (Figure 4*a*), although the orientation of the methylene group stays the same, that of the methyl groups changes to ~55°. The result can also be understood by having the polymer chains lie more or less flat on the surface.

SFG was used to probe the glass transition at the surface of APP and IPP in a 10^{-5} Torr vacuum (35). An increase in the ratio of the symmetric stretch of the CH_2 group to that of the CH_3 group in the spectra of both APP and IPP was observed on cooling through the glass transition between 0 and 20°C. It indicates that the CH_2 groups (backbone) become more polar-ordered at the surface below T_g. Above T_g, the chains are more disordered and the CH_2 groups more randomly oriented, leading to a reduced CH_2 peak intensity. The spectral change is more prominent for APP than IPP. This can be attributed to the higher amorphous content of APP because the glass transition only involves the amorphous component of the polymer. There is no evidence that the surface glass transition of PP occurs at a different temperature from that of the bulk glass transition.

A series of ethylene-propylene rubber copolymers of varying bulk ethylene concentration (0 to 42 mol%) has been studied (A. Opdahl, R.A. Philips, G.A. Somorjai, submitted). All of the copolymers preferentially orient methyl side groups out of the surface.

PS

PS materials are widely used in packaging, construction, electronics, and appliances, and their surface properties have also been extensively studied (69–71). SFG has been applied to study surface structure of polystyrene thin films on various substrates to obtain molecular level understanding of the PS surface structures.

The phenyl side groups in PS exhibit five vibrational modes of aromatic C–H stretches, which are both IR and Raman active (22). Figure 5 shows the ssp and sps SFG spectra of a 370-nm PS film on a 2-nm SiO_2 layer on Si, which are quite similar to the SFG spectra of a phenylsiloxane self-assembled monolayer (SAM) formed on a 340-nm CVD SiO_2 layer on a Si substrate (22). The similarities in spectra clearly indicate that the phenyl groups at the PS surface point in the same direction as those in the SAM. In the SAM, the phenyl ring is directed away from the substrate owing to the silane coupling chemistry. Therefore, on the PS surface, phenyl groups are also directed away from the substrate. The alignment of the phenyl groups at the PS surface can be obtained by quantitatively analyzing the ratios of the vibrational peaks observed in sps and ssp spectra (22). The phenyl groups at the PS free surface were shown to be tilted away from the surface normal in an angular range around 57°.

The surface structure of PS on sapphire has been studied using SFG in a total internal reflection geometry (23). The reflected SFG signal can be enhanced by 1

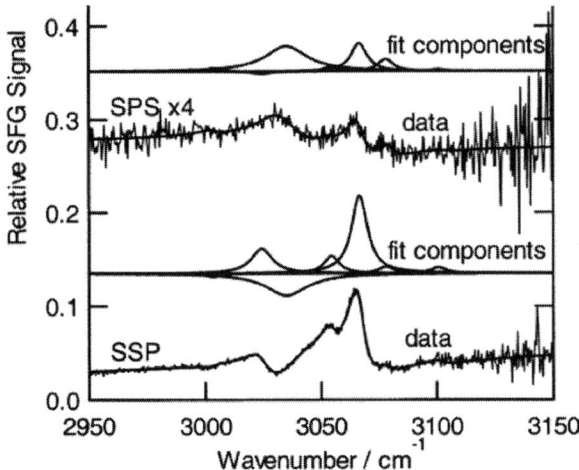

Figure 5 SFG spectra of a 370-nm PS film on a 2-nm SiO_2 layer on Si.

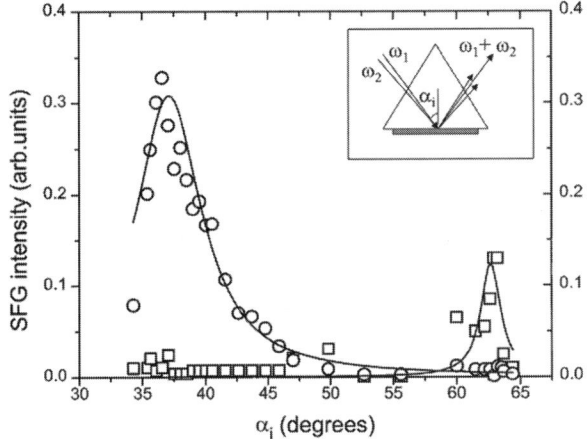

Figure 6 The ssp SFG intensity of PS films on a sapphire prism as a function of incident angles for two infrared input frequencies, 3027 (*squares*) and 3067 cm^{-1} (*circles*). The data represented by squares are scaled by a factor of 5.

to 2 orders of magnitude when the incident angles of the input beams are close to the critical angle for total internal reflection due to the higher Fresnel coefficients. By carefully choosing the incident angles of the input beams, the PS/air surface as well as the "hidden" PS/sapphire interface can be explored separately. Figure 6 shows the incident angle dependence of ssp SFG intensity at two infrared frequencies, 3028 and 3067 cm^{-1} (23). These two frequencies were chosen because the strongest SFG signals were observed at 3028 and 3067 cm^{-1} for the PS/sapphire and PS/air interfaces, respectively. The SFG intensity is enhanced near both critical angles corresponding to the PS/air (36°) and PS/sapphire (64°) interfaces.

The ssp and ppp SFG spectra of the PS/air and PS/sapphire interfaces were collected with the incident angles of the input beams at 36° and 64°, respectively. The SFG spectra of PS/air are dominated by the peak at 3069 cm^{-1}. The SFG spectra of PS/sapphire are dominated by the peaks at 3023 and 3059 cm^{-1}. Analysis of the spectra shows that the phenyl rings are orientated approximately parallel to the surface normal at the PS/air interface in comparison to nearly perpendicular to the surface normal at the PS/sapphire interface. The result also indicates that the phenyl groups of PS on sapphire orient differently than PS on silicon at the PS/air interfaces.

PMMA

PMMA is one of the most important polymers, extensively used as material in lithography, biomedical implants, barriers, membranes, and microcapillary electrophoresis devices (74–78).

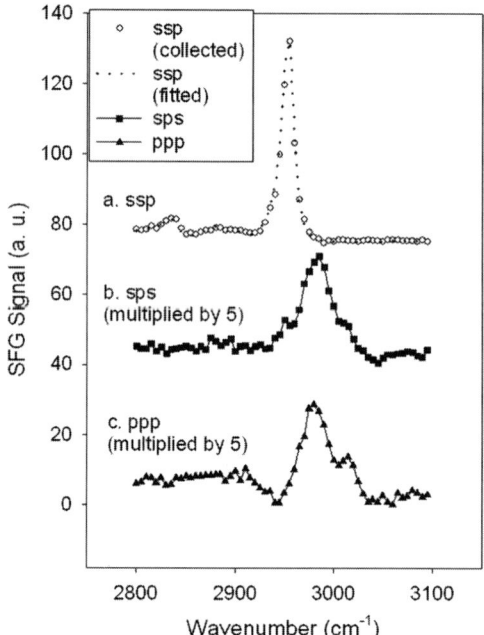

Figure 7 SFG spectra of PMMA: (*a*) ssp; (*b*) sps; (*c*) ppp.

Both IR and Raman spectra of PMMA have been extensively studied. Some arguments still exist over peak assignments (79–81). SFG spectra of PMMA in the C–H stretching region are presented in Figure 7 (39). The strong peak at 2955 cm^{-1} in the ssp spectrum is due to the symmetric stretch of the ester methyl group. Two peaks have been observed in both sps and ppp spectra. The peak at 2990 cm^{-1} comes from the asymmetric stretch of the ester methyl group as well as that of the alpha methyl group. The peak at 3016 cm^{-1} is another asymmetric stretch of the ester methyl group. The SFG spectra show that the ester methyl groups dominate on the PMMA surface. Analysis of these spectra indicates that the ester methyl groups tend to stand up on the surface, whereas the alpha methyl groups appear to lie down on the surface. No resonance features of the methylene groups can be detected in the spectra. The results illustrate that the high surface tension and hydrophilic property of the PMMA surface are due to the high coverage of the ester methyl groups on the surface.

Polyimides

Polyimide is one of the most widely used polymers in the electronics industry (82), suitable as a packaging and insulating material for electronic devices and circuit boards. SFG has been applied to study surfaces of polyimides with different alkyl side chains. Figure 8 shows the SFG spectra of polyimides with alkyl side chain

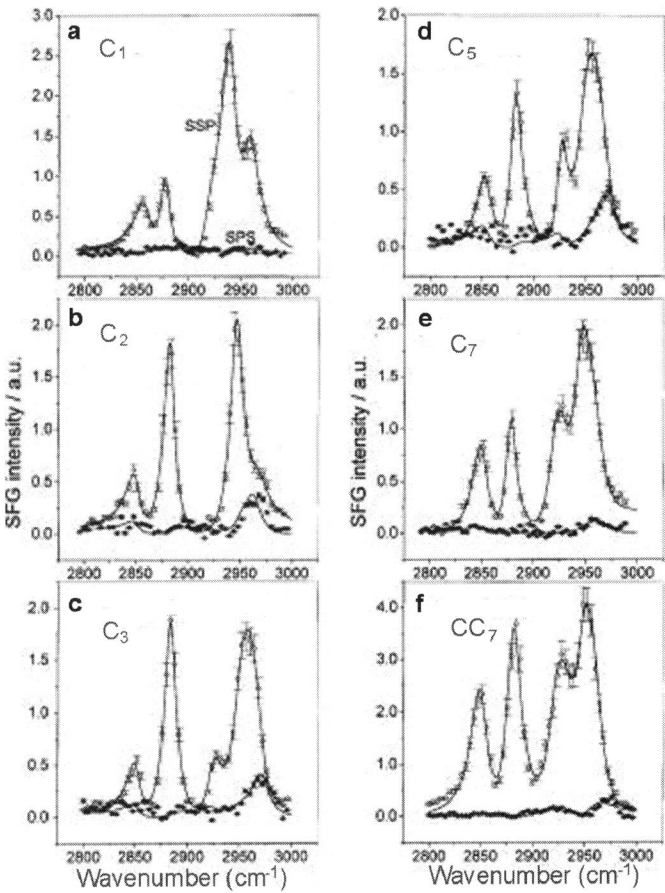

Figure 8 SFG spectra of polyimides with alkyl side chain lengths varying from C_1 to C_7: (*a*) C_1; (*b*) C_2; (*c*) C_3; (*d*) C_5; (*e*) C_7; (*f*) CC_7.

lengths varying from C_1 to C_7. The CH_3 of C_1 polyimide is deduced to have a very broad orientational distribution, with its average along the surface normal, similar to the case of methanol. The spectra of C_2 and C_3 are qualitatively similar to those of a surface monolayer of all-trans alkyl chains oriented more or less along the surface normal. Here, the average orientation of the CH_3 groups has a 35° tilt from the surface normal. Therefore, the conclusion can be drawn that for the C_1–C_3 polyimides, the orientations of the alkyl chains at the surface are similar to those of the C_1–C_3 alcohols, with probably broader orientational distributions. The same appears to be true for polyimides with longer alkyl chains. However, gauche defects set in with the longer chain lengths to enhance the CH_2 modes in the ssp spectra, but the number of defects seems small as judged from their relatively weak spectral peaks. A schematic model of side alkyl chains at the air-polyimide interface can be deduced (37).

END GROUP EFFECTS ON POLYMER SURFACE STRUCTURES

Polymer end groups occupy the termini of polymer molecules and are therefore more mobile than backbone groups. If surface energies of polymer end groups are lower than those of polymer backbone groups, they will segregate to the polymer surface to lower the energy of the system. Since end groups are usually quite small compared to the polymer backbone, they can be used to effectively modify the polymer surface structure and properties, leaving the polymer bulk structure and properties unchanged. In this section, we summarize the SFG studies of the end group effects on surface structures of a common polymer, poly(ethylene glycol) (PEG) and a more complicated biomedical polymer, polyurethane.

PEG

PEG is commonly used in many areas of medicine and biological science (83). Surface modification with PEG to provide protein and cell-rejecting properties is the focus of much research (84, 85). PEG polymers are widely available with different kinds of end groups, but their surfaces have not been extensively studied.

FTIR spectra of three PEG samples, PEG diol, PEG methyl ether, and PEG dimethyl ether, are identical (18), but contact angle measurements show that their surface tensions are different (86). The SFG spectra of different PEG films given in Figure 9 are also markedly dissimilar, showing that their molecular surface structures are distinct (18). The SFG spectrum of PEG diol in Figure 9a exhibits a strong C–H symmetric stretch peak of OCH_2 at 2865 cm^{-1}, whereas no O–H stretch signal can be detected in the frequency range from 3000 to 3800 cm^{-1} (not shown), indicating that the surface is covered by the backbone. The SFG spectrum (Figure 9b) of PEG methyl ether has two strong peaks: The stronger peak at 2820 cm^{-1} is due to the C–H symmetric stretch of the OCH_3 end group, and the peak at 2865 cm^{-1} is due to the backbone. This shows that the hydrophobic methoxy end group covers a significant fraction of the surface, and the backbone covers the rest. The stronger peak at 2820 cm^{-1} from PEG dimethyl ether (Figure 9c) shows that its surface coverage of hydrophobic OCH_3 end groups is even higher. Correspondingly, the peak at 2865 cm^{-1} due to the backbone appears much weaker. The conclusion can be drawn that the hydrophobic methoxy end groups tend to partition to the surface to lower surface tension or energy. For PEG diol with the hydrophilic hydroxyl end groups, the backbones like to cover the surface to lower surface tension.

Biomedical Polymers: BS and BP

Polyurethanes have been used for a number of blood contacting applications because of their favorable tensile and fatigue properties. To develop polyurethanes with surfaces that will not cause blood coagulation, a novel method has been developed to control surface properties without significantly changing the polymer

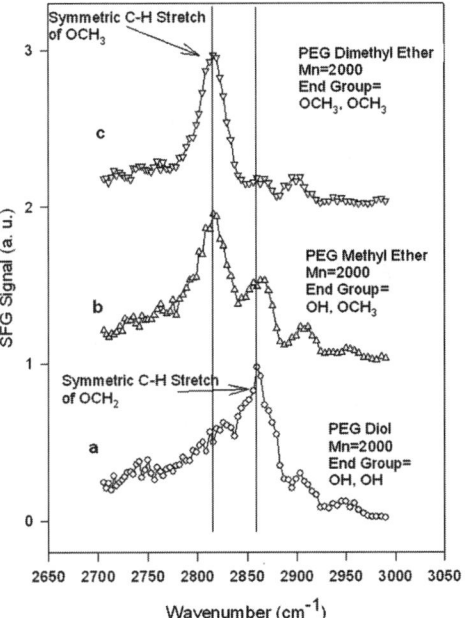

Figure 9 SFG spectra of poly (ethylene glycol) with different end groups: (*a*) PEG diol; (*b*) PEG methyl ether; (*c*) PEG dimethyl ether.

bulk properties: to append to polymer molecules certain monofunctional surface-modifying end groups (SME) such as poly(dimethyl siloxane) (PDMS) and PEG. Again, SFG can be used to probe surface structures of polyurethane with various end groups: BN (without SME), BS (with PDMS as end groups), and BP (with PEG as end groups).

Attenuated total reflection infrared (ATR-IR) spectra of BS, BP, and BN in the C–H stretching region are essentially identical (18). The bulk compositions of these three samples are very similar, because the end groups (PDMS and PEG, respectively) of BS and BP are only 3 wt%. The surface sensitivity of the ATR-IR technique is not sufficient to detect segregation of PDMS or PEG on BS or BP surfaces.

Unlike ATR spectra, SFG spectra of BN, BS, and BP are different (Figure 10). There are four peaks in the ssp spectrum of BN (Figure 10*a*). The peaks at $\sim$2850 and $\sim$2920 cm^{-1} are the symmetric and asymmetric stretches of aliphatic CH$_2$ groups, respectively. The $\sim$2940 cm^{-1} peak is due to Fermi resonance. The peak at 2787 cm^{-1} can be attributed to the symmetric stretch of the CH$_2$ connected to an oxygen atom of the carbamate group (NCOO). The BS spectrum (Figure 10*c*) is very different from that of BN and has a strong peak at $\sim$2915 cm^{-1}. This peak is from the C–H symmetric stretch of Si-CH$_3$ in PDMS (Figure 10*b*), showing that the hydrophobic PDMS end groups largely cover the BS surface. The surface

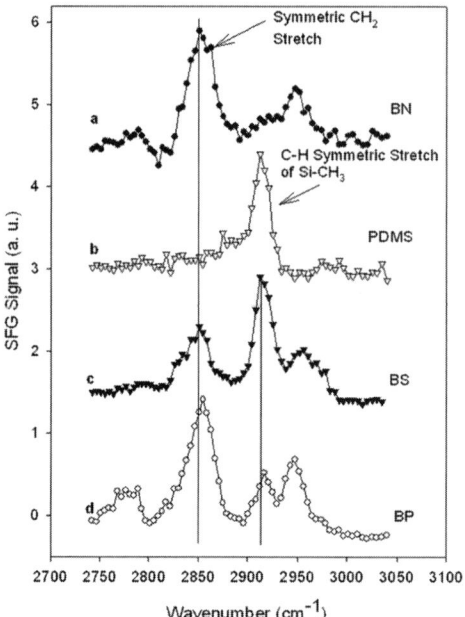

Figure 10 SFG spectra of polyurethane with or without polymer end groups: (*a*) BN, polyurethane without polymer end groups; (*b*) PDMS; (*c*) BS, polyurethane with PDMS as end groups; (*d*) BP, polyurethane with PEG as end groups.

is, however, not completely covered by PDMS because the 2850 cm^{-1} backbone peak is still visible in the spectrum. The spectrum indicates that although there is only 3 wt% PDMS in BS, it tends to segregate to the surface. The SFG spectrum of BP is similar to that of BN, the surface of which is covered by the polyurethane backbones. No hydrophilic PEG end groups can be detected in the spectrum, showing that the BP surface is also mainly covered by the backbones. The results illustrate, in general, that hydrophobic end groups, such as PDMS and methoxy, like to segregate to the polymer/air interface, and the hydrophilic end groups, such as PEG and hydroxyl, like to remain in the bulk so that the interface is covered by the more hydrophobic backbones.

MODIFICATION OF POLYMER SURFACES

Modification of the chemical composition and structure of a polymer surface is important to many applications, ranging from adhesion, friction, and wear to biomedical utilization (87, 88). These applications require special surface properties such as bondability, lubricity, or hydrophilicity. Although polymers possess excellent bulk physical and chemical properties, they may sometimes not have the surface properties needed for these applications. Hence, UV irradiation (38), plasma treatment (38), rubbing (40), wet treatment (36), oxygen ion and radical irradiation (87) and other methods have been used to modify polymer surfaces to obtain the desired surface properties. For example, as mentioned above, polyimide is widely used in the electronics industry as packaging and insulating material for electronic devices and circuit boards. These applications usually involve polyimide/metal interfaces. However, adhesion of polyimide to metals such as copper is not sufficiently strong. To improve the adhesion, therefore, the polyimide surface is commonly wet treated in a strong base solution (36). Modification of PS surface by plasma treatment and UV light irradiation can also enhance the PS surface adhesion, improve its biocompatibility, and activate its ability to align adsorbed liquid crystals (LC) (89–92). Rubbing of a polymer surface can align the surface polymer chains, which in turn align the liquid crystal film deposited on it—an effective method used in industry to manufacture liquid crystal display devices.

The mechanism of surface modification has been subject to intensive investigation in order to better control surface properties (27). SFG has been successfully applied to study surface modification of various polymer materials. In this section, we summarize the SFG studies of polymer surface modification by several different methods.

Wet Treatment

The polyimide surface can be modified by reaction in a base solution (36). Figure 11 describes how the SFG spectrum of polyimides in the C=O stretch region changes

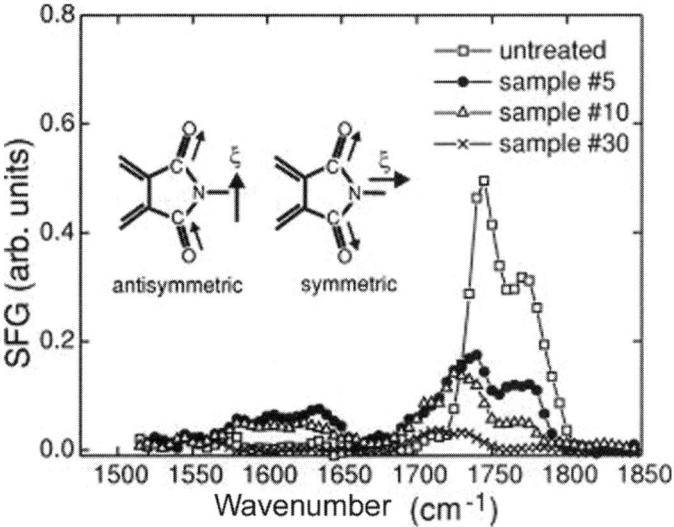

Figure 11 SFG spectra of poly-n-alkyl-pyromellitic imide without and with 5, 10, and 30 seconds of wet treatment.

when the surface is immersed in a strong NaOH solution at 25°C for different lengths of time: 5, 10, and 30 seconds. Both the antisymmetric (a, 1739 cm^{-1}) and symmetric (s, 1775 cm^{-1}) C=O stretching modes of the imide (sketched in the inset of Figure 11) decrease in intensity with reaction time, with the s mode disappearing first. A new broad band appears around 1600–1660 cm^{-1} and can be assigned to C=O stretches from amide groups. It also disappears after 30 s in the solution. Analysis of the SFG spectra in Figure 11 indicates that reaction in the NaOH solution converts the imide rings at the surface to amides and then eventually etches away all the polymer. Etching seems to be more effective on the amorphous than on the crystalline part of the polyimide.

Plasma Treatment, UV Irradiation, Oxygen-Ion Treatment, and Oxygen-Radical Treatment

Modification of surface structures of PS thin films deposited on silica using UV light and plasma treatment has been investigated by SFG (38). The observed spectral changes provide direct support to the proposed surface reactions involving aromatic groups, which cannot be concluded from studies by other techniques. The SFG spectrum shows the dramatic decrease of the aromatic CH stretches after UV irradiation and plasma treatment, due to oxidation of the PS surface. However, the levels of oxidation are different. The oxidation of PS to a higher oxidation state, forming carbonyl/carboxyl species, was observed by SFG on the PS surface after plasma treatment, but not after UV radiation. Along with XPS findings, SFG results suggest that the two modification processes may take different pathways. The ring opening, which results in the formation of aldehyde/carboxylate species, is the main pathway for plasma treatment. Oxygen uptake, forming species of low oxidation states, is the main mechanism in UV irradiation. SFG studies also indicate that kinetics of photooxidation induced by UV irradiation at the PS surface is much faster than that in the bulk, due to the short penetration depth of UV light and limited oxygen diffusion into the bulk.

Surface modifications of PET by reaction with oxygen ions and oxygen radicals have also been investigated by SFG (87). The results indicate that an impinging oxygen ion beam induces degradation of the polymer chains and converts the ester moiety to −COOH terminal species and others on the surface. On the other hand, an oxygen radical beam induces a ring-opening reaction of the surface phenyl groups, leading to new oxygen-containing species formation accompanied by a corresponding loss of aromatic character of the polymer surface.

Rubbing

Mechanical rubbing of a polymer surface makes the surface structure anisotropic. SFG vibrational spectroscopy allows the detection of the structure change by rubbing (40). Study on a rubbed PVA surface shows that the PVA main chains at the surface are aligned by rubbing along the rubbing direction. Detailed analysis of the spectra yields a quantitative distribution for the chains that appear to lie, on average, nearly parallel to the surface along the rubbing direction, with a slight upward tilt. The spreads of the distribution in all angles are around 30°. Similar results have been observed on rubbed surfaces of polyimides and PS.

POLYMER BLENDS

Polymer blends provide ways of enhancing the properties of a pure polymer. Low surface tension components in polymer blends tend to aggregate to the surface to lower the free energy of the system (68, 94). Unlike copolymers, components in polymer blends are more mobile, making surface segregation easy. SFG can be used to study surface structures of polymer blends. The polymer blends studied by SFG include biomedical polymer blends of phenoxy (PHE) and polyurethane with different end groups (42, 43), blends of PS and poly(n-butyl methacrylate) (PBMA) (44), and blends of APP and poly(ethylene-co-propylene) rubber (PE-co-PP) (A. Opdahl, R.A. Phillips, G.A. Somorjai, submitted).

Blends of Phenoxy with Different Polyurethanes

Polymer blends of PHE and polyurethane are ideal candidates for various biopolymer applications, such as intravenous catheter tubing, which will soften after insertion into veins. As the surfaces of these blends are to be in contact with blood after insertion, the ability to control their surface properties is of great significance to prevent possible complications related to thrombosis and embolization caused by the contact of blood with a foreign surface. Because the surface properties are controlled by surface structures, understanding of surface structures of these polymer blends at the molecular level is very important.

Three groups of PHE polymer blends, BS/PHE, BSP/PHE, and BF/PHE, have been studied by SFG (42, 43). The blend in each group has two thermodynamically compatible components: One is PHE, and the other is a modified polyurethane BS, BSP, or BF. The molecular structures of all four polymers are shown in Figure 1b. As seen from their chemical structures, BS, BSP, and BF all contain both hydrophobic [PDMS or $(-(-CF_2-)_n-)$], and hydrophilic (polyurethane) components. The surface tensions of PHE, BS, BSP, and BF are 45, 22, 26, and 16 dyne/cm, respectively.

The SFG spectra (Figure 12) of BS/PHE blend surfaces show that BS likes to segregate to the surface. For a blend with 0.17 wt% BS bulk concentration, the SFG spectrum is very similar to that of pure PHE. With increasing BS bulk concentration, the spectrum changes rapidly: The prominent methyl resonance of PHE at 2875 cm^{-1} weakens, and the prominent 2915 cm^{-1} peak of BS strengthens. This signifies an enrichment of the BS component at the free surface of the BS/PHE polymer blend. When the BS concentration reaches above 1.7 wt%, the blend spectrum becomes nearly identical to that of pure BS, signifying that the blend surface is mainly covered by BS.

SFG spectra of BF/PHE and BSP/PHE blends also show that BF and BSP tend to segregate to the blend surface. BF and BSP appear to completely cover the blend

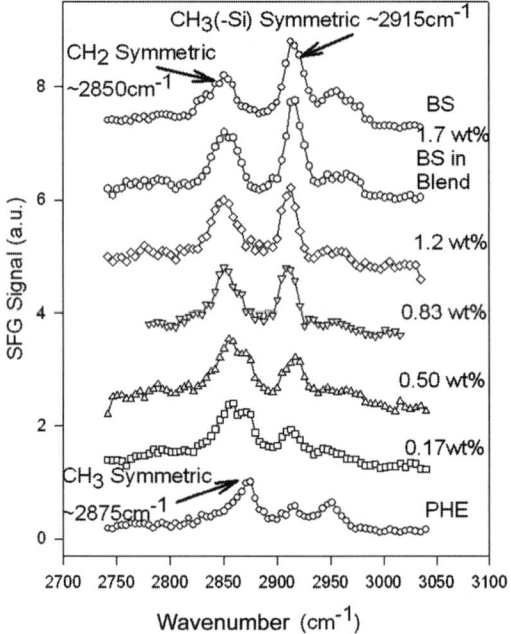

Figure 12 SFG spectra of BS/PHE polymer blends.

surface when their bulk concentrations reach 1 wt% and 3.5 wt%, respectively. The very small bulk concentrations required for complete coverages indicate that the surface energies of the polyurethane components are much smaller than that of PHE. They also suggest that BSP, BS, and BF have, respectively, small, smaller, and smallest surface energies, as substantiated by surface tension measurements. Reduction of surface tension of a blend surface as compared to the pure PHE surface was also observed. There is a roughly linear correlation between these required bulk concentrations and the surface free-energy differences between the PHE base polymer and the three surface-active polymer components, BF, BS, and BSP. This relation may be used to predict the wt% of other surface-active polymers that may be added to PHE to fully cover the surface of the polymer blend.

PS/PBMA

The PS/PBMA blend system provides a good model to investigate the surface structure of blends composed of partially compatible components. From the SFG spectra of PS/PBMA blends, the conclusion can be made that PBMA likes to segregate to the blend surface (44). When the bulk concentration of PBMA in the blend exceeds 60 wt%, PBMA nearly covers the entire surface. As the PBMA concentration decreases from 60 wt%, signals from PS begin to show up, but even with only 4 wt% PBMA, the spectrum indicates that PBMA can still be easily detected on the surface. Note that no PS resonances could be detected in the ppp SFG spectra of the blends even with 96 wt% PS in the bulk. Analysis of SFG spectra for the blends with different input/output polarization combinations shows that orientation of the phenyl groups is different from that of the pure PS surface: Phenyl groups tend to tilt more towards the surface (44). This explains the disappearance of the PS signal in the ppp SFG spectra of the blends.

POLYMER SURFACE RESTRUCTURING IN WATER

Since Langmuir described the reorientation of hydrophilic groups on a solid surface in contact with a water drop (95), much research has been undertaken to understand polymer surface dynamics and structures in water. For many applications, it is important to know how the structure of a polymer surface would change in an aqueous environment. For example, biomedical polymers implanted in a body would be in contact with blood and tissue. Their surfaces need to be characterized in situ to understand how the surface structures are related to biocompatibility. Most surface science techniques are operable only with the sample under ultrahigh vacuum (1, 2). Freeze-drying XPS has been developed to study polymer surface restructuring in water, but it cannot provide orientational information of surface functional groups (96). Contact angle measurements show that many polymer surfaces reconstruct after they are in contact with water (97–100), but they cannot provide a picture of the surface changes at the molecular level. Theoretical models have long been developed to simulate details of polymer surface restructuring in water (26), but have not yet been tested by molecular level experiments. As mentioned earlier, SFG is a powerful technique for probing solid/liquid interfaces. To avoid strong absorption of IR input by water, special sample preparation and beam geometries must be adopted: for example, a thin layer of water sandwiched between a CaF_2 window and a polymer film on a substrate. The input beams can then access the polymer/water interface through water from the water side (45). The input beams can also be sent through the substrate and the thin polymer film to reach the polymer/water interface (46). In this section, we summarize the SFG studies of surface restructuring of BS, PMMA, PBMA, poly(n-octyl methacrylate) (POMA), poly(2-hydroxyethyl methacrylate) (PHEMA) and PEG in water.

Contact angle measurements of BS and various polymethacrylates show that the hydrophobicity of these polymer surfaces changes after contacting water (26, 45). Assumptions have been made that these changes might be induced by reorientation of the surface functional groups or by diffusion between various surface and bulk functional groups. These motions are possible only if the polymer molecules are mobile. The mobility of the polymer backbone is related to the glass transition temperature (T_g) of the molecules. The reorientation of side groups may occur more easily, even at temperatures below T_g. SFG studies of various polymers show that surface restructurings of polymers in water are related to their glass transition temperatures (46).

Different polymer surfaces respond differently in water. For POMA (46), PHEMA (47), and PEG (48), their polymer backbones change immediately after they contact water. For example, dramatic changes of PEG surfaces in water have been detected by SFG (48). The surface arrangement of PEG molecules, originally relatively well-ordered in air, becomes disorganized in the presence of water. For PBMA (46), the orientation of the ester side group on the surface alters in water, but that of the polymer backbone does not. For BS (45), the surface coverage of the hydrophobic PDMS end group decreases, and more of the hydrophilic polyurethane backbone covers the surface in water. For PMMA (46), which has a high glass transition temperature of 105°C, no substantial surface restructuring has been observed in water.

Restructuring of Polymer Backbone Immediately After Contacting Water

Cross-linked PHEMA is a widely used contact lens material—hydrogel (47). The SFG spectra of hydrated PHEMA in water, wet PHEMA just removed from water, and dehydrated PHEMA in air are presented in Figure 13 (47). The spectrum of the dry PHEMA in air indicates a preponderance of methyl groups at the surface,

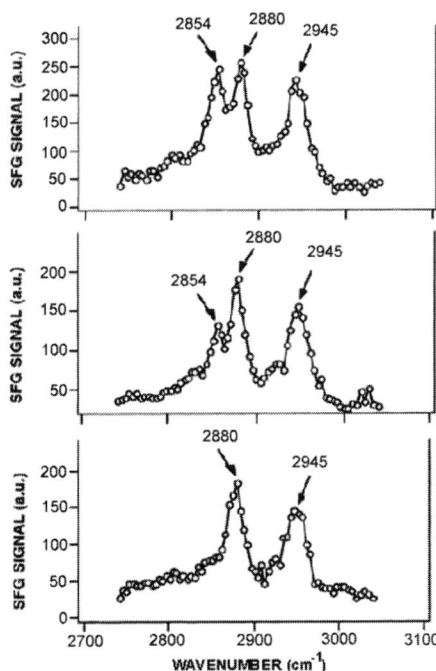

Figure 13 SFG (ssp) spectra of PHEMA: (*top spectrum*) hydrated PHEMA at the water/polymer interface; (*middle spectrum*) hydrated PHEMA at the air/polymer interface; (*bottom spectrum*) dry PHEMA at the air/polymer interface.

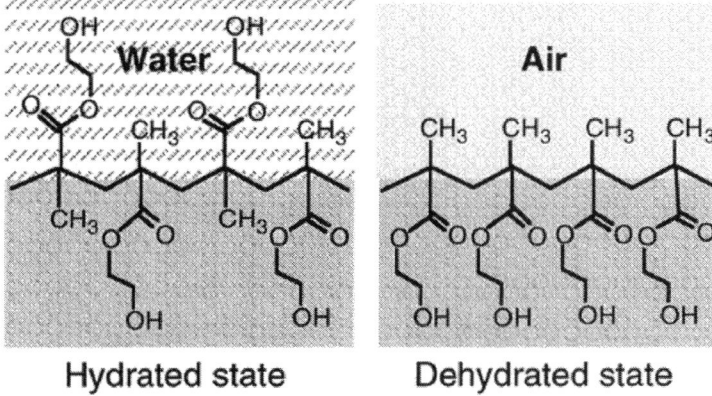

Figure 14 Schematics describing surface structures of PHEMA in air (*right*) and in water (*left*).

and that of the hydrated PHEMA in water revealed a peak of the symmetric CH_2 stretch adjacent to the oxygen atoms in the ethylene glycol group (-OCH_2CH_2OH), the hydrophilic component of the polymer. The intensity of this peak varied from moderate to weak as the hydrogel transformed from the hydrated state to the dehydrated state. This provides direct evidence that PHEMA can adopt two discrete surface states depending on the environment, as illustrated in Figure 14. When PHEMA is exposed to air, polar side chains turn into the bulk, leaving the nonpolar methyl groups to project out of the surface to form a "hydrophobic conformation." In water, polar ethylene glycol groups migrate to the surface and coexist with methyl groups at the surface, creating a "hydrophilic conformation."

The SFG spectra of POMA show that the POMA/air interface is dominated by the methyl and methylene groups (46). When POMA makes contact with water, its SFG spectra immediately disappear. With the sample out of water and dried, the spectra reappear, but are very different from those before contacting water. We believe that polymer backbones of POMA change immediately after contacting water, because POMA has a very low T_g ($-20°C$) and thus the backbones are very mobile. The molecules on the surface in water become randomly oriented so that no SFG signal can be detected. The SFG signal may also disappear because the POMA surface is too rough. AFM images show that the POMA surface is flat in air, but rough in water.

Side Group Reorientation on Polymer Surface in Water

The SFG spectra of PBMA before, during, and after contact with water are displayed in Figure 15. The figure shows that the surface structure of PBMA changes after the sample contacts water and recovers after the sample is removed from water. Analysis of the spectra indicates that, in air, the methyl group in the ester chain tilts more towards the surface normal, whereas in water it lies closer to the surface

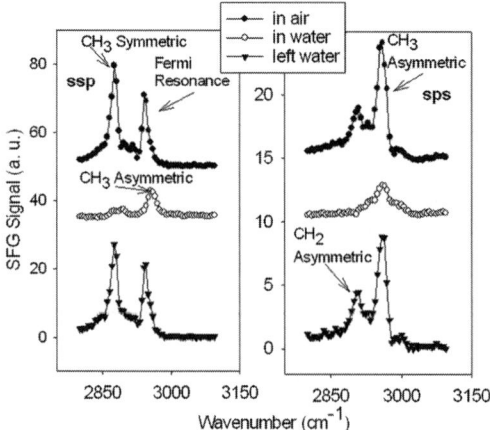

Figure 15 SFG spectra of PBMA before, during, and after being in contact with water. (*left*) ssp; (*right*) sps.

(46). The spectra of PBMA in water for 12 h are similar to those taken immediately after it is immersed in water. The recovery of the spectra immediately after the sample is removed from water and the absence of additional detectable changes of spectra after prolonged exposure to water show that the spectral changes come mainly from reorientation of side chains.

Slow Polymer Surface Change in Water

SFG spectra show that when BS is in contact with water, a significant change of the polymer surface structure takes place in response to alteration of environment (45). Figure 16 displays the time evolution of the ssp SFG spectra of BS in water at 300°K. The surface structural transformation of BS in response to the environmental change from air to water is slow, and it takes about 25 h at 300°K. The key features of the observed spectral changes of BS in water are weakening of the methyl resonance of PDMS at 2915 cm^{-1} and strengthening of the 2850 and 2785 cm^{-1} bands for the CH_2 groups of the polyurethane backbone. Upon hydration, hydrophobic interaction between water and BS drives the hydrophobic PDMS segments away from the polymer surface, reorients them to minimize contact with water, and attracts more hydrophilic polyurethane backbones to the polymer/water interface. The result is a decrease of the CH_3 peaks of PDMS and an increase of the CH_2 peaks of BS in the SFG spectra.

The polarization dependence of the SFG spectra also reflects a surface restructuring process. The average tilt of the CH_3 symmetric axis from the surface normal was found to be 35° at the polymer-air interface. The same angle obtained from analysis of the spectra of BS after contact with water for 51 hours was 60°.

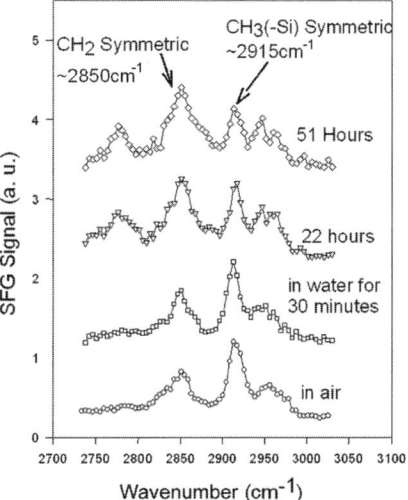

Figure 16 Time evolution of the ssp SFG spectra of BS in water.

Absence of Surface Restructuring of Polymers in Water

Both ssp and sps SFG spectra of PMMA in contact with water are similar to those of the PMMA/air interface (46), although they are much weaker. Analysis of the spectra shows that the weaker spectra are mainly due to the changes of the Fresnel coefficients for the two interfaces. This indicates that there is little structural change for the ester methyl, alpha methyl, and methylene groups at the PMMA surface after contacting water. The observed difference between the advancing and receding water contact angles on PMMA is not due to surface restructuring of PMMA in water but rather to the small amount of water absorption at the polymer water interface, as suggested by Andrade et al. (101).

SUMMARY

SFG has been successfully developed into a powerful technique to study surface structures of polymer materials. Various tests show that SFG can probe polymer surface structures with submonolayer surface sensitivity. Orientations and orientational distributions of functional groups on polymer surfaces can be deduced by SFG. Understanding of surface structures of different polymers ranging from simple polymers such as PE and PP to very complicated polymer systems such as biomedical polymer blends has been achieved at the molecular level by SFG studies. These studies show that surface energy minimization is an important driving force to determine the polymer surface structure. For example, hydrophobic end groups, which have a lower surface energy than polymer backbones, tend to segregate to polymer surfaces in air, but in water they would diffuse into the bulk. On the contrary, hydrophilic end groups, which have higher surface energies than polymer backbones, would face the polymer bulk when the surface is exposed to air, but turn toward the water when immersed in water. In some cases, such a driving force can be overridden by other effects such as lack of mobility of the low surface energy components in the polymers. Phenyl side groups in PS, and incomplete surface restructuring of complex biopolymers in water are examples.

From SFG studies, some information on polymer surfaces has already been obtained. Surface structures of polymers (PE, PP, PS, and PMMA) can be determined and correlated to other surface properties such as surface tension. Surface structural change of polymers (PP) with temperature can be probed, providing insight into the polymer surface glass transition. Observation of segregation of small hydrophobic end groups to air/polymer interfaces supports the idea that small end groups on polymer units can effectively modify the polymer surface structures. Modification of polymer surfaces by other methods such as UV irradiation, plasma treatment, wet etching, oxygen ion and radical treatments, and rubbing can all be monitored at the molecular level, and their mechanisms deduced. How the surface structures vary with the bulk composition in polymer blends can also be systematically investigated and understood. Structural information on the "buried" polymer/liquid and polymer/solid interfaces can also be obtained. Detection of polymer surface restructuring in water is now possible, an important step to understand biocompatibility in biomedical polymer applications. Understanding of surface structures of polymer materials through SFG studies at the molecular level will help in defining correlations between polymer surface structures and properties, and assist in the design of polymeric materials with desired surface properties.

ACKNOWLEDGMENTS

We are grateful to all the coworkers quoted in this review for their invaluable contributions. We would like to acknowledge support from the National Science Foundation, Department of Energy, University of California, and University of Michigan. We want to thank Drs. L.J. Richter, A. Dhinojwala, and M.S. Yeganeh for providing figures.

Visit the Annual Reviews home page at www.annualreviews.org

LITERATURE CITED

1. Woodruff D, Delchar T. 1986. *Modern Techniques of Surface Science*. Cambridge: Cambridge Univ. Press
2. Somorjai GA. 1994. *Introduction to Surface Chemistry and Catalysis*. New York: Wiley
3. Shen YR. 1984. *The Principle of Nonlinear Optics*. New York: Wiley

4. Shen YR. 1989. *Nature* 337:519–25
5. Eisenthal KB. 1996. *Chem. Rev.* 96:1343–60
6. Binnig G, Quate CF, Gerber C. 1986. *Phys. Rev. Lett.* 56:930–33
7. Cohen SH, Lightbody ML. 1999. *Atomic Force Microscopy/Scanning Tunneling Microscopy 3*. New York: Kluwer
8. Rider KB, Hwang KS, Salmeron M, Somorjai GA. 2001. *Phys. Rev. Lett.* 86:4330–33
9. Lindsay SM, Nagahara LA, Thundat T, Knipping U, Rill RL, et al. 1989. *J. Biomol. Struct. Dyn.* 7:279–87
10. Shen YR. 1994. *Surf. Sci.* 299:551–62
11. Miranda PB, Shen YR. 1999. *J. Phys. Chem. B* 103:3292–307
12. Bain CD. 1995. *J. Chem. Soc. Faraday Trans.* 91:1281–96
13. Gragson DE, Richmond GL. 1998. *J. Phys. Chem. B* 102:3847–61
14. Walker RA, Gruetzmacher JA, Richmond GL. 1998. *J. Am. Chem. Soc.* 120:6991–7003
15. Conboy JC, Messmer MC, Richmond GL. 1996. *J. Phys. Chem. B* 100:7617–22
16. Chen Z, Gracias DH, Somorjai GA. 1999. *Appl. Phys. B* 68:549–57
17. Gracias DH, Chen Z, Shen YR, Somorjai GA. 1999. *Acc. Chem. Res.* 320:930–40
18. Chen Z, Ward R, Tian Y, Baldelli S, Opdahl A, et al. 2000. *J. Am. Chem. Soc.* 122:10615–20
19. Shultz MJ, Schnitzer C, Simonelli D, Baldelli S. 2000. *Int. Rev. Phys. Chem.* 19:123–53
20. Pizzolatto RL, Yang YJ, Wolf LK, Messmer MC. 1999. *Anal. Chim. Acta* 397:81–92
21. Kim J, Cremer PS. 2000. *J. Am. Chem. Soc.* 122:12371–72
22. Briggman KA, Stephenson JC, Wallace WE, Richter LJ. 2001. *J. Phys. Chem. B* 105:2785–91
23. Gautam KS, Schwab AD, Dhinojwala A, Zhang D, Dougal SM, Yeganeh MS. 2000. *Phys. Rev. Lett.* 85:3854–57
24. Löbau J, Wolfrum K. 1997. *J. Opt. Soc. Am. B* 14:2505–12
25. Scatena LF, Brown MG, Richmond GL. 2001. *Science* 292:908–12
26. Andrade JD. 1988. *Polymer Surface Dynamics*. New York: Plenum
27. Carbassi F, Morra M, Occhiello E. 1994. *Polymer Surfaces: From Physics to Technology*. Chichester: Wiley
28. Ratner BD, Castner DG. 1996. *Surface Modification of Polymeric Biomaterials*. New York: Plenum
29. Feast WJ, Munro HS. 1987. *Polymer Surfaces and Interfaces*. New York: Wiley
30. Feast WJ, Munro HS, Richards RW. 1992. *Polymer Surfaces and Interfaces II*. New York: Wiley
31. Jones RAL, Richards RW. 1999. *Polymers at Surfaces and Interfaces*. Cambridge: Cambridge Univ. Press
32. Recum AF. 1986. *Handbook of Biomaterials Evaluation: Scientific, Technical, and Clinical Testing of Implant Materials*. New York: Macmillan
33. Park JB, Lakes RS. 1992. *Biomaterials: An Introduction*. New York: Plenum
34. Zhang D, Shen YR, Somorjai GA. 1997. *Chem. Phys. Lett.* 281:394–400
35. Gracias DH, Zhang D, Lianos L, Ibach W, Shen YR, Somorjai GA. 1999. *Chem. Phys.* 245:277–84
36. Kim D, Shen YR. 1999. *Appl. Phys. Lett.* 74:3314–16
37. Oh-e M, Lvovsky AI, Wei X, Shen YR. 2000. *J. Chem. Phys.* 113:8827–32
38. Zhang D, Dougal SM, Yeganeh MS. 2000. *Langmuir* 16:4528–32
39. Wang J, Chen CY, Buck SM, Chen Z. 2001. *J. Phys. Chem. B* 105: In press
40. Wei X, Zhuang X, Hong SC, Goto T, Shen YR. 1999. *Phys. Rev. Lett.* 82:4256–59
41. Gautam KS, Dhinojwala A. 2001. *Macromolecules* 34:1137–39
42. Zhang D, Gracias DH, Ward R, Gauckler M, Tian Y, et al. 1998. *J. Phys. Chem.* 102:6225–30
43. Chen Z, Ward R, Tian Y, Eppler AA,

Shen YR, Somorjai GA. 1999. *J. Phys. Chem. B* 103:2935–42
44. Chen CY, Wang J, Woodcock SE, Chen Z. 2001. *Langmuir.* In press
45. Zhang D, Ward RS, Shen YR, Somorjai GA. 1997. *J. Phys. Chem. B* 101:9060–64
46. Wang J, Woodcock SE, Buck SM, Chen CY, Chen Z. 2001. *J. Am. Chem. Soc.* 123:9470–71
47. Chen Q, Zhang D, Somorjai GA, Bertozzi CR. 1999. *J. Am. Chem. Soc.* 121:446–47
48. Dreesen L, Humbert C, Hollander P, Mani AA, Ataka K, et al. 2001. *Chem. Phys. Lett.* 333:327–31
49. Fowkes FM. 1964. *Contact Angle, Wettability, and Adhesion.* Washington, DC: Am. Chem. Soc. Adv. Chem. Ser. 43
50. Kwok DY, Lam CNS, Li A, Neumann AW. 1999. *J. Adhes.* 68:229–55
51. Castner DG, Ratner BD. 1990. *Surf. Interface Anal.* 15:479–86
52. Briggs D. 1989. *Br. Polym. J.* 21:3–15
53. Brown A, Vickerman JC. 1986. *Surf. Interface Anal.* 8:75–81
54. Stohr J. 1992. *NEXAFS Spectroscopy.* Berlin: Spring-Verlag
55. Hirosawa I. 1996. *Jpn. J. Appl. Phys. Lett.* 35:5873–75
56. Russell TP. 1996. *Physica B* 221:267–83
57. Cole KC, Guevremont J, Ajji A, Dumoulin MM. 1994. *Appl. Spectrosc.* 48:1513–21
58. Yui N, Suzuki Y, Mori H, Terano M. 1995. *Polym. J.* 27:614–22
59. Kip BJ, van Eijk MCP. 1991. *J. Polym. Sci. B* 29:99–108
60. Guyot-Sionnest P, Hunt JH, Shen YR. 1987. *Phys. Rev. Lett.* 59:1597–600
61. Hirose C, Akamatsu N, Domen K. 1992. *J. Chem. Phys.* 96:997–1004
62. Hirose C, Yamamto H, Akamatsu N, Domen K. 1993. *J. Phys. Chem.* 97:10064–69
63. Hirose C, Akamatsu N, Domen K. 1992. *Appl. Spectrosc.* 46:1051–72
64. Zhuang X, Miranda PB, Kim D, Shen YR. 1999. *Phys. Rev. B* 59:12632–40
65. Zhang JY, Huang JY, Shen YR. 1993. *J. Opt. Soc. Am. B* 10:1758–64
66. Wei X, Hong SC, Lvovsky AI, Held H, Shen YR. 2000. *J. Phys. Chem. B* 104:3349–54
67. Deleted in proof
68. Wu SH. 1982. *Polymer Interface and Adhesion.* New York: Marcel Dekker
69. Zhang XM, Tasaka S, Inagaki N. 2000. *J. Polym. Sci. B Polym. Phys.* 38:654–58
70. Tanaka K, Taura A, Ge SR, Takahara A, Kajiyamas T. 1996. *Macromolecules* 29:3040–42
71. Meyers GF, Dekoven BM, Seitz JT. 1992. *Langmuir* 8:2330–35
72. Young RJ, Lovell PA. 1994. *Introduction to Polymers.* London: Chapman & Hall
73. Johnson KL, Kendall K, Roberts AD. 1971. *Proc. R. Soc. London Ser. A* 324:301–13
74. Stroeve P, Franses EI. 1987. *Molecular Engineering of Ultrathin Films.* London: Elsevier
75. Roberts GG. 1990. *Langmuir-Blodgett Films.* New York: Plenum
76. Ulman A. 1991. *An Introduction to Ultrathin Organic Films: From Langmuir-Blodgett to Self Assembly.* New York: Academic
77. Kuan SWJ, Frank CW, Fu CC, Allee DR, Maccagno P, Pease RFW. 1988. *J. Vac. Sci. Technol. B* 6:2274–79
78. Ford SM, Kar B, McWhorter S, Davies J, Soper SA, et al. 1998. *J. Microcolumn Sep.* 10:413–22
79. Dirlikov S, Koenig JL. 1979. *Appl. Spectrosc.* 33:555–61
80. Lipschitz I. 1982. *Polym. Plast. Technol. Eng.* 19:53–106
81. Dybal J, Krimm S. 1990. *Macromolecules* 23:1301–8
82. Patel JS. 1993. *Annu. Rev. Mater. Sci.* 23:269–94
83. Harris JM, Zalipsky S, eds. 1997. *Poly(Ethylene Glycol) Chemistry and*

Biological Applications. Washington, DC: Am. Chem. Soc. Symp. Ser. 680
84. Malmsten M, Emoto K, VanAlstine JM. 1998. *J. Colloid Interface Sci.* 202:507–17
85. McPherson T, Kidane A, Szleifer I, Park K. 1998. *Langmuir* 14:176–86
86. Wu S. 1975. In *Polymer Handbook*, ed. J Brandrup, EH Immergut, pp. VI/411–34. New York: Wiley
87. Miyamae T, Yamada Y, Uyama H, Nozoye H. 2001. *Appl. Surf. Sci.* 180:126–37
88. Chan CM. 1993. *Polymer Surface Modification and Characterization.* New York: Hanser
89. Wild S, Kesmodel LL. 2001. *J. Vac. Sci. Technol. A* 19:856–60
90. Dupont-Gillain CC, Adriaensen Y, Derclaye S, Rouxhet PG. 2000. *Langmuir* 16:8194–200
91. Rossier JS, Girault HH. 1999. *Phys. Chem. Chem. Phys.* 1:3647–52
92. Hasegawa M. 1999. *Jpn. J. Appl. Phys. Pt 2.* 38:255–57
93. Deleted in proof
94. Dee GT, Sauer BB. 1998. *Adv. Phys.* 47:161–205
95. Langmuir I. 1938. *Science* 87:493–500
96. Lewis KB, Ratner BD. 1993. *J. Colloid Interface Sci.* 159:77–85
97. Ruckenstein E, Gourisankar SV. 1986. *J. Colloid Interface Sci.* 109:557–66
98. Yasuda H, Charlson EJ, Charlson EM, Yasuda T, Miyama M, Okuno T. 1991. *Langmuir* 7:2394–400
99. Yasuda T, Okuno T, Yasuda H. 1994. *Langmuir* 10:2435–39
100. Hogt AH, Gregonis DE, Andrade JD, Kim SW, Dankert J, Feijen J. 1985. *J. Colloid Interface Sci.* 106:289–98
101. Andrade JD, Gregonis DE, Smith LM. 1983. In *Physico-Chemical Aspects of Polymer Surfaces*, ed. KL Mittal, 2:911–22. New York: Plenum

QUANTUM MECHANICAL METHODS FOR ENZYME KINETICS

Jiali Gao and Donald G. Truhlar
Department of Chemistry and Supercomputer Institute, University of Minnesota, 207 Pleasant Street S.E., Minneapolis, Minnesota 55455-0431;
e-mail: gao@chem.umn.edu; truhlar@umn.edu

Key Words catalysis, dynamics, free energy, potential of mean force, QM/MM, rate constant, tunneling, variational transition state theory

■ **Abstract** This review discusses methods for the incorporation of quantum mechanical effects into enzyme kinetics simulations in which the enzyme is an explicit part of the model. We emphasize three aspects: (*a*) use of quantum mechanical electronic structure methods such as molecular orbital theory and density functional theory, usually in conjunction with molecular mechanics; (*b*) treating vibrational motions quantum mechanically, either in an instantaneous harmonic approximation, or by path integrals, or by a three-dimensional wave function coupled to classical nuclear motion; (*c*) incorporation of multidimensional tunneling approximations into reaction rate calculations.

INTRODUCTION

The past few years have seen significant progress on the accurate molecular modeling of enzyme reactions. This review summarizes recent progress on incorporating quantum mechanical effects into such modeling.

Almost all enzyme reactions can be well described by the Born-Oppenheimer approximation, in which the sum of the electronic energy and the nuclear repulsion provides a potential energy function, or potential energy surface (PES), governing the interatomic motions. Therefore the molecular modeling problem breaks into two parts: the PES and the dynamics. After a brief overview in the next section, we consider these in turn in the following section. The final section surveys applications.

THEORETICAL FRAMEWORK OF ENZYME KINETICS

The key to elucidating the mechanism of enzymatic reactions is an understanding of the specific interactions between the substrate and its enzyme environment and their effect on the reaction rate. Although our emphasis in this review is quantum mechanical effects in enzymology, we begin by considering classical mechanical rate theory and then proceed to discuss the way in which quantum mechanical contributions can be included. The starting point of our discussion is transition state theory (TST) (1–12), which is based on a hypersurface (the transition state) that divides reactants from products in phase space. TST can be derived based on two fundamental assumptions: (*a*) Reactant states are in local equilibrium along a progress coordinate, which is the reaction coordinate, and (*b*) trajectories that cross the transition state hypersurface do not recross it before becoming thermalized on the reactant or product side (7, 13). When these assumptions are satisfied, the TST rate constant, which is the local equilibrium one-way flux coefficient for crossing the hypersurface in the direction of products, is equal to the net reactive flux coefficient, which is the observed rate constant. Because it neglects recrossing, the classical TST rate gives an upper limit to the true classical reaction rate for the case in which reactant states are in equilibrium with each other even though they are not in equilibrium with the product. The dynamic recrossing can be taken into account by a transmission coefficient (4, 8, 14–16). Thus, the exact classical rate constant at temperature T can be written as

$$k(T) = \gamma(T) \frac{1}{\beta h} \exp[-\beta \Delta G^{\neq}(T)], \qquad 1.$$

where $\gamma(T)$ is the transmission coefficient, $\beta = 1/k_B T$, k_B is Boltzmann's constant, h is Planck's constant, and $\Delta G^{\neq}(T)$ is the molar free energy of activation. Note that both $\Delta G^{\neq}(T)$ and $\gamma(T)$ depend on the choice of the transition state hypersurface, and thus they have a complementary nature. A "better" transition state hypersurface leads to a higher $\Delta G^{\neq}(T)$ and a larger $\gamma(T)$, at least in classical mechanics. The theory may be applied to gas-phase reactions (1–12) and to reactions in condensed phases including enzymes (8, 10–13, 17–21). Experience has shown that TST provides accurate estimates of rate constant, provided that one chooses the location of the dividing surface along the reaction coordinate to minimize the calculated rate constant; this is called variational TST (VTST) (5) to emphasize the variation or TST for simplicity. For the purposes of this review, TST is especially important because it provides a clear route to the incorporation of quantum effects (9, 11, 22), which we consider below.

Equation 1 shows that there are two critical quantities to be evaluated: the equilibrium free energy of activation $\Delta G^{\neq}(T)$ and the dynamical transmission coefficient $\gamma(T)$. In many cases the more important of these quantities is the former because the rate constant depends exponentially on the free energy of activation. Thus, it is not surprising that major efforts have been devoted to developing methods for the accurate description of PESs that allow one to estimate free energies of activation. In classical mechanics one can always, in principle, find a hypersurface for gas phase reaction such that there are no recrossings, and hence $\gamma(T)$ has a value of unity. However, such a hypersurface may be very intricate, and finding it may be equivalent to solving the full dynamic problem.

Let z be the reaction coordinate or progress variable. For multidimensional condensed-phase systems such as reactions in enzymes, a one-dimensional optimization of the transition state location along z is usually carried out. In such cases it is not always clear how best to choose the reaction coordinate. Most work uses a simple function of valence coordinates (e.g., making and breaking bond distances) (23, 24), whereas other studies employ a collective bath coordinate, as in Marcus theory (25–28). In practice, a good choice of the reaction coordinate, z, plus optimization of the location of the transition state hypersurface along z and sometimes of its orientation relative to z is expected to result in a transmission coefficient with a value close to 1, making it quantitatively unimportant. On the other hand, a poor choice of the reaction coordinate or the transition state will require more effort in computing $\gamma(T)$, which might have a significantly reduced value. In VTST one optimizes the transition state hypersurface to minimize the need for a recrossing correction (1, 3, 5, 6, 8–12, 29, 30).

The equilibrium assumption is normally expected to be well satisfied except for very fast reactions (31). The word "equilibrium" is sometimes used in a second way to describe TST. If a set of coordinates **x** are not involved in the definition of the reaction coordinate z, the TST one-way flux expression can be derived by assuming that coordinates **x** are equilibrated instantaneously with respect to motion along z, although TST does not require this. Nevertheless, in this sense the statement that coordinates are not in equilibrium is the same as saying that they should be included in the reaction coordinate by including them in the definition of the transition state. Thus, an effect that is a nonequlibrium contribution for one choice of transition state hypersurface may be an equilibrium contribution for another. If mode m of the bath is not in equilibrium with motion along the reaction coordinate, it exerts a frictional effect on it. In such a case the calculation can be improved in two ways: (*a*) including mode m in the definition of the reaction coordinate to optimize the definition of the transition state hypersurface (because the reaction coordinate is defined as motion orthogonal to that hypersurface) or (*b*) including these "nonequilibrium" or frictional effects in a transmission coefficient that is less than unity to account for recrossing.

Note that the transition state is not a single structure. It is an ensemble of structures, and the free energy of activation is related to the potential of mean force (PMF), $W(T, z)$ (8, 11, 15, 24, 32–34), that is obtained by averaging over protein, substrate, and solvent configurations along z. Thus, in classical mechanics, denoted by CM,

$$\Delta G^{\neq}(T) = W_{CM}(T, z^{\neq}) - \left[W_{CM}(T, z_R) + G^R_{CM}(z) \right] + C(T, z), \qquad 2.$$

where z_R and $z^{\neq}$ are values of the reaction coordinate at the reactant state and at the transition state, $G^R_{CM}(z)$ corresponds to the free energy of the mode in the reactant (R) state that correlates with the progress coordinate z, and $C(T, z)$ is a small correction term that is due to the Jacobian of the transformation from a locally rectilinear reaction coordinate to the curvilinear reaction coordinate z.

The CM potential of mean force is defined by

$$W_{CM}(T, z) = -RT \ln \rho_{CM}(T, z) + W^o_{CM}, \qquad 3.$$

where $\rho_{CM}(T, z)$ represents the classical mechanical probability density as a function of z, which can be evaluated by carrying out molecular dynamics or Monte Carlo simulations using umbrella sampling (35) or free energy perturbation techniques (36).

Quantum mechanics enters the computation of the PMF and free energy of activation in two major ways. First, the electronic structure of the atoms that are involved in the bond-making and bond-breaking process should be treated quantum mechanically in calculating the PES that governs the motions that determine $W(T, z)$. This is necessary because although traditional molecular mechanics force fields that describe the PES of substrates by a set of parameterized analytical functions have been very successful in optimizing equilibrium structure and give reasonably accurate results for the energy contours for small-amplitude vibrations around such structures (37), they do not have the correct functional form for describing chemical reactions. The reader is referred to key references (38–41) and a recent review (42) for further discussion of molecular mechanics models, the most widely used of which, for enzyme kinetics (including interactions with aqueous solvent), are AMBER, CHARMM, GROMOS, OPLS, and TIP3P. In the section entitled "Potential Energy Surface" we summarize techniques for incorporating electronic structural methods in modeling enzymatic reactions.

The second aspect of quantum mechanics that affects the PMF is the quantum mechanical nature of nuclear motion. In particular, because a reacting system is bound in its transition state as the reaction coordinate is fixed and missing as a degree of freedom, the most important quantum effects on the motions within the transition state hypersurface can be incorporated by the usual methods for bound vibrations (43). As we review in "Potential of Mean Force," it is necessary to incorporate such quantum vibrational effects in evaluation of the free energy of activation. The transmission coefficient, though, may be considered as arising from motions normal to the transition state hypersurface, i.e., dynamical recrossing and tunneling, and from the nonseparability of the reaction coordinate.

The recrossing contribution to the transmission coefficient, $\gamma(T)$, in Equation 1 can be approximated classically by propagating trajectories forward and backward from an ensemble of transition state configurations and counting the number of crossings of the transition state hypersurface. The general formalism for such calculations was first developed systematically by Keck (14) and elaborated by Anderson (4) and others (13, 15, 44). These workers used classical methods to define the transition state ensemble and calculated the transmission coefficient by propagating trajectories forward and backward from phase points in that ensemble for both gas-phase and liquid-phase reactions. This procedure has been extended to an enzyme reaction by Neria & Karplus (21). The next level of sophistication is to select transition state phase points from a semiclassical treatment of the transition state, i.e., from a distribution of transition states with quantized vibrations, but still estimate the recrossing factor from purely classical trajectories (45, 46). Even better, one can both select from a semiclassical distribution of transition states and include quantum effects in the dynamical calculation used to estimate recrossing. Schenter et al. (47) presented one method for doing this, and we have presented a method that is specifically designed for enzyme reactions to include the effect of quantization transverse to the reaction coordinate on motion along a minimum-energy reaction path (48). Our method leads to the $\Gamma(T)$ factor discussed in "Dynamics and Tunneling." Billeter et al. (49) have presented an alternative approach, also discussed in that section.

The inclusion of the quantum nature of the motion that advances the reaction coordinate, chief among which is tunneling, is the third aspect of quantum mechanical contributions in enzyme kinetics modeling and is also addressed in "Dynamics and Tunneling." A critical aspect is that tunneling requires a multi-dimensional treatment because reaction coordinate motion is not separable (9).

METHODS

From the above discussion, we can see that there are three main ways in which quantum mechanics should be included in enzyme kinetics modeling. First, the electronic structure of the atoms directly involved in the chemical step must be treated quantum mechanically in order to describe the bond-making and bond-breaking processes. Second, the discrete nature of quantum mechanical vibrational energies should be incorporated in the description of nuclear motion for computing the PMF and the transmission coefficient. Finally, consideration of multidimensional tunneling contributions is required in calculating the transmission coefficient, particularly for reactions involving hydrogen (H, H^+, and H^-) transfer. Although quantum mechanical approaches have been used in various ways for studying condensed-phase systems (24, 50), it was only recently that computational studies have fully included all three of these aspects of quantum mechanical effects in calculations including an explicit enzyme environment (48, 51–59). In what follows, we summarize methods that have been developed by our groups and others for including quantum mechanical contributions in enzyme kinetics modeling.

Potential Energy Surface

As explained above, $\Delta G^{\neq}(T)$ results from averaging the PES appropriately over an ensemble of structures, and therefore its accuracy depends on the accuracy of the PES. To compute the PES for macromolecular processes such as enzymatic reactions, it is necessary to use a chemical model that is capable of describing the formation and breaking of chemical bonds and also is suitable for capturing the complexity of the system. A fully quantum mechanical treatment of the entire enzyme system, in principle, satisfies these criteria, and quantum mechanical algorithms designed to involve an effort that scales linearly with system size have been developed and applied to protein systems in energy calculations (60–69). Although this approach has many attractive features, it is very expensive, and this has limited its application to biological problems.

Fortunately, in most enzymatic reactions it is not necessary to treat the electronic structure of the entire enzyme-solvent system quantum mechanically. The most promising approaches for modeling enzymatic reactions are combined quantum mechanical and molecular mechanical (QM/MM) methods, in which a system is divided into a quantum mechanical region and a molecular mechanical region (28, 70–83). The quantum mechanical region typically includes atoms that are directly involved in the chemical step and they are treated explicitly by a quantum mechanical electronic structure method, whereas the molecular mechanical region consists of the rest of the system and is approximated by a molecular mechanical force field. This way of combining quantum mechanics with molecular mechanics was initially developed for gas-phase calculations (84) and was first applied to enzyme systems by Warshel & Levitt (71). A combined QM/MM potential has the advantage of both the computational efficiency offered by molecular mechanics, allowing for free energy simulations of macromolecules, and the computational accuracy provided by quantum mechanics such that the method can be systematically improved (85). The past decade has seen a rapid increase in number and diversity of applications, including reactions in solution (24) and in enzymes (59, 86). The method has been implemented in combination with semiempirical and ab initio molecular orbital theory and with density functional theory (DFT). In fact, combined QM/MM techniques have emerged as the method of choice for modeling enzymatic reactions.

The most popular methods for treating the quantum mechanical subsystem in enzyme reactions have been semiempirical molecular orbital methods (87), such as AM1 (88), PM3 (89), MNDO (90), Hartree-Fock (HF) theory (91), Møller-Plesset perturbation theory (MP2) (91), configuration interaction with single excitations (91), the Becke-Lee-Yang-Parr density functional (92, 93), and the three-parameter hybrid version (B3LYP) (94) of this DFT method. Basis sets have included 3-21G (91) 6-31G* (in which a d shell is added to the 6-31G basis) (91), basis sets designed for use with effective core potentials (95), and plane waves (96).

Combined QM/MM methods have been reviewed in several articles (70, 74, 80, 97, 98) and a full review is not repeated in this article. Here, we focus on the most critical issue in application to enzyme systems, i.e., the treatment of the boundary that separates the quantum mechanical and molecular mechanical regions. Gao et al. identified criteria that they considered to be important for the treatment of a covalent bond at the QM/MM interface (99). First, without introducing extra degrees of freedom, any boundary method should reproduce the correct geometry in comparison with that predicted by the corresponding quantum mechanical calculation on the entire system. Second, the electronic structure, including charge distribution and electrostatic potential, of the full quantum mechanical system should be retained in the quantum mechanical fragment, e.g., the boundary atom should have the same electronegativity as that in the quantum mechanical molecule. Third, the torsional PES about the bond connecting the quantum mechanical region to the molecular mechanical region should be consistent with results from pure quantum mechanical and molecular mechanical calculations.

An early and still widely used scheme is the "link-atom" approach, in which extra, unphysical atoms are added to the quantum mechanical region, one for each bond at the interface to the molecular mechanical region (72, 74). These link atoms, typically hydrogen atoms, are placed at appropriate distances along the bond, and they can be constrained in the direction of the molecular mechanical boundary atom or allowed to freely move. The forces between these link atoms and molecular mechanical atoms are unphysical, so all or some of the forces of the molecular mechanical subsystem on the quantum mechanical subsystem are sometimes omitted. When this is done, the quantum mechanical fragment does not experience the real electrostatic environment of the enzyme active site (100, 101). To circumvent the problem of omitting forces on the link atom, a scaling scheme (scaled-position link-atom method) has been proposed to transfer forces on link atoms to other "physical" atoms (102). Rather than using hydrogen as the link atom, Ostlund (102a) suggested using pseudo-halogen atoms parameterized to mimic the missing valence bond between the quantum mechanical and molecular mechanical subsystems (80), which would also include hyperconjugation interactions owing to the presence of p orbitals. A similar idea has been implemented as a pseudopotential in DFT by Zhang et al. (103).

Other methods dispense with the unphysical link atoms and employ hybrid orbitals. Warshel & Levitt suggested that a hybrid oribital could be used to treat the QM/MM bond, though the approach was not elaborated further (71). Two formalisms that have proved successful for treating the boundary region are strictly localized bond orbitals (104–108) and generalized hybrid orbitals (GHOs) (99, 109). Rivail and coworkers (104–108) called their scheme the local self-consistent field (LSCF) method. In this method the valency of the quantum mechanical boundary atom is saturated by a frozen hybrid orbital with a predetermined density from calculations of a smaller model system. The charge densities of the frozen hybrid orbitals are kept fixed in an SCF calculation of the quantum mechanical system. The LSCF scheme has been developed both for semiempirical (104) and ab initio molecular orbital QM/MM methods (106, 107) and recently it has been extended to DFT (110).

The GHO method, which was initially developed for semiempirical Hamiltonians (99, 109), differs from the LSCF approach in orbital optimization and system partition as illustrated by Figure 1. In the LSCF method the boundary atom is a quantum mechanical atom with a frozen density for the hybrid orbital pointing toward the molecular mechanical fragment, whereas the boundary atom in the GHO scheme is treated as both a quantum mechanical and molecular mechanical atom. The hybridization of the hybrid orbitals on the boundary atom in the GHO method is determined by the molecular geometry and is dynamically varied during the simulation. Furthermore, the charge density of the hybrid orbital that is bonded to the quantum mechanical fragment is optimized self-consistently with all other atomic orbitals in the SCF calculation. An advantage of the GHO method is that it uses fully transferable semiempirical parameters for the boundary atoms in precisely the same way as the standard semiempirical algorithm (99). There are

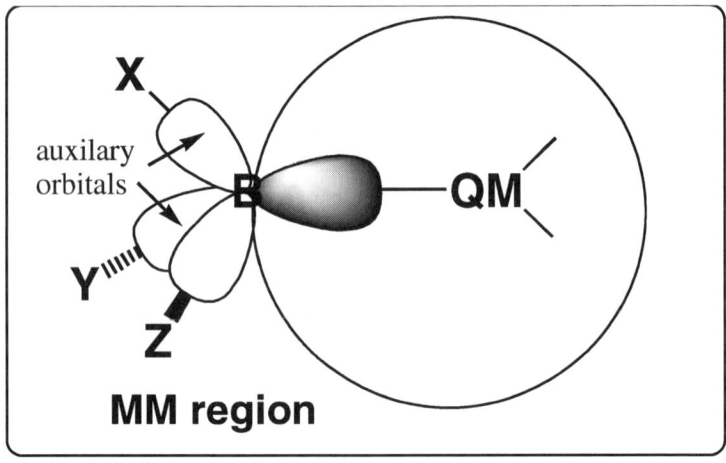

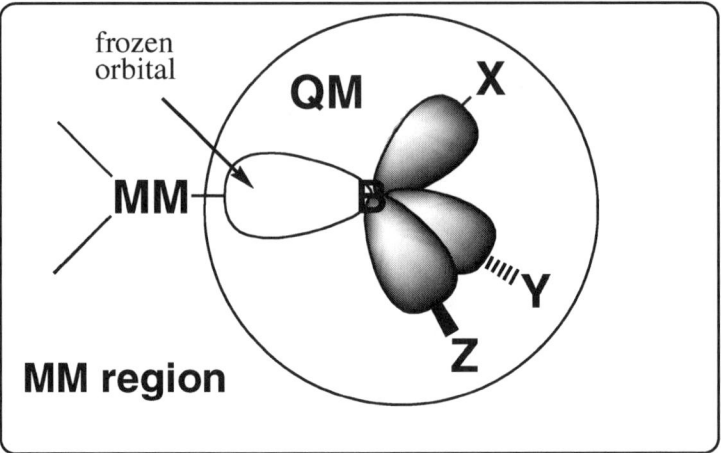

Figure 1 Schematic representation of the local self-consistent field method, (*bottom*) and the generalized hybrid orbital approach, (*top*) for the treatment of a covalent bond across the boundary between the quantum and classical regions. QM, quantum mechanical; MM, molecular mechanical.

no extra degrees of freedom introduced into the system, and all molecular mechanical partial charges are fully retained in SCF calculations. Thus, it provides a better treatment of electrostatic interactions than the link-atom method does. This method has been successfully used to study a number of enzymatic reactions and is being extended to ab initio QM/MM calculations.

Antes & Thiel described a similar approach (111, 112), in which the boundary atom is represented by a hydrogen-like atom with its semiempirical Hamiltonian parameters optimized to reproduce the geometry and energy of a set of model compounds. Bersuker et al. described a double self-consistent procedure to allow charge transfer (113), as in the GHO method, between the quantum mechanical and molecular mechanical fragments.

The development of satisfactory boundary methods allows combined QM/MM methods to be used effectively in free energy calculations on enzymatic reactions. However, when ab initio or DFT methods are used for the quantum mechanical part, one must use at least a polarized split valence basis set and include electron correlation to achieve the accuracy needed for predicting the enzyme reaction rate. Consequently, it is expensive to perform free energy calculations using ab initio QM/MM methods. Semiempirical molecular orbital methods are computationally more efficient and they offer the opportunity to develop specific reaction parameters (SRPs) to improve the accuracy for specific systems (58, 114–116). Because the shape of the PES is usually well represented, an even simpler approach is to add a semiempirical valence bond term such as the London-Eyring-Polanyi-Sato function (54, 117) to correct the errors in the predicted free energies of reaction and activation. This approach has been used in the study of hydride transfer reactions in liver alcohol dehydrogenase (48, 54).

Semiempirical QM/MM methods can be used in what are called dual-level simulations (118–122). In this approach QM/MM energies are separated into a solvent-independent (gas phase) term and a solute-solvent interaction component such that a high-level (HL) method, typically MP2 or DFT, can be used to determine the intrinsic energy of the reactants in the gas phase while the time-consuming free energy calculations are carried out with a lower-level (LL) model. A similar approach is the integrated molecular orbital molecular mechanics (IMOMM) method (123), in which the lower-level calculation is performed by a pure molecular mechanical method rather than by using a combined QM/MM potential. In dual-level calculations, the total potential energy (and free energy) of the system can be separated as follows (118, 122):

$$E_{tot} = E_{qm}^{HL}(gas) + \Delta E_{qm/mm}^{LL} + E_{mm}, \qquad 4.$$

where $E_{qm}^{HL}(gas)$ is the energy of reactants computed at a high level of theory in the gas phase, typically ab initio calculations that include electron correlations, E_{mm} is the molecular mechanical energy of the protein, and $\Delta E_{qm/mm}^{LL}$ is the QM/MM interaction energy modeled with a lower-level quantum mechanical method, e.g., the AM1 model (88) combined with the CHARMM force field (38). In this approach, the free energy calculation is carried out using a lower-level QM/MM potential, whereas the high-level quantum mechanical result is made as a posterior correction to the intrinsic (i.e., gas-phase) energy of the quantum mechanical part from the lower-level calculation (118, 122). This energy separation is made possible by using the QM/MM energy decomposition scheme developed previously (76), and the method has been applied to the solvation of organic molecules (118, 119, 122, 124).

Warshel and others have used a valence bond formalism to approximate the electronic structure of the reactants in the enzyme active site (125). In the most widely used valence bond approach (20, 49, 97, 125–135), computational efficiency is obtained by treating the energies of valence bond configurations with molecular mechanical force fields and using a constant or one-variable exponential function for the resonance integrals (H_{12}); such calculations are denoted as empirical valence bond (EVB) (127, 128). Unlike combined QM/MM methods, the electronic wave function is implicit in the EVB model, and thus the overlap integrals between valence bond states are approximated as zero; the effect of neglecting overlap integrals on the ground-state potential energy surface (PES) is assumed to be absorbed in the parameter-fitting step. The inclusion of a solvation energy term in the diagonal EVB matrix elements provides a way to model reactions in solution and in enzymes (127). The off-diagonal matrix element, H_{12}, is assumed to be environment independent. The EVB method provides a fast analytical procedure for obtaining approximate reactive PESs and has been used extensively (97).

Effective diabatic states such as those used in EVB-type models are not pure valence bond states and they are mixtures of valence bond configurations, which include both covalent and ionic contributions. Although the charge distribution of pure valence bond configurations in ab initio valence bond theory can be described by Lewis resonance structures with charges localized in specific valence bond orbitals, the charge distribution of mixed or partially mixed valence bond states (or effective diabatic states) depends on molecular geometry and environment (28, 136). Although a fixed-charge diabatic valence bond state can be parameterized to model reactions, it is dangerous to infer energetic insights using such diabatic states. It would be possible to develop a semiempirical valence bond model in which empirical charges for individual effective valence bond states vary as a function of reactant geometry, but the functional form and calibration procedure would be complicated.

The EVB method when used as described above is on the border between a classical and a quantum mechanical electronic structure method, but it is considered as a classical method (like molecular mechanics) in this review. It is, however, possible to use valence bonds as a quantum mechanical method, as discussed next.

A mixed molecular orbital and valence bond (MOVB) method was recently introduced (28, 136); it combines and extends many features of combined QM/MM techniques as well as ab initio valence bond theory. In the MOVB method, the effective diabatic state is treated by a block-localized wave function theory (137, 138), corresponding to the traditional Lewis resonance structures. Because the block-localized wave function approach is based on an ab initio molecular orbital wave function that corresponds to a mixed valence bond state, the geometrical dependence of gas-phase charge distribution and solute-solvent electronic polarization are naturally included in the MOVB method's Hamiltonian matrix (28, 136). Thus, the significance of omitting all or some of these factors can be systematically investigated without the need to empirically adjust parameters (28, 136).

Schmitt & Voth described a multistate empirical valence bond model for proton transport in water (139, 140). The approach extends the two-state EVB method by introducing an exchange charge distribution in the off-diagonal terms to mimic transition dipole moments. Inclusion of these exchange (or transition) charges, which is also present in MOVB (28), led to qualitatively different results in modeling proton transport in water in comparison with a similar approach that excluded such explicit off-diagonal solute-solvent interactions (139, 140).

In another valence bond approach a multiconfiguration molecular mechanics method was proposed with a systematic parameterization scheme for the off-diagonal element (141, 142). This method can accurately reproduce ab initio energies and gradients and does not suffer from the limitation of the formalism to matrices of order 2 in that it can be systematically converged to reproduce high-level calculations. In this respect, it is a systematic fitting tool based on the valence bond formalism rather than an approximate valence bond calculation per se.

Potential of Mean Force

Next we turn our attention from the electronic structure considerations that yield the PES to the nuclear dynamics. The first challenging goal is to obtain an accurate estimate of $\Delta G^{\neq}(T)$ by computing the potential of mean force (PMF) along the reaction coordinate. Two methods are generally used in the calculation of PMFs for reactions in solution and in enzymes, the umbrella sampling (35) technique and the free energy perturbation theory (36); these are reviewed elsewhere (143). Because these calculations are often carried out using purely classical mechanics, zero-point energy and effects of the quantization of vibrational motion are neglected. It has been shown in several studies (48, 49, 52, 54, 58, 144, 145) that omission of these quantum effects can lead to significant errors in computed free energies of activation, particularly in systems involving hydrogen transfer. Thus, it is desirable to make quantal corrections to the classical mechanical-PMF or to perform simulations that directly include nuclear quantum mechanical effects.

To this end, our groups (48, 145) have developed a convenient procedure to include the effect of quantization of molecular vibrations in free energy calculations. In this method, the quantal PMF is related to the classical PMF of Equation 3 by

$$W_{\text{QC}}(T, z) = W_{\text{CM}}(T, z) + \Delta W_{\text{vib}}(T, z), \qquad 5.$$

where $\Delta W_{\text{vib}}(T, z)$ is an ensemble average of the instantaneous harmonic approximation to the quantal correction to the classical vibrational free energy for vibrational modes orthogonal to the reaction coordinate, and the subscript QC denotes "quasiclassical" and specifies the fact that tunneling and other quantum effects on the reaction coordinate, z, are excluded. The procedure involves instantaneous normal mode analysis based on local harmonic approximations, and this should be more accurate than using true normal mode analysis at a single local minimum. In applying Equation 5, the lowest vibrational modes corresponding to low-frequency vibrational motion, for which the classical approximation should be adequate, are omitted in the calculation of $\Delta W_{\text{vib}}(T, z)$. The constraint that the modes included are orthogonal to z can be achieved by a projection operator (48, 145). To evaluate the QC free energy of activation, quantum corrections must also be made to the

$G_{CM}^R(z)$ term in Equation 2. A procedure for estimating the required frequency ω_R of the reaction-coordinate mode of the reactant has been presented (48).

Free energy simulations of model proton shifts in water as well as enzymatic hydrogen transfer reactions in a number of enzymes indicate that inclusion of quantum vibrational free energy contributions reduces the classical barrier height by 2–4 kcal/mol and that more degrees of freedom than those associated with the migrating atoms are needed in these calculations (49, 52, 54, 145, 146).

Another method for including quantum effects on vibrations is the path integral technique (147). In this approach the ensemble averages for the quantum system can be obtained by carrying out a classical simulation in which the quantized particles are represented by ring polymers of classical particles. To determine the PMF, it has been proposed to define the positions of the quantized particles and hence the value of the reaction coordinate using the average positions (148–154), or centroids, of the ring polymers, leading to a method called path integral quantum transition state theory (TST). Although this approach has been successful for some problems and was applied to an enzyme reaction (53), it has been noted that difficulty exists for asymmetric reactions at low temperature (154–158). An algorithm has been proposed in which classical free energy calculations are performed first, and then quantum corrections are added by a path integral calculation with the centroids constrained at the classical atom positions (146, 153). This approach has been called the quantum-classical path (QCP) method and has also been applied to enzymes (133, 135, 146, 153). A path integral approach that does not use the centroid concept has been shown to overcome the deficiencies of the centroid-based approach but has not so far been applied to enzymes (157). A path integral method designed to reduce the variance in sampling has also been proposed (158a). It is interesting to note that Tuckerman & Marx (159), who employed a centroid path integral method in the study of the proton transfer reaction of malonic acid, found that the reaction rate is underestimated by a factor of two if only the transfer proton is treated quantum mechanically in the framework of classical heavy atoms, compared with a full path integral quantum mechanics treatment.

Recently, Billeter et al. (49) used a hybrid approach to study enzymatic hydrogen transfer reactions. In this approach the free energy of activation is first calculated by a mixed quantum/classical molecular dynamics (MQCMD) method, and then a transmission coefficient is added by molecular dynamics with quantum transitions (MDQT). We discuss the MQCMD step here and the MDQT step in the next section. In the MQCMD step the nuclear degrees of freedom are divided into a quantal subset and a classical subset; the quantal subset is treated by using the Fourier grid Hamiltonian multiconfigurational self-consistent field method (160, 161) to calculate a three-dimensional hydrogen vibrational wave function, which is expressed as a linear combination of products of one-dimensional wave functions. Since the one-dimensional basis functions are directly represented on a cubic grid, the numerical calculation of the vibrational wave function requires knowledge of the potential energy and forces at each grid point for each dynamics integration step. The latter calculations are the computational bottleneck in the Fourier grid Hamiltonian multiconfigurational self-consistent field approach because a large number of grid points and memory space are needed (49). To generate the quantum PMF from classical simulations, a free energy perturbation approach analogous to the QCP algorithm (153) is used to yield the free energy difference between the quantum (Fourier grid Hamiltonian multiconfigurational self-consistent field) and classical EVB-mapping potential along the reaction coordinate.

In an approach called the quantum mechanical–free energy method (162–167) or the "solvated gas-phase" approach (135), free energy perturbation theory is used to add environmental free energies (i.e., bath free energies) along gas-phase reaction paths (162–168). This involves free energy perturbation on the molecular mechanics subsystem for a fixed quantum mechanical subsystem. An advantage of this method is that it allows high-level quantum mechanical calculations for the gas phase portion of the problem, but a serious disadvantage is that it does not treat the effect of the bath on the primary subsystem self-consistently. This effect is included to some extent by fixing certain anchor atoms of the quantum mechanical model to the protein coordinate frame (162, 169). However, we do not consider this method further because it does not introduce quantum mechanics into a full enzymic simulation.

Dynamics and Tunneling

Whereas the previous section was concerned with quantum effects within the transition state hypersurface, in which the reaction coordinate is missing, this section focuses on the reaction coordinate. However, we note that the calculation of free energy of activation and dynamical effects are not independent of each other. There are two major factors contributing to the transmission coefficient, $\gamma(T)$: (a) dynamical recrossing, which can be estimated classically and is discussed above, and (b) nuclear tunneling and other quantum effects on the reaction coordinate that are missing in the calculation of the PMF. It can be difficult, however, to separate quantum effects on the reaction coordinate from those on other degrees of freedom if path integral or quantum-classical molecular dynamics calculations are carried out. To a certain degree, the effect of the nuclear tunneling contributions is absorbed into the computed PMF in these methods. Thus, methods discussed in the previous section are often related to techniques reviewed below. From a more general perspective, we note that any reliable approach to calculating tunneling effects must be multidimensional to take account of the nonseparability of the reaction coordinate.

We can distinguish three ways to include quantum mechanical effects on reaction-coordinate motion. The first category involves dynamical calculations that do not involve transition state theory; an example of such an approach that has been applied to enzyme dynamics is the quantum-classical molecular dynamics method reviewed below. The second category includes quantal TST methods that recognize the finite spatial extension of a quantum mechanical transition state along the reaction coordinate (the infinitesimal thickness of a classical transition state hypersurface violates the Heisenberg uncertainty principle). In such methods, tunneling contributes to the free energy of activation. An example of this approach is path-integral–quantum TST, including the QCP algorithm, discussed in the previous section. The third way to include quantum effects involves first calculating a free energy of activation in which reaction coordinate motion is separated from all other degrees of freedom normal to it. In this technique, introduced above, the free energy of activation is quasiclassical because the progression coordinate is classical and the other degrees of freedom are quantal. The dynamical recrossing and quantum effects on the reaction coordinate, including the effects of nonseparability, are treated by means of a transmission coefficient. An example of this approach is ensemble-averaged variational transition state theory with multidimensional tunneling (EA-VTST/MT) (48, 59) for estimating transmission coefficients of enzymatic reactions. The MQCMD/MDQT method (49) lies between categories 2 and 3.

Bala et al. (55, 170) described a mean-field quantum-classical molecular dynamics model in which the dynamics of the quantum subsystem is described by the time-dependent Schrodinger equation, whereas the rest of the system is treated by Newtonian dynamics. The coupling between the quantum subsystem and the classical atoms is modeled by a mean-field method through extended Hellman-Feynman forces. The method has been used to study a proton tunneling process in the active site of phospholipase A_2 (55). Other mean-field approaches include the time-dependent self-consistent field method (171, 172) and the density matrix evolution technique (173, 174).

The MQCMD/MDQT technique of Billeter et al. (49, 175) uses reactive flux for computing the transmission coefficient for enzyme reactions. The MQCMD step is discussed in the previous section. In the second step the MDQT surface hopping method (160, 175–178) is used to calculate a transmission coefficient. Although MDQT was originally developed for treating electronically nonadiabatic transitions, it is the nuclear motion of hydrogen that is treated quantum mechanically in applications to enzymes. The method includes both vibrationally adiabatic and vibrationally nonadiabatic motions of the hydrogen, as does the EA-VTST/MT method described below. For calculating rate constants, the MDQT method uses a fictitious surface hopping algorithm that does not depend on the quantum amplitude for backward trajectories. Then, forward trajectories are propagated with the true surface hopping method by integrating the quantum amplitude of the same trajectories. This approach assumes a Boltzmann distribution of vibrationally adiabatic states at the transition state dividing surface for both the fictitious and true surface hopping algorithms (175).

In our own approach, which is EA-VTST/MT, the net transmission coefficient is approximated as (48, 58)

$$\gamma(T) = \langle \kappa(T)\Gamma(T)\rangle_{\neq} \cong \frac{1}{M}\sum_{i=1}^{M}\kappa_i(T)\Gamma_i(T), \qquad 6.$$

where the brackets $\langle \cdots \rangle_{\neq}$ represent an ensemble average over transition state configurations at temperature T, $\Gamma(T)$ accounts for dynamical recrossing, $\kappa(T)$ accounts for tunneling through the effective barrier and nonclassical diffractive reflection from the barrier top, i denotes a particular member of the quasiclassical transition state ensemble, and M is the total number of configurations that have been sampled in the calculation. The ensemble of transition state configurations can be generated during the PMF calculation using umbrella sampling (58; M. Garcia-Viloca, C. Alhambra, D.G. Truhlar, J. Gao, unpublished manuscript) or in a separate molecular dynamics simulation with the constraint that the reaction coordinate corresponds to the transition state value (52, 54). The progress coordinate z that defines this ensemble can be either a combination of internal coordinates or a diabatic energy-gap coordinate representing collective solvent motions. In our work so far, we have used the difference of the breaking and making bond distances. For each configuration i in the transition state ensemble, the recrossing factor $\Gamma_i(T)$ is evaluated by an embedded cluster (19, 21, 48, 179) formalism in which an improved choice of the transition state hypersurface is determined using quantized variational TST (VTST) for that individual configuration, and the "tunneling factor" $\kappa_i(T)$ is also computed in the same calculation by a semiclassical multidimensional tunneling (MT) approximation, either the small-curvature tunneling approximation (180) or the microcanonically optimized MT approximation (181). The latter involves choosing at each tunneling energy the better of the small-curvature (180) and large-curvature (182, 183) tunneling approximations.

For each configuration i in the transition state ensemble, the system is partitioned into a primary zone and a secondary zone. The primary zone consists of N_1 atoms, which can be the same quantum mechanical fragment as that used in the combined QM/MM potential or a different set of atoms, and they have kinetic energy. The secondary zone represents the rest of the system, which is kept frozen in the calculation of an isoinertial minimum energy path (MEP) for the embedded primary zone (48). The progress variable along the MEP becomes the reaction coordinate, denoted as s to be distinguished from z used in the PMF calculation. The calculation yields a generalized transition-state free energy of activation $\Delta G^{\text{GT}}(T, s_i)$ and a semiclassical transmission coefficient $\kappa(T)$ by methods fully described in previous work (5, 180–190). In a canonical ensemble of primary systems interacting with its frozen environment the minimum-flux dividing hypersurface is the one with maximum $\Delta G^{\text{GT}}(T, s_i)$, which is located at $s_i = s_i^{\neq}$ (this defines $s_i^{\neq}$). The equilibrium classical flux is reduced relative to the dividing hypersurface that corresponds to $\Delta G_{\text{QC}}^{\neq}(T)$, which is the quasiclassical free energy of activation described in the previous section, by a factor of (19, 21, 48)

$$\Gamma_i(T, z) = \exp\{-\beta[\Delta G^{\text{GT}}(T, s^{\neq}) - \Delta G_{\text{QC}}^{\neq}(T)]\}. \qquad 7.$$

Note that $\Delta G^{\text{GT}}(T, s_i)$ is normalized so that it equals $\Delta G_{\text{QC}}^{\neq}(T)$ when $z(s_i) = z_i^{\neq}$. Equation 7 is called the quasiclassical transmission factor (48).

The net effect of quantum contributions on the reaction coordinate motion, relative to value reduced by Equation 7, is given by the semiclassical transmission coefficient $\kappa_i(T)$. The methods used for tunneling approximate the quantum mechanical transmission probabilities by the Wentzel-Kramers-Brillouin approximation (191) in terms of imaginary action integrals; such a treatment is called semiclassical because it uses classical variables (or their analytic continuation to complex momenta) to approximate quantum mechanical dynamics (192). The net effect of $\kappa_i(T)$ is usually to increase the reactive flux because the tunneling has a larger Boltzmann weight than does the nonclassical reflection. Note that $\kappa_i(T)$ is an implicit function of $z^{\neq}$, the location of the maximum PMF along the reaction coordinate z, because it is computed in the framework of the transition state configurations chosen at $z^{\neq}$, and it is an implicit function of $s_i^{\neq}$ because it gives a net factor relative to the rate constant reduced by Equation 7.

Calculations employing Equation 6 may be carried out using either the static-secondary-zone (SSZ) approximation or the equilibrium-secondary-zone (ESZ) approximation. In the SSZ approximation the effective free energy profile for the calculation of $\Gamma(T)$ and the effective tunneling potential for the calculation of $\kappa(T)$ are evaluated with a frozen secondary zone. Although the frozen bath approximation is reasonable for the short timescale of barrier crossing for enzymatic reactions, especially for hydrogen transfer reactions, the SSZ approximation does not include dynamical and entropic contributions from the surrounding protein in the transmission coefficient. This is corrected in an ESZ approach (48), in which these effective quantities also include thermal energetic and entropic contributions from the secondary zone; the latter are obtained by additional secondary-zone free energy calculations along the MEP for each configuration in the transition state ensemble (48).

The SSZ and ESZ versions of the EA-VTST/MT method were first presented in a paper on the study of liver alcohol dehydrogense (LADH) (48). The SSZ model was subsequently applied to methyl amine dehydrogenase (58) and xylose isomerase (M. Garcia-Viloca, C. Alhambra, D.G. Truhlar, J. Gao, unpublished manuscript). We also employed a preliminary version of this approach in our earlier papers on enolase (52) and LADH (54), but this review covers details only for the latter more satisfactory EA-VTST/MT method.

It is useful to compare the MQCMD/MDQT (49) and EA-VTST/MT (48) approaches to estimating the free energy of activation and transmission coefficient because both appear to be very promising. Both methods are based on Equation 1 and incorporate the desirable feature that the free energy difference between the transition state and reactant is evaluated by computing the potential of mean force along a progress variable z. Both methods include recrossing and quantum effects in the transmission coefficient; this is very efficient because such effects are more localized in time and space than the semiglobal dynamics that determine $\Delta G^{\neq}(T)$. Because the inclusion of a recrossing correction can make up for a nonoptimal choice of a transition state hypersurface, such methods can in principle be very accurate. The accuracy ultimately depends on the quality of the transmission coefficient and the accuracy of the PES. The recrossing estimated using the MDQT method has the advantage of involving full dynamics but the disadvantage of not quantizing all particles owing to computational cost (so far only one particle has been quantized). Furthermore, in weakly coupled cases the MDQT algorithm does not treat the quantized and classical particles fully self-consistently owing to frustrated hops (178, 193). The dynamical recrossing correction estimated by using EA-VTST/MT in the embedded cluster approximation quantizes all but one degree of freedom in the primary zone (which may therefore involve ∼100 or more quantized degrees of freedom) but involves extra approximations for the secondary zone. Both approaches include tunneling, although the MDQT technique includes tunneling effects in the calculation of the PMF. The EA-VTST/MT method incorporates tunneling contributions in the transmission coefficient. The VTST/MT method has been very extensively validated against accurate quantum dynamics for smaller systems (5, 9, 10, 187, 194, 195).

In practice, the MQCMD/MDQT approach has been applied with a discrete energy gap coordinate for z, whereas EA-VTST/MT has used internal coordinates. However, these choices are not intrinsic restrictions to the methods; these two choices for the reaction coordinate have recently been compared for nonenzymatic reactions in polar media (28, 136, 196). In another study MDQT and VTST/MT were compared for a proton transfer in a polar medium (196a).

APPLICATIONS

We divide the discussion of applications into two sections: those that have quantum effects in the nuclear dynamics and those that do not. In the latter we include only studies involving a quantum mechanical treatment of electronic structure, and we further divide these studies into those that use statistical mechanics to compute free energies and those that do not. The latter deal only with potential energies. Simulations involving only classical mechanical dynamics with molecular mechanics potential functions are not included in this review, even though—taking a broader view—most molecular mechanics potentials used for biochemical simulations are originally fit using at least some input from quantum mechanical electronic structure calculations. Furthermore, although much can be learned from analytic few-dimensional models and general discussions or calculations on model systems that do not explicitly include the specific enzyme (or most of it), such models are not reviewed here. One problem with such models is that they often make unrealistic assumptions motivated by the goal of simplifying the problem. Our subject here is the emerging field of quantum mechanical methods for modeling detailed interactions and the resulting dynamics in specific enzymes with the atoms of the enzyme explicit. For lack of space, we omit discussion of electron transfer reactions.

Quantum Treatments of Nuclear Dynamics

Systems are discussed in this subsection in roughly chronological order.

LACTATE DEHYDROGENASE Hwang et al. (197) used the quantum-classical path (QCP) method to calculate the free energy of activation profile as a function of an energy-gap reaction coordinate (125) for the hydride transfer from nicotinamide adenine dinucleotide hydride (NADH) to pyruvate catalyzed by lactate dehydrogenase. The potential energy function was modeled by empirical valence bond (EVB) with an empirical exponential function coupling the valence bond states. Quantum effects on nuclear motion lowered the free energy of activation from 12 to 9 kcal/mol.

CARBONIC ANHYDRASE   Hwang & Warshel (146) used the QCP method to calculate the free energy of activation profile for proton transfer between water molecules catalyzed by the zinc metalloenzyme carbonic anhydrase. The potential was modeled by EVB. They treated six atoms as quantal and found that quantum nuclear motion effects lower the free energy of activation from 12 to 10.5 kcal/mol (146). They calculated a kinetic isotope effect (KIE) of 3.9 (146, 198).

PHOSPHOLIPASE A_2 (PLA_2)   Bala et al. (55) applied the quantum-classical molecular dynamics method to study proton transfer from water to histidine with nearly simultaneous nucleophilic attack of the resulting OH^- on the carbonyl carbon atom of a phospholipid substrate, as catalyzed by PLA_2. The PES was modeled by an approximate valence bond method that is similar to EVB but uses density functional theory (DFT) calculations instead of empirical data to parameterize the valence bond coupling. Quantum nuclear motion effects lowered the average barrier from 2.5 kcal/mol to 0.5.

ASPARTIC TRANSCARBAMOYLASE   Pawlak et al. (199) studied the reaction between carbonyl phosphate and aspartate catalyzed by aspartic transcarbamoylase. The reactants were treated by the PM3 model in the rigid field of an optimized structure of the active site residues, whose interaction with the quantum mechanical subsystem was modeled by atomic potential charges determined in a supermolecular calculation involving 490 atoms. Proton transfer was found to be slightly more advanced in the flexible model of the active site. Kinetic isotope effects were calculated by conventional TST with quantized harmonic vibrations and unit transmission coefficient using the rigid active site. Both wild-type enzyme and a mutant were studied.

GLUTATHIONE REDUCTASE   Transition states were optimized for hydride transfer from nicotinamide adenine dinucleotide phosphate hydride (NADPH) to a flavin cofactor as catalyzed by glutathione reductase (51). The quantum mechanical system was treated by AM1, and the molecular mechanical part of the calculation employed CHARMM and TIP3P; the subsystems were joined by link atoms. KIEs were calculated by conventional TST with quantized harmonic vibrations and a Wigner transmission coefficient. The latter is as large as 4.2, and the authors noted that Wigner approximation is not reliable for such a large transmission coefficient (51).

ENOLASE   Alhambra et al. (52) studied the proton transfer from 2-phospho-D-glycerate to the ε-nitrogen of Lys345 in the active site of the homodimeric enzyme enolase. The mechanism of the enolase reaction is particularly interesting because the rate-limiting step involves the abstraction of a proton from a weak carbon acid through electrostatic stabilization of the carbanion intermediate by two magnesium ions in the active site. The PES was modeled with 25 AM1 atoms connected to an 8863-atom CHARMM22/TIP3P environment through generalized hybrid orbital (GHO) boundary atoms (99). Umbrella sampling simulations were carried out using molecular dynamics to determine the potential of mean force (PMF) along a reaction coordinate defined as the difference between the making and breaking bond distances of the transferring proton. Then, a molecular mechanical simulation with strong harmonic restraints of the reaction coordinate at the transition state value was performed to generate a representative enzyme structure for dynamics calculations. Subsequently, the system was partitioned into a primary zone and secondary zone to yield an isoinertial minimum energy path (MEP) for 25 quantum mechanical atoms in a bath of 8863 classical atoms. The reactant, transition state, and product were optimized, and the MEP of the proton transfer reaction was calculated and used as the basis for VTST/MT calculations of the rate constant and the deuteron KIE. Quantum effects on nuclear motion decreased the free energy of activation from 16.8 kcal/mol to 14.4 and increased the rate constant by a factor of 56 (a factor of 34 owing to quantizing vibrational motion orthogonal to the MEP and an additional factor of 1.7 from adding quantum effects—primarily tunneling—to the reaction coordinate) (52). Good agreement with the experimental KIE was achieved only by including the isotope dependence of the dynamical bottleneck location.

DIHYDROFOLATE REDUCTASE Castillo et al. (200) studied hydride transfer from NADPH to 7,8-dihydrofolate catalyzed by dihydrofolate reductase. The PES was modeled by QM/MM with link atoms between a 53-atom AM1 region and a CHARMM24 region. A saddle point was optimized, and KIEs were calculated by conventional TST with harmonic quantized vibrations and no tunneling. The hydrogen/deuterium (H/D) KIE was calculated to be 3.7–3.9, compared with an experimental value of 3.0.

GLYOXALASE I Feierberg et al. (133) studied the proton abstraction of a hemithioacetal by Glu172 of the zinc metalloenzyme glyoxalase I. The hemiacetal results from binding of methylglyoxal substrate to a glutathione cofactor. The PES was modeled by the EVB method. The free energy of activation as a function of an energy-gap reaction coordinate was calculated by the QCP method, and quantum effects on nuclear motion were found to lower the free energy of activation from $\sim$14.8 to $\sim$12.3 kcal/mol. H/D and hydrogen/tritium (H/T) KIEs of 5 ± 1 and 8 ± 3 were calculated, compared with gas-phase values of 5–6 and 9–12, respectively.

LIVER ALCOHOL DEHYDROGENASE (LADH) One of the most extensively studied enzymes is LADH (48, 49, 54, 135, 201), which catalyzes the reversible conversion of an alcohol to an aldehyde in the active site of a ternary complex that requires a nicotinamide adenine dinucleotide (NAD^+) coenzyme. The chemical step is hydride transfer from the alcoholate anion to NAD^+. A catalytic Zn ion is present in the active site, which is coordinated to the alcoholate form of the substrate and one His and two Cys residues. The work of Klinman and coworkers provides strong experimental evidence for hydrogen tunneling in the chemical step in alcohol dehydrogenation by wild-type LADH, mutant LADH, and yeast alcohol dehydrogenase (202, 203). Two important questions to be addressed are (*a*) Can kinetic isotope effects be accurately predicted for such enzymatic reactions? and (*b*) To what extent does quantum mechanical tunneling contribute to the rate enhancement in LADH? LADH catalysis has now been studied with explicit enzyme and nuclear-motion quantum effects by three groups (48, 49, 54, 135).

In our first study of LADH (54) the hydride transfer reaction was successfully studied with a multidimensional semiclassical treatment of the KIEs using the same theory that was applied to the enolase reaction described above (52), but for LADH we also included statistical averaging over several secondary-zone configurations. Subsequently, we developed the more complete method outlined above and in an account article (59) and described in detail elsewhere (48), and we applied the new theory to the LADH reaction. We summarize the latter simulation (48) here. [We note that the inclusion of protein motion effects on the transmission coefficient in the new calculation (48) makes a quantitative but not a qualitative change as compared with the results of the earlier (54) calculation.]

Unlike the enolase case, in which the semiempirical AM1 model was adequate, the hydride transfer in LADH directly involves a Zn ion. As a result, energetic quantities obtained from semiempirical (AM1 or PM3) calculations are not in good accord with available experimental data and high-level ab initio results for model reactions. To correct this, we added a semiempirical valence bond term to a QM/MM calculation based on 21 AM1 atoms joined to 5506 CHARMM22/TIP3P atoms by GHO boundary atoms (48). The semiempirical valence bond parameters were adjusted so that the combined potential reproduces the $MP2/6-31+G^*$ potential energy surface for the model reaction between Zn-complexed benzyl alcoholate and NAD^+ in the gas phase. The resulting PES shows that the active site of LADH is characterized by a delicate hydrogen-bonding network as well as hydrophobic interactions, which all make important contributions to the dynamics of the hydride transfer process. We obtained a quasiclassical free energy of activation of 14.7 kcal/mol for oxidation of benzyl alcoholate, whereas the calculated maximum of the classical mechanical free energy of activation curve as a function of the difference between the making and breaking hydride bond distances is 16.5 kcal/mol; the difference shows the importance of quantum mechanical vibrational free energies. The experimental phenomenological free energy of activation is 15.6 kcal/mol.

The transmission coefficient γ was averaged using the EA-VTST/MT method in two ways over 18 equilibrium configurations in the transition state, for each of which we determined isoinertial MEPs with various isotopic substitutions at the primary and secondary positions of the hydride transfer reaction in LADH (48). In the SSZ approximation the dynamic contributions of the surrounding protein-solvent bath are excluded. These effects on the shape of the potential surface and the position of the transition state are incorporated in our ESZ algorithm by computing the free energy change of the secondary zone system along the reaction paths determined in the SSZ stage. Average values of $\gamma^{(SSZ)} = 2.5$ and 2.4 were obtained for the hydride transfer from the microcanonically optimized multidimensional-tunneling and small Λ-curvature-tunneling approximations in the SSZ approach; the average transmission coefficient was increased to $\gamma^{(ESZ)} = 4.1$ by including the protein dynamics contributions. These results indicate a modest increase in the rate constant owing to quantum mechanical tunneling.

It is essential to include quantum mechanical tunneling effects to obtain accurate KIEs. The result is particularly striking for the secondary H/T KIE, which is much larger than the value one would expect from the H/D secondary KIE on the basis of vibrational partition functions. The calculated H/T secondary KIE is 1.36, in good agreement with the experimental value of 1.33.

What is the origin of the exaltation of the secondary H/T KIE in the LADH reaction? It results from tunneling effects that increase the H/T secondary KIE by 17% in stage 2 calculations. (Incorporating the dynamic contributions of the surrounding protein-solvent bath in the ESZ calculation did not have a large effect.) Analysis (54) of the vibrationally adiabatic ground-state potential barrier for one of the SSZ configurations shows that it is not well approximated by a parabola and that tunneling along the MEP would give rise to an 8% effect in the ratio of the computed transmission coefficients; another 8% comes from reaction-path curvature, which is larger for hydrogen than for tritium in the region of the dominant tunneling paths. Inclusion of nuclear motion tunneling effects for the perprotio case lowers the free energy of activation from 14.7 kcal/mol to 13.9.

For LADH we compared small-curvature and large-curvature tunneling algorithms. Although the large-curvature algorithm (and hence microcanonically optimized MT) is often required for accurate results in gas-phase bimolecular reactions involving the transfer of a light particle between two heavy moieties (9, 142, 186, 194, 204), we have previously pointed out (205) that the curvature for heavy-light-heavy unimolecular reactions (such as the reactions of enzyme-substrate complexes) need not be large. Thus, it is not known how important large-curvature tunneling will turn out to be for enzymatic reactions. We found that large-curvature tunneling is not important for the LADH reaction in this study.

Hammes-Schiffer and coworkers (49, 134, 206) studied the reaction of benzyl alcoholate anion using the MDQT approach with one nucleus-treated quantum mechanical. The PES was an EVB potential based on a modified GROMOS united-atom force field with four explicit hydrogen atoms and with a constant value for the coupling matrix element H_{12}. The transition state ensemble was identified by classical mechanical and MDQT calculations of the free energy profile as a function of an energy gap. Quantum mechanical nuclear-motion effects lower the free energy of activation from 17.2 kcal/mol to 15.4. Two sample MDQT trajectories were presented to illustrate how one could include recrossing effects as a transmission coefficient (49).

Villà & Warshel applied the QCP method to these problems (135), assuming that the classical and quantum mechanical transmission coefficients are identical. They presented a preliminary calculation of the free energy of activation profile along an energy-gap coordinate. Quantum effects on nuclear motion lowered the free energy of activation from 14.5 to 12 kcal/mol.

TRIOSE PHOSPHATE ISOMERASE   Karplus and coworkers studied the conversion of dihydroxyacetone phosphate to D-glyceraldehyde-3-phosphate by triose phosphate isomerase. QM/MM was used to calculate free energy barriers by adding quantized vibrational free energies to B3LYP single-point potential energies at structures optimized with a smaller quantum mechanical region (207).

METHYL AMINE DEHYDROGENASE   Methyl amine dehydrogenase oxidizes primary amines to an aldehyde and ammonia by using the tryptophan tryptophylquinone prosthetic group as a cofactor. The rate-limiting step for the half reaction in which tryptophan tryptophylquinone is reduced corresponds to the breakdown of an iminoquinone intermediate by proton transfer from the carbon atom of a methylimino moiety on the prosthetic group to an Asp residue of the enzyme. Experimental studies (208, 209) have found large primary KIEs in comparison of CH_3NH_2 to CD_3NH_2.

Faulder et al. (57) studied this reaction with a QM/MM PES in which a 31-atom region described by PM3 is joined by link atoms to an environment described by AMBER. The reactant structure was optimized to a local minimum; then the molecular mechanical region was frozen and a transition state was optimized in the reactant environment. This was used as a starting point for calculating a 31-atom MEP in the frozen environment. This was used in a canonical variational theory/small-curvature tunneling calculation that yielded a KIE of 11.1 at room temperature, compared with a conventional TST value (without tunneling) of 6 and an experimental value of 18. The overbarrier mechanism was predicted to contribute about 4% of the observed rate, with 96% owing to tunneling.

Our own study included a subset of 48 atoms in the quantum mechanical region that is treated by the semiempirical PM3 model with specific reaction parameters (SRPs) determined for the proton transfer reaction in methyl amine dehydrogenase (58). The SRPs were only used for the methyl carbon of the substrate, methyliminoquinone, and their values were adjusted to yield good agreement with ab initio MP2/6-31G* and B3LYP/6-31G* results for two model reactions. An additional 10,977 atoms, of which 7248 were protein atoms and 3729 were from 1243 solvent water molecules, were included in the molecular mechanical part of the system, which was joined to the quantum mechanical part by GHO boundary atoms (99). Atoms within 24 Å of the reaction center were included in the dynamics calculations, and the transmission coefficients were determined by using the static-secondary zone (SSZ) approximation.

From the classical mechanical PMF, the estimated free energy of activation is 20.3 kcal/mol for the proton transfer reaction from methyliminoquinone to Asp-76.

Inclusion of quantum mechanical vibrational free energies averaged over 400 instantaneous configurations corresponding to the reactant and transition region and the effect of the average Γ (the dynamic recrossing transmission coefficient) reduces the effective barrier height to a value of 17.1 kcal/mol (58). The latter result, which does not include quantum mechanical tunneling contributions, is the quasiclassical free energy of activation. The net effect of including tunneling and recrossing is to further lower the energy of activation by 2.5 kcal/mol to 14.6. Thus, the overall lowering of the free energy of activation by including quantum effects in the nuclear motion is 5.7 kcal/mol.

Our calculations indicate that overbarrier processes contribute only about 1% to the rate, with tunneling accounting for the rest. We calculated a KIE of 17 in excellent agreement with the experimental value of 18 (208, 209). Conventional TST without tunneling yielded a KIE of only 6.

XYLOSE ISOMERASE   Xylose isomerase converts the aldose D-xylose to the ketose D-xylulose and, as summarized by Hu et al. (210), it is believed to operate by catalyzing a 1,2-hydride shift in the ring-opened aldose. Two Mg^{2+} cofactors play a role (211, 212). One of the two magnesium ions (Mg-1) in the active site is tightly held by the protein, whereas the second magnesium ion (Mg-2) is mobile; Mg-2 occupies two possible positions that are 5.4 and 3.8 Å from Mg-1, respectively, in the active site of xylose isomerase complexed with D-glucose (213). It is the latter configuration (shorter Mg-Mg separation) that corresponds to the active form of the enzyme for the hydride transfer step in xylose isomerase. We (M. Garcia-Viloca, C. Alhambra, D.G. Truhlar, J. Gao, unpublished manuscript) found by molecular dynamics simulations that the two magnesium ions move apart again as the hydride transfer reaction takes place. Thus, there is a breathing motion of Mg^{2+} ions that mediates the proton and hydride transfer reactions in xylose isomerase. This reaction is a prototypical example for understanding the role of metal ions in catalysis. Two groups have simulated the reaction including quantum effects on nuclear motion.

Nicoll et al. (56) used a QM/MM PES in which a 42-atom system described by PM3 is joined by link atoms to an environment described by AMBER. Kinetic isotope effects were calculated by the canonical VTST/small-curvature tunneling approximation for a single 42-atom MEP in a fixed environment. The room temperature H/D KIE was found to be 6.3, compared with a nontunneling result of 3. Adding corrections from a transition state optimized with B3LYP/6-31G changed the canonical VTST/small-curvature tunneling KIE to 5.8.

The results from our own preliminary molecular dynamics simulations (59) of the Michaelis complex between xylose isomerase and xylose were consistent with the mechanism proposed by Whitlow et al. (214). In this mechanism the 2-hydroxy proton of the substrate is first transferred to a hydroxide ion that is ligated to Mg-2, which leads to the replacement of one of the bidenate interactions between Mg-2 and Asp254 by an interaction with the 2-alkoxide anion. Concomitantly, Mg-2 migrates about 1 Å toward Mg-1 to resume its second position described above; this is the active configuration for the hydride transfer step, and the carbonyl group of the substrate becomes a ligand to Mg-2, replacing the neutral Asp256. We proposed that Asp256 is protonated after abstracting a proton from the water molecule, which is the base for abstracting a proton from the 2-hydroxy group of the xylose.

We used a potential that includes a simple valence bond potential adjusted to match the MP2/6-31 + G* results for a model hydride transfer reaction in the gas phase and then further refined on the basis of the experimental overall k_{cat} for the enzymatic reaction. Tests showed that QM/MM calculations with a 19-atom quantum mechanical subsystem agreed well with calculations based on a 79-atom quantum mechanical subsystem that includes the magnesium ions in the quantum mechanical region, so the final calculations were based on the former. The quantum mechanical subsystem was joined to a CHARMM/TIP3P environment (38, 215) by GHO boundary atoms.

Inclusion of quantum effects in the nuclear motion lowered the free energy of activation from 25.0 kcal/mol to 22.6. The calculated KIE is 3.8 at 298 K, in general agreement with experimental values in the range of 2.7–4.0 at 333 K (216–218). Tunneling accounts for about 90% of the rate.

VIRAL NEURAMIDASE   Thomas et al. (53) carried out PMF calculations for the formation of the covalent intermediate catalyzed by viral neuramidase (53). They chose the bond length of one of the two bonds being made as the reaction coordinate. They also used path integral methods to estimate the quantum effects on this step. In the path integral calculations two oxygen atoms and one hydrogen atom were treated as quantum particles; this reduces the free energy of activation from 5.3 kcal/mol to 3.8.

Other Applications with Quantum Electronic Structure and Statistical Mechanics

Much useful information can be obtained from studies that use quantum mechanical electronic structure calculations to calculate PMFs along enzyme reaction paths, transition state theory rate constants, or other dynamical variables based on ensemble averaging, even when quantum effects are not included in the nuclear motion. This section reviews such studies, in alphabetical order of enzyme.

ACETOHYDROXY ACID ISOMEROREDUCTASE   Proust-De Martin et al. (219) used link-atom QM/MM to calculate free energy profiles for the alkyl migration/proton transfer step catalyzed by the Mg-enzyme acetohydroxy acid isomeroreductase. Models with 18–63 quantum mechanical atoms were explored. The quantum mechanical method was AM1. Electrostatic effects were found to be important. The calculated ΔG_{act} is 38 kcal/mol, which is 2–4 times higher than experiment.

CATECHOL O-METHYLTRANSFERASE   Kollman and coworkers (163, 164, 167) applied the quantum mechanical–free energy method with a cratic term to the S_N2

methyl cation transfer catalyzed by catechol O-methyltransferase and obtained good agreement with experiment.

CITRATE SYNTHASE   This system has also been studied by QM/MM methods by two groups using different approaches: potential energy profiles with AM1 and MP2/6-31G* for the quantum mechanical subsystem (98, 220, 221) and quantum mechanical–free energy with restrained electrostatic potential charges from HF/6-31 + G* calculations (163, 167).

HUMAN IMMUNODEFICIENCY VIRUS TYPE-1 (HIV-1) PROTEASE   Liu et al. (222) used link-atom PM3/GROMOS87 calculations in molecular dynamics simulations to calculate a PMF as a function of the making C-O bond distance.

OROTIDINE 5′-MONOPHOSPHATE DECARBOXYLASE (ODCase)   ODCase is involved in the last step of pyrimidine biosynthesis, which catalyzes the decarboxylation of orotidine 5′-monophosphate at a remarkable acceleration rate of 17 orders of magnitude over the uncatalyzed aqueous process. ODCase has been extensively studied both experimentally and computationally. In a joint X-ray crystallographic and computational study, Wu et al. (223) solved the structure of the apo enzyme and the structure complexed with a transition state analog and proposed a mechanism on the basis of free energy calculations of the reaction profile using a combined AM1/CHARMM potential. The computed $\Delta G^{\neq}$ of 37.2 and 14.8 kcal/mol for the uncatalyzed and catalyzed reactions were in excellent agreement with the corresponding experimental values (38.5 and 15.2 kcal/mol, respectively). Wu et al. analyzed the free energy of transfer for the reactive part, orotate ion, from aqueous solution into the enzyme active site using free energy perturbation theory (75, 76) and found that orotate is destabilized by 17.8 kcal/mol owing to electrostatic stress from an Asp residue and a change of the polar aqueous environment to a largely hydrophobic binding pocket (223) surrounding the orotate ion. The reduction in free energy of activation by ODCase was attributed to reactant destabilization, an idea originally proposed by Jencks (224). The destabilization effect is compensated by binding contributions from interactions with the rest of the ribose and phosphate groups of the substrate, which form extended hydrogen bonding and ion-pair interactions with the enzyme.

Warshel and coworkers, on the other hand, carried out a calculation using an EVB potential and suggested that transition state stabilization, rather than reactant destabilization, is responsible for the enormous rate enhancement of 23 kcal/mol by ODCase (225, 226). They interpreted the catalytic enhancement as due to the dipolar environment of the enzyme active site being preorganized such that there is a lower reorganization penalty for the enzymatic reaction than the aqueous one (225). Although the authors did not compute reorganization energy for this reaction and the concept of preorganized dipoles of the enzyme environment may be reasonable to account for transition state stabilization in some enzymes, the application of this explanation to the ODCase reaction neglects the effect of conformational changes of key residues in ODCase's active site (see Figure 11 in Reference 225). To make this argument work for ODCase, the residue Lys72 has to be deleted from the enzyme and included as part of the substrate. However, it is not clear if this choice of reference state is reasonable for comparison with the reaction catalyzed under physiological conditions by the real enzyme.

PARA-HYDROXY BONZOATE HYDROXYLASE   The 3-hydroxylation of p-hydroxy benzoate is catalyzed by para-hydroxy bonzoate hydroxylase, with the chemical step being electrophilic aromatic attack of a C4a-peroxyflavin intermediate formed from the cofactor. Billeter et al. (227) carried out studies with a 102- or 103-atom AM1 subsystem and 6902 GROMOS96 molecular mechanical atoms. The QM/MM interactions were treated by either the mechanical embedding model or the electronic embedding model (228), and free energy profiles were calculated as functions of the making bond distance. They also optimized a transition state structure in the presence of enzyme.

PROTEIN TYROSINE PHOSPHATASE   Alhambra et al. (100) carried out free energy calculations and obtained the PMF for the reaction in a low-molecular-weight protein tyrosine phosphatase. They attributed the lowering of the free energy of activation to Walden-inversion enforced transition state stabilization. The protein tyrosine phosphatase reaction has also been investigated by Aqvist and coworkers using an EVB model (229). They found that a protonated Asp129 residue is favored in the catalytic step, which was used in the study by Alhambra et al. (100).

SUBTILISIN   Colombo et al. (230, 231) used classical molecular dynamics with free energy perturbation to study the enantioselectivity of the serine protease subtilisin in water, dimethylformamide, and hexane. They used the PM3 and MNDO quantum mechanical methods and the AMBER force field for the enzyme and substrate, the OPLS force field for organic solvents, and the TIP3P force field for water.

TRIOSE PHOSPHATE ISOMERASE   Studies (207) involving quantized vibrations are summarized above. There have also been studies with quantal electronic structure and classical nuclear motion. First, Karplus and coworkers used a QM/MM method to determine the potential energy along a reaction path and to provide a decomposition of the energy lowering into contributions from specific residues (232, 233). Later, Cui et al. (234) used a self-consistent-charge density-functional tight-binding quantum mechanical (235) subsystem or an AM1 or AM1-SRP quantum mechanical subsystem with a CHARMM molecular mechanical subsystem and optimized the active site structure with a fixed secondary zone.

Zhang et al. (169) also studied the triose phosphate isomerase reaction using QM/MM methods. They used an HF/3-21G description of a 37-atom quantum mechanical system joined by the pseudobond approach to a 6014-atom molecular mechanical subsystem described by the AMBER and TIP3P force fields. They first optimized a 3008-atom subsystem reactant geometry and then calculated a forward reaction path all the way to products by a series of restrained minimizations along a distinguished coordinate. They then followed the reverse path back from the products to reactants, obtaining agreement of the potential energy curves within 1 kcal/mol. Then, using the forward reaction path, they used the quantum mechanical–free energy method to obtain the change in free energy of interaction of the quantum mechanical and molecular mechanical subsystems as a function of the reaction coordinate.

Quantum Treatments of Electronic Structure Without Statistical Mechanics

Quantum mechanical electronic structure calculations for an active site in the field of the enzyme environment can be a powerful technique for elucidating the factors underlying enzyme catalysis even when statistical averaging is not employed, provided one proceeds carefully and cautiously and uses good judgment. In fact, this approach has already become so powerful that the amount of work being carried out has exploded, and a complete summary would require more pages than have been allotted for this review. Because it is a very exciting area, though, we summarize selected applications from various groups to illustrate the state of the art and the variety of approaches employed. The following paragraphs are in alphabetical order of enzyme. [Calculations without explicit enzymes are not included, but some of these were reviewed recently (236, 237).]

ALDOSE REDUCTASE   Two studies were carried out using link-atom QM/MM to elucidate the mechanism of the reduction of D-glyceraldehyde by aldose reductase. Lee et al. (238) optimized the geometry of active site residues and constructed reaction paths and reduced-dimensionality potential surfaces. Várnai et al. (239) examined a chain of points on hypothetical proton transfer and hydride transfer reaction paths.

BACTERIORHODOPSIN   Ben-Nun et al. (240) applied the full multiple spawning wave packet dynamics method to study the coupled-electronic-state all-*trans* → 13-*cis* photoisomerization of retinal in bacteriorhodopsin. Their multi-state potential function was an electronically diabatic one with molecular mechanical diagonal elements and constant off-diagonal elements. Rajamani & Gao carried out molecular dynamics simulation of bacteriorhodopsin in a lipid bilayer and the protonated Schiff base in solution using combined AM1/CHARMM and configuration interaction with single excitations/3-21G/CHARMM potentials to compute the opsin shift of bacteriorhodopsin (241). The ensemble-averaged opsin shift of 4200 cm^{-1} without dispersion contributions is in reasonable agreement with the experimental value of 5100 cm^{-1}. When dispersion effects are included, the estimated opsin shift is 5200 cm^{-1}.

CARBONIC ANHYDRASE   Studies (146) involving quantized vibrations are summarized above. Toba et al. (242) investigated carbonic anhydrase using a three-stage approach. First they used QM/MM calculations with a PM3 quantum mechanical subsystem to partially optimize a reactant geometry. At this geometry they used an MNDO quantum mechanical treatment to calculate partial atomic charges by electrostatic potential fitting. These charges were then used with the AMBER force field for classical molecular dynamics simulations.

CHORISMATE MUTASE   The transformation of chorismate to prephenate by chorismate mutase has been studied independently by several groups using QM/MM. Lyne et al. (243) carried out simulations to study the origin of the catalytic effect. One study (244) focused on comparative evaluation of methodological choices (with quantum mechanical partial atomic charges from electrostatic fitting of a B3LYP/6-31G* treatment of 24 atoms in the environment of 4117 MM atoms) and another (245, 246) (based on 24 AM1 atoms and 6245 CHARMM and TIP3P atoms) focused on the origin of catalytic efficiency. We have also carried out free energy perturbation-VTST/MT calculations to estimate KIEs using an AM1-SRP potential for the chorismate reaction (C. Alhambra & J. Gao, unpublished manuscript).

COLD SHOCK PROTEIN A   Merz and coworkers (62, 63) carried out PM3 quantum mechanical calculations on selected geometries of the entire cold shock protein A in the presence of water. This allowed them to study charge transfer from the protein to water, which was found to be nonnegligible. This raises important questions about the location of QM/MM boundaries because current QM/MM algorithms do not allow charge transfer across the boundary.

CYTIDINE DEAMINASE   Lewis et al. (66, 67) applied PM3 quantum mechanical methods to optimize geometries of a 1330-atom subsystem of cytidine aminase with the active site occupied by cytidine, by a transition state analog, or by uridine (the product). The Zn-S distance at the transition state is largest for the transition state analog.

DIHYDROFOLATE REDUCTASE   Dynamics calculations (21) on dihydrofolate reductase are discussed above. Gready and coworkers (69, 248–250) carried out an interesting series of studies that led to a comparison of QM/MM calculations with quantum mechanical electronic structure calculations on the whole enzyme-substrate system. The quantum mechanical level for the comparison was PM3, and the molecular mechanical subsystem was treated by AMBER94 and TIP3P force fields. They discussed the polarization of the quantum mechanical subsystem by the molecular mechanical one in light of these results, warning against possible pitfalls in the QM/MM calculations, which were 500 times less expensive than the fully quantum mechanical calculations.

ELASTASE   Topf et al. (251) reported link-atom QM/MM energy minimizations on three stationary points along the deacylation step in the elastase reaction. A total of 49 atoms were treated by quantum mechanics using HF/3-21G and HF/6-31G*, and 12,010 atoms were treated molecular mechanically by CHARMM and TIP3P.

ENOLASE Dynamics calculations (52) on the enolase reaction are discussed above. More recently Liu et al. (252) have carried out QM/MM calculations of free energy profiles as functions of distance along a restrained-optimization reaction path. They used the pseudobond approach to the QM/MM interface with HF/3-21G and B3LYP/6-31G* quantum mechanical calculations.

GALACTOSE OXIDASE Röthlisberger et al. (253) applied QM/MM calculations to the H abstraction step catalyzed by galactose oxidase. The quantum region of ~ 70 atoms was treated by the Becke-Lee-Yang-Parr density functional with a plane wave basis (80 Ry cutoff), and the molecular mechanical region of $\sim 16,000$ atoms was treated by CHARMM. They used the scaled-position link-atom method for linking quantum mechanics to molecular mechanics. They optimized structures by energy minimization to serve as typical thermally accessible configurations for further analysis.

HALOALKANE DEHALOGENASE Lau et al. (254) used QM/MM calculations to study the $S_N 2$ displacement of chloride from 1,2-dichloroethane catalyzed by haloalkane dehalogenase. They used AM1-SRP for the quantum mechanical subsystem (15 atoms) and a combination of CHARMM parameters and MM-SRP parameters optimized for this system for the molecular mechanical subsystem (4237 atoms). The conclusions were based on a two-dimensional grid of energies for two choices of the other coordinates.

HIV-1 PROLEASE A free energy calculation on this system is mentioned above (222). Chatfield et al. used an ab initio QM/MM approach with CHARMM for the MM part to determine a minimum energy path for the reaction in the enzyme (255, 256). Trylska et al. (257) presented a parameterization of a 23 × 23 approximate valence bond description of the complete enzymatic reaction catalyzed by HIV-1 protease.

HYDROGENASE Amara et al. (258) used a B3LYP/effective core potential treatment of a 30-atom quantum subsystem joined by link atoms to a CHARMM subsystem of 10,000 atoms in residues within 27 Å of the active site to optimize reactant structures of Ni-Fe hydrogenase.

β-LACTAMASES Both class A (259–261) and class B (262) β-lactamases have been studied with QM/MM, in both cases employing energy minimizations, either with a fixed molecular mechanical region or with the molecular mechanical region also energy optimized (261).

LACTATE DEHYDROGENASE The lactate dehydrogenase system presents an informative case history in which two seemingly similar QM/MM calculations led to different mechanistic conclusions about the order of proton transfer and hydride transfer and the degree of synchronicity of these steps (263–266). Furthermore, transition state optimizations starting at various points lead to different structures. Although this result is not surprising in itself (this is one reason why reliable dynamics procedures for enzyme reactions should be based on an ensemble of transition structures such as those in the highest window of a free energy profile), in this case the variation in disposition of active site residues was very large (265). Poetically, the authors commented, "depending upon how a gentle breeze may gust at the crucial moment, a raindrop falling on the Andes may flow either to the Atlantic or the Pacific. So it is with geometry optimization in a large and flexible system" (266). Although the classification of transition structures into the two mechanisms was not as sensitive to initial conditions of the optimization, the energetics of the two mechanisms are similar enough that it will be hard to tell which is preferred by nature.

MALATE DEHYDROGENASE   Bash and coworkers (267, 268) calculated two-dimensional energy surfaces for the conversion of malate to oxaloacetate by malate dehydrogenase.

MANDELATE RACEMATE   A QM/MM study of potential energies for the racemization of vinylglycolate catalyzed by mandelate racemate found three competitive parallel mechanisms (269). The authors commented that an exploration of the PES to locate possibly unexpected critical points should precede the choice of reaction coordinate for calculating a free energy of activation profile.

PAPAIN   Han et al. (270) scanned the PES for proton transfer catalyzed by papain with HF and B3LYP in the active site surrounded by fixed enzymes with AMBER charges. Harrison et al. (271) used AM1/CHARMM to find a transition structure for amide hydrolysis by papain.

PARA-HYDROXY BONZOATE HYDROXYLASE   Free energy profiles (227) are discussed above. Whereas those calculations indicated a free energy of activation of 12 kcal/mol, energy optimizations by the same authors with the same QM/MM Hamiltonian yielded 21 kcal/mol, showing the danger of the latter type of calculation for predicting rate constants. Using the difference of two bond lengths as a reaction coordinate, Ridder et al. (272–274) used link-atom QM/MM with AM1 and CHARMM22 for the subsystems to find a distinguished-coordinate reaction path, which involves optimizing other coordinates for fixed values of the reaction coordinate. This allowed them to estimate a barrier height and study the role of the cofactor and active site amino acids.

PHENOL HYDROXYLASE   Ridder et al. (275) used AM1/CHARMM to calculate a two-dimensional energy surface for hydroxylation of phenol by phenol hydroxylase. Starting at the transition structure so identified, they performed energy minimizations for a series of substituted phenols to examine the variation in catalytic activity.

PLA₂ Dynamics calculations (55) on PLA$_2$ are discussed above. Hillier and coworkers (276, 277) applied the QM/MM method to the transition state for amide hydrolysis by PLA$_2$. Schürer et al. (278) also studied PLA$_2$, a coenzyme that hydrolyzes an ester bond of L-glycerophospholipids. They treated an active-site model of 156 atoms with PM3 and surrounded this with a fixed molecular mechanical region described by the PARAM94 force field of AMBER. Coulomb interactions between the quantum mechanical subsystem and the molecular mechanical region were modeled with dielectric constant 4, but this was varied from 1 to 78 to study the effect of this assumption on the calculated potential energy profiles.

PROTEIN TYROSINE PHOSPHATASE Free energy calculations are mentioned above (100, 229). Hart et al. (279, 280) computed the transition structure using AM1/MM and PM3/MM potentials and predicted a dissociative mechanism for the dephosphorylation reaction in protein tyrosine phosphatase.

RUBISCO Moliner et al. (281) calculated AM1/CHARMM transition state structures for the carboxylation of D-ribulose-1,5-diphosphate catalyzed by rubisco.

THERMOLYSIN Antonczak et al. (282, 283) studied two mechanisms for the hydrolysis of formamide and a tripeptide by the Zn protease thermolysin. They used AM1 with the LSCF formalism with parameters extracted from a model system, and they used the AMBER force field for the molecular mechanical part. They optimized reactant and transition state structures of the quantum mechanical subsystem with the fixed molecular mechanical subsystem.

THYMIDINE PHOSPHORYLASE Burton et al. (280) used a PM3/AMBER hybrid with a charge redistribution algorithm for minimizing adverse effects from link atoms to calculate potential energy barriers for phosphorolysis in thymidine phosphorylase. They optimized reactant and transition state structures of the quantum mechanical subsystem with fixed molecular mechanical regions and then energy minimized the molecular mechanical region with a fixed quantum mechanical subsystem to learn about enzyme structure changes during the reaction. They also explored the effects of moving residues from the molecular mechanical region into the quantum mechanical subsystem.

CONCLUDING REMARKS

In this article we review methods for enzyme kinetics modeling and applications that incorporate quantum mechanical effects in the computation. Three aspects are emphasized, namely the treatment of the electronic structure of atoms involved in the chemical process in an enzyme, the incorporation of quantum vibrational free energies into the calculation of the potential of mean force, and the use of multidimensional tunneling methods for estimating the transmission coefficient. It is evident from the applications we have summarized here that the incorporation of quantum mechanical effects is essential for enzyme kinetics simulations, and the computational accuracy has been tremendously increased in the past several years in comparison with experiment. These computational approaches provide insights and help interpret experimental data such as kinetic isotope effects and the significance of such factors as dynamics and tunneling in enzymatic processes. Still, there is much to improve in computational techniques, sampling, and time and length scales relevant to physiological conditions.

Visit the Annual Reviews home page at www.annualreviews.org

LITERATURE CITED

1. Wigner EP. 1938. *Trans. Faraday Soc.* 34:29–41
2. Glasstone S, Laidler KJ, Eyring H. 1941. *The Theory of Rate Processes.* New York: McGraw-Hill
3. Keck JC. 1967. *Adv. Chem. Phys.* 13:85–121
4. Anderson JB. 1973. *J. Chem. Phys.* 58:4684–92
5. Truhlar DG, Garrett BC. 1980. *Acc. Chem. Res.* 13:440–48
6. Pechukas P. 1981. *Annu. Rev. Phys. Chem.* 32:159–72
7. Pechukas P. 1982. *Ber. Bunsenges. Phys. Chem.* 86:372–78
8. Truhlar DG, Hase WL, Hynes JT. 1983. *J. Phys. Chem.* 87:2664–82. Erratum. 1983. *J. Phys. Chem.* 87:5523
9. Truhlar DG, Garrett BC. 1984. *Annu. Rev. Phys. Chem.* 35:159–89
10. Tucker SC, Truhlar DG. 1989. In *New Theoretical Concepts for Understanding Organic Reactions*, ed. J Bertran, IG Csizmadia, pp. 291–346. Dordrecht, The Netherlands: Kluwer
11. Truhlar DG, Garrett BC, Klippenstein SJ. 1996. *J. Phys. Chem.* 100:12771–800
12. Garrett BC, Truhlar DG. 1998. In *Encyclopedia of Computational Chemistry*, ed. PvR Schleyer, NL Allinger, T Clark, J Gasteiger, PA Kollman, HF Schaefer III. 5:3094–104. Chichester, UK: Wiley
13. Grimmelmann EK, Tully JC, Helfand E. 1981. *J. Chem. Phys.* 74:5300–10
14. Keck JC. 1962. *Discuss. Faraday Soc.* 33:173–82
15. Chandler D. 1978. *J. Chem. Phys.* 68:2959–70
16. Montgomery JA Jr, Chandler D, Berne BJ. 1979. *J. Chem. Phys.* 70:4056–66
17. Grote RF, Hynes JT. 1980. *J. Chem. Phys.* 73:2715–32
18. Northrup SH, Pear MR, Lee CY, McCammon JA, Karplus M. 1982. *Proc. Natl. Acad. Sci. USA* 79:4035–39
19. Bergsma JP, Gertner BJ, Wilson KR, Hynes JT. 1987. *J. Chem. Phys.* 86:1356–76
20. Hwang JK, King G, Creighton S, Warshel A. 1988. *J. Am. Chem. Soc.* 110:5297–311
21. Neria E, Karplus M. 1997. *Chem. Phys. Lett.* 267:26–30
22. Miller WH. 1996. In *Dynamics of Molecules and Chemical Reactions*, ed. RE Wyatt, JZH Zhang, pp. 387–410. New York: Marcel Dekker
23. Natanson GA, Garrett BC, Truong TN, Joseph T, Truhlar DG. 1991. *J. Chem. Phys.* 94:7875–92
24. Gao J. 1996. *Acc. Chem. Res.* 29:298–305
25. Marcus RA. 1956. *J. Chem. Phys.* 24:966–78

26. Warshel A. 1982. *J. Phys. Chem.* 86: 2218–24
27. Schenter GK, Garrett BC, Truhlar DG. 2001. *J. Phys. Chem. B* 105:9672–85
28. Mo Y, Gao J. 2000. *J. Phys. Chem. A* 104:3012–20
29. Garrett BC. 2000. *Theor. Chem. Acc.* 103:200–4
30. Villa J, Truhlar DG. 1997. *Theor. Chem. Acc.* 97:317–23
31. Boyd RK. 1977. *Chem. Rev.* 77:93–119
32. Dellago C, Bolhuis PG, Chandler D. 1998. *J. Chem. Phys.* 108:9236–45
33. Dellago C, Bolhuis PG, Csajka FS, Chandler D. 1998. *J. Chem. Phys.* 108:1964–77
34. Dellago C, Bolhuis PG, Chandler D. 1999. *J. Chem. Phys.* 110:6617–25
35. Valleau JP, Torrie GM. 1977. In *Modern Theoretical Chemistry*, ed. BJ Berne, 5:169–94. New York: Plenum
36. Zwanzig R. 1954. *J. Chem. Phys.* 22:1420–26
37. Brooks CL III, Karplus M, Pettitt BM. 1988. *Adv. Chem. Phys.* 71:1–259
38. MacKerell AD Jr, Wiorkiewicz-Kuczera J, Karplus M. 1995. *J. Am. Chem. Soc.* 117:11946–75
39. Cornell WD, Cieplak P, Bayly CI, Gould IR, Merz KM Jr, et al. 1995. *J. Am. Chem. Soc.* 117:5179–97
40. MacKerell AD Jr, Bashford D, Bellott M, Dunbrack RL, Evanseck JD, et al. 1998. *J. Phys. Chem. B* 102:3586–616
41. Jorgensen WL, Tirado-Rives J. 1988. *J. Am. Chem. Soc.* 110:1657–66
42. Wang W, Donini O, Reyes CM, Kollman PA. 2001. *Annu. Rev. Biophys. Biomol. Struct.* 30:211–43
43. Johnston HS. 1966. *Gas Phase Reaction Rate Theory.* New York: Ronald
44. Bennett CH. 1977. *ACS Symp. Ser.* 46:63–97
45. Smith IWM. 1981. *J. Chem. Soc. Faraday Trans. II* 77:747–59
46. Truhlar DG, Garrett BC. 1987. *Faraday Discuss. Chem. Soc.* 84:464
47. Schenter GK, Messina M, Garrett BC. 1993. *J. Chem. Phys.* 99:1674–84
48. Alhambra C, Corchado JC, Sanchez ML, Garcia-Viloca M, Gao J, Truhlar DG. 2001. *J. Phys. Chem. B* 105:11326–40
49. Billeter SR, Webb SP, Iordanov T, Agarwal PK, Hammes-Schiffer S. 2001. *J. Chem. Phys.* 114:6925–36
50. Cramer CJ, Truhlar DG. 1999. *Chem. Rev.* 99:2161–200
51. Burton NA, Harrison MJ, Hillier IH, Jones NR, Tantanak D, Vincent MA. 1999. *ACS Symp. Ser.* 721:401–10
52. Alhambra C, Gao J, Corchado JC, Villa J, Truhlar DG. 1999. *J. Am. Chem. Soc.* 121:2253–58
53. Thomas A, Jourand D, Bret C, Amara P, Field MJ. 1999. *J. Am. Chem. Soc.* 121:9693–702
54. Alhambra C, Corchado JC, Sanchez ML, Gao J, Truhlar DG. 2000. *J. Am. Chem. Soc.* 122:8197–203
55. Bala P, Grochowski P, Nowinski K, Lesyng B, McCammon JA. 2000. *Biophys. J.* 79:1253–62
56. Nicoll RM, Hindle SA, MacKenzie G, Hillier IH, Burton NA. 2001. *Theor. Chem. Acc.* 106:105–12
57. Faulder PF, Tresadern G, Chohan KK, Scrutton NS, Sutcliffe MJ, et al. 2001. *J. Am. Chem. Soc.* 123:8604–5
58. Alhambra C, Sanchez ML, Corchado J, Gao J, Truhlar DG. 2001. *Chem. Phys. Lett.* 347:512–18; Erratum. In press
59. Truhlar DG, Gao J, Alhambra C, Garcia-Viloca M, Corchado J, et al. 2002. *Acc. Chem. Res.* In press
60. Dixon SL, Merz KM Jr. 1996. *J. Chem. Phys.* 104:6643–49
61. Van der Vaart A, Merz KM Jr. 1999. *J. Phys. Chem. A.* 103:3321–29
62. Nadig G, Van Zant LC, Dixon SL, Merz KM Jr. 1999. *ACS Symp. Ser.* 721:439–47
63. Van der Vaart A, Merz KM Jr. 1999. *J. Am. Chem. Soc.* 121:9182–90
64. Lee T-S, York DM, Yang WT. 1996. *J. Chem. Phys.* 105:2744–50

65. York DM, Lee T-S, Yang WT. 1998. *Phys. Rev. Lett.* 80:5011–14
66. Lewis JP, Carter CW Jr, Hermans J, Pan W, Lee T-S, Yang WT. 1998. *J. Am. Chem. Soc.* 120:5407–10
67. Lewis JP, Liu SB, Lee T-S, Yang WT. 1999. *J. Comput. Phys.* 151:242–63
68. Stewart JJP. 1997. *Theochem* 401:195–205
69. Titmuss SJ, Cummins PL, Bliznyuk AA, Rendell AP, Gready JE. 2000. *Chem. Phys. Lett.* 320:169–76
70. Gao J, Thompson MA. 1998. *Combined Quantum Mechanical and Molecular Mechanical Methods.* Washington, DC: Am. Chem. Soc.
71. Warshel A, Levitt M. 1976. *J. Mol. Biol.* 103:227–49
72. Singh UC, Kollman PA. 1986. *J. Comput. Chem.* 7:718–30
73. Tapia O, Lluch JM, Cardenas R, Andres J. 1989. *J. Am. Chem. Soc.* 111:829–35
74. Field MJ, Bash PA, Karplus M. 1990. *J. Comput. Chem.* 11:700–33
75. Gao J. 1992. *J. Phys. Chem.* 96:537–40
76. Gao J, Xia XF. 1992. *Science* 258:631–35
77. Stanton RV, Hartsough DS, Merz KM Jr. 1993. *J. Phys. Chem.* 97:11868–70
78. Gao J. 1994. *ACS Symp. Ser.* 569:8–21
79. Gao J. 1994. *J. Am. Chem. Soc.* 116:9324–28
80. Gao J. 1995. In *Reviews in Computational Chemistry*, ed. KB Lipkowitz, DB Boyd, 7:119–85. New York: VCH
81. Freindorf M, Gao J. 1996. *J. Comput. Chem.* 17:386–95
82. Gao J. 1997. *J. Comput. Chem.* 18:1062–971
83. Alhambra C, Byun K, Gao J. 1998. *ACS Symp. Ser.* 712:35–49
84. Warshel A, Karplus M. 1972. *J. Am. Chem. Soc.* 94:5612–25
85. Gao J, Furlani TR. 1995. *IEEE Comput. Sci. Eng.* 2(3):24–33
86. Monard G, Merz KM Jr. 1999. *Acc. Chem. Res.* 32:904–11
87. Pople JA, Santry DP, Segal GA. 1965. *J. Chem. Phys.* 43:S129–35
88. Dewar MJS, Zoebisch EG, Healy EF, Stewart JJP. 1985. *J. Am. Chem. Soc.* 107:3902–9
89. Stewart JJP. 1990. *J. Comput.-Aided Mol. Des.* 4:1–105
90. Dewar MJS, Thiel W. 1977. *J. Am. Chem. Soc.* 99:4899–907
91. Hehre WJ, Radom L, Schleyer PvR, Pople JA. 1986. *Ab Initio Molecular Orbital Theory.* New York: Wiley
92. Becke AD. 1988. *Phys. Rev. A* 38:3098–100
93. Lee CT, Yang WT, Parr RG. 1988. *Phys. Rev. B* 37:785–89
94. Stephens PJ, Devlin FJ, Chabalowski CF, Frisch MJ. 1994. *J. Phys. Chem.* 98:11623–27
95. Krauss M, Stevens WJ. 1984. *Annu. Rev. Phys. Chem.* 35:357–85
96. Car R, Parrinello M. 1985. *Phys. Rev. Lett.* 55:2471–74
97. Aqvist J, Warshel A. 1993. *Chem. Rev.* 93:2523–44
98. Mulholland AJ. 2001. *Theor. Comput. Chem.* 9:597–653
99. Gao J, Amara P, Alhambra C, Field MJ. 1998. *J. Phys. Chem. A* 102:4714–21
100. Alhambra C, Wu L, Zhang Z-Y, Gao J. 1998. *J. Am. Chem. Soc.* 120:3858–66
101. Reuter N, Dejaegere A, Maigret B, Karplus M. 2000. *J. Phys. Chem. A* 104:1720–35
102. Eichinger M, Tavan P, Hutter J, Parrinello M. 1999. *J. Chem. Phys.* 110:10452–67
102a. Ostlund NS. Personal communication. See also: 1994. *HyperChem, Getting Started*, pp. 195–96 Hypercube, Inc.
103. Zhang YK, Lee T-S, Yang WT. 1999. *J. Chem. Phys.* 110:46–54
104. Thery V, Rinaldi D, Rivail JL, Maigret B, Ferenczy GG. 1994. *J. Comput. Chem.* 15:269–82
105. Monard G, Loos M, Thery V, Baka K, Rivail J-L. 1996. *Int. J. Quantum Chem.* 58:153–59

106. Assfeld X, Rivail J-L. 1996. *Chem. Phys. Lett.* 263:100–6
107. Assfeld X, Ferre N, Rivail J-L. 1998. *ACS Symp. Ser.* 712:234–49
108. Ferre N, Assfeld X, Rivail J-L. 2002. Preprint. In press
109. Amara P, Field MJ, Alhambra C, Gao J. 2000. *Theor. Chem. Acc.* 104:336–43
110. Murphy RB, Philipp DM, Friesner RA. 2000. *Chem. Phys. Lett.* 321:113–20
111. Antes I, Thiel W. 1998. *ACS Symp. Ser.* 712:50–65
112. Antes I, Thiel W. 1999. *J. Phys. Chem.* 103:9290–95
113. Bersuker IB, Leong MK, Boggs JE, Pearlman RS. 1998. *ACS Symp. Ser.* 712:66–91
114. Gonzalez-Lafont A, Truong TN, Truhlar DG. 1991. *J. Phys. Chem.* 95:4618–27
115. Rossi I, Truhlar DG. 1995. *Chem. Phys. Lett.* 233:231–36
116. Bash PA, Ho L, MacKerell AD Jr, Levine D, Hallstrom P. 1996. *Proc. Natl. Acad. Sci. USA* 93:3698–703
117. Truhlar DG, Parr CA. 1971. *J. Phys. Chem.* 75:1844–60
118. Gao J. 1993. *J. Am. Chem. Soc.* 115:2930–35
119. Gao J, Xia X. 1994. *ACS Symp. Ser.* 568:212–28
120. Gao J. 1995. *J. Am. Chem. Soc.* 117:8600–7
121. Sehgal A, Shao L, Gao J. 1995. *J. Am. Chem. Soc.* 117:11337–40
122. Byun K, Mo Y, Gao J. 2001. *J. Am. Chem. Soc.* 123:3974–79
123. Maseras F, Morokuma K. 1995. *J. Comput. Chem.* 16:1170–79
124. Gao J. 1994. *Proc. Indiana Acad. Sci.* 106:507–19
125. Warshel A. 1991. *Computer Modeling of Chemical Reactions in Enzymes and Solutions.* New York: Wiley
126. Coulson CA, Danielson U. 1954. *Ark. Fys.* 8:245–55
127. Warshel A, Weiss RM. 1980. *J. Am. Chem. Soc.* 102:6218–26
128. Warshel A, Weiss RM. 1981. *Ann. NY Acad. Sci.* 367:370–82
129. Warshel A, Russell ST, Weiss RM. 1982. *Stud. Org. Chem.* 10:267–74
130. Warshel A, Russell S. 1986. *J. Am. Chem. Soc.* 108:6569–79
131. Aqvist J, Fothergill M, Warshel A. 1993. *J. Am. Chem. Soc.* 115:631–35
132. Aqvist J. 1997. *Underst. Chem. React.* 19:341–62
133. Feierberg I, Luzhkov V, Aqvist J. 2000. *J. Biol. Chem.* 275:22657–62
134. Agarwal PK, Webb SP, Hammes-Schiffer S. 2000. *J. Am. Chem. Soc.* 122:4803–12
135. Villà J, Warshel A. 2001. *J. Phys. Chem. B* 105:7887–907
136. Mo Y, Gao J. 2000. *J. Comput. Chem.* 21:1458–69
137. Mo Y, Peyerimhoff SD. 1998. *J. Chem. Phys.* 109:1687–97
138. Mo Y, Zhang YQ, Gao J. 1999. *J. Am. Chem. Soc.* 121:5737–42
139. Schmitt UW, Voth GA. 1998. *J. Phys. Chem. B* 102:5547–51
140. Schmitt UW, Voth GA. 1999. *J. Chem. Phys.* 111:9361–81
141. Kim Y, Corchado JC, Villa J, Xing J, Truhlar DG. 2000. *J. Chem. Phys.* 112:2718–35
142. Albu TV, Corchado JC, Truhlar DG. 2001. *J. Phys. Chem. A* 105:8465–87
143. Field MJ. 1993. In *Computer Simulation of Biomolecular Systems*, ed. WF van Gunsteren, PK Weiner, AJ Wilkinson, 2:82–123. Leiden, The Netherlands: ESCOM
144. Espinosa-Garcia J, Corchado JC, Truhlar DG. 1997. *J. Am. Chem. Soc.* 119:9891–96
145. Garcia-Viloca M, Alhambra C, Truhlar DG, Gao J. 2001. *J. Chem. Phys.* 114:9953–58
146. Hwang J-K, Warshel A. 1996. *J. Am. Chem. Soc.* 118:11745–51
147. Feynman RP, Hibbs AR. 1965. *Quantum Mechanics and Path Integrals.* New York: McGraw-Hill

148. Gillan MJ. 1988. *Philos. Mag. A* 58:257–83
149. Voth GA, Chandler D, Miller WH. 1989. *J. Chem. Phys.* 91:7749–60
150. Messina M, Schenter GK, Garrett BC. 1993. *J. Chem. Phys.* 98:8525–36
151. Cao J, Voth GA. 1994. *J. Chem. Phys.* 101:6184–92
152. Lobaugh J, Voth GA. 1992. *Chem. Phys. Lett.* 198:311–15
153. Hwang JK, Warshel A. 1993. *J. Phys. Chem.* 97:10053–58
154. Jang S, Voth GA. 2000. *J. Chem. Phys.* 112:8747–57. Erratum. 2001. *J. Chem. Phys.* 114:1944
155. Makarov DE, Topaler M. 1995. *Phys. Rev. E* 52:178–88
156. Messina M, Schenter GK, Garrett BC. 1995. *J. Chem. Phys.* 103:3430–35
157. Mills G, Schenter GK, Makarov DE, Jonsson H. 1997. *Chem. Phys. Lett.* 278:91–96
158. Jang S, Schwieters CD, Voth GA. 1999. *J. Phys. Chem. A* 103:9527–38
158a. Mielke SL, Truhlar DG. 2001. *J. Chem. Phys.* 115:652–62
159. Tuckerman ME, Marx D. 2001. *Phys. Rev. Lett.* 86:4946–49
160. Hammes-Schiffer S. 1998. *J. Phys. Chem. A* 102:10443–54
161. Webb SP, Hammes-Schiffer S. 2000. *J. Chem. Phys.* 113:5214–27
162. Stanton RV, Peraekylae M, Bakowies D, Kollman PA. 1998. *J. Am. Chem. Soc.* 120:3448–57
163. Kuhn B, Kollman PA. 2000. *J. Am. Chem. Soc.* 122:2586–96
164. Lee TS, Massova I, Kuhn B, Kollman PA. 2000. *J. Chem. Soc. Perkin Trans. II* 3:409–15
165. Donini O, Darden T, Kollman PA. 2000. *J. Am. Chem. Soc.* 122:12270–80
166. Peraekylae M, Kollman PA. 2000. *J. Am. Chem. Soc.* 122:3436–44
167. Kollman PA, Kuhn B, Donini O, Peraekylae M, Stanton R, Bakowies D. 2001. *Acc. Chem. Res.* 34:72–79
168. Chandrasekhar J, Smith SF, Jorgensen WL. 1984. *J. Am. Chem. Soc.* 106:3049–50
169. Zhang YK, Liu HY, Yang WT. 2000. *J. Chem. Phys.* 112:3483–92
170. Bala P, Grochowski P, Lesyng B, McCammon JA. 1996. *J. Phys. Chem.* 100:2535–45
171. Alimi R, Gerber RB, Hammerich AD, Kosloff R, Ratner MA. 1990. *J. Chem. Phys.* 93:6484–90
172. Gerber RB, Alimi R. 1991. *Isr. J. Chem.* 31:383–93
173. Mavri J, Berendsen HJC. 1995. *J. Phys. Chem.* 99:12711–17
174. Berendsen HJC, Mavri J. 1996. *Int. J. Quantum Chem.* 57:975–83
175. Hammes-Schiffer S, Tully JC. 1995. *J. Chem. Phys.* 103:8528–37
176. Tully JC. 1990. *J. Chem. Phys.* 93:1061–71
177. Hammes-Schiffer S, Tully JC. 1994. *J. Chem. Phys.* 101:4657–67
178. Fang J-Y, Hammes-Schiffer S. 1999. *J. Phys. Chem. A* 103:9399–407
179. Lauderdale JG, Truhlar DG. 1986. *J. Chem. Phys.* 84:1843–49
180. Liu YP, Lynch GC, Truong TN, Lu DH, Truhlar DG, Garrett BC. 1993. *J. Am. Chem. Soc.* 115:2408–15
181. Liu YP, Lu DH, Gonzalez-Lafont A, Truhlar DG, Garrett BC. 1993. *J. Am. Chem. Soc.* 115:7806–17
182. Garrett BC, Truhlar DG, Wagner AF, Dunning TH Jr. 1983. *J. Chem. Phys.* 78:4400–13
183. Fernandez-Ramos A, Truhlar DG. 2001. *J. Chem. Phys.* 114:1491–96
184. Garrett BC, Truhlar DG, Grev RS, Magnuson AW. 1980. *J. Phys. Chem.* 84:1730–48
185. Truhlar DG, Isaacson AD, Garrett BC. 1985. In *Theory of Chemical Reaction Dynamics*, ed. M Baer, pp. 65–137. Boca Raton, FL: CRC Press
186. Kreevoy MM, Ostovic D, Truhlar DG, Garrett BC. 1986. *J. Phys. Chem.* 90:3766–74

187. Truhlar DG, Garrett BC. 1987. *J. Chim. Phys.* 84:365–69
188. Truong TN, Lu D-h, Lynch GL, Liu Y-P, Melissas VS, et al. 1993. *Comput. Phys. Commun.* 75:143–59
189. Truhlar DG, Liu Y-P, Schenter GK, Garrett BC. 1994. *J. Phys. Chem.* 98:8396–405
190. Chuang Y-Y, Truhlar DG. 1999. *J. Am. Chem. Soc.* 121:10157–67
191. Schatz GC, Ratner MA. 1993. In *Quantum Mechanics in Chemistry*, pp. 167–72. Englewoods Cliffs, NJ: Prentice-Hall
192. Miller WH. 1974. *Adv. Chem. Phys.* 25:69–177
193. Jasper AW, Hack MD, Truhlar DG. 2001. *J. Chem. Phys.* 115:1804–16
194. Allison TC, Truhlar DG. 1998. In *Modern Methods for Multidimensional Dynamics Computations in Chemistry*, ed. DL Thompson, pp. 618–712. Singapore: World Sci.
195. Pu J, Corchado JC, Truhlar DG. 2001. *J. Chem. Phys.* 115:6266–67
196. Schenter GK, Garrett BC, Truhlar DG. 2001. *J. Phys. Chem. B* 105:9672–85
196a. McRae RP, Schenter GK, Garrett BC, Svetlicic Z, Truhlar DG. 2001. *J. Chem. Phys.* 115:8460–80
197. Hwang JK, Chu ZT, Yadav A, Warshel A. 1991. *J. Phys. Chem.* 95:8445–48
198. Warshel A, Hwang JK. 1992. *Faraday Discuss.* 93:225–38
199. Pawlak J, O'Leary MH, Paneth P. 1999. *ACS Symp. Ser.* 721:462–72
200. Castillo R, Andres J, Moliner V. 1999. *J. Am. Chem. Soc.* 121:12140–47
201. Antoniou D, Schwartz SD. 2001. *J. Phys. Chem. B* 105:5553–58
202. Bahnson BJ, Park DH, Kim K, Plapp BV, Klinman JP. 1993. *Biochemistry* 32:5503–7
203. Bahnson BJ, Klinman JP. 1995. *Methods Enzymol.* 249:373–97
204. Bondi DK, Connor JNL, Garrett BC, Truhlar DG. 1983. *J. Chem. Phys.* 78:5981–89
205. Truhlar DG. 1994. *J. Chem. Soc. Faraday Trans.* 90:1740–43
206. Webb SP, Agarwal PK, Hammes-Schiffer S. 2000. *J. Phys. Chem. B* 104:8884–94
207. Cui Q, Karplus M. 2001. *J. Am. Chem. Soc.* 123:2284–90
208. Brooks HB, Jones LH, Davidson VL. 1993. *Biochemistry* 32:2725–29
209. Basran J, Sutcliffe MJ, Scrutton NS. 1999. *Biochemistry* 38:3218–22
210. Hu H, Liu HY, Shi YY. 1997. *Proteins* 27:545–55
211. Lavie A, Allen KN, Petsko GA, Ringe D. 1994. *Biochemistry* 33:5469–80
212. Allen KN, Lavie A, Glasfeld A, Tanada TN, Gerrity DP, et al. 1994. *Biochemistry* 33:1488–94
213. Bhosale SH, Rao MB, Deshpande VV. 1996. *Microbiol. Rev.* 60:280–300
214. Whitlow M, Howard AJ, Finzel BC, Poulos TL, Winborne E, Gilliland GL. 1991. *Proteins Struct. Funct. Genet.* 9:153–73
215. Jorgensen WL, Chandrasekhar J, Madura JD, Impey RW, Klein ML. 1983. *J. Chem. Phys.* 79:926–35
216. Lee C, Bagdasarian M, Meng M, Zeikus JG. 1990. *J. Biol. Chem.* 265:19082–90
217. Van Tilbeurgh H, Jenkins J, Chiadmi M, Janin J, Wodak SJ, et al. 1992. *Biochemistry* 31:5467–71
218. van Bastelaere PBM, Kersters-Hilderson HLM, Lambeir A-M. 1995. *Biochem. J.* 307:135–42
219. Proust-De Martin F, Dumas R, Field MJ. 2000. *J. Am. Chem. Soc.* 122:7688–97
220. Mulholland AJ, Richards WG. 1999. *ACS Symp. Ser.* 721:448–61
221. Mulholland AJ, Lyne PD, Karplus M. 2000. *J. Am. Chem. Soc.* 122:534–35
222. Liu H, Mueller-Plathe F, van Gunsteren WF. 1996. *J. Mol. Biol.* 261:454–69
223. Wu N, Mo Y, Gao J, Pai EF. 2000. *Proc. Natl. Acad. Sci. USA* 97:2017–22
224. Jencks WP. 1975. *Adv. Enzymol. Relat. Areas Mol. Biol.* 43:219–410

225. Warshel A, Strajbl M, Villa J, Florian J. 2000. *Biochemistry* 39:14728–38
226. Warshel A, Florian J, Strajbl M, Villa J. 2001. *ChemBioChem* 2:109–11
227. Billeter SR, Hanser CFW, Mordasini TZ, Scholten M, Thiel W, van Gunsteren WF. 2001. *Phys. Chem. Chem. Phys.* 3:688–95
228. Bakowies D, Thiel W. 1996. *J. Phys. Chem.* 100:10580–94
229. Hansson T, Nordlund P, Aqvist J. 1997. *J. Mol. Biol.* 265:118–27
230. Columbo G, Ottolina G, Carrea G, Merz KM Jr. 2000. *Chem. Commun.* pp. 559–60
231. Colombo G, Toba S, Merz KM Jr. 1999. *J. Am. Chem. Soc.* 121:3486–93
232. Bash PA, Field MJ, Davenport RC, Petsko GA, Ringe D, Karplus M. 1991. *Biochemistry* 30:5826–32
233. Karplus M, Evanseck JD, Joseph D, Bash PA, Field MJ. 1992. *Faraday Discuss.* 93:239–48
234. Cui Q, Elstner M, Kaxiras E, Frauenheim T, Karplus M. 2001. *J. Phys. Chem. B* 105:569–85
235. Elstner M, Porezag D, Juugnickel G, Elsner J, Haugk M, et al. 1998. *Phys. Rev. B* 58:7260–68
236. Clark T, Gedeck P, Lanig H, Schurer G. 1997. In *Molecular Modeling and Dynamics of Bioinorganic Systems*, ed. L Bauci, P Camba, pp. 307–17. Dordrecht, The Netherlands: Kluwer
237. Siegbahn PEM, Blomberg MRA. 1999. *Annu. Rev. Phys. Chem.* 50:221–49
238. Lee YS, Hodoscek M, Brooks BR, Kador PF. 1998. *Biophys. Chem.* 70:203–16
239. Várnai P, Richards WG, Lyne PD. 1999. *Proteins* 37:218–27
240. Ben-Nun M, Molnar F, Lu H, Phillips JC, Martinez TJ, Schulten K. 1998. *Faraday Discuss.* 110:447–62
241. Rajamani R, Gao J. 2002. *J. Comput. Chem.* 23:96–105
242. Toba S, Colombo G, Merz KM Jr. 1999. *J. Am. Chem. Soc.* 121:2290–302
243. Lyne PD, Mulholland AJ, Richards WG. 1995. *J. Am. Chem. Soc.* 117:11345–50
244. Hall RJ, Hindle SA, Burton NA, Hillier IH. 2000. *J. Comput. Chem.* 21:1433–41
245. Marti S, Andres J, Moliner V, Silla E, Tunon I, et al. 2001. *J. Am. Chem. Soc.* 123:1709–12
246. Marti S, Andres J, Moliner V, Silla E, Tunon I, Bertran J. 2000. *J. Phys. Chem. B* 104:11308–15
247. Deleted in proof
248. Cummins PL, Gready JE. 1998. *ACS Symp. Ser.* 712:250–63
249. Cummins PL, Gready JE. 2000. *J. Phys. Chem. B* 104:4503–10
250. Greatbanks SP, Gready JE, Limaye AC, Rendell AP. 2000. *J. Comput. Chem.* 21:788–811
251. Topf M, Várnai P, Richards WG. 2001. *Theor. Chem. Acc.* 106:146–51
252. Liu H, Zhang Y, Yang W. 2000. *J. Am. Chem. Soc.* 122:6560–70
253. Röthlisberger U, Carloni P, Doclo K, Parrinello M. 2000. *J. Biol. Inorg. Chem.* 5:236–50
254. Lau EY, Kahn K, Bash PA, Bruice TC. 2000. *Proc. Natl. Acad. Sci. USA* 97:9937–42
255. Chatfield DC, Brooks BR. 1995. *J. Am. Chem. Soc.* 117:5561–72
256. Chatfield DC, Eurenius KP, Brooks BR. 1998. *Theochem J. Mol. Struct.* 423:79–92
257. Trylska J, Grochowski P, Geller M. 2001. *Int. J. Quantum Chem.* 82:86–103
258. Amara P, Volbeda A, Fontecilla-Camps JC, Field MJ. 1999. *J. Am. Chem. Soc.* 121:4468–77
259. Pitarch J, Pascual-Ahuir J-L, Silla E, Tunon I, Ruiz-Lopez MF. 1999. *J. Comput. Chem.* 20:1401–11
260. Pitarch J, Pascual-Ahuir J-L, Silla E, Tunon I, Moliner V. 1999. *J. Chem. Soc. Perkin Trans. 2*, pp. 1351–56
261. Pitarch J, Pascual-Ahuir J-L, Silla E, Tunon I. 2000. *J. Chem. Soc. Perkin Trans. 2*, pp. 761–67

262. Suarez D, Merz KM Jr. 2001. *J. Am. Chem. Soc.* 123:7687–90
263. Ranganathan S, Gready JE. 1997. *J. Phys. Chem. B* 101:5614–18
264. Moliner V, Turner AJ, Williams IH. 1997. *Chem. Commun.*, pp. 1271–72
265. Turner AJ, Moliner V, Williams IH. 1999. *Phys. Chem. Chem. Phys.* 1:1323–31
266. Moliner V, Williams IH. 2000. *Chem. Commun.*, pp.1843–44
267. Cunningham MA, Ho LL, Nguyen DT, Gillilan RE, Bash PA. 1997. *Biochemistry* 36:4800–16
268. Cunningham MA, Bash PA. 1999. *ACS Symp. Ser.* 721:384–400
269. Garcia-Viloca M, Gonzalez-Lafont A, Lluch JM. 2001. *J. Am. Chem. Soc.* 123:709–21
270. Han W-G, Tajkhorshid E, Suhai S. 1999. *J. Biomol. Struct. Dyn.* 16:1019–32
271. Harrison MJ, Burton NA, Hillier IH. 1997. *J. Am. Chem. Soc.* 119:12285–91
272. Ridder L, Mulholland AJ, Vervoort J, Rietjens IMCM. 1998. *J. Am. Chem. Soc.* 120:7641–42
273. Ridder L, Mulholland AJ, Rietjens IMCM, Vervoort J. 2000. *J. Am. Chem. Soc.* 122:8728–38
274. Ridder L, Palfey BA, Vervoort J, Rietjens IMCM. 2000. *FEBS Lett.* 478:197–201
275. Ridder L, Mulholland AJ, Rietjens IMCM, Vervoort J. 1999. *J. Mol. Graph. Model.* 17:163–75
276. Waszkowycz B, Hillier IH, Gensmantel N, Payling DW. 1991. *J. Chem. Soc. Perkin Trans. 2*, pp. 2025–32
277. Waszkowycz B, Hillier IH, Gensmantel N, Payling DW. 1991. *J. Chem. Soc. Perkin Trans. 2*, pp. 225–31
278. Schürer G, Lanig H, Clark T. 2000. *J. Phys. Chem. B* 104:1349–61
279. Hart JC, Burton NA, Hillier IH, Harrison MJ, Jewsbury P. 1997. *Chem. Commun.*, pp. 1431–32
280. Burton NA, Harrison MJ, Hart JC, Hillier IH, Sheppard DW. 1998. *Faraday Discuss.* 110:463–73
281. Moliner V, Andres J, Oliva M, Safont VS, Tapia O. 1999. *Theor. Chem. Acc.* 101:228–33
282. Antonczak S, Monard G, Ruiz-Lopez MF, Rivail J-L. 1998. *J. Am. Chem. Soc.* 120:8825–33
283. Antonczak S, Monard G, Lopez MR, Rivail J-L. 2000. *J. Mol. Model.* 6:527–38

SURFACE FEMTOCHEMISTRY: Observation and Quantum Control of Frustrated Desorption of Alkali Atoms from Noble Metals

Hrvoje Petek[1] and Susumu Ogawa[2]

[1]*Department of Physics and Astronomy, University of Pittsburgh, Pittsburgh, Pennsylvania 15260; e-mail: petek@pitt.edu*
[2]*Advanced Research Laboratory, Hitachi Ltd., Hatoyama, Saitama 350-0395, Japan; e-mail: ogawa@harl.hitachi.co.jp*

Key Words surface photochemistry, time-resolved photoemission, alkali chemisorption, coherent control

■ **Abstract** This review presents a case study of the direct, real-time observation of a surface photochemical reaction, namely the frustrated photodesorption of alkali atoms from noble metal surfaces. Charge transfer excitation of an electron from the metal substrate into an unoccupied resonance of the alkali atom instantaneously turns on the repulsive Coulomb force inducing the nuclear motion of both the adsorbate and substrate atoms. The incipient nuclear wave packet dynamics are documented for the case of Cs/Cu(111) through the accompanying change in the surface electronic structure. The intimate view of atoms attempting to escape the surface bond highlights the unique role of the substrate in the electronic and nuclear dynamics that ultimately determine the product yields. Moreover, slow dephasing of the coherent polarization is exploited to demonstrate the control of nuclear wave packets through the phase of the excitation light.

INTRODUCTION

The application of lasers to the study of chemical reaction dynamics has led to enormous advances in our understanding of the fundamental photophysical and photochemical processes governing the course of a broad range of prototypical gas and solution phase reactions (1). Motivated by the aspiration to control matter with light, chemists have been exploring the fundamental forces that are created and dissipated upon the electronic excitation of molecules (2). The development of femtosecond lasers with substantially shorter pulse durations than typical periods of molecular vibrations allows us to view chemical reactions in terms of coherent nuclear wave packet motion and even to design optical fields that can alter their outcome (1, 3, 4). Surface photochemical reactions also are of great interest for fundamental and practical reasons, because the electronic excitation leads to substantially different reaction products and product energy distributions than can be obtained through conventional thermal activation. Whereas it has become possible to view, induce reactions of, and synthesize single molecules on surfaces (5–7), because of exceedingly fast electronic relaxation, the dynamics of photochemical reactions on metal surfaces could only be gleaned indirectly through the analysis of stable reaction products that remain on the surface or are desorbed into the gas phase (8, 9). Borrowing from the arsenal of gas-phase techniques, surface chemists have resorted to extracting the information on nuclear wave packet dynamics from the disposal of energy into translation, rotation, and vibration of products. These product distributions are then related to several classes of simple, typically one-dimensional potential energy surfaces that may account for the experimental results (9, 10).

In the standard scenario for surface photochemistry on metal surfaces, the electronic and nuclear dynamics in desorption induced by electronic transition (DIET) can be broken up into several fundamental steps. As illustrated in Figure 1 (see color insert), these processes include (10–13) (a) the photoexcitation of hot electrons with $0 - h\nu$ energy within the skin depth (10–20 nm) of the metal substrate; (b) the electronic excitation via transport and scattering of hot electrons into adsorbate-induced surface resonances; (c) the ensuing nuclear wave packet motion on dissociative (Menzel-Gomer-Redhead) (14, 15) or attractive (Antoniewicz) (16) excited state potential energy surfaces (PES); and (d) the electronic relaxation of the adsorbate (10–13). In both scenarios, photodesorption occurs from the ground state PES when the excited state lifetime is sufficiently long for the desorbing particle to acquire kinetic energy on the excited state PES to overcome the chemisorption potential on the ground PES. This nuclear kinetic energy is conserved in the nonadiabatic transition to the ground state, making desorption energetically feasible. Thus, the reaction yields and the product energy distributions depend on the excited state lifetime and the shapes of the ground and excited PES. Analyses of DIET yields and distributions, along with the observed unoccupied adsorbate resonance widths, suggest that the lifetimes of electronically excited adsorbates are $\ll 10$ fs (9–11), precluding direct, real-time observation of the nuclear wave packet motion on metal surfaces.

Undaunted by such pessimistic estimates, in the past decade several laboratories have initiated research programs on the ultrafast electron and nuclear dynamics of clean and adsorbate-covered surfaces by means of time-resolved photoemission techniques (17–22). Because it provides electron energy and momentum-resolved spectroscopic information, time-resolved two-photon photoemission (TR-2PP) is a general and sensitive technique for investigating the dynamics of hot electrons in metals (19, 20, 23, 24), as well as the phase and energy relaxation of intrinsic surface states, which are confined by the crystal and image potentials (22, 25, 26). TR-2PP is particularly useful for studying adsorbate-covered metal surfaces, where it is possible to observe enhanced two-photon photoemission (2PP) signals owing to transitions through the intermediate unoccupied surface resonances of the adsorbate-surface complex. This makes it possible to probe the electronic excitation and relaxation dynamics directly in the time domain (18), or indirectly through the linewidth and laser polarization analysis of 2PP spectra (27–31). Although short-lived reaction intermediates can be detected through time-resolved vibrational spectroscopy (32–34) or time-resolved valance or core level photoemission (35), TR-2PP is uniquely suitable because it is possible interact directly with the electronic excitations that drive the surface photochemistry.

The two chemisorption systems most intensively studied by TR-2PP are CO/Cu(111) and Cs/Cu(111) (21, 27–29, 36–42). Because of the extensively studied electronic interaction of CO with transition and noble metals, and the encyclopedic interest CO holds in surface science, Wolf and coworkers have investigated the spectroscopy and dynamics of CO/Cu(111) (28). Consistent with the current understanding of DIET, both the frequency and time domain measurements revealed that the lifetime of the unoccupied $2\pi^*$ resonance of CO is in the range of 1–5 fs, precluding a real-time observation of the nuclear dynamics. Although Cs/Cu(111) is a less obvious model system for surface photochemistry, alkali atom chemisorption is also of fundamental and practical interest on account of the dramatic effects the alkali atoms have on the surface electronic structure of metals that impart useful properties in both the thermionic emission and catalysis (43). Moreover, the photon- and electron-induced desorption of alkali atoms from oxide surfaces is thought to be responsible for their large abundance in the planetary atmospheres (44–46). Furthermore, the ionic interaction between alkali atoms and metal surfaces is common to a broader class of chemisorption systems on metals, as well as to gas-phase alkali halide ion-pair molecules that have well-understood femtochemistry (47, 48). Most importantly, TR-2PP measurements show that the alkali atom lifetimes on metal surfaces range between 1 and 50 fs, which makes it possible both to study their nuclear dynamics and, just as importantly, to critically examine the factors that influence the electronic relaxation rates of adsorbates on metals. The remainder of this review is devoted to presenting the experimental results on the alkali atom dynamics on noble metals, and more generally, a perspective for surface femtochemistry.

EXPERIMENTAL TECHNIQUE

Comprehensive reviews of TR-2PP can be found in References (17) and (18). TR-2PP experiments are typically performed with <100-fs laser pulses that excite 2PP from conductive surfaces. Two-photon excitation lifts electrons from below the Fermi level E_F, through unoccupied states in the 0 to $h\nu$ energy range, to final states in the $h\nu$–$2h\nu$ range, as shown in Figure 2a (see color insert). Electrons that are excited above the vacuum level Φ are emitted from the surface into the vacuum. The photoelectrons are detected with energy and momentum resolution by employing, for instance, time-of-flight or conventional electrostatic electron energy analyzers. The energy- and momentum-resolved detection represents a significant advantage over all optical methods, which provide only integrated information, for studying ultrafast charge carrier dynamics in solid state materials. Because in the absence of elastic and inelastic scattering, the energy and parallel momentum $k_\parallel$ are conserved in photoemission, 2PP spectra contain spectroscopic information on the initial and intermediate states in the excitation process. This makes it possible to identify in 2PP spectra the occupied and unoccupied bands involved in the 2PP process that can be intrinsic to the substrate or the adsorbate, or belong to the substrate-adsorbate complex. Further information regarding the orbital symmetries can be obtained from the dependence of 2PP spectra on the polarization of the exciting light (29).

When the 2PP process is excited with an ultrafast pump-probe pulse sequence that has a variable delay, TR-2PP provides further information regarding the charge carrier and even nuclear dynamics (18). The photoemission current as a function of the pump-probe delay provides a two-pulse correlation measurement of the charge carrier phase and energy dynamics involved in the two-photon excitation process (49, 50). For instance, the excitation to an unoccupied adsorbate resonance can occur coherently through a direct optical transition or through a variety of indirect, and therefore incoherent, processes involving the transport and scattering of hot electrons and holes. In surface photochemical studies the excitation of an adsorbate resonance via substrate electrons is thought to be more probable than the direct optical transition owing to a larger optical density and relaxed optical selection rules (8, 10). The intermediate state population dynamics will also be influenced by decay processes, such as the resonant charge or energy transfer to the bulk, or the inelastic scattering of the adsorbate electron with the Fermi sea. Finally, the 2PP signal can reveal the nuclear dynamics of the adsorbate through the coupling of the electronic and nuclear motion if the excited state lifetime is sufficiently long to achieve significant nuclear displacements. In the simplest case of photodesorption following the Menzel-Gomer-Redhead scenario, the transfer of electronic to nuclear energy will result in a corresponding decrease in the electronic energy of the unoccupied adsorbate resonance. This will appear as a decay of the intermediate state population from the observation window set by the electron energy analyzer (27, 41). The nuclear motion can also affect the photoemission cross section from the adsorbate resonance, as well as the electronic relaxation rates, which are predicted to decrease exponentially with the internuclear distance (51). Many of these dynamical processes occur on timescales that can be comparable to or faster than the excitation pulse. Therefore, accurate pulse characterization and realistic modeling of the 2PP process including the electron and nuclear dynamics are essential for extracting meaningful data from TR-2PP measurements (52–54).

Owing to the extremely short timescales probed by a 2PP experiment, the signal can strongly depend on the optically induced coherence in the excitation process (22, 55–57). The incoming radiation field coherently couples all transitions that can be excited within the bandwidth or the inverse interaction time with the external field. The coherently induced polarization evolves through free induction decay creating an outgoing radiation field, i.e., reflection. In an energy- and momentum-integrated measurement this decay is instantaneous because the inhomogeneous width of the metal response function exceeds that of the laser pulse. However, in an energy- and momentum-resolved measurement the homogeneous or inhomogeneous width can be less than the transform-limited width of the laser pulse or the energy resolution of the analyzer. In this case the coherent response of a metal can be resolved in a TR-2PP measurement (25, 57).

A graphic method for probing coherence in the 2PP process is through measurement of interferometric two-pulse correlations (I2PC) of 2PP using identical pump and probe pulses (25). In an I2PC measurement, such as that simulated in Figure 2b, the pump and probe pulses excite the sample collinearly, and the optical delay is scanned interferometrically with subwavelength resolution (39). Because the delay specifies a well-defined phase between the pulses, the I2PC measurement is composed of an interference signal with the period of the carrier frequency at zero delay. At later delays the interference may evolve to other frequencies spanned by the excitation pulse as a consequence of the coherent response of the sample. The observed fringes are a manifestation of the optical interference between the excitation fields in the interferometer used for generating the delay and more importantly of quantum interference between the pump- and probe-induced polarizations in the sample. Moreover, because the 2PP signal has a quartic dependence on the external field, the I2PC signal also contains oscillations at twice the carrier frequency corresponding to the nonlinear polarization excited between the initial and final states in the 2PP process (25).

Decay of the coherent oscillations in an I2PC trace results from both the phase relaxation of the coherent polarization and the inhomogeneous spectral broadening (53). Decoherence can occur through quasi-elastic momentum scattering of carriers with phonons or defects or through inelastic carrier-carrier scattering. The carrier-phonon interaction increases linearly with temperature, whereas the carrier-carrier interaction increases quadratically with the carrier energy above the Fermi level (58, 59). The inhomogeneous broadening leads to a loss of fringe contrast, because different frequencies excited within the inhomogeneous width dephase as the delay increases. Likewise, the fringe contrast also depends on the energy width subtended by the analyzer. Phase relaxation times in metals can extend to $\sim$100 fs for occupied surface states (60, 61) and even to >1 ps for the image potential states (22). The carrier-carrier and carrier-phonon interactions are generally stronger in the bulk metals, but for localized bands, such as those formed by d-electrons in copper, the phase relaxation time can exceed 35 fs (63). The effective phase relaxation times can be deduced from I2PC scans from the coherent fringes either by fitting the data to a model that describes the 2PP process in terms of optical Bloch equations (53) or, more qualitatively, by decomposing the I2PC signals into envelopes of the oscillations or their phase average (18), as shown in Figure 2b.

SURFACE ELECTRONIC STRUCTURE

Alkali Atom Chemisorption

Before discussing the experimental results on TR-2PP measurements, it is important to introduce the pertinent aspects of the bonding and electronic structure of alkali atoms on noble metals that relate to the experimental results (64–70).

The leading interaction experienced by an alkali atom near a metal surface is the Coulomb repulsion between its ns valance electron and the negative image charge induced in the metal by screening of the positively charged alkali atom core. This interaction will polarize the electronic cloud leading in the lowest order to the hybridization of the alkali atom ns and np orbitals into a bonding and antibonding pair (71). In addition to the hybridization, according to the Langmuir-Gurney model for the chemisorption of alkali atoms (65), the ns resonance shifts upward in energy and broadens in proportion to the electron transfer rate between the atom and the surface. At $\sim$10 Å from the surface the ns electron is destabilized with respect to the Fermi level, making the electron transfer to the substrate energetically favorable (51, 66, 67). The delocalization of charge leads to the formation of a strong ($\sim$2 eV) alkali atom–metal surface chemisorption bond. Nevertheless, the charge transfer is not complete as judged by the bond length and other characteristics of ionic bonding (68, 72).

In this scenario the ns resonance is predicted to have maximum density of states (DOS) at >2 eV above E_F and a width of several eV (all energies are given relative to E_F) (51, 73, 74). However, the unequivocal characterization of the density of states associated with the chemisorption bond by standard photoemission and inverse photoemission techniques has been elusive (75, 76). The clearest spectroscopic signature of the alkali atom interaction with metal substrates is a ubiquitous unoccupied resonance, which appears at $\sim$3 eV in the limit of zero alkali atom coverage (37, 76, 77). Although this pronounced feature is sometimes attributed to the bonding density of states (78–80), its width and energy are too narrow to allow for significant density below E_F and, therefore, to be consistent with incomplete charge transfer of the ns electron to the substrate (72, 74). A more consistent assignment is to the ns + np antibonding density of states. In this resonance the electron charge cloud is strongly polarized away from the surface and into the vacuum, where the Coulomb repulsion is minimized (69, 81, 82). Thus, the electronic excitation of this antibonding resonance (AR) corresponds to an electron transfer from the substrate to the vacuum side of the alkali atom.

Based on this simple model, it is possible to construct the set of PES for the chemisorption and photodesorption of alkali atoms on metals shown in Figure 1b. The surfaces are constructed as prescribed by Hellsing et al. for the potassium photodesorption from graphite (45). The asymptotic energies of Cs and Cs$^+$ are given exactly from the difference between the measured work function of Cs/Cu(111) at $\sim$0.1 ML Cs coverage and the ionization potential of Cs atoms. Note that because the work function of Cu(111) and the chemisorption energy of Cs can decrease by as much as 3.4 and 2 eV, respectively, by increasing the Cs coverage Θ_{Cs} (83), the PES for photodesorption also strongly depend on the coverage. The Cs coverage can be used even to control the occurrence and the internuclear distance for the crossing between the neutral and ionic PES. Furthermore, the concept of PES may be questionable because of strong nonadiabatic effects that might arise owing to the changes in the local work function (84) following the forward or backward electron transfer.

Disregarding these subtleties, the ground state PES is constructed by splicing a long-range Coulomb potential with a short-range Morse potential such that the chemisorption energy is 1.9 eV, which is estimated from temperature-programmed desorption (TPD) of Cs/Ag(100), and the bond length is 2.97 eV, as predicted by theory (81, 85, 86). The excited state PES is constructed by assuming a long-range van der Waals interaction appropriate for the polarizability of Cs (87) and a short-range Morse repulsion that gives the correct vertical transition energy for the excitation of the AR at $\Theta_{Cs} = 0.1$ ML. PES constructed in this manner predict that the optically induced charge transfer excitation of the AR will initiate the dissociative wave packet motion on the excited state PES. Note that quite similar PES describe the optically induced charge transfer and subsequent wave packet dynamics for gas-phase NaI and NaBr (48, 88). In general, qualitatively similar potential energy surfaces may be appropriate for describing the surface femtochemistry of other chemisorption systems when the work function of the substrate is larger than the ionization potential of the adsorbate.

Electronic Structure of Noble Metals

The dynamical properties of alkali atoms are significantly affected by the electronic structure of the metal substrate (27). Figure 3 shows schematic band structures for momentum parallel to the low index (110), (100), and (111) planes of copper. The defining feature of the band structure that strongly influences the electronic relaxation rates of alkali atoms is the presence of projected band gaps that at the Γ point ($k_{\parallel} = 0$) extend between $-0.85 \sim 4.2$ eV for Cu(111) and $1.6 \sim 8.0$ eV for Cu(100) (27). These band gaps define a range of momenta that are purely imaginary, implying that the electron propagation into the metal is exponentially damped. The band gaps support several intrinsic surface states including the well-known occupied Shockley surface state (SS) at -0.4 eV on Cu(111) and the image potential states that converge to the work function (89, 90). These states correspond to intrinsic quantum wells that are bound by the image and vacuum potentials at the metal-vacuum interface. Their wavefunctions are concentrated at the surface and penetrate the bulk only as evanescent waves when resonant with the band gaps. This confinement of electrons to the metal-vacuum interface often leads to increased lifetimes of surface states with respect to the bulk bands at the same energy, because of a reduced spatial overlap with the bulk bands and restricted phase space for electron-electron scattering at surfaces (59, 90, 91).

Upon deposition of alkali atoms, the surface electronic structure changes in several significant aspects, including a substantial decrease in the work function and introduction of the Cs antibonding resonance. Because decreasing the work function increases the width of the confining potential at the surface, surface states are stabilized according to their proximity to the vacuum level. Because the AR is stabilized at a faster rate than SS, tuning of the SS → AR transition frequency into resonance with a fixed frequency excitation laser is possible by varying the alkali atom coverage (40).

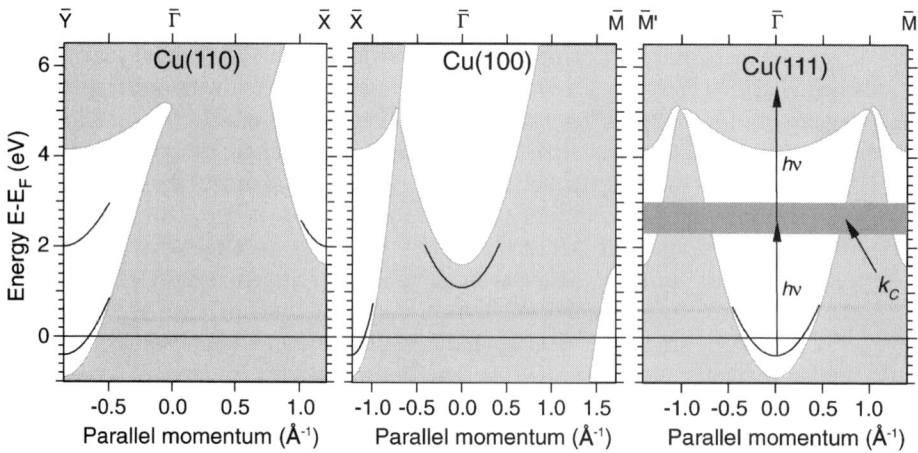

Figure 3 Projected band structure for the parallel momentum along the three low index surfaces of copper. Projected band gaps shown as white regions support several intrinsic surface states (*black curves*), which form resonances outside of the band gaps. The Cs overlayer at low coverage is disordered, and therefore the antibonding resonance (AR), which is observed for all three surfaces at approximately the same energy, does not disperse, implying localization of the excitation on single Cs atoms. The shaded area for AR conveys the energy range spanned by AR through variations in the Cs coverage, temperature, and nuclear motion. k_c represents the minimum parallel momentum required for the decay of AR by resonant charge transfer into propagating Bloch states.

Although the AR exists within the same band gaps as the intrinsic surface states, the conditions for its confinement at the surface are significantly different. At low alkali atom coverages (<0.5 ML) the distribution of alkali atoms is dominated by the dipole-dipole repulsion rather than by the corrugation of the surface potential, resulting in formation of a disordered hexactic liquid phase (92, 93). Owing to disorder and weak overlap between the alkali atom electronic charge densities, the excitation is localized on single alkali atoms, and therefore, the AR does not form a dispersive band. Because $k_\parallel$ of the AR is undefined, momentum conservation does not prevent the resonant charge transfer from the AR to the propagating Bloch states at $k_\parallel \neq 0$. However, the tunneling rates into the bulk bands can vary significantly by reason of the momentum-dependence of the barrier widths (81), as is discussed below.

ELECTRONIC STRUCTURE AND SPECTRA

Figure 4*a* (see color insert) presents several 2PP spectra taken at $k_\parallel = 0$ and 300 K for Cs/Cu(111) with progressively higher Θ_{Cs} starting with the clean Cu(111) surface. The 2PP spectrum of the clean surface shows a single peak arising from

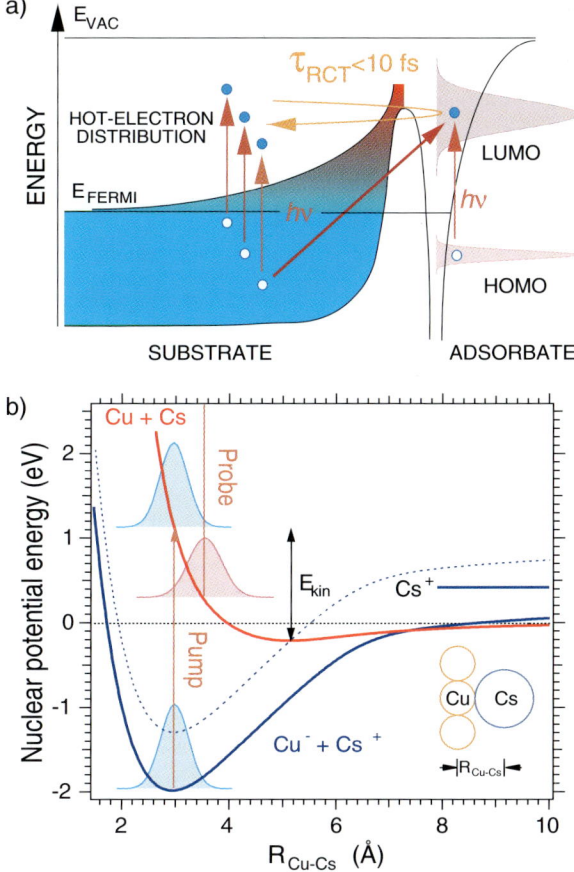

Figure 1 (*a*) Excitation pathways in surface photochemistry. Photochemistry is initiated through the excitation of an unoccupied state of an adsorbate either indirectly by the scattering of photogenerated hot electrons in the bulk, or directly by optical transition from the occupied states of the bulk or the surface. Resonant charge transfer (RCT) is thought to be the dominant mechanism for the rapid electronic quenching. (*b*) The potential energy surfaces for Cs photodesorption that are estimated from the chemisorption energy, van der Waals interaction, ionization potential, and excitation energy of Cs atoms (27, 41, 45). Photoexcitation projects the ground state probability distribution to the excited state, where it evolves as a coherent nuclear wave packet under the influence of the dissociative potential. In the desorption induced by electronic transition (DIET) scenario (10) desorption occurs from the ground state potential energy surface (PES), following a nonadiabatic transition via RCT when the kinetic energy gained on the excited state PES exceeds the chemisorption energy. The *dashed line* represents the shifted ground state potential, required by energy conservation, which accounts for the electron-hole pair generation in the substrate. Because at low coverage Cs does not appear to have a well defined bonding site, R_{Cu-Cs} is defined as the distance of Cs from the topmost Cu layer.

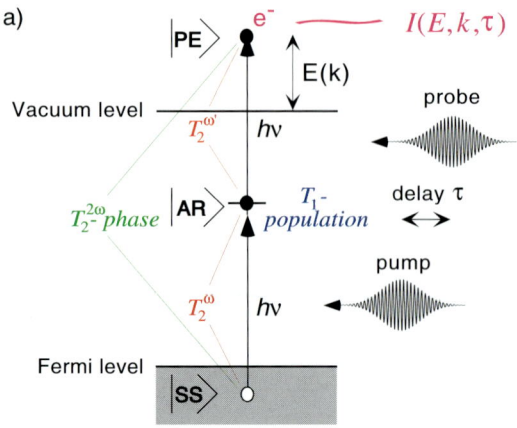
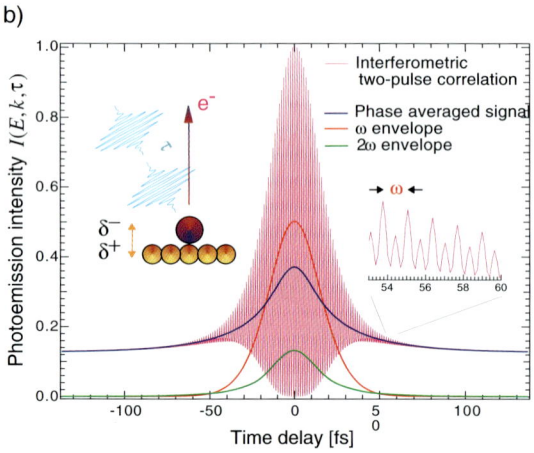

Figure 2 (*a*) Excitation scheme for two-photon photoemission. A single pulse or a pump-probe sequence lifts an electron from below the Fermi level, through an unoccupied intermediate state, to above the vacuum level. To obtain two-photon photoemission (2PP) spectra, the photoemitted electron current is recorded with energy and momentum resolution, thereby capturing the information regarding the occupied and unoccupied band structure. SS, occupied Shockley surface state; AR, Cs antibonding resonance; PE, the slice of the photoemission continuum detected by the electron energy analyzer. (*b*) Interferometric two-pulse correlation (I2PC) scans are recorded by scanning the pump-probe delay while recording the 2PP current at a fixed energy and momentum, thereby providing information on the electronic excitation and relaxation processes. Oscillations at approximately the excitation frequency and its second harmonic provide information on the phase relaxation times T_2 of the coherent polarization induced in the sample. Long time decay is dominated by the population dynamics including decay T_1 through resonant charge transfer or inelastic electron-electron scattering. Decomposition of I2PC scans into component envelopes helps to visualize the individual dynamical processes.

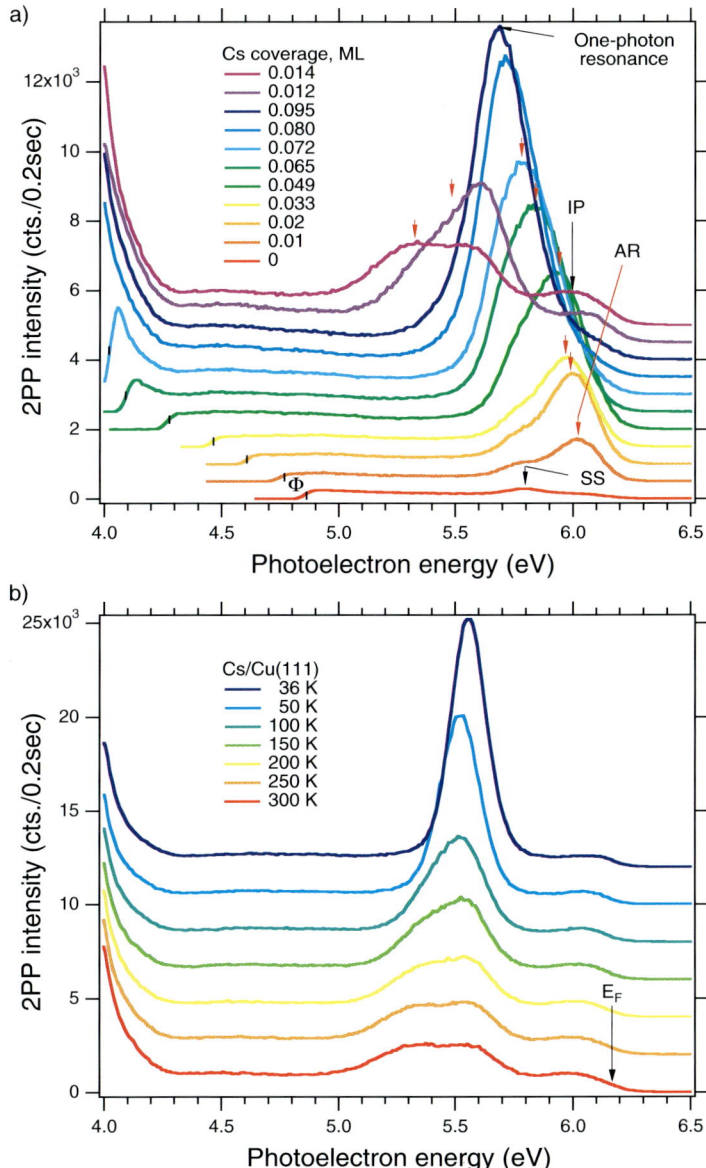

Figure 4 (*a*) A series of two-photon photoemission (2PP) spectra of Cs/Cu(111) at 300° K for various Cs coverages starting with the clean surface. One monolayer corresponds to $4.41 * 10^{14}$ atoms/cm^2. The antibonding resonance (AR) is indicated with a *red arrow*. SS, Shockley surface state; IP, image potential state. (*b*) Temperature dependence of 2PP spectra for the 0.14-ML sample in (*a*). The spectra are obtained by progressively cooling the sample from 300° to 36° K.

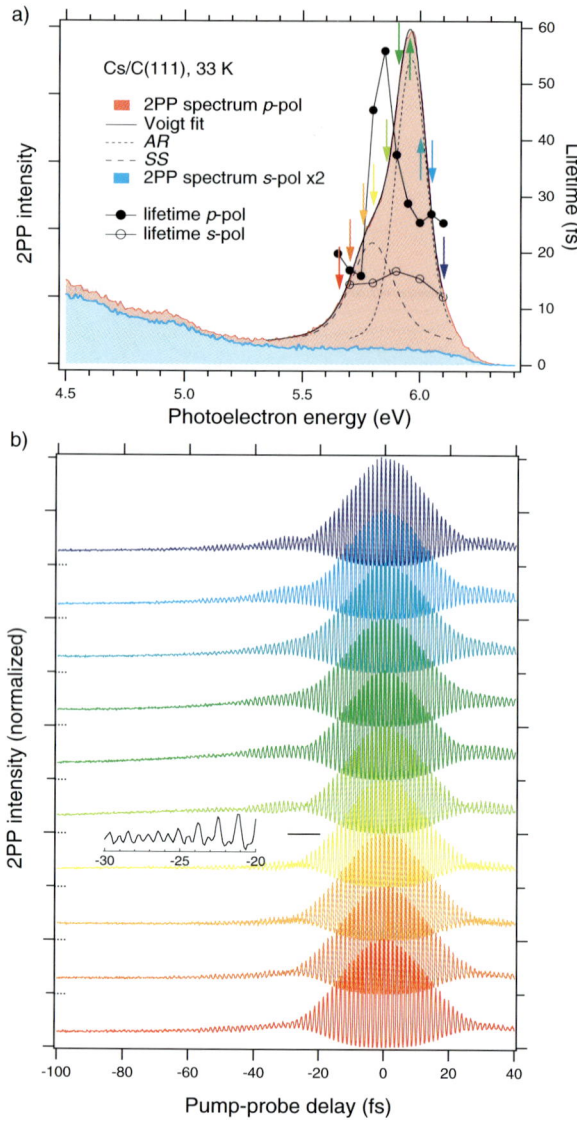

Figure 5 (*a*) Two-photon photoemission (2PP) spectrum for near resonant Shockley surface state (SS) → antibonding resonance (AR) excitation. Voigt lineshape fit of the spectrum (*solid line*) is used to deconvolute SS and AR, which are indicated by *long* and *short dashed lines*, respectively. Lifetimes from forced exponential fits are provided to contrast the surface and bulk dynamics. (*b*) Interferometric two-pulse correlation scans for several energies (indicated by corresponding arrows in (*a*) near the SS → AR resonance. The expanded detail shows the coherent oscillations, which result from the interference between the pump- and probe-induced polarizations.

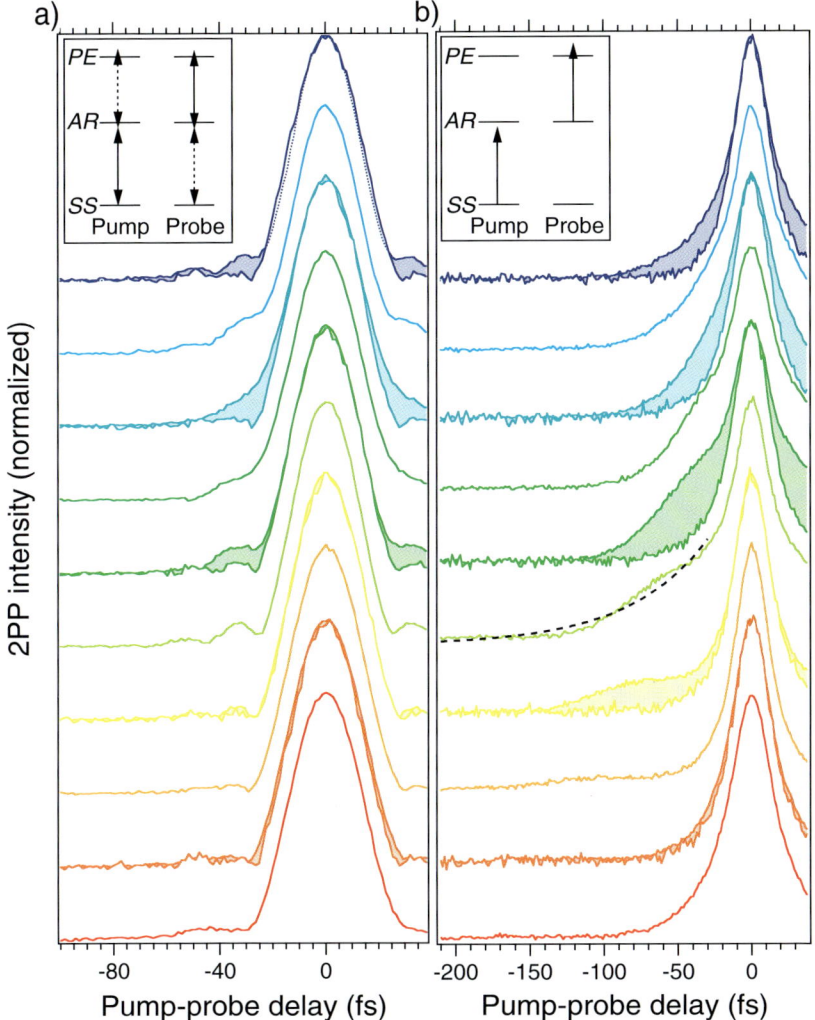

Figure 6 Decomposition of the interferometric two-pulse correlation (I2PC) scans in Figure 5 for p-polarization and representative scans taken with s-polarization into envelopes of oscillations at (*a*) ω_i and (*b*) the I2PC phase average. As indicated in the insets, the data in (*a*) probe the polarization dynamics, whereas the data in (*b*) also include the longer time scale population dynamics (note the abscissa scale change). Measurements taken with p-polarization are *shaded*, to contrast with the faster dynamics of bulk electrons taken with s-polarization. *Dashed line* shows a fit to an exponential decay, to emphasize the nonexponential nature of the population dynamics.

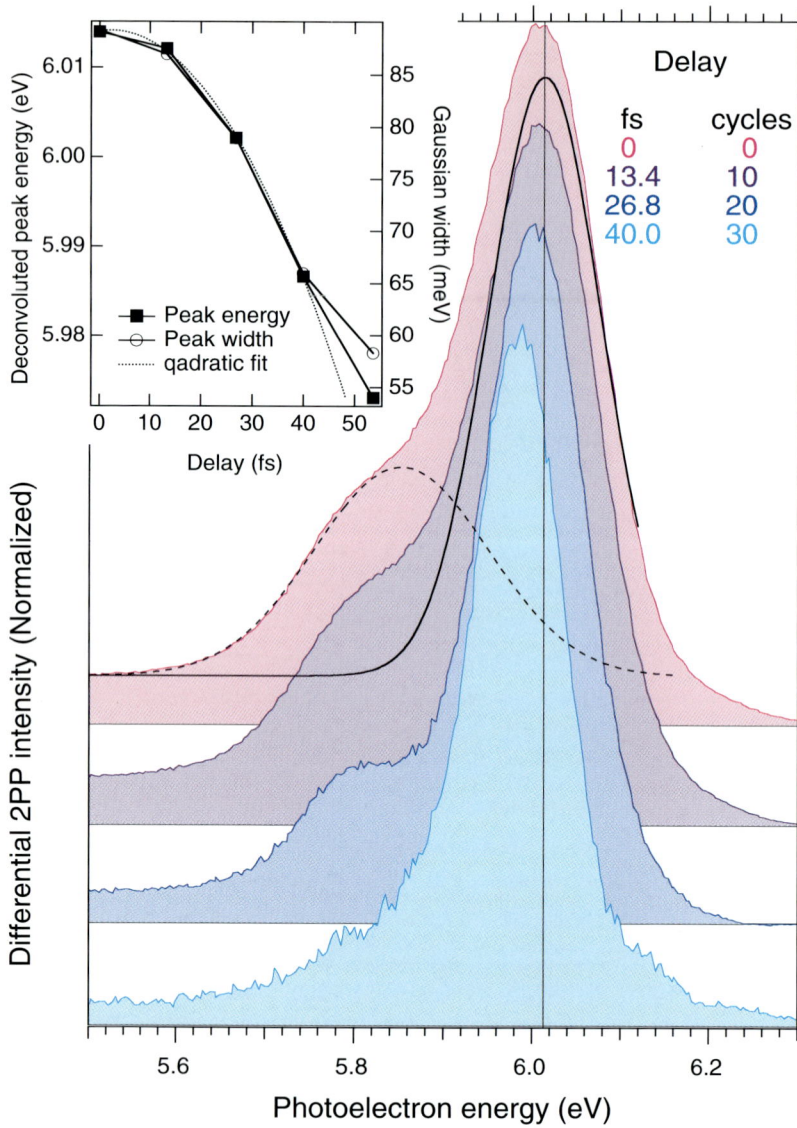

Figure 7 Normalized two-photon photoemission (2PP) difference spectra for several pump-probe delays. Spectra are displaced vertically for clarity. Gaussian deconvolution of Shockley surface state (SS) and antibonding resonance (AR) peaks is shown with *dashed* and *solid curves*, respectively. The *vertical line* indicates the energy of AR at zero pump-probe delay. The *inset* shows the peak energy and width from the fits to AR for each delay and a second-order polynomial fit to the peak position for delays ≤40 fs.

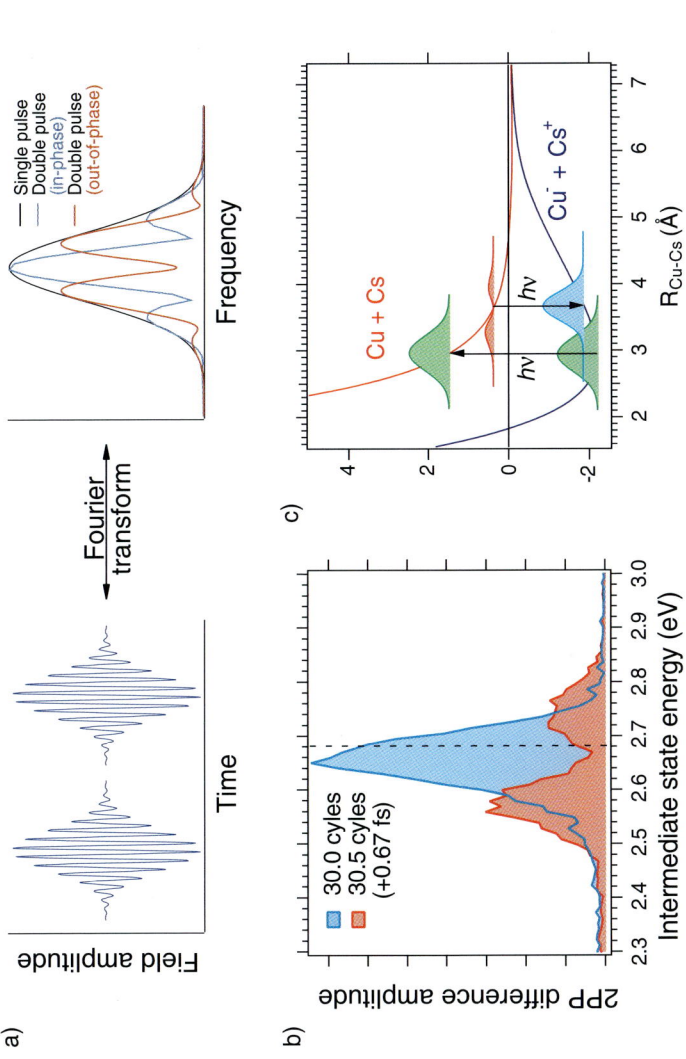

Figure 9 (*a*) Electric field amplitudes in time and frequency domains for a pulse pair. The spectra depend on the precise phase delay between the pulses, when the coherence length of the measuring instrument exceeds the delay. (*b*) Two-photon photoemission (2PP) difference spectra for pulse delays of 30.0 and 30.5 cycles. *Vertical dashed line* indicates the energy of the antibonding resonance at zero delay. (*c*) The quantum control scheme appropriate for the out-of-phase excitation. Destructive interference between the pump and probe excited wave packets generates a displaced ground state wave packet.

the two-photon excitation from the occupied Shockley surface state, which is on a broad pedestal corresponding to the momentum-integrated DOS feature. Upon addition of Cs atoms, a new peak appears with a final state energy above that of the SS. Based on previous studies and measurements of the excitation light wavelength dependence of 2PP spectra, this feature can be assigned to the Cs-induced AR (21, 30, 76). Increasing the Cs coverage decreases the work function Φ, and correspondingly, the energies of the AR, and to a lesser extent those of the SS, as explained in the previous section. Significantly, integrated intensities of the AR and SS peaks increase to a maximum as the two features blend into a single peak, indicating that the 2PP yield is enhanced as the SS→AR transition is tuned into resonance with the excitation laser.

It should be noted, however, that the resonance condition exists even at the lowest coverages for $k_\parallel \neq 0$ as a consequence of different dispersions of the SS and AR bands, as shown in Figure 3. This signal can appear at $k_\parallel = 0$, because the excitation to the AR is localized on single Cs atoms, and therefore, $k_\parallel$ is undefined. However, this signal probably only contributes to the 2PP signal at $k_\parallel = 0$ for the incoherent two-photon absorption excitation. As Θ_{Cs} increases beyond the resonant condition, separate SS and AR peaks reappear, but with a higher intensity than for the preresonant condition, probably because of a larger Cs coverage. In addition, new peaks belonging to the d-bands (below 4.0 eV) and image potential (IP) states emerge at the work function and Fermi edges, respectively, owing to the work function reduction (30, 52, 94). Similar spectra also have been observed for Na/Cu(111), K/Cu(111), Na/Cu(100), Cs/Cu(100), Cs/Cu(110), Cs/Ag(111), and Rb/Cu(111) (27, 37, 40, 77). However, only the (111) surfaces of noble metals have the occupied SS at $k_\parallel = 0$, which is a prerequisite for observing the distinct resonance-enhanced excitation of the AR (27).

The 2PP spectra of Cs/Cu(111) also exhibit an interesting temperature dependence that is reproduced in Figure 4b for the highest coverage sample of Figure 4a (38). As the sample is cooled, the AR undergoes a linear shift to higher energy while its spectral width decreases. These changes probably are associated with the cooling of the thermally excited Cu-Cs vibration, which has an estimated frequency of 8 meV (93). Cooling of the anharmonic Cu-Cs vibration should narrow the distribution of the vibrationally excited states, thereby reducing the average Cu-Cs distance R_{Cu-Cs} as well as its distribution. Because at low temperatures the R_{Cu-Cs} distance is shorter, the resonant SS → AR transition occurs higher up the repulsive wall of the excited state leading to the upward shift of the AR peak. The shorter R_{Cu-Cs} distance also implies a smaller surface dipole, which is probably responsible for an upward shift in the work function with temperature (38). Since the AR energy is related to the work function, the magnitude of surface dipole will also contribute to the upward shift with cooling. The linewidth increase with temperature is probably related to the spread in the R_{Cu-Cs} distance at elevated temperatures, as well as the electron-phonon scattering (63, 95). The intensity changes in Figure 4b are observed at a constant Cs atom coverage, providing further evidence of the resonance enhancement of the SS → AR transition.

The coverage and temperature dependence of the 2PP spectra demonstrate that the SS→AR excitation occurs directly by an optical transition from an intrinsic state of the metal to a state of the adsorbate-surface complex. This conclusion is consistent with the polarization dependence of 2PP spectra in Figure 5a (see color insert), which shows that the AR cannot be populated with s-polarized excitation, because in a direct transition this is forbidden by symmetry. The direct excitation of Cs atoms differs from many surface photochemical studies, which attribute the photochemistry to the excitation of unoccupied states of adsorbates via the hot electron mechanism (8, 10, 96).

PHASE AND ENERGY RELAXATION

The most compelling evidence for the direct charge transfer excitation comes from the I2PC scans taken with p-polarized light, such as those shown in Figure 5b, for selected observation energies in the region for observation of the near resonant SS → AR transition. The I2PC signal consists of rapid oscillations at approximately the excitation frequency ω_l, which is most prominent when the pump and probe pulses overlap in space and time near the zero delay, and a more slowly evolving component, which persists on longer (>200 fs) timescales. As shown in Figure 6 (see color insert), the two components can be visualized separately by decomposing the I2PC signal into separate envelopes of the coherent oscillations at ω_l and the slowly varying component representing an average over the rapid coherent oscillations, as explained above in "Experimental Technique." The former represents the coherent response of the surface to the external driving field, and the latter also includes the longer timescale population dynamics in the intermediate state. Also shown are several I2PC envelopes taken with s-polarization to contrast the surface dynamics with that of bulk hot electrons. The difference between p- and s-polarized scans can be attributed to the relatively slowly evolving polarization and population dynamics associated with the SS → AR transition.

Coherent Phase Dynamics

An ultrafast laser pump pulse interacting with a metal surface excites a broad range of electron-hole pairs with different initial and final energies that can be coupled within the optical bandwidth. Although each oscillator (*e-h* pair) is driven at ω_l, its response will occur at its specific resonance frequency, which may be time dependent on account of the many-body screening effects or the nuclear dynamics (97). The phase evolution of this polarization can be interrogated with the probe pulse, which interacts with the sample at a precisely defined delay, resulting in the coherent interference signal.

The I2PC scans in Figures 5 and 6 demonstrate the coherent response of Cs/Cu(111) for observation energies in the range of the SS → AR transition. The excitation pulses induce coherent charge oscillations corresponding to the electron transfer between the adsorbate and the substrate, which persist for a significantly longer time than the duration of each excitation pulse. Were Cs atoms excited indirectly through the hot electron mechanism, the I2PC scans would not show the coherent oscillations, other than those arising from the trivial effect of the optical interference of light when the two pulses overlap, because the optical transition moment coupling the ground and excited states would not exist. Thus, the I2PC scans provide unassailable evidence for the direct optical excitation of the adsorbate. The coherent nature of the excitation process is essential for the quantum control of photodesorption dynamics that is demonstrated below in "Quantum Control."

In addition to elucidating the excitation process, Figures 5 and 6 measure the phase relaxation processes operating at the Cs/Cu(111) surface. At 33 K, the I2PC measurements are consistent with 15–20 fs electronic phase relaxation time for the Cs antibonding resonance. The decay rate of the coherent oscillations increases for higher substrate temperatures, as can be anticipated from the temperature dependence of the AR linewidth in Figure 4b. This temperature-dependent contribution to the dephasing may be due to both an increasing inhomogeneous broadening of the AR peak and increasing participation of the electron-phonon scattering in the phase and energy relaxation. The precise role of phonons is not understood, but both the thermally induced surface-lateral and normal motion of Cs atoms can induce local perturbations in the electronic structure arising from fluctuations in the local Cs density or the Cu-Cs internuclear distance. Such fluctuations can lead to both inhomogeneous broadening and the electronic relaxation of the AR via elastic and inelastic scattering processes. The increased spectral widths resulting from the inhomogeneous alkali atom distributions or from vibrationally induced decoherence lead to the loss of the fringe contrast (53).

Nonexponential Population Dynamics

After suppressing the coherent fringes, the phase-averaged signal in Figure 6b reveals that in addition to the coherent polarization decay, the I2PC scans contain a longer time-scale component that measures the more slowly evolving population dynamics. The most striking aspect is the nonexponential decay of the 2PP signal, which is most prominent for observation energies at 0.1–0.2 eV below the AR resonance. The deviation from exponential decay is sensitive to the surface coverage, becoming most prominent when the SS $\rightarrow$ AR transition frequency is 0.1–0.2 eV above the laser frequency. Interpretation of the measured kinetics in terms of the AR lifetime would lead to the highly improbable conclusion that the AR decay rate is a nonmonotonic function of time that, furthermore, depends on the observation energy (27). Complex population kinetics could arise from other dynamical processes such as the energy migration in the adsorbate overlayer through the dipole-dipole interaction (98); however, in known examples such as the CO stretching vibration on Ru(111), this occurs on a considerably longer timescale and does not lead to similar kinetic behavior. The more likely scenario is that the unusual kinetics arise from the effect of the Cs atom motion on the 2PP process.

According to Figure 6b, the decay of the AR can be monitored for up to 200 fs, which is sufficiently long for substantial nuclear motion to occur even for heavy atoms such as Cs. The nuclear motion can affect the 2PP signal in several ways. Because the resonant charge transfer (RCT) rates can have a strong dependence on R_{Cu-Cs}, the lifetime of a desorbing particle could vary with the time after excitation leading to the nonexponential kinetics. Although it may be possible to calculate the lifetime of an adsorbate at a fixed internuclear distance, experimentally it is a challenge to measure an exponential decay for a desorbing particle if the internuclear distance, and therefore the coupling to the substrate, varies with time. Theory predicts an RCT tunneling rate that increases exponentially with the distance of alkali atoms from a jellium surface (51). However, when band structure effects are included, factors affecting the lifetimes are significantly more complex. The AR lifetimes for Cs/Cu(110) and Cs/Cu(100) are too short to deduce from I2PC measurements, and the AR state linewidth analysis only can set a lower limit of 2–3 fs (27, 40). These crystal face–dependent differences have been attributed to the band structure anisotropy (37, 38, 81). Specifically, for Cs/Cu(111), RCT can only occur in an off-normal direction, because resonant bulk bands only exist at large $k_\parallel$. Because the effective tunneling barrier is thicker for the off-normal direction, RCT is strongly suppressed (99). In fact, theoretical simulations indicate that the electronic relaxation of AR on Cs/Cu(111) is dominated by inelastic electron-electron scattering (82). Because both off-normal RCT and electron-electron scattering could have a complex dependence on the distance of Cs above the Cu(111) surface, it is difficult a priori to exclude the distance-dependent lifetimes as the cause of the nonexponential kinetics.

Another mechanism for the effect of nuclear motion on the I2PC measurements is the dependence of 2PP transition moments on R_{Cu-Cs}. The transition moment for the excitation of an electron from AR to the photoemission continuum exists only near the surface because the ionization potential of free Cs atoms is 3.9 eV. Whereas both the dependence of the lifetime and the 2PP transition moments on R_{Cu-Cs} can contribute to the complex time evolution of the AR population signal, it is not obvious how to attribute the specific nonexponential dynamics in Figure 6b to these effects.

The most intuitive and simple way to reproduce the nonexponential decays is through the dependence of the surface electronic structure on R_{Cu-Cs} (27, 41). In "Electronic Structure and Spectra," above, the temperature dependence of 2PP spectra already suggested that the energy of the AR is sensitive to changes in R_{Cu-Cs}. Upon electronic excitation, the Coulomb repulsion between the Cs atom and the metal substrate might induce dissociative wave packet motion, as suggested in Figure 1b. Because of the stretching of the Cu-Cs bond, the Coulomb interaction should decrease, leading to the stabilization of the AR. Thus, the simplest explanation for the nonexponential dynamics is that the dissociative wave packet motion leads to the change of the AR energy and, therefore, the tuning of transition frequencies. For observation energies 0.1–0.2 eV below the AR peak, a delay of 100–200 fs appears to be required for the wave packet motion such that the two photon excitation via the AR becomes a fully resonant process. This tuning effect increases the transition probability for energies below the AR peak maximum, resulting in a higher photoemission yield at longer delays than if the I2PC scans were primarily affected by the true electronic relaxation rates.

Although the tuning of the final excitation step can explain in part the unusual distortion of the I2PC scans on the low energy wing, the discussion of the unusual kinetics is incomplete without considering the initial excitation step. Because of the coherence in the excitation process, the I2PC scans are also affected by the tuning of the SS → AR transition frequency on account of the wave packet motion. This frequency chirp of the coherent polarization will be exploited in "Quantum Control," below, to demonstrate the quantum control of the nuclear wave packet motion. The nonexponential decays are most pronounced when the wave packet motion converts a near-resonant three-level system into a resonant one, as suggested in Figure 2. In recording the I2PC scans in Figure 5, the Cs coverage is adjusted to achieve such transient tuning of the electronic structure into resonance with the laser, thereby maximizing the distortion of population dynamics from the exponential behavior.

This scenario for the coupling of the nuclear and electronic degrees, and how it leads to the nonexponential population dynamics, has been confirmed (41) through simulations of the I2PC scans with a three-level optical Bloch equation model. The optical Bloch equation calculation of the 2PP yield includes both the energy and phase dynamics. In order to model the change in the electronic structure, the intermediate state energy is described by a quadratic function of the delay. Such dependence is expected if both the slope of the repulsive potential that drives the wave packet motion and the energy of the AR are approximated by linear functions of R_{Cu-Cs}. The linear approximation should hold in the Franck-Condon region but must deviate at larger distances owing to the curvature in the dissociative potential. Simulations with the optical Bloch equation scheme in Reference (41) reproduce the data well for the AR lifetime of 50 fs, the phase relaxation times for the coherence among the three levels of 15 fs, and the quadratic decrease of the AR energy by 0.12 eV in 100 fs. This scenario suggests that it should be possible to detect the change in the surface electronic structure and even to quantify the wave packet motion by recording the time dependence in the 2PP spectra, as is demonstrated in the next section.

NUCLEAR WAVE PACKET DYNAMICS

Time-Dependent Electronic Structure

The time dependence of the surface electronic structure induced by the wave packet motion can be extracted when the 2PP process is excited by the pump-probe pulse pair with a delay that is comparable to the excited state dynamics. Considering initially just the population dynamics, and ignoring the interference that arises from the existence of two indistinguishable excitation pathways, the 2PP signal with two-pulse excitation will have components in which each pulse excites 2PP independently and a component in which the two pulses act jointly: the pump pulse creating the intermediate state population, and the delayed probe pulse testing its evolution. 2PP spectra in which each pulse acts only individually are obtained for long delays (240 fs), providing a reference for isolating the time-dependent component in which the pulses act jointly. 2PP difference spectra in Figure 7 (see color insert) are obtained from 2PP measurements at several delays of precisely $2\pi n$ integer number of optical cycles, in which the reference spectrum has been subtracted. Because in addition to the population dynamics, these two-pulse spectra also probe the coherent polarization created in the sample for delays of as long as 60 fs, the relative phase for each measurement must be set precisely in order to be meaningful.

The difference spectra in Figure 7 show that the energy and the width of the AR decrease systematically with the delay for the first 40 fs after the excitation. These changes are quantified by a Gaussian deconvolution of the AR peak and summarized in the inset of Figure 7. Both the peak position and width of the AR decrease quadratically with the delay during the initial decay. For longer delays, the difference spectra acquire additional structure and increase in width, probably as a result of the interaction of the Cs atom with the substrate, as reported in Reference (41). The quadratic frequency shift has been anticipated in the previous section, and is discussed further to extract the wave packet motion; the spectral narrowing corresponds to quantum control of the nuclear wave packet motion.

Estimation of the Excited State Potential Energy Surface

That the energy of the Cs antibonding resonance $E_{AR}(R_{Cu\text{-}Cs})$ depends on the distance of Cs atoms above the Cu surface is the essential postulate for interpreting the temperature and pump-probe delay dependence of the AR peak in 2PP difference spectra. However, it is not implicit that the electronic stabilization of AR corresponds to the gain in nuclear kinetic energy in the reaction coordinate, which would allow the direct experimental measurement of the wave packet motion. The spectral changes in Figure 7 are constrained by energy conservation. The two-photon excitation process generates the final state consisting of a hole in the metal and a free electron propagating in the vacuum. In addition, the photoemission process terminates the nuclear motion on the excited state potential energy surfaces (PES), generating a Cs cation (presumably the ground state), which is displaced from its equilibrium position while retaining the nuclear kinetic energy gained on the excited state PES (100). The energy balance for the two-photon excitation process can thus be written as $2h\nu = E^e_{kin} + E^n_{kin} + E^h + \Phi + E^{e,n}_{dis}$, where $E^e_{kin} = E_{AR}(R_{Cu\text{-}Cs}) + h\nu - \Phi$ is the kinetic energy of electrons detected by the analyzer, E^n_{kin} is the conserved nuclear kinetic energy in the reaction coordinate, E^h is the hole energy, and $E^{e,n}_{dis}$ is the remaining electronic and nuclear energy that resides neither in the outgoing electron nor in the reaction coordinate. This remaining energy includes dissipation arising from the multiple scattering of Cu substrate atoms initiated by the half-collision that is not directly associated with the reaction coordinate and the vibrational excitation retained by the cation following the photoemission, as well as electron-hole pair excitations in the metal associated with the screening charge redistribution in the optically induced transfer of an electron from the substrate onto the Cs atom and into the vacuum.

Because $E_{dis}^{e,n}$ is unknown, the wave packet motion is extracted from the observed 2PP spectral changes and the theoretical $E_{AR}(R_{Cu\text{-}Cs})$ dependence calculated by Borisov et al. (82). The wave packet propagation for the first 50 fs presented in Figure 8a is derived from the observed electronic energy as a function of the delay $E_{AR}(\tau)$ from Reference (41), and the theoretical $E_{AR}(R_{Cu\text{-}Cs})$ dependence reproduced in Figure 8b. Based on this transformation, the nonexponential I2PC decays result from changes in the surface electronic structure induced by lifting of Cs atoms by ∼0.15 Å above the Cu surface 50 fs after the excitation (41).

Having obtained $R_{Cu\text{-}Cs}(\tau)$, the dissociative potential is obtained from classical mechanics as follows: The wave packet trajectory in Figure 8a is differentiated to obtain the acceleration $\frac{\partial^2 R_{Cu\text{-}Cs}(\tau)}{\partial \tau^2} = a$. Then the force $F = ma$ can be integrated, i.e., $U = -\int F dR$, to derive the potential curve near the Franck-Condon region. However, one unknown aspect, namely how the momentum is distributed among the recoiling particles, still needs to be considered. One limiting case is that of elastic desorption from a massive hard wall (HW), in which the momentum transfer to Cs atoms is maximum and the entire 1.1 eV potential energy is available to be converted into the translation of outgoing Cs atoms. This limit is plausible considering the large lateral extent of the calculated Cs electronic cloud in AR (82), which suggests that a desorbing Cs atom can simultaneously scatter from several Cu atoms, implying a large effective mass for the substrate. In the HW limit the appropriate mass for calculating the force is that of a Cs atom. As shown in Figure 8b, the derived slope of the excited state PES at the point of vertical excitation is only -0.75 eV/Å, as compared with the predicted slope of -1.9 eV/Å for the PES in Figure 1b and -2.55 eV/Å for the photodesorption of K from graphite (45). However, the derived slope is nearly the same as for $E_{AR}(R_{Cu\text{-}Cs})$ from theory, implying that if the HW model applies, at least for the first 50 fs the electronic energy of AR is converted quantitatively into E_{kin}^n, and $E_{dis}^{e,n}$ is negligibly small.

The main difficulty with the HW model is reconciling the discrepancy between the predicted and derived excited state PES slopes. One possibility is that the binding energy of Cs on Cu(111) is larger than estimated from the temperature-programmed desorption data for Cs/Ag(100) (86). Substantial uncertainty in chemisorption energy can arise through the choice of the prefactor for modeling temperature-programmed desorption data. Stolz et al. used a prefactor of $3 * 10^{11}$ s^{-1} for Cs/Ag(100), whereas $1.6 * 10^{13}$ s^{-1} is appropriate for a noninteracting two-dimensional lattice gas of Cs atoms (86; F.M. Zimmermann, personal communication). The larger prefactor is consistent with the theoretical modeling of temperature-programmed desorption data for desorption of K from graphite (74). With the revised prefactor, the binding energy of Cs atoms is 2.3 eV, implying a shallower slope of the PES of -1.1 eV/Å and 0.7 eV energy available for the photodesorption. This shows that the slope of the potential is quite sensitive to

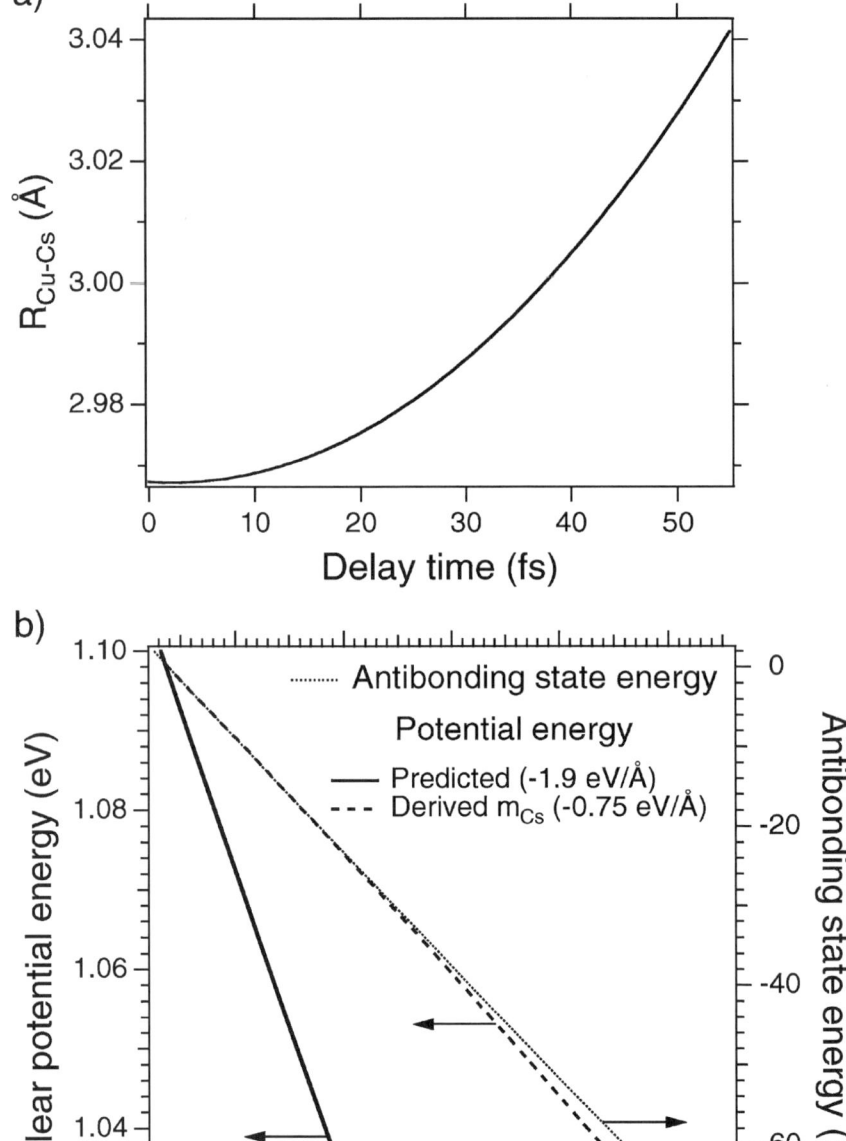

parameters such as the chemisorption energy and the choice of how to describe the repulsion in the excited state.

The large difference between the derived and predicted potentials suggests that the acceleration of Cs atoms is modest either because the actual potential is not as steep as predicted in Figure 1b or because the surface is not so rigid as implied by the HW model. If the repulsive force results in the recoil of the substrate atoms, the available energy for the ejection of Cs will be less than in the HW model. The effect of the energy transfer to the surface can be estimated from the binary collision (BC) model, which has been used successfully to reproduce the energy transfer and angular distributions of heavy hyperthermal particles such as Xe, scattering with lighter semiconductor and metal surfaces (101). In the modeling of scattering of 1–10 eV Xe atoms from the Ag(100) surface, ~100% of the kinetic energy component corresponding to the propagation along the local surface normal is transferred to the recoil of a single Ag atom. Considering the mass and the energy range for the half-collision energy, the BC model could also be appropriate for describing the photodesorption of Cs from Cu(111); however, the substantially higher occupation of the diffuse 6s state of Cs probably implies a larger effective number of surface collision partners than for Xe.

In the BC limit the Cs desorption is described as the dissociation of a Cu-Cs diatomic molecule, where the Cu and Cs particles recoil with opposite and equal momentum and an ~2:1 ratio of kinetic energies. The BC model is likely to be most valid during the instant of the excitation and progressively to fail once the internuclear distance increases and both Cu and Cs start interacting with other particles. Because the reduced mass for the motion is less than in the HW model on account of the recoil of the lighter Cu atom, the force required to achieve the observed acceleration $\frac{\partial^2 R_{Cu\text{-}Cs}(\tau)}{\partial \tau^2} = a$ is reduced by a factor of 3.09. At the instant of the excitation the motion of the Cu atom may be ballistic and it will displace the local image plane. However, upon encountering the repulsive wall of second-layer Cu atoms, it can transfer its momentum as in the recoil of balls in a Newton's cradle. [According to a low-energy electron diffraction (LEED) analysis of Cs coverlayer structure at monolayer Cs coverage, the surface is already buckled by ~0.1 Å at the atop site in the ground state (Y.K. Kim, J. Rundgren, H. Over, unpublished results).] Such impulsive transfer of momentum has been postulated to produce a soliton-like shock wave in the scattering of heavy atoms and molecules from lighter surface atoms (102). Because the lighter Cu atom recoils, accelerating the Cs atom with respect to the static image plane by the same amount as in the HW

Figure 8 (a) The wave packet position as a function of time after excitation as deduced from antibonding resonance peak shifts, such as in Figure 7, and the $E_{AR}(R_{Cu\text{-}Cs})$ curve from theory (82). (b) The $E_{AR}(R_{Cu\text{-}Cs})$ curve from theory and slopes of the excited state potential energy surfaces in the Franck-Condon region from Figure 1b, and derived from the wave packet motion based on the hard wall collision model.

model would require more work than for an HW collision. This led to the conclusion that the slope of the dissociative potential in the BC model could be 3.09 times steeper than in the HW model in Reference (27). However, the reference plane for determining $E_{AR}(R_{Cu-Cs})$ is likely to be dynamic during the initial recoil and only reach the static limit when Cs is no longer interacting with only a single Cu atom, i.e., when the BC model is no longer valid. Thus, unless the local image plane is fixed throughout the half-collision, the HW model gives the maximum acceleration of Cs atoms.

Frustrated Desorption

The coherent wave packet motion for the first 40 fs after the excitation provides a unique glimpse of photodesorption dynamics at a metal surface. Although the empirical slopes of the excited state PES derived from asymptotic chemical parameters and the time-resolved dynamical behavior in Figure 8 are far from agreement, the analysis of the wave packet motion implies a quantitative conversion of the electronic energy into the nuclear motion along the reaction coordinate, which is consistent with the gradual loss of coherence. At least for the initial 40 fs, Cs atoms appear to be desorbing from an HW surface with minimal energy dissipation. The gradual loss of coherence with a T_2 time of 15–20 fs is consistent with the excitation of phonons for a surface with a Debye energy of 35 meV.

Despite having up to ∼1.1 eV available potential energy for desorption on the excited state PES, the desorption yield of Cs atoms is too small to quantify by conventional or even by hot-wire ionization mass spectroscopic methods (27). Estimates from depletion of Cs atoms from the surface yield an upper limit for the photodesorption cross section of $\sigma_{Cs/Cu} < 2.4 * 10^{-23}$ cm^2, which is to be compared with $\sigma_{K/C} = 2 * 10^{-20}$ cm^2 for the desorption of K from graphite (45). The large difference in cross sections may reflect the substantial difference in slopes for the photodesorption, and perhaps different probabilities for the excitation of phonons in the substrate. Assuming that the HW model is appropriate for both systems, the larger repulsive slope and smaller mass of K implies ∼12 times faster acceleration from the surface. Because the estimated excited state lifetime of K is comparable to that of Cs/Cu(111), K is able to travel on the excited state PES more than 10 times further than Cs before relaxing to the ground state. This should dramatically affect the desorption yield if the electronic quenching depends exponentially on the distance from the surface.

Although Cs atoms appear to be dissociating from a rigid surface for the initial 40 fs, the 2PP difference spectra in Reference (41) suggest that on a longer timescale there could be substantial transfer of energy to the surface phonons. After the initial narrowing, for delays longer than 40 fs, the AR peak in the 2PP difference spectra acquires substantial width and perhaps even structure. The transition from initial ballistic motion into a multiple substrate-atom scattering regime can be seen in the difference spectrum in Figure 7 for 40 fs delay, which shows a narrow AR peak on top of a broad pedestal. The excitation of surface phonons is expected for a heavy particle, because it occurs with a substantial probability even in the scattering of the lightest atoms and molecules with <100 meV kinetic energies (103). Because of the electron momentum constraint on RCT, the quenching rate may even accelerate during the course of the half-collision, because of the recoil-induced local heating of the lattice. Thus, the substantial energy transfer to the surface phonons can frustrate the desorption both by decreasing the available energy for the nuclear motion and by promoting the electronic quenching of Cs atoms.

Considering the evolution of the 2PP difference spectra, the following picture emerges for the frustrated desorption of Cs from the Cu(111) surface. Photoinduced charge transfer creates strong Coulomb repulsion between the Cs atom and its nearest neighbor substrate atoms. The initial motion appears to be mainly ballistic with Cs atoms pushing off an elastic surface. The loss of coherence in the excitation process gives the timescale for dissipation, which probably occurs through the phonon generation in the substrate. The modeling of the dissociative wave packet motion suggests that the Cu-Cs bond is elongated by only $\sim$0.15 Å in 50 fs after the excitation (27, 41). Considering that the elongation corresponds to a mere 5% of the bond length, electronic quenching by the RCT and electron-electron scattering can efficiently induce the nonadiabatic transition to the ground state terminating the photodesorption dynamics.

QUANTUM CONTROL

Because the SS $\rightarrow$ AR excitation occurs by a direct optical transition and the phase relaxation of the induced coherence persists for up to 60 fs, it is possible to use the optical phase to control the nuclear wave packet motion on both the ground and excited state PES (42). The clue that quantum-mechanical interference effects can influence the photodesorption process is the narrowing of the difference spectra with the pump probe delay in Figure 7. This effect is not intrinsic to the photodesorption process; rather, it is a consequence of the specific choice of the relative phase for these two-pulse measurements.

The spectral narrowing is a manifestation of optical Ramsey fringe effect—a well-known coherent phenomenon in gas-phase spectroscopy when two phase-related optical fields interact coherently in a sample (104). Ramsey fringe–like effects are expected in interferometric TR-2PP when the coherence excited in the sample decays on a longer time scale than the delay between the identical pump and probe pulses. Characteristic spectral features related to the specific phase between the excitation fields were first observed in 2PP spectra of SS on the clean Cu(111) surface (56). Figure 9a (see color insert) schematically represents the relationship between optical excitation fields and the observed spectra. A 2PP spectrum of a homogeneously broadened sample is a convolution of the Fourier transform of the electric field of the excitation pulse (raised to the appropriate power) and the spectral lineshape of the sample (53). When the duration of the electric field pulse is much shorter than the relaxation time of the coherence excited in the sample, the spectral lineshape is dominated by that of the excitation field. The spectrum of a two-pulse sequence, as described in Figure 9a, depends on the precise phase delay between the pulses. If the spectrum of such a sequence is measured with an instrument with a longer coherence length than the delay, phase-dependent interference structure will be resolved. Specifically, when the delay is set to a $2\pi n$ integer multiple of the optical cycle, the spectra will constructively interfere at the center frequency ω_l and at side-bands (Ramsey fringes), for which the delay corresponds to $2\pi m$ ($m = n \pm 1, 2 \ldots$) cycles. Likewise, destructive interference occurs for the frequency components in which the delay corresponds to $\pi(2n \pm 1)$ cycles. Such phase-dependent spectral structure can also be resolved when the excitation field excites a sample with a dephasing time comparable to or longer than the pulse delay. In this case the coherence length for resolving Ramsey fringes is given by the dephasing time. Thus, in 2PP spectra of SS reported in Reference (56) in-phase excitation results in spectral "squeezing," giving a sharper SS peak than in single-pulse 2PP spectrum, whereas out-of-phase excitation results in spectral "dilation."

Because both the SS and AR have dephasing times comparable to or longer than the excitation pulse, analogous coherent effects are expected in the two-pulse difference spectra of Figure 7. A novel twist of the Cs/Cu(111) system is the coherent nuclear wave packet motion and the consequent time-dependent electronic structure. Because of the wave packet motion, the Ramsey fringe effect is influenced not only by the dephasing rate but also by the frequency shift (chirp) of the SS $\rightarrow$ AR transition. Thus, the difference spectra reflect the coherent nature of the excitation process in both the narrowing of the AR peak and its delay-dependent shift. The narrowing and shifting of the peak results from the constructive interference between the polarization induced by the probe and the chirped polarization induced by the pump when the delay corresponds to $2\pi n$ cycles of ω_l. The constructive interference occurs at ω_l only for zero delay, whereas at longer delays it shifts to a lower frequency corresponding to the chirp in the polarization. Because the regions of constructive and destructive interference are determined by the delay, meaningful analysis of the difference spectra can only be performed when the relative phase is precisely known.

The above interpretation of the evolution of the 2PP difference spectra with the delay demonstrates that there is an intimate relationship between the coherence induced by the excitation field and the nuclear wave packet motion that can be exploited to control the nuclear dynamics. The quantum control of the desorptive motion is demonstrated in Figure 9b, which shows two 2PP difference spectra, in which the delay is set to 30 and 30.5 cycles of the excitation frequency. With in-phase excitation, a single redshifted and narrowed peak is observed, as in Figure 7. Changing the delay by precisely +0.67 fs results in a dramatic change in the difference spectrum that can only be accounted for by the interference between the coherent polarizations. In the case of the in-phase excitation the target state in this quantum interference experiment is a displaced and spectrally narrowed excited state wave packet. Such a state can only be generated by quantum interference rather than by excitation with a single transform-limited pulse, because the sample is not resonant at the frequency of the propagated wave packet. In the case of the out-of-phase excitation, the target state is a displaced and spectrally narrowed ground state wave packet. The 2PP spectrum for the delay of 30.5 cycles shows a hole at the energy of the excited state wave packet for the in-phase excitation. The spectral hole results from the π-phase shift between the pump-induced and probe-induced polarizations, which leads to enhanced reflection and, therefore, suppressed absorption at this antiresonant frequency. With the out-of-phase excitation, the pump creates an excited state population, which evolves for the duration of the delay to a new internuclear position, whence it is transferred to the ground state through quantum interference. Rotational and vibrational wave packet manipulation through quantum interference has also been demonstrated in gas-phase diatomic molecules (105–107). The novel feature described for Cs/Cu(111) is the manipulation of atomic motion on a metal surface with sub-femtosecond and sub-Angstrom resolution.

PERSPECTIVES

This review describes several unique aspects of the electronic and nuclear dynamics of Cs atoms chemisorbed on Cu(111) that have been deduced from the interferometric time-resolved two-photon photoemission measurements. The direct measurement of the surface electronic structure during the process of photodesorption sheds new light on old problems that previously could only be addressed by less direct means. To conclude the review, this section considers what general implications can be drawn from Cs/Cu(111) on photon- and electron-stimulated dynamics at surfaces.

Whereas most studies of surface photochemistry invoke the hot electron excitation mechanism, this study provides unimpeachable evidence for the direct excitation of the Cs antibonding resonance. Time-resolved photoemission studies on adsorbate-covered surfaces, which can directly probe the excitation mechanisms of adsorbates on surfaces, have been too few to allow us to draw general conclusions. Yet, other 2PP studies consistently show evidence for the direct excitation process (29, 96). In the future, the issue of the excitation mechanism should be addressed by 2PP techniques for systems that also have well-characterized surface photochemistry.

Surface photochemistry and linewidth analysis studies provide considerable evidence that the electronic relaxation on metal surfaces occurs on <10 fs timescales. This is also supported by direct TR-2PP studies of CO/Cu(111), Cs/Cu(100), and Cs/Cu(110) (27, 28). Yet, alkali atoms on (111) surfaces of noble metals (27, 40), and in particular the Cs/Cu(111) system, demonstrate that the lifetimes can be considerably longer if the RCT decay channel is attenuated through a judicious choice of the surface and bulk electronic structure. In the case of Cs/Cu(111) the projected band gap of Cu(111) and the polarizability of the Cs atom have the combined effect of suppressing the off-normal RCT process (27, 82). This is in stark contrast to CO/Cu(111), in which the adsorbate interacts with the surface through its $2p_x$ and $2p_y$ orbitals, which do not allow efficient hybridization of the electronic cloud away from the surface and also have the appropriate symmetry to interact with the off-normal bands of the substrate (38, 108). The example of alkali atoms chemisorbed on (111) faces of noble metal surfaces provides simple considerations for seeking out other chemisorption systems that may exhibit similar properties. In particular, other molecular chemisorption systems are able to efficiently displace charge away from the surface and may achieve similar stabilization to Cs/Cu(111).

Furthermore, the example of Cs/Cu(111) reminds us that only time-domain techniques can measure electronic state lifetimes on surfaces. Although the spectroscopic linewidths that are available from photoemission and related electronic spectroscopies contain a contribution from lifetimes, it is incorrect to equate linewidths with lifetimes. Even intrinsic, homogeneous linewidths have contributions form other phase relaxation processes such as the electron-phonon interaction and defect scattering, which need to be quantified before the true lifetime can be extracted (109). The need to perform time-domain measurements to extract surface state lifetimes has been established for image-potential states (95), and Cs/Cu(111) provides another important example for the unoccupied states of chemisorbates.

Cs/Cu(111) also provides the first chance to follow the first stages of a surface photodesorption process directly on a surface. The initial wave packet evolution can be described in terms of elastic scattering of Cs from a hard wall. However, the wave packet broadening suggests that at later stages there is a substantial energy transfer to the surface phonons owing to the recoil of Cu atoms from the heavy adsorbate. Although the recoil effects are pronounced for heavy chemisorbates such as Cs, the inelastic scattering of adsorbates from surfaces during the photodesorption should be a general phenomenon that also affects the dynamics of lighter scatterers. Because the energy lost to recoil is unavailable for the propagation along the desorption coordinate, the inelastic scattering should strongly affect the desorption yield. If the recoil effect is not considered in analyses of photodesorption yields, it could lead to incorrect conclusions regarding the slopes of desorbing potentials and excited state lifetimes.

Finally, despite extremely efficient phase and energy relaxation of electronically excited states of adsorbates, the example of Cs/Cu(111) demonstrates that it is possible to control the photodesorption dynamics through the phase of the electric field of the excitation pulse. Although the wave packet dynamics can be monitored for only a modest extension of the Cu-Cs bond in the excited state, the ability to prepare a displaced wave packet in the ground state should make it possible to study the coupling of the adsorbate and substrate nuclear motions on the time scale of the Cu-Cs vibration. Furthermore, the Cs desorption process can deposit as much as 1.1 kcal of energy to the lattice. It may be possible to use this local heating of the lattice to drive other chemical processes, and therefore to use the phase of light to alter the chemical properties of a metal surface.

ACKNOWLEDGMENTS

It is a pleasure to acknowledge important contributions from past collaborators who have made this work possible, including A. Heberle, S. Matsunami, M. Moriya, W. Nessler, and M.J. Weida. Partial funding for this research has been generously provided by a NEDO Joint International Research Grant. H. Petek is grateful to A.F. Heinz, R. Coalson, and especially F.M. Zimmermann for insightful discussions of billiard ball dynamics, the Alexander von Humboldt Foundation for financial support, and J. Kirschner and M. Nyvlt for hospitality while writing this manuscript at the Max-Planck Institute for Microstructure Physics.

Visit the Annual Reviews home page at www.annualreviews.org

LITERATURE CITED

1. Polanyi JC, Zewail AH. 1995. *Acct. Chem. Res.* 2:119–32
2. Zare RN. 1998. *Science* 279:1875–79
3. Assion A, Baumert T, Bergt M, Brixner T, Kiefer B, et al. 1998. *Science* 282:919–22
4. Zewail AH. 2000. *J. Phys. Chem. A* 104:5660–94
5. Ho W. 1998. *Acc. Chem. Res.* 31:567–73
6. Stipe BC, Rezaei MA, Ho W. 1998. *Science* 279:1907–9
7. Hla S-W, Bartels L, Meyer G, Rieder K-H. 2000. *Phys. Rev. Lett.* 85:2777–80
8. Zhu X-Y. 1994. *Annu. Rev. Phys. Chem.* 45:113–44
9. Zimmermann FM, Ho W. 1995. *Surf. Sci. Rep.* 22:127–247
10. Dai H-L, Ho W, eds. 1995. *Laser Spectroscopy and Photochemistry on Metal Surfaces*, Vol. 5. Singapore: World Sci.
11. Gadzuk JW. 1995. *Surf. Sci.* 342:345–58
12. Cavanagh RR, King DS, Stephenson JC, Heinz TF. 1993. *J. Phys. Chem.* 97:786
13. Ho W. 1996. *J. Phys. Chem.* 100:13050–60
14. Menzel D, Gomer R. 1964. *J. Chem. Phys.* 41:3311–28
15. Redhead PA. 1964. *Can. J. Phys.* 42:886–905
16. Antoniewicz PR. 1980. *Phys. Rev. B* 21:3811–15
17. Haight R. 1995. *Surf. Sci. Rep.* 21:275–325
18. Petek H, Ogawa S. 1997. *Prog. Surf. Sci.* 56:239–310
19. Schmuttenmaer CA, Aeschlimann M, Elsayed-Ali HE, Miller RJD, Mantel DA, et al. 1994. *Phys. Rev. B* 50:8957–60
20. Hertel T, Knoesel E, Wolf M, Ertl G. 1996. *Phys. Rev. Lett.* 76:535–38
21. Bauer M, Pawlik S, Aeschlimann M. 1997. *Phys. Rev. B* 55:10040–43
22. Höfer U, Shumay IL, Reuβ C, Thomann U, Wallauer W, Fauster T. 1997. *Science* 277:1480–82
23. Fann WS, Storz R, Tom HWK, Bokor J. 1992. *Phys. Rev. B* 46:13592–95
24. Ogawa S, Nagano H, Petek H. 1997. *Phys. Rev. B* 55:10869–77
25. Ogawa S, Nagano H, Petek H, Heberle A. 1997. *Phys. Rev. Lett.* 78:1339–42
26. Harris CB, Ge N-H, Lingle RL Jr, McNeill JD, Wong CM. 1997. *Annu. Rev. Phys. Chem.* 48:711–44
27. Petek H, Nagano H, Weida MJ, Ogawa S. 2001. *J. Phys. Chem. B* 105:6767–79
28. Bartels L, Meyer G, Rieder K-H, Velic D, Knoesel E, et al. 1998. *Phys. Rev. Lett.* 80:2004–7
29. Wolf M, Hotzl A, Knoesel E, Velic D. 1999. *Phys. Rev. B* 59:5926–35
30. Fischer N, Schuppler S, Fauster T, Steinmann W. 1994. *Surf. Sci.* 314:89–96
31. Munakata T, Sakashita T, Tsukakoshi M, Nakamura J. 1996. *Surf. Sci.* 357/358:629–33
32. Bandara A, Kubota J, Onda K, Wada A, Kano SS, et al. 1998. *J. Phys. Chem. B* 102:5951–54
33. Bonn M, Funk S, Hess C, Denzler DN, Stampfl C, et al. 1999. *Science* 285:1042–45

34. Hess C, Wolf M, Bonn M. 2000. *Phys. Rev. Lett.* 85:4341–44
35. Haight R, Seidler PF. 1994. *Appl. Phys. Lett.* 65:517–19
36. Velic D, Knoesel E, Wolf M. 1999. *Surf. Sci.* 424:1–6
37. Bauer M, Pawlik S, Aeschlimann M. 1999. *Phys. Rev. B* 60:5016–28
38. Ogawa S, Nagano H, Petek H. 1999. *Phys. Rev. Lett.* 82:1931–34
39. Ogawa S, Nagano H, Petek H. 1999. *Surf. Sci.* 427/428:34–38
40. Petek H, Weida MJ, Nagano H, Ogawa S. 2000. *Surf. Sci.* 451:22–30
41. Petek H, Nagano H, Weida MJ, Ogawa S. 2000. *Science* 288:1402–4
42. Petek H, Weida MJ, Nagano H, Ogawa S. 2000. *J. Phys. Chem. A* 104:10234–39
43. Bradshaw AM, Bonzel HP, Ertl G. 1989. *Physics and Chemistry of Alkali Metal Adsorption.* Amsterdam: Elsevier
44. Yakshinskiy BV, Madey TE. 1999. *Nature* 400:642–44
45. Hellsing B, Chakarov DV, Österlund L, Zhadanov VP, Kasemo B. 1996. *J. Chem. Phys.* 106:982–1002
46. Wilde M, Beauport I, Stuhl F, Al-Shamery K, Freund H-J. 1999. *Phys. Rev. B* 59:13401–12
47. Engel V, Metiu H, Almeida R, Marcus RA, Zewail AH. 1988. *Chem. Phys. Lett.* 152:1–7
48. Cong P, Roberts G, Herek JL, Mohktari A, Zewail AH. 1996. *J. Phys. Chem.* 100:7832–48
49. Knoesel E, Hotzel A, Wolf M. 1998. *Phys. Rev. B* 57:12812–24
50. Petek H, Nagano H, Ogawa S. 1999. *Appl. Phys. B* 68:369–76
51. Nordlander P, Tully JC. 1990. *Phys. Rev. B* 42:5564–78
52. Petek H, Nagano H, Weida M, Ogawa S. 2000. *Chem. Phys.* 251:71–86
53. Weida MJ, Ogawa S, Nagano H, Petek H. 2000. *J. Opt. Soc. Am. B* 17:1443–51
54. Weida MJ, Ogawa S, Nagano H, Petek H. 2000. *Appl. Phys. A* 71:553–59
55. Ogawa S, Nagano H, Petek H, Heberle AP. 1997. *Phys. Rev. Lett.* 78:1339–42
56. Petek H, Heberle AP, Nessler W, Nagano H, Kubota S, et al. 1997. *Phys. Rev. Lett.* 79:4649–52
57. Hertel T, Knoesel E, Hotzel A, Wolf M, Ertl G. 1997. *J. Vac. Sci. Technol. A* 15:1503–9
58. Grimvall G. 1976. *Phys. Scr.* 14:63–78
59. Echenique PM, Pitarke JM, Chulkov EV, Rubio A. 2000. *Chem. Phys.* 251:1–35
60. Reinert F, Nicolay G, Schmidt S, Ehm D, Hüfner S. 2001. *Phys. Rev. B* 63:115415
61. Purdie D, Hengsberger M, Garnier M, Baer Y. 1998. *Surf. Sci.* 407:L671–75
62. Deleted in proof
63. Petek H, Nagano H, Ogawa S. 1999. *Phys. Rev. Lett.* 83:832–35
64. Langmuir I. 1932. *J. Am. Chem. Soc.* 54:2798–827
65. Gurney RW. 1935. *Phys. Rev.* 47:479–82
66. Muscat JP, Newns DM. 1978. *Prog. Surf. Sci.* 9:1–43
67. Lang ND, Williams AR. 1978. *Phys. Rev. B* 18:616–36
68. Ishida H. 1988. *Phys. Rev. B* 38:8006–21
69. Ishida H, Liebsch A. 1990. *Phys. Rev. B* 42:5505–15
70. Wertheim GK, Riffe DM, Citrin PH. 1994. *Phys. Rev. B* 49:4834–41
71. Muscat JP, Newns DM. 1979. *Surf. Sci.* 84:262–74
72. Diehl RD, McGrath R. 1997. *J. Phys. C* 9:951–68
73. Ishida H, Liebsch A. 1992. *Phys. Rev. B* 45:6171–87
74. Lou L, Österlund L, Hellsing B. 2000. *J. Chem. Phys.* 112:4788–96
75. Woratschek B, Sesselmann W, Küppers J, Ertl G, Haberland H. 1985. *Phys. Rev. Lett.* 55:1231–34
76. Arena DA, Curti FG, Bartynski RA. 1997. *Phys. Rev. B* 56:15404–11
77. Fischer N, Schuppler S, Fischer R,

Fauster T, Steinmann W. 1993. *Phys. Rev. B* 47:4705–13
78. Sandell A, Hjortstam O, Nilsson A, Brühwiler PA, Eriksson O, et al. 1997. *Phys. Rev. Lett.* 78:4994–97
79. Sandell A, Brühwiler PA, Nilsson A, Bennich P, Rudolf P, Mårtensson N. 1999. *Surf. Sci.* 429:309–19
80. Keller C, Stichler M, Comelli G, Esch F, Lizzit S, et al. 1998. *Phys. Rev. B* 57: 11951–54
81. Borisov AG, Kazansky AK, Gauyacq JP. 1999. *Surf. Sci.* 430:165–75
82. Borisov AG, Gauyacq JP, Kazansky AK, Chulkov EV, Silkin VM, Echenique PM. 2001. *Phys. Rev. Lett.* 86:488–91
83. Lindgren SÅ, Walldén L. 1978. *Solid State Commun.* 25:13–15
84. Wandelt K. 1997. *Appl. Surf. Sci.* 111:1–10
85. Lindgren SÅ, Walldén L, Rundgren J, Westrin P, Neve J. 1983. *Phys. Rev. B* 28:6707–12
86. Stolz H, Hüfer M, Wassmuth H-W. 1993. *Surf. Sci.* 287/288:564–67
87. Persson BNJ, Dubois LH. 1989. *Phys. Rev. B* 39:8220–35
88. Rose TS, Rosker MJ, Zewail AH. 1998. *J. Chem. Phys.* 88:6672–73
89. Matzdorf R. 1997. *Surf. Sci. Rep.* 30:153–206
90. Fauster T, Steinmann W. 1994. In *Electromagnetic Waves: Recent Developments in Research*, ed. P Halevi, pp. 1–33. Amsterdam: Elsevier
91. Kliewer J, Berndt R, Chulkov EV, Silkin VM, Echenique PM, Crampin S. 2000. *Science* 288:1399–1402
92. Fan WC, Ignatiev A. 1988. *J. Vac. Sci. Technol. A* 6:735–38
93. Senet P, Toennies JP, Witte G. 1999. *Chem. Phys. Lett.* 299:389–94
94. Lindgren SÅ, Walldén L. 1992. *Phys. Rev. B* 45:6345–47
95. Knoesel E, Hotzel A, Wolf M. 1998. *J. Electron Spectrosc. Relat. Phenom.* 88–91:577–84
96. Bauer M, Pawlik S, Burgermeister R, Aeschlimann M. 1998. *Surf. Sci.* 402–404:62–65
97. Gumhalter B, Petek H. 2000. *Surf. Sci.* 445:195–208
98. Bonn M, Hess C, Wolf M. 2002. *J. Chem. Phys.* 115:7725–35
99. Tersoff J, Hamann DR. 1985. *Phys. Rev. B* 31:805–13
100. Ramakrishna R, Willig F, May V. 2000. *Phys. Rev. B* 62:R16330–33
101. Amirav A, Cardillo MJ, Trevor PL, Lim C, Tully JC. 1987. *J. Chem. Phys.* 87: 1796–1807
102. Gerber RB, Amirav A. 1986. *J. Phys. Chem.* 90:4483–91
103. Gumhalter B, Langreth DC. 1999. *Phys. Rev. B* 60:2789–809
104. Salour MM, Cohen-Tannoudji C. 1977. *Phys. Rev. Lett.* 38:757–60
105. Scherer NF, Carlson RJ, Matro A, Du M, Ruggiero AJ, et al. 1991. *J. Chem. Phys.* 95:1487–1511
106. Uberna R, Amitay Z, Loomis RA, Leone SR. 1999. *Faraday Discuss. Chem. Soc.* 113:385–400
107. Blanchet V, Nicole C, Bouchène MA, Girard B. 1997. *Phys. Rev. Lett.* 78:2716–19
108. Gauyacq JP, Borisov AG, Raseev G, Kazansky AK. 2000. *Faraday Discuss. Chem. Soc.* 117:15–25
109. Weinelt M. 2000. *Appl. Phys. A* 71:493–502

CONNECTING LOCAL STRUCTURE TO INTERFACE FORMATION: A Molecular Scale van der Waals Theory of Nonuniform Liquids

John D. Weeks
Institute for Physical Science and Technology and Department of Chemistry and Biochemistry, University of Maryland, College Park, Maryland 20742; e-mail: jdw@ipst.umd.edu

Key Words molecular field theory, Gaussian fluctuations, wetting and drying transitions, linear response theory, hydrophobic interactions

■ **Abstract** This article reviews a new and general theory of nonuniform fluids that naturally incorporates molecular scale information into the classical van der Waals theory of slowly varying interfaces. The method optimally combines two standard approximations, molecular (mean) field theory to describe interface formation and linear response (or Gaussian fluctuation) theory to describe local structure. Accurate results have been found in many different applications in nonuniform simple fluids and these ideas may have important implications for the theory of hydrophobic interactions in water.

INTRODUCTION

This article reviews recent progress we and our coworkers have made in developing a new and general theory of nonuniform fluids (1–7), based on a reexamination of the ideas that lead to the classic van der Waals (VDW) theory (8, 9) of the liquid-vapor interface. The VDW interface theory, developed twenty years after the VDW equation of state, is equally far-reaching and it has a great many virtues that merit further consideration in the light of modern developments in statistical mechanics (9). It is physically motivated, treating separately the excluded volume effects associated with the short-ranged and harshly repulsive intermolecular forces and the averaged effects of the longer-ranged attractive interactions. Both thermodynamic and structural features are connected together naturally in an elegant and self-consistent approach.

In principle, the VDW theory can be applied to a very general problem: the structure and thermodynamics of a fluid in the presence of a general external field. However, the usual theory makes a crucial assumption that the fluid density in the presence of the field is in some sense slowly varying. Although this assumption seems appropriate for the liquid-vapor interface in zero field where the VDW theory had its original and spectacularly successful application (9), it fails badly when applied to the more general and often rapidly varying fields associated with a number of problems of current interest. By trying to understand precisely where and why the classical theory failed, we have been able to develop a new perspective that allows us to address these more general problems.

For example, the field can describe the interaction of fixed solutes with a solvent liquid. The solvation free energy is related to the changes in free energy as the interaction field is turned on from zero to full strength (10). A significant part of the solvation free energy arises from the required expulsion of solvent molecules from the region occupied by the solute. This involves very strong excluded volume interactions that can significantly perturb the density around the solutes. These considerations also play a major role in the theory of hydrophobic interactions (4, 11–29). We will later review and clarify work (4) by Lum, Chandler, and Weeks (LCW) based on this perspective. (As emphasized by Lawrence Pratt in his review of hydrophobic interactions in this volume (30), the general theory of hydrophobic interactions is incredibly complicated, and we only deal with certain important but limited aspects here.) Similar issues arise in trying to understand effective interactions between solutes arising from depletion forces (31).

An external field representing one or more fixed solvent particles leads to theories for multiparticle correlations functions describing the molecular scale structure of the bulk solvent liquid (32). Nonuniform fluids confined by walls, slits, pores, etc. can be described using appropriate external fields. Particular fields can enhance and alter features seen in the local structure of the uniform fluid. Confining fields can also produce shifts and rounding of bulk liquid-vapor and critical point phase transitions and can change the effective dimensionality of bulk fluid correlations, e.g., from three- to two-dimensional behavior in sufficiently narrow slits. In addition the field can induce new density correlations associated with interface formation, leading to wetting or drying transitions and related phenomena-like capillary condensation (33–42). Thus, in an example we will consider in some detail later, a vapor-like drying region of lower density can form near a hard wall in a Lennard-Jones (LJ) fluid over a range of thermodynamic states near coexistence. We refer to the recent review by Gelb et al. (33) for a detailed discussion of many of these possibilities.

This possible mixing of local structural components along with interfacial components in the density response to an external field leads to the great variety of behavior seen in nonuniform fluids. The classical VDW theory gives a qualitatively accurate description of the smooth interfacial components but fails for the local structure features that also must be taken into account for even a qualitatively accurate description of problems like those discussed above.

Of course, it has long been recognized that the VDW assumption of a slowly varying interface is at fault. But that realization alone does not give us much insight into how to improve things. Most modern attempts (43–60) to address these problems have used density functional theory (DFT). Here one tries to express the free energy as a (generally nonlocal) functional of the density, and the

VDW theory emerges when a local density approximation, appropriate for a slowly varying interface, is made. But what is that functional when the local density is rapidly varying and even gradient corrections are inadequate? Most workers have used a weighted density approximation in which one averages in some way over the rapidly varying local structure, and this approach has had great success in some applications. However, there is no theory of theories (61) for how to choose suitable weighting functions, and a host of different and often highly formal schemes have been proposed. These complications stand in strong contrast to the simplicity and physical appeal of the original VDW theory. We will discuss DFT further below.

We show here that there is another and very simple way to derive the VDW interface equation directly without first approximating the free energy. This allows one to think about the local density or slowly varying profile approximation in a new way where the theory itself suggests what is needed to correct it. Our work can be viewed as simply implementing these corrections to the classical VDW interface equation, and it seems appropriate to refer to it for the purposes of this review as a molecular scale van der Waals (MVDW) theory.

More precisely, we show that there are two key approximations in this interpretation of the VDW theory: (a) the introduction of an effective single particle potential or molecular field to describe the locally averaged effects of the attractive interactions in the nonuniform fluid and (b) the use of a hydrostatic approximation to determine the density response to the effective field that takes into account only the local value of the field. The latter approximation is accurate only for very slowly varying fields, where it reduces to the local density approximation. This is what limits the utility of the VDW theory in most of the applications discussed above. But from this perspective, it is easy to see how the theory can be corrected by considering nonlocal effects from the effective field. Our new MVDW theory does this using what is probably the simplest possible method, linear response theory (10, 62), which we argue (5, 6) is especially accurate when used to correct the hydrostatic approximation.

In the following, we will present this interpretation of the VDW theory first for the nonuniform LJ fluid. Then we will describe the corrections and generalizations that lead to the MVDW theory, and we review results for several applications to nonuniform LJ and hard sphere fluids and to water. Not all the simplifications that arise from this viewpoint were realized originally, so the present review can serve as a simpler and more concise guide to the theory. In particular, we have tried to clarify certain aspects of the LCW theory for hydrophobic interactions (4). A different view of the LCW theory that may go beyond the MVDW picture considered originally and discussed here is given in Reference (63).

We first consider the LJ fluid, where the physics is particularly clear. One reason why the LJ fluid is aptly called a simple liquid is that its local structure at typical liquid densities can be well approximated by an even simpler model, the hard sphere fluid. As pointed out by Widom (64, 65), in most typical configurations of the uniform LJ fluid the vector sum of the longer-ranged attractive forces on a given molecule from pairs of oppositely situated neighbors tends to cancel. This leaves only the excluded volume correlations induced by the harshly repulsive molecular cores, well described by hard spheres of an appropriate size (10, 66). The hard sphere model is thus of fundamental importance in the theory of liquids as the simplest model that can give a realistic description of the structural correlations arising from excluded volume effects.

As emphasized by VDW himself (9), this cancellation argument must fail for nonuniform fluids, and there will exist unbalanced attractive forces, whose effect on the structure can be quite significant, leading in some cases to interface formation. In general, there is a competition between these two sources of structural correlations that a proper theory of nonuniform fluids should account for (1–7). This physical picture will help us understand what is missing in the classical VDW theory and how it might be improved. The insights gained from a careful study of this simple system may suggest physically well-grounded approximations that could be applied more generally. We believe this is the case for the MVDW theory.

We start by defining the interactions in the nonuniform LJ fluid. Particles interact with a known external field $\phi(\mathbf{r})$, which we will initially assume is nonzero only in a local region, e.g., the potential arising when a LJ or hard core solute is fixed at the origin. We describe the system using a grand ensemble with fixed chemical potential μ^B, which determines ρ^B, the uniform bulk fluid density far from the perturbation where $\phi(\mathbf{r}) = 0$. The LJ pair potential $w(r) \equiv u_0(r) + u_1(r)$ is separated into rapidly and slowly varying parts associated with the intermolecular forces (67–69) so that all the harshly repulsive forces arise from u_0 and all the attractive forces from u_1. Thus, $u_0(r) = w(r) + \epsilon$ for $r \leq r_0$, where $r_0 \equiv 2^{1/6}\sigma$ is the distance to the potential minimum where the LJ force changes sign, and is zero otherwise, while $u_1(r) = -\epsilon$ for $r \leq r_0$ and equals $w(r)$ otherwise. Here σ and ϵ are the usual length and energy parameters in the LJ potential. With this separation $u_1(r)$ is relatively slowly varying and smooth, with a continuous derivative even at r_0. We will make use of these features in the theory described below.

MOLECULAR FIELD APPROXIMATION

Structure of Nonuniform Reference Fluid

We now begin our derivation of the VDW interface equation. The fundamental approximation in this interpretation of the VDW theory is the introduction of an effective single particle potential or molecular field that describes the locally averaged effects of the attractive interactions (8) in the nonuniform LJ fluid. Since the attractive interactions are relatively slowly varying, such an averaged treatment seems physically reasonable. The structure of the resulting reference or mimic system, where the attractive intermolecular interactions have been replaced by the effective single particle potential, is supposed to accurately approximate that of the original system. Thus the structure of the nonuniform LJ fluid is assumed to be given by that of the simpler nonuniform reference fluid (1–3), with only repulsive intermolecular pair interactions $u_0(r)$ (equal to the LJ repulsions) and a chemical potential μ_0^B corresponding to the same bulk density ρ^B but in a different renormalized or effective reference field (ERF) $\phi_R(\mathbf{r})$. Because u_0 is harshly repulsive, many properties of the reference fluid can be accurately approximated by those of a fluid of hard spheres with a diameter chosen by the usual blip function expansion (10, 66), as described in detail in References (5 and 6). Although this approximation is not essential, it is numerically very convenient, and for most purposes we will treat the reference system as a hard sphere fluid in the presence of the ERF.

Before we discuss the specific molecular field equation (Equation 3) usually used to determine the ERF, let us consider some general physical consequences arising from the use of any ERF to describe the effects of attractive interactions that will be important in what follows. Since the goal is to produce structure in the reference fluid approximating that of the full fluid to the extent possible, it is natural to choose $\phi_R(\mathbf{r})$ in principle so that the local (singlet) densities (70) at every point $\mathbf{r}$ in the two fluids are equal:

$$\rho_0(\mathbf{r}; [\phi_R], \mu_0^B) = \rho(\mathbf{r}; [\phi], \mu^B). \qquad 1.$$

Of course this density is not known in advance, so in practice we will make approximate choices for ϕ_R motivated by molecular field ideas. Here the subscript 0 denotes the reference fluid, the absence of a subscript denotes the LJ fluid, and the notation $[\phi]$ indicates that the correlation functions are functionals of the external field. Unless we want to emphasize this point, we will suppress this functional dependence, e.g., writing Equation 1 as $\rho_0(\mathbf{r}) = \rho(\mathbf{r})$.

But the reference fluid can also describe certain features of higher order correlation functions. We expect that when ϕ_R is chosen so that Equation 1 holds, this will produce similar local environments for the (identical) repulsive cores in the two fluids, which at high density will mainly determine higher order density correlations through excluded volume effects. Thus, when Equation 1 is satisfied, it seems plausible that pair correlations are approximately equal too (1, 2):

$$\rho^{(2)}(\mathbf{r}_1, \mathbf{r}_2) \simeq \rho_0^{(2)}(\mathbf{r}_1, \mathbf{r}_2). \qquad 2.$$

In the dense uniform LJ fluid it is well known that correlations are dominated by the repulsive forces, and this near equality lies behind the success of perturbation theories of liquids (10, 68, 69). [However, this structural assumption is rigorously true only in the artificial Kac limit where the attractive interactions are infinitely weak and long-ranged (64, 71).] The most general use of this idea asserts that all structural effects of attractive forces in the nonuniform LJ fluid can be approximately described in terms of the structure of the reference fluid in the appropriately chosen ERF.

There are many advantages in using the simpler reference system to define structure. In particular, the uniform reference fluid is well defined for all densities from dilute vapor to dense fluid and there are no conceptual problems that arise from densities in the two-phase region, as would be the case in the traditional theories requiring such densities in the original LJ fluid. (See, e.g., "Hydrostatic

Approximation and the Classical VDW Theory" below.) Moreover, as we will discuss in detail later, there exist simple and accurate theories for the reference fluid structure based on linear response theory.

However, there are inherent limitations in such an approximation, most notably in describing long wavelength correlations such as those found near the critical point. Some other shortcomings of the molecular field approximation when applied to realistic fluid models are easily understood. For example, the effects of long wavelength capillary wave fluctuations (72) of the free liquid-vapor interface clearly cannot be described using such reference system correlation functions however ϕ_R is chosen.

Simple Molecular Field Equation

We now turn to the choice of ϕ_R. In our interpretation, the classical VDW theory uses the simple molecular field (MF) equation (Equation 3) for the ERF. This is just a transcription of the usual molecular field equation for the Ising model to a continuum fluid with attractive interactions $u_1(r)$ and can be arrived at in a number of different ways (8, 44). Particularly relevant for our purposes here is the derivation discussed in detail in References (1–3 and 7) that starts from the balance of forces in the reference and LJ fluids as described by the exact Yvon-Born-Green (YBG) hierarchy (10) and uses Equations 1 and 2 to arrive at the MF equation by an approximate integration. This can be looked on as a modern version of the closely related calculation VDW originally carried out (9). The final result is well known:

$$\phi_R(\mathbf{r}_1) = \phi(\mathbf{r}_1) + \int d\mathbf{r}_2\, \rho_0(\mathbf{r}_2; [\phi_R], \mu_0^B) u_1(r_{12}) + 2a\rho^B. \qquad 3.$$

Here

$$a \equiv -\frac{1}{2} \int d\mathbf{r}_2 u_1(r_{12}) \qquad 4.$$

corresponds to the attractive interaction parameter a in the uniform fluid VDW equation, as discussed below. The last term in Equation 3 represents a constant of integration in the derivation in References (1–3 and 7) and is chosen so that ϕ_R vanishes far from a localized perturbation where the density becomes equal to ρ^B.

Because of the integration over the slowly varying attractive component $u_1(r_{12})$ of the intermolecular potential, the second term on the right in Equation 3 is smooth and relatively slowly varying even when ρ_0 itself has discontinuities and oscillations, as could arise from a ϕ with a hard core. Thus, the ERF ϕ_R is quite generally comprised of the original field ϕ and an additional slowly varying term

$$\phi_s(\mathbf{r}_1) \equiv \int d\mathbf{r}_2\, \rho_0(\mathbf{r}_2) u_1(r_{12}) + 2a\rho^B \qquad 5.$$

that takes account of spatial variations of the attractive interactions, i.e., the unbalanced attractive forces in the nonuniform LJ fluid.

Some thermodynamic implications of the MF approximation can be seen when we consider Equation 3 in the limit of a constant field. In the grand ensemble, this is equivalent to a shift of the chemical potential, producing a shift in the uniform density, as discussed in more detail in the next section. Thus, we can use Equation 3 to relate chemical potentials in the reference and LJ systems. Let $\mu(\rho)$ and $\mu_0(\rho)$ denote the chemical potential as a function of density in the uniform LJ system and reference system, respectively. (These also depend on the temperature, but we are usually interested in density variations along particular isotherms, so we will not indicate the temperature dependence explicitly.) Then $\phi^\rho = \mu^B - \mu(\rho)$ is the exact value of that constant field such that the density in the LJ fluid changes from ρ^B to ρ, and $\phi_R^\rho = \mu_0^B - \mu_0(\rho)$ is the analogous field in the reference fluid producing the same density change. Using Equation 1, Equation 3 gives the MF approximation relating ϕ^ρ and ϕ_R^ρ, or equivalently, the MF approximation relating $\mu(\rho)$ and $\mu_0(\rho)$. This can be written in the familiar VDW form (64):

$$\mu(\rho) = \mu_0(\rho) - 2a\rho. \qquad 6.$$

In the limit of a uniform system, Equation 3 describes all effects of attractive interactions in terms of the constant parameter a. Indeed, the theory then reduces to the generalized uniform fluid VDW theory of Longuet-Higgins & Widom (64, 65), where one combines an accurate description of the uniform (hard sphere) reference system with the simple treatment of the attractive interactions in terms of a constant background potential a. In the MF approximation, $\mu(\rho)$ is defined by the right side of Equation 6 and has meaning even for values of ρ in the two-phase region.

In general, to determine ϕ_R a self-consistent solution of Equation 3 is required, since ϕ_R appears explicitly on the left side and implicitly on the right side through $\rho_0(\mathbf{r}; [\phi_R], \mu_0^B)$. In principle, since a unique density $\rho_0(\mathbf{r}; [\phi_R], \mu_0^B)$ is associated with a given external field ϕ_R through the partition function, Equation 3 is self-contained and hence self-consistent values for both ϕ_R and ρ_0 can be found, by iteration, for example. Such solutions were found (1, 2), using computer simulations to accurately determine the associated density $\rho_0(\mathbf{r}; [\phi_R], \mu_0^B)$ for a variety of external fields.

In practice, one must make additional approximations beyond the MF assumption to obtain $\rho_0(\mathbf{r}; [\phi_R], \mu_0^B)$ in an accurate and computationally practical way. It is here that the main limitation of the classical VDW theory arises.

In our derivation the classical VDW theory results when a second approximation, appropriate for a slowly varying density field (8), is used to determine the density $\rho_0(\mathbf{r}; [\phi_R], \mu_0^B)$ induced by ϕ_R. This approximation takes account only of the local value of the field through a shifted chemical potential and can be used for a general $\phi_R(\mathbf{r})$. When $\phi_R(\mathbf{r})$ is very slowly varying it is exact. But in more general cases it gives inaccurate results, spoiling the predictions of the VDW theory. To see how this comes about, we first define this local hydrostatic density response for a general field, and then show how its use transforms Equation 3 into the VDW interface equation as it is usually presented.

HYDROSTATIC DENSITY

Local Response to General Field

Consider first a given general external field, ϕ_R. Because the chemical potential acts like a constant field in the grand partition function, the associated density $\rho_0(\mathbf{r}; [\phi_R], \mu_0^B) \equiv \rho_0(\mathbf{r})$ is a functional of ϕ_R and a function of μ_0^B and depends only on the difference between these quantities. Thus for any fixed position $\mathbf{r}_1$, we can define a shifted chemical potential $\mu_0^{\mathbf{r}_1} \equiv \mu_0^B - \phi_R(\mathbf{r}_1)$ and a shifted field $\phi_R^{\mathbf{r}_1}(\mathbf{r}) \equiv \phi_R(\mathbf{r}) - \phi_R(\mathbf{r}_1)$, whose parametric dependence on $\mathbf{r}_1$ is denoted by a superscript, and we have for all $\mathbf{r}$ the exact relation $\rho_0(\mathbf{r}; [\phi_R], \mu_0^B) = \rho_0(\mathbf{r}; [\phi_R^{\mathbf{r}_1}], \mu_0^{\mathbf{r}_1})$.

By construction, the shifted field $\phi_R^{\mathbf{r}_1}(\mathbf{r})$ vanishes at $\mathbf{r} = \mathbf{r}_1$. If ϕ_R is very slowly varying, then it remains very small for $\mathbf{r}$ near $\mathbf{r}_1$. In that case, to determine the density at $\mathbf{r}_1$ we can approximate $\rho_0(\mathbf{r}_1; [\phi_R^{\mathbf{r}_1}], \mu_0^{\mathbf{r}_1})$ by $\rho_0(\mathbf{r}_1; [0], \mu_0^{\mathbf{r}_1}) \equiv \rho_0^{\mathbf{r}_1}$, the density of the uniform fluid (in zero field) at the shifted chemical potential $\mu_0^{\mathbf{r}_1}$. This defines $\rho_0^{\mathbf{r}_1}$, the hydrostatic density response to the field ϕ_R. Note that $\rho_0^{\mathbf{r}_1}$ depends only on the local value of the field ϕ_R at $\mathbf{r}_1$ through its dependence on $\mu_0^{\mathbf{r}_1}$. When the field varies slowly enough, the hydrostatic approximation, where $\rho_0(\mathbf{r}_1)$ at each $\mathbf{r}_1$ is replaced by the corresponding uniform fluid density $\rho_0^{\mathbf{r}_1}$, is very accurate (5).

One can equivalently define the hydrostatic density $\rho_0^{\mathbf{r}_1}$ using $\mu_0(\rho)$, the chemical potential of the uniform reference fluid as a function of density; it satisfies

$$\mu_0(\rho_0^{\mathbf{r}_1}) = \mu_0^B - \phi_R(\mathbf{r}_1). \qquad 7.$$

This equation plays a central role in all that follows. The hydrostatic density is easy to determine for a general ϕ_R. In particular, if ϕ_R has a hard core with $\phi_R = \infty$ in a certain region of space, then the hydrostatic density $\rho_0^{\mathbf{r}_1}$ correctly vanishes in that same region, corresponding to the vanishing density of the uniform fluid at the chemical potential $\mu_0^{\mathbf{r}_1} = -\infty$. However, because of the strictly local response to the field, any nonlocal excluded volume correlations induced by the hard core potential outside the hard core region are not properly described by the hydrostatic approximation. Still, it remains well defined even in this limit, whereas approximations based on using properties of a uniform fluid evaluated at the local density (which can easily exceed close packing) break down completely (43, 44).

Local Response to ERF

To obtain the VDW and MVDW theories we now determine the local hydrostatic response to the ERF in Equation 3. This arises after Equation 3 for $\phi_R(\mathbf{r}_1)$ is substituted into Equation 7:

$$\mu_0(\rho_0^{\mathbf{r}_1}) = \mu_0^B - \phi(\mathbf{r}_1) - \int d\mathbf{r}_2\, \rho_0(\mathbf{r}_2)\, u_1(r_{12}) - 2a\rho^B. \qquad 8.$$

As written, Equation 8 just defines the hydrostatic density $\rho_0^{\mathbf{r}_1}$ in terms of the local value of the ERF at $\mathbf{r}_1$, which itself involves an integral over the full density

$\rho_0(\mathbf{r}_2; [\phi_R], \mu_0^B)$ at all other points $\mathbf{r}_2$. To explicitly determine $\rho_0^{\mathbf{r}_1}$ we need to specify $\Delta\rho_0(\mathbf{r}_2) \equiv \rho_0(\mathbf{r}_2) - \rho_0^{\mathbf{r}_2}$, as is clear when Equation 8 is exactly rewritten in the following form:

$$\mu_0(\rho_0^{\mathbf{r}_1}) - 2a\rho_0^{\mathbf{r}_1} = \mu_0^B - 2a\rho^B - \phi(\mathbf{r}_1) - \int d\mathbf{r}_2 [\rho_0^{\mathbf{r}_2} - \rho_0^{\mathbf{r}_1}] u_1(r_{12})$$
$$- \int d\mathbf{r}_2 \Delta\rho_0(\mathbf{r}_2) u_1(r_{12}). \qquad 9.$$

As we now show, different approximations for $\Delta\rho_0(\mathbf{r}_2)$ immediately give the classical VDW and MVDW interface equations.

HYDROSTATIC APPROXIMATION AND THE CLASSICAL VDW THEORY

The classical VDW interface equation arises when one assumes that the full density response to the ERF, $\rho_0(\mathbf{r}_2)$ for all $\mathbf{r}_2$, is accurately approximated by $\rho_0^{\mathbf{r}_2}$, the local hydrostatic response. This hydrostatic approximation is consistent when the ERF and the induced density are varying slowly enough. Thus, in this derivation of the classical VDW theory the last term in Equation 9 is ignored and Equation 8 or 9 is approximated by an integral equation involving only the hydrostatic density:

$$\mu(\rho_0^{\mathbf{r}_1}) = \mu^B - \phi(\mathbf{r}_1) - \int d\mathbf{r}_2 [\rho_0^{\mathbf{r}_2} - \rho_0^{\mathbf{r}_1}] u_1(r_{12}), \qquad 10.$$

along with the assumption that $\rho_0(\mathbf{r}_2) = \rho_0^{\mathbf{r}_2}$ for all $\mathbf{r}_2$. Here we have also used Equation 6 to replace $\mu_0(\rho)$ by $\mu(\rho)$, with the understanding that the latter is really defined by the right side of Equation 6. Consistent with the assumption of a slowly varying profile and the fact that u_1 is reasonably short-ranged, $\rho_0^{\mathbf{r}_2}$ in the last term on the right is often expanded to second order in a Taylor series about $\rho_0^{\mathbf{r}_1}$, yielding, in the simple case of a liquid-vapor interface with planar symmetry and $\phi = 0$, a differential equation for the interface profile ρ_0^z:

$$\mu(\rho_0^z) = \mu^B + md^2\rho_0^z/dz^2 \qquad 11.$$

where

$$m \equiv -\frac{1}{6}\int d\mathbf{r}\, r^2 u_1(r). \qquad 12.$$

Whereas there is a strictly local response to the value of the ERF at $\mathbf{r}_1$ in Equation 10 yielding $\rho_0^{\mathbf{r}_1}$, the ERF itself at $\mathbf{r}_1$ depends nonlocally on $\rho_0^{\mathbf{r}_2}$ through the integration over the attractive interactions. In the classical VDW theory, this provides the only source of nonlocality.

Equations 10 and 11 are completely equivalent to the VDW theory for the density profile as it is usually presented (8). In our derivation the theory describes hydrostatic densities in the reference system, and $\mu(\rho)$ is also defined in terms of reference system quantities given on the right side of Equation 6. This provides a simple and consistent interpretation that avoids all problems associated with densities in the two-phase region of the LJ fluid.

However, in view of Equation 1, one can replace $\rho_0^{r_1}$ by ρ^{r_1} in Equations 10 and 11 and formally eliminate all explicit mention of the reference system. This is the way the standard theory is usually presented. This formal rewriting of Equations 10 and 11 seems to require only properties of the original LJ system and could be useful when applying the theory to other fluids where the appropriate reference system is not so apparent. Work for water (4) along these lines will be reviewed later. However, this obscures some of the physical underpinnings of the theory in terms of the basic MF approximation, and it is not clear from the form of these equations how they should be corrected in cases where the density profile is more rapidly varying.

Of course, one might hope that there could exist a more general formulation of the theory where one does not need the MF approximation at all. Despite some formal results from density functional theory, which we will discuss later, we believe in practice this is not likely to be the case. Long-standing conceptual problems (8) arise in interpreting what is meant by $\mu(\rho^{r_1})$ and ρ^{r_1} itself in Equations 10 and 11 for density values in the two-phase region of the uniform LJ fluid, unless they are defined by using a MF approximation, either explicitly, as in the right-hand side of Equation 6, or implicitly, by assuming some kind of analytic continuation of values from the stable phases. See "Correcting the Molecular Field Approximation" below for further discussion.

Thus, to improve the classical VDW theory, we return to Equations 8 and 9. Here the essential physics involving the interplay between the MF approximation for the ERF and the hydrostatic density response is clear. If we can understand in detail how to improve the VDW theory for the LJ system, we may gain insights that could apply more generally.

CORRECTING THE HYDROSTATIC APPROXIMATION

Optimized Linear Response and the HLR Equation

As emphasized in "Local Response to ERF" above, in our derivation of the VDW theory one first determines the local density response $\rho_0^{r_1}$ to the ERF $\phi_R(\mathbf{r}_1)$ in Equations 8 and 9. This immediately reduces to the classical VDW interface equation if one makes the further approximation of ignoring the density difference $\Delta\rho_0(\mathbf{r}_2) \equiv \rho_0(\mathbf{r}_2) - \rho_0^{r_2}$, produced by definition from nonzero values of the shifted field $\phi_R^{r_2}(\mathbf{r})$ away from $\mathbf{r}_2$. Thus, to correct the VDW theory, we should take the nonlocal response to the shifted ERF into account.

But one can calculate $\Delta\rho_0$ in a very simple way by using linear response theory, which exactly relates small changes in the density and field. This approach is clearly correct when $\Delta\rho_0$ and $\phi_R^{r_2}$ are uniformly small, and in this sense is analogous to a gradient correction to the local density approximation in DFT. However, in contrast to the gradient correction, we will see that there are good physical reasons to believe that our linear response treatment could remain accurate even for large perturbations (3, 5, 6). This will allow us to develop a new and generally accurate theory for the nonuniform reference fluid in an arbitrary external field and also will help us correct one of the major shortcomings of the classical VDW theory for fluids with attractive interactions.

We start from the general linear response equation (Equation 10) for a nonuniform system in a general field ϕ_R, with chemical potential μ_0^B, inverse temperature $\beta = (k_B T)^{-1}$, and associated density $\rho_0(\mathbf{r})$:

$$-\beta \delta \phi_R(\mathbf{r}_1) = \int d\mathbf{r}_2 \, \chi_0^{-1}(\mathbf{r}_1, \mathbf{r}_2; [\rho_0]) \, \delta \rho_0(\mathbf{r}_2), \qquad 13.$$

which relates small perturbations in the density and potential through the (inverse) linear response function

$$\chi_0^{-1}(\mathbf{r}_1, \mathbf{r}_2; [\rho_0]) \equiv \delta(\mathbf{r}_1 - \mathbf{r}_2)/\rho_0(\mathbf{r}_1) - c_0(\mathbf{r}_1, \mathbf{r}_2; [\rho_0]). \qquad 14.$$

Here, c_0 is the direct correlation function of the system with density $\rho_0(\mathbf{r})$. In most cases, we will consider perturbations about a uniform system, so c_0 will take the simple form $c_0(r_{12}; \rho)$. Because we want to focus on effects of the perturbing field, we have used the inverse form of linear response theory, where the field appears explicitly only on the left-hand side of Equation 13, evaluated at $\mathbf{r}_1$.

Could Equation 13 also be used to determine the finite density response $\Delta \rho_0$ to a large field perturbation $\Delta \phi_R$? Such a linear relation between a (possibly infinite) external field perturbation on the left-hand side and the finite induced density change on the right must certainly fail for values of $\mathbf{r}_1$ where the field is very large. Conversely, Equation 13 should be most accurate for those values of $\mathbf{r}_1$ where the field is small—in particular where the field vanishes—and then through the integration over all $\mathbf{r}_2$, it relates density changes in the region where the field vanishes to density changes in the regions where the field is nonzero (3, 5).

This optimal condition for the validity of linear response theory holds true automatically when we use Equation 13 to determine the change at each $\mathbf{r}_1$ from the uniform fluid with density $\rho_0^{\mathbf{r}_1}$ induced by the shifted field. Thus, we set $\chi_0^{-1} = \chi_0^{-1}(r_{12}; \rho_0^{\mathbf{r}_1})$ in Equation 13 and take $\delta \phi_R = \phi_R^{\mathbf{r}_1}$ and $\delta \rho_0(\mathbf{r}_2) = \rho_0(\mathbf{r}_2) - \rho_0^{\mathbf{r}_1}$. Because $\phi_0^{\mathbf{r}_1}(\mathbf{r})$ is zero at $\mathbf{r}_1$ by construction, the left side of Equation 13 vanishes, and we have

$$0 = \int d\mathbf{r}_2 \, \chi_0^{-1}(r_{12}; \rho_0^{\mathbf{r}_1}) \left[\rho_0(\mathbf{r}_2) - \rho_0^{\mathbf{r}_1} \right], \qquad 15.$$

which can be rewritten using Equation 14 as

$$\rho_0(\mathbf{r}_1) = \rho_0^{\mathbf{r}_1} + \rho_0^{\mathbf{r}_1} \int d\mathbf{r}_2 \, c_0(r_{12}; \rho_0^{\mathbf{r}_1}) \left[\rho_0(\mathbf{r}_2) - \rho_0^{\mathbf{r}_1} \right]. \qquad 16.$$

This is our final result, which we refer to as the hydrostatic linear response (HLR) equation. A more formal derivation involving a coupling parameter integration and yielding another related equation for $\rho_0(\mathbf{r}_1)$ is given elsewhere (5).

Note that the field appears only implicitly through its local effect on $\rho_0^{\mathbf{r}_1}$. The HLR equation is useful by itself as a way to determine the density change induced by a known external field $\phi_R(\mathbf{r})$. It builds on properties of the uniform reference system, requiring in particular $c_0(r_{12}; \rho)$ and $\mu_0(\rho)$. Quantitatively accurate and computationally convenient approximations for these functions are known from the GMSA theory of Waisman (73, 74); in many cases the simpler Percus-Yevick (PY) approximation (10, 62), discussed below, gives sufficient accuracy. Given these functions and a known external field $\phi_R(\mathbf{r})$, Equation 16 is a linear integral equation that can be solved self-consistently to determine the induced density $\rho_0(\mathbf{r}_1)$ for all $\mathbf{r}_1$. We will first discuss some general properties of the HLR equation and examine its accuracy in several test cases where the external field $\phi_R(\mathbf{r}_1)$ is known. Then we will discuss its use in correcting the classical VDW theory where $\phi_R(\mathbf{r}_1)$ is determined self-consistently.

Properties of the HLR Equation for Fixed ϕ_R

Equation 16 relates the density $\rho_0(\mathbf{r}_1)$ at each $\mathbf{r}_1$ to an integral involving the density, $\rho_0(\mathbf{r}_2)$, at all other points and a locally optimal uniform fluid kernel, $c_0(r_{12}; \rho_0^{\mathbf{r}_1})$, that depends implicitly on $\mathbf{r}_1$ through $\rho_0^{\mathbf{r}_1}$. This $\mathbf{r}_1$ dependence is the most important new feature of the HLR equation and it represents the main reason for improved results over conventional methods, which generally use only $c_0(r_{12}; \rho_0^B)$ along with various nonlinear closures that try to directly relate the field and the density in regions where the field is nonzero (10). See, e.g., Equation 17 below.

HARD SPHERE SOLUTE AND THE PY EQUATION   Consider, in particular, the important test case where ϕ_R represents the potential of a hard sphere solute fixed at the origin. The local hydrostatic response $\rho_0^{\mathbf{r}_1}$ is simply a step function in this case, equal to zero for $\mathbf{r}_1$ inside the core region and to ρ_0^B outside. In reality, at high density there are very large nonlocal oscillatory excluded volume correlations in the true $\rho_0(\mathbf{r}_1)$. As we will see, the HLR equation gives results in this special case equivalent to the generally accurate PY approximation. In addition, we can use results from recent computer simulations (75) and related work on the Gaussian field model (76) to provide us with further insights into why this simple linear response treatment can remain surprisingly accurate even for hard cores.

We can connect the HLR equation with conventional integral equation theory leading to the PY equation by noting that the usual solute-solvent direct correlation function $C_S(\mathbf{r}_1)$ can be defined by

$$C_S(\mathbf{r}_1) = \int d r_2 \, \chi_0^{-1}(r_{12}; \rho_0^B)[\rho_0(\mathbf{r}_2) - \rho_0^B], \qquad 17.$$

where exact values for all functions are used on the right side. Thus, $-C_S(\mathbf{r}_1)/\beta$ is that function that must replace the perturbing solute-solvent potential to give exact results when the full induced density change from ρ_0^B is used in the linear response equation (Equation 13). [Using Equation 14, we see that Equation 17 is equivalent to the standard solute-solvent Ornstein-Zernike equation, which is the usual way (10) of defining $C_S(\mathbf{r}_1)$.]

Similarly, in the derivation of the HLR equation, we can replace the zero on the left side of Equation 15 by a new function $\widetilde{C}_S(\mathbf{r}_1)$ that exactly satisfies:

$$\widetilde{C}_S(\mathbf{r}_1) = \int d\mathbf{r}_2 \, \chi_0^{-1}(r_{12}; \rho_0^{\mathbf{r}_1})[\rho_0(\mathbf{r}_2) - \rho_0^{\mathbf{r}_1}]. \qquad 18.$$

Note that the local value of the perturbing field $\phi_R^{\mathbf{r}_1}$ associated with $\widetilde{C}_S(\mathbf{r}_1)$ always vanishes at $\mathbf{r}_1$, unlike the case for $C_S(\mathbf{r}_1)$ above. The HLR equation (Equation 16) follows from Equation 18 by setting $\rho_0^{\mathbf{r}_1}\widetilde{C}_S(\mathbf{r}_1)$ equal to zero everywhere, and to the extent that the HLR equation is accurate, we expect $\rho_0^{\mathbf{r}_1}\widetilde{C}_S(\mathbf{r}_1)$ to be generally very small. In particular, the HLR equation automatically satisfies the core condition $\rho_0 = 0$ inside the hard core region where both $\rho_0^{\mathbf{r}_1}$ and $\rho_0(\mathbf{r}_1)$ vanish. Corrections to the HLR equation arise from nonzero values of $\rho_0^{\mathbf{r}_1}\widetilde{C}_S(\mathbf{r}_1)$ outside the core.

If we compare Equation 17, with the core condition imposed, and Equation 18 for $\mathbf{r}_1$ outside the core region where $\rho_0^{\mathbf{r}_1} = \rho_0^B$, we see they are identical. Thus $\widetilde{C}_S(\mathbf{r}_1)$ outside the core exactly equals $C_S(\mathbf{r}_1)$, the tail of the hard sphere solute-solvent direct correlation. This is generally believed to be small and in the very accurate GMSA equation it is approximated by a rapidly decaying Yukawa function (73, 74). In the HLR approximation this tail is set equal to zero. Thus, the HLR equation predicts that the hard sphere solute-solvent direct correlation vanishes identically outside the core. As is well known, this is equivalent to the PY closure for hard core systems (10). A GMSA treatment could be used in Equation 18 to correct the HLR results if more accuracy is needed. A self-consistent application of these ideas when the solute is the same size as the solvent, requiring that $C_S = c_0 = 0$ outside the core, yields the standard PY equation for hard spheres as interpreted by Stell (77), with $c_0(r)$ equal to zero outside the core and $\rho_0(r) = \rho_0^B g_0(r)$ equal to zero inside. Similarly, the HLR equation reduces to the PY wall-particle equation when the radius of the solute tends to infinity and the PY c_0 is used.

COMPUTER SIMULATIONS AND GAUSSIAN FLUCTUATIONS   Further insight into the surprising accuracy of linear response theory for uniform hard sphere fluids comes from recent computer simulations by Crooks & Chandler (75), following related work on water (17). They have shown that even large spontaneous density fluctuations in a uniform hard sphere fluid can be accurately described using the same Gaussian probability distribution that describes small fluctuations. Small density changes induced by a small perturbing field are described by the linear response equation (Equation 13). By the fluctuation-dissipation theorem, the same linear response function $\chi_0^{-1}(r_{12}; \rho)$ controls the spontaneous Gaussian density fluctuations in the uniform fluid (10, 76). In particular, they considered the probability distribution for finding N hard spheres in a spherical volume of the fluid. They found that even cavity or void formation with $N = 0$ was reasonably well described by the Gaussian theory. Because a cavity influences the rest of the fluid in exactly the same way as a hard sphere solute of the same size, the density response of the hard sphere fluid to such a solute (or imposed region of zero density) should indeed be well described using linear response theory with the uniform fluid response function, as in Equation 15. Thus, a key feature exploited in Equation 16 is that density fluctuations in the reference fluid are to a remarkable extent Gaussian. The simulation results pertain to fluctuations in a uniform fluid. Similarly, to describe effects of a general external field, the HLR equation considers perturbations at each r_1 to the uniform hydrostatic fluid density $\rho_0^{r_1}$ induced by the shifted field $\phi_R^{r_1}$, which vanishes at least locally at r_1.

OTHER RAPIDLY VARYING MODEL POTENTIALS Thus, we see that the HLR equation (Equation 16) satisfies the following limits: (*a*) It is exact when $\phi_R(\mathbf{r})$ is very slowly varying, and it exactly describes to linear order small corrections to the hydrostatic approximation; (*b*) It is exact for any $\phi(\mathbf{r})$ at low enough density, where there is a local relation between the potential and induced density; (*c*) For a field $\phi(\mathbf{r})$ from a general hard core solute, Equation 16 reduces to the PY approximation, as discussed above. To give more indications of the accuracy of Equation 16, we review solutions (5) for some model potentials designed to show both the strengths and the weaknesses of the present methods and compare with computer simulations.

Figure 1 shows the correlation functions $g_0(r) \equiv \rho_0(r)/\rho_0^B$ for a hard sphere system at a moderately high bulk density $\rho_0^B = 0.49$ induced by the deep attractive spherical parabolic model potentials shown in the inset. (Reduced units, with distances measured in units of the hard sphere diameter are used.) The HLR equation reproduces the increased density inside the well, and the nonlocal oscillatory excluded volume correlations, which show a local density minimum at the center of the well due to packing effects.

Since the external field enters Equation 16 only locally through its effect on $\rho_0^{r_1}$, it is also easy to use it for the inverse problem of determining the field associated with a given density profile. As an example, the crosses in the inset give the potentials predicted by Equation 16 given the simulation data for *g(r)*.

Although these results are qualitatively very satisfactory for the most part, there are some problems. Note, in particular, the slightly negative density at the center of the well; a positive density is not guaranteed in this linear theory. (The related HM equation derived in Reference (5) always yields a nonnegative density and does slightly better here, but it performs less well in the hard wall limit.) One would expect methods based on expanding about the uniform hydrostatic fluid to be least accurate for potentials with very steep gradients. Thus, poor results at high density were found for a repulsive planar triangular barrier potential that rose from 0 to 10 k_BT over a distance of one hard sphere diameter. Results actually improve as the barrier height increases and the potential approaches a hard wall potential. The theory does better for hard cores because there is no region of space where there is a large gradient in the external field while, at the same time, the local density is nonzero. A GMSA-like treatment based on Equation 18 may improve matters here. But large repulsive potentials with very steep gradients are better treated by blip function methods (66) or other expansions about a hard core system.

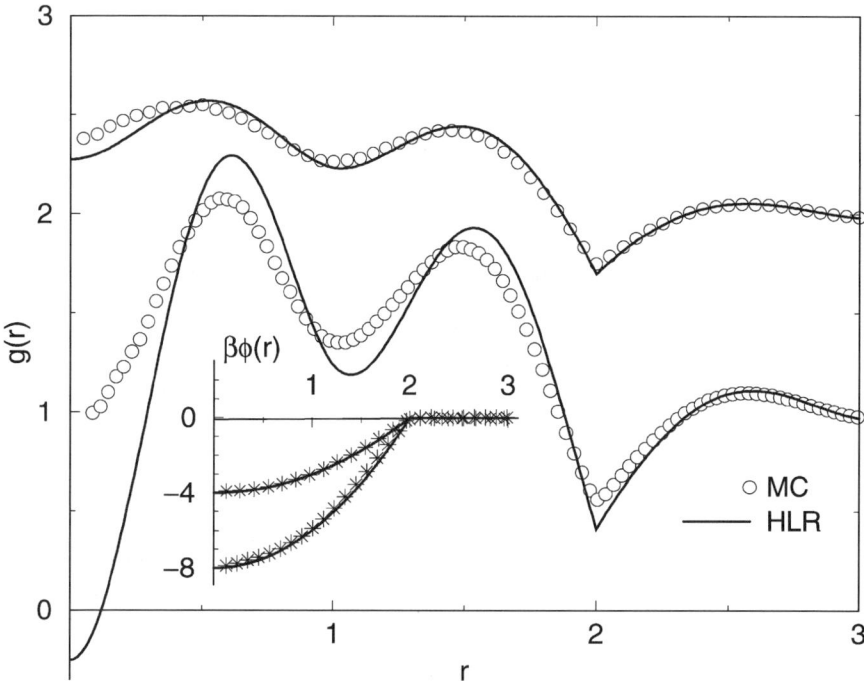

Figure 1 Correlation functions for hard spheres in the presence of spherical parabolic potentials shown in the inset (solid lines) as given by theory and simulation. The upper curve corresponds to the smaller potential and has been displaced upward by one unit. Also shown in the inset (crosses) are the potentials predicted by Equation 16 given the simulation data.

CORRECTING THE MOLECULAR FIELD APPROXIMATION

We now return to fluids with attractive interactions, described using a self-consistently determined ERF. Thinking about the hydrostatic limit also can help us correct some quantitatively inadequate features of the simple MF approximation (Equation 3) for the ERF (7). In the limit of a uniform system, this describes all effects of attractive interactions in terms of the constant VDW parameter a. Although this very simple approximation captures much essential physics and gives a qualitative description of the uniform fluid thermodynamic properties, it certainly is not quantitatively accurate. For example, when used to describe a slowly varying liquid-vapor interface, it will predict shifted MF values for the densities of the coexisting bulk liquid and vapor phases. The main problem with the theory is not so much its description of the local density gradients, but its predictions for the thermodynamic properties of the coexisting bulk phases (60).

To achieve quantitative agreement with bulk thermodynamic properties, one can replace the constant a by a function α that depends (weakly, one hopes, to the extent the classical VDW theory is reasonably accurate) on temperature and density. This has been suggested many times in the literature, usually in the context of perturbation theories for uniform fluids (10), where approximate expressions giving such functions have been derived (78). We review here a simpler and more empirical approach that can be used to determine α if an accurate analytic equation of state for the bulk fluid is known. This approach results in a simple modification of the MF expression for the ERF that insures that exact thermodynamic results (consistent with the given uniform fluid equation of state) are found in the hydrostatic limit of a very slowly varying ERF.

To that end, let us consider again the MF equation in the limit of a constant field. Instead of using the MF approximation for $\mu(\rho)$ as in Equation 6, we assume that $\mu(\rho)$ is known from an accurate bulk equation of state. In particular, we determine $\mu(\rho)$ from the 33-parameter equation of state for the LJ fluid given by Johnson et al. (79). This provides a very good global description of the stable liquid and vapor phases in the LJ fluid and provides a smooth interpolation in between by using analytic fitting functions. Thus, it naturally produces a modified VDW loop in the two-phase region, and it seems quite appropriate for our use here in improving the simplest MF description of the uniform fluid.

Using known properties of the hard sphere fluid, we also have essentially exact expressions for $\mu_0(\rho)$. Using these we can relate $\mu(\rho)$ and $\mu_0(\rho)$ in analogy to Equation 6:

$$\mu(\rho) = \mu_0(\rho) - 2\alpha(\rho)\rho \qquad 19.$$

but with the constant a replaced by a function $\alpha(\rho)$ of density and temperature chosen so that Equation 19 holds. Because even the simplest MF approximation is qualitatively accurate, we expect that the ratio $\alpha(\rho)/a$ will be of order unity and rather weakly dependent on density and temperature.

We indeed found that the constant a was a good overall compromise (7). However, the true $\alpha(\rho)$ fit to the equation of state exhibits variations of up to about 15% as a function of density and temperature, illustrating the need for an accurate equation of state for quantitative accuracy.

Because of the strictly local response, these results for a constant field can be used to determine exact results in the hydrostatic limit of very slowly varying fields. We want to modify Equation 3 so that in the hydrostatic limit it will reproduce these exact values, while still giving reasonable results for more rapidly varying fields.

There is no unique way to do this, but the following simple prescription seems very natural and gives our final result, which we will call the modified molecular field (MMF) approximation for the ERF:

$$\phi_R(\mathbf{r}_1) - \phi(\mathbf{r}_1) = \frac{\alpha(\rho_0^{\mathbf{r}_1})}{a} \int d\mathbf{r}_2\, \rho_0(\mathbf{r}_2; [\phi_R], \mu_0^B) u_1(r_{12}) + 2\alpha(\rho^B)\rho^B. \qquad 20.$$

Thus, the MF integral in Equation 3 is multiplied by a factor $\alpha(\rho_0^{r_1})/a$ of order unity that depends on r_1 through the dependence of the hydrostatic density $\rho_0^{r_1}$ on the local value of the field $\phi_R(r_1)$, and the constant of integration $2a\rho^B$ is replaced by the appropriate limiting value of the modified integral. Note that the hydrostatic density $\rho_0^{r_1}$ remains smooth and relatively slowly varying outside the core even when $\phi_R(r_1)$ contains a hard core. The nonlocal oscillatory excluded volume correlations that can exist in the full density $\rho_0(r_1)$ do not appear in $\rho_0^{r_1}$ because of the strictly local response to ϕ_R. Results using Equation 20 will be discussed below.

MVDW THEORY OF NONUNIFORM FLUIDS WITH ATTRACTIVE INTERACTIONS

Two-Step Method

The ERF must be calculated self-consistently for fluids with attractive interactions. Our new MVDW theory corrects the classical VDW theory by using Equation 16 to accurately determine $\rho_0(r_2)$ in Equation 8. Thus the MVDW theory requires the simultaneous solution of two equations: Equation 8 (or the closely related equation that arises if Equation 20 is used for the ERF) and Equation 16. Equation 8 determines $\rho_0^{r_1}$, the local hydrostatic response to the ERF as in the VDW theory described above, and Equation 16 determines the full nonlocal response $\rho_0(r_1)$. [Similarly, the VDW theory can be viewed as replacing Equation 16 by the hydrostatic approximation $\rho_0(r_1) = \rho_0^{r_1}$.]

One can think about solving these equations by a two-step iterative method (3). For a given approximation to the ERF, determine in a first step the associated smooth hydrostatic density $\rho_0^{r_1}$ from Equation 8. Then in a second step, take account of nonlocal and usually oscillatory corrections to this profile, generally induced by rapidly varying features in the external field ϕ, using a locally optimal application of linear response theory, Equation 16. This new density $\rho_0(r_1)$ is then used to compute a new approximation for the ERF, and the two steps are iterated to self consistency. This process is easy to implement numerically and rapid convergence has been found in most cases we have examined (3, 6, 7).

We noted in "Simple Molecular Field Equation" above that the simple MF ERF can quite generally be written as $\phi_R = \phi + \phi_s$, where ϕ_s in Equation 5 takes account of the unbalanced attractive interactions and is smooth and relatively slowly varying. We are often interested in cases where ϕ is a hard core potential. [If ϕ also has a slowly varying part, say from weak attractive interactions between a solute and solvent, the latter should be added (6) to ϕ_s in the discussion that follows.] In this case, one can implement the two-step process in a slightly different way, which is physically suggestive and was, in fact, how we first thought about the problem (3, 4, 6).

Let us consider in the first step the hydrostatic response $\rho_s^{r_1}$ to the slowly varying part ϕ_s of the ERF alone. Because there is a strictly local response to the ERF, $\rho_s^{r_1}$

differs from $\rho_0^{r_1}$ for the full ERF only inside the core region, where $\rho_0^{r_1}$ vanishes while $\rho_s^{r_1}$ remains continuous and smooth. This smoothness allows us to use the gradient approximation in the next to last term of Equation 9 in determining $\rho_s^{r_1}$ if we wish. Because ϕ_s describes the unbalanced attractive interactions associated with interface formation, we can interpret the smooth $\rho_s^{r_1}$ as an interfacial component of the full density response $\rho_0(\mathbf{r}_1)$.

Then in the second step, we take into account the response to the remaining hard core part of ϕ_R and all nonlocal effects. This will cause $\rho_0(\mathbf{r}_1)$ to vanish inside the core and, depending on the extent to which the density $\rho_s^{r_1}$ near the core has been reduced, can induce nonlocal oscillatory excluded volume corrections to $\rho_s^{r_1}$ outside the core, which we again calculate using linear response theory. Thus, speaking pictorially, the second step takes into account the nonlocal Gaussian fluctuations induced by the hard core potential about the smoothly varying interfacial component of the density profile. The strongly non-Gaussian component associated with interface formation and arising from the unbalanced attractive interactions is taken into account in the first step. The final profile results from the self-consistent interplay between the components described in each step.

Because we are using linear response theory in all cases to correct a hydrostatic response, these alternate ways of implementing the two-step procedure are completely equivalent. In particular, in the first interpretation described above, we calculate the hydrostatic response $\rho_0^{r_1}$ to the full ERF and correct it for all $\mathbf{r}_1$ using Equation 15. The hard core condition $\rho_0(\mathbf{r}) = 0$ for all $\mathbf{r}$ inside the core region follows automatically from perturbing about the local hydrostatic density. In the second interpretation, we calculate in the first step the hydrostatic response $\rho_s^{r_1}$ to the interfacial component ϕ_s alone and correct it by using linear response only for $\mathbf{r}_1$ outside the core, while imposing $\rho_0(\mathbf{r}_2) = 0$ in the core region:

$$0 = \int d\mathbf{r}_2 \, \chi_0^{-1}(r_{12}; \rho_s^{r_1}) [\rho_0(\mathbf{r}_2) - \rho_s^{r_1}]. \qquad 21.$$

Because $\rho_s^{r_1}$ equals $\rho_0^{r_1}$ outside the core, the solution $\rho_0(\mathbf{r})$ to Equation 21 with the core condition imposed is identical to that given by Equation 15.

Hard Sphere Solute in LJ Fluid Near Liquid-Vapor Coexistence

These ideas have been applied to the nonuniform LJ fluid in a variety of different situations, including fluids near a hard wall, fluids confined in slits and tubes, and the liquid-vapor interface, with generally very good results (1–3, 5–7, 80; K. Katsov, J.D. Weeks, unpublished manuscript; K. Vollmayr-Lee, K. Katsov, J.D. Weeks, unpublished manuscript). We will review here results (7) for a system studied by Huang & Chandler (HC) that combines many of these limits (81), and is physically very relevant for our later discussion about hydrophobic interactions.

HC carried out extensive computer simulations to determine properties of the LJ liquid at a state very near the triple point with $\rho^B = 0.70$ and $T = 0.85$ when the radius of a hard sphere solute is varied, and compared the results to the LCW theory reviewed below. (We use the standard LJ reduced units.) By definition, the solute centered at the origin interacts with the LJ particles through the hard core potential:

$$\phi(r; S) = \begin{cases} \infty, & r \leq S, \\ 0, & r > S. \end{cases} \qquad 22.$$

The MMF theory discussed above allows us to reduce this problem to that of the reference fluid in the presence of the effective field $\phi_R(r; S)$ satisfying Equation 20. We have calculated self-consistently the ERF $\phi_R(r; S)$ and the associated density response $\rho_0(r; S)$ of the reference fluid, solving Equations 20 and 16 by iteration. In Figure 2, we compare these results for the density profiles in the presence of the hard sphere solutes with S equal to 1.0, 2.0, 3.0, and 4.0 in reduced units with the simulation results of the same LJ system by HC. There is very good agreement between theory and simulation.

Figure 3 shows the corresponding ERFs obtained in these calculations. For small solutes with S less than about 0.7, attractive interactions do not give rise

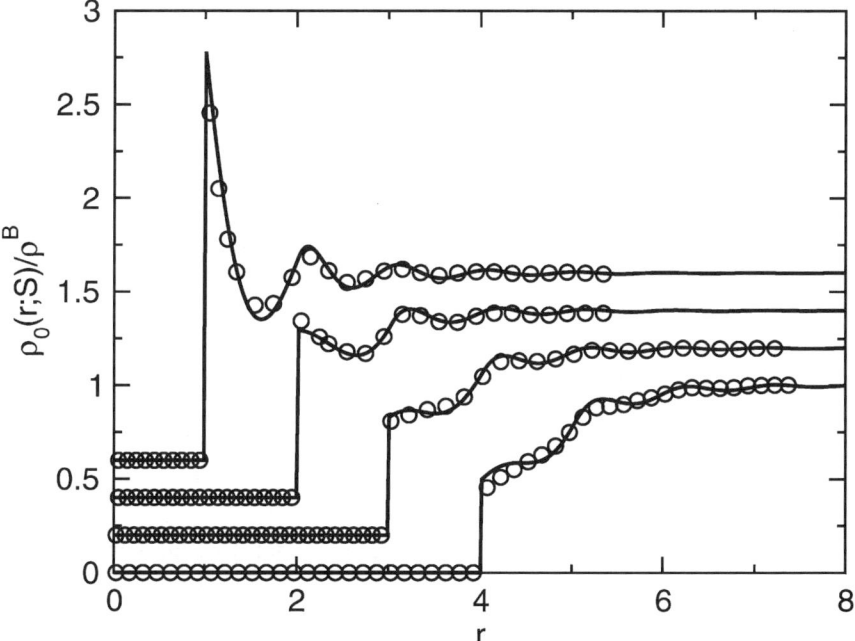

Figure 2 Density profiles of the LJ fluid ($T = 0.85$, $\rho^B = 0.70$) in the presence of the hard sphere solute with $S = 1.0, 2.0, 3.0$, and 4.0. Circles denote simulation results (81). Lines are results of the self-consistent approach based on the modified molecular field determined from Equation 20. For ease of viewing, the density profiles for $S = 1.0, 2.0$, and 3.0 have been shifted vertically by 0.6, 0.4, and 0.2 units, respectively.

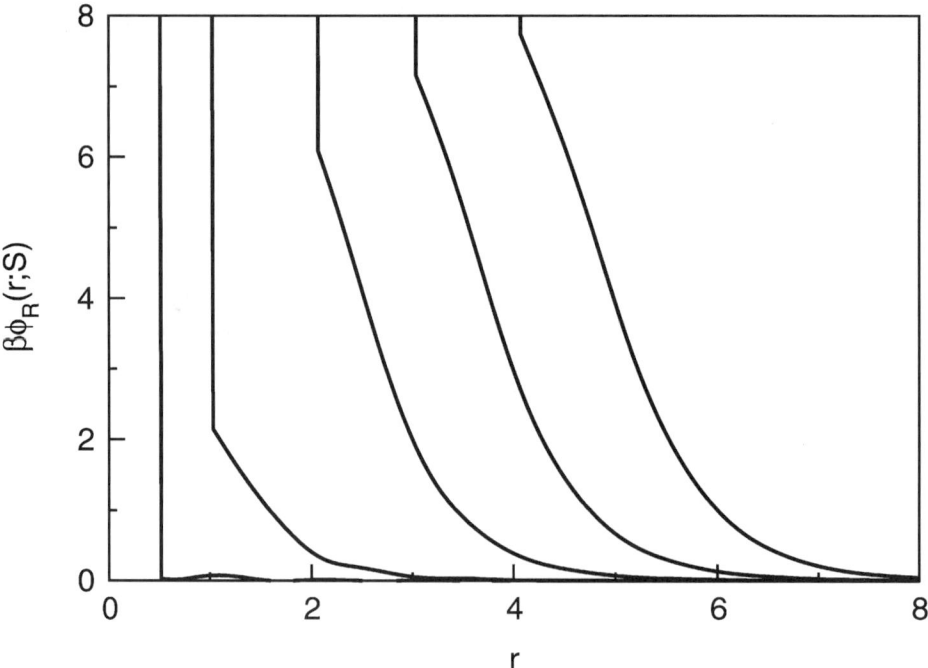

Figure 3 Self-consistent molecular field of the LJ fluid for the solute with $S = 0.5$, 1.0, 2.0, 3.0, and 4.0, obtained from Equation 20.

to any substantial modification of the bare external field, as can be seen from the plot of $\phi_R(r; S)$ for $S = 0.5$. (Clearly, for $S = 0$ there are no solute-induced interactions of any kind and the profile reduces to the constant ρ^B.) However, the effects due to unbalanced attractions become important even for $S = 1.0$, which is about the same size as the LJ core, and all larger sizes give rise to a very strong and relatively soft repulsion in $\phi_R(r; S)$. The corresponding density profiles show pronounced depletion near the surface of the solute, characteristic of surface-induced drying.

Solvation Free Energy

Another quantity of great interest is the free energy of the nonuniform system. This is the main focus of attention in density functional theory, briefly discussed in "Density Functional Theory" below. In contrast, the MVDW approach focuses first on the liquid structure. We believe this permits physical insight to play a more direct role. However, since we can determine the density response to an arbitrary external field, the free energy can be easily calculated from a coupling-parameter-type integration that connects some initial state, e.g., the bulk fluid, whose free energy is known, to the final state as the field is varied. In the present case, there is a very simple route to the free energy of the nonuniform LJ system that uses structural features that we know from simulations are accurately determined.

In particular, let us consider the change in free energy of the LJ fluid as the range of the external field representing the hard core solute is varied from zero to its full extent S. This construction is the basis of scaled particle theory (82), and it is well known that the free energy change takes a particularly simple form:

$$\beta \Delta \Omega_S = 4\pi S^3 \int_0^1 d\lambda \, \lambda^2 \rho_\lambda(\lambda S^+), \qquad 23.$$

which requires only the contact value $\rho_\lambda(\lambda S^+)$ of the density profile. This is very accurately given by the theory described above. To use this virial route, we can replace the λ-integration by a sum and calculate the density profile for several values of λ at the fixed bulk chemical potential μ^B.

Using Equation 23, we obtain the dependence of solvation free energy on the size of the hard sphere solute. The free energy per unit surface area of the solute $\Delta \Omega_S / 4\pi S^2$ we obtain is shown in Figure 4. For small solutes unbalanced attractive

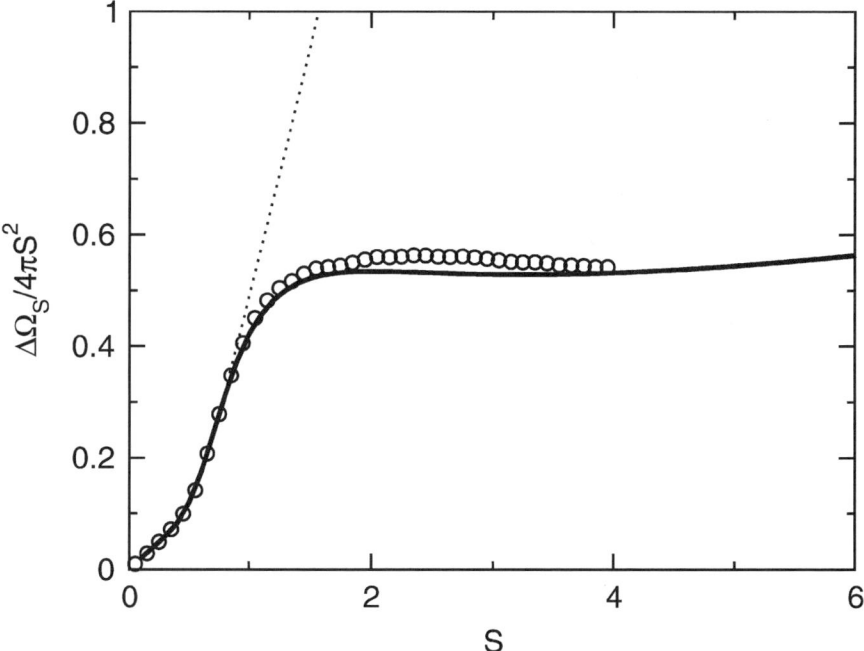

Figure 4 Dependence of the solvation free energy on the cavity size S. Circles denote results of simulations (81). Lines are obtained from Equation 23 by using the results of the molecular field equation (Equation 20) (solid), and by neglecting the molecular field (dotted).

interactions do not play an important role, and the solvation free energy agrees well with a pure hard sphere model, which completely neglects attractions by using the bare hard core solute potential, as shown by the dotted line. At the solute size of about 0.7, the behavior changes drastically and the reduced free energy rapidly crosses over to the practically constant plateau in agreement with the simulation results. The small slope of the curves in Figure 4 for large S can be understood by separating the free energy into volume ($V_S = 4\pi S^3/3$) and surface ($A_S = 4\pi S^2$) contributions as discussed by HC:

$$\Delta\Omega_S \approx V_S p^B + A_S \gamma_S. \qquad 24.$$

The first term in this expression corresponds to the work required to remove liquid particles from the volume occupied by the solute, where p^B is the bulk liquid pressure, and is very small for the values of S considered here. The second term determines the cost of forming the liquid-solute interface and is proportional to the interface tension γ_S, which is essentially independent of the solute size for large solutes.

HYDROPHOBIC INTERACTIONS IN WATER

We now discuss an important extension of these ideas to hydrophobic interactions in water (4), as described by LCW. We first consider water at ambient conditions as the radius of a hard sphere solute at the origin is varied. Because this state is very close to the liquid-vapor phase boundary, phenomena involving interfaces can be very important. This system serves as a simple model of a hydrophobic object in water—the solute does not participate in hydrogen bonding and it creates an excluded volume region where the density of water molecules vanishes. Weak VDW attractions between the solute and solvent do not change the qualitative nature of the phenomena we will discuss and can easily be taken into account (6, 83).

But how can one sensibly apply the theory to water? The local structure of water is certainly very different from that of the LJ fluid and anything relying on the detailed properties of a hard sphere reference system cannot be trusted. However, the two systems do have certain essential features in common that can be exploited in a properly generalized MVDW theory.

First, small fluctuations in liquid water are Gaussian, and computer simulations had earlier shown that even relatively large fluctuations leading to the formation of molecular sized cavities can be well described using the same Gaussian distribution (17). Indeed, Pratt & Chandler (13) developed a quantitative theory for the solvation of small apolar molecules using the experimental linear response function for water that takes into account the structural and free energy changes induced by the excluded volume of the solute. These ideas have been significantly extended and placed on a firmer conceptual basis in recent work by Pratt et al. (17, 19, 20, 22, 30, 84). [See also Reference (76).] This suggests that the Gaussian/linear response ideas used in the second step of the two-step method in the

MVDW theory, using response functions appropriate for water, could be modified to apply to water.

Second, just as for the LJ fluid above, the main nonGaussian feature to be expected in this application is associated with interface formation. Long ago, Stillinger (11) gave a qualitative and physically very suggestive description of what would be expected as the radius of the hard sphere solute is increased. Whereas small solutes should not significantly disturb the hydrogen bond network, which can simply go around the solute, in the vicinity of a sufficiently large solute or wall the network must be completely disrupted. In the latter case, Stillinger argued, the arrangement of molecules near the solute and the interface free energy should resemble that of the liquid-vapor interface, which optimally solves the similar problem of going from a complete hydrogen-bond network in the liquid to no hydrogen bonds in the dilute vapor.

Thus, Stillinger envisioned a drying transition very like that studied in the last section for the LJ fluid as the solute size is increased. This phenomenon is very general. The differences in local structure should have little effect on the generic and qualitative physics leading to interface formation. Thus, it seems plausible that the interfacial component determined in the first step of the MVDW theory could be described, at least qualitatively, by a MF treatment similar to that used for the LJ fluid provided appropriate thermodynamic parameters for water are used.

We see that key features of both steps of the MVDW theory have some analogues for water. The hard sphere reference fluid picture for the LJ fluid was incisive in developing the general ideas leading to the MVDW theory. Given that understanding, we may be able to develop an analogous approach for water and other fluids that does not rely on the details of the reference fluid, or indeed explicitly introduce a reference fluid at all.

Following LCW, let us examine the essential features of the MVDW theory and see how they can be modified to apply to water. As in the last part of "Hydrostatic Approximation and the Classical VDW Theory" above, we will try to describe everything formally in terms of the properties of water itself, and not a reference system, though MF ideas will be introduced to define what is meant in two-phase regions, etc. A crucial part of the physics of interface formation in the MVDW theory is the description of the unbalanced attractive interactions in Equation 5. In the LJ fluid, this can be rewritten and reinterpreted in terms of an averaged or coarse-grained density $\bar{\rho}(\mathbf{r}_1)$ with a normalized weighting function proportional to u_1:

$$-2a\bar{\rho}(\mathbf{r}_1) \equiv \int d\mathbf{r}_2 \, \rho(\mathbf{r}_2) u_1(r_{12}). \qquad 25.$$

LCW argue that the unbalanced attractive interactions in water can be described by a similar coarse graining of the water density, with the coarse graining carried out over the appropriate range λ of the attractive interactions in water.

Now consider the first step of the two-step method, as described in the last part of "Two-Step Method" above, where the smooth interfacial component $\rho_s^{\mathbf{r}_1}$

associated with the field ϕ_s in Equation 5 is determined from Equation 9. Using the notation of Equation 25 and expanding the next to last term, Equation 9 becomes:

$$\mu\left(\rho_s^{r_1}\right) = \mu^B + m\nabla^2 \rho_s^{r_1} + 2a\left[\bar{\rho}(\mathbf{r}_1) - \bar{\rho}_s^{r_1}\right].\qquad 26.$$

With appropriate changes of notation, this is exactly Equation 5 of LCW. However, LCW did not consistently interpret $\rho_s^{r_1} [= n_s(\mathbf{r}_1)$ in LCW] as the hydrostatic density, and some later simplifications that arise from this were not exploited. Because the slowly varying and generic interfacial component $\rho_s^{r_1}$ in Equation 26 should be essentially independent of local structure, LCW used a simple VDW form for $\mu(\rho)$, which automatically interpolates in the two-phase region, but with VDW parameters a and b chosen to reproduce the density and compressibility of liquid water at phase coexistence and $T = 298$ K. The parameter m was fit to the surface tension of water and the VDW relationship between a and m determined the coarse-graining scale λ for $\bar{\rho}(\mathbf{r})$, which LCW carried out using a simple Gaussian weight. Although the details of this fitting procedure are rather arbitrary, and the VDW equation is probably inadequate to describe the quantitative relation between the energy density and surface tension in water (63), LCW showed that small variations in these parameter values did not change the qualitative picture that emerged.

Now turn to the second linear response step, described for the LJ fluid by Equation 21. What is the analogue of the LJ reference $\chi_0^{-1}(r_{12}; \rho)$ for water? In view of Equation 2, one can formally consider fluctuations in the full fluid instead. For ρ in the stable liquid phase, small fluctuations are Gaussian and the uniform fluid $\chi^{-1}(r_{12}; \rho)$ can be used directly. However, when interfaces form, we need an approximation for χ^{-1} in our MF treatment that remains well defined for all values of ρ in the two-phase region as well as in the vapor phase, as is the case for the reference χ_0^{-1} for the LJ fluid. In effect, LCW devised an interpolation scheme for a χ^{-1} for water that has these properties.

LCW essentially considered an alternate but equivalent version of Equation 21:

$$\rho(\mathbf{r}_1) - \rho_s^{r_1} = \int d\mathbf{r}_2\, \chi\left(r_{12}; \rho_s^{r_1}\right) C_S(\mathbf{r}_2),\qquad 27.$$

which involves the standard linear response function $\chi(r_{12}; \rho) \equiv \rho\delta(\mathbf{r}_1-\mathbf{r}_2) + \rho^2 h(r_{12}; \rho)$, the inverse to $\chi^{-1}(r_{12}; \rho)$. Here, $h + 1$ is the radial distribution function for water, and C_S is a generalized solute-solvent direct correlation function, similar to that in Equation 17. This is nonzero only inside the core (PY closure), and is completely determined by the imposed requirement that $\rho(\mathbf{r})$ equals zero for all $\mathbf{r}$ in the core region. LCW replaced $h(r_{12}; \rho)$ by $h(r_{12}; \rho^B)$ in the definition of χ, and used experimental values for the latter. This approximation is exact in the bulk liquid phase, reduces to the correct ideal gas value at very low density, and smoothly interpolates in between. Equation 6 in LCW reduces to our Equation 27 when it is realized that $\rho_s^{r_1}$ is the hydrostatic density. Because $\rho_s^{r_1}$ is slowly varying, no essential differences should result from using either equation.

Thus LCW implemented in an approximate, but plausible, way both steps of the MVDW theory, making use of experimental thermodynamic and structural data. They obtained results for the solute-solvent density distribution function qualitatively very similar to the results in Figure 2 for the analogous LJ-hard sphere system. As the radius of the solute increased, they found a crossover on the biologically relevant length scale of nanometers from wetting with peak densities greater than the bulk to drying with peak densities less than the bulk. (The unbalanced attractive interactions cause significant density perturbations in both regimes, so this terminology is somewhat misleading.) Weak attractive VDW attractions between the solute and water can shift inward the position of the interface when partial drying occurs and suppress complete drying, but should not have an important effect on the basic interface structure and free energy changes or the length scale for the crossover (83).

LCW also calculated solvation free energies, using a Gaussian approximation that is quantitatively somewhat less accurate than the coupling parameter method discussed above for the LJ system, but quite sufficient for qualitative purposes. Again they found behavior very similar to Figure 4, with free energies scaling with surface area only on large length scales of order nanometers. LCW also studied assemblies of extended idealized hydrophobic objects (plates and rods) and found that the drying can lead to strong attractions between sufficiently close pairs of such surfaces as the intervening water is expelled. Thus, there is a length scale dependence of hydrophobic interactions. LCW suggest that such phase transitions could play an important role in aspects of protein folding where extended mostly hydrophobic regions approach one another.

It is beyond the scope of this article (and the expertise of this author) to assess the validity of that last conjecture, given the many complications occurring in nature. Clearly, much more theoretical and experimental effort is called for on all aspects of the theory. Some interesting recent work along these lines can be found in References (21–29 and 85–91). Our purpose is merely to argue that interface formation is a fundamental piece of physics that almost certainly occurs for the idealized models discussed herein and that LCW have developed a qualitatively reasonable MF treatment of that process based on sound statistical mechanical principles. From that perspective, the scaling of hydrophobic solvation energies with exposed surface area with a value close to the surface tension of water can be justified only on large length scales (4, 91).

DENSITY FUNCTIONAL THEORY

As discussed in the "Introduction," the most commonly used theory for the structure and thermodynamics of nonuniform fluids is density functional theory (DFT). Space limitations permit only a few general remarks here focusing on possible relations to the MVDW theory and the advantages and disadvantages of each approach. Two standard reviews of DFT are found in References (43 and 44).

A main focus of DFT is the intrinsic free energy density functional $F([\rho], [w])$, which arises from the usual grand canonical free energy Ω by a Legendre transform (10), where the functional dependence on the chemical potential and external field ϕ is replaced by a functional dependence on the (uniquely associated) singlet density ρ. Our notation emphasizes that F, just like Ω, remains a functional of the intermolecular pair (and any higher order) potentials w. In principle, this formalism is exact and F contains all the information in Ω. An exact hierarchy of (direct) correlation functions can be derived by successive functional derivatives of F with respect to ρ.

Although it must be as hard to determine F exactly as it is to determine Ω, the hope is that practically useful approximations may suggest themselves more naturally in this representation, and the sound theoretical basis of DFT may provide a means for systematic corrections. By starting from the free energy, certain exact sum rules relating integrals of those correlation functions to the thermodynamic properties are automatically and consistently satisfied in any DFT (45, 59, 60); this is not the case for structurally based methods such as the MVDW theory. Particularly suggestive is the fact that the classical VDW theory can be viewed as a DFT for F where a MF approximation is made for the functional dependence on w and a local density approximation is made for the functional dependence on ρ (43, 44). Although this certainly seems promising, it turns out to be rather difficult to do significantly better from this starting point.

Can the DFT formalism help us improve on the MF approximation? Consider the first functional derivative of F, which from basic properties of the Legendre transform is easily seen to satisfy

$$\delta F([\rho], [w])/\delta\rho(\mathbf{r}) = \mu - \phi(\mathbf{r}). \qquad 28.$$

In principle, if we knew the exact F we could solve this equation, determining the density $\rho(\mathbf{r})$ associated with a given potential $\phi(\mathbf{r})$. Whereas most discussions have focused on the density dependence, the main problem in determining an accurate F, from a fundamental point of view, is its dependence on w. When there are attractive interactions, the exact F must describe critical phenomena, capillary waves, and a host of other properties for which we have essentially no idea what the true functional dependence on w should be. A treatment incorporating even the simplest capillary wave correlations for the liquid-vapor interface produced a density functional very different from conventional ideas (44, 72, 92), and this does not begin to address the range of phenomena arising from the general functional dependence on w. So as a practical matter, we are essentially forced to accept the MF treatment of the attractive interactions, perhaps modified slightly as in "Correcting the Molecular Field Approximation" above, to give a better description of the uniform fluid (60).

The true F for a system with repulsive forces only must be significantly simpler. Even here, the dependence on a general repulsive w_0 can cause problems; any density functional must also explicitly or implicitly approximate the dependence on w_0. It is not obvious, from a fundamental point of view, how to modify a functional that gives a good description for say, hard spheres, so that it can describe much softer repulsions. [See References (53 and 54) for interesting work on this question.] Unlike most quantum-mechanical applications of DFT, where there is a single Coulomb interaction potential, we must deal with a range of interactions and hence potentially many different functionals.

Most workers have quite reasonably focused on the basic hard sphere system, and here some significant progress has been made in determining accurate approximations for F_0, particularly in the development of fundamental measure theory (50–52, 58). However, the latter is very complicated, and one might hope for a simpler and perhaps more physically motivated approach.

Unfortunately, the DFT formalism itself gives few indications of how to proceed in a practical manner. The basic problem can be seen when we consider Equation 28 for a system of hard spheres and a general ϕ_R, i.e.,

$$\delta F_0([\rho_0], [u_0])/\delta\rho_0(\mathbf{r}) = \mu_0^B - \phi_R(\mathbf{r}). \qquad 29.$$

This equation can be viewed as a generalization of our hydrostatic equation (Equation 7), which, in principle, could be used to determine the exact density if a proper F_0 could be found. Indeed, Equation 7 follows from Equation 29 for a very slowly varying profile when a local density approximation is made. But how should this be corrected for rapidly varying densities? To go beyond the local density approximation, various weighted density approximations have been proposed. However, from a fundamental point of view, except in special low-dimensional cases (58), it is not even clear that F_0 can be written as a simple functional of weighted densities; certainly the original Ω_0 is not a functional of a weighted $\phi_R(\mathbf{r})$. This contrasts with the development of the MVDW theory from the VDW theory. There the simplest correction (linear response) is clearly correct for small perturbing potentials, and it remains reasonably accurate even in the hard core limit, reducing naturally to the PY approximation. Usually in DFT, this limit is imposed by hand.

Because many treatments of DFT have discussed the formal advantages of the method, we have mostly focused here on some of the difficulties as we see them. Despite these (perhaps pedantic) objections, there have been impressive successes arising from DFT, and there clearly are close connections between some of the ideas of DFT and the MVDW theory. In particular, although we do think about molecular fields, the HLR equation (Equation 16) explicitly involves only densities. Indeed, the reason for its success is the removal of any explicit dependence on the field through the expansion about the hydrostatic density. However, a deeper connection to DFT has so far escaped us; this is an active topic of our current research.

FINAL REMARKS

We conclude with a few general remarks. The MVDW theory combines in a self-consistent way two standard and widely used ingredients: molecular (or mean) field theory and linear response (or Gaussian field) theory. The MF equation for the ERF takes account of the unbalanced attractive forces discussed in the original VDW theory (9); all that is required is to determine the molecular field and the induced structure accurately. Although the hydrostatic approximation used in the classical VDW theory is not generally accurate, it serves as an optimal starting point for corrections based on linear response theory. The resulting theory can handle problems involving fluctuations on a variety of length scales where each ingredient alone would fail. Because of its sound physical basis, we are hopeful that the MVDW theory will prove useful in many different applications. In particular, assessing and further developing both the physical and biophysical implications of the LCW theory in realistic environments seems an important topic for future research. One new direction we are thinking about involves fluids with long-ranged Coulomb forces. Here there are a wide range of phenomena, such as charge ordering and double layer formation, arising from very strong and competing interactions on many length scales. This will certainly put our current ideas to a most severe test.

There are also many basic theoretical questions left open. Can the MVDW theory be understood as some kind of weighted DFT? Is it possible to extend these ideas to solid-fluid interfaces? Can one go significantly beyond the MF picture in describing the effects of attractive interactions while still maintaining a tractable theory? The LCW theory, despite its use of properties of water only, is definitely a MF theory. But the idea of short-wavelength Gaussian fluctuations related to local structure and long wavelength slowly varying fluctuations related to interface formation seems more general. A reinterpretation of LCW theory from this perspective is found in Reference (63). We close with what is probably the most basic question: Is it is generally correct to imagine that fluctuations are essentially Gaussian in most noncritical liquids except when interfaces form? A deeper understanding of whether and why this is true is called for. If this physically suggestive picture remains valid, then the MVDW theory has captured in a surprisingly simple way much of the noncritical physics of the liquid state.

ACKNOWLEDGMENTS

This work was supported by the National Science Foundation through grants CHE-9528915 and CHE-0111104. Significant parts of the work described here were carried out by Jeremy Broughton, Robin Selinger, Katharina Vollmayr-Lee, and Ka Lum. Kirill Katsov played the key role in the conceptual development of the HLR equation and its numerical solution, and David Chandler's ideas and insights shaped the LCW theory. We are grateful to Michael Fisher and Jim Henderson for many helpful discussions.

Visit the Annual Reviews home page at www.annualreviews.org

LITERATURE CITED

1. Weeks JD, Selinger RLB, Broughton JQ. 1995. *Phys. Rev. Lett.* 75:2694–97
2. Weeks JD, Vollmayr K, Katsov K. 1997. *Physica A* 244:461–75
3. Weeks JD, Katsov K, Vollmayr K. 1998. *Phys. Rev. Lett.* 81:4400–3
4. Lum K, Chandler D, Weeks JD. 1999. *J. Phys. Chem. B* 103:4570–77
5. Katsov K, Weeks JD. 2001. *Phys. Rev. Lett.* 86:440–43
6. Vollmayr-Lee K, Katsov K, Weeks JD. 2001. *J. Chem. Phys.* 114:416–25

7. Katsov K, Weeks JD. 2001. *J. Phys. Chem. B* 105:6738–44
8. Rowlinson JS, Widom B. 1989. *Molecular Theory of Capillarity*. Oxford: Clarendon
9. Rowlinson JS. 1979. *J. Stat. Phys.* 20:197–244
10. Hansen JP, McDonald IR. 1986. *Theory of Simple Liquids*. London: Academic
11. Stillinger FH. 1973. *J. Solut. Chem.* 2:141–58
12. Tansford C. 1973. *The Hydrophobic Effect—Formation of Micelles and Biological Membranes*. New York: Wiley-Intersci.
13. Pratt LR, Chandler D. 1977. *J. Chem. Phys.* 67:3683–704
14. Lee CY, McCammon JA, Rossky PJ. 1984. *J. Chem. Phys.* 80:4448–55
15. Rossky PJ, Lee SH. 1989. *Chem. Scr.* 29A:93–95
16. Wallqvist A. 1990. *Chem. Phys. Lett.* 165:437–42
17. Hummer G, Garde S, Garcia AE, Pohorille A, Pratt LR. 1996. *Proc. Natl. Acad. Sci. USA* 93:8951–55
18. Lum K, Luzar A. 1997. *Phys. Rev. E* 56:R6283–86
19. Hummer G, Garde S. 1998. *Phys. Rev. Lett.* 80:4193–96
20. Hummer G, Garde S, Garcia AE, Paulaitis ME, Pratt LR. 1998. *J. Phys. Chem. B* 102:10469–82
21. Silverstein KAT, Haymet ADJ, Dill KA. 1999. *J. Chem. Phys.* 111:8000–9
22. Hummer G, Garde S, Garcia AE, Pratt LR. 2000. *Chem. Phys.* 258:349–70
23. Leung K, Luzar A. 2000. *J. Chem. Phys.* 113:5845–52
24. Luzar A, Leung K. 2000. *J. Chem. Phys.* 113:5836–44
25. Southall NT, Dill KA. 2000. *J. Phys. Chem. B* 104:1326–31
26. Bratko D, Curtis RA, Blanch HW, Prausnitz JM. 2001. *J. Chem. Phys.* 115:3873–77
27. Wallqvist A, Gallicchio E, Levy RM. 2001. *J. Phys. Chem. B* 105:6745–53
28. Truskett TM, Debenedetti PG, Torquato S. 2001. *J. Chem. Phys.* 114:2401–18
29. Silverstein KAT, Dill KA, Haymet ADJ. 2001. *J. Chem. Phys.* 114:6303–14
30. Pratt LR. 2002. *Annu. Rev. Phys. Chem.* 53:409–36
31. Roth R, Evans R, Dietrich S. 2000. *Phys. Rev. E* 62:5360–77
32. Percus JK. 1962. *Phys. Rev. Lett.* 8:462–64
33. Gelb LD, Gubbins KE, Radhakrishnan R, Sliwinska-Bartkowiak M. 1999. *Rep. Prog. Phys.* 62:1573–659
34. Evans R, Tarazona P. 1984. *Phys. Rev. Lett.* 52:557–60
35. Evans R, Marconi UMB, Tarazona P. 1986. *J. Chem. Phys.* 84:2376–99
36. Evans R, Marconi UMB, Tarazona P. 1986. *J. Chem. Soc. Faraday Trans.* II 82:1763–87
37. Tarazona P, Marconi UMB, Evans R. 1987. *Mol. Phys.* 60:573–95
38. Ball PC, Evans R. 1988. *J. Chem. Phys.* 89:4412–23
39. Gotzelmann B, Dietrich S. 1997. *Phys. Rev. E* 55:2993–3005
40. Gotzelmann B, Evans R, Dietrich S. 1998. *Phys. Rev. E* 57:6785–800
41. Rocken P, Somoza A, Tarazona P, Findenegg G. 1998. *J. Chem. Phys.* 108:8689–97
42. Neimark AV, Ravikovitch PI, Vishnyakov A. 2000. *Phys. Rev. E* 62:R1493–96
43. Evans R. 1990. In *Liquids at Interfaces*, ed. J Charvolin, JF Joanny, J Zinn-Justin, pp. 1–98. Amsterdam: North Holland
44. Evans R. 1992. In *Fundamentals of Inhomogeneous Fluids*, ed. D Henderson, pp. 85–175. New York: Dekker
45. Henderson JR, Tarazona P, van Swol F, Velasco E. 1992. *J. Chem. Phys.* 96:4633–38
46. Henderson JR. 1994. *Phys. Rev. E* 50:4836–46
47. Meister TF, Kroll DM. 1985. *Phys. Rev. A* 31:4055–57
48. Percus JK. 1991. *Physica A* 172:1–16
49. Rosenfeld Y. 1994. *Phys. Rev. E* 50:R3318–21
50. Rosenfeld Y. 1996. *J. Phys. Condens. Matter* 8:9289–92

51. Rosenfeld Y, Schmidt M, Lowen H, Tarazona P. 1997. *Phys. Rev. E* 55:4245–63
52. Tarazona P, Rosenfeld Y. 1997. *Phys. Rev. E* 55:R4873–76
53. Schmidt M. 1999. *Phys. Rev. E* 60:R6291–94
54. Schmidt M. 2000. *Phys. Rev. E* 62:4976–81
55. Tarazona P, Evans R. 1984. *Mol. Phys.* 52:847–57
56. Tarazona P. 1984. *Mol. Phys.* 52:81–96
57. Tarazona P. 1985. *Phys. Rev. A* 31:2672–79
58. Tarazona P. 2000. *Phys. Rev. Lett.* 84:694–97
59. van Swol F, Henderson JR. 1989. *Phys. Rev. A* 40:2567–78
60. van Swol F, Henderson JR. 1991. *Phys. Rev. A* 43:2932–42
61. Andersen HC. 1975. *Annu. Rev. Phys. Chem.* 26:145–66
62. Percus JK. 1964. In *The Equilibrium Theory of Classical Fluids*, ed. HL Frisch, JL Lebowitz, pp. II-33–II-170. New York: Benjamin
63. TenWolde PR, Sun SX, Chandler D. 2001. *Phys. Rev. E.* In press
64. Widom B. 1967. *Science* 157:375–79
65. Longuet-Higgins HC, Widom B. 1964. *Mol. Phys.* 8:549–56
66. Andersen HC, Weeks JD, Chandler D. 1971. *Phys. Rev. A* 4:1597–607
67. Chandler D, Weeks JD. 1970. *Phys. Rev. Lett.* 25:149–52
68. Weeks JD, Chandler D, Andersen HC. 1971. *J. Chem. Phys.* 54:5237–47
69. Chandler D, Weeks JD, Andersen HC. 1983. *Science* 220:787–94
70. Sullivan D, Stell G. 1978. *J. Chem. Phys.* 69:5450–57
71. Lebowitz JL. 1974. *Physica* 73:48–60
72. Weeks JD. 1977. *J. Chem. Phys.* 67:3106–21
73. Waisman E. 1973. *Mol. Phys.* 25:45–48
74. Waisman E, Lebowitz JL, Henderson D. 1976. *Mol. Phys.* 32:1373–81
75. Crooks GE, Chandler D. 1997. *Phys. Rev. E* 56:4217–21
76. Chandler D. 1993. *Phys. Rev. E* 48:2898–905
77. Stell G. 1963. *Physica* 29:517
78. Weeks JD, Broughton JQ. 1983. *J. Chem. Phys.* 78:4197–205
79. Johnson JK, Zollweg JA, Gubbins KE. 1993. *Mol. Phys.* 78:591–618
80. Katsov K. 2000. *From uniform to nonuniform fluids: a journey with density fluctuation theory*. PhD thesis. College Park: Univ. Maryland. 113 pp.
81. Huang DM, Chandler D. 2000. *Phys. Rev. E* 61:1501–6
82. Reiss H, Frisch HL, Lebowitz JL. 1959. *J. Chem. Phys.* 31:369–80
83. Huang DM, Chandler D. 2001. *J. Phys. Chem. B.* In press
84. Garde S, Hummer G, Garcia AE, Paulaitis ME, Pratt LR. 1996. *Phys. Rev. Lett.* 77:4966–68
85. Brown MG, Raymond EA, Allen HC, Scatena LF, Richmond GL. 2000. *J. Phys. Chem. A* 104:10220–26
86. Gragson DE, Richmond GL. 1998. *J. Phys. Chem. B* 102:3847–61
87. Kuhn PS, Barbosa MC, Levin Y. 2000. *Physica A* 283:113–18
88. Kargupta K, Konnur R, Sharma A. 2001. *Langmuir* 17:1294–305
89. Carey C, Yuen-Kit C, Rossky PJ. 2000. *Chem. Phys.* 258:415–25
90. Yuen-Kit C, Wen-Shyan S, Rossky PJ. 1999. *Biophys. J.* 76:1734–43
91. Huang DM, Geissler PL, Chandler D. 2001. *J. Phys. Chem. B* 105:6704–9
92. Bedeaux D, Weeks JD. 1985. *J. Chem. Phys.* 82:972–79

AUTHOR INDEX

Abbet S, 402
Abella ID, 21
Aberth W, 91
Abrikosov IA, 336
Abseher R, 423, 424
Achibi Y, 101, 102
Ada ET, 401
Adebodun F, 350, 369
Adelman DE, 71, 84, 85, 90
Adriaensen Y, 453
Aeschlimann M, 224, 508, 509, 512, 515, 516, 518, 527
Agarwal PK, 471, 476, 482, 485, 487
Agarwal R, 25
Agmon N, 313
Ahlrichs R, 275
Ahmed M, 109, 136
Aizawa M, 382, 402
Ajji A, 440
Akamatsu N, 440
Akasaka K, 425
Akazawa H, 390
Alavi A, 328
Alber F, 277
Albrecht AC, 23, 24, 31
Albrecht AW, 27
Albu TV, 477, 487
Aldén M, 336
Alderman DW, 354
Alexander AJ, 90
Alexander MH, 159
Alhambra C, 4, 471–73, 475, 477, 478, 480–82, 484–86, 488, 489, 495
Aliev MR, 141, 153, 154
Alimi R, 480
Allan DC, 250, 252, 319
Allee DR, 447
Allen HC, 557

Allen KN, 489
Allen L, 18, 19
Allerhand A, 349, 350
Allison TC, 79, 81, 82, 86, 87, 483, 487
Almdal K, 179, 185
Almeida R, 509
Al-Shamery K, 509
Althorpe SC, 42–44, 46, 52, 56, 58, 59, 84, 88
Alvarenga AD, 187
Amaldi E, 67
Amara P, 471–73, 478, 485, 488, 490, 495
Amat G, 154
Ambroise JP, 190
Amirav A, 523
Amitay Z, 527
Amos LA, 190
Amos WB, 190
Anand R, 135
Anbar M, 91
Andaloro G, 410
Andersen HC, 410, 419, 421, 423, 424, 535–37, 546
Andersen LH, 43, 62, 91
Anderson EM, 187, 188, 190
Anderson JB, 72, 84, 292, 468, 470
Anderson JR, 331
Anderson SL, 382, 402
Andersson DR, 390
Andrade JD, 438, 457, 461
Andreoni WA, 250, 254, 271, 276, 277
Andres J, 472, 485, 494, 497
Andreson A, 329
Andrews AP, 178–80, 183–86
Andrews KP, 180, 183, 184
Andrist M, 112
Angelico VJ, 398, 399

Anisimov MA, 187, 190
Anlauf S, 190
Antes I, 474
Antonczak S, 497
Antoniewicz PR, 508
Antoniou D, 485
Aoiz FJ, 71–73, 83, 84, 86, 87, 90
Aquilanti V, 73
Aqvist J, 472, 476, 478, 485, 492, 497
Arasaki Y, 46
Arena DA, 512, 515
Arias TA, 250, 252, 319
Arisaka F, 195
Arndt R, 29
Arnett DC, 23
Arnold J, 71
Arnold W, 358, 359, 370
Arnold WD, 368, 372, 373, 375
Arthur J, 410, 422
Arthur JW, 410
Asakura K, 175
Asakura S, 175, 190, 191, 195
Ashbaugh HS, 410, 412, 414, 418, 420, 422, 425, 427, 428
Ashcroft NW, 204
Ashfold MNR, 82, 83
Asplund M, 33
Asplund MC, 34
Assfeld X, 473
Assion A, 43, 507
Ast T, 398
Astinov V, 31
Atabek O, 76, 79
Ataka K, 439, 458
Atkins PW, 264
Attach E, 214
Auger P, 68

563

Augeri DJ, 376
Augspurger JD, 350, 366, 369
Auzinsh M, 72, 84
Avaldi L, 110
Avouris Ph, 209, 214
Ayers JD, 73, 86, 87, 90
Azinovic D, 91

Baba Y, 390
Bachelet GB, 323
Backx P, 423, 424
Bacon RF, 187, 190
Bader RFW, 283, 372, 373
Baek DH, 401
Baer M, 82
Baer T, 102, 103, 108, 109, 112, 124, 126, 128, 130, 132, 133
Baer Y, 511
Baerends EJ, 274, 276
Bagchi B, 21, 314
Bagdasarian M, 490
Bahn S, 343, 344
Bahnson BJ, 486
Bai YS, 20, 22, 24
Bain CD, 438, 442
Baka K, 473
Baker AD, 101, 102, 109
Baker C, 101, 102, 109
Baker D, 423
Baker J, 108
Bakowies D, 479, 490–92
Bala P, 471, 480, 484, 497
Baldelli S, 438, 441, 450
Baldwin DP, 108
Baldwin RL, 420
Balents L, 217
Balint-Kurti GG, 159
Ball PC, 534
Ballone P, 178
Balsara NP, 185
Balucani N, 83
Bañares L, 71, 73, 83, 84, 86, 87, 90
Bandara A, 509
Bandow S, 202
Baranger E, 68

Barbara PF, 21
Barbosa LAMM, 331, 332
Barbosa MC, 557
Bard AJ, 222
Bardo RD, 159
Barkema GT, 293, 300
Barnett RN, 252, 286
Baroni S, 281
Barstis TLO, 389, 394, 397–99
Barteau MA, 324, 325
Bartels L, 237, 508, 509, 527
Bartunik HD, 370
Bartynski RA, 512, 515
Bash PA, 472, 473, 475, 492, 495, 496
Bashford D, 470
Basran J, 488, 489
Bates FS, 179
Batista ER, 262
Batlogg B, 242
Bauer CJ, 368
Bauer M, 224, 508, 509, 512, 515, 516, 518, 527
Bauernschmitt R, 275
Baumert T, 41, 42, 43, 507
Baumgärtel H, 128
Bax A, 350, 354, 355, 358, 364, 365
Bayly CI, 470
Bean BD, 73, 85–87, 90
Beauport I, 509
Beck RD, 384, 394, 395
Beck WF, 25
Becke AD, 278, 472
Becker ED, 19, 21
Becker PC, 23
Beckman RA, 361
Bedeaux D, 558
Bedrov D, 425
Begeron DJ, 209
Beglov D, 410
Beguelin DR, 410, 412
Beijersbergen JHM, 383, 395
Belafhal A, 156
Belch AC, 307
Bell AT, 301

Bellissent R, 190
Bello I, 390, 401
Bellott M, 470
Ben-Amotz D, 413
Benedict LX, 204, 217
Bengaard HS, 339
Benigno A, 26, 27
Benn R, 21, 27
Ben-Naim A, 411, 421, 429
Bennemann K-H, 196
Bennett AMA, 330
Bennett CH, 292, 470
Bennich P, 512
Ben-Nun M, 493
Ben-Shaul A, 196
Berendsen HJC, 410, 412, 413, 480
Berg BA, 312
Berg M, 22, 24–27, 29
Berghold G, 279
Bergman DL, 410
Bergsma JP, 468, 481
Bergt M, 507
Berkowitz J, 101, 102, 104, 107, 108
Berkowitz M, 307
Berleb S, 221
Bernasconi L, 278, 279
Bernasconi M, 262, 263, 265, 266, 279, 281
Berndt R, 215, 513
Berne BJ, 410, 423, 425, 427, 468
Bernhardt TM, 402
Bernier P, 202
Bernstein RB, 69, 73
Berry RS, 44, 45
Bersuker IB, 475
Bertolo M, 324
Bertozzi CR, 439, 458
Bertran J, 494
Besenbacher F, 343
Bethe HA, 68
Bettermann G, 29
Beutler TC, 410, 412
Beveridge DL, 410, 423
Beylen MV, 179

Bhosale SH, 489
Bibby JM, 315
Bickelhaupt FM, 276
Biesecker JP, 382
Bigot B, 410
Bigot JY, 23
Billeter SR, 471, 476, 482, 485, 492, 496
Binder K, 293, 300
Binnig G, 438
Biondi MA, 68
Bird DM, 322, 323
Birdsall B, 368
Bjerre N, 61
Blackman GS, 326
Blackwell M, 43
Blair W, 178, 185, 186
Blais NC, 72, 79, 82
Blake JF, 410
Blanch HW, 534, 557
Blanchet V, 43, 44, 59, 60, 62, 527
Blank DA, 29, 31, 32
Blasé X, 204
Blatt JM, 68
Bliznyuk AA, 471, 494
Blochl PE, 253, 267–70, 339
Blomberg MRA, 276, 493
Bloomfield VA, 411
Blumenstein M, 365
Blush JA, 128, 133, 136
Bockmann A, 359
Bockrath M, 210, 215
Bode J, 365
Boden N, 235
Bodensteiner T, 190
Boehlen J-M, 375
Boero M, 263, 329, 330
Boggs JE, 475
Bohm A, 75
Bohr N, 68
Bokor J, 508
BOLHUIS PG, 291–318; 264, 291, 295, 299, 300, 306, 309, 312, 313, 469
Bonad J, 213
Bonder EM, 191

Bondi DK, 487
Bonn M, 509, 517
Bonzel HP, 224, 509
Boo DW, 43, 62, 257
Boothroyd AI, 72
Booze J, 102, 124
Bordas J, 191
Borejdo J, 195
Borer PN, 7
Borg HJ, 326
Borisov AG, 512, 518, 521, 523, 527
Borkowski J, 83, 86
Born M, 142
Borne TB, 109
Borodin O, 425
Bosch W, 177
Bouchaud JP, 178
Bouchéne MA, 527
Boudaïffa B, 69
Boudon S, 423
Boué F, 180, 184, 190
Boulougouris GC, 410, 421, 422
Bower JE, 402
Bowman JM, 70, 71, 81, 87
Bowron DT, 422, 426
Boyd KJ, 382, 402
Boyd RK, 469
Boys SF, 278
Bozovic D, 210
Bradforth SE, 91
Bradshaw AM, 509
Bragin J, 215
Bramley MJ, 159
Bratko D, 534, 557
Bratos S, 410
Bräuchle G, 384
Braun E, 190, 194, 195
Braun J, 239
Braun M, 91
Brédas JL, 221, 234
Breit G, 68
Bret C, 471, 478, 490
Bridgeman CH, 410
Briggman KA, 438, 441, 443, 446

Briggs D, 440
Brixner T, 507
Brocks G, 142, 159
Bronikowski MJ, 212
Brookhart M, 330
Brooks BR, 493, 495
Brooks CL, 410, 423, 425
Brooks CL III, 470
Brooks HB, 488, 489
Brouard M, 90
Broughton JQ, 533, 536, 548, 550
Brovchenko I, 428
Brown A, 440
Brown FB, 72, 79
Brown MG, 438, 557
Brown WA, 322, 324, 326
Brown WB, 178
Brown WL, 402
Bruckmeier R, 91
Bruge F, 266
Brühwiler PA, 512
Bruice TC, 495
Brundle CR, 101, 102, 109
Brune H, 211
Brutting W, 221
Bryant MA, 399
Bryant Z, 303, 304, 312
Buchenau H, 71
Buck SM, 439, 441, 443, 448, 457
Buckingham AD, 410, 422
Buckman SJ, 68
Buckner JK, 410, 423, 425
Buda F, 252
Buhl H, 29
Bulusheva LG, 214
Buma WJ, 43
Bumm LA, 243
Bunker PR, 141, 151, 257
Buntine MA, 82
Buontempo JT, 30
Burgermeister R, 516, 527
Burgi L, 211
Burke K, 322
Burke PG, 68
Burlacu S, 195

Burleigh DC, 159
Bursing H, 25
Burton NA, 471, 484, 489, 494, 496, 497
Burum DP, 284
Bushuev YG, 410
Bustamante C, 8, 11
Butcher EA, 81
Butcher MJ, 209
Butler LJ, 157
Buttet J, 402
Byers Brown W, 281
Byun K, 472, 475

Calbert JPh, 234
Callahan J, 394
Candau SJ, 177, 196
Canuto S, 410
Cao J, 252, 478
Cao M, 360
Capecchi G, 73
Capek V, 233, 234, 238, 239
Car R, 250, 254, 262, 281, 284, 310, 472
Carbassi F, 438, 453
Cardenas R, 472
Cardillo MJ, 523
Carey C, 557
Carlier M-F, 191, 195
Carloni P, 276, 277, 285, 495
Carlson RJ, 527
Carlson TA, 41, 44
Carney GD, 159
Carolini P, 277
Carrea G, 492
Carrington T Jr, 141, 159, 160, 163, 165
Carroll SJ, 402
Carter CW Jr, 471, 494
Carter EA, 253, 254, 268
Carter S, 159
Casavecchia P, 83
Case DA, 90, 351, 358, 359
Casimir HBG, 142
Castillo JF, 72, 73, 84, 86, 87, 90
Castillo R, 185, 485

Castner DG, 438–40
Castner EW Jr, 21
Cates ME, 174, 177, 178, 196
Cavalli S, 73
Cavanagh RR, 508
Cederbaum LS, 43, 44, 59, 84
Ceperley DM, 321
Cerjan CJ, 293
Ceyer ST, 326, 328, 343
Chabalowski CF, 472
Chakarov DV, 509, 512, 521, 524
Chakrabarti A, 196
Chambaud G, 115
Chambers SA, 326
Chamorro M, 10
Chan CM, 326, 452
Chan DYC, 409
Chan HS, 423
Chan MSW, 331
CHANDLER D, 291–318; 29, 264, 291–93, 295, 299, 300, 306, 307, 309–14, 409, 413, 419, 424, 428, 429
Chandler DW, 46, 62, 85
Chandra N, 44
Chandrasekhar J, 479, 490
Chang WH, 401
Chao SD, 72, 73, 84
Chapman FM, 73
Chapuisat X, 141, 156, 159, 160, 164
Charlier JC, 217
Charlson EJ, 457
Charlson EM, 457
Chase B, 202
Chatelain A, 213
Chatfield DC, 81, 82, 86, 495
Chau P, 410
Chau PL, 410, 422
Chelli R, 423, 427, 428
Chelmick T, 130, 133
Chen C, 267
Chen CY, 439, 441, 443, 448, 455, 457
Chen J, 221, 243

Chen JG, 324
Chen P, 108, 128, 133, 136
Chen PJ, 324
Chen Q, 439, 458
Chen S, 32, 33
Chen SH, 196
Chen WW, 103, 115, 126, 129, 132, 215
Chen XY, 10
Chen Y, 43
Chen YJ, 106, 115
CHEN Z, 437–65; 438–41, 443, 448, 450, 455, 457
Cheng HP, 252
Cheng K, 257, 258
Cheng Y, 423
Cheo PK, 68
Chernyak V, 30
Chesnut DB, 356
Cheung CL, 201, 202, 205, 209, 212, 215, 217
Cheung YS, 106–8, 115
Chiadmi M, 490
Chico L, 217
Chidsey CED, 398
Chien EYT, 368
Chien RL, 44, 45
Chieux P, 190
Child MS, 70, 73, 141, 167
Childers RF, 349
Chingas GC, 352
Chipot C, 423, 427
Chiu ML, 350, 369
Chiu SW, 108
Cho M, 23, 25, 30, 31, 34
Cho WR, 387, 394
Chohan KK, 471, 488
Choi HJ, 208, 210
Chong DP, 249, 251
Chorkendorff I, 326, 327, 343
Chou JJ, 364
Christensen A, 336
Christophe Y, 27
Chu ZT, 483
Chuang Y-Y, 481
Chulkov EV, 511–13, 518, 521, 523, 527

AUTHOR INDEX 567

Chung J, 370
Chung JW, 401
Chupka WA, 104, 111
Ciccotti G, 253, 254, 268
Cieplak P, 470
Ciobîca IM, 339
Citrin PH, 511
Clark CW, 68
Clark T, 493, 497
Clarke JS, 108
Clary DC, 79, 90
Clausen BS, 343, 344
Clauss W, 209
Clayton J, 190
Cloutier P, 69
Cobden DH, 215
Coburn JW, 381
Cochran HD, 410
Cockett MCR, 108
Cohen JS, 349
Cohen ML, 208, 210, 217, 250, 252, 322
Cohen SH, 438
Cohen-Tannoudji C, 525
Coker DF, 276
Colbert DT, 201
Cole HB, 359
Cole KC, 440
Cole RH, 265
Collins PG, 213
Colombo G, 492, 493
Colton MC, 71, 81
Columbo G, 492
Colwell SM, 167
Comelli G, 512
Conboy JC, 438
Conde L, 185
Condon EU, 67
Cong P, 509, 513
Connor JNL, 79, 487
Continetti RE, 42, 46, 62
Cooks RG, 381, 382, 385, 387, 394, 396, 398, 401, 402
Cooper WC, 186
Corchado JC, 471, 472, 475, 477, 478, 480–89, 495

Cordery R, 176
Cordes HG, 329
Cornell WD, 470
Cornil J, 234
Cornilescu G, 364, 365
Corr D, 384
Corrales LR, 177, 180, 183, 184
Cosby PC, 91, 119
Costas ME, 196
Coulson CA, 262, 476
Covell DG, 410, 412, 425
Cowburn D, 361
Cowin JP, 382, 402
Cramer CJ, 245, 471
Cramer SM, 410
Crampin S, 513
Creer G, 331
Creighton S, 468, 476
Cremer PS, 438
Crofton MW, 257
Crommie MF, 209, 215
Crooks GE, 291, 413, 544, 545
Cross PC, 167
Crothers DM, 7
Cruz CHB, 23, 26
Csajka FS, 291, 309, 469
Csaszar AG, 159
Cuccaro SA, 81
Cui Q, 488, 492
Cummings PT, 410
Cummins PL, 471, 494
Cunningham MA, 496
Curioni A, 250, 271, 276, 277
Curti FG, 512, 515
Curtis RA, 534, 557
Curtiss LA, 103, 108
Cyr DR, 43, 44, 59, 60, 62
Cyrier M, 25

D'Agostino O, 67
Dahl S, 326, 327, 343, 344
Dai HJ, 201, 202
Dai H-L, 508, 516
Dainton FS, 175
Dal Peraro M, 277

Daley SP, 326, 328
Dalton BJ, 44
Dang LX, 423, 424
Danielson U, 476
Dankert J, 457
Darakjian Z, 81
Darden T, 479
Darling BT, 141, 150
Das SS, 178–80, 183–86
Dashevsky VG, 410, 423
Datz S, 91
Davenport RC, 492
Davidson VL, 488, 489
Davies JA, 46, 62, 447
Davis AV, 43
Davis PP, 235
Dawber G, 110
Day TJF, 303, 308
Debenedetti PG, 534, 557
de Boeij WP, 18, 23, 25
Debrunner PG, 370
Decatur S, 33
Decius JC, 167
deDios AC, 352, 353, 356–58, 360, 365–67, 369–71
Dee GT, 455
de Gennes PG, 175
DeGrado WF, 33
de Heer WA, 213
Dehmer JL, 121
Dehmer PM, 104, 121
Dejaegere A, 473
DeJong PHK, 410, 422
Dekker C, 201, 202, 204, 206, 209, 212, 214, 217
Dekoven BM, 443, 446
Delaglio F, 364, 365
Delaney P, 208
de Lange CA, 43, 382
Delchar TA, 385, 437, 457
DELLAGO C, 291–318; 264, 291, 295, 299, 300, 307, 309–13, 469
Delley B, 215
DeLucia FC, 163
De Maeyer L, 263, 313

D'Mello M, 82, 83
D'Mello MJ, 83, 86
de Miranda MP, 90
Demirdoven N, 31, 34
Demontis P, 410
Dengler B, 7
Dennison DM, 141, 150
Denny J, 361
Denny RA, 30
Denzler DN, 509
Derclaye S, 453
des Cloizeaux J, 175
Descotes L, 190
Deshpande VV, 489
de Silva V, 315
De Silvestri S, 24
Devlin FJ, 472
Devoret MH, 202
de Vries MS, 135
Dewar MJS, 472, 475
Deyerl HJ, 108
Dhinojwala A, 438, 439, 441, 443, 446, 447
Dianoux AJ, 190
Dickenson H, 130, 133
Dickerson RE, 9
Diehl H, 159
Diehl RD, 512
Dietrich S, 534
Diffey WM, 25
Dill D, 41, 44
Dill KA, 410, 428, 534, 557
Dimov SS, 112
Dirlikov S, 448
Ditchfield R, 283
Dixon RN, 83
Dixon SL, 471, 494
Dlott DD, 33
Dobber MR, 43
Doclo K, 495
Doege G, 29
Doll JD, 312
Domcke W, 41, 43, 44, 59
Domen K, 440
Dong F, 72
Dong W, 328
Donhauser ZJ, 243

Doniach S, 301, 422, 425
Donini O, 470, 479, 490, 491
Dou Y, 384, 385
Dougal SM, 438, 439, 441, 443, 446, 447, 453, 454
Douglas JF, 176, 183, 184, 191, 192, 194, 196
Drabe KE, 382
Draxl MK, 103, 106, 123
Dresselhaus G, 201, 203, 212–14, 217
Dresselhaus M, 201, 203, 212, 214
Dresselhaus MS, 213, 217
Dreesen L, 439, 458
Dressler R, 115
Dreyfuss MP, 184
Dreyfuss P, 184
Dronskowski R, 339
Du M, 527
Du R, 303, 304, 312
Dubois LH, 513
Dudde R, 224, 233
Dudowicz J, 176, 183, 184, 191, 192, 194, 196
Duffy EM, 410
Dumas R, 490
Dumesic JA, 339
Dumoulin MM, 440
Dunbrack RL, 470
Dunlap BI, 201, 217
Dunning F, 111
Dunning TH Jr, 481
Dupont-Gillain CC, 453
Duppen K, 18, 23, 29, 30
Durell SR, 410
Durup J, 119
Dutton G, 229, 230, 233, 240
Duxon SP, 45
Dwyer JP, 70
Dybal J, 448
Dyke JM, 108
Dykstra CE, 350, 366, 369

Earle EM, 371
East ALL, 257, 258
Eastman MA, 352

Eastman P, 301
Ebata T, 135
Ebbesen TW, 201, 217
Eberly JH, 18, 19
Echenique PM, 511–13, 518, 521, 523, 527
Eckart C, 75, 142, 146
Eckert T, 91
Economou IG, 410, 421, 422
Edgecombe KE, 278
Edington MD, 25
Egeberg RC, 326, 327
Egelstaff PA, 190
Eggert S, 205
Ehbrecht M, 402
Ehm D, 511
Eichinger M, 271, 277, 473
Eichler A, 337
Eigen M, 263, 264, 313
Eigler DM, 209, 215
Eisenberg A, 175, 178, 183, 186, 187
Eisenberg D, 262
Eisenmesser E, 410
Eisenthal KB, 34, 438
Eklund PC, 202, 213
Elber R, 301
Ellis K, 110
Ellison GB, 382, 402
Elsayed-Ali HE, 508
Elsner J, 492
Elstner M, 492
Emoto K, 450
Engberts JBFN, 410, 412
Engel T, 328
Engel V, 43, 509
Engholm JR, 33
Enriquez PA, 90
Epa VC, 151
Eppler AA, 439, 440, 455
Epstein ST, 281
Eriksson O, 512
Ernzerhof M, 322
Errington JR, 410, 422
Ertl G, 230, 328, 336, 343, 508–12
Esch F, 512

Escribano R, 257
Espinosa-Garcia J, 477
Esry BD, 91
Estes JE, 191
Eu BC, 69
Eurenius KP, 495
Evans C, 381, 382, 385, 394, 396, 401, 402
Evans M, 102, 103, 105–7, 109, 110, 112, 113, 115, 116, 118, 119, 121, 129
Evans MW, 422
Evans R, 534, 538, 540, 557, 558
Evanseck JD, 277, 470, 492
Even U, 112, 394
Eyring H, 69, 187, 468
Ezzell R, 191

Faeder SMG, 27
Faglioni F, 339
Fairbrother F, 186
Faivre G, 186
Falke JJ, 366
Fan L, 322
Fan WC, 514
Fanelli R, 187, 190
Fang J-Y, 480, 483
Fann WS, 508
Fano U, 68
Farmanara P, 43
Farrow RL, 85
Faulder PF, 471, 488
Faulkner LR, 222
Fauster T, 43, 223, 224, 227, 508, 510–13, 515
Fauth K, 213
Fayer MD, 20, 22–25, 32–34
Fazio DD, 73
Fèlix C, 402
Feast WJ, 438
Fedorov D, 115, 116
Fedorov DG, 115, 116
Fedrigo S, 402
Feeney J, 368
Feher F, 187, 188
Feierberg I, 476, 478, 485

Feijen J, 457
Feldstein MJ, 23
Feng B, 402
Fenn PT, 115
Feranchak BT, 392
Ferenczy GG, 473
Fermi E, 67
FERNÁNDEZ-ALONSO F, 67–99; 73, 85–87, 90
Fernández-Prini R, 410, 412
Fernandez-Ramos A, 481
Ferrante C, 23, 33
Ferre N, 473
Feshbach H, 68, 70
Fesik SW, 375
Fetters LJ, 179, 185
Feynman RP, 18, 478
Field MJ, 253, 471–73, 477, 478, 485, 488, 490, 492, 495
Field RW, 152
Field TA, 44, 46, 56
Fielding HH, 42
Figger H, 91
Filipponi A, 422, 426
Findenegg G, 534
Finger D, 191
Finney JL, 422, 426
Finzel BC, 489
Fischer HE, 422
Fischer I, 43, 108, 136
Fischer N, 508, 512, 515
Fischer R, 512, 515
Fischer SF, 29
Fishkind DJ, 191
Fitts DD, 4
Flaig R, 372
Fleischman SH, 410
Fleming GR, 21, 23, 25, 29–32, 314
Fleniken J, 163
Florian J, 491, 492
Floris F, 410, 412
Flory PJ, 176
Flugge S, 159
Fois ES, 410, 423
Fonseca A, 217

Fonseca T, 21
Fontecilla-Camps JC, 495
Forbes JG, 176, 191, 192, 194, 195
Ford SM, 447
Forester T, 410
Fork RL, 23, 26
Forro L, 213
Forsman J, 410
Fossum ER, 399
Foster JS, 212
Fothergill M, 476
Foulkes WMC, 321, 322
FOURKAS JT, 17–40; 26, 27, 29, 30, 32
Fowkes FM, 439
Fox RE, 68
Fragnito HL, 23
Francis RS, 23, 33
Frank CW, 447
Frank HS, 422
Frank I, 274–76
Frank KH, 224, 233
Franke H, 397
Franses EI, 447
Frantz DD, 312
Frauenfelder H, 68
Frauenheim T, 492
Frechard F, 339
Frederick JH, 141, 160
Freed KF, 159, 176, 183, 184, 191, 192, 194, 196
Freeman DL, 312
Freeman MP, 5
Freindorf M, 472
Freitag M, 209
Freitas JE, 29
Frenkel D, 292, 293, 300, 314
Freudenberg T, 43, 44, 59, 60, 62
Freund H-J, 509
Friedel J, 336
Friedel P, 213
Frieden C, 191
Friedman HL, 411, 421, 424, 427, 429, 430

Friedman RS, 75, 81, 82, 86, 87
Friesner RA, 473
Frisbie CD, 244
Frisch MJ, 472
Frischkorn C, 43
Frohnmeyer T, 43
Frydman L, 352
Fu CC, 447
Fujii A, 135
Fujimura Y, 44, 56
Fujita M, 201
Fukui K, 268
Funk S, 509
Furlani TR, 472
Fushman D, 361
Fygenson DK, 190, 194, 195

Gador D, 387
Gadzuk JW, 224, 508
Gaffney KJ, 230, 233, 240
Gahl C, 230, 232–35, 237, 240
Gale JD, 332
Gallagher T, 111
Galli G, 265
Gallicchio E, 410, 425, 428, 534, 557
Gamba A, 410, 423
Ganapathy S, 352
Ganduglia-Pirovano MV, 336
GAO J, 437–65; 410, 427, 469, 471, 472, 475–78, 480–86, 488, 489, 491, 495
Garbis S, 381, 392
García AE, 409, 410, 413–15, 417, 420–27, 429, 430, 534, 545, 554, 557
Garcia-Viloca M, 471, 472, 475, 477, 478, 480, 485, 486, 489, 496
Garde S, 4, 409, 411, 413–15, 417, 418, 420, 423, 425, 427, 429, 430
Gardecki JA, 21
Gardissat J-L, 186

Garnier F, 221
Garnier M, 511
Garrett BC, 70, 79, 81, 307, 468, 469, 471, 478, 481, 483, 487
Garrett RH, 369
Gasser U, 309
Gatti F, 159, 160, 164
Gauckler M, 439, 440, 455
Gautam KS, 438, 439, 441, 443, 446, 447
Gauyacq JP, 385, 512, 518, 521, 523, 527
Ge N-H, 42, 43, 55, 224, 227, 229, 238, 240, 508
Ge Q, 337, 338
Ge SR, 443, 446
Gebhard W, 190
Gedeck P, 493
Gee G, 186
Geerlings JJC, 386, 389
Geiger A, 410, 428
Geisler M, 43
GEISSLER PL, 291–318; 264, 291, 307, 310, 311, 313, 314, 428, 557
Gelb LD, 534
Gelbart WM, 196
Geller M, 495
Gensmantel N, 497
Gentile ME, 409, 412, 414, 419, 430
Georges AT, 45
Gerber C, 438
Gerber G, 41–43
Gerber RB, 480, 523
Gerfin T, 375
Gerhards M, 135
Gerjouy E, 68
Gerrity DP, 489
Gershman LC, 191
Gertner BJ, 468, 481
Gervazio FL, 423, 427, 428
Geyer CJ, 312
Ghosh T, 422–24, 426
Giannozzi P, 277, 281
Gibson JM, 201

Gilbert T, 108, 136
Gill SJ, 427
Gillan MJ, 250, 252, 478
Gillilan RE, 301, 496
Gilliland GL, 489
Gingrich NS, 190
Girard B, 527
Girault HH, 453
Gislason EA, 390
Glasfeld A, 489
Glassey WV, 339
Glasstone S, 468
Gleeson MA, 395
Godbout N, 358, 359, 363, 369–71
Goddard WA, 339
Godefroid M, 163
Goerler GP, 187, 188
Goldby IM, 402
Golden DE, 68, 75
Goldman S, 423, 424
Golonzka O, 31, 34
Gomer R, 508
Gomez MA, 409, 412, 414, 415, 419, 429, 430
Gomez PC, 257
Gomez-Llorente J, 76
Gonzalez-Lafont A, 475, 481, 496
Gonze X, 281
Goode JG, 135
Goodman DW, 324, 328, 379
Gordon M, 115, 116
Gordon MS, 115, 116
Gostein M, 85
Goto T, 439, 441, 453, 454
Gotzelmann B, 534
Gould IR, 470
Gourisankar SV, 457
Graber T, 257
Gracias DH, 438–40, 443, 445, 455
Graener H, 33
Gragson DE, 438, 557
Graham RL, 108
Gralla J, 7
Grande S, 410

Grandinetti PJ, 352
Granstrom EL, 244
Grant DM, 354
Grant ER, 115
Gratzel M, 222
Graybeal JD, 67
Gready JE, 471, 494, 496
Greatbanks SP, 494
GREELEY J, 319–48
Greeley JN, 381, 389, 390, 392
Green CD, 262
Green DC, 27
Green WH, 75
Green WH Jr, 159
Greenblatt BJ, 43
Greene CH, 91
Greenfield SR, 32, 33
GREER SC, 173–200; 174, 176–80, 183–88, 190–92, 194
Gregonis DE, 457, 461
Gregor T, 284
Grev RS, 481
Griffiths DV, 368
Grill V, 381, 382, 385, 401, 402
Grimley JS, 358, 359
Grimmelmann EK, 468, 470
Grimsditch M, 187
Grimvall G, 511
Grisham CM, 369
Grochowski P, 471, 480, 484, 495, 497
Grönbech-Jensen N, 301, 422
Groot M-L, 25
Grosberg AY, 175, 303, 304, 312
Grote RF, 468
Grubmüller H, 301
Gruetzmacher JA, 438
Gruner K, 180, 183
Gu X, 180, 183
Guardia E, 307
Gubbins KE, 428, 534, 548
Guerra CF, 276
Guevremont J, 440

Guillot B, 410
Guisoni N, 410
Guissani Y, 410
Gujrati PD, 176
Gülseren O, 323
Gumhalter B, 516, 525
Gunthard HsH, 159, 163
Gunther H, 21, 27
Guo H, 277
Guo K, 350, 369
Gurney RW, 511, 512
Güthe F, 110
Guttman CM, 178, 185, 186
Guyon PM, 102, 109
Guyot-Sionnest P, 440
Gygi F, 253, 265

Haak JR, 410, 413
Haas R, 68
Habenicht W, 102, 111
Haberland H, 512
Häberli M, 375
Hack MD, 483
Haddon RC, 242
Hafner JH, 201, 210, 215, 331, 332, 337, 339
Hagfeldt A, 222
Hahn JR, 401
Hahn S, 30
Haight R, 224, 238, 508, 509
Hajduk PJ, 375, 376
Hajlaoui R, 221
Hald K, 61
Hall GE, 90
Hall RI, 110
Hall RJ, 494
Hallstrom P, 475
Halvick P, 81
Hamada N, 201
Hamann DR, 323, 518
Hamm P, 33
Hammer B, 319, 322, 332–35, 343
Hammerich AD, 480
Hammes-Schiffer S, 471, 476, 478, 480, 482, 483, 485, 487

Hammick DL, 187
Hammond BL, 321
Han J, 205
Han W-G, 496
Handy NC, 141, 159, 164, 167
Hanley L, 381, 382, 392, 393, 398, 401
Hansen J, 44, 45
Hansen JP, 534, 544, 545, 548, 557
Hansen LB, 322, 333
Hanser CFW, 492, 496
Hansson T, 492, 497
Harbich W, 402
Harder R, 394
Harrell S, 365, 367
Harris CB, 42, 43, 55, 224, 227, 229, 230, 233, 235, 238, 240, 508
Harris JM, 450
Harris RA, 9
Harris RE, 186
Harrison MJ, 471, 484, 496, 497
Harrison SW, 410
Hart JC, 497
Hartke B, 253
Hartmann SR, 21, 28
Hartsough DS, 472
Hase WL, 468, 469
Hasegawa M, 453
Hasegawa Y, 209
Haslett TL, 402
Hassanien A, 209
Hasselbrink E, 230
Haug K, 81
Haugk M, 492
Havlin RH, 358, 363, 370, 371
Hayaishi T, 110
Hayden CC, 41, 43, 44, 46, 59, 60, 62, 72
Hayes EF, 73, 81
Hayes TL, 8
Haymet ADJ, 410, 422–24, 534, 557

He GZ, 108
Head-Gordon T, 410, 423
Healy EF, 472, 475
Heberle A, 508, 511
Heberle AP, 510, 525, 526
Heck AJR, 46
Heermann DW, 293, 300
Hehre WJ, 472
Heiland W, 394, 397
Heimann P, 102, 103, 105, 106, 109, 110, 112, 113, 115, 121, 129
Heinz TF, 508
Heiz U, 402
Helbing J, 41, 43
Held H, 441
Helfand E, 468, 470
Helfrich W, 176
Heller J, 358
Hellsing B, 509, 512, 521, 524
Hellwarth RW, 18
Helm H, 91
Henderson D, 544
Henderson JR, 534, 547, 558
Hengsberger M, 511
Henkelman G, 301
Henley EM, 68
Hennen PC, 105
Henriques VB, 410
Henry L, 154
Hensel F, 190
Heo J, 31
Hepburn JW, 102–5, 112, 115, 118, 121, 122, 134
Herbst E, 163
Herek JL, 509, 513
Herman GS, 326
Hermans J, 471, 494
Hernández-Cobos J, 410
Herrero VJ, 71, 72, 83, 86
Herrmann G, 394
Herron JT, 103, 106, 123
Herschbach DR, 69, 90
Hertel IV, 43, 44, 59, 60, 62, 105

Hertel T, 230, 242, 508, 510, 511
Herwig J, 329
Herzberg G, 141
Herzenberg A, 68
Hess C, 509, 517
Hibbs AR, 478
Hilber R, 104
Hill JR, 33
Hillier IH, 471, 484, 489, 494, 496, 497
Hindle SA, 471, 489, 494
Hindman JC, 284
Hines JV, 10
Hinton JF, 352
Hipes PG, 71, 81
Hirata Y, 27
Hirayama Y, 209
Hirosawa I, 440
Hirose C, 440
Hirschfelder JO, 69, 281
Hjortstam O, 512
Hla S-W, 508
Ho JT, 176
Ho L, 475
Ho LL, 496
Ho W, 508, 516
Hochlaf M, 115
Hochstrasser RM, 33, 34
Hodoscek M, 493
Hofer U, 43, 223, 224, 508, 510, 511
Hoffman GA, 263
Hoffmann FM, 328
Hoffmann R, 339
Hofmann H, 43
Hofstein JD, 135
Hogt AH, 457
Hohenberg P, 249, 322
Hollander P, 439, 458
Hollenstein U, 104, 112
Holmes KC, 190
Holt WE, 187
Holthausen M, 249, 251
Homoelle BJ, 25
Hong M, 359
Hong SC, 439, 441, 453, 454

Hood RQ, 265, 322
Horng ML, 21
Horowitz CJ, 72
Horowitz G, 221
Hosokawa S, 187
Hotta K, 175
Hotzel A, 223, 230, 232, 234, 235, 237, 240, 510, 511, 515, 528
Hotzl A, 508, 527
Hougen JT, 141, 142, 154, 156, 159, 163
Houston JE, 244
Houston PL, 90
Howard AJ, 489
Howard BJ, 156, 157
Howard JB, 141, 147, 150
Howells WS, 190
Hsu CW, 102, 103, 105, 106, 108–10, 112, 113, 115, 116, 118, 119, 121, 129, 136
Hu H, 489
Hu P, 328, 340, 341
Hu QJ, 102, 121
Hu R, 33
Hu XK, 102, 121
Huang DM, 428, 550, 553, 554, 557
Huang JC, 106, 107, 115
HUANG J-L, 201–20; 201, 202, 204–7, 209, 212, 213, 215–17
Huang JS, 185
Huang JY, 440
Huber M, 343
Huels MA, 69
Hüfer M, 513, 521
Hüffer S, 329
Hüfner S, 511
Hui KM, 185
Huisken F, 402
Hullinger VD, 75
Humbert C, 439, 458
Hummer G, 4, 409, 411, 413–15, 417, 418, 420, 423, 425, 427, 429, 430

Humphreys SE, 323
Hunt JH, 440
Hunting D, 69
Hunziker HE, 135
Huo WM, 82
Hush NS, 410
Huster D, 359
Hutschka F, 332
Hutson JM, 159
Hutter J, 250, 253, 254, 262, 264, 271, 274–77, 279, 291, 303, 311, 313, 473
Hwang JK, 468, 476, 478, 479, 483, 484, 493
Hwang KS, 438
Hybl JD, 27
Hynes JT, 253, 254, 268, 468, 469, 481

Ibach W, 439, 443, 445
Iedema MJ, 382, 402
Igarashi K, 402
Ignatiev A, 514
Ihm H, 324
Ihm J, 208, 210, 250, 252
Ikeguchi M, 410, 412
Ikeshoji T, 263
Imai N, 175
Imke U, 390, 394
Impey RW, 490
Inaba R, 29
Inagaki N, 443, 446
in'tVeld PJ, 413
Iordanov T, 471, 476, 482, 485
Ippen EP, 24
Isaacson AD, 481
Ishida H, 511, 512
Ishioka K, 230, 232–35, 237, 240
Itoh H, 27
Iung C, 141, 159, 160, 164
Ivin KJ, 175
Ivkov R, 191, 194, 195
Iwasaki A, 135
Iwata S, 101, 102, 115

Jacobi K, 324
JACOBS DC, 379–407; 381, 384, 389–92, 394, 397–99
Jacobs DT, 180, 183
Jacobsen CJH, 343, 344
Jacobsen KW, 336
Jaffe R, 423, 427
Jagod M, 257
James A, 322
Jamneala T, 215
Janes N, 352
Jang S, 252, 478
Janin J, 490
Janmey PA, 195
Jansen TIC, 30
Janssen JW, 202, 212, 214
Jardetzky O, 349
Jarrold MF, 402
Jarvis GK, 102, 103, 106, 112, 115, 116, 118, 119, 124, 128–30, 132–34
Jarzeba W, 21
Jasper AW, 483
Jayaram B, 423
Jayasinghe L, 72, 84
Jeandupeux O, 211
Jencks WP, 491
Jenkins J, 490
Jenkins SJ, 337
Jensen E, 108
Jensen MJ, 91
Jensen P, 141
Jensen PJ, 196
Jeon HS, 185
Jessup RS, 183
Jewsbury P, 497
Ji Y, 401
Jia Y, 25
Jishi RA, 215
Joannopoulos JD, 250, 252, 319
Johannesson G, 301
Johansson B, 336
Johns JWC, 141
Johnson AD, 326, 328
Johnson AT, 209

Johnson BR, 69
Johnson JK, 548
Johnson KL, 444
Johnson P, 102, 103, 134
Johnson PM, 135
Johnston HS, 470
Jonas DM, 25, 27, 30
Jonas J, 263, 413
Jones GS, 324, 325
Jones LH, 488, 489
Jones NR, 471, 484
Jones RAL, 438
Jones RO, 178
Jonsson B, 410
Jonsson H, 262, 301, 478
Joo T, 23–25
Jordan RE, 229, 235
Jordanides XJ, 25, 30
Jorgensen WL, 410, 423, 425, 427, 470, 479, 490
Joseph D, 492
Joseph T, 469
Joubert D, 253
Jourand D, 471, 478, 490
Journet C, 202
Jouvet C, 43, 44, 59, 60, 62
Jucknischke O, 185
Judson RS, 85, 90
Jungen Ch, 44
Junker BR, 75
Justum Y, 159, 164
Juugnickel G, 492

Kabsch W, 190
Kachalova GS, 370
Kador PF, 493
Kahlow MA, 21
Kahn K, 495
Kaira A, 410
Kaiser B, 402
Kajiyamas T, 443, 446
Kakitani K, 276
Kaler EW, 422, 425
Kalyuzhnyi Y, 410
Kaminski G, 410
Kaminsky W, 329
Kandel SA, 85, 90

Kane CL, 205, 208, 209
Kang H, 385, 387, 394, 401
Kang TJ, 21, 26, 27, 29
Kanisawa K, 209
Kano SS, 509
Kanter EP, 257
Kao CT, 326
Kappes MM, 394, 395
Kapral R, 253, 254, 268
Kar B, 447
Karaborni S, 410
Kargupta K, 557
Karim OA, 307
Karlstrom G, 410
Karplus M, 69, 72, 301, 312, 468, 470, 472, 473, 475, 481, 488, 490–92, 494
Kas J, 191
Kasai M, 175, 191
Kasemo B, 509, 512, 521, 524
Kasi SR, 385
Kassel LS, 311
Katsov K, 533, 535, 536, 538, 540, 543, 546, 549, 550, 554
Katsumata S, 101, 102
Katz J, 187, 188
Kaufman DJ, 86, 87
Kaufman LJ, 29, 31, 32
Kauzmann W, 425
Kawamura M, 195
Kawashima H, 32
Kaxiras E, 492
Kaye JA, 70
Kazansky AK, 512, 518, 521, 523, 527
Keck JC, 468–70
Keith TA, 283
Keller C, 512
Keller JS, 108
Kelley TW, 244
Kelly KF, 243
Kelly PS, 322
Kemble EC, 153
Kempter V, 387, 389
Kendall K, 444

Kendrick BK, 72, 73, 84, 85, 87
Kennedy SJ, 173, 175–77, 184, 186, 187, 196
Kenny S, 322
Kent JT, 315
Keogh GP, 30
Keogh WJ, 72
Kern CW, 159
Kern K, 211
Kersters-Hilderson HLM, 490
Kesmodel LL, 453
Keyes T, 30
Khalil M, 31, 34
Khare R, 413
Khawaja E, 44, 45
Khidekel V, 29, 30
Khokhlov AR, 175
Kidane A, 450
Kiefer B, 507
Kim B, 43, 44, 59, 60, 62
Kim BC, 401
Kim D, 30, 265, 439–41, 443, 453
Kim G, 389, 394, 397, 398
Kim HI, 244
Kim J, 438
Kim K, 486
Kim MS, 132, 134
Kim P, 201, 202, 204, 206, 207, 209, 212, 213, 216
Kim SJ, 255, 361
Kim SW, 457
Kim SY, 132, 134
Kim Y, 158, 163, 477
Kim YJ, 326, 413
Kim ZH, 90
Kimber BJ, 368
Kimura K, 101, 102
Kincaid RH, 410
King DA, 322, 324, 326, 337, 338
King DS, 358, 508
King GC, 115, 468, 476
King-Smith RD, 263
Kinoshita S, 27
Kinosian HJ, 191

Kioke M, 102, 103, 105, 106, 112
Kip BJ, 440
Kirkwood JC, 31
Kirkwood JG, 4
Kistemaker PG, 383, 395
Kittlemann O, 105
Klein ML, 252, 258, 260, 265, 266, 490
Kleiner A, 205
Kleiner I, 163
Kleinermanns K, 135
Kleint C, 224
Klement W, 187, 188
Kleyn AW, 339, 383, 390, 394, 395
Kliewer J, 513
Kliner DAV, 71, 82, 84, 85, 90
Klinman JP, 486
Klippenstein SJ, 81, 468, 469
Kloc C, 242
Klopper W, 255
Klosek MM, 304
Klug DD, 255, 257, 260
Knapp EW, 29
Knee JL, 43, 53
Knight AEW, 153
Knipping U, 438
Knobler CM, 177
Knoesel E, 223, 230, 232, 237, 508–11, 515, 527, 528
Knox WH, 26
Koch EE, 224, 233
Koch L, 186
Koch MHJ, 191
Koch W, 249, 251
Kocher JP, 410, 412, 413
Koenders BG, 382
Koenig JL, 448
Koh JC, 187, 188
Kohguchi H, 46, 56
Kohn B, 402
Kohn W, 249, 322
Kokoouline V, 91
Kolbuszewski M, 257
Kollman P, 410, 423, 427

Kollman PA, 410, 470, 472, 473, 479, 490, 491
Kondo J, 215
Koneshan S, 410
Konnur R, 557
Koop OY, 410
Köppel H, 43, 44, 59, 76, 84, 91, 92
Koppers WR, 383, 395
Koritsanszky T, 372
Korn ED, 191, 195
Korn G, 105
Kornienko O, 401
Kose R, 322
Kosloff R, 480
Kouki F, 221
Kouri DJ, 81, 82, 85, 90
Koyano I, 102, 136
Kozhevnikov VF, 187, 188, 190
Krack M, 286
Kramer GM, 255
Krane H-G, 372
Krause H, 134
Krause JL, 76, 91, 92
Krauss M, 472
Kreckel H, 91
Kreevoy MM, 481, 487
Kress JD, 81, 409, 430
Kresse G, 253
Krimm S, 448
Krischok S, 387, 389
Kroenke CD, 361
Krohn S, 91
Kroll DM, 534
Kronig RdeL, 141
Kropholler M, 395
Kroto HW, 141
Krynicki K, 262
Kua J, 339
Kuan SWJ, 447
Kubarych KJ, 31
Kubo MM, 410, 425
Kubo R, 20, 294
Kubota J, 509
Kubota S, 510, 525, 526
Kugel KI, 187, 190

Kuhn B, 479, 490, 491
Kuhn PS, 557
Kuik GJ, 382
Kuipers L, 402
Kuk Y, 402
Kulander KC, 91
Kumar SK, 196
Kunc K, 277
Kundu S, 25
Kung AH, 104
Kuppermann A, 69–72, 81–84
Küppers J, 512
Kurland RJ, 361
Kurnit NA, 21
Kusalik P, 410
Kussell E, 413
Kutzelnigg W, 255, 283
Kuyatt CE, 68
Kwak K, 30
Kwok AS, 33
Kwok DY, 439
Kwok RWM, 401
Kwon YK, 208

Laaksonen A, 410, 423
Laasonen K, 255, 257, 262, 264–266, 313
Laenen R, 33, 34
Laguna MA, 402
Laidler KJ, 468
Lakes RS, 439, 443
Lakshminarayan C, 43, 53
Lam CNS, 439
Lamb WJ, 263
Lambeir A-M, 490
Lambin PH, 217
Lambropoulos P, 45
Lammich L, 91
Lamy de la Chapelle M, 202
Landau DP, 177, 178
Landau LD, 76
Landman U, 252, 286
Lang MJ, 25, 30
Lang ND, 511, 512
Lange M, 91
Langevin D, 178

Langford JC, 315
Langmuir I, 457, 511
Langreth DC, 525
Lanig H, 493, 497
Lannoo M, 213
Lanziserra DV, 90
Lapicki A, 382, 402
Largo A, 104
Laria D, 311, 410, 412
Larkin JA, 187, 188
Larsen DS, 25, 30
Larsen JH, 326, 327
Larsen JJ, 41, 43, 44, 59–62
Lasksonen A, 265, 266
Lau EY, 495
Lau KC, 103, 126, 129, 132, 133
Lau WM, 390, 401
Laubereau A, 33, 34
Lauderdale JG, 481
Launay JM, 70, 71
Lavie A, 489
LaViolette RA, 186, 409, 412, 414–16, 419, 430
Lawrence WD, 153
Laws DD, 352, 353, 358, 369, 370
Lazaridis T, 410
Le H, 358, 360, 361, 363–65, 367, 368
Lebowitz JL, 537, 544
LeClaire JE, 46, 135
Ledentu V, 328
LeDourneuf M, 70, 71
Lee B, 409–13, 420, 421, 429
Lee CT, 472, 490
Lee CY, 468, 534
Lee FS, 367
Lee HC, 370, 401
Lee HW, 387, 394
Lee MB, 343
Lee SB, 343
Lee SH, 72, 534
Lee SI, 10
Lee SM, 394, 401
Lee TJ, 257

Lee T-S, 471, 473, 479, 490, 494
Lee YK, 352
Lee YS, 493
Lee YT, 72, 104, 112, 257, 258
Lefebvre R, 76, 79
Lefebvre-Brion H, 115
Leforestier C, 81
Le Guillou JC, 178
Lehmann J, 43
Lehmann KK, 152
Lehr L, 43
Leiderer P, 177
Lemay SG, 212
Leone SR, 49, 52, 527
Leong MK, 475
Lepetit B, 82
Lesemann M, 413
Lester WA, 321
Lesyng B, 471, 480, 484, 497
Leuchs G, 44, 45
Leung K, 534, 557
Levene HB, 71, 85
Levin Y, 557
Levine D, 475
Levine MD, 7
Levine RD, 69, 73, 78, 86, 394
Levitt M, 410, 472, 473
Lev-On T, 90
Levy M, 179
Levy RM, 410, 423, 425, 428, 534, 557
Lewis JP, 471, 494
Lewis KB, 457
Li A, 439
Li H, 425
Li J, 215
Li S, 364
Li TS, 371
Li WK, 108
Lian C, 365, 367
Liang W, 210
Lianos L, 439, 443, 445
Liao CL, 106, 108, 115, 136
Liao CX, 106, 107, 115

Libchaber A, 190, 194, 195
LIEBER CM, 201–20; 201, 202, 204–7, 209, 210, 212, 213, 215–17
Liebsch A, 511, 512
Liew CC, 263
Lifshitz EM, 76
Light JC, 71, 73, 82, 91
Lightbody ML, 438
Lightstone FC, 265
Lim C, 523
Lim H, 381, 392
Lim M, 33, 34
Limaye AC, 494
Lin CC, 141
Lin CL, 410, 422
Lin ST, 267
Lin MY, 185
Lindgren SÅ, 512, 513, 515
Lindh R, 257
Lindner P, 185
Lindsay SM, 438
Lineberger WC, 43, 62
Lingle RL Jr, 42, 43, 55, 224, 227, 229, 235, 238, 240, 508
Linse P, 410, 423, 427
Liphardt J, 11
Lipschitz I, 448
Lipson RH, 112
Lisovskaya TY, 187, 190
Litster JD, 176
Littau KA, 22, 24, 25
Littlejohn RG, 141
Liu B, 72
Liu HY, 479, 489, 491, 492, 495
Liu JB, 103, 115, 126, 129, 132
Liu K, 72, 83
Liu SB, 471, 494
Liu SH, 230, 233
Liu Y-P, 481
Liu ZF, 257, 258, 266, 267
Lizzit S, 512
Lluch JM, 472, 496
Löbau J, 438

Lobaugh J, 478
Lobban C, 422
Lochbrunner S, 41, 43, 44, 59, 60, 62
Loesch HJ, 90
Loffreda D, 326
Logadottir A, 326, 327, 343, 344
Lohrenz JCW, 253, 267
Loiseau A, 202
Lombardi JR, 159
London F, 69
Long P, 43
Longuet-Higgins HC, 535, 539
Loomis RA, 527
Loos M, 473
Lopes MCA, 115
Lopez MR, 497
Lopez-Martens R, 43
Lorente N, 385
Loring RF, 28
Los J, 386, 389
Lotshaw WT, 30
Lou L, 215, 512, 521
Louck JD, 151
Louie SG, 204, 208, 210, 215, 217, 283, 284
Lovell PA, 443
Lowen H, 534, 559
Lu DH, 481
Lu H, 493
Lu KT, 102, 103, 105, 106, 112
Lu Y, 110, 111
Lucas AA, 217
Lucas M, 411, 429
Lucia J, 43
Luck LA, 366
Lüdemann S, 423, 424
Ludowise P, 43
Luger P, 372
Lukka TJ, 141, 143, 167
Lum K, 428, 429, 533, 534, 542, 549, 554, 557
Lumbroso H, 186
Lundqvist BI, 319

Lutz CP, 209, 215
Lutz HD, 187, 188
Luzar A, 534, 557
Luzhkov V, 476, 478, 485
Lvovsky AI, 439, 441, 443, 449
Lynch GC, 79, 81, 82, 481
Lynch GL, 481
Lynden-Bell R, 410
Lynden-Bell RM, 410
Lyne PD, 491, 493, 494
Lyubartsev AP, 265, 266

Ma A, 30
Ma J, 26, 27
Ma NL, 108
Ma ZX, 108, 136
Maarouf AA, 208
Maazouz M, 384, 399
Maazouz PL, 399
Maccagno P, 447
Mack J, 376
MacKenzie G, 471, 489
Mackenzie S, 136
MacKerell AD Jr, 470, 475, 490
Mackie AD, 410
MacKnight WJ, 186
Madan B, 410, 412
Madden PA, 250, 278, 279
Madey TE, 509
Madhavan V, 215
Madura JD, 490
Maekawa H, 34
Maestre MF, 8
Magnuson AW, 481
Mahapatra S, 43, 44, 59, 84, 91
Mahnert J, 110, 128
Maigret B, 423, 427, 473
Makarewicz J, 141, 160
Makarov DE, 478
Malmsten M, 450
Malow M, 102, 103, 112, 124, 128, 130, 132, 133
Mancera RL, 410, 422
Mandl F, 68

Mani AA, 439, 458
Mannherz HG, 190
Manoharan HC, 215
Manolopoulos DE, 71, 72, 82, 83, 90
Mantel DA, 508
Mantooth BA, 243
Manz J, 42, 43, 70, 79
Mao DM, 112
Mao J, 368
Marassi FM, 361
Marchi M, 276, 423, 427, 428
Marconi UMB, 534
Marcus RA, 73, 311, 469, 509
Mardia KV, 315
Marek D, 375
Margl P, 253, 267, 268
Marinero EE, 104
Markley JL, 349, 361, 362, 364
Maroncelli M, 21
Marston CC, 70
Mårtensson N, 512
Marti S, 494
Martin JDD, 102, 121, 122
Martin JML, 363
Martin JS, 381, 389, 390, 392
Martin PG, 72
Martin RL, 409, 430
Martin RM, 322
Martinez TJ, 493
Martins JL, 323
Martinu L, 381
Martyna GJ, 252, 286
Maruyama Y, 394, 402
Marx D, 250, 252–55, 257, 258, 260, 264, 274, 275, 278, 303, 313, 478
Marzari N, 262, 278, 281
Maser WK, 202
Maseras F, 475
Massobrio C, 402
Massova I, 479, 490
Mataga N, 27
Matkowsky BJ, 304
Matousek P, 44, 46, 56
Matro A, 527

Matsudaira P, 191
Matsui T, 110, 111, 410
Matsuoka T, 187
Matubayasi N, 410, 423, 425
Matzdorf R, 513
Mauri F, 283, 284
Mavri J, 480
MAVRIKAKIS M, 319–48; 323, 324, 325, 333, 337
May K, 394, 395
May V, 520
McCague C, 257, 258
McCammon JA, 307, 468, 471, 480, 484, 497, 534
McClelland GM, 90
McConkey A, 110
McCormack EF, 121
McCormick JE, 368
McCoy AB, 159
McDermott A, 359
McDonald IR, 534, 544, 545, 548, 557
McEachran RP, 322
McEuen PL, 201, 215
McGarvey BR, 361
McGlashan ML, 175
McGrath R, 512
McGreevey RL, 190
McKoy V, 46
McLendon GL, 222
McMahon MT, 369, 370, 372
McNeill JD, 42, 43, 55, 224, 227, 229, 230, 235, 238, 240, 508
McPherson T, 450
McQuarrie DA, 320
McRae RP, 483
McWeeny R, 321
McWhorter S, 447
Mead CA, 72, 79, 82
Meadows DH, 349
Meadows RP, 375
Medlin JW, 324
Meerschdorf M, 43
Mei HS, 265
Meier C, 43
Meier W, 82, 83

Meijer EJ, 273
Meister TF, 534
Mele EJ, 205, 208, 209
Melissas VS, 481
Meller J, 301
Mendoza R, 376
Meng EC, 410
Meng M, 490
Menou M, 159, 164
Menzel D, 508
Merchant KA, 33, 34
Merck A, 329
Merkt F, 82, 104, 112, 113, 128, 130, 133, 136
Mermin ND, 204
Merrall GT, 186
Merz KM Jr, 276, 470–72, 492–95
Meserole CA, 384, 385
Mesmer RE, 410
Messer JK, 163
Messina M, 471, 478
Messmer MC, 438
Metiu H, 509
Metz RB, 91
Meyer B, 186
Meyer G, 237, 508, 509, 527
MEYER H, 141–72; 158, 163
Meyer R, 79, 159, 163
Meyers GF, 443, 446
Mezei M, 423
Micha DA, 69
Michaelides A, 328, 340, 341
Michalak A, 268, 331
Michels MAJ, 192
Mielke SL, 81, 82, 478
Mihailovic D, 209
Mikami N, 135
Mikkilineni R, 425
Milchev A, 177, 178
Milkovich R, 179
Miller AD, 230, 233
Miller MA, 293
Miller RJD, 31, 222, 508
Miller SA, 387, 394
Miller WB, 69

Miller WH, 69, 71, 75, 76, 81, 85, 90, 293, 316, 468, 478, 482
Millie Ph, 43
Mills G, 478
Milne CJ, 31
Minn FL, 69
Mintmire JW, 201, 204, 205, 208, 212
Miranda PB, 438, 440, 441
Mistrík I, 91
Mitas L, 321, 322
Mitchell DJ, 409, 413
Mitchell KA, 141
Mitchell SA, 398, 399
Mitra R, 389, 394, 397, 398
Mitsuke K, 106, 115
Mittendorfer F, 337
Miura S, 260
Miyake A, 178
Miyama M, 457
Miyamae T, 452–54
Miyasaka H, 27
Mizes HA, 212
Mladenovic M, 81, 141, 160, 164
Mo Y, 361, 469, 472, 475–77, 483, 491
Moffitt W, 4
Mohktari A, 509, 513
Moiseyev N, 75, 76
Mole T, 331
Molenbroek AM, 343
Moliner V, 485, 494–97
Molnar F, 493
Molteni C, 266, 276
Monard G, 276, 472, 473, 497
Monnell JD, 243
Monot R, 402
Montez B, 365, 367, 368
Montgomery JA Jr, 468
Moody CM, 368
Mooij M, 212
Moore CB, 75, 263
Moore D, 5
Moore KD, 356
Moos G, 242

Mooseker MS, 191
Mordasini T, 276, 277
Mordasini TZ, 492, 496
Mordaunt DH, 83
Morgenroth W, 372
Mori H, 440
Morikawa Y, 333, 334
Morioka Y, 110, 111, 124
Morokuma K, 475
Morosi G, 410
Morra M, 438, 453
Morris JR, 381, 384, 389, 390, 392, 394, 397, 398
Morris MR, 398
Morrison MA, 68
Mortensen JJ, 322
Moseley JT, 119
Moser S, 72, 84
Moskovits M, 402
Moss RE, 156, 157
Mountain RD, 410, 413, 423
Mrzel A, 209
Muckerman JT, 69
Muckl AG, 221
Mueller-Plathe F, 491, 495
Muga JG, 86
Muhlpfordt A, 112
Mukamel S, 28–30, 32, 34
Mulholland AJ, 472, 491, 494, 496
Müller EA, 428
Müller H, 255, 387, 389
Muller LJ, 29
Muller M, 28
Müller U, 91
Muller W, 176
Müller-Dethlefs K, 102, 111
Mulliken RS, 278
Munakata T, 230, 508
Mundy CJ, 279
Munoz C, 160
Munro HS, 438
Murad E, 381
Murakami H, 27
Murakami J, 394
Murata Y, 390
Murayama K, 195

Murphy LR, 423, 425
Murphy RB, 473
Murry RL, 30
Muruyama K, 195
Muscat JP, 511, 512
Myers DJ, 33

Nadig G, 471, 494
Nagahara LA, 438
Nagano H, 230, 232, 233, 237, 508, 510, 511, 513, 515, 517, 518, 524–27
Nagasawa Y, 23, 25
Nagy JB, 217
Naitoh Y, 29, 30
Nakamura J, 230, 508
Nakamura S, 410, 412
Nakanishi K, 410, 423
Nakano E, 191
Nalewajski RF, 249, 251
Narasimhan LR, 22, 24, 25
Narmann CHA, 397
Natanson GA, 159, 469
Natta G, 329
Natzle WC, 263
Nauts A, 156, 159, 164
Nazarkin A, 105
Ndoni S, 179
Needs RJ, 321, 322
Neilson GW, 410, 422
Neimark AV, 534
Nelson KA, 29, 32
Neria E, 468, 470, 481, 494
Nesbitt DJ, 152
Nessler W, 510, 525, 526
Neuhaus T, 312
Neuhauser D, 85, 90
Neumann AW, 439
Neumark DM, 41, 43, 62, 72, 91
Neurock M, 319, 321, 322, 333
Neusser H, 134
Neve J, 513
New MH, 423, 425
Newman MEJ, 293, 300

Newns DM, 511, 512
Newton RG, 74
Newton SQ, 360
NG C-Y, 101–40; 41, 59, 62, 101–3, 106, 108–10, 112, 113, 115, 121, 124, 126, 128–30, 132–34
Ng TL, 185
Nguyen DT, 496
Ni B, 401
Nibbering ETJ, 18, 23
Nicolay G, 511
Nicole C, 527
Nicoll RM, 471, 489
Niederjohann B, 83, 86
Nieh J-C, 71, 85
Nielsen HH, 141
Niemantsverdriet JW, 326
Nifosi R, 277
Nilsson A, 512
Ninham BW, 409, 413
Nip AM, 365
Nir E, 135
Niranjan PS, 176, 191, 192, 194
NØRSKOV JK, 319–48; 319, 322, 323, 332–36, 339, 343, 344
Noga J, 255
Noguchi T, 44, 56
Nordlander P, 239, 510, 512, 518
Nordlund P, 492, 497
Norris JR, 25
Northrup SH, 468
Norwood K, 108
Nourbakhsh S, 108
Nowakowski M, 185
Nowinski K, 471, 480, 484, 497
Noworyta JP, 410
Nozik AJ, 222
Nozoye H, 452–54
Nusair M, 322

Occhiello E, 438, 453

Odom TW, 201, 202, 204, 206, 207, 209, 212, 213, 215, 216
O'Driscoll KF, 196
OGAWA S, 507–31; 42, 43, 224, 227, 230, 232, 233, 237, 238, 242, 508, 510, 511, 513, 515, 517, 518, 524, 525, 527
Oh-e M, 439, 441, 443, 449
Ohmine I, 30
Ohta K, 25
Oka T, 255, 257
Okada K, 115
Okada T, 27
Okamoto H, 29
Okamoto M, 209
Okazaki S, 410
Okotrub AV, 214
Okumura K, 27, 30
Okuno T, 457
OLDFIELD E, 349–78; 349, 350, 352, 353, 356–61, 363–71, 373, 375
O'Leary MH, 484
Oleinikova A, 428
Olender R, 301
Oliva M, 497
Oliveira IT, 410, 412
Olson JS, 371
O'Malley TF, 69
Onda K, 509
Ono Y, 331
Onoa B, 11
Ooi T, 175
Oosawa F, 175, 190, 191, 195
Opdahl A, 438, 441, 450
Opella SJ, 361
Oppenheimer R, 142
Ordoejon P, 214
Orel AE, 91
Orr-Ewing AJ, 73, 85, 90
Osapay K, 351
Osgood RM Jr, 224, 227
Oshiyama A, 201
Østerlund L, 509, 512, 521, 524

Ostlund NS, 473
Ostovic D, 481, 487
Ott A, 178
Ottolina G, 492
Ouw D, 25
OUYANG M, 201–20; 201, 202, 205, 209, 212, 217
Over H, 328, 336
Owen GDT, 196
Owicki JC, 410
Oxtoby DW, 314
Ozaki Y, 43, 62
Ozenne JB, 119

Paal Z, 343
Pack RT, 81, 159
Padowitz DF, 229, 235
Padró JA, 307
Pai EF, 190, 491
Paillard V, 402
Pailthorpe BA, 409, 413
Palese S, 30
Palfey BA, 496
Pallassana V, 333
Palm H, 104, 112
Palmer AG III, 361
Palmer RE, 402
Palmieri P, 73
Pan W, 471, 494
Panagiotopoulos AZ, 196, 410, 422
Pande VS, 303, 304, 312
Paneth P, 484
Pangali C, 410, 423
Panhuis M, 410
Pantaloni D, 191, 195
Pantano S, 277
Papadakis CM, 179
Papanikolas JM, 49, 52
Papazyan A, 21
Papoian G, 339
Papoian GA, 339
Papousek D, 141, 153
Pargas RE, 360
Parhikhteh H, 85
Park DH, 486
Park H, 44

Park JB, 439, 443
Park K, 30, 34, 450
Park KD, 350, 369
Park KH, 401
Park ST, 132, 134
Parker DH, 136
Parker GA, 81
Parker GT, 81
Parr CA, 475
Parr RG, 249, 251, 472
Parrinello M, 250, 252, 254, 255, 257, 258, 262–66, 271, 274–77, 279, 281
Parsegian VA, 410
Parthasathy A, 188
Paschek D, 428
Pascual-Ahuir J-L, 495
Pasquarello A, 281
Passino SA, 23, 25
Pate BH, 152
Patel CKN, 68
Patel JS, 448
Patore G, 252
Patterson C, 410
Patterson CH, 410
Patterson J, 365, 367
Paulaitis ME, 409, 410, 413, 414, 418, 420, 422, 423, 425, 427, 428, 429, 534, 554
Pauls S, 43
Paulsen EK, 359
Pavese M, 252
Pawlak J, 484
Pawlik S, 224, 508, 509, 512, 515, 516, 518, 527
Payling DW, 497
Payne MC, 250, 252, 278, 319, 322, 332
Payne VA, 423, 425
Pear MR, 468
Pearlman DA, 410, 423, 427
Pearlman RS, 475
Pearson JG, 356, 363, 364, 366–68
Pease RFW, 447
Peatman WB, 109

Pechukas P, 70, 468, 469
Peden CHF, 326, 328
Pedersen HB, 91
Peebles LH Jr, 178
Peeters A, 360, 363
Pellegrini M, 422, 425
Pemberton JE, 399
Pendergast P, 81
Pendyala K, 180, 183
Penno M, 110
Pepi F, 394, 396
Peraekylae M, 479
Perakyla M, 479, 490, 491
Percus JK, 413, 534, 535, 544
Perdew JP, 322
Perelygin IS, 410
Perrin CL, 196
Perry DS, 152
Persson BNJ, 513
Petchek RG, 176
PETEK H, 507–31; 42, 43, 224, 227, 230, 232, 233, 237, 238, 242, 508, 510, 511, 513, 515–18, 524–27
Peterka DS, 109, 136
Peterson KA, 33
Peterson MR, 72
Petsalakis ID, 91
Petsko GA, 489, 492
Pettitt BM, 470
Peyerimhoff SD, 476
Pfab R, 108, 136
Pfeiffer W, 43
Pfeuty P, 173, 175–77, 180, 184, 186, 187, 190
Pfrommer BG, 283, 284
Phelps AV, 68
Philipp DM, 473
Phillips DL, 71, 85
Phillips GN Jr, 371
Phillips JC, 493
Piana S, 285
Pickard CJ, 278
Pickett HM, 159
Pierotti RA, 411, 429
Pilgrim W-C, 190

Pines A, 358
Piryatinski A, 30, 34
Pitarch J, 495
Pitarke JM, 511, 513
Pizzolatto RL, 438
Plapp BV, 486
Podolsky B, 142, 148
Pohorille A, 409, 411, 413, 417, 423, 425–27, 429, 430, 534, 545, 554
Polanyi JC, 41, 507
Polanyi M, 69
Polenova T, 359
Polik WF, 75
Pollack GL, 420
Pollak E, 70, 71, 79
Poltavtsev YG, 190
Poltras D, 381
Pomerantz AE, 73, 86, 87, 90
Pomes R, 410
Pontecorvo B, 67
Pople JA, 103, 108, 472
Popov AN, 370
Popp D, 190
Porezag D, 492
Porter RN, 69, 72
Post CB, 410
Postawa Z, 384, 385
Postma HWC, 217
Postma JPM, 410, 412, 413
Poulos TL, 489
Powell RE, 187
PRATT LR, 409–36; 293, 409, 411–15, 417, 418, 420, 423–25, 427, 429, 430
Pratt ST, 112, 121
Prausnitz JM, 534, 557
Prevost M, 413
Price JM, 258
Privalov PL, 410, 427
Procacci P, 423, 427, 428
Proust-De Martin F, 490
Pshenichnikov MS, 18, 23, 25
Pu J, 483
Pulay P, 352
Purdie D, 511

Pusey PN, 309
Putrino A, 281, 283

Qi PX, 361
Qian J, 391, 392
Qian XM, 103, 126, 129, 132, 133
Qin X, 399
Quate CF, 438
Quinteros CL, 389, 398, 399

Rabalais JW, 380, 385, 401
Rábanos VS, 71, 72
Rademann K, 402
Radhakrishnan R, 534
Radloff W, 43, 44, 59, 60, 62
Radmer RJ, 410
Radom L, 472
Radzicka A, 413
Raftery MA, 365
Raghavachari K, 103, 108
Rahman A, 410
Rahman M, 196
Rajagopal G, 321, 322
Rajamani R, 493
Rakitzis TP, 85, 90
Ramakrishna R, 520
Ramaniah LM, 265
Ramsperger HC, 311
Ramswell JA, 42
Rance M, 361
Rand RP, 410
Ranganathan S, 496
Rank JA, 423
Rao AM, 202
Rao BG, 410, 423
Rao M, 410, 423
Rao MB, 489
Rapaport DC, 410
Rasaiah J, 410
Rasaiah JC, 410
Raschke TM, 410
Raseev G, 527
Rasetti F, 67
Rassolov V, 103, 108
Rathje J, 49, 52
Ratner BD, 438–40, 457

Ratner MA, 480, 482
Rau DC, 410
Rauk A, 274
Rauscher C, 33, 34
Ravikovitch PI, 534
Ravimohan C, 410
Ravishanker G, 423
Rawlett AM, 221, 243
Raymond EA, 557
Raz T, 394
Re M, 410, 412
Rechtien JH, 394
Rector KD, 33
Recum AF, 439
Redaelli C, 410, 423
Redfern PC, 103, 108
Redhead PA, 508
Redondo A, 409, 430
Reed LH, 410
Reed MA, 221, 243
Rehfus BD, 257
Reichle R, 91
Reichman DR, 30
Reid KL, 44–46, 56
Reid SA, 75
Reijerse JFCJM, 326
Reijnen PHF, 390, 394
Reilly JP, 43
Reimers JR, 410
Reinert F, 511
Reiser G, 102, 111
Reisler H, 75
Reiss H, 553
Rella CW, 33
Rembaum A, 175
Remerie K, 410, 412
Remler DK, 250
Rempe SB, 159, 409, 415, 416, 428, 430
Rendell AP, 471, 494
Resta R, 263, 278
Rettner CT, 104
Reuβ C, 508, 510, 511
Reuss C, 43, 223, 224
Reuter N, 473
Rey R, 307
Reyes CM, 470

Reynolds PJ, 321
Rezaei MA, 508
Rhim WK, 284
Riande E, 185
Rice JE, 257
Rice OK, 311
Rice SA, 82
Richards FM, 365, 426
Richards G, 494
Richards RW, 438
Richards WG, 491, 493, 494
Richmond GL, 438, 557
Richter D, 185
Richter LJ, 438, 441, 443, 446
Rick SW, 423
Ridder L, 496
Rider KB, 438
Rieder K-H, 237, 508, 509, 527
Riederer DE, 387, 394, 398
Rietjens IMCM, 496
Riffe DM, 511
Rill RL, 438
Rinaldi D, 473
Ringe D, 489, 492
Ringling J, 105
Rinnen K-D, 82
Rinzler AG, 201, 202, 204, 206, 209, 212, 214
Ritze HH, 43
Rivail J-L, 473, 497
Roberts AD, 444
Roberts DE, 183
Roberts GCK, 368, 509, 513
Roberts GG, 447
Roberts MA, 422, 426
Robertson DH, 204, 208, 212
Robinson GN, 72
Rochefort A, 214
Rocken P, 534
Rockenberger J, 384
Rodriguez J, 311
Rokhsar DS, 303, 304, 312
Rolland D, 136
Romanov DA, 214
Römelt J, 70, 79
Romero AH, 279

Rose TS, 513
Rosenberg RO, 425
Rosenfeld Y, 534, 559
Rosengaard NM, 336
Rosenstock HM, 103, 106, 123
Rosker MJ, 513
Rosmus P, 115
Ross J, 69
Rossi I, 475
Rossier JS, 453
Rossky PJ, 410, 423, 424, 427, 430, 534, 557
Rostrup-Nielsen JR, 339
Roth R, 534
Röthig C, 394
Rothlisberger U, 265, 495
Rouault Y, 177, 178
Rouch J, 196
Roush W, 190
Rousseau R, 257
Roux B, 410
Rouxhet PG, 453
Rovira C, 277
Roweis ST, 315
Rowland SC, 187
Rowlinson JS, 412, 533, 536, 538, 539, 541, 542
Rozanska X, 332
Ruban A, 335
Ruban AV, 336
Rubio A, 204, 208, 214, 511, 513
Rubio MA, 185
Ruckenstein E, 457
Rudolf P, 512
Ruggiero AJ, 21, 527
Ruiz-Garcia J, 180, 183–85, 187, 190
Ruiz-Lopez MF, 495, 497
Rundgren J, 513
Ruscic B, 107, 108
Russell ST, 476
Russell TP, 440
Ryabov V, 75, 76

Sadeghi R, 76, 91, 92
Sáez Rábanos V, 90

Safont VS, 497
Safron SA, 69
Safvan CP, 91
Sagnella DE, 265
Saito N, 294
Saito R, 201, 203, 212, 214, 217
Saito S, 30, 208
Saitta AM, 258
Sakashita T, 230, 508
Salahub DR, 214, 277
Salaneck WR, 221
Salmeron M, 438
Salour MM, 525
Salverda JM, 25
Salzmann R, 369–71
Samson JAR, 104
Sanche L, 69, 224
Sanchez A, 402
Sanchez I, 413
Sanchez ML, 471, 475, 477, 478, 480–82, 485, 486, 488, 489
Sanchez-Portal D, 214
Sandell A, 512
Sander M, 102, 111
Sanders LK, 358, 359, 363, 369, 371, 372
Santry DP, 472
Sarkisov GN, 410, 423
Sasaki TA, 390
Sass CS, 385
Sato S, 44, 56
Sauer BB, 455
Saul LK, 315
Sauter B, 33
Sautet P, 326, 328
Savin A, 278
Sawada H, 175
Sawada S, 201
Sawyer DW, 262
Sayvetz A, 156
Scatena LF, 438, 557
Scettino V, 423, 427, 428
Schaefer HF III, 255
Schaeffer L, 178
Schäfer L, 179, 360, 363

Schatz GC, 69–71, 81, 482
Scheffler M, 328, 336
Schegler E, 265
Schenter GK, 469, 471, 478, 481, 483
Scheraga HA, 410
Scherer NF, 23, 527
Scheurer C, 34
Schick CP, 43, 44, 59, 60, 62
Schiffer H, 271
Schilling L, 30
Schiwek M, 135
Schlag EW, 102, 109, 111
Schlathölter T, 397
Schlesinger MJ, 365
Schleyer PvR, 255, 472
Schlüter M, 323
Schmickler W, 222
Schmid R, 330, 331
Schmidt K, 397
Schmidt M, 534, 558, 559
Schmidt S, 511
Schmitt M, 41, 43, 44, 59, 60, 62
Schmitt UW, 303, 308, 476, 477
Schmuttenmaer CA, 508
Schmutz H, 112
Schneider WD, 215, 402
Schnieder L, 72, 82, 83, 86
Schnitzer C, 438
Schnitzer R, 91
Schofield A, 309
Schofield P, 176
Scholten M, 492, 496
Schon JH, 242
Schor HHR, 79
Schreiber H, 423, 424
Schreiner PR, 255
Schroder U, 159
Schulten K, 493
Schultz DG, 381, 392, 393, 398
Schultz T, 41, 59, 62, 108
Schulz GJ, 68
Schulze B, 277

Schuppler S, 508, 512, 515
Schürer G, 493, 497
Schurhammer R, 410
Schuss Z, 304
Schwab AD, 438, 443, 446, 447
Schwartz D, 383
Schwartz SD, 485
Schwegmann S, 328, 336
Schweizer KS, 29
Schwenke DW, 70, 79, 81, 82, 86
Schwettman HA, 33
Schwieters CD, 478
Sciortino F, 196
Scoles G, 152
Scott RL, 176, 177, 187, 188
Scrutton NS, 471, 488, 489
Scuseria G, 255
Sebastiani D, 281, 283, 284, 285
Secrest D, 69
Seekamp-Rahn K, 83, 86
Seel M, 43, 44, 59
Segal GA, 472
Segall MD, 278
Segré E, 67
Sehgal A, 475
SEIDEMAN T, 41–65; 41–44, 46, 48, 51, 52, 56–62, 75, 76
Seidler PF, 509
Seiersen K, 91
Seifert G, 33
Seigbahn PEM, 276
Seiler R, 112
Seitz JT, 443, 446
Selden LA, 191
Selinger RLB, 533, 536, 550
Selmi M, 410, 412
Senet P, 514, 515
Sesselmann W, 512
Severance DL, 423, 427
Sevick EM, 301
Seyfried V, 43
Shafer NE, 73, 85, 90
Shafer-Ray N, 72, 84

Shafer-Ray NE, 73, 85, 90
Shaffer JP, 41, 43, 44, 59, 60, 62
Shah R, 278, 332
Shakhnovich EI, 303, 304, 312, 413
Sham LJ, 249, 322
Shank CV, 23, 26
Shao L, 475
Shapiro M, 69, 159
Sharafeddin O, 81
Sharma A, 557
Sharp K, 410
Shavitt I, 69
Sheehan PE, 201
Shen J, 381, 382, 385, 401, 402
Shen LX, 10
Shen YG, 390
SHEN YR, 437–65; 438–41, 443, 445, 449, 453–55, 457, 458, 460
Sheppard DW, 497
Sherrill AB, 324
Sherwood MH, 354
Sheterline P, 190, 191
Shi YJ, 112
Shi YY, 489
Shiell RC, 102, 103, 112, 115, 118, 121, 134
Shimada J, 413
Shimizu K, 410, 412
Shimizu S, 410, 412, 423
Shirley EL, 204
Shobatake K, 72
Shortley GH, 67
Shubin AA, 331
Shuker SB, 375
Shultz MJ, 438
Shumay IL, 43, 223, 224, 508, 510, 511
Sibert EL III, 159, 167
Siegbahn P, 72
Siegbahn PEM, 493
Siegert AJF, 75
Signorell R, 112, 113, 128, 130, 133

Silinsh EA, 233, 234, 238, 239
Silkin VM, 512, 513, 518, 521, 523, 527
Silla E, 494, 495
Silverman PJ, 402
Silverstein KAT, 410, 534, 557
Silverstein TP, 412
Silvestrelli PL, 262, 263, 279, 281
Silvi B, 278
Simeonidis K, 33, 34
Simon D, 326
Simonelli D, 438
Simons JP, 75, 90
Simonson JM, 410
Simpson WR, 73, 85, 90
Sims D, 184
Singh UC, 410, 423, 472, 473
Sinnott SB, 401
Sintes T, 196
Sitkoff D, 358, 359
Sitz GO, 85
Siu CK, 266
Skalozub A, 141, 160
Skipper NT, 410, 423, 424
Skodje RT, 70, 72, 73, 76, 79, 84, 91, 92
Skouteris D, 72
Skriver HL, 335, 336
Slaten BL, 188
Sligar SG, 33, 350, 368, 369, 371
Sliwinska-Bartkowiak M, 534
Slusher JT, 410
Small BL, 330
Smalley RE, 201, 202, 204, 206, 209, 212, 214
Smargriass E, 252
Smit B, 291–93, 295, 296, 300, 311, 312, 314
Smith B, 371
Smith D, 410
Smith DE, 423, 424
Smith FT, 74
Smith GD, 425

Smith IWM, 470
Smith JM, 43, 53
Smith LM, 461
Smith PE, 410
Smith SB, 11
Smith SF, 479
Smith SJ, 44, 45
Smith W, 410
Snijders JG, 30, 276
Snowdon KJ, 390, 394
Soep B, 43
Softley TP, 130, 133, 136
Sokolovski D, 72
Soler JM, 214
Solgadi D, 43
Somasundaram T, 410
Somogyi A, 394
SOMORJAI GA, 437–65; 319, 326, 329, 437–40, 443, 445, 455, 457, 458, 460
Somoza A, 534
Son KA, 244
Song X, 25
Song XB, 43
Song Y, 102, 103, 106, 112, 115, 116, 118, 124, 126, 128–30, 132–34
Soper PD, 258
Soper SA, 447
Sostarecz A, 384, 385
Southall N, 410
Southall NT, 410, 428, 534, 557
Sparks SW, 359
Sparrow JC, 190, 191
Spellmeyer D, 410
Spera S, 350, 354, 355
Sperandeo-Mineo RM, 410
Sprandel LL, 159
Sprik M, 253, 262, 264, 268, 271, 273, 313
Springer BA, 371
Springer M, 85
Srinivasan J, 81
Stam CH, 235
Stamatopoulou A, 413

Stampfl C, 328, 336, 509
Stampfuss P, 43, 44, 59
Stanton RV, 472, 479, 490, 491
Stapelfeldt H, 61
Stark K, 72
Stebbings R, 111
Steffen T, 29, 30
Steiner BW, 103, 106, 123
Steinhauser O, 423, 424
Steinmann W, 224, 227, 508, 512, 513, 515
Stell G, 414, 419, 430, 537, 545
Stellbrink J, 185
Stephens PJ, 472
Stephenson JC, 438, 441, 443, 446, 508
Stert V, 43, 44, 59, 60, 62
Steudel R, 186
Stevens JE, 108
Stevens RM, 69
Stevens WJ, 472
Stewart JJP, 471, 472, 475
Stich I, 322
Stichler M, 512
Stilbs P, 410, 423
Stillinger FH, 186, 410, 411, 418, 427, 534, 555
Stimson S, 102, 106, 112, 113, 115, 118, 119, 121, 129
Stipe BC, 508
Stock G, 41, 59
Stockmayer WH, 178
Stohr J, 440
Stolow A, 41, 43, 44, 59, 60, 62
Stoltze P, 323, 335, 336
Stolz H, 513, 521
Stolz M, 190
Stone M, 413
Storz R, 508
Stossel TP, 190
Straatsma TP, 410
Stracke P, 387
Strajbl M, 491, 492

AUTHOR INDEX 585

Strand MP, 44, 45
Strasser D, 91
Stratt RM, 30
Strauss R, 186
Street SC, 379
Strehel M, 43
Strey H, 191
Stroeve P, 447
Stryer L, 369
Stuhl F, 509
Suarez D, 495
Subirana JA, 8
Suck D, 190
Sucre MG, 76
Suffritti GB, 410
Suhai S, 496
Suits AG, 102, 105, 109, 136, 257
Sullivan DM, 422, 537
Sun H, 368
Sun SX, 428, 535, 556, 560
Sun Y, 410
Sun YX, 410
Sung J, 30, 31
Suslick KS, 33, 369
Susman S, 187
Sutcliffe BT, 141, 142, 159, 160, 162, 164, 165, 321
Sutcliffe MJ, 471, 488, 489
Suzuki T, 42, 43, 46, 56
Suzuki Y, 440
Svetlicic Z, 483
Svishchev I, 410
Swalen JD, 141
Swaminathan S, 410, 423
Swinton FL, 412
Swope W, 410, 421
Syage J, 43
Sykes BD, 365
Szabo CM, 358, 359
Szleifer I, 450
Szornel C, 190
Szwarc M, 174, 178, 179, 184, 196
Szymanski P, 233
Szymanski R, 196

Tachikawa H, 253
Taganov NG, 178
Tai Y, 394, 402
Tajkhorshid E, 496
Takahara A, 443, 446
Takatsuka K, 46
Takayanagi K, 209
Tamura K, 187
Tanada TN, 489
Tanaka H, 410, 412, 413, 423
Tanaka K, 102, 136, 443, 446
Tanaka T, 110, 111, 303, 304, 312
Tanford C, 190
Tang JX, 191, 255
Tang KES, 411
Tani A, 410, 412
Tanimura Y, 27, 29, 30
Tannor DJ, 391, 392
Tans SJ, 202, 214
Tansford C, 534
Tantanak D, 471, 484
Tapia O, 472, 497
Tarazona P, 534, 558, 559
Tartaglia P, 196
Tasaka S, 443, 446
Tassios DP, 421
Tasumi M, 29
Taura A, 443, 446
Tavan P, 271, 473
Taylor DP, 135
Taylor H, 76
Taylor JR, 74
Taylor PC, 187, 188, 190
Teillet-Billy D, 385
Tenenbaum JB, 315
Tennyson J, 142, 159, 162
TenWolde PR, 535, 556, 560
Teodoro M, 371
Terakura K, 263, 329, 330
Terano M, 440
Tersoff J, 518
Testa A, 281
Teter MP, 250, 252, 319
Thalweiser R, 42, 43
Thantu N, 43, 44, 59, 60, 62
Theil MH, 196

Theodorakopoulos G, 91
Theodorou DN, 301, 410, 421, 422
Thery V, 473
Thess A, 202
Thevuthasan S, 326
Thiel W, 472, 474, 492, 496
Thimm JK, 215
Thirumalai D, 70, 79, 410, 423
Thirumoorthy K, 413
Thomann U, 43, 223, 224, 508, 510, 511
Thomas A, 471, 478, 490
Thompson DE, 33, 34
Thompson MA, 472
Thompson TC, 70, 79
Thon A, 43
Thundat T, 438
Tian C, 361
Tian Y, 438–41, 450, 455
TINOCO I Jr, 1–15; 5–11
Tirado-Rives J, 423, 470
Titenko YV, 190
Titmuss SJ, 471, 494
Tjandra N, 358, 361, 364
Toba S, 492, 493
Tobias DJ, 425
Tobolsky AV, 175, 178, 183, 186, 187, 196
Toda M, 294
Todorov SS, 399
Toennies JP, 71, 514, 515
Tokmakoff A, 27, 30–34
Tokumoto M, 209
Tolbert CA, 27
Tom HWK, 508
Tomanek D, 208, 214
Tomasi J, 410, 412
Tomas-Oliveira I, 410, 412
Tominaga K, 29, 30, 34
Tompson CW, 190
Tonhara H, 410
Tonkyn RG, 104
Topaler M, 478
Topf M, 494
Topley B, 69

Toral R, 196
Torchia DA, 359
Tornaghi E, 277
Torquato S, 534, 557
Torrie GM, 418, 470, 477
Touhara H, 410
Tour JM, 221, 243
Touro FJ, 190
Townes CH, 6
Towrie M, 44–46, 56
Tozzini V, 277
Treacy MMJ, 201
Trento L, 402
Tresadern G, 471, 488
Trevor PL, 523
Troullier N, 323
Trout BI, 263
Truck GW, 103, 108
TRUHLAR DG, 467–505;
    69, 70, 72, 75, 79, 81,
    82, 86, 87, 307, 468,
    469, 471, 481, 483, 487
Truong TN, 469, 475, 481
Truskett T, 413
Truskett TM, 534, 557
Trylska J, 495
Tsai J, 410
Tschinke V, 322
TSE JS, 249–90; 255, 257,
    258, 260, 266
Tsekouras AA, 382, 402
Tsubouchi M, 46, 56
Tsukakoshi M, 230, 508
Tsumori K, 383, 395
Tucker SC, 468, 469, 483
Tuckerman ME, 252, 258,
    260, 264–66, 286, 303,
    313, 478
Tuckett RP, 85
Tugcu N, 410
Tuinstra HLJT, 202, 214
Tull CE, 322
Tully JC, 44, 157, 468, 470,
    480, 510, 512, 518, 523
Tully-Smith DM, 553
Tunon I, 494, 495
Turner AJ, 496

Turner DH, 8
Turner DW, 101, 102, 109
Tzvetkov T, 399

Uberna R, 527
Ugarte D, 213
Uhlenbeck OC, 7
Ulman A, 447
Ulness DJ, 31
Umek P, 209
Underwood JG, 46
Unterberg C, 135
Urahata S, 410
Urbach W, 178
Urbain X, 91
Urbic T, 410
Urdahl RS, 33
Utz AL, 326, 328
Uyama H, 452–54

Vager Z, 257
Valentini JJ, 71, 85, 90
Valleau JP, 418, 470, 477
Van Alsenoy C, 360, 363
Van Alstine JM, 450
van Bastelaere PBM, 490
van Belle D, 410, 423, 424
Vanden Bout D, 26, 27, 29
van den Hoek PJ, 390, 394
van den Hout M, 212
van der Avoird A, 142, 159
Vanderbilt D, 253, 262, 263,
    278, 281, 323
van der Schoot P, 178, 184,
    192
Van der Vaart A, 471, 494
van der Zwan G, 25
Vandeweert E, 384, 385
Vandoni G, 402
van Eijk MCP, 440
van Gestel J, 192
van Grondelle R, 25
van Gunsteren WF, 410, 412,
    491, 492, 495, 496
van Hove MA, 326
Vanka K, 331
van Mourik F, 25

Vanolli F, 402
van Santen RA, 319, 321,
    322, 326, 331, 332, 339
van Slooten U, 390
van Swol F, 534, 547, 558
Van Tilbeurgh H, 490
Van Vleck JH, 148
Van Voorhis T, 310
Van Voorst JDW, 28
Van Zant LC, 471, 494
Varandas AJC, 72, 77, 79, 84
Varmus HE, 10
Várnai P, 493, 494
Vega LF, 410
Velasco E, 534, 558
Velic D, 223, 230, 232, 237,
    508, 509, 527
Venema LC, 201, 202, 204,
    206, 209, 212, 214
Vernon FL Jr, 18
Vervoort J, 496
Vickerman JC, 440
Vidal CR, 104
Viel A, 159
Vigneron JP, 217
Villà J, 469, 471, 476–79,
    481, 482, 484–86, 488,
    491, 492, 495
Villeneuve DM, 43
Vincent MA, 471, 484
Viner JM, 187, 188, 190
Vishnyakov A, 534
Vlachy V, 410
Vlugt TJH, 291, 295, 296,
    311, 312
Vohringer P, 23, 25
Vohs JM, 324, 325
Volbeda A, 495
Volkel A, 159
Volkov AV, 372
Voll S, 43
Vollmayr K, 533, 536, 538,
    543, 549, 550
Vollmayr-Lee K, 533, 535,
    543, 549, 550, 554
Volpi GG, 83
von Delft J, 215

Vondrak T, 230, 236, 245
von Issendorff B, 402
von Rosenvinge T, 265
Vosko SH, 322
Voter AF, 301
Voth GA, 252, 303, 308, 476–78
Voutsas EC, 421
Vrakking MJJ, 43, 112

Wada A, 509
Wade N, 394, 396
Wagner AF, 481
Waisman E, 544
Wales D, 293
Walker RA, 438
Walker RB, 81
Wallace PR, 203
Wallace WE, 438, 441, 443, 446
Wallauer W, 43, 223, 224, 508, 510, 511
Walldén L, 512, 513, 515
Wallensteinm R, 104
Walling AE, 360
Wallqvist A, 410, 412, 423, 425, 427, 428, 534, 557
Walsh CA, 22, 24, 25
Walsh TR, 293
Walther H, 44, 45
Wand AJ, 361
Wandelt K, 512
Wang H, 229, 230, 233, 245, 382
Wang J, 361, 439, 441, 443, 448, 455, 457
Wang J-F, 364
Wang K, 46
Wang L, 46, 56
Wang P, 112
Wang W, 470
Wang XG, 141, 160, 224, 227
Wang XGN, 167
Wang YM, 267
Wang Z-G, 196
Ward R, 438–41, 450, 455

Ward RS, 439, 440, 457, 458, 460
Warren TD, 41, 44
Warshel A, 367, 468, 469, 472, 473, 475, 476, 478, 479, 483–86, 488, 491–93
Warth C, 394, 395
Waschewsky GCG, 108
Waseda H, 276
Wasserman E, 258
Wassmuth H-W, 513, 521
Waszkowycz B, 497
Watanabe K, 423, 424
Watanabe N, 410
Watson JKG, 141, 143, 146, 148, 150, 151, 153, 154, 156
Watts E, 85
Watts RO, 159
Watwe RM, 339
Weale KE, 196
Webb SP, 21, 471, 476, 478, 482, 485, 487
Weber PM, 43, 44, 59, 60, 62
Weber TA, 186
Webster F, 71, 82
Webster OW, 174
Weeding TL, 383, 395
Weeks ER, 309
WEEKS JD, 533–62; 419, 428, 429, 533, 535–38, 540, 542, 543, 546, 549, 550, 554, 557
Wei D, 277
Wei H, 159, 163, 165
Wei X, 439, 441, 443, 449, 453, 454
Weibel P, 402
Weida MJ, 230, 232, 233, 237, 508, 513, 515, 517, 524, 525, 527
Weigel R, 422, 426
Weiguny A, 159
Weinelt M, 528
Weiner AM, 24
Weingarten HI, 365

Weinhold F, 361, 362
Weinkauf R, 43
Weis P, 384
Weisel MD, 328
Weiss H, 329
Weiss M, 343
Weiss RM, 476
Weiss V, 42
Weisskopf VF, 68
Weitz DA, 309
Weitzel KM, 102, 103, 110, 112, 124, 128, 130, 132, 133
Welge KH, 82, 83, 86
Wemmer DE, 358
Wen-Shyan S, 557
Wenzel W, 43, 44, 59
Werner H-J, 72
Wertheim GK, 511
Wesselink GA, 235
West ED, 187, 188
Western CM, 82, 83
Westervelt RA, 23
Westler WH, 362
Westler WM, 361, 362
Westrin P, 513
Wheeler JC, 173, 175–77, 184, 186–88, 196
Whitaker BJ, 42, 43, 46, 56
White CT, 201, 204, 205, 208, 212
White ET, 255, 257
White JA, 322
White JM, 324
White MG, 104, 112, 128, 133, 136
Whitlow M, 489
Whitten TA, 178
Widom B, 417, 430, 533–39, 541, 542, 559
Wiedmann RT, 104, 112, 128, 133, 136
Wiegershaus F, 387
Wiersma DA, 18, 23, 25
Wiewiorowski TK, 188, 190
Wigner EP, 68, 468, 469
Wijesundara MB, 401

Wild S, 453
Wilde M, 509
Wildöer JWG, 201, 202, 204, 206, 209, 212, 214
Wilk L, 322
Wilkens SJ, 361, 362
Wilkerson W, 43
Willerding B, 394
Williams AR, 511, 512
Williams IH, 496
Williams KW, 202
Williams RM, 49, 52
Williamson AJ, 322
Willig F, 222, 520
Willner L, 185
Wilson EB Jr, 141, 147, 150, 167
Wilson JE, 410, 422
Wilson KR, 301, 468, 481
Wilson MA, 413, 429
Winborne E, 489
Winger BE, 398
Winget P, 245
Wingreen NS, 215
Winograd N, 384, 385
Winter R, 190
Winters HF, 381
Wiorkiewicz-Kuczera J, 470, 475, 490
Wipff G, 410
Wishart DS, 365
Witte G, 514, 515
Wittmer JP, 177, 178
Wodak SJ, 410, 412, 413, 423, 424, 490
Wodtke AM, 72
Wojdelski M, 369–71
Wold DJ, 244
Wolf LK, 438
Wolf M, 223, 230, 232–35, 237, 240, 508–11, 515, 517, 527, 528
Wolfenden R, 413
Wolff B, 104
Wolfgang B, 215
Wolfrum J, 71
Wolfrum K, 438

Wolfsberg M, 159
Wolinski K, 352
Wolter HH, 45
Wong CM, 42, 43, 55, 224, 227, 229, 230, 238, 240, 508
Wong EW, 201
Woo TK, 267, 268
Wood RH, 410, 422
Woodcock SE, 439, 441, 455, 457
Woodruff DP, 385, 437, 457
Woody RW, 5
Woolf TB, 301
Woratschek B, 512
Wöste L, 42, 43
Woywood C, 141, 160
Wrede E, 72, 83, 86
Wu G, 372
Wu L, 473, 492, 497
Wu N, 491
Wu Q, 382, 398
Wu S-F, 69, 78, 450
Wu SH, 443, 455
Wu Y-SM, 72, 82–84
Wuilleumier FJ, 41, 44
Wunderlich C, 91
Wyatt RE, 69–71, 82, 83
Wynne K, 28
Wysocki VH, 394, 398, 399

Xantheas SS, 160, 262
Xia B, 361, 362
Xia XF, 472, 475, 491
Xie J, 382
Xing J, 477
Xu C, 379
Xu H, 73, 82, 85
Xu Q-H, 25
Xu Y, 337
Xu ZR, 84

Yadav A, 483
Yakovlev AL, 331
Yakshinskiy BV, 509
Yamada Y, 452–54
Yamaguchi H, 209

Yamaguchi S, 359
Yamaguchi W, 394, 402
Yamamoto H, 390
Yamamoto S, 276
Yamamto H, 440
Yamato T, 276
Yamazaki T, 101, 102
Yang J, 376
Yang L, 205
Yang M, 25
Yang QY, 343
Yang T-S, 23
Yang W, 249, 251, 322, 495
Yang WT, 471–73, 479, 492, 494
Yang XM, 102, 105
Yang YJ, 438
Yannoulis P, 224, 233
Yao Z, 217
Yarkony DR, 82, 83
Yarne DA, 266
Yasuda H, 457
Yasuda T, 457
Yeganeh MS, 438, 439, 441, 443, 446, 447, 453, 454
Yeh LI, 258
Yencha A, 115
Yim W-L, 266, 267
Yim YH, 402
Yin HM, 108
Yokoyama T, 209
Yoon YG, 215
York DM, 471
Yoshihara K, 29, 30
Yoshii H, 110, 111
Yoshimura K, 402
Yoshmura K, 394
Young RJ, 443
Young WS, 423
Yu C-h, 81, 363, 381, 392
Yu HG, 77
Yu J-Y, 23, 25–27
Yuen-Kit C, 557
Yui N, 440

Zajfman D, 257
Zalipsky S, 450

Zamampour HP, 108
Zanni MT, 34, 43
ZARE RN, 67–99; 44, 54, 55, 71, 73, 82, 84–87, 90, 104, 143, 148, 382, 507
Zassetsky A, 410
Zeigler T, 267, 268
Zeikus JG, 490
Zeng J, 410
Zettle A, 213
Zewail AH, 41, 507, 509, 513
Zgierski M, 43, 44, 59, 60, 62
Zhadanov VP, 509, 512, 521, 524
Zhang DH, 81, 438–41, 443, 445–47, 453–55, 457, 458, 460
Zhang JY, 440
Zhang JZH, 71, 81, 85, 90, 141
Zhang L, 423, 424
Zhang XM, 443, 446
Zhang Y, 81, 322, 495
Zhang YK, 473, 479, 492
Zhang YQ, 476
Zhang Z-Y, 473, 492, 497
Zhao M, 81, 82
Zheltikov AM, 28
Zheng KM, 180, 183, 184, 186
Zhidomirov GM, 331
Zhong D, 41–43
Zhong Q, 230, 232, 234, 235, 237, 240
Zhu L, 102, 103, 134
ZHU X-Y, 221–47; 222, 229, 230, 233, 236, 240, 241, 245, 508, 516
Zhuang J, 180, 184, 185
Zhuang X, 439–41, 453, 454
Zichi DA, 410
Ziegler CJ, 33, 369
Ziegler T, 253, 268, 274, 319, 322, 330, 331
Zilm KW, 359
Zimdars D, 23, 32, 33
Zimmerle CT, 191
Zimmermann FM, 508
Zingaro RA, 186
Zinn-Justin J, 178
Zobel D, 372
Zoebisch EG, 472, 475
Zollweg JA, 548
Zuckerman DM, 301
Zundel G, 264
Zunger A, 250, 252, 322
Zwanzig R, 294, 470, 477
Zygan-Maus R, 45

SUBJECT INDEX

A
Ab initio Hartree-Fock quantum chemical techniques, 352
Ab initio molecular dynamics (AIMD), 250–53, 257–58, 260, 262, 286
ABCRATE program, 79
Abstraction reactions, 398–400
 in modulating scattered ion product signals, 399
Acetohydroxy acid isomeroreductase, 490
Acetylene
 protonated, 257
Acids, 264–65
Actin model, 190–92
 chemical kinetics and molecular weight distribution, 195
 experimental considerations, 192
 structure, 195
 thermodynamics, 192–95
 transport properties, 195
Adenosine triphosphate (ATP), 190–92
Adiabatic potentials, 76, 80–81
 of Feshbach resonances, 80–81
Adsorption calorimetry
 single-crystal, 322
Advanced Light Source (ALS), 102, 106
Aggregations, 173–200
 biomolecular, 190–95
 chemical kinetics and molecular weight distribution, 178–79

statistical mechanics, 175–78
terminology, 173–74
theoretical framework, 174–79
thermodynamics, 174–75
AIM
 See Atoms in molecules theory
AIMD
 See Ab initio molecular dynamics
Air
 polymer surfaces in, 443–49
Aldose reductase, 493
Alignment
 pump-induced, 61
Alkali atoms
 chemisorption, 511–13
 frustrated desorption from noble metals, 507–31
Alkane conformational equilibrium in water, 425
Alkanethiol SAMs, 244
Alkyl side chain lengths
 polyimides with, 449
Alloy formation trends
 surface, 334–36
ALS
 See Advanced Light Source
AMBER force field, 269, 356, 363, 470, 488–89, 492, 494, 496–97
 PARAM field of, 497
Amino acids
 carbon-13 and nitrogen-15 NMR, 351–61
 chemical shifts in, 349–78
Ammonia synthesis

calculation of turnover frequencies for, 344
catalyst developments, 343–44
Angular correlations
 time-resolved techniques of measuring, 61–62
Angular structure
 time-resolved, 52–56
Anhydrases
 carbonic, 484
Anions
 pentafluorobenzyl alcoholate, 487
Antibonding resonance (AR), 513, 527
Antoniewicz model, 241
APP
 See Atactic PP
Applications
 industrial, 342–44
Applications for enzyme kinetics, 483–97
 with quantum electronic structure and statistical mechanics, 490–93
 quantum treatments of electronic structure without statistical mechanics, 493–97
 quantum treatments of nuclear dynamics, 483–90
Applications of time-resolved photoelectron angular distributions, 56–61
 time-evolving electronic symmetry, 59–61
 time-evolving rotational composition, 56–59
Applications of transition path sampling, 309–16

591

biomolecule isomerization, 312–13
computing quantum dynamics, 315–16
diffusion of isobutane in silicalite, 311
harvesting long trajectories, 314–15
heptamer of cold Leonard-Jones disks, 309
isomerization of a solvated model dimer, 309–10
parallel tempering, 311–12
recognizing patterns in stable states, metastable states, and transition states, 315
solvation dynamics, 313–14
water autoionization, 313
water clusters, 310–11
Approximations
Born-Oppenheimer, 142
equilibrium-secondary-zone, 482
frozen core, 322
generalized gradient, 322
hydrostatic, 542–47
local density, 275, 322
modified molecular field, 548
molecular field, 536–39
static-secondary-zone, 482
AR
See Antibonding resonance
Armchair SWNTs, 207–9
Fermi points of, 206
Aspartic transcarbamoylase, 484
Atactic PP (APP), 443, 445, 455
Atomic ionization
two-pulse, 45
Atoms in molecules (AIM) theory, 372
Attachment/detachment experiments, 90–102

Attraction
distinguishing basins of, 302–3
Attractive interactions
nonuniform fluids with, 549–54
Auger neutralization, 68, 395
Autoionization
water, 313

B
β-lactamases, 495
Backbones
restructuring of polymer, 458–59
Bacteriorphodopsin, 493
Band formation
and intermolecular interaction, 233–36
Barrier resonances, 75–77
Barriers
potential, 75
Basins of attraction
distinguishing, 302–3
Beams
See Crossed-beam techniques
Becke, Lee, Yang, Parr (BLYP) gradient-corrected functional, 262, 275, 472
B3LYP single-point potential energies, 488–89
Benzene-benzene PMF, 427–28
Berry phase method, 281
Biological polymers, 2
Biological systems, 276–78
Biomedical polymers, 450–52
poly (ethylene glycol), 451
polyurethane, 452
Biomolecular aggregations, 190–95
actin model, 190–92
Biomolecular reactions
gas-phase, 487

Biomolecule isomerization, 312–13
Biopolymer-liquid interfaces, 438–39
BKMP
See Boothroyd-Keogh-Martin-Peterson PES
Blends
phenoxy, 455–56
polymer, 455–57
Blip function methods, 546
Bloch equations, 18, 511, 519
Bloch-vector diagrams, optical, 19–25
for fifth-order Raman spectroscopy, 31
for three-pulse photon echoes, 24
BLYP
See Becke, Lee, Yang, Parr gradient-corrected functional
B3LYP single-point potential energies, 488–89
Boltzmann distributions, 49, 480
Boothroyd-Keogh-Martin-Peterson (BKMP) PES, 72, 83–84, 86
Born-Oppenheimer approximations, 142, 147, 156–57
Born-Oppenheimer surfaces, 252
Bovine pancreatic trypsin inhibitor (BPTI) studies, 359
BP polymers, 450–52, 455–56
Breakthroughs in hydrophobic effects, 411–14
good theories as Gaussian, 414
modeling occupancy probabilities, 413
nonequivalence with

SUBJECT INDEX 593

Pratt-Chandler theory,
413–14
the small size hypothesis,
411–12
transient cavities probing
packing and fluctuations,
412–13
Breit-Wigner formula, 68
Brillouin zones, 203
Broken symmetry, 207–9
BS polymers, 450–52,
455–57

C

Calculation of spectroscopic
properties, 278–85
analysis of valence charge
distribution, 278–79
infrared spectrum, 279–81
vibration modes, Raman
scattering and NMR
chemical shift, 281–85
Calculations
DFT cluster, 332, 337–40,
344
Hartree-Fock, 255
molecular orbital, 362
multi-reference doubles
configuration interactions,
275
of potential energy
surfaces, for MeOH
decomposition, 342
quasiclassical trajectory,
71, 86–88
spin-unrestricted, 362
of turnover frequencies for
ammonia synthesis, 344
See also Approximations
Calorimetry
single-crystal adsorption,
322
Car & Parrinello molecular
dynamics (CPMD), 250,
258–60, 263–67,
270–72, 276–77, 285,
310, 313

Carbon cluster cations
mass distribution of
scattered, 395
Carbon hydrogenation
transition state structures
for intermediates of, 341
Carbon nanotubes
See Single-walled carbon
nanotubes
Carbon-13 NMR chemical
shifts
in amino acids, peptides,
and proteins, 351–61
Carbonic anhydrase, 484,
493–94
Carbonmonoxymyoglobin,
370
CARS
See Coherent anti-Stokes
Raman spectroscopy
Catalysis
in polymerization, 329–31
with zeolites, 331–32
Catalysis on metal surfaces,
319–48
chemisorption trends,
332–38
first principles techniques,
320–23
reactivity trends, 338–42
synergy between theory
and experiment, 323–32
from theory to industrial
applications, 342–44
Catalyst developments
in ammonia synthesis,
343–44
steam-reforming, 343
Catalysts
Ziegler-Natta, 329–30
Catechol
O-methyltransferase,
490–91
Cations
scattered carbon cluster,
mass distribution of, 395
Cavities

transient, probing packing
and fluctuations, 412–13
$C_6F_6/Cu(111)$
molecular resonances and
interfacial electron
transfer in, 230–38
CH_4^+
in high-n Rydberg states,
dissociation mechanism
for, 128–34
Chain lengths
See Side chain lengths
Chain rule approach, 162
Channeltrons, 383
Charge transfers
nonadiabatic, 385–89
resonant rates, 518
See also Valence charge
distribution
CHARMM force field, 358,
470, 475, 484–86,
490–97
Chemical equilibrium
approach, 176
Chemical kinetics, 178–79
of actin, 195
of inorganic
polymerizations, 188–89
of organic polymerizations,
184–85
Chemical reactions, 266–76
direct approach to, 266–67
dynamic simulation of,
267–68
hybrid quantum
mechanics-molecular
mechanics methods,
268–71, 472–76
reactions in solution,
271–73
reactions in the excited
state, 274–76
Chemical shift anisotropy
(CSA) values, 359
Chemical shifts in amino
acids, peptides, and
proteins, 349–78

carbon-13 and nitrogen-15
NMR chemical shifts in
amino acids, peptides,
and proteins, 351–61
drug design based on,
373–75
fluorine-19 NMR and
electrostatic field effects,
365–68
hemeproteins and model
systems, 369–71
hydrogen bonding and
electrostatics, 371–73
paramagnetic systems,
361–63
predicting/refining aspects
of peptide and protein
structure, 363–65
Chemisorption trends,
332–38
alkali atom, 511–13
CO on transition metals,
337–38
oxygen on transition
metals, 336–37
simple model for periodic
trends, 332–34
trends in segregation
energies and surface alloy
formation, 334–36
Chorismate mutase, 494
CI
See Configuration
interaction methods
Circular dichromism
measuring, 8
Citrate synthase, 490–91
Classical VDW theory
hydrostatic approximations
and, 541–42
Clathrate-style hydration
structures, 422–23
Clostridium pasteurianum,
362
Cluster calculations
DFT, 332, 337–40, 344
Clusters

water, 310–11
CO on transition metals,
337–38
potential energy surface for
diffusion of CO, 338
Coherent anti-Stokes Raman
spectroscopy (CARS),
28–29, 71, 85
Coherent phase dynamics,
516–17
Coincidence spectroscopies,
42
threshold photoelectron-
photoion, 124
Cold ion storage rings, 91
Cold Lennard-Jones disks
heptamer of, 309
Cold shock protein A, 494
Collisions
See Reactive collisions of
hyperthermal energy
molecular ions
Committor distributions
versus order parameters,
305–8
potential energies and
corresponding free energy
functions, 306
potential versus free energy
landscapes, 308
in transition path sampling,
305–8
Competition
between delocalization and
localization, 238–44
Complex systems
difficulty of identifying
transition state surfaces,
292–93
events in, 292–93
transition path sampling,
292–93
transition state theory,
292
and transition state theory,
292
Composition

rotational, time-evolving,
56–59
Compressibilities
isothermal, 412
Computation
cost of transition path
sampling, 296
of quantum dynamics,
315–16
Computer simulations
and Gaussian fluctuations,
545–46
Configuration interaction (CI)
methods, 321–22
calculating multi-reference
doubles, 275
Configuration space
random walks in, 294–95
Conformational equilibrium
in water
alkane, 425
Constants
spin-orbit splitting, 118
Contact
densities, 426–27
hydrophobic interactions,
423
problems with, 244–45
Contours
of a free energy surface,
302
Control
quantum, 507–31
Coordinates
See Internal coordinates;
Jacobi coordinates;
Normal mode
coordinates; Radau
coordinates; Reaction
coordinates; Valence
coordinates
Core electron representations,
322–23
Coriolis coupling, 56, 59,
146, 152–53, 155
Correction
of the hydrostatic

SUBJECT INDEX 595

approximation, 542–47
 of the molecular field
  approximation, 547–49
Correlations
 interferometric two-pulse,
  511, 516–19, 521
 vector, in scattering
  experiments, 88–90
Coulomb explosion imaging,
 257
Coulomb potential
 long-range, 513
 repulsion, 512, 518, 525
Coupling
 Coriolis, 56
 hyperfine, 44
 rotation-vibration, 58
CPMD
 See Car & Parrinello
  molecular dynamics
Cross sections
 See Differential cross
  sections
Crossed-beam techniques
 photoinitiated, 72
 in search of resonances,
  83–84
Crystallography
 ultra-high-resolution X-ray,
  372
CS_2^+
 photoionization efficiency
  spectrum for, 106–7
CSA
 See Chemical shift
  anisotropy values
Curvature
 See Finite curvature effect
  of SWNTs
Cytidine deaminase, 494

D

DABCO molecules, 59–60
Dainton & Ivin equation, 175
DCM
 See Dielectric continuum
  model

DCSs
 See Differential cross
  sections
Deaminase
 cytidine, 494
Debye energy, 524
Decarboxylase
 orotidine
  5′-monophosphate,
  491–92
Decomposition
 of MeOH, calculation of
  potential energy surfaces
  for, 342
Defects
 See Surface defects
Dehalogenases
 haloalkane, 495
Dehydrogenases
 lactate, 483–84, 495–96
 liver alcohol, 485–88
 malate, 496
 methyl amine, 488–89
Dehydrogenation-
  hydrogenation reactions,
  339–40
Delocalization
 competing with
  localization, 238–44
Densities
 contact, 426–27
 hydrostatic, 540–41
Density functional
 perturbation theory
  (DFPT), 281
 generalized, 281–85
 vibration modes, Raman
  scattering and NMR
  chemical shift, 281–85
Density functional theory
 (DFT) techniques, 322,
  326, 328, 350, 359, 472,
  534, 557–59
 biological systems, 276–78
 calculation of spectroscopic
  properties, 278–85
 chemical reactions, 266–76

cluster calculations, 332,
 337–40, 344, 484
isolated systems, 254–60
liquid and solution, 260–66
molecular dynamics with,
 249–90
time-dependent, 275
Density matrices, 19
Department of Energy
 (DOE), 8
Dephasing spectroscopy,
 31–32
Desorption
 frustrated, 507–31
Desorption induced by
 electronic transition
 (DIET), 508–9
Detachment
 energy-resolved, of
  negative ions, 62
Detachment/attachment
 experiments, 90–92
Detection
 ion/surface scattering
  chamber for, 384
 product, 383–85
 steradiancy, 110
DFD
 See
  Dudowicz-Freed-Douglas
  lattice model
DFPT
 See Density functional
  perturbation theory
DFT
 See Density functional
  theory
Dielectric continuum model
 image states and, 227–30
Dielectric continuum model
 (DCM), 229
DIET
 See Desorption induced by
  electronic transition
Differential cross sections
 (DCSs), 71, 73, 88–92
 direct observation of the

H_3 transition-state region, 90–92
  electron attachment/ detachment experiments, 90–92
  state-resolved, 73, 88–92
  vector correlations in scattering experiments, 88–90
Diffraction studies
  photoelectron X-ray, 326
Diffusion
  of isobutane in silicalite, 311
Diffusion Monte Carlo techniques, 321
Dihydrofolate reductase, 485, 494
Dimer
  solvated model, 309–10
Dipole operator
  matrix elements of transition, 62
Discharge lamps
  laboratory, 103–4
Dispersion
  one-dimensional energy, 209–12
  parallel curve, 234
Dissociation
  of CD_4^+, 130–32
  of CH_4^+, 126–28
  of CH_4^+ in high-n Rydberg states, 128–34
  of CH_3X^+ (X = Br AND I), 132
  of N_2, 327
  of O_2 in high-n Rydberg states, 119–21
Dissociative scattering, 389–98
  experimental kinetic energy distribution for the dominant scattered ions, 393
  mass distribution of scattered carbon cluster cations in, 395
  relative product yields from, 391, 397
  scattered ion mass spectra recorded for, 396
Distributions
  Flory-Schulz, 178
  molecular weight, 178–79
  See also Kinetic energy distribution
Dividing surfaces
  repulsive periodic orbit, 76
Dominant scattered ions
  experimental kinetic energy distribution for, 393
Double-many-body- expansion (DMBE) PES, 72, 79
Drug design, 373–75
  nonbonding van der Waals shells in, 374–75
Dudowicz-Freed-Douglas (DFD) lattice model, 176–78, 183, 191–92, 194
Dynamics
  nonexponential population, 517–19
  nuclear wave packet, 519–25
  simulation of chemical reactions, 267–68
  and tunneling, 479–83
  See also Ion dynamics; Molecular dynamics; Phase dynamics; Quantum dynamics; Thermodynamics; Unbiased dynamics

E

E-R
  See Eley-Rideal mechanism
EA-VTST/MT
  See Ensemble-averaged variational transition state theory with multidimensional tunneling
Eckart barriers, 142, 146
  potential, 75–76
Effective Hamiltonians, 151–54
  the Watsonian, 153–54
Effective reference fields (ERFs), 537–38, 547–51
  local response to, 540–41
Efficiency sampling
  photoionization, of photofragments, 107–9
Eigenvalues
  Siegert, 75
Elastase, 494–95
Electric field gradients
  Mössbauer, 370
Electron
  attachment/detachment experiments, 90–92
Electron kinetic-energy spectra
  experimental studies of, 388–89
Electron localization function (ELF), 278
Electron transfer (ET) at molecule-to-metal interfaces, 221–47
  experimental techniques in two-photon photoemission, 225–27
  image states and the dielectric continuum model, 227–30
  implications for molecule electronic devices, 242–45
  lifetimes and, 236–38
  molecular resonances and interfacial electron transfer in C_6F_6/Cu(111), 230–38
  polarization and

localization, 238–42
Electronic devices, 242–45
  contact problem, 244–45
  delocalization or
    localization, 243–44
  energetics and dynamics,
    242–43
Electronic spectroscopies,
    21–27
  ion impact, 387
  three-pulse photon echoes,
    23–25
  transient hole burning,
    25–27
  two-dimensional
    Fourier-transform
    electronic spectroscopy,
    27
  two-pulse photon echoes,
    22–23
Electronic structures, 511–14
  alkali atom chemisorption,
    511–13
  and catalysis on metal
    surfaces, 319–48
  chemisorption trends,
    332–38
  electronic structure of
    noble metals, 513–14
  first principles techniques,
    320–23
  methods in, 320–22
  modifications due to,
    156–58
  of noble metals, 513–14
  quantum treatments of,
    with statistical mechanics,
    490–93
  quantum treatments of,
    without statistical
    mechanics, 493–97
  reactivity trends, 338–42
  in surface femtochemistry,
    514–16
  synergy between theory
    and experiment, 323–32
  from theory to industrial

applications, 342–44
time-dependent,
  519–20
Electronic symmetry
  time-evolving, 59–61
Electrons
  See Core electron
    representations
Electrostatic field effects
  fluorine-19 NMR and,
    365–68
  hydrogen bonding and,
    371–73
Eley-Rideal (E-R)
  mechanism, 328, 399
ELF
  See Electron localization
    function
Ellipsometry, 440
Empirical valence bond
  (EVB) modeling, 484,
  491
End group effects on polymer
  surface structures,
  450–52
  biomedical polymers,
    450–52
  PEG, 450
Energetics
  and dynamics,
    242–43
Energies
  See Debye energy;
    Exchange-correlation
    energy; Excitation
    energy; Free energy;
    Hyperthermal energy;
    Kinetic energy;
    Kohn-Sham energy;
    Minimum energy paths;
    Potential energies;
    Relaxation energy;
    Segregation energy
    trends; Vibrational
    energies
Energy- or state-selected
  studies of ion

dynamics, 123–34
pulsed field
  ionization-photoion
  measurements, 134
synchrotron-based pulsed
  field ionization-
  photoelectron-photoion
  coincidence
  measurements, 124–33
threshold
  photoelectron-photoion
  coincidence studies, 124
Energy dispersion
  one-dimensional, 209–12
Energy relaxation, 516–19
  coherent phase dynamics,
    516–17
  nonexponential population
    dynamics, 517–19
  surface femtochemistry,
    516–19
Energy-resolved detachment
  of negative ions, 62
Energy-selected dissociation
  studies
  of CD_4^+, 130–32
  of CH_4^+, 126–28
  of CH_3X^+ (X = Br
    AND I), 132
Energy surfaces
  See Free energy surfaces;
    Potential energy surfaces
Enolase, 484–85, 495
Ensemble-averaged
  variational transition
  state theory with
  multidimensional
  tunneling
  (EA-VTST/MT), 480,
  482–83
Enzyme kinetics, 483–97
  applications with quantum
    electronic structure and
    statistical mechanics,
    490–93
  dynamics and tunneling,
    479–83

methods for, 471–83
potential energy surface,
   471–77
potential of mean force,
   477–79
quantum mechanical
   methods for, 467–505
quantum treatments of
   electronic structure
   without statistical
   mechanics, 493–97
quantum treatments of
   nuclear dynamics, 483–90
theoretical framework of,
   467–71
EQMC
   See Exact-Quantum-
   Monte-Carlo PES
Equations
   See Bloch equations;
   Dainton & Ivin equation;
   GMSA equation;
   Hydrostatic linear
   response equation;
   Ornstein-Zernike
   equations; Percus-Yevick
   equation; Simple
   molecular field equation;
   Time-independent
   Schrödinger equation;
   Van't Hoff equation
Equilibrium-secondary-zone
   (ESZ) approximation,
   482, 487
ERFs
   See Effective reference
   fields
Escherichia coli, 366
Estimation
   of the excited state
   potential energy surface,
   520–24
ESZ
   See Equilibrium-
   secondary-zone
   approximation
Ethylene glycol, 23–25, 451

HITCI in, 25
Ethylene insertion, 330
Euler angles, 49, 142, 145,
   151, 166
EVB
   See Empirical valence
   bond modeling
Evolution
   temporal, of polymer
   chains, 259
   See also Time-evolution
Exact-Quantum-Monte-Carlo
   (EQMC) PES, 72, 84
Exchange-correlation (XC)
   energy, 322
Excitation energy
   low-spin singlet, 275
Excitation mechanisms,
   231–33
   photoelectron kinetic
   energy for molecular
   resonance, 232
Excited state
   estimation of potential
   energy surfaces,
   520–24
   pyrazine, 56
   reactions in, 274–76
Experimental considerations
   in actin, 192
   in inorganic
   polymerizations, 187
   in organic polymerizations,
   179
   in reactive collisions of
   hyperthermal energy
   molecular ions, 381–85
Experimental kinetic energy
   distribution
   for dominant scattered
   ions, 393
Experimental PADs from
   nanosecond ionization
   time-resolved
   photoelectron angular
   distributions of, 45
Experimental studies

of curvature-induced gaps,
   206–7
of detachment/attachment,
   90–92
of electron kinetic-energy
   spectra, 388–89
Experimental studies of
   hydration structure,
   422–23
   clathrate-style, 422–23
   pressure dependence of
   hydrophobic hydration,
   422
Experimental techniques
   in surface femtochemistry,
   509–11
   in two-photon
   photoemission, 225–27
Extension towards N atoms
   Jacobi coordinates of,
   163–64
   of valence coordinates, 166

F
F-actin
   See Filamentary actin
Fast atom reactions
   nascent product structures
   formed in, 108
FCA
   See Frozen Core
   Approximations
FDCD
   See Fluorescence-detected
   circular dichromism
FEL
   See Free-electron lasers
Femtochemistry
   surface, 509–11, 514–16
Femtosecond two-photon
   photoemission
   spectrometers, 227
Femtosecond VUV radiation
   tunable, 105
Fermi contact, 361–62
Fermi electrons, 238
Fermi points, 206

SUBJECT INDEX 599

Feshbach resonances, 76–82
  adiabatic potentials of, 80–81
Fibrinogen
  polymerization of, 2
Field effect transistors (FETs), 221
Field effects
  electrostatic, 365–68, 371–73
  local self-consistent, 474
  See also Effective reference fields; Force fields; General fields; Hartree-Fock self-consistent field; Molecular field approximation
Field-free Hamiltonians, 57
Field gradients
  Mössbauer electric, 370
Fifth-order Raman spectroscopy, 29–31
  optical Bloch-vector diagram for, 31
Filamentary actin (F-actin), 190–91
Filters
  quadrupole mass, 383–84
Fine structure spectroscopy
  near-edge X-ray absorption, 440
Finite curvature effect of SWNTs, 204–7
  experimental studies of curvature-induced gaps, 206–7
  theoretical models, 204–6
Finite size effects, 213–15
First principles techniques, 320–23
  core electron representations, 322–23
  electronic structure methods, 320–22
  models of surface structure, 323

time-independent Schrödinger equation, 320
Fischer-Tropsch reactions, 337
Fixed Φ_P
  computer simulations and Gaussian fluctuations for, 545–46
  hard sphere solute and the PY equation for, 544–45
  HLR equation properties for, 544–47
  other rapidly varying model potentials for, 546–47
Flory-Huggins lattice models, 176–77
Flory-Schulz distributions, 178
Fluctuations
  Gaussian, computer simulations of, 545–46
  in transient cavities, 412–13
Fluorescence-detected circular dichroism (FDCD), 8
Fluorine-19 NMR
  and electrostatic field effects, 365–68
Fluorobenzene, 366
Folded RNA
  free energy functions of, 7
Force fields
  AMBER, 269, 356, 363, 470, 488–89, 494, 496–97
  CHARMM, 358, 470, 475, 484–86, 490–97
  GROMOS, 470, 487, 491
  OPLS, 470, 492
  PARAM, 497
  TIP3P, 470, 484–86, 492, 494–95
Formulas
  Breit-Wigner, 68
  molecular, 438–39
Forward scattering

in the H + D_2 HD(v' = 3, j') + D reaction, 85–88
4-by-4 Mueller matrix, 8
Frameshifting, 10
Franck-Condon factors, 21, 59, 91, 519
Free-electron lasers (FELs), 32
Free energy functions
  and corresponding potential energies, 306
  of folded RNA, 7
Free energy landscapes
  versus potential, 308
Free energy surfaces
  contours of, 302
Fringe effects
  Ramsey, 525–26
Frozen core approximations (FCA), 322
Frustrated desorption of alkali atoms from noble metals
  observation and quantum control of, 507–31
Full width at half maximum (FWHM) resolution, 103–5
Functionals
  See Becke, Lee, Yang, Parr gradient-corrected functional; Density functional perturbation theory; Density functional theory; Local density approximation functional
Functions
  free energy, 306
  maximally-localized Wannier, 278–79
  partition, ratio of, 301

G

G-matrix technique, 159, 167–68
Galactose oxidase (GO), 495
Gallus domesticus, 364
Gas-phase biomolecular

reactions, 487
Gauge-including-atomic-orbital (GIAO) methods, 352, 354, 356, 365–67
Gaussian charges, 269
Gaussian convolution, 206
Gaussian field model, 544
Gaussian fluctuations
 computer simulations, 545–46
Gaussian theories, 414
General fields
 local response to, 540
Generalized density
 functional perturbation theory
 vibration modes, Raman scattering and NMR chemical shift using, 281–85
Generalized gradient approximations (GGA), 322
Generalized hybrid orbital (GHO) approach, 473–75, 490
Geometric phase (GP) effects
 role of, 82–84
GIAO
 See Gauge-including-atomic-orbital methods
Globular actin (G-actin), 190–91
Glutathione reductase, 484
Glyoxalase I, 485
GMSA equation, 544–46
Gradient approximations
 generalized, 322
Gradient-corrected functional
 Becke, Lee, Yang, Parr, 262
Green's function form, 321
GROMOS force field, 470, 491
 united-atom, 487

H

$H + D_2$ reaction, 72–73
 resonances in, 82–88

$H + D_2$ $HD(v' = 3, j') + D$ reaction
 forward scattering in, 85–88
H_3 transition-state region
 direct observation of, 90–92
Haloalkane dehalogenase, 495
Hamiltonians
 field-free, 57
 isomorphic, 156
 reduced, 154
 simple, 159
 translational invariance of, 142
 Watson, 143–58
Hamiltonians in internal coordinates, 158–69
 Jacobi coordinates, 160–64
 rovibrational S-vectors, 166–69
 valence coordinates, 164–66
Hamiltonians in normal mode coordinates, 143–58
 effective Hamiltonians, 151–54
 modifications due to electronic structure, 156–58
 modifications for linear molecules, 154–56
 nonlinear molecules, 143–51
Hammer-Nørskov model
 analyses of, 333–35
Hard sphere solutes
 benzene-benzene PMF, 427–28
 contact densities, 426–27
 in LJ fluid near liquid-vapor coexistence, 550–52
 PFM for stacked plates in water, 427

and the PY equation, 544–45
size dependence of hydrophobic hydration for, 426–29
theory of interface formation, 428–29
Hard wall (HW) model, 521, 523–24
Hartree-Fock (HF) quantum chemical techniques, 366, 372
 ab initio, 352
Hartree-Fock (HF) theory, 255, 472
 geometry optimization, 363
Hartree-Fock self-consistent field (HFSCF), 320–22
Hellman-Feynman forces, 253
Hemeproteins
 and model systems, 369–71
Heptamers
 of cold Lennard-Jones disks, 309
Hermitian operators, 156
High-n Rydberg states
 dissociation mechanism for CH_4^+ in, 128–34
 dissociation mechanism for O_2 in, 119–21
 lifetime measurements of, 116–19
High-pressure reactivity, 328–29
Higher-order optical correlation spectroscopy in liquids, 17–40
 background, 18–21
 electronic spectroscopies, 21–27
 hybrid vibrational spectroscopies, 34
 infrared spectroscopies, 32–34
 Raman spectroscopy, 27–32

SUBJECT INDEX   601

Highly ordered pyrolytic
 graphite (HOPG)
 surfaces, 394–95,
 401–2
HITCI
 in ethylene glycol, 25
HLR equation
 See Hydrostatic linear
 response equation
Hole burning
 transient, 25–27
Homogeneity, 29
HREELS spectrum, 324, 326
Human immunodeficiency
 virus type 1 (HIV-1)
 prolease, 495
 protease, 491
HW
 See Hard wall model
Hybrid orbital approach
 generalized, 474
Hybrid quantum
 mechanics-molecular
 mechanics (QM/MM)
 methods, 268–71,
 472–76
Hybrid vibrational
 spectroscopies, 34
Hydration structures
 clathrate-style, 422–23
 recent experimental studies
 of, 422–23
Hydrogen bonding
 and electrostatics,
 371–73
Hydrogenase, 495
Hydrogenation-
 dehydrogenation
 reactions, 339–40
Hydrophobic effects
 breakthroughs in, 411–14
 molecular theory of,
 409–36
 potentials of mean forces
 among primitive
 hydrophobic species in
 water, 423–25

pressure dependence of
 hydrophobic interactions,
 425–26
recent experimental studies
 of hydration structure,
 422–23
technical observations and
 extensions, 414–20
Hydrophobic hydration and
 size dependencies
 for hard sphere solutes,
 426–29
Hydrophobic hydration and
 temperature dependencies,
 420–22
 model explanation, 420–22
 solubilities, 420
Hydrophobic interactions
 contact, 423
 noncontact, 423–24
 in water, 554–57
Hydrostatic approximations,
 542–47
 and the classical VDW
 theory, 541–42
 optimized linear response
 and the HLR equation,
 542–44
 properties of the HLR
 equation for fixed Φ_P,
 544–47
Hydrostatic density, 540–41
 local response to ERF,
 540–41
 local response to general
 field, 540
Hydrostatic linear response
 (HLR) equation
 computer simulations and
 Gaussian fluctuations,
 545–46
 hard sphere solute and the
 PY equation, 544–45
 optimized linear response
 and, 542–44
 other rapidly varying
 model potentials, 546–47

properties for fixed Φ_P,
 544–47
Hydroxylase
 para-hydroxy bonzoate,
 492, 496
 phenol, 496–97
Hyperfine coupling, 44
Hyperthermal energy
 molecular ions
 ion sources and transport,
 381–82
 product detection, 383–85
 reactive collisions of,
 381–85
 surface preparation,
 382–83
Hypochromism, 6–7

I

Image states
 and the dielectric
 continuum model, 227–30
Imides
 poly-n-alkyl-pyromellitic,
 453
IMJs
 See Intramolecular
 junctions
Importance sampling
 to correct occupancies,
 417–19
 of trajectory space, 293–96
Industrial applications,
 342–44
 new ammonia synthesis
 catalysts, 343–44
 new steam-reforming
 catalysts, 343
Inelastic neutron-scattering,
 190
Infrared spectroscopies,
 32–34
 two- and three-pulse
 infrared echo
 spectroscopy, 32–33
 two-dimensional infrared
 echo spectroscopy, 34

Infrared spectrum, 279–81
Infrared transient hole
  burning, 33–34
Inhomogeneity, 29
Initial trajectory
  stepwise route to, 299–301
  transition path sampling of,
    296–97
Initiated monomers, 174
Inorganic polymerizations,
    186–90
  chemical kinetics and
    molecular weight
    distribution, 188–89
  experimental
    considerations, 187
  structure, 190
  sulfur model, 186–87
  thermodynamics, 187–88
  transport properties, 190
Insertion
  of ethylene, 330
Instantaneous structure
  of liquid sulfur at different
    temperatures, 261
Integral cross sections
  resonance structures in,
    84–85
Integrated molecular orbital
  molecular mechanics
  (IMOMM), 475
Integrated photoelectron
  intensity, 402–3
Intensity
  integrated photoelectron,
    402–3
Interactions
  configuration, 321–22
  contact hydrophobic, 423
  intermolecular, 233–36
  noncontact hydrophobic,
    423–24
  nonuniform fluids with
    attractive, 549–54
Interface formation
  theory of, 428–29
  See also Molecule-to-metal
    interfaces
Interface formation,
  connecting local
  structure to, 533–62
  correcting the molecular
    field approximation,
    547–49
  density functional theory,
    557–59
  hydrophobic interactions in
    water, 554–57
  hydrostatic approximation
    and the classical VDW
    theory, 541–42
  hydrostatic density, 540–41
  molecular field
    approximation, 536–39
  MVDW theory of
    nonuniform fluids with
    attractive interactions,
    549–54
Interfacial electron transfer in
  C_6F_6/Cu(111), 230–38
  excitation mechanisms,
    231–33
  intermolecular interaction
    and band formation,
    233–36
  lifetimes and
    molecule-to-metal
    electron transfer, 236–38
Interferometric two-pulse
  correlations (I2PC), 511,
    516–19, 521
Intermediates of C, N, and O
  hydrogenation
  transition state structures
    for, 341
Intermolecular interaction
  and band formation,
    233–36
  parallel dispersion curve,
    234
Internal coordinates
  Hamiltonians in, 158–69
Intramolecular junctions
  (IMJs), 216–18
Invariance
  translational, of
    Hamiltonians, 142
Ion deposition and surface
  modification, 400–4
  integrated photoelectron
    intensity, 402–3
Ion dynamics
  state- or energy-selected
    studies of, 123–34
Ion impact electron
  spectroscopy, 387
Ion mass spectroscopies
  temperature programmed
  static secondary, 326
Ion/surface scattering
  chamber, 384
Ionization
  nanosecond, experimental
    PADs from, 45
  resonance enhanced
    multiphoton, 71, 82, 85,
    381–82
  two-color photoinduced
    Rydberg, 135
  two-pulse atomic, 45
Ionization spectroscopy
  photo-induced Rydberg,
    136
Ions
  formation of negative, near
    a metal surface, 390
  molecular, 381–85
  neutralization of positive,
    near a metal surface, 386
  scattered, 393, 396, 399
  sources and transport of,
    381–82
I2PC
  See Interferometric
    two-pulse correlations
IPP
  See Isotactic PP
Iso-enthalpy solubility
  minimum, 421
Isobutane
  diffusion in silicalite, 311

Isolated systems, 254–60
  proton sharing in water complexes, 258–60
  proton transfer in malonaldehyde, 260
  temporal evolution of polymer chains, 259
Isomerases
  triose phosphate, 492–93
  xylose, 489–90
Isomerization
  biomolecule, 312–13
  of a solvated model dimer, 309–10
Isomeroreductases
  acetohydroxy acid, 490
Isomorphic Hamiltonians, 156
Isotactic PP (IPP), 443, 445
Isothermal compressibilities, 412

J
Jacobi coordinates, 70, 159–64
  extension towards N atoms, 163–64
  S-vectors for the triatomic system in, 169
Jahn Teller sheets, 91
Jahn-Teller-symmetrydistortion effect, 91
Junctions
  intramolecular, 216–18

K
Kinetic energy
  detection, zero electron, 43
  photoelectron, for molecular resonance, 232
Kinetic energy distribution
  experimental, for dominant scattered ions, 393
Kinetic energy spectra
  experimental studies of electron, 388–89

Kinetic isotope effects (KIEs), 484–85, 488–89
  secondary, 487
Kinetics
  chemical, 178–79
  enzyme, 483–97
  of single-molecules, 11–12
Kohn-Sham energy, 252
Kondo phenomena in 1D, 215–16

L
Laboratory discharge lamps, 103–4
Lactamases
  β, 495
Lactate dehydrogenase, 483–84, 495–96
Lactobacillus casei, 368
LADH
  See Liver alcohol dehydrogenase
Lamps
  laboratory discharge, 103–4
Laser-based detection methods, 385
Laser-based pulsed field ionization-photoelectron (PFI-PE) studies, 111–12
Lasers
  free-electron, 32–33
  vacuum ultraviolet, 104–5
Lattice models
  Dudowicz-Freed-Douglas, 176–78, 183
  Flory-Huggins, 176–77
Lawrence Berkeley Lab, 8
LDA
  See Local density approximation functional
LDPE
  See Low-density PE
LEED
  See Low energy electron diffraction studies
Lennard-Jones (LJ) disks

cold, heptamer of, 309
Lennard-Jones (LJ) fluid, 534–39
  near liquid-vapor coexistence, hard sphere solute in, 550–52
Lennard-Jones (LJ) model, 422, 425
Lifetime measurements
  of high-n Rydberg states, 116–19
Lifetimes and
  molecule-to-metal electron transfer, 236–38
  in time-resolved two-photon photoemission spectroscopy, 237
Lights
  See Advanced Light Source; Laboratory discharge lamps
Linear molecules
  modifications for, 154–56
Linear response
  optimized, and the HLR equation, 542–44
Liquid and solution, 260–66
  acid and superacid, 264–65
  liquid sulfur, 260–62
  liquid water, 262–64
  other systems, 265–66
Liquid-vapor coexistence
  hard sphere solute in LJ fluid, 550–52
Liquids
  higher-order optical correlation spectroscopy in, 17–40
  nonuniform, 533–62
Liu-Siegbahn-Truhlar-Horowitz (LSTH) PES, 72, 83–84
Liver alcohol dehydrogenase (LADH), 482, 485–88
Living polymers, 179
LJ fluid

See Lennard-Jones fluid
Local density approximation (LDA) functional, 275, 322
Local response
  to ERFs, 540–41
  to general fields, 540
Local self-consistent field (LSCF) method, 473–74
Local structure, connecting to interface formation, 533–62
  correcting the molecular field approximation, 547–49
  density functional theory, 557–59
  hydrophobic interactions in water, 554–57
  hydrostatic approximation and the classical VDW theory, 541–42
  hydrostatic density, 540–41
  molecular field approximation, 536–39
  MVDW theory of nonuniform fluids with attractive interactions, 549–54
Localization, 238–42
  competing with delocalization, 238–44
  competition between delocalization and, 238–41
  molecular polarons and surface photochemistry, 241–42
London-Eyring-Polanyi-Sato function, 475
Long-range Coulomb potential, 513
Long trajectories harvesting, 314–15
Loops, 9–10
  tetraloops, 9
Lorentzian affinity level, 386

Low-density PE (LDPE), 443–44
Low energy electron diffraction (LEED) studies, 324, 326, 523
Low-pressure reactivity, 328–29
Low-spin singlet excitation energy, 275
Lowest unoccupied molecular orbital (LUMO), 221–22, 229–30, 232–37, 242–43
LSCF
  See Local self-consistent field method
LSTH
  See Liu-Siegbahn-Truhlar-Horowitz PES

M

Malate dehydrogenase, 496
Malonaldehyde
  proton transfer in, 260
Mandelate racemate, 496
Mass distribution
  of scattered carbon cluster cations, 395
Mass spectra
  scattered ion, 396
Mass spectroscopies
  photoionization measurements, 106–9
  temperature programmed static secondary ion, 326
Matrix elements
  of the transition dipole operator, 62
Mauri, Pfrommer, and Louie (MPL) method, 284
Maximally-localized Wannier functions, 278–79
Mayer-Montroll series, 418
MDQT
  See Molecular dynamics with quantum transitions
Mean force

potential, 469–70
Measurement
  of angular correlations, time-resolved techniques of, 61–62
  of circular dichroism, 8
  of photoionization efficiency, 101
  photoionization mass spectrometric, 106–9
  of viscosities, 1
Mechanical methods
  See Quantum mechanical methods for enzyme kinetics; Quantum mechanics-molecular mechanics methods; Quantum treatments of electronic structure without statistical mechanics; Statistical mechanics
Menzel-Gomer-Redhead (MGR) model, 241
MeOH decomposition
  calculation of potential energy surfaces for, 342
MEPs
  See Minimum energy paths
Metal surfaces
  catalysis on, 319–48
  negative ion formation near, 390
  neutralization of a positive ion near, 386
  single-crystal transition, 324
  See also Noble metals
Metallic-metallic (M-M) IMJs, 217
Metallic-semiconductor (M-S) IMJs, 217
Metastable states
  pattern recognition in, 315
Methyl amine dehydrogenase, 488–89

SUBJECT INDEX   605

Methylcyclohexane (MCH), 183
Methyltransferases
  catechol O, 490–91
MF
  See Molecular field approximation
Microchannel plates (MCPs), 383
Microscopy
  See Scanning tunneling microscopy studies
Minimum energy paths (MEPs), 73–75, 79, 481–82, 485–89
Mixed quantum/classical molecular dynamics (MQCMD) method, 478, 480
MMF
  See Modified molecular field approximation
Model dimer
  isomerization of solvated, 309–10
Model potentials
  rapidly varying, 546–47
Model systems
  hemeproteins and, 369–71
Models
  actin, 190–95
  Antoniewicz, 241
  dielectric continuum, 227–30
  Dudowicz-Freed-Douglas lattice, 176–78, 183
  empirical valence bond, 484
  Flory-Huggins lattice, 176–77
  Gaussian field, 544
  Hammer-Nørskov, 333–35
  hard wall, 521, 523–24
  Lennard-Jones, 425
  Menzel-Gomer-Redhead, 241
  moment, 418

occupancy probability, 413
organic polymerization, poly(α-methylstyrene), 179
surface structure, 323
Modifications
  due to electronic structure, 156–58
  for linear molecules, 154–56
Modifications of polymer surfaces, 452–55
  plasma treatment, UV irradiation, oxygen-ion treatment, and oxygen-radical treatment, 454
  poly-n-alkyl-pyromellitic imide for, 453
  rubbing, 454–55
  wet treatment, 453–54
Modified molecular field (MMF) approximation, 548, 551
Modulation
  of scattered ion product signal, 399
Molecular dynamics
  ab initio, 250–53, 257–58, 260, 262, 286
  Car & Parrinello, 250, 258–60, 263–67, 270–72, 276–77, 285, 310, 313
  mixed quantum/classical, 478, 480
  pump-probe approach to, 42–43
  simulations, 392
Molecular dynamics with density functional theory, 249–90
  biological systems, 276–78
  calculation of spectroscopic properties, 278–85
  chemical reactions, 266–76
  isolated systems, 254–60

liquid and solution, 260–66
Molecular dynamics with quantum transitions (MDQT) approach, 478
Molecular field (MF) approximation, 536–39, 548–49, 555, 558
  modified, 548, 551
  simple molecular field equation, 538–39
  structure of nonuniform reference fluid, 536–38
Molecular field (MF) equation
  simple, 538–39
Molecular formulas, 438–39
Molecular Hamiltonians, 141–72
  in internal coordinates, 158–69
  in normal mode coordinates, 143–58
  the Watson Hamiltonian, 143–58
Molecular ions
  hyperthermal energy, reactive collisions of, 381–85
Molecular mechanics methods
  See Quantum mechanics-molecular mechanics methods
Molecular orbital and valence bond (MOVB) method
  mixed, 476
Molecular orbital (MO) calculation, 362
Molecular polarons
  and surface photochemistry, 241–43
Molecular resonances in C_6F_6/Cu(111), 230–38
  excitation mechanisms, 231–33
  intermolecular interaction

and band formation,
233–36
lifetimes and
molecule-to-metal
electron transfer, 236–38
photoelectron kinetic
energy for, 232
Molecular theory of
hydrophobic effects,
409–36
breakthroughs in, 411–14
hydrophobic hydration and
temperature dependencies,
420–22
potentials of mean forces
among primitive
hydrophobic species in
water, 423–25
pressure dependence of
hydrophobic interactions,
425–26
recent experimental studies
of hydration structure,
422–23
size dependence of
hydrophobic hydration for
hard sphere solutes,
426–29
technical observations and
extensions, 414–20
Molecular van der Waals
(MVDW) theory of
nonuniform liquids,
533–62
with attractive interactions,
549–54
correcting the molecular
field approximation,
547–49
density functional theory,
557–59
hard sphere solute in LJ
fluid near liquid-vapor
coexistence, 550–52
hydrophobic interactions in
water, 554–57
hydrostatic approximation
and the classical VDW
theory, 541–42
hydrostatic density, 540–41
molecular field
approximation, 536–39
solvation free energy,
552–54
two-step method, 549–50
Molecular weight distribution
(MWD), 178–79,
184–86, 195–96
of actin, 195
of inorganic
polymerizations, 188–89
of organic polymerizations,
184–85
Molecule electronic devices,
242–45
contact problem, 244–45
delocalization or
localization, 243–44
energetics and dynamics,
242–43
Molecule/surface collisions
fundamental processes
associated with, 380
Molecule-to-metal interfaces,
electron transfer at,
221–47
and lifetimes, 236–38
Molecules
DABCO, 59–60
linear, modifications for,
154–56
nonlinear, quantization of,
148–51
See also Single-molecule
thermodynamics and
kinetics
Møller-Plesset perturbation
theory (MP2), 472
Moment modeling, 418
Monolayers
self-assembled, 244, 383,
394
Monomers
initiated versus
uninitiated, 174
pure liquid, 175
Monophosphates
See Orotidine
$5'$-monophosphate
decarboxylase
Mössbauer spectroscopy, 351,
371
electric field gradients, 370
MOVB
See Molecular orbital and
valence bond method
MP2
See Møller-Plesset
perturbation theory
MQCMD
See Mixed quantum/
classical molecular
dynamics method
MRDCI
See Multi-reference
doubles configuration
interactions calculations
MSP
See Multipole shielding
polarizabilities
Mueller matrix
4-by-4, 8
Multi-reference doubles
configuration
interactions (MRDCI)
calculations, 275
Multibunch synchrotron
radiation, 110
Multidimensional tunneling
(MT)
variational transition state
theory with, 480, 482–83
Multiphoton ionization
resonance enhanced, 71,
82, 85, 381–82
Multipole shielding
polarizabilities (MSP)
of fluorobenzene, 366
local reaction field
technique, 367
Mutases

SUBJECT INDEX 607

chorismate, 494
MVDW
 See Molecular scale van
  der Waals theory of
  nonuniform liquids
MWD
 See Molecular weight
  distribution

N

Nanosecond ionization
 experimental PADs from,
 45
Nascent product structures
 formed in
 photodissociation or fast
 atom reactions, 108
National Institutes of Health
 (NIH), 2
Near-edge X-ray absorption
 fine structure (NEXAFS)
 spectroscopy, 440
Negative ions
 energy-resolved
  detachment of, 62
 formation of near a metal
  surface, 390
Neutralization
 Auger, 395
 of a positive ion near a
  metal surface, 386
Neutron reflection, 440
Neutron scattering
 inelastic, 190
Nicotinamide adenine
 dinucleotide hydride
 (NADH), 483–84
Nicotinamide adenine
 dinucleotide phosphate
 hydride (NADPH),
 483–84
Nitrogen dissociation, 327
Nitrogen hydrogenation
 transition state structures
 for intermediates of, 341
Nitrogen-15 NMR chemical
 shifts

in amino acids, peptides,
 and proteins, 351–61
Noble metals
 electronic structure of,
  513–14
 frustrated desorption of
  alkali atoms from, 507–31
Nonadiabatic charge transfer,
 385–89
 experimental electron
  kinetic-energy spectra,
  388–89
 negative ion formation near
  a metal surface, 390
 neutralization of a positive
  ion near a metal surface,
  386
Nonbonding van der Waals
 shells, 374–75
Noncontact hydrophobic
 interactions, 423–24
Nonequivalence
 with Pratt-Chandler theory,
  413–14
Nonexponential population
 dynamics, 517–19
Nonlinear molecules,
 143–51
 quantization, 148–51
Nonlinear optical process
 second-order, 440–41
Nonuniform liquids
 molecular scale van der
  Waals theory of, 533–62
Normal mode coordinates
 Hamiltonians in, 143–58
Nuclear dynamics
 quantum treatments of,
  483–90
Nuclear magnetic resonance
 (NMR) studies, 18, 266,
 349–53
 chemical shifts using
  generalized density
  functional perturbation
  theory, 281–85
 fluorine-19, and

electrostatic field effects,
 365–68
nitrogen-15 chemical
 shifts, 351–61
Nuclear Overhauser effect
 (NOE) violations, 364
Nuclear wave packet
 dynamics, 519–25
 estimation of the excited
  state potential energy
  surface, 520–24
 frustrated desorption,
  524–25
 surface femtochemistry,
  519–25
 time-dependent electronic
  structure, 519–20
Nucleic acid studies
 physical chemistry, 1–15
 RNA structure, 9–11
 single-molecule
  thermodynamics and
  kinetics, 11–12
 spectroscopy, 8–9
 thermodynamics, 7
 University of California,
  Berkeley, 4–7, 13–14
 University of Wisconsin,
  Madison, 1–3
 Yale University, 3–4

O

Occupancy probabilities
 models of, 413
One-dimensional electronic
 properties of
 single-walled carbon
 nanotubes, 201–20
 armchair SWNTs, 207–9
 broken symmetry, 207–9
 energy dispersion, 209–12
 finite curvature effect of
  SWNTs, 204–7
 finite size effects, 213–15
 initial STM studies, 204
 intramolecular junctions,
  216–18

Kondo phenomena in 1D,
 215–16
prospects, 218
SWNT end states, 212–13
theoretical background,
 202–4
OPLS force field, 470, 492
Optical Bloch-vector
 diagrams, 19–25, 511,
 519
 for fifth-order Raman
  spectroscopy, 31
 for three-pulse photon
  echoes, 24
Optical correlation
 spectroscopy in liquids,
 17–40
 background, 18–21
 electronic spectroscopies,
  21–27
 hybrid vibrational
  spectroscopies, 34
 infrared spectroscopies,
  32–34
 Raman spectroscopy,
  27–32
Optical process
 second-order, nonlinear,
  440–41
Optical rotation
 of polypeptides, 4
Optimized linear response
 and the HLR equation,
 542–44
Orbital approach
 generalized hybrid, 474
Order parameters versus
  committor distributions
 potential energies and
  corresponding free energy
  functions, 306
 potential versus free energy
  landscapes, 308
 in transition path sampling,
  305–8
Organic polymerizations,
 179–90

chemical kinetics and
 molecular weight
 distribution, 184–85
experimental
 considerations, 179
poly(α-methylstyrene)
 model, 179
structure, 186
thermodynamics, 180–84
transport properties,
 185–86
Ornstein-Zernike equations,
 414, 545
Orotidine 5′-monophosphate
 decarboxylase
 (ODCase), 491–92
Overtone dephasing
 spectroscopy, 31–32
Oxametallacycles
 surface, 325
Oxidases
 galactose, 495
Oxidation reactions
 partial, 340–42
Oxygen
 in high-n Rydberg states,
  dissociation mechanism
  for, 119–21
 on transition metals,
  336–37
Oxygen hydrogenation
 transition state structures
 for intermediates of, 341
Oxygen-ion treatment
 for modification of
 polymer surfaces, 454
O-methyltransferase
 catechol, 490–91
Oxygen-radical treatment
 for modification of
 polymer surfaces, 454

P

Packets
 See Wave packet dynamics
Packing
 probes of, in transient

cavities, 412–13
PADs
 See Photoelectron angular
 distributions
Papain, 496
Para-hydroxy bonzoate
 hydroxylase, 492, 496
Parabolic barriers
 potential, 75
Parallel dispersion curve, 234
Parallel tempering, 311–12
PARAM force field, 497
Paramagnetic systems,
 361–63
Partial oxidation reactions,
 340–42
 calculated potential energy
  surfaces for MeOH
  decomposition, 342
 transition state structures
  for intermediates of C, N,
  and O hydrogenation, 341
Partial-wave summation,
 84–85
Partition functions
 ratio of, 301
Path-integral quantum
 transition state theory,
 478
 See also Minimum energy
 paths; Quantum-classical
 path method; Transition
 path sampling
Pattern recognition
 in stable states, metastable
 states, and transition
 states, 315
PAW
 See Projected
 augmented-wave method
PBMA
 See Poly(n-butyl
 methacrylate)
PDMS
 See Poly(dimethyl siloxane)
PE
 See Polyethylenes

SUBJECT INDEX

PEG
  See Poly(ethylene glycol)
Pentafluorobenzyl alcoholate anion, 487
PEPICO
  See Photoelectron-photoion coincidence measurements scheme
Peptides
  carbon-13 and nitrogen-15 NMR chemical shifts in, 351–61
  chemical shifts in, 349–78
  predicting/refining aspects of structure, 363–65
Percus-Yevick (PY) equation
  analog type, 413, 419
  hard sphere solute and, 544–45
Periodic orbit dividing surfaces (PODS)
  repulsive, 76
Periodic trends
  Hammer-Nørskov model analyses, 334–35
  simple model, 332–34
Perturbation theory
  generalized density functional, 281–85
  Møller-Plesset, 472
  Van Vleck, 159
PES
  See Photoelectron spectroscopy
PESs
  See Potential energy surfaces
PFI-IP
  See Pulsed field ionization-ion pair spectroscopy
PFI-PE
  See Pulsed field ionization-photoelectron spectroscopy

PFI-PEPICO
  See Pulsed field ionization-photoelectron-photoion coincidence measurements
Phase
  role of geometric, 83–84
Phase dynamics
  coherent, 516–17
  in nonexponential population dynamics, 517–19
  surface femtochemistry, 516–19
PHEMA
  See Poly(2-hydroxyethyl methacrylate)
Phenol hydroxylase, 496–97
Phenoxy (PHE)
  blended with different polyurethanes, 455–56
Phosphatases
  protein tyrosine, 492, 497
Phosphate isomerases
  triose, 492–93
Phospholipase A_2 (PLA_2), 484, 487
Phosphorylases
  thymidine, 497
Photo-induced Rydberg ionization spectroscopy (PIRI), 136
Photochemistry
  surface, 241–42
Photodissociation
  nascent product structures formed in, 108
Photoelectron angular distributions (PADs), time-evolving, 56–61
  electronic symmetry, 59–61
  rotational composition, 56–59
Photoelectron angular distributions (PADs), time-resolved, 41–65
  applications, 56–61

background, 42–26
energy-resolved detachment of negative ions, 62
experimental PADs from nanosecond ionization, 45
matrix elements of the transition dipole operator, 62
role of pump-induced alignment, 61
theory, 46–56
Photoelectron diffraction studies
  X-ray, 326
Photoelectron intensity
  integrated, 402–3
Photoelectron kinetic energy
  for molecular resonance, 232
Photoelectron-photoion coincidence (PEPICO) measurements, 102, 123–24
  threshold, 124
Photoelectron spectroscopy (PES)
  threshold, 109–11
  X-ray, 440
Photoemission spectroscopy, two-photon, 221–47
  femtosecond, 227
  time-resolved, 237
Photoinduced Rydberg ionization (PIRI) measurements
  two-color, 135
Photoinitiated crossed-beam techniques, 72
Photoion-photoelectron studies, 106–23
  nascent product structures formed in photodissociation or fast atom reactions, 108
Photoionization efficiency (PIE)

measuring, 101
sampling of
  photofragments, 107–9
spectrum for CS_2^+, 106–7
Photoionization mass
  spectrometric
  measurements, 106–9
laser-based pulsed field
  ionization-photoelectron
  studies, 111–12
pulsed field ionization-ion
  pair spectroscopy, 121–23
synchrotron-based
  pulsed field
  ionization-photoelectron
  studies, 112–21
threshold photoelectron
  spectroscopy, 109–11
Photoloc approach, 85–88
Photon echoes
  three-pulse, 23–25
  two-pulse, 22–23
Physical chemistry of nucleic
  acid, 1–15
RNA structure, 9–11
single-molecule
  thermodynamics and
  kinetics, 11–12
spectroscopy of nucleic
  acids, 8–9
thermodynamics of nucleic
  acids, 7
University of California,
  Berkeley, 4–7, 13–14
University of Wisconsin,
  Madison, 1–3
Yale University, 3–4
Physical interpretation of
  time-resolved
  photoelectron angular
  distributions
  angular structure, 52–56
  temporal structure, 50–52
PIE
  See Photoionization
    efficiency
PLA_2

See Phospholipase A_2
Plasma treatment
  for modification of
  polymer surfaces, 454
PMF
  See Potential mean force
PMMA
  See Poly(methyl
    methacrylate)
Podolsky transformation,
  142, 155–56
Polarizabilities, 424–25
Polarization, 238–42
  competition between
    delocalization and
    localization, 238–41
  molecular polarons and
    surface photochemistry,
    241–42
Polarons
  in molecular, and surface
    photochemistry, 241–42
Poly(2-hydroxyethyl
  methacrylate)
  (PHEMA), 457–59
Poly-n-alkyl-pyromellitic
  imide
  for modification of
    polymer surfaces, 453
Polyatomic radicals, 108
Poly(n-butyl methacrylate)
  (PBMA), 455–60
Poly(dimethyl siloxane)
  (PDMS), 451–52, 460
Polyenes
  trans-linear, 60
Poly(ethylene glycol) (PEG),
  450–51, 457–58
Polyethylenes (PE), 443–6
  low-density, 443–44
  SFG spectra of various, 444
  ultrahigh molecular
    weight, 443–44
Polyimides, 448–49
  with alkyl side chain
    lengths, 449
Polymer backbones

restructuring of,
  immediately after
  contacting water, 458–59
Polymer blends, 455–57
  of phenoxy with different
    polyurethanes, 455–56
  PS/PBMA, 456–57
Polymer chains
  temporal evolution of, 259
Polymer surfaces
  analysis of SFG spectra,
    441–43
  end group effects on
    structures of, 450–52
  modification of, 452–55
  molecular formulas,
    438–39
  SFG technique, 440–41
  studied by sum frequency
    generation vibrational
    spectroscopy, 437–65
Polymer surfaces in air,
  443–49
  PE and PP, 443–46
  PMMA, 447–48
  polyimides, 448–49
  PS, 446–47
  SFG spectra of various
    polyethylenes, 444
  SFG spectra of various
    polypropylenes, 445
Polymer surfaces in water,
  457–61
  absence of restructuring,
    461
  side group reorientation,
    459–60
  slow change, 460
Polymerization catalysis,
  329–31
  ethylene insertion, 330
Polymerizations, 173–200
  biomolecular aggregations,
    190–95
  chemical kinetics and
    molecular weight
    distribution, 178–79

of fibrinogen, 2
inorganic, 186–90
organic, 179–90
questions remaining, 196
statistical mechanics,
175–78
terminology, 173–74
theoretical framework,
174–79
thermodynamics, 174–75
Polymers
biological, 2
biomedical, 450–52
blends, 455–57
living, 179
synthetic, 1
Poly(α-methylstyrene) model
of organic polymerizations,
179
Poly(methyl methacrylate)
(PMMA), 447–48,
457–58, 461
Poly(n-octyl methacrylate)
(POMA), 457–58
Polypeptides
optical rotation of, 4
synthetic, 5
Polypropylenes (PP), 443–46
atactic, 443, 445, 455
isotactic, 443, 445
SFG spectra of various, 445
Polyproteins, 10
Polystyrenes (PS), 446–47,
456–57
Polyurethanes, 452
phenoxy blended with
different, 455–56
POMA
See Poly(n-octyl
methacrylate)
Population dynamics
nonexponential, 517–19
Porter-Karplus PES, 72
Positive ions
neutralization of near a
metal surface, 386
Potential barriers

parabolic and Eckart, 75
Potential energies
B3LYP single-point, 488
and corresponding free
energy functions, 306
long-range Coulomb, 513
Potential energy landscapes
versus free, 308
Potential energy surfaces
(PESs), 69–72, 338,
471–77, 487–89, 512–13
Boothroyd-Keogh-Martin-
Peterson, 72, 83–84,
86
calculation of, for MeOH
decomposition, 342
for diffusion of CO, 338
double-many-body-
expansion, 72,
79
estimation of the excited
state, 520–24
Exact-Quantum-Monte-
Carlo, 72, 84
generalized hybrid orbital
approach, 474
Liu-Siegbahn-Truhlar-
Horowitz, 72,
83–84
local self-consistent field
method, 474
Porter-Karplus, 72
Potential mean force (PMF),
469–70, 485
benzene-benzene, 427–28
quantum, 479
for stacked plates in water,
427
Potentials
adiabatic, of Feshbach
resonances, 80–81
alkane conformational
equilibrium in water, 425
among primitive
hydrophobic species in
water, 423–25
contact hydrophobic

interactions, 423
of mean forces, 477–79
noncontact hydrophobic
interactions, 423–24
question of polarizabilities,
424–25
simulation results, 424
PP
See Polypropylenes
Pratt-Chandler theory, 409,
426–30
nonequivalence with,
413–14
Pressure dependence
of hydrophobic hydration,
422
of hydrophobic
interactions, 425–26
Probabilities
modeling occupancy, 413
Probes
of packing in transient
cavities, 412–13
Product detection, 383–85
ion/surface scattering
chamber for, 384
Product signals
scattered ion, modulation
of, 399
Product structures
nascent, formed in
photodissociation or fast
atom reactions, 108
Product yields
relative, 397
Projected augmented-wave
(PAW) method, 253,
268–70
Prolease
HIV-1, 495
Protease
HIV-1, 491
Protein A
cold shock, 494
Protein tyrosine phosphatase,
492, 497
Proteins

carbon-13 and nitrogen-15
NMR chemical shifts in,
351–61
chemical shifts in, 349–78
predicting/refining aspects
of structure, 363–65
Proton sharing
in water complexes,
258–60
Proton transfer
in malonaldehyde, 260
mechanism for, 308
Protonated acetylene, 257
PS
See Polystyrenes
PS/PBMA blends, 456–57
Pseudoknots, 10
Pulsed field ionization-ion
pair (PFI-IP)
spectroscopy, 121–23
spectrum for H_2S, 122–23
spectrum for O_2, 121–22
Pulsed field
ionization-photoelectron
(PFI-PE) spectroscopy
dissociation mechanism for
O_2 in high-n Rydberg
states, 119–21
laser-based, 111–12
lifetime measurements of
high-n Rydberg states,
116–19
spectrum for O_2^+ ($X^2\Pi_g$,
$v^+ = 0$–38), 115–16
spin-orbit splitting
constants, 118
synchrotron-based, 112–21
Pulsed field ionization-
photoelectron-photoion
coincidence
measurements
(PFI-PEPICO)
dissociation mechanism for
CH_4^+ in high-n Rydberg
states, 128–34
energy-selected
dissociation study of

CD_4^+, 130–32
energy-selected
dissociation study of
CH_3X^+ (X = Br AND I),
132
energy-selected
dissociation study of
CH_4^+, 126–28
synchrotron-based, 124–33
zero point vibrational
energies, 133
Pulsed field
ionization-photoion
measurements, 134
synchrotron-based, 134
Pump-induced alignment
role of, 61
Pump-probe approach
to molecular dynamics,
42–43
Pure liquid monomers,
175
PY equation
See Percus-Yevick
equation
Pyrazine
excited state, 56

Q

QCP
See Quantum-classical
path method
QCT
See Quasiclassical
trajectory calculations
QM/MM
See Quantum
mechanics-molecular
mechanics methods
Quadrupole mass filters,
383–84
Quantum-classical path
(QCP) method, 478–80,
483–85, 488
Quantum control
of frustrated desorption of
alkali atoms from noble

metals, 507–31
of surface femtochemistry,
525–27
Quantum dynamics
computation of, 315–16
Quantum electronic structure
with statistical
mechanics, 490–93
acetohydroxy acid
isomeroreductase, 490
catechol
O-methyltransferase,
490–91
citrate synthase, 490–91
HIV-1 protease, 491
orotidine
5'-monophosphate
decarboxylase, 491–92
para-hydroxy bonzoate
hydroxylase, 492
protein tyrosine
phosphatase, 492
subtilisin, 492
triose phosphate isomerase,
492–93
Quantum electronic structure
without statistical
mechanics, 493–97
aldose reductase, 493
β-lactamases, 495
bacteriorphodopsin, 493
carbonic anhydrase,
493–94
chorismate mutase, 494
cold shock protein A, 494
cytidine deaminase, 494
dihydrofolate reductase,
494
elastase, 494–95
enolase, 495
galactose oxidase, 495
haloalkane dehalogenase,
495
HIV-1 prolease, 495
hydrogenase, 495
lactate dehydrogenase,
495–96

malate dehydrogenase, 496
mandelate racemate, 496
papain, 496
para-hydroxy bonzoate
  hydroxylase, 496
phenol hydroxylase,
  496–97
PLA$_2$, 497
protein tyrosine
  phosphatase, 497
rubisco, 497
thermolysin, 497
thymidine phosphorylase,
  497
Quantum mechanical
  methods for enzyme
  kinetics, 467–505
  applications, 483–97
  methods, 471–83
  theoretical framework,
  467–71
Quantum
  mechanics-molecular
  mechanics (QM/MM)
  methods, 472–76,
  485–97
  hybrid, 268–71, 472–76
Quantum Monte Carlo
  (QMC) techniques, 321
Quantum PMF, 479
Quantum transitions
  molecular dynamics with,
  478, 487
Quantum treatments of
  nuclear dynamics,
  483–90
  aspartic transcarbamoylase,
  484
  carbonic anhydrase, 484
  dihydrofolate reductase,
  485
  enolase, 484–85
  glutathione reductase, 484
  glyoxalase I, 485
  lactate dehydrogenase,
  483–84
  liver alcohol
    dehydrogenase, 485–88
  methyl amine
    dehydrogenase, 488–89
  phospholipase A$_2$, 484
  xylose isomerase, 489–90
Quasichemical theory,
  415–17
Quasiclassical trajectory
  (QCT) calculations, 71,
  86–88

R

Rabi-type cycling, 53
Racemate
  mandelate, 496
Radau coordinates, 159
Radiation
  multibunch synchrotron,
  110
  tunable femtosecond VUV,
  105
Radicals
  polyatomic, 108
Ramachandran shielding
  surfaces, 357–58, 370
Raman scattering
  using generalized density
  functional perturbation
  theory, 281–85
Raman spectroscopies,
  27–32, 440
  coherent anti-Stokes,
  28–29, 71
  fifth-order, 29–31
  overtone dephasing, 31–32
  Raman-echo, 28–29
Ramsey fringe effect, 525–26
Random walks in
  configuration space
  transition path sampling,
  294–95
Ratios
  of partition functions, 301
Reaction coordinates
  versus order parameters,
  305–8
Reaction processes, 385–404

dissociative scattering,
  389–98
  in the excited state, 274–76
  fast atom, nascent product
  structures formed in, 108
  Fischer-Tropsch, 337
  gas-phase biomolecular,
  487
  H + D$_2$, 72–73
  H + D$_2$ HD(v′ = 3, j′) +
  D, forward scattering in,
  85–88
  hydrogenation-
  dehydrogenation,
  339–40
  low-spin singlet excitation
  energy, 275
  N$_2$ dissociation, 327
  nonadiabatic charge
  transfer, 385–89
  partial oxidation, 340–42
  in solution, 271–73
  with subsurface species,
  326–28
  See also Abstraction
  reactions; Chemical
  reactions
Reactive collisions of
  hyperthermal energy
  molecular ions, 379–407
  experimental methodology,
  381–85
  fundamental processes
  associated with
  molecule/surface
  collisions, 380
  ion sources and transport,
  381–82
  processes, 385–404
  product detection, 383–85
  surface preparation,
  382–83
Reactive scattering
  resonances, 73–82
  barrier resonances, 75–77
  Feshbach resonances,
  77–82

shape resonances, 73–75
Reactivity trends, 338–42
  high- and low-pressure, 328–29
  hydrogenation-dehydrogenation reactions, 339–40
  partial oxidation reactions, 340–42
Reduced Hamiltonians, 154
Reductases
  aldose, 493
  dihydrofolate, 485, 494
  glutathione, 484
Reference fields
  effective, local response to, 540–41
Relative yields
  of product, 397
  of scattered O^- fragments, 391
Relaxation energy, 516–19
REMPI
  See Resonance enhanced multiphoton ionization
Reorientation
  side group, on polymer surface in water, 459–60
Representations
  of a core electron, 322–23
  of semiclassical initial values, 315
Repulsion
  Coulomb, 512, 518, 525
  on periodic orbit dividing surfaces, 76
Resolution
  full width at half maximum, 103–5
Resonance enhanced
  multiphoton ionization (REMPI), 71, 82, 85, 381–82
Resonances
  antibonding, 513, 527
  barrier, 75–77
  Feshbach, 77–82

molecular, in
  C_6F_6/Cu(111), 230–38
  reactive scattering, 73–82
  shape, 73–75
  signatures, 72
Resonances in the $H + D_2$ reaction, 82–88
  crossed-beam experiments in search of resonances, 83–84
  forward scattering in the $H + D_2$ HD($v' = 3, j'$) + D reaction, 85–88
  partial-wave summation, 84–85
  resonance structure in the integral cross section, 84–85
  role of geometric phase, 83–84
Resonant charge transfer (RCT) rates, 518
Resonant periodic orbits (RPOs), 70
Restructuring of polymer backbones
  immediately after contacting water, 458–59
Reversible polymerizations and aggregations, 173–200
  biomolecular aggregations, 190–95
  chemical kinetics and molecular weight distribution, 178–79
  inorganic polymerizations, 186–90
  organic polymerizations, 179–90
  questions remaining, 196
  statistical mechanics, 175–78
  terminology, 173–74
  theoretical framework, 174–79
  thermodynamics, 174–75

Reversible work, 12, 297–303
  for changing ensembles of trajectories, 298–99
  contours of a free energy surface, 302
  distinguishing basins of attraction, 302–3
  in equilibrium statistical mechanics, 297–98
  ratio of partition functions, 301
  sequence of trajectory ensembles, 300
  stepwise route to the initial trajectory, 299–301
  transition path sampling, 297–303
  unbiased dynamics, 301
Ribozomes, 11
RNA
  folded, free energy functions of, 7
  structure of, 9–11
Rotation
  optical, of polypeptides, 4
Rotation-vibration coupling
  effect of, 58
Rotational composition
  time-evolving, 56–59
Rovibrational S-vectors, 166–69
  for the triatomic system in Jacobi coordinates, 169
Rubbing
  for modification of polymer surfaces, 454–55
Rubisco, 497
Rydberg ionization spectroscopy
  photo-induced, 136
  two-color photoinduced, 135
Rydberg states, high-n
  dissociation mechanism for CH_4^+ in, 128–34
  dissociation mechanism for

O_2 in, 119–21
  lifetime measurements of, 116–19
Rydberg-tagging, 82

S

S-vectors
  rovibrational, 166–69
  for the triatomic system in Jacobi coordinates, 169
Sample trajectories
  MDQT approach, 487
Sampling
  to correct occupancies, importance of, 417–19
  photoionization efficiency, 107–9
  trajectory space, importance of, 293–96
  transition path, 291–318
SAMs
  See Self-assembled monolayers
Scanning tunneling microscopy (STM) studies
  armchair SWNTs, 207–9
  broken symmetry, 207–9
  1D energy dispersion, 209–12
  finite curvature effect of SWNTs, 204–7
  finite size effects, 213–15
  initial studies, 204
  intramolecular junctions, 216–18
  Kondo phenomena in 1D, 215–16
  one-dimensional electronic properties of single-walled carbon nanotubes, 201–20
  prospects, 218
  SWNT end states, 212–13
  theoretical background, 202–4
Scattered carbon cluster cations

  mass distribution of, 395
Scattered ions
  experimental kinetic energy distribution for dominant, 393
  mass spectra recorded, 396
  modulation of product signal, 399
  See also Inelastic neutron-scattering
Scattering
  angles, 50
  forward, in the $H + D_2$ $HD(v' = 3, j') + D$ reaction, 85–88
  ion/surface chambers, 384
  vector correlations in, 88–90
Scattering resonances, 67–99
  barrier, 75–77
  beyond state-resolved differential cross sections, 88–92
  Feshbach, 77–82
  in the $H + D_2$ reaction, 82–88
  from nuclear physics to chemical reactions, 67–73
  shape, 73–5
  wave packet snapshots, 89
  See also Reactive scattering resonances
Schrödinger equation, 73, 77, 171, 321
  time-independent, 320
Second-order nonlinear optical process, 440–41
Secondary ion mass spectroscopy (SIMS), 440
  temperature programmed static, 326
Segregation energy trends, 334–36
Self-assembled monolayers (SAMs), 244, 383, 394
Self-consistent field method

  Hartree-Fock, 320–22
  local, 474
Semiclassical initial value representation (SC-IVR), 315
Separatrix
  in transition path sampling, 303–5
Sequence
  of trajectory ensembles, 300
SFG
  See Sum frequency generation vibrational spectroscopy
Shape resonances, 73–75
Shared protons
  in water complexes, 258–60
Shells
  van der Waals nonbonding, 374–75
Shielding surfaces
  Ramachandran, 357–58
Shockley surface state, 513
Side chain lengths
  alkyl, polyimides with, 449
Side group reorientation
  on polymer surface in water, 459–60
Siegert eigenvalues, 75
Silicalite
  diffusion of isobutane in, 311
Simple Hamiltonians, 159
Simple molecular field equation, 538–39
SIMS
  See Secondary ion mass spectroscopy
Simulations
  computer, and Gaussian fluctuations, 545–46
  dynamic, of chemical reactions, 267–68
  molecular dynamic, 392
  results of, 424

Single-crystal adsorption calorimetry, 322
Single-crystal transition metal surfaces, 324
Single-molecule thermodynamics and kinetics, 11–12
Single-walled carbon nanotubes (SWNTs)
armchair, 207–9
broken symmetry, 207–9
1D energy dispersion, 209–12
end states, 212–13
finite curvature effect of, 204–7
finite size effects, 213–15
initial STM studies, 204
intramolecular junctions, 216–18
Kondo phenomena in 1D, 215–16
prospects, 218
scanning tunneling microscopy studies of the one-dimensional electronic properties of, 201–20
theoretical background, 202–4
Singlet excitation energy low-spin, 275
Site preferences and vibrational spectroscopies, 324–26
surface oxametallacycles, 325
Size dependence of hydrophobic hydration for hard sphere solutes, 426–29
benzene-benzene PMF, 427–28
contact densities, 426–27
PFM for stacked plates in water, 427
theory of interface formation, 428–29
See also Finite size effects
Slow polymer surface change in water, 460
Small size hypothesis, 411–12
isothermal compressibilities, 412
Solubilities, 420
Solubility minimum iso-enthalpy, 421
Solutions
See Hard sphere solute; Liquid and solution
Solvated model dimer isomerization of, 309–10
Solvation
dynamics of, 313–14
free energy from, 552–54
Spectra
and electronic structure, 514–16
photoionization efficiency, for CS_2^+, 106–7
in surface femtochemistry, 514–16
See also Kinetic energy spectra; Mass spectra
Spectrometries
femtosecond two-photon photoemission, 227
photoionization mass measurements, 106–9
Spectroscopic properties calculation of, 278–85
Spectroscopies, 351, 440
coincidence, 42, 124
dephasing, 31–32
electronic, 21–27
fifth-order Raman, 29–31
higher-order optical correlation, in liquids, 17–40
infrared, 32–34
ion impact electron, 387
Mössbauer, 351, 370–71
near-edge X-ray absorption fine structure, 440
NMR, 349–53
of nucleic acids, 8–9
optical, 18
overtone dephasing, 31–32
photo-induced Rydberg ionization, 136
photoelectron, 109–11, 440
photoemission, two-photon, 227, 237
Raman-echo, 28–29
sum frequency generation vibrational, 437–65
temperature programmed static secondary ion mass, 326
threshold photoelectron, 109–11
time-resolved two-photon photoemission, 237, 508–11
two- and three-pulse infrared echo, 32–33
vibrational, 34, 324–26
X-ray photoelectron, 440
Spin-coating, 443
Spin-orbit splitting constants, 118
Spin-unrestricted calculations, 362
Splitting constants spin-orbit, 118
SSZ
See Static-secondary-zone approximation
Stable states
pattern recognition in, 315
Stacked plates
in water, PMF for, 427
Stanford Magnetic Resonance Laboratory, 9
State- or energy-selected studies of ion dynamics, 123–34
pulsed field ionization-photoion measurements, 134

synchrotron-based pulsed
  field ionization-
  photoelectron-photoion
  coincidence
  measurements, 124–33
  threshold
  photoelectron-photoion
  coincidence studies, 124
State-resolved differential
  cross sections, 73, 88–92
  direct observation of the
  H_3 transition-state region,
  90–92
  electron attachment/
  detachment experiments,
  90–92
  vector correlations in
  scattering experiments,
  88–90
Static-secondary-zone (SSZ)
  approximation, 482, 487
Statistical mechanics, 175–78
  quantum electronic
  structure and, 490–93
  quantum treatments of
  electronic structure with,
  490–93
  quantum treatments of
  electronic structure
  without, 493–97
Steam-reforming catalyst
  developments, 343
Stepwise route
  to the initial trajectory,
  299–301
Steradiancy detection, 110
STM
  See Scanning tunneling
  microscopy studies
Stokes shift
  time-resolved fluorescence,
  21–22
Stokes vector, 8
Structures
  of actin, 195
  hydration, 422–23
  of inorganic

polymerizations, 190
  local, 533–62
  nascent product, 108
  of nonuniform reference
  fluids, 536–38
  of organic polymerizations,
  186
  predicting/refining aspects
  of peptide and protein,
  363–65
  See also Electronic
  structures
Subtilisin, 492
Successive Monte Carlo
  sampling, 321
Sulfur, liquid, 260–62
  instantaneous structure at
  different temperatures,
  261
Sulfur model
  of inorganic
  polymerizations, 186–87
Sum frequency generation
  (SFG) vibrational
  spectroscopy
  second-order, nonlinear
  optical process technique,
  440–41
  spectra of various
  polyethylenes, 444
  spectra of various
  polypropylenes, 445
  studying polymer surfaces
  by, 437–65
Summation
  partial-wave, 84–85
Superacids, 264–65
Surface alloy formation
  trends, 334–36
Surface defects
  effect of, 326
Surface electronic structures,
  511–14
  alkali atom chemisorption,
  511–13
  electronic structure of
  noble metals, 513–14

Surface femtochemistry
  electronic structure and
  spectra, 514–16
  experimental technique,
  509–11
  nuclear wave packet
  dynamics, 519–25
  observation and quantum
  control of frustrated
  desorption of alkali atoms
  from noble metals,
  507–31
  phase and energy
  relaxation, 516–19
  quantum control, 525–27
Surface modification
  ion deposition and, 400–4
Surface oxametallacycles,
  325
Surface photochemistry
  molecular polarons and,
  241–42
Surface preparation, 382–83
Surface restructuring of
  polymers in water,
  457–61
  absence of, 461
Surface states (SS)
  Shockley, 513
Surface structure
  models of, 323
Surfaces
  Born-Oppenheimer, 252
  dividing, repulsive periodic
  orbit, 76
  HOPG, 394–95, 401–2
  negative ion formation near
  metal, 390
  positive ion neutralization
  near metal, 386
  Ramachandran shielding,
  357–58
  single-crystal transition
  metal, 324
  See also Free energy
  surfaces; Interface
  formation;

Molecule/surface
collisions; Potential
energy surfaces;
Transition state surfaces
SWNTs
See Single-walled carbon
nanotubes
Symmetry
broken, 207–9
electronic, time-evolving,
59–61
Symmetrydistortion effect
Jahn-Teller, 91
Synchrotron-based
pulsed field
ionization-photoelectron
(PFI-PE) studies,
112–21
dissociation mechanism for
O_2 in high-n Rydberg
states, 119–21
lifetime measurements of
high-n Rydberg states,
116–19
pulsed field
ionization-photoelectron
spectrum for O_2^+ ($X^2\Pi_g$,
$v^+ = 0$–38), 115–16
spin-orbit splitting
constants, 118
Synchrotron-based pulsed
field ionization-
photoelectron-photoion
coincidence
measurements
(PFI-PEPICO), 124–33
dissociation mechanism for
CH_4^+ in high-n Rydberg
states, 128–34
energy-selected
dissociation study of
CD_4^+, 130–32
energy-selected
dissociation study of
CH_4^+, 126–28
energy-selected
dissociation study of

CH_3X^+ (X = Br AND I),
132
zero point vibrational
energies, 133
Synchrotron-based pulsed
field ionization-photoion
measurements, 134
Synchrotron radiation
multibunch, 110
vacuum ultraviolet, 105–6
Synergy of experiment with
theory, 323–32
catalysis with zeolites,
331–32
effect of surface defects,
326
high- and low-pressure
reactivity, 328–29
polymerization catalysis,
329–31
reactions with subsurface
species, 326–28
site preferences and
vibrational
spectroscopies, 324–26
Synthases
citrate, 490–91
Synthetic polypeptides, 5

T

TALOS program, 364–65
Temperature dependencies
in hydrophobic hydration,
420–22
Temperature programmed
static secondary ion
mass spectroscopy
(TPSSIMS), 326
Tempering
parallel, 311–12
Temporal evolution
of polymer chains, 259
Temporal structure
time-resolved, 50–52
Tetrahydrofuran (THF), 180,
183–84
Tetraloops, 9

TEXAS90 program, 352
Theoretical framework
of enzyme kinetics, 467–71
Theories
of interface formation,
428–29
quasichemical, 415–17
See also Atoms in
molecules theory; Density
functional perturbation
theory; Density functional
theory; Hartree-Fock
theory; Molecular theory
of hydrophobic effects;
Møller-Plesset
perturbation theory;
Pratt-Chandler theory;
Time-dependent density
functional theory;
Time-resolved
photoelectron angular
distributions theory;
Transition states theory;
Van Vleck perturbation
theory; Variational
transition states theory
Theory in synergy with
experiment, 323–32
catalysis with zeolites,
331–32
effect of surface defects,
326
high- and low-pressure
reactivity, 328–29
polymerization catalysis,
329–31
reactions with subsurface
species, 326–28
site preferences and
vibrational spectroscopies,
324–26
Theory of time-resolved
photoelectron angular
distributions, 46–56
physical interpretation,
50–56
time-evolution and

observables, 47–50
Thermodynamics, 174–75
    of actin, 192–95
    of inorganic
        polymerizations, 187–88
    of nucleic acids, 7
    of organic polymerizations, 180–84
    of single-molecules, 11–12
Thermolysin, 497
Three-pulse infrared echo
    spectroscopy, 32–33
Three-pulse photon echoes, 23–25
    optical Bloch-vector
        diagram for, 24
Threshold-photoelectron-
    photoion coincidence
    (TPEPICO) studies, 102, 124, 128
Threshold photoelectron
    (TPE) spectroscopy, 109–11, 128
Thymidine phosphorylase, 497
Time-dependent density
    functional theory
    (TDDFT), 275
Time-dependent electronic
    structure, 519–20
Time-evolution
    and observables, 47–50
Time-evolving electronic
    symmetry, 59–61
Time-evolving rotational
    composition, 56–59
    effect of rotation-vibration
        coupling, 58
Time-independent
    Schrödinger equation
    (TISE), 320–22
Time-of-flight (TOF)
    selection, 110, 113, 126–27
Time-resolved fluorescence
    Stokes shift (TRFSS), 21–22

Time-resolved photoelectron
    angular distributions, 41–65
    applications, 56–61
    background, 42–46
    energy-resolved
        detachment of negative
        ions, 62
    experimental PADs from
        nanosecond ionization, 45
    matrix elements of the
        transition dipole operator, 62
    role of pump-induced
        alignment, 61
    theory, 46–56
Time-resolved techniques
    for measuring angular
    correlations, 61–62
Time-resolved temporal
    structure, 50–52
Time-resolved two-photon
    photoemission (TR-2PP)
    spectroscopy, 237, 508–11
TIP3P force field, 470, 484–86, 492, 494–95
TISE
    See Time-independent
    Schrödinger equation
Tobacco mosaic virus, 190
TOF
    See Time-of-flight
    selection
TPE
    See Threshold
    photoelectron
    spectroscopy
TPEPICO
    See Threshold-
    photoelectron-photoion
    coincidence
    measurements method
TR-2PP
    See Time-resolved
    two-photon
    photoemission

spectroscopy
Trajectories
    harvesting long, 314–15
    sample, MDQT approach, 487
    See also Initial trajectory
Trajectory calculations
    quasiclassical, 71, 86–88
Trajectory ensembles
    sequence of, 300
Trajectory space
    importance sampling of, 293–96
Trans-linear polyenes, 60
Transcarbamoylase
    aspartic, 484
Transfer rates
    resonant charge, 518
Transfers
    proton, in malonaldehyde, 260
Transformations
    Podolsky, 142
Transient cavities
    probing packing and
    fluctuations, 412–13
Transient hole burning, 25–27
    infrared, 33–34
Transition dipole operator
    matrix elements of, 62
Transition metal surfaces
    single-crystal, 324
Transition metals
    CO on, 337–38
    oxygen on, 336–37
Transition path sampling, 291–318
    applications of, 309–14
    biomolecule isomerization, 312–13
    committors, the separatrix
        and the transition state
        ensemble in, 303–5
    computational cost of, 296
    diffusion of isobutane in
        silicalite, 311

events in complex systems, 292–93
future applications, 314–16
heptamer of cold Leonard-Jones disks, 309
importance sampling of trajectory space, 293–96
initial trajectory, 296–97
isomerization of a solvated model dimer, 309–10
order parameters versus reaction coordinates and committor distributions, 305–8
parallel tempering, 311–12
random walks in trajectory space, 294–95
reversible work, 297–303
solvation dynamics, 313–14
transition state ensemble in, 303–5
water autoionization, 313
water clusters, 310–11
Transition state ensemble in transition path sampling, 303–5
Transition state region direct observation of H_3, 90–92
Transition state structures for intermediates of C, N, and O hydrogenation, 341
Transition state surfaces difficulty of identifying, 292–93
Transition states pattern recognition in, 315
Transition states theory (TST), 292, 468–69, 478–81, 485, 488
conventional, 489
path-integral quantum, 478
Translational invariance of Hamiltonians, 142
Transport properties of actin, 195

of inorganic polymerizations, 190
ion sources, 381–82
of organic polymerizations, 185–86
Trends
See Chemisorption trends; Periodic trends; Reactivity trends; Segregation energy trends; Surface alloy formation trends
Triatomic system in Jacobi coordinates, S-vectors for, 169
Triethylenediamine, 57
Triose phosphate isomerase, 492–93
TST
See Transition states theory
Tunable femtosecond VUV radiation, 105
Tunneling and dynamics, 479–83
variational transition state theory with multidimensional, 480, 482–83
Tunneling microscopy studies, 201–20
armchair SWNTs, 207–9
broken symmetry, 207–9
1D energy dispersion, 209–12
finite curvature effect of SWNTs, 204–7
finite size effects, 213–15
initial STM studies, 204
intramolecular junctions, 216–18
Kondo phenomena in 1D, 215–16
prospects, 218
SWNT end states, 212–13
theoretical background, 202–4
Turnover frequencies

calculation of, for ammonia synthesis, 344
Two-color photoinduced Rydberg ionization measurements, 135
Two-dimensional spectroscopy Fourier-transform electronic, 27
infrared echo, 34
Two-photon photoemission (2PPE) spectroscopy, 221–47
experimental techniques in, 225–27
femtosecond, 227
image states and the dielectric continuum model, 227–30
implications for molecule electronic devices, 242–45
molecular resonances and interfacial electron transfer in C_6F_6/Cu(111), 230–38
polarization and localization, 238–42
time-resolved, 237
Two-pulse atomic ionization, 45
Two-pulse correlations interferometric, 511, 516–19, 521
Two-pulse infrared echo spectroscopy, 32–33
Two-pulse photon echoes, 22–23
Tyrosine phosphatase protein, 492, 497

U

Ultra-high-resolution X-ray crystallography, 372
Ultra-high vacuum (UHV) conditions, 328, 340, 379

Ultrahigh molecular weight
    PE (UHMWPE),
    443–44
Ultraviolet lasers
    vacuum, 104–5
Ultraviolet synchrotron
    radiation
    vacuum, 105–6
Unbiased dynamics, 301
Uninitiated monomers, 174
United-atom force field
    GROMOS, 487
University of California,
    Berkeley, 4–7, 13–14
University of Wisconsin,
    Madison, 1–3
UV irradiation
    for modification of
    polymer surfaces, 454

V

Vacuum ultraviolet (VUV)
    light sources, 103–6
    laboratory discharge lamps,
    103–4
    vacuum ultraviolet lasers,
    104–5
    vacuum ultraviolet
    synchrotron radiation,
    105–6
Vacuum ultraviolet (VUV)
    spectroscopy and
    chemistry, 101–40
    outlook, 135–36
    photoion-photoelectron
    studies, 106–23
    state- or energy-selected
    studies of ion dynamics,
    123–34
    two-color photoinduced
    Rydberg ionization
    measurements, 135
    vacuum ultraviolet light
    sources, 103–6
Vacuums
    See Ultra-high vacuum
    conditions
Valence charge distribution,
    278–79
    maximally-localized
    Wannier functions,
    278–79
    Wannier function centers,
    280
Valence coordinates, 164–66
    extension towards N atoms,
    166
Van der Waals (VDW) shells
    nonbonding, 374–75
Van der Waals (VDW) theory,
    419–20
    classical, and hydrostatic
    approximations, 541–42
    of nonuniform liquids,
    molecular scale, 533–62
Van Vleck perturbation
    theory, 159
Van't Hoff equation, 175
Variational Monte Carlo
    techniques, 321
Variational transition states
    theory (VTST), 468–69,
    489, 494
    with multidimensional
    tunneling,
    ensemble-averaged, 480,
    482–83
VDW
    See Van der Waals
Vector correlations
    in scattering experiments,
    88–90
Vibration modes
    using generalized density
    functional perturbation
    theory, 281–85
    See also Rotation-vibration
    coupling; Rovibrational
    S-vectors
Vibrational energies
    zero point, 133
Vibrational spectroscopies,
    324–26
    hybrid, 34
surface oxametallacycles,
    325
Violations
    of nuclear Overhauser
    effect, 364
Viscosities
    measuring, 1
VTST
    See Variational transition
    states theory
VUV
    See Vacuum ultraviolet

W

Wannier function centers
    (WFCs), 262, 279–80
Wannier functions
    maximally-localized,
    278–79
Water
    alkane conformational
    equilibrium in, 425
    hydrophobic interactions
    in, 554–57
    liquid, 262–64
    PFM for stacked plates in,
    427
    polymer surface
    restructuring in, 457–61
    restructuring of polymer
    backbone immediately
    after contacting, 458–59
    slow polymer surface
    change in, 460
Water autoionization, 313
Water clusters, 310–11
Water complexes
    proton sharing in, 258–60
Watson Hamiltonians, 143–58
Watsonians, 153–54
Wave packet dynamics
    nuclear, 519–25
Wave packet snapshots, 89
Weight
    See Molecular weight
    distribution
Wet treatment

for modification of polymer surfaces, 453–54
WFCs
  See Wannier function centers
Wigner cusps, 76

X

X-ray absorption fine structure spectroscopy
  near-edge, 440
X-ray crystallography
  ultra-high-resolution, 372
X-ray photoelectron diffraction (XPD) studies, 326
X-ray photoelectron spectroscopy (XPS), 440
XC
  See Exchange-correlation energy
Xylose isomerase, 489–90

Y

Yale University, 3–4
Yukawa function, 545
Yvon-Born-Green (YBG) hierarchy, 538

Z

Zeolites
  catalysis with, 331–32
Zero electron kinetic energy (ZEKE) detection, 43
Zero point vibrational energies (ZPVEs), 130–32, 133
Ziegler-Natta (ZN) catalysts, 329

Cumulative Indexes

CONTRIBUTING AUTHORS, VOLUMES 49–53

A
Abramson EH, 50:279–313
Aksay IA, 51:601–22
Anderson JB, 51:501–26
Andreoni W, 49:405–39
Armentrout PB, 52:423–61
Armstrong NR, 52:391–422

B
Balucani N, 50:347–76
Bartell LS, 49:43–72
Bashford D, 51:129–52
Beck C, 50:443–84
Bergmann K, 52:763–809
Blomberg MRA, 50:221–49
Boato G, 50:23–50
Bolhuis PG, 53:291–318
Brockman JM, 51:41–63
Brooks CL III, 52:499–535
Brown JM, 50:279–313
Buckingham AD, 49:xiii–xxxv
Bürgi HB, 51:275–96
Butler LJ, 49:125–71

C
Callender RH, 49:173–202
Campbell EEB, 51:65–98
Carrington A, 52:1–13
Casavecchia P, 50:347–76
Case DA, 51:129–52
Chakraborty AK, 52:537–73
Chandler D, 53:291–318
Chapovsky PL, 50:315–45
Cheatham TE III, 51:435–71
Chemla DS, 52:233–53
Chen Z, 53:437–65
Collier CP, 49:371–404

Continetti RE, 52:165–92
Corn RM, 51:41–63
Cukier RI, 49:337–69

D
Dabbs DM, 51:601–22
Dahan M, 52:233–53
Dantus M, 52:639–79
de Boeij WP, 49:99–123
Dellago C, 53:291–318
Deniz AA, 52:233–53
de Pablo JJ, 50:377–411
Dlott DD, 50:251–78
Dyer RB, 49:173–202

E
Eachus RS, 50:117–44
Ediger MD, 51:99–128
Emmett MR, 50:517–36
Escobedo FA, 50:377–411

F
Farantos SC, 50:443–84
Fayer MD, 52:315–56
Fernández–Alonso F, 53:67–99
Field RW, 50:443–84; 52:811–52
Flynn GW, 49:297–336
Fourkas JT, 53:17–40
Freed JH, 51:655–89
Frydman L, 52:463–98

G
Gallagher SC, 51:355–80
Gallicchio E, 49:531–67
Gao J, 53:467–505
Garashchuk S, 51:553–600

Geissler PL, 53:291–318
Giancarlo LC, 49:297–336
Gilmanshin R, 49:173–202
Golumbfskie AJ, 52:537–73
Gordon MS, 49:233–66
Greeley J, 53:319–48
Greer SC, 53:173–200
Gross EM, 52:391–422
Gruebele M, 50:485–516

H
Halberstadt N, 51:405–33
Halfmann T, 52:763–809
Hall GE, 51:243–74
Hansen J-P, 51:209–42
Hansma HG, 52:71–92
Harata A, 50:193–219
Heath JR, 49:371–404
Hemley RJ, 51:763–800
Hendrickson CL, 50:517–36
Hermans LJF, 50:315–45
Herschbach D, 51:1–39
Hershberger JF, 52:41–70
Ho T-S, 50:537–70
Hollebeek T, 50:537–70
Huang J-L, 53:201–20
Hudson PK, 51:473–99

I
Ishikawa H, 50:443–84

J
Jacobs DC, 53:379–407
Janda KC, 51:405–33
Jarrold MF, 51:179–207
Johnson MR, 51:297–321
Jongma R, 52:811–52
Jónsson H, 51:623–53
Joyeux M, 50:443–84

623

K
Kamins TI, 51:527–51
Kearley GJ, 51:297–321
Keske JC, 51:323–53
Klimov DK, 52:751–62
Knickelbein MB, 50:79–115
Knoll W, 49:569–638
Kollman PA, 51:435–71
Kondow T, 51:731–61
Koput J, 50:443–84

L
Laurence TA, 52:233–53
Lee N, 52:751–62
Levine RD, 51:65–98
Levy RM, 49:531–67
Lieber CM, 53:201–20
Liu K, 52:139–64
Löwen H, 51:209–42
Lüchow A, 51:501–26

M
MacMillan F, 52:279–313
Mafuné F, 51:731–61
Makri N, 50:167–91
Marchetti AP, 50:117–44
Mavrikakis M, 53:319–48
McLaughlin L, 52:93–106
Medeiros-Ribeiro G, 51:527–51
Meyer H, 53:141–72
Miller RE, 52:607–37
Mohanty U, 52:93–106
Muenter AA, 50:117–44
Mukamel S, 51:691–729
Murad E, 49:73–98
Myers AB, 49:267–95

N
Nelson BP, 51:41–63
Neumark DM, 52:255–77
Ng C-Y, 53:101–40
Nikitin EE, 50:1–22
Nitzan A, 52:681–750
Nocera DG, 49:337–69
Nørskov JK, 53:319–48

North SW, 51:243–74
Nozik AJ, 52:193–231

O
Ogawa S, 53:507–31
Ohlberg DAA, 51:527–51
Oldfield E, 53:349–78
Oudejans L, 52:607–37
Ouyang M, 53:201–20

P
Pate BH, 51:323–53
Petek H, 53:507–31
Pratt DW, 49:481–530
Pratt LR, 53:409–36
Prenni AJ, 51:473–99
Price DL, 50:571–601
Prisner T, 52:279–313
Pshenichnikov MS, 49:99–123

R
Rabitz H, 50:537–70
Richmond GL, 52:357–89
Rohrbacher A, 51:405–33
Rohrer M, 52:279–313
Ross J, 50:51–78

S
Sawada T, 50:193–219
Schinke R, 50:443–84
Schmidt MW, 49:233–66
Schultz PG, 52:233–53
Schwartz DK, 52:107–37
Seideman T, 53:41–65
Shea J-E, 52:499–535
Shen Q, 50:193–219
Shen YR, 53:437–65
Shore BW, 52:763–809
Siegbahn PEM, 50:221–49
Silva M, 52:811–52
Slutsky LJ, 50:279–313
Somorjai GA, 53:437–65
Steinfeld JI, 49:203–32

T
Taatjes CA, 52:41–70
Tannor DJ, 51:553–600

Thirumalai D, 52:751–62
Tinoco I Jr, 53:1–15
Toennies JP, 49:1–41
Tolbert MA, 51:473–99
Trautman JK, 49:441–80
Trenary M, 51:381–403
Trewhella J, 51:355–80
Trouw FR, 50:571–601
Truhlar DG, 53:467–505
Tse JS, 53:249–90
Tully JC, 51:153–78
Tycko R, 52:575–606

V
Valentini JJ, 52:15–39
Vilesov AF, 49:1–41
Vitanov NV, 52:763–809
Vlachy V, 50:145–65
Vlad MO, 50:51–78
Volpi GG, 50:23–50, 347–76
Vossmeyer T, 49:371–404

W
Wall ME, 51:355–80
Weeks JD, 53:533–62
Weiss S, 52:233–53
Wiersma DA, 49:99–123
Wightman RM, 52:391–422
Williams RS, 51:527–51
Wodtke AM, 52:811–52
Wolkow RA, 50:413–41
Woodruff WH, 49:173–202
Woodson SA, 52:751–62
Wormhoudt J, 49:203–32

X
Xie XS, 49:441–80

Y
Yan Q, 50:377–411

Z
Zare RN, 53:67–99
Zhu X-Y, 53:221–47
Zondlo MA, 51:473–99

CHAPTER TITLES, VOLUMES 49–53

Biophysical Chemistry

Fast Events in Protein Folding: The Time Evolution of Primary Processes	RH Callender, RB Dyer, R Gilmanshin, WH Woodruff	49:173–202
Molecular Electronic Spectral Broadening in Liquids and Glasses	AB Myers	49:267–95
Computer Simulations with Explicit Solvent: Recent Progress in the Thermodynamic Decomposition of Free Energies, and in Modeling Electrostatic Effects	RM Levy, E Gallicchio	49:531–67
Density Functional Theory of Biologically Relevant Metal Centers	PEM Siegbahn, MRA Blomberg	50:221–49
The Fast Protein Folding Problem	M Gruebele	50:485–516
Peptides and Proteins in the Vapor Phase	MF Jarrold	51:179–207
Large-Scale Shape Changes in Proteins and Macromolecular Complexes	ME Wall, SC Gallagher, J Trewhella	51:355–80
Molecular Dynamics Simulation of Nucleic Acids	TE Cheatham III, PA Kollman	51:435–71
On the Characteristics of Migration of Oligomeric DNA in Polyacrylamide Gels and in Free Solution	U Mohanty, L McLaughlin	52:93–106
From Folding Theories to Folding Proteins: A Review and Assessment of Simulation Studies of Protein Folding and Unfolding	J-E Shea, CL Brooks III	52:499–535
Biomolecular Solid State NMR: Advances in Structural Methodology and Applications to Peptide and Protein Fibrils	R Tycko	52:575–606
Early Events in RNA Folding	D Thirumalai, N Lee, SA Woodson, DK Klimov	52:751–62

Chemical Kinetics–Reactions

Chemical Reaction Dynamics Beyond the Born-Oppenheimer Approximation	LJ Butler	49:125–71
Nonlinear Kinetics and New Approaches to Complex Reaction Mechanisms	J Ross, MO Vlad	50:51–78
Reactions of Transition Metal Clusters with Small Molecules	MB Knickelbein	50:79–115
Constructing Multi-Dimensional Molecular Potential Energy Surfaces from Ab Initio Data	T Hollebeek, T-S Ho, H Rabitz	50:537–70
Delayed Ionization and Fragmentation En Route to Thermionic Emission: Statistics and Dynamics	EEB Campbell, RD Levine	51:65–98
The Dynamics of Noble Gas-Halogen Molecules and Clusters	A Rohrbacher, N Halberstadt, KC Janda	51:405–33
Semiclassical Calculation of Chemical Reaction Dynamics via Wavepacket Correlation Functions	DJ Tannor, S Garashchuk	51:553–600

Chemical Kinetics–State-to-State

Proton-Coupled Electron Transfer	RI Cukier, DG Nocera	49:337–69
Crossed-Beam Studies of Reaction Dynamics	P Casavecchia, N Balucani, GG Volpi	50:347–76
HCP ↔ CPH Isomerization: Caught in the Act	H Ishikawa, RW Field, SC Farantos, M Joyeux, J Koput, C Beck, R Schinke	50:443–84
State-to-State Chemical Reaction Dynamics in Polyatomic Systems: Case Studies	JJ Valentini	52:15–39
Recent Progress in Infrared Absorption Techniques for Elementary Gas-Phase Reaction Kinetics	CA Taatjes, JF Hershberger	52:41–70
Crossed-Beam Studies of Neutral Reactions: State-Specific Differential Cross Sections	K Liu	52:139–64

Photofragment Translational Spectroscopy
of Weakly Bound Complexes: Probing
the Interfragment Correlated Final
State Distributions   L Oudejans,   52:607–37
    RE Miller

The Dynamics of "Stretched Molecules":
Experimental Studies of Highly
Vibrationally Excited Molecules with
Stimulated Emission Pumping   M Silva, R Jongma,   52:811–52
    RW Field,
    AM Wodtke

Colloids

Effective Interactions Between Electric
Double Layers   J-P Hansen,   51:209–42
    H Löwen

Electrochemistry

Light-Emitting Electrochemical Processes   NR Armstrong,   52:391–422
    RM Wightman,
    EM Gross

Geochemistry and Cosmochemistry

The Shuttle Glow Phenomenon   E Murad   49:73–98

Chemistry and Microphysics of Polar
Stratospheric Clouds and Cirrus Clouds   MA Zondlo,   51:473–99
    PK Hudson,
    AJ Prenni,
    MA Tolbert

Laser Chemistry

Photothermal Applications of Lasers:
Study of Fast and Ultrafast Photothermal
Phenomena at Metal-Liquid Interfaces   A Harata, Q Shen,   50:193–219
    T Sawada

Ultrafast Spectroscopy of Shock
Waves in Molecular Materials   DD Dlott   50:251–78

Transient Laser Frequency Modulation
Spectroscopy   GE Hall,   51:243–74
    SW North

Liquid State

Simulation of Phase Transitions in Fluids   JJ de Pablo,   50:377–411
    Q Yan,
    FA Escobedo

Generalized Born Models of Macromolecular Solvation Effects	D Bashford, DA Case	51:129–52
Structures and Dynamics of Molecules on Liquid Beam Surfaces	T Kondow, F Mafuné	51:731–61
Coherent Nonlinear Spectroscopy: From Femtosecond Dynamics to Control	M Dantus	52:639–79
Higher-Order Optical Correlation Spectroscopy in Liquids	JT Fourkas	53:17–40
Molecular Theory of Hydrophobic Effects: "She is Too Mean to Have Her Name Repeated."	LR Pratt	53:409–36

Magnetic Resonance

Nuclear Spin Conversion in Polyatomic Molecules	PL Chapovsky, LJF Hermans	50:315–45
New Technologies in Electron Spin Resonance	JH Freed	51:655–89
Pulsed EPR Spectroscopy: Biological Applications	T Prisner, M Rohrer, F MacMillan	52:279–313
Spin-1/2 and Beyond: A Perspective in Solid State NMR Spectroscopy	L Frydman	52:463–98

Miscellaneous

Explosives Detection: A Challenge for Physical Chemistry	JI Steinfeld, J Wormhoudt	49:203–32
Electrospray Ionization Fourier Transform Ion Cyclotron Resonance Mass Spectrometry	CL Hendrickson, MR Emmett	50:517–36
Effects of High Pressure on Molecules	RJ Hemley	51:763–800

Polymers and Macromolecules

Interfaces and Thin Films as Seen by Bound Electromagnetic Waves	W Knoll	49:569–638
Spatially Heterogeneous Dynamics in Supercooled Liquids	MD Ediger	51:99–128
Generalized Born Models of Macromolecular Solvation Effects	D Bashford, DA Case	51:129–52
Self-Assembled Ceramics Produced by Complex-Fluid Templation	DM Dabbs, IA Aksay	51:601–22

Polymer Adsorption-Driven Self-Assembly of Nanostructures	AK Chakraborty, AJ Golumbfskie	52:537–73
Reversible Polymerizations and Aggregations	SC Greer	53:173–200
Scanning Tunneling Microscopy Studies of the One-Dimensional Electronic Properties of Single-Walled Carbon Nanotubes	M Ouyang, J-L Huang, CM Lieber	53:201–20

Prefatory Chapters

Molecules in Optical, Electric, and Magnetic Fields: A Personal Perspective	AD Buckingham	49:xiii–xxxv
Nonadiabatic Transitions: What We Learned from Old Masters and How Much We Owe Them	EE Nikitin	50:1–21
Experiments on the Dynamics of Molecular Processes: A Chronicle of Fifty Years	G Boato, GG Volpi	50:23–50
Fifty Years in Physical Chemistry: Homage to Mentors, Methods, and Molecules	D Herschbach	51:1–39
A Free Radical	A Carrington	52:1–13
Physical Chemistry of Nucleic Acids	I Tinoco Jr	53:1–15

Quantum Chemistry

The Construction and Interpretation of MCSCF Wavefunctions	MW Schmidt, MS Gordon	49:233–66
Computational Approach to the Physical Chemistry of Fullerenes and Their Derivatives	W Andreoni	49:405–35
Time-Dependent Quantum Methods for Large Systems	N Makri	50:167–91
Chemical Dynamics at Metal Surfaces	JC Tully	51:153–78
Monte Carlo Methods in Electronic Structures for Large Systems	A Lüchow, JB Anderson	51:501–26
Semiclassical Calculation of Chemical Reaction Dynamics via Wavepacket Correlation Functions	DJ Tannor, S Garashchuk	51:553–600

Laser-Induced Population Transfer by Adiabatic Passage Techniques	NV Vitanov, T Halfmann, BW Shore, K Bergmann	52:763–809
AB Initio Molecular Dynamics With Density Functional Theory	JS Tse	53:249–90
Quantum Mechanical Methods for Enzyme Kinetics	J Gao, DG Truhlar	53:467–505

Scattering–Elastic and Inelastic

Structure and Transformation: Large Molecular Clusters as Models of Condensed Matter	LS Bartell	49:43–72
Chemical Applications of Neutron Scattering	FR Trouw, DL Price	50:571–601
Motion and Disorder in Crystal Structure Analysis: Measuring and Distinguishing Them	HB Bürgi	51:275–96
Quantitative Atom-Atom Potentials from Rotational Tunneling: Their Extraction and Their Use	MR Johnson, GJ Kearley	51:297–321
Scattering Resonances in the Simplest Chemical Reaction	F Fernández-Alonso, RN Zare	53:67–99
Reactive Collisions of Hyperthermal Energy Molecular Ions with Solid Surfaces	DC Jacobs	53:379–407

Solids and Ordered Arrays

Nanocrystal Superlattices	CP Collier, T Vossmeyer, JR Heath	49:371–404
The Photophysics of Silver Halide Imaging Materials	RS Eachus, AP Marchetti, AA Muenter	50:117–44
Applications of Impulsive Stimulated Scattering in the Earth and Planetary Sciences	EH Abramson, JM Brown, LJ Slutsky	50:279–313
Surface Plasmon Resonance Imaging Measurements of Ultrathin Organic Films	JM Brockman, BP Nelson, RM Corn	51:41–63

Thermodynamics of the Size and Shape of Nanocrystals: Epitaxial Ge on Si(001)	RS Williams, G Medeiros-Ribeiro, TI Kamins, DAA Ohlberg	51:527–51
Self-Assembled Ceramics Produced by Complex-Fluid Templation	DM Dabbs, IA Aksay	51:601–22
Theoretical Studies of Atomic-Scale Processes Relevant to Crystal Growth	H Jónsson	51:623–53
Spectroscopy and Hot Electron Relaxation Dynamics in Semiconductor Quantum Wells and Quantum Dots	AJ Nozik	52:193–231

Spectroscopy–Infrared, Raman, and Electronic

Spectroscopy of Atoms and Molecules in Liquid Helium	JP Toennies, AF Vilesov	49:1–41
Ultrafast Solvation Dynamics Explored by Femtosecond Photon Echo Spectroscopies	WP de Boeij, MS Pshenichnikov, DA Wiersma	49:99–123
Explosives Detection: A Challenge for Physical Chemistry	JI Steinfeld, J Wormhoudt	49:203–32
Molecular Electronic Spectral Broadening in Liquids and Glasses	AB Myers	49:267–95
Optical Studies of Single Molecules at Room Temperature	XS Xie, JK Trautman	49:441–80
High Resolution Spectroscopy in the Gas Phase: Even Large Molecules Have Well-Defined Shapes	DW Pratt	49:481–530
Transient Lasar Frequency Modulation Spectroscopy	GE Hall, SW North	51:243–74
Quantitative Atom-Atom Potentials from Rotational Tunneling: Their Extraction and Their Use	MR Johnson, GJ Kearley	51:297–321
Decoding the Dynamical Information Embedded in Highly Mixed Quantum States	JC Keske, BH Pate	51:323–53
Reflection Absorption Infrared Spectroscopy and the Structure of Molecular Adsorbates on Metal Surfaces	M Trenary	51:381–403

Multidimensional Femtosecond Correlation Spectroscopies of Electronic and Vibrational Excitations	S Mukamel	51:691–729
Coincidence Spectroscopy	RE Continetti	52:165–92
Ratiometric Single-Molecule Studies of Freely Diffusing Biomolecules	AA Deniz, TA Laurence, M Dahan, DS Chemla, PG Schultz, S Weiss	52:233–53
Time-Resolved Photoelectron Spectroscopy of Molecules and Clusters	DM Neumark	52:255–77
Fast Protein Dynamics Probed with Infrared Vibrational Echo Experiments	MD Fayer	52:315–56
Time-Resolved Photoelectron Angular Distributions: Concepts, Applications, and Directions	T Seideman	53:41–65
Vacuum Ultraviolet Spectroscopy and Chemistry by Photoionization and Photoelectron Methods	C-Y Ng	53:101–40
The Molecular Hamiltonian	H Meyer	53:141–72
Electron Transfer at Molecule-Metal Interfaces: A Two-Photon Photoemission Study	X-Y Zhu	53:221–47
Studies of Polymer Surfaces by Sum Frequency Generation Vibrational Specctroscopy	Z Chen, YR Shen, GA Somorjai	53:437–65

Statistical Mechanics

Structure and Transformation: Large Molecular Clusters as Models of Condensed Matter	LS Bartell	49:43–72
Ionic Effects Beyond Poisson-Boltzmann Theory	V Vlachy	50:145–65
Transition Path Sampling: Throwing Ropes Over Rough Mountain Passes, in the Dark	PG Bolhuis, D Chandler, C Dellago, PL Geissler	53:291–318
Connecting Local Structure to Interface Formation: A Molecular Scale van der Waals Theory of Nonuniform Liquids	JD Weeks	53:533–62

Surfaces

Scanning Tunneling and Atomic Force Microscopy Probes of Self-Assembled, Physisorbed Monolayers: Peeking at the Peaks	LC Giancarlo, GW Flynn	49:297–336
Interfaces and Thin Films as Seen by Bound Electromagnetic Waves	W Knoll	49:569–638
Controlled Molecular Adsorption on Silicon: Laying a Foundation for Molecular Devices	RA Wolkow	50:413–41
Surface Plasmon Resonance Imaging Measurements of Ultrathin Organic Films	JM Brockman, BP Nelson, RM Corn	51:41–63
Chemical Dynamics at Metal Surfaces	JC Tully	51:153–78
Reflection Absorption Infrared Spectroscopy and the Structure of Molecular Adsorbates on Metal Surfaces	M Trenary	51:381–403
Structures and Dynamics of Molecules on Liquid Beam Surfaces	T Kondow, F Mafuné	51:731–61
Surface Biology of DNA by Atomic Force Microscopy	HG Hansma	52:71–92
Mechanisms and Kinetics of Self-Assembled Monolayer Formation	DK Schwartz	52:107–37
Structure and Bonding of Molecules at Aqueous Surface	GL Richmond	52:357–89
Electron Transmission through Molecules and Molecular Interfaces	A Nitzan	52:681–750
Electronic Structure and Catalysis on Metal Surfaces	J Greeley, JK Nørskov, M Mavrikakis	53:319–48
Surface Femtochemistry: Observation and Quantum Control of Frustrated Desorption of Alkali Atoms from Noble Metals	H Petek, S Ogawa	53:507–31

Thermochemistry and Thermodynamics

Computer Simulations with Explicit Solvent: Recent Progress in the Thermodynamic Decomposition of Free Energies and in Modeling Electrostatic Effects	RM Levy, E Gallicchio	49:531–67
Reactions and Thermochemistry of Small Transition Metal Cluster Ions	PB Armentrout	52:423–61